Frontiers of Quantum Chemistry

Marek J. Wójcik · Hiroshi Nakatsuji
Bernard Kirtman · Yukihiro Ozaki
Editors

# Frontiers of Quantum Chemistry

Springer

*Editors*
Marek J. Wójcik
Jagiellonian University
Kraków
Poland

Bernard Kirtman
University of California
Santa Barbara, CA
USA

Hiroshi Nakatsuji
Quantum Chemistry Research Institute
Nishikyo-ku, Kyoto
Japan

Yukihiro Ozaki
School of Science and Technology
Kwansei Gakuin University
Sanda
Japan

ISBN 978-981-10-5650-5 ISBN 978-981-10-5651-2 (eBook)
https://doi.org/10.1007/978-981-10-5651-2

Library of Congress Control Number: 2017946658



Printed on acid-free paper

This Springer imprint is published by Springer Nature
The registered company is Springer Nature Singapore Pte Ltd.
The registered company address is: 152 Beach Road, #21-01/04 Gateway East, Singapore 189721, Singapore

# Preface

Quantum chemistry plays a crucial role in many areas of chemistry, physics and biology. There is significant research ongoing in this area of science. International conferences in quantum chemistry are frequently organized in many countries. This book came about as a consequence of the International Symposium "Frontiers of Quantum Chemistry" on November 11, 2015 at Kwansei Gakuin University in Japan.

This book consists of twenty chapters. It presents the frontiers of current research in quantum chemistry in various areas of theory. The chapter topics include: development of correlated single particle theories; density-functional theory; relativistic time-dependent density-functional theory; quantum many-body methods; response theory; SACCI methodology; electron communications theory; vibrational linear and nonlinear optical properties theory; and molecular dynamics. Other chapter topics include: applications in kinetics; description of molecular properties and chemical reactions; nonlinear optical properties in ultraviolet spectroscopy; circular dichroism; nuclear quantum effect and H/D isotope effect in hydrogen-bonded systems; properties of periodic materials subjected to external electric and magnetic fields; photoisomerization reactions; chemical enhancement of surface-enhanced Raman spectroscopy; NIR spectroscopy; and spectroscopy of hydrogen-bonded systems.

The first fourteen chapters, representing the first part of the book, present newly developed methods which might play important roles in future spectroscopic studies. The presented methods elucidate state-of-the-art modern quantum chemistry. It should be pointed out that these chapters revolve around different approaches. Concepts based on the post-Hartree-Fock methods are presented as well as those based on density functional theory. Several chapters are concerned with vibrational linear and nonlinear optical properties which represent modern spectroscopic analysis. Some of the chapters go beyond currently used methods, such as electron communications theory. The second part of the book is devoted to discussions about the applications of current methods. These six chapters cover the most significant and important applications across several fields of science. The application of molecular dynamics simulations for large systems looks very promising. Determination of isotope effects, as well as reaction mechanism studies, are also essential applications, especially in computational enzymology and biology. These chapters represent a complete source of quantum chemistry knowledge.

We hope readers of this book will learn a great deal about the current state of quantum chemistry and its applications and importance in basic and applied science.

We would like to thank Dr. Shinichi Koizumi and Ms. Taeko Sato of Springer Japan for their continuous efforts toward publishing this book.

Kraków, Poland — Marek J. Wójcik
Kyoto, Japan — Hiroshi Nakatsuji
Santa Barbara, USA — Bernard Kirtman
Sanda, Japan — Yukihiro Ozaki
March 2017

# Contents

1 **Rigorous and Empirical Approaches to Correlated Single-Particle Theories** . . . . . 1
Johannes T. Margraf and Rodney J. Bartlett

2 **Circular Dichroism Spectroscopy with the SAC-CI Methodology: A ChiraSac Study** . . . . . 21
Tomoo Miyahara and Hiroshi Nakatsuji

3 **Frontiers of Coupled Cluster Chiroptical Response Theory** . . . . . 49
T. Daniel Crawford

4 **Response Theory and Molecular Properties** . . . . . 69
Shane M. Parker and Filipp Furche

5 **Response Properties of Periodic Materials Subjected to External Electric and Magnetic Fields** . . . . . 87
Bernard Kirtman, Lorenzo Maschio, Michel Rérat and Michael Springborg

6 **Quantum Chemical Methods for Predicting and Interpreting Second-Order Nonlinear Optical Properties: From Small to Extended π-Conjugated Molecules** . . . . . 117
Benoît Champagne, Pierre Beaujean, Marc de Wergifosse, Marcelo Hidalgo Cardenuto, Vincent Liégeois and Frédéric Castet

7 **Embedding Methods in Quantum Chemistry** . . . . . 139
Albrecht Goez and Johannes Neugebauer

8 **Calculation of Vibrational Spectra of Large Molecules from Their Fragments** . . . . . 181
Shigeki Yamamoto and Petr Bouř

9 **Describing Molecules in Motion by Quantum Many-Body Methods** . . . . . 199
Ove Christiansen

**10 Relativistic Time-Dependent Density Functional Theory for Molecular Properties** ........ 223
Muneaki Kamiya and Takahito Nakajima

**11 Warming Up Density Functional Theory** ........ 249
Justin C. Smith, Francisca Sagredo and Kieron Burke

**12 Quantum Chemistry at the High Pressures: The eXtreme Pressure Polarizable Continuum Model (XP-PCM)** ........ 273
Roberto Cammi

**13 Transition States of Spin-State Crossing Reactions from Organometallics to Biomolecular Excited States** ........ 289
Naoki Nakatani, Akira Nakayama and Jun-ya Hasegawa

**14 Electron Communications and Chemical Bonds** ........ 315
Roman F. Nalewajski

**15 Molecular Dynamics Simulations of Vibrational Spectra of Hydrogen-Bonded Systems** ........ 353
Mateusz Z. Brela, Marek Boczar, Łukasz Boda and Marek J. Wójcik

**16 Nuclear Quantum Effect and H/D Isotope Effect on Hydrogen-Bonded Systems with Path Integral Simulation** ........ 377
Kimichi Suzuki, Yukio Kawashima and Masanori Tachikawa

**17 Vibrational Linear and Nonlinear Optical Properties: Theory, Methods, and Application** ........ 401
Heribert Reis, David Lauvergnat, Josep M. Luis and Robert Zaleśny

**18 Ab Initio Molecular Dynamics Study on Photoisomerization Reactions: Applications to Azobenzene and Stilbene** ........ 431
Tetsuya Taketsugu and Yu Harabuchi

**19 Density Functional Theoretical Studies on Chemical Enhancement of Surface-Enhanced Raman Spectroscopy in Electrochemical Interfaces** ........ 455
Ran Pang, De-Yin Wu and Zhong-Qun Tian

**20 Advances in Anharmonic Methods and Their Applications to Vibrational Spectroscopies** ........ 483
Krzysztof B. Beć, Justyna Grabska and Yukihiro Ozaki

# Chapter 1
# Rigorous and Empirical Approaches to Correlated Single-Particle Theories

**Johannes T. Margraf and Rodney J. Bartlett**

**Abstract** In this contribution, we will discuss how electron correlation is described in different types of single-particle approaches, namely those based on many-body theory, density functional theory, and semiempirical approximations. Our goal is to provide some insight into how the term "correlation" is used in different communities and what the physical and computational advantages and drawbacks of the different approaches are. This interdisciplinary perspective will illuminate how researchers with different backgrounds think about electron correlation in single-particle methods. In this context, the main focus lies on MO-like theories such as KS-DFT and semiempirical methods, since these are the most common "general-purpose" approaches in quantum chemistry, which allow predicting ground state and excited state energies and properties.

**Keywords** Electron correlation • Wavefunction theory • Density functional theory • Semiempirical molecular orbital theory • Correlated orbital theory

## 1.1 Introduction

The electrostatic interactions between nuclei and electrons determine the structure and properties of molecules and solids. Electronic structure theory is, therefore, a critical part of modern research in chemistry and materials science. For systems of chemical interest, theoreticians are faced with the many-electron problem, and usually have to resort to approximations to numerically describe the highly complex electron–electron interactions. One of the most popular classes of these approxi-

---

J.T. Margraf · R J. Bartlett (✉)
Quantum Theory Project, University of Florida, Gainesville, FL 32611, USA
e-mail: bartlett@qtp.ufl.edu

J.T. Margraf
Chair for Theoretical Chemistry, Technische Universität München, Lichtenbergstr. 4, 85747 Garching, Germany

 
M.J. Wójcik et al. (eds.), *Frontiers of Quantum Chemistry*,
https://doi.org/10.1007/978-981-10-5651-2_1

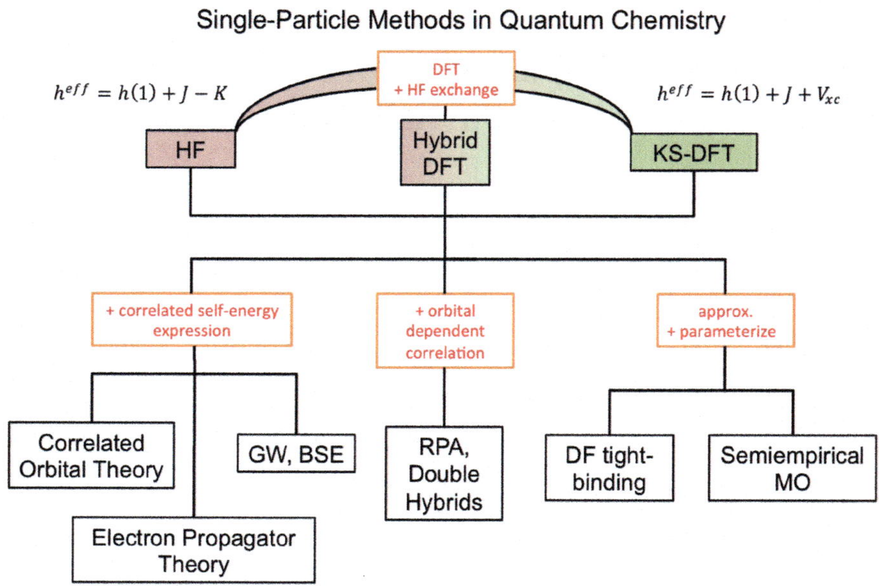


**Fig. 1.1** Relationship between different single-particle methods in quantum chemistry. The prototypical effective single-particle theories are HF (equation in the *top left*) and KS-DFT (*top right*). Other single-particle methods augment (e.g., via correlated self-energy expressions), combine (hybrid DFT), or approximate (semiempirical methods) these approaches

mations is formed by effective single-particle theories (see Fig. 1.1 for a schematic overview of single-particle methods in quantum chemistry) [1].

In single-particle theories, the interaction of N electrons is mathematically described via N one-electron equations. This dramatically reduces the computational complexity of the problem, but also introduces an approximation, namely that each electron interacts with the average field of all other electrons, via an effective Hamiltonian $h^{eff}$. For this reason, such approaches are known as mean field theories and have long been familiar to chemists due to *Hartree-Fock* (HF) theory (see Fig. 1.1) [1–4]. The HF wavefunction is the variationally optimal representation of the electronic structure within the mean field approximation [5].

The HF one-particle equations have the following form:

$$\left(h(i) + \sum_j \left(J_j(i) - K_j(i)\right)\right)\varphi_i(i) = \varepsilon_i \varphi_i(i), \tag{1.1}$$

where $h(i)$ is the one-electron "core" Hamiltonian, $\varphi_i(i)$ is the $i$th molecular orbital, $\varepsilon_i$ is the orbital eigenvalue, and $J_j(i)$ and $K_j(i)$ are the Coulomb and exchange operators,

$$J_j(i) - K_j(i) = \int \varphi_j(i)^* \left(1 - P_{ij}\right)\varphi_j(i)dj. \quad (1.2)$$

By only considering the interaction of each electron with the average field of all other electrons, the "correlated motions" between the movements of individual electron pairs are neglected. This electron correlation effect only amounts to a small percentage (typically ca. 1%) of the total energy. Unfortunately, it often translates to very large errors for the energy differences critical to chemistry. In order to make quantitative predictions (e.g., to determine which possible product of a reaction is more stable or to accurately describe spectroscopy), we must, therefore, take electron correlation into account.

The most commonly accepted quantification of electron correlation is attributed to Löwdin, though the concept was also mentioned even earlier by Wigner [6, 7]. Löwdin defined the correlation energy as the difference between the full CI and restricted HF (RHF) energies in a given basis [8, 9]:

$$E_C = E_{FCI} - E_{HF}. \quad (1.3)$$

In this context, electron correlation is mainly understood as the instantaneous electrostatic repulsion between individual electrons, which is missing from the mean-field description in RHF. In modern usage, electron correlation is typically measured as the difference between the nonrelativistic "right" answer and any single determinant approximation including unrestricted HF (UHF), although in Löwdin's days UHF would be considered correlated since different spin–orbitals are allowed different spatial extent.

There is also a broader, statistical definition of electron correlation. In this sense, electrons are correlated if there is a difference between the ordinary and conditional probability density of finding an electron at a given point in space [10]. This includes the instantaneous repulsion effect described above (Coulomb correlation), as well as the fact that electrons obey Fermi statistics, i.e., that the wavefunction changes its sign when indices are permuted (Fermi correlation). In both cases, the physical consequence of correlation is that it becomes less likely to find two electrons close to each other. This depletion of probability density around an electron is often referred to as a "hole" (e.g., the "Fermi hole") [11, 12].

It is important to note that HF does describe Fermi correlation through the exchange operator, so the term electron correlation is typically used as in Löwdin's definition, i.e., as Coulomb correlation. Up to this point, there is a clear separation between exchange and correlation. Exchange is treated exactly via the HF reference, and the remaining goal is to approximate the correlation energy and wavefunction as accurately as possible. As we shall see, the case is more complicated in DFT, where typically both exchange and correlation are approximated via density-dependent functionals. In particular, in DFT, one speaks of a "many-electron self-interaction," which has to mean in rigorous wavefunction theory, the higher-order couplings between the exchange operator and the Coulomb operator as would occur in MBPT and coupled cluster theory.

The correlation energy from Eq. 1.3 is missing from HF due to the mean field approximation, which comes from representing the wavefunction as a single Slater determinant [1]:

$$\Psi^{HF} = 1/\sqrt{N!}\,\det(\chi_1, \chi_2, \ldots, \chi_N). \tag{1.4}$$

Correlation can be introduced in a formally simple way by approximating the wavefunction as a linear combination of Slater determinants (or symmetry-adapted configuration state functions) as is done in configuration interaction and multi-configurational SCF methods [13]:

$$\Psi^{CI} = \sum_i c_i \Phi_i. \tag{1.5}$$

Unfortunately, using the full set of possible excited configurations (i.e., full CI) is computationally intractable for all but the smallest systems. Truncating the number of determinants to a certain degree of excitations (e.g., all singles and doubles, CISD) leads to a method that introduces correlation in finite systems but violates size extensivity, a critical failing [14, 15].

These problems can be overcome to a prominent extent by using coupled cluster (CC) theory, where the wavefunction has the correct exponential form [15–17]:

$$\Psi^{CC} = e^T \Phi_0, \tag{1.6}$$

where $\Phi_0$ is the mean-field single determinant.

CC can also be considered as an infinite-order many-body perturbation theory (MBPT):

$$e^T \Phi_0 = \Phi_0 + \sum_{k=1}^{\infty} (R_0 V)_L^k \Phi_0 \tag{1.7}$$

where

$$R_0 = (E_0 - H_0)^{-1} Q, \tag{1.8}$$

$$Q = 1 - |0\rangle\langle 0|, \tag{1.9}$$

and

$$V = H - H_0. \tag{1.10}$$

CC and MBPT allow the accurate calculation of the correlation energy and have been extended to predict the vibrational, optical, and electronic properties of molecules. These "post-HF" many-body methods are the state of the art in most modern applications of quantum chemistry.

As Mulliken famously noted, the high sophistication of these methods has the consequence that many concepts that are dear to the heart of chemists and physicists "vanish into thin air" [18]. Molecular orbitals (MOs) are one such endangered species, but an extremely useful one. For instance, the spatial extent of a molecular orbital, its eigenvalue, and the energy gap to other orbitals can be used to understand chemical reactions and ionization processes in molecules. Indeed, scanning tunneling microscopy and photoemission tomography experiments show that observable electronic states have many of the characteristics typically associated with MOs (nodal structure, energetic ordering, etc.) [19, 20]. Similarly, in condensed matter physics, the band structure is an essential concept for understanding the electronic and optical properties of materials. In this context, correlated single-particle methods are very appealing because they promise to combine the quantitative accuracy of wavefunction theory with the conceptual advantages of molecular orbital theory.

In the following, we will discuss how electron correlation is described in different types of single-particle approaches, namely those based on many-body theory, density functional theory, and semiempirical approximations. Our goal is to provide some insight into how the term "correlation" is used in different communities and what the physical and computational advantages and drawbacks of the different approaches are. This interdisciplinary perspective will illuminate how researchers with different backgrounds think about electron correlation in single-particle methods. In this context, the main focus lies on MO-like theories such as KS-DFT and semiempirical methods, since these are the most common "general-purpose" approaches in quantum chemistry, which allow predicting ground state and excited state energies and properties. We will begin with rigorous many-body approaches, however, to define the framework used to build future simplified and semiempirical methods.

## 1.2 Correlation in Single-Particle Many-Body Theories

The formally exact way of formulating a correlated single-particle method begins with the Dyson equation, which can be written (in an "inverse" form) as [21–23]:

$$\left(\hat{h}(i) + \sum_j \left(J_j(i) - K_j(i)\right) + \Sigma(i,\omega)\right)\varphi_i(i) = \omega_i \varphi_i(i), \tag{1.11}$$

where $\Sigma(i,\omega)$ is the frequency-dependent self-energy and $\omega_i$ is the $i$th pole of the Green's function.

Written in this way, Eq. 1.11 resembles the Schrödinger equation and illustrates how HF theory can be understood as an approximation to Dyson theory (or Dyson theory as a generalized HF theory) [21].

A common feature of all Green's function and propagator methods (such as the GW approximation, the outer-valence Green's function, or electron propagator theory) is the frequency-dependent self-energy operator. The concept of self-energy is essential for describing correlation in a single-particle framework. The idea is that the presence of a particle instantaneously polarizes its environment. Instead of treating correlation within or between pairs or groups of particles (as in WFT), there is a mutual interaction between each particle and the surrounding field. This conceptual difference to WFT allows retaining a single-particle description while still describing correlation. This approach is exact in the limit of an exact self-energy operator, just like WFT is exact in the full CI limit. In practice, calculating the self-energy has similar computational demands as calculating the correlation energy in WFT and similar perturbational approximations are made.

The eigenvalues of Eq. 1.11 correspond to ionization potentials and electron affinities. As a consequence, propagator methods are mainly used to calculate energy differences and transition probabilities of photoelectron spectra and related processes [23, 24]. They are also increasingly relevant in the context of molecular electronics, where electron affinities and ionization potentials define the driving forces for the fundamental processes in device operation.

The analogy between Dyson theory and HF is reflected in Koopmans' theorem, as HF orbital energies can be understood as approximations to the poles of the Dyson equation. However, the frequency dependence of the self-energy leads to an important distinction to HF: the set of Dyson amplitudes,

$$\varphi_i(x_1) = \int \Phi_i^{N-1*}(x_2, \ldots, x_n)\Phi^N(x_1, \ldots, x_n)dx_2 \ldots dx_N, \tag{1.12}$$

which would be a single orbital $\varphi_i(x_1)$ in a single determinant approximation, now consists of as many amplitudes as could occur in a full CI calculation. Thus, this set is overdetermined (and non-orthogonal), meaning that there are vastly more ionization amplitudes than electrons. On one hand this is advantageous, because it allows describing complicated electronic situations, such as shake-up ionizations. On the other hand, it means that Dyson theory in a correlated form is not an MO-like theory that is the objective of this contribution.

In an effective one-particle theory, one wants the eigenvalues of the occupied orbitals to correspond to *principal* IPs, meaning those that measure the energy required to remove a single electron from any orbital in a molecule with no other orbital transitions. Furthermore, whereas one can evaluate the one-particle density matrix and the total energy from the Greens' function, the contour integrals invoked to do so are subject to *all* the poles of the Greens' function, i.e., the same number as the Dyson amplitudes. This feature is drastically changed once the electron propagator is replaced by its coupled cluster form [25, 26]. This is the essence of a *correlated orbital theory* (COT) which is both MO-like and rigorously includes electron correlation [27]. The transition from the Greens' function to CC theory also provides the total energy and density matrices in the far more immediate and tractable form familiar to all CC work [28].

As we have discussed, the treatment of correlation in WFT is equivalent to the exact frequency-dependent self-energy expression in the limit of full CI. It is, therefore, possible to define frequency-independent self-energy operators based on

CC theory. This leads to a kind of "quantitative" MO theory whose eigenvalues must be the exact *principal* IPs and EAs in the basis and correlation limit,

$$\left(\hat{h}(i) + \sum_j \left(J_j(i) - K_j(i)\right) + \Sigma_{CC}(i)\right)\varphi_i(i) = \varepsilon_i\varphi_i(i). \quad (1.13)$$

The above COT follows from manipulations of the IP/EA-EOM-CC equations [22],

$$\left(\bar{\mathrm{H}}R_k\right)_c = \omega_k R_k \quad (1.14)$$

and

$$L_k\bar{\mathrm{H}} = L_k\omega_k, \quad (1.15)$$

where $\bar{\mathrm{H}}$ indicates a projection of the operator, $e^{-T}He^{T}$, onto some set of determinants. When those determinants correspond to one hole (1h), 2h1p, 3h2p, etc., the eigenvalues are the principal ionization potentials; while for 1p, 2p1h, 3p2h, etc., the eigenvalues are electron affinities. $\Sigma_{CC}$ follows because of the manipulations that allow these many-electron expressions to be reduced to an effective one-particle and frequency-independent form. It can always be written in terms of the CC ground state amplitudes, $T$, and the EOM-CC amplitudes, $R, L$. By extending the equations to double ionization or double attachment and beyond, one can provide frequency-independent equations for any other sector of Fock space and its appropriate self-energy operator. So there is no restriction to the above except choosing to focus on the 1h and 1p sectors to define a correlated orbital theory.

## 1.3 Correlation in Kohn-Sham Density Functional Theory

Density functional theory (DFT) was not originally conceived as a single-particle method, since it describes electronic structure in terms of the electron density $\rho(\vec{r})$ and not via individual electrons [29]. This idea is extremely appealing, because it reduces the dimensionality of the electronic structure problem from 3 * N variables (three spatial coordinates per electron) to the three variables necessary to describe the spatial distribution of the electron density. In practice, however, orbital-free DFT methods suffer from a poor description of the kinetic energy of inhomogeneous electron distributions, which is particularly critical for covalently bonded systems [30, 31]. Since according to the virial theorem,

$$2\langle T\rangle = -\langle V\rangle \quad (1.16)$$

for an atom, substantial errors in the kinetic energy propagate to the total energy and cannot be tolerated in electronic structure calculations.

The great popularity of DFT in quantum chemistry is due to the method of Kohn and Sham (KS-DFT), which sacrifices a purely density-based description in favor of a MO-like framework [32]. This allows a more accurate evaluation of the kinetic energy, while exchange and correlation energies are determined via an unknown density-dependent functional $E_{xc}[\rho]$.

The KS-DFT approach leads to a set of familiar single-particle equations:

$$\left(\hat{h}(i) + \sum_j J_j(i) + V_{xc}(i)\right)\varphi_i(i) = \varepsilon_i \varphi_i(i), \tag{1.17}$$

where $V_{xc}$ is the exchange-correlation potential:

$$V_{xc} = \frac{\delta E_{xc}[\rho]}{\delta \rho}. \tag{1.18}$$

Like Dyson theory or WFT, DFT is often considered to be formally exact, although this is not true, since there is no general mechanism to describe two-particle quantities like $S^2$. Furthermore, all results suffer from the significant caveat that the exact form of the exchange-correlation functional is in general unknown (and probably unknowable) [33]. There are many approximate functionals, both derived from theoretical arguments and empirical fitting (or some superposition of these approaches) [34–36].

Since the beginnings of KS-DFT, there has been much debate about the meaning of KS orbitals [36–44]. As mentioned above, DFT was originally devised as an orbital-free method. In this tradition, model systems like the homogeneous electron gas have played an important role in defining exact conditions for the design of density functional approximations [45].

Many researchers subscribe to the view that KS orbitals are a physically meaningless construct that is merely convenient for calculating the electronic kinetic energy and ensuring the N-representability of the density [34, 46]. At the same time, KS band structures are commonly employed to study the electronic properties of solids, implicitly acknowledging that KS eigenfunctions possess some significance (although the band theory approximations are poor). Furthermore, it can be shown that for the exact functional the eigenvalue of the highest occupied KS-MO is equal to the negative principal IP of the system:

$$\varepsilon_{HOMO} = -IP. \tag{1.19}$$

The topic of this chapter is the inclusion of correlation in a single-particle picture, so we will not discuss correlation from an "orbital-free" standpoint. This has been extensively done elsewhere [11, 35, 47–51]. Instead, we want to look at the KS equations in light of the previous section. $V_{xc}$ is usually further separated into the exchange and correlation potentials $V_x$ and $V_c$. When comparing Eq. 1.17 with the single-particle equations of Dyson theory or COT (Eqs. 1.11 and 1.13), one is tempted to see $V_x$ as a (local) approximation to the HF exchange operator $K$ and $V_c$ as a (local, frequency independent) approximation to the self-energy $\Sigma(\omega)$ [38].

The analogy between $V_x$ and $K$ should not be taken too literally, however. Handy in particular has promoted the view that DFT exchange functionals also describe left-right correlation effects, while the correlation functional exclusively describes dynamic correlation [52–54]. Similarly, many have argued that exchange and correlation should not be handled separately in DFT [55]. Empirically, this idea is reflected by the generally poor performance of functionals that combine 100% HF exchange with DFT correlation (e.g., HFLYP vs. BLYP). In that sense, a more fitting analogy is that [43]

$$V_{xc} \approx K + \Sigma. \tag{1.20}$$

Independent of whether exchange and correlation are considered separately or not, this implies that all KS orbital eigenvalues should be equal to negative vertical IPs. A formal proof of this is given by Bartlett's IP theorem, in the context of adiabatic TD-DFT [56]. If TD-DFT is formally exact for single excitations, then it must be able to correctly describe ionizations as excitations into continuum orbitals. It can be shown that in this case the excitation energy reduces to the energy difference between the occupied KS orbital eigenvalue and the continuum orbital eigenvalue (which is zero). It follows that the occupied KS eigenvalue must be the negative vertical IP.

Gritsenko, Baerends, and coworkers have developed a similar "Koopmans' theorem of DFT" independently [39, 41, 57, 58]. Their work draws from the analysis of KS potentials obtained from reverse engineering accurate electron densities, which represent good approximations to the exact KS potential for that particular system. They find the "exact" KS eigenvalues for valence IPs to be accurate to within 0.1 eV. A related approximate $V_{xc}$, based on the statistical averaging of orbital potentials (SAOP), shows mean absolute deviations of 0.4 eV for valence IPs [41].

Knowing what the eigenvalues should be, it is disheartening that the *eigenvalues* of conventional DFT functionals like BLYP and PBE are typically several eV in error, even though they can provide quite accurate total energies and ΔKS-DFT IPs. In other words, accurate exchange-correlation functionals do not automatically lead to accurate potentials [59, 60]. This follows from the obvious consequence that a typical KS-DFT functional will be correct to second order in $\rho$ (because it is variational), while the potential, being its derivative (Eq. 1.18), is only correct to first order up to a constant. In many cases, DFT functionals provide significantly improved answers when used non-self-consistently with more accurate HF densities (particularly when self-interaction problems are pathological) [59–62]. Obviously, HF densities are not perfect either, because they completely neglect correlation and tend to be too diffuse. But thanks to the approximate validity of Koopman's theorem, the HF potential tends to be significantly better than one iteratively generated by a typical (local) density functional approximation.

From this point of view, it appears that the search for improved density functional approximations should actually be the search for improved KS *potentials*, i.e., ones that comply with the IP theorem. A good place to start is ab initio DFT, which is the closest DFT analogue to the many-body methods discussed in the previous section [56, 63–66]. Here, the optimized effective potential method (OEP) is used to generate local exchange and correlation functionals based on the corresponding orbital-dependent ab initio expressions [64, 67]. Perhaps unsurprisingly, the OEP exact exchange (EXX) potential provides eigenvalues that are fairly good predictions of IPs as eigenvalues, comparable with HF. When a correlation potential based on MBPT(2) (semicanonical second-order OEP, OEP2-sc) is added, the accuracy of the eigenvalues is further improved. For core ionizations, it proves beneficial to use a 50:50 combination of local OEP EXX and nonlocal HF exchange. Using this approach, all KS eigenvalues correspond to negative IPs to within 1 eV [68]. Furthermore, such OEP functionals are self-interaction free.

The calculation of the orbital-dependent exchange and correlation potentials in OEP is computationally demanding. In this sense, the OEP methods lack the appeal of pure or hybrid DFT approaches like PBE or B3LYP. As a consequence, considerable effort has been put into developing conventional DFT methods that can provide an accurate eigenvalue spectrum. A major improvement in this context was brought about by the introduction of range-separated hybrid (RSH) exchange functionals. Here, the $r_{12}^{-1}$ operator is split into a short- and long-range component [69–71]:

$$\frac{1}{r_{12}} = \frac{1 - \mathrm{erf}(\mu r_{12})}{r_{12}} + \frac{\mathrm{erf}(\mu r_{12})}{r_{12}}. \tag{1.21}$$

This allows using the asymptotically correct HF exchange integrals at long range, while still benefiting from the description of static correlation effects from the exchange functional at short range. RSH functionals like CAM-B3LYP, ωB97-XD, and LC-BOP are especially useful for describing charge-transfer states with TD-DFT [69–72]. They also display improved orbital energies compared to their non-range-separated counterparts, although they are still far from the accuracy of OEP-based functionals.

A commonly used remedy for this is the use of "tuned" RSH functionals [73–76]. Here, the range-separation parameter $\mu$ is self-consistently adjusted so that the HOMO eigenvalue is equal to the ΔSCF ionization potential (which is generally a good approximation to the experimental one for finite systems, but not possible for solids). This technique is useful when one is interested in excitation energies or band structures, but it cannot be applied to general thermochemistry, since each molecule effectively uses a different functional in this approach.

Bartlett and coworkers have developed several density functional approximations that aim to fulfill the IP theorem for all orbitals, and are therefore 'consistent'. The CAM-QTP functionals are reparameterizations of CAM-B3LYP. CAM-QTP00 was solely parameterized to reproduce all orbital energies of water as vertical IPs [77]. It was shown that this parameterization leads to generally accurate eigenvalues and unprecedented accuracy for core-excitation energies [78]. In CAM-QTP01, the focus was concentrated on the valence orbitals, in order to improve valence excitation energies and general thermochemistry [79].

In a similar vein, the B2IP-PLYP functional is an IP-optimized double-hybrid (DH) functional [80]. Here, the intrinsic inconsistency between potential and functional in DHs was investigated using the OEP method. It was shown that the consistent potential to Grimme's B2-PLYP [81] can display similar eigenvalues as the semicanonical OEP2-sc method. By using an IP-optimized potential, the consistent DH potential is emulated without the increased computational cost of an OEP2-sc calculation. This approach leads to an improved performance of B2IP-PLYP for a wide range of thermochemical benchmarks, compared to B2-PLYP.

All of these attempts of IP optimization have in common that they focus on the exchange potential, while the correlation potential in conventional KS-DFT appears to have a lesser influence on the orbital eigenvalues. This is in stark contrast to what is observed in many-body theory (and OEP2-sc), where correlation corrections to Koopmans' IPs can be sizeable and even change the order of orbitals [82]. This underscores the fact that in conventional DFAs, the exchange potential and functionals clearly also contribute to the correlation energy as defined in Eq. 1.3. Replacing DFT exchange with HF exchange, therefore, means losing some long-range correlation effects. This may explain why RSHs like CAM-B3LYP are generally not more accurate than their parent functionals in thermochemical benchmarks. In particular, RSHs are significantly more sensitive to multireference cases. This could possibly be avoided by including some other form of long-range correlation.

In this section, we have seen that KS-DFT can be understood as a correlated single-particle method with an exchange-correlation potential $V_{xc}$ that approximates HF exchange and the self-energy. From this point of view, it is evident that $V_{xc}$ should be chosen such that the eigenvalues correspond to vertical IPs, as they do in the many-body approaches discussed above. This is the case for accurate potentials derived from exact densities or from first principles via the OEP approach. To a degree, acceptable eigenvalues can be achieved with conventional functionals by modifying the exchange potential, in particular by using range separation to admix large fractions of HF exchange at long range. Since DFT exchange potentials are also responsible for correlation, this unfortunately leads to some loss of accuracy in terms of thermochemistry. It, therefore, appears that a better solution would be to also obtain accurate correlation potentials in KS-DFT. These potentials should improve the quality of KS eigenvalues beyond what is possible by modifying the exchange potentials and will hopefully allow avoiding the unsatisfactory trade-off between obtaining good eigenvalues and accurate thermochemistry that is often observed.

## 1.4 Correlation in Semiempirical Methods

Semiempirical MO theory (SEMO) uses simple quantum mechanical expressions that contain adjustable parameters, the idea being that all the physics necessary to qualitatively describe the problem can be included in a simple model. Any neglected effects are approximately treated by adjusting the parameters to reproduce suitable reference data.

The earliest quantum chemical approach of this type is probably Erich Hückel's theory of π-electrons from 1931 [83]. This theory was instrumental in understanding the properties of unsaturated and aromatic compounds and was later extended by Mulliken, Parr, and Coulson [84, 85]. By the 1950s, the model of Pariser and Parr (independently also developed by Pople, hence the acronym PPP) could be applied to the accurate prediction of absorption band energies and intensities [86, 87]. Roald Hoffmann's "extended Hückel theory" from 1963 further included σ-electrons and allowed studying isomerization energies in saturated hydrocarbons and orbital symmetries in a wide variety of chemical reactions [88]. All these early methods were developed to allow the interpretation of experiments, and often directly used experimental data (such as atomic ionization potentials) as parameters.

In parallel, rigorous ab initio MO theory (i.e., HF) was developed by Roothaan, Slater, Löwdin, and others [1, 5, 8, 89]. In this context, Pople proposed several simplified SCF procedures, by analyzing which types of two-electron integrals could be neglected in minimal basis HF without destroying rotational invariance. This led to the theories with complete neglect of differential overlap (CNDO), intermediate neglect of differential overlap (INDO), and neglect of diatomic differential overlap (NDDO) [90–92]. Apart from the neglect of certain integrals, these methods also include further approximations, namely that the orbitals are assumed to be orthonormal (i.e., no Löwdin orthogonalization is performed in the secular equations), and that core electrons are not included in the MO treatment. Due to the harsh approximations made, these SCF procedures were basically useless in an ab initio framework. By introducing semiempirical expressions for the remaining integrals, the errors could be strongly reduced, however. In this manner, close to minimal basis HF results could be recovered at a fraction of the cost.

The hierarchy of methods proposed by Pople still lies at the foundation of modern SEMO. However, it quickly became obvious that minimal basis set HF calculations are not accurate enough for predictive quantum chemistry. This goal can only be rigorously achieved by methods that are converged with respect to the correlation and basis set limits. However, the computational efficiency and parametric flexibility of the SEMO equation remained highly appealing. This led several researchers to explore how accuracies that go beyond the parent minimal basis HF theory could be achieved in a semiempirical framework. This new trend in semiempirical theory was pursued, e.g., by Del Bene and Jaffé [93, 94], Dewar and Zerner [95–98].

Zerner adapted the INDO approach to spectroscopy, with the goal of reproducing experimental excitation energies from CIS calculations. This INDO/S method was much in the spirit of the earlier PPP method, with the benefit of including σ and lone-pair electrons. It was thus able to account for inductive effects and n-π*-type excitations. For many years, this method (distributed in the ZINDO program) was among the only methods available for accurately calculating excitation energies and oscillator strengths in large (bio)molecules, and is still used in applications where computational efficiency is critical (e.g., for very large systems or excited state dynamics).

In order to predict ground state properties of molecules, Dewar proposed a different modification of INDO (MINDO) [99]. The target properties in the parameterization of this method were experimental heats of formation, geometries, dipole moments, and ionization potentials. MINDO/3 (the original MOPAC method) was very attractive for chemical applications at the time, since it offered comparable accuracy to the available ab initio methods (small basis HF) at a fraction of the cost [100]. Subsequently, Dewar and Thiel moved to the more elaborate NDDO approximation, which formed the basis of his successful MNDO method [95, 100, 101]. The MNDO framework has since been further developed and improved by various researchers, resulting in a wide array of methods like AM1 (Dewar et al.) [95], PMx (Stewart) [102–106], MNDO/d (Thiel and Voityuk) [107, 108], OMx (Thiel et al.) [109–113], AM1* (Clark et al.) [114–121], PDDG (Jorgensen et al.) [122], PMO (Truhlar et al.) [123–125], and RM1 (Simas et al.) [126]. Concurrently, similar "general-purpose" methods were developed within the INDO framework, initially by Karl Jug and later by Bredow et al. (MSINDO) [127–129].

Methods like INDO/S and MNDO need to include a wide range of effects to reproduce experimental data. As alluded to above, the most prominent issues relate to the minimal basis set and missing electron correlation, but experimental data further contains thermal, relativistic, non-Born-Oppenheimer and zero-point energy (ZPE) effects. This cluttering of physically distinct phenomena into the parameterization was necessary in the 1970s, but it can be avoided in modern methods by using high-quality ab initio reference data. It was, e.g., shown that NDDO-based "transfer Hamiltonians" can achieve coupled cluster accuracy for specific systems [130–134]. In this case, electron correlation and finite basis set errors are treated implicitly.

Simply speaking, dynamical correlation makes it less likely to find two electrons close to each other. This is often illustrated with simple examples, using the model of helium-like two-electron system [135]. If one electron is close to the core, the other electron will likely be more distant from it (radial correlation). Similarly, if one electron is on one side of the nucleus, the other electron will more likely be on the opposite side (angular correlation). The HF mean-field description is, therefore, too repulsive. The difficulty with describing this effect in an ab initio method is that the exact wavefunction has a cusp at zero interelectronic distance, while approximate wavefunctions are analytically differentiable. To accurately model the cusp, either a large basis set with high angular momentum functions must be used, or the interelectronic distance must explicitly be included in the Hamiltonian (see below) [10, 28, 135–137].

How is this effect incorporated into SEMO methods? All one-center integrals in MNDO-like methods are fixed parameters. They were originally fit to atomic experimental ionization energies by Oleari et al., and partially or fully reparameterized in MNDO and subsequent methods [96, 138]. We recently reported new values for these integrals, based on accurate EOM-CCSD(T) calculations [139].

Oleari's and our integrals are smaller than they would be in an ab initio HF framework. In this way, the over-repulsion of HF is corrected in an average way, leading to accurate results for atomic IPs and EAs.

In the case of molecules, additional two-electron, two-center integrals need to be computed in the NDDO approximation. It would clearly be inconsistent to use the ab initio integrals here, since the two-center integrals should converge to the one-center ones in the zero-distance limit. In MNDO and subsequent methods, the repulsion integrals are computed via a classical multipole approximation, which is accurate at long distances. The integrals are then scaled to achieve the correct one-center limit (Klopman-Ohno scaling) [140, 141]. The two-center integrals are thus smaller than their ab initio counterparts in the short range, but equivalent in the long-range limit. In fact, some modern implementations have abandoned the MNDO multipole approximation in favor of directly using scaled ab initio integrals with Gaussian basis sets [112, 113].

Integral scaling can by no means describe electron correlation fully. For instance, long-range dynamic correlation effects like dispersion (often attributed to simultaneous excitations/charge fluctuations on different molecules) cannot be included in this average manner [142]. There have been some attempts to combine the implicit treatment of SEMO with explicit WFT correlation expressions. The potential danger of this approach is that there can be double counting of correlation contributions by the implicit and explicit expressions.

Danovich combined the outer-valence Green's function (OVGF) method with semiempirical Hamiltonians. This allowed the accurate prediction of ionization potentials in systems of unprecedented size [143–145]. To avoid double counting effects, Thiel reparameterized MNDO using second-order Brillouin-Wigner perturbation theory with Epstein-Nesbet denominators (MNDO/c) [146]. Overall, this treatment brought little advantage compared to regular MNDO for ground state properties, although it proved beneficial for transition states [147]. Unfortunately, the description of dispersion in SEMO remains poor even at the full CI level, due to the limitations of the minimal basis set. In particular, it is not diffuse enough to meaningfully describe the wavefunction between two molecules and contains no high angular momentum functions.

As in DFT, there has been some success with dispersion corrections (e.g., Grimme's D3 method), which use atomic polarizabilities to determine averaged isotropic dispersion coefficients [148]. The dispersion correction can then be evaluated via an atom pairwise potential:

$$E^{D3} = \sum_{AB} \sum_{n=6,8} \frac{C_n^{AB}}{r_{AB}^n} f(r_{AB}). \tag{1.22}$$

Such corrections clearly fit the general spirit of semiempirical approaches. Indeed, methods like AM1 already included additional Gaussian potentials to this end. Unfortunately, this can lead to unexpected problems, as methods like AM1 and

PM3 tend to be far too attractive at very short ranges for some functional groups (a problem that is then made worse by the also attractive dispersion correction) [149].

As a side note, it should be mentioned that non-covalent interactions are generally rather poorly described by semiempirical methods, not just due to missing long-range correlation. Purely electrostatic interactions like hydrogen or halogen bonds are also typically underestimated. Partially, this is because electrostatics are usually not strongly valued in the parameterization procedure. Additionally, there are limits to what is possible within a minimal basis set, as the polarization of hydrogen atoms can play an important role in this type of interaction [150, 151]. A common fix for these problems is the use of additional atom pairwise corrections, e.g., for hydrogen and halogen bonds [149, 152–154]. Unlike for the dispersion corrections, there is no physical motivation for these corrections, however. It should in principle be possible to achieve fairly accurate electrostatics at the SEMO-SCF level. Indeed, the OMx methods appear to be quite promising in this context [111].

For cases where static correlation is critical (i.e., when several determinants have significant weights in the CI expansion), it is necessary to go beyond HF-like calculations. This can, for example, be achieved via UNO-CAS or MR-CI treatments [110, 155, 156]. Static correlation is also significant when describing excited states, in particular (open-shell) singlets [110, 156]. As mentioned above, correlated semiempirical calculations may suffer from double counting of dynamic correlation. This is, however, not critical, due to the small active spaces typically used in the CI calculations.

## 1.5 Basis Set Aspect of a COT

Clearly, in a viable semiempirical theory one has to address both correlation and basis set effects simultaneously, and much of the latter occurs due to the above-mentioned "cusp" effect. In the previous sections, the issue of correlation in effective one-electron theories was addressed. One can also offer a basis set complement. In explicitly correlated CC theory, one inserts the detailed effects of $|r_{12}|^{-1}$ into the wavefunction, today, normally in a Slater germinal form, called F12; the CC wavefunction is then written as

$$\Psi^{F12-CC} = e^{T^B} e^{T} \Phi_0, \tag{1.23}$$

where $e^{T^B}$ accounts for the basis correction. The two-electron amplitudes $T_2^B$ consist of occupied orbital indices, $t(ij,kl)$, unlike those in the usual $T_2$ that involve $t(ij,ab)$ with $a$ and $b$ being virtual orbitals. The $t(ij,kl)$ amplitudes are fixed solely by the known, rigorous cusp conditions [157]. The two operators commute $[T^B, T] = 0$. This enables building a new cusp-corrected Hamiltonian, $H_{cusp}$, that is a component of the usual correlated Hamiltonian of CC theory,

$$\bar{H} = e^{-T} e^{-T^B} H e^{T^B} e^T = e^{-T} H_{cusp} e^T. \tag{1.24}$$

Further, $H_{cusp}$ will be universal, in the sense that the cusp conditions pertain to any molecule. The effect of $H_{cusp}$ will be to modify the one, two, three,... electron integrals that occur in the second-quantized form of $\bar{H}$ to account for short-range correlation. Because this feature occurs at the Hamiltonian level, all properties including analytical gradients should be available from differentiation of the Hamiltonian, a property that so far has resisted treatment in F12-CC implementations. Going through the same steps that led to Eq. 1.13, one now has a basis set corrected COT template for future semiempirical and KS-DFT approximations.

Please note that the main effect of using explicitly correlated wavefunctions is that it reduces the need for high angular momentum basis function to describe the interelectronic cusp. This type of effect can reasonably be expected to be included in a semiempirical fashion (i.e., within an effective Hamiltonian). The basis set still needs to be flexible enough to adequately describe the electron density, however. This should not be expected from a minimal basis set, particularly when no polarization functions are used. An accurate semiempirical implementation of COT will, therefore, likely require a moderately sized basis set of at least split-valence + polarization quality.

## 1.6 Conclusions

This chapter presents an overview on how electron correlation effects can be incorporated into single-particle theories. We have seen that many-body self-energy expressions allow modifying the HF-like equations to obtain "correlated orbitals," both in a frequency-dependent (Dyson) and independent (COT) fashion.

In KS-DFT, the exchange-correlation potential $V_{xc}$ can be seen as a local, frequency-independent approximation to the self-energy $\Sigma$. In this context, it can be shown that KS orbital eigenvalues should correspond to vertical ionization potentials, as they do in Dyson theory and COT. The concepts of exchange and correlation are, however, not as clearly separated in DFT as in WFT, so the direct analogy between $V_{xc}$ and $\Sigma$ should not be taken too literally.

In SEMO, scaling the semiempirical two-electron integrals implicitly treats some dynamic short-range correlation effects. Further correlation effects like static or long-range correlation (dispersion) are missing from this description. They can, however, be included by post-HF treatments or empirical corrections. Since these effects are (to a large extent) distinct, double counting of implicitly and explicitly included correlation is typically not a large problem. Ideally, this problem is avoided by including all corrections and post-HF treatments during the parameterization.

The viewpoint presented in this chapter is that both KS-DFT and SEMO can be understood as approximations to the exact correlated single-particle methods represented by Dyson theory and correlated orbital theory. In this sense, the exchange-correlation potential (in DFT) and the semiempirical integral approximations (in SEMO) can be understood as approximate self-energy expressions. How well the orbital eigenvalues of these approximate methods correspond to vertical ionization potentials is an important yardstick for how good the approximation is [158].

**Acknowledgements** This work was supported by the Alexander von Humboldt Foundation (via a Feodor Lynen Research Fellowship for JTM and a Humboldt Research Award for RJB) and by the United States Air Force Office of Scientific Research (AFOSR, Grant No. FA9550–11-1-0065).

## References

1. J.C. Slater, Phys. Rev. **81**, 385 (1951)
2. J.C. Slater, Phys. Rev. **32**, 339 (1928)
3. V. Fock, Zeitschrift Für Phys. **61**, 126 (1930)
4. D. Hartree, W. Hartree, Proc. R. Soc. A Math. Phys. Eng. Sci. **150**, 9 (1935)
5. C.C.J. Roothaan, Rev. Mod. Phys. **23**, 69 (1951)
6. E. Wigner, Trans. Faraday Soc. **34**, 678 (1938)
7. E. Wigner, Phys. Rev. **46**, 1002 (1934)
8. P.O. Löwdin, Phys. Rev. **97**, 1490 (1955)
9. P.-O. Löwdin, Phys. Rev. **97**, 1474 (1955)
10. L. Kong, F.A. Bischoff, E.F. Valeev, Chem. Rev. **112**, 75 (2012)
11. A.D. Becke, J. Chem. Phys. **119**, 2972 (2003)
12. A.D. Becke, J. Chem. Phys. **88**, 1053 (1988)
13. D. Cremer, Wiley Interdiscip. Rev. Comput. Mol. Sci. **3**, 482 (2013)
14. R.J. Bartlett, J. Phys. Chem. **93**, 1697 (1989)
15. R.J. Bartlett, G.D. Purvis, Phys. Scr. **21**, 255 (1980)
16. G.D. Purvis, R.J. Bartlett, J. Chem. Phys. **76**, 1910 (1982)
17. J. Čížek, J. Chem. Phys. **45**, 4256 (1966)
18. R.S. Mulliken, J. Chem. Phys. **43**, S2 (1965)
19. P. Puschnig, A.D. Boese, M. Willenbockel, M. Meyer, D. Lüftner, E.M. Reinisch, T. Ules, G. Koller, S. Soubatch, M.G. Ramsey, F.S. Tautz, J. Phys. Chem. Lett. 208 (2016)
20. J. Krüger, F. Eisenhut, J.M. Alonso, T. Lehmann, E. Guitián, D. Pérez, D. Skidin, F. Gamaleja, D.A. Ryndyk, C. Joachim, D. Peña, F. Moresco, G. Cuniberti, Chem. Commun. 1 (2017)
21. J.V. Ortiz, Wiley Interdiscip. Rev. Comput. Mol. Sci. **3**, 123 (2013)
22. Y. Öhrn, G. Born, Adv. Quantum Chem. **13**, 1 (1981)
23. L.S. Cederbaum, W. Domck, Adv. Chem. Phys. **36**, 205 (1977)
24. T. Körzdörfer, N. Marom, Phys. Rev. B—Condens. Matter Mater. Phys. **86**, 1 (2012)
25. M. Nooijen, J. Snijders, Int. J. Quantum Chem. **83**, 55 (1992)
26. L. Meissner, R.J. Bartlett, Int. J. Quantum Chem. **48**, 67 (1993)
27. R.J. Bartlett, Chem. Phys. Lett. **484**, 1 (2009)
28. R.J. Bartlett, M. Musiał, Rev. Mod. Phys. **79**, 291 (2007)
29. P. Hohenberg, W. Kohn, Phys. Rev. **136**, B864 (1964)
30. V.V. Karasiev, S.B. Trickey, Comput. Phys. Commun. **183**, 2519 (2012)
31. V.V. Karasiev, R.S. Jones, S.B. Trickey, F.E. Harris, New Dev. Quantum Chem. **661**, 25 (2009)
32. W. Kohn, L.J. Sham, Phys. Rev. **140**, A1133 (1965)

33. N. Schuch, F. Verstraete, Nat. Phys. **5**, 8 (2007)
34. A.D. Becke, J. Chem. Phys. **140** (2014)
35. C. Lee, W. Yang, R.G. Parr, Phys. Rev. B **37**, 785 (1988)
36. W. Kohn, A.D. Becke, R.G. Parr, J. Phys. Chem. **100**, 12974 (1996)
37. W. Yang, A.J. Cohen, P. Mori-Sánchez, J. Chem. Phys. **136**, 204111 (2012)
38. P. Duffy, D.P. Chong, M.E. Casida, D.R. Salahub, Phys. Rev. A **50**, 4707 (1994)
39. O.V. Gritsenko, P.R.T. Schipper, E.J. Baerends, Chem. Phys. Lett. **302**, 199 (1999)
40. Y. Wang, R.G. Parr, Phys. Rev. A **47**, R1591 (1993)
41. D.P. Chong, O.V. Gritsenko, E.J. Baerends, J. Chem. Phys. **116**, 1760 (2002)
42. E.R. Davidson, AIP Conf. Proc. **1504**, 5 (2012)
43. S. Hamel, P. Duffy, M.E. Casida, D.R. Salahub, J. Electron Spectrosc. Relat. Phenom. **123**, 345 (2002)
44. R. Van Meer, O.V. Gritsenko, E.J. Baerends, J. Chem. Theory Comput. **10**, 4432 (2014)
45. J.P. Perdew, J. Sun, R.M. Martin, B. Delley, Int. J. Quantum Chem. **116**, 847 (2016)
46. A. Savin, F. Colonna, R. Pollet, Int. J. Quantum Chem. **93**, 166 (2003)
47. T. Chachiyo, J. Chem. Phys. **145**, 21101 (2016)
48. P.-F. Loos, P.M.W. Gill, Int. J. Quantum Chem. **112**, 1712 (2012)
49. K. Burke, A. Cancio, T. Gould, S. Pittalis, J. Chem. Phys. **145**, 54112 (2016)
50. J.P. Perdew, K. Burke, Int. J. Quantum Chem. **57**, 309 (1996)
51. E. Prodan, W. Kohn, Proc. Natl. Acad. Sci. **102**, 11635 (2005)
52. N.C. Handy, A.J. Cohen, Mol. Phys. **99**, 403 (2001)
53. A.J. Cohen, N.C. Handy, Mol. Phys. **99**, 607 (2001)
54. K. Molawi, A.J. Cohen, N.C. Handy, Int. J. Quantum Chem. **89**, 86 (2002)
55. R. Peverati, D.G. Truhlar, J. Chem. Theory Comput. **8**, 2310 (2012)
56. R.J. Bartlett, V.F. Lotrich, I.V. Schweigert, J. Chem. Phys. **123**, 62205 (2005)
57. O.V. Gritsenko, Ł.M. Mentel, E.J. Baerends, J. Chem. Phys. **144**, 204114 (2016)
58. E.J. Baerends, O.V. Gritsenko, J. Chem. Phys. **123** (2005)
59. P. Verma, A. Perera, R.J. Bartlett, Chem. Phys. Lett. **524**, 10 (2012)
60. N. Oliphant, R.J. Bartlett, J. Chem. Phys. **100**, 6550 (1994)
61. M.C. Kim, E. Sim, K. Burke, Phys. Rev. Lett. **111**, 1 (2013)
62. M.C. Kim, H. Park, S. Son, E. Sim, K. Burke, J. Phys. Chem. Lett. **6**, 3802 (2015)
63. A. Görling, M. Levy, Phys. Rev. A **50**, 196 (1994)
64. S. Ivanov, S. Hirata, R.J. Bartlett, Phys. Rev. Lett. **83**, 5455 (1999)
65. I. Grabowski, V. Lotrich, S. Hirata, Mol. Phys. **108**, 3313 (2010)
66. R.J. Bartlett, I. Grabowski, S. Hirata, S. Ivanov, J. Chem. Phys. **122**, 34104 (2005)
67. S. Hirata, S. Ivanov, I. Grabowski, R.J. Bartlett, K. Burke, J.D. Talman, J. Chem. Phys. **115**, 1635 (2001)
68. P. Verma, R.J. Bartlett, J. Chem. Phys. **137** (2012)
69. H. Iikura, T. Tsuneda, T. Yanai, K. Hirao, J. Chem. Phys. **115**, 3540 (2001)
70. Y. Tawada, T. Tsuneda, S. Yanagisawa, T. Yanai, K. Hirao, J. Chem. Phys. **120**, 8425 (2004)
71. T. Yanai, D.P. Tew, N.C. Handy, Chem. Phys. Lett. **393**, 51 (2004)
72. J.-D. Chai, M. Head-Gordon, Phys. Chem. Chem. Phys. **10**, 6615 (2008)
73. S. Refaely-Abramson, M. Jain, S. Sharifzadeh, J.B. Neaton, L. Kronik, Phys. Rev. B **92**, 1 (2015)
74. S. Refaely-Abramson, S. Sharifzadeh, N. Govind, J. Autschbach, J.B. Neaton, R. Baer, L. Kronik, Phys. Rev. Lett. **109**, 1 (2012)
75. T. Stein, L. Kronik, R. Baer, J. Am. Chem. Soc. **131**, 2818 (2009)
76. T. Körzdörfer, R.M. Parrish, N. Marom, J.S. Sears, C.D. Sherrill, J.-L. Brédas, Phys. Rev. B **86**, 205110 (2012)
77. P. Verma, R.J. Bartlett, J. Chem. Phys. **140** (2014)
78. P. Verma, R.J. Bartlett, J. Chem. Phys. **145** (2016)
79. Y. Jin, R.J. Bartlett, J. Chem. Phys. **145**, 34107 (2016)
80. J.T. Margraf, P. Verma, R.J. Bartlett, J. Chem. Phys. **145**, 104106 (2016)

81. S. Grimme, J. Chem. Phys. **124**, 34108 (2006)
82. L.S. Cederbaum, Chem. Phys. Lett. **25**, 562 (1974)
83. E. Hückel, Zeit. Phys. **70**, 204 (1931)
84. R.G. Parr, R.S. Mulliken, J. Chem. Phys. **18**, 1338 (1950)
85. C.A. Coulson, D.P. Craig, J. Jacobs, Proc. R. Soc. A **206**, 297 (1951)
86. R. Pariser, R.G. Parr, J. Chem. Phys. **21**, 466 (1953)
87. J.A. Pople, Trans. Faraday Soc. **49**, 1375 (1953)
88. R. Hoffmann, J. Chem. Phys. **39**, 1397 (1963)
89. P.-O. Löwdin, J. Chem. Phys. **18**, 365 (1950)
90. J.A. Pople, D.P. Santry, G.A. Segal, J. Chem. Phys. **43**, S129 (1965)
91. J.A. Pople, G.A. Segal, J. Chem. Phys. **43**, S136 (1965)
92. J.A. Pople, J. Chem. Phys. **47**, 2026 (1967)
93. J. Del Bene, H.H. Jaffé, J. Chem. Phys. **48**, 4050 (1968)
94. J. Del Bene, H.H. Jaffé, J. Chem. Phys. **48**, 1807 (1968)
95. M.J.S. Dewar, E.G. Zoebisch, E.F. Healy, J.J.P. Stewart, J. Am. Chem. Soc. **107**, 3902 (1985)
96. M.J.S. Dewar, D.H. Lo, J. Am. Chem. Soc. **94**, 5296 (1972)
97. M. Dewar, W. Thiel, J. Am. Chem. Soc. **99**, 4899 (1977)
98. J. Ridley, M. Zerner, Theor. Chim. Acta **32**, 111 (1973)
99. R.C. Bingham, M.J.S. Dewar, D.H. Lo, J. Am. Chem. Soc. **97**, 1285 (1975)
100. M.J.S. Dewar, J. Mol. Struct. **100**, 41 (1983)
101. M. Dewar, W. Thiel, J. Am. Chem. Soc. **99**, 4907 (1977)
102. J.J.P. Stewart, J. Mol. Model. **13**, 1173 (2007)
103. J.J.P. Stewart, J. Mol. Model. **19**, 1 (2013)
104. J.J.P. Stewart, J. Comput. Chem. **10**, 209 (1989)
105. J.J.P. Stewart, J. Mol. Model. **10**, 155 (2004)
106. J.J.P. Stewart, J. Comput. Chem. **11**, 543 (1990)
107. W. Thiel, A.A. Voityuk, Theor. Chim. Acta **81**, 391 (1992)
108. W. Thiel, A. Voityuk, J. Phys. Chem. **100**, 616 (1996)
109. P.O. Dral, X. Wu, L. Spörkel, A. Koslowski, W. Thiel, J. Chem. Theory Comput. **12**, 1097 (2016)
110. D. Tuna, Y. Lu, A. Koslowski, W. Thiel, J. Chem. Theory Comput. **12**, 4400 (2016)
111. T. Tuttle, W. Thiel, Phys. Chem. Chem. Phys. **10**, 2159 (2008)
112. P.O. Dral, X. Wu, L. Spörkel, A. Koslowski, W. Weber, R. Steiger, M. Scholten, W. Thiel, J. Chem. Theory Comput. **12**, 1082 (2016)
113. M. Kolb, W. Thiel, J. Comput. Chem. **14**, 775 (1993)
114. P. Winget, A.C. Horn, C. Selçuki, B. Martin, T. Clark, J. Mol. Model. **9**, 408 (2003)
115. H. Kayi, T. Clark, J. Mol. Model. **16**, 29 (2010)
116. H. Kayi, T. Clark, J. Mol. Model. **13**, 965 (2007)
117. H. Kayi, T. Clark, J. Mol. Model. **17**, 2585 (2011)
118. P. Winget, T. Clark, J. Mol. Model. **11**, 439 (2005)
119. H. Kayi, T. Clark, J. Mol. Model. **16**, 1109 (2010)
120. H. Kayi, T. Clark, J. Mol. Model. **15**, 295 (2009)
121. H. Kayi, T. Clark, J. Mol. Model. **15**, 1253 (2009)
122. M.P. Repasky, J. Chandrasekhar, W.L. Jorgensen, J. Comput. Chem. **23**, 1601 (2002)
123. L. Fiedler, J. Gao, D.G. Truhlar, J. Chem. Theory Comput. **7**, 852 (2011)
124. M. Isegawa, L. Fiedler, H.R. Leverentz, Y. Wang, S. Nachimuthu, J. Gao, D.G. Truhlar, J. Chem. Theory Comput. **9**, 33 (2013)
125. P. Zhang, L. Fiedler, H.R. Leverentz, D.G. Truhlar, J. Gao, J. Chem. Theory Comput. **8**, 2983 (2012)
126. G.B. Rocha, R.O. Freire, A.M. Simas, J.J.P. Stewart, J. Comput. Chem. **27**, 1101 (2006)
127. T. Bredow, G. Geudtner, K. Jug, J. Comput. Chem. **22**, 861 (2001)
128. B. Ahlswede, K. Jug, J. Comput. Chem. **20**, 572 (1999)
129. B. Ahlswede, K. Jug, J. Comput. Chem. **20**, 563 (1998)

130. R.J. Bartlett, J. McClellan, J.C. Greer, S. Monaghan, J. Comput. Mater. Des. **13**, 89 (2006)
131. R. Bartlett, C. De Taylor, A. Korkin, in *Handbook of Materials Modeling SE - 4*, ed. by S. Yip (Springer, Netherlands, 2005), pp. 27–57
132. C.E. Taylor, M.G. Cory, R.J. Bartlett, W. Thiel, Comput. Mater. Sci. **27**, 204 (2003)
133. D.E. Taylor, K. Runge, R.J. Bartlett, Mol. Phys. **103**, 2019 (2005)
134. J.J. McClellan, T.F. Hughes, R.J. Bartlett, Int. J. Quantum Chem. **105**, 914 (2005)
135. D.P. Tew, W. Klopper, T. Helgaker, J. Comput. Chem. **28**, 1307 (2007)
136. F. Neese, E.F. Valeev, J. Chem. Theory Comput. **7**, 33 (2011)
137. M.B. Ruiz, J.T. Margraf, A.M. Frolov, Phys. Rev. A **88**, 1 (2013)
138. L. Oleari, L. Di Sipio, G. De Michelis, Mol. Phys. **10**, 97 (1966)
139. J.T. Margraf, D. Claudino, R.J. Bartlett, Mol. Phys. **8976**, 1 (2016)
140. K. Ohno, Theor. Chim. Acta **2**, 219 (1964)
141. G. Klopman, J. Am. Chem. Soc. **88**, 4550 (1964)
142. S. Grimme, Wiley Interdiscip. Rev. Comput. Mol. Sci. **1**, 211 (2011)
143. D.K. Danovich, V.K. Turchaninov, V.G. Zakrzewski, J. Mol. Struct. THEOCHEM **209**, 77 (1990)
144. D. Danovich, V. Zakrzewski, E. Domnina, J. Mol. Struct. THEOCHEM **188**, 159 (1989)
145. D. Danovich, V. Zakrzewski, E. Domnina, J. Mol. Struct. THEOCHEM **187**, 297 (1989)
146. W. Thiel, J. Am. Chem. Soc. **103**, 1413 (1981)
147. S. Schroder, W. Thiel, J. Am. Chem. Soc. **108**, 7985 (1986)
148. S. Grimme, J. Antony, S. Ehrlich, H. Krieg, J. Chem. Phys. **132**, 0 (2010)
149. J. Řezáč, P. Hobza, J. Chem. Theory Comput. **8**, 141 (2012)
150. P. Winget, C. Selçuki, A.H.C. Horn, B. Martin, T. Clark, Theor. Chem. Acc. **110**, 254 (2003)
151. B. Martin, T. Clark, Int. J. Quantum Chem. **106**, 1208 (2006)
152. J. Řezáč, P. Hobza, Chem. Phys. Lett. **506**, 286 (2011)
153. N.D. Yilmazer, M. Korth, Comput. Struct. Biotechnol. J. **13**, 169 (2015)
154. M. Korth, M. Pitonak, J. Rezac, P. Hobza, J. Chem. Theory Comput. **6**, 344 (2010)
155. P.O. Dral, T. Clark, J. Phys. Chem. A **115**, 11303 (2011)
156. M.R. Silva-Junior, W. Thiel, J. Chem. Theory Comput. **6**, 1546 (2010)
157. D. Bokhan, S. Ten-no, J. Noga, Phys. Chem. Chem. Phys. **10**, 3320 (2008)
158. R.J. Bartlett, D.S. Ranasinghe, Chem. Phys. Lett. (2017)

# Chapter 2
# Circular Dichroism Spectroscopy with the SAC-CI Methodology: A ChiraSac Study

**Tomoo Miyahara and Hiroshi Nakatsuji**

**Abstract** Circular dichroism (CD) spectroscopy reflects sensitively the various chemistries involved in chiral molecules and molecular systems. The CD spectra are very sensitive to the conformational changes of molecules: the rotation around the single bond including a chiral atom. It is also sensitive to the changes in the stacking interactions in the chiral DNA and RNA. Since these changes are low-energy processes, we expect that the CD spectra include a lot of information of the chiral molecular systems. On the other hand, the SAC-CI method is a highly reliable excited-state theory and gives very reliable theoretical CD spectra. Therefore, by comparing the experimental CD spectra with the theoretical SAC-CI spectra calculated for various chemical situations, one can study various chemistries such as the nature of the weak interactions involved in chiral molecular systems and biology. Based on these facts, we are developing a new molecular technology called ChiraSac, a term combining chirality and SAC-CI, to study chiral molecular systems and the chemistry involved thereof. We utilize highly reliable SAC-CI method together with many useful quantum chemical methods involved in Gaussian suite of programs. In this chapter, we review our ChiraSac studies carried out to clarify the chemistries of some chiral molecules and molecular systems: large dependences of the CD spectra on the conformations of several chiral molecules in solutions and the effects of the stacking interactions of the nuclear-acid bases in DNA and RNA on the shapes of their CD spectra. The results of our several studies show that the ChiraSac is a useful tool for studying the detailed chemistry involved in chiral molecular and biological systems.

**Keywords** Chirality · Circular dichroism spectra · SAC-CI theory · Sensitivity on low-energy processes · Conformational dependence · Stacking interaction · DNA · RNA

---

T. Miyahara · H Nakatsuji (✉)
Quantum Chemistry Research Institute, Kyoto Technoscience Center 16,
14 Yoshida-Kawara-Machi, Sakyou-ku, Kyoto 606-8305, Japan
e-mail: h.nakatsuji@qcri.or.jp

 
M.J. Wójcik et al. (eds.), *Frontiers of Quantum Chemistry*,
https://doi.org/10.1007/978-981-10-5651-2_2

## 2.1 Introduction

The symmetry adapted cluster (SAC)/SAC—configuration interaction (SAC-CI) [1–6] method proposed by Nakatsuji and Hirao in 1978 is one of the most reliable excited-state theories. Figure 2.1 explains the subjects of the SAC/SAC-CI program available in Gaussian program. SAC is an accurate coupled-cluster theory for ground state, and SAC-CI is the theory for the excited, ionized, and electron-attached states produced from the ground state calculated by the SAC theory. The energy gradient method that calculates the forces acting on the nuclei of molecules facilitates the studies of the molecular equilibrium geometries and dynamics among various electronic states [7–13]. The SAC/SAC-CI code was released in 2003 through Gaussian suite of programs [14] and has been applied to various chemistry and physics involving various electronic sates, ground and excited states from singlet to septet spin multiplicities as shown in Fig. 2.1 [15–20].

In this review, we focus on the application of the SAC/SAC-CI theory and program to the chiral world of chemistry and biology. Chirality is caused by an asymmetric carbon atom that is bonded with four different atoms or groups as shown in Fig. 2.2. In the chiral molecule of Fig. 2.2, since all bonds are single bond and then their rotations around the bond are low-energy processes, the groups ($R_1$ to $R_4$) can rotate easily by an external stimulus. Namely, even a weak interaction causes a different conformation. As such weak interaction, solute-solvent interaction, hydrogen-bonding interaction and/or stacking interaction in biological system, etc., are very interesting subject.

The circular dichroism (CD) spectroscopy [21, 22] is one of the most powerful techniques to study chemistry caused by the chirality. Furthermore, the CD spectra are very sensitive to the conformational changes of the chiral molecules [23–35]. Therefore, the CD spectra are used not only for distinguishing the absolute chiral geometries of amino acids, sugars, etc., but also for understanding the nature of the

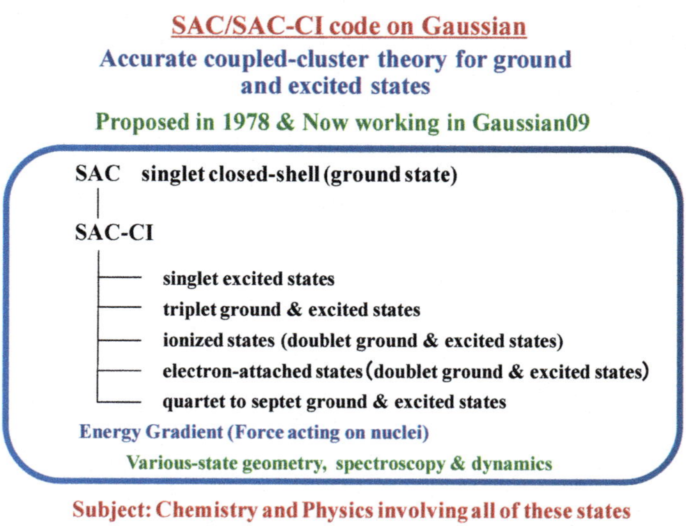


**Fig. 2.1** Overview of the SAC-CI program in Gaussian

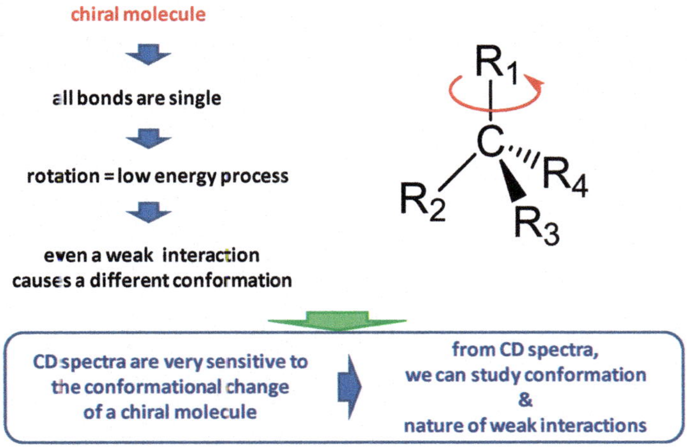


**Fig. 2.2** Chiral molecule, conformation and weak interactions

interactions that leads to the observed structures and conformations of the chiral molecules in solution or proteins. Though the UV spectra change only slightly by the change in conformations and/or weak interactions, the CD spectra sometimes change drastically with such changes. Therefore, from the CD spectra, we can obtain much information concerning the nature of weak interactions that are involved in the observed structures and conformations.

Recently, we have applied the SAC/SAC-CI method to the analysis of the CD spectra of various molecules and systems [36–44]. It was confirmed that the CD spectra are very sensitive to the rotations around the single bond of the chiral atom [40–42] and to the weak interactions involved in the chiral DNA and RNA helical structures [40, 43, 44]. Thus, through the CD spectra, we can study the natures of the weak interactions involved in these subjects. However, for such studies, we need a highly reliable theory like the SAC-CI theory to analyze the information involved in the CD spectra. We have confirmed that the SAC-CI method can provide reliable theoretical CD spectra of molecules and molecular systems [36–44]. Based on this fact, we are developing a theoretical methodology, ChiraSac [41–44], to study chemistry involved in chiral molecular systems. Figure 2.3 is an overview of the ChiraSac project. There, we use the SAC-CI method as a central reliable methodology, as well as other useful methods included in the Gaussian suite of program [14] to study the nature of chemistry provided by the CD spectral information on the structure, conformation, interaction with protein environment, solvent effect, etc.

In this review, we introduce some recent studies of the CD spectroscopy with the SAC-CI method. The computational details are described in the reference of each study. First, we discuss the gauge-origin independent method required for the CD spectral calculations [38]. Second, we discuss the conformational dependence of the CD spectra on the rotation around the single bonds involving the chiral atom: the molecule we deal with here is α-hydroxyphenylacetic acid (HPAA) [41] and deoxyguanosine [40]. Third, we discuss the CD spectral change with the substituent

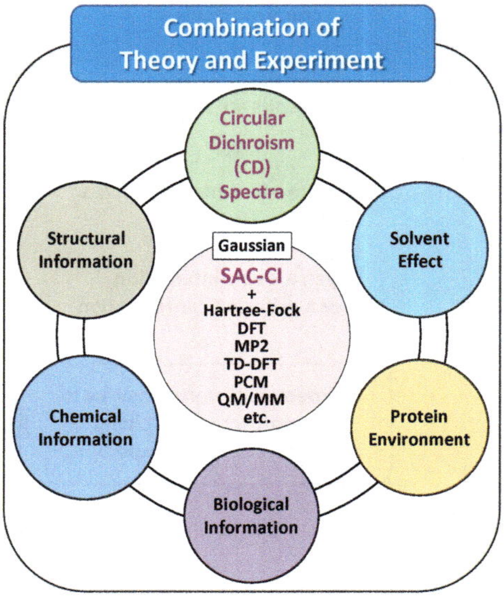


**Fig. 2.3** Overview of a ChiraSac project. (Reprinted from Ref. [42] with permission from *The Journal of Physical Chemistry A* **2014**, *118*, 2931–2941. Copyright 2014 American Chemical Society.)

change for the uridine derivatives [42]. Finally, we discuss the weak interactions involved in the CD spectra of the double-helical DNA and RNA [40, 43, 44].

## 2.2 Rotatory Strength

The intensity of the CD spectra is calculated as the rotatory strength ($R_{0a}$). It is the imaginary part of the scalar product of the electric and magnetic transition moments between the ground state ($\Psi_0$) and the excited state ($\Psi_a$) [45]

$$R_{0a} = \mathrm{Im}\{\langle\Psi_0|\hat{\mu}|\Psi_a\rangle\langle\Psi_a|\hat{m}|\Psi_0\rangle\} \quad (2.1)$$

where $\hat{\mu}$ is an electric dipole moment operator and $\hat{m}$ is a magnetic dipole moment operator. This formula is called the length-gauge expression. The electric transition dipole moment is gauge-origin independent if the ground and excited-state wave functions are orthogonal to each other. On the other hand, the magnetic transition dipole moment is gauge-origin dependent. Using the Hypervirial theorem [46] given by

$$\langle a|\nabla|0\rangle = (E_a - E_0)\langle a|\hat{\mu}|0\rangle, \quad (2.2)$$

Equation (2.1) can be readily transformed to

$$R_{0a} = \mathrm{Im}\left\{\frac{\langle\Psi_0|\nabla|\Psi_a\rangle\langle\Psi_a|\hat{m}|\Psi_0\rangle}{E_a - E_0}\right\}. \quad (2.3)$$

called a velocity form [47], in which the gauge-origin dependent terms cancel each other. Therefore, Eq. (2.3) is gauge-origin independent.

## 2.3 Gauge-Origin Dependency

The gauge-origin dependency of the rotatory strength was studied for R-methyloxirane calculated by both length and velocity forms (Eqs. (2.1) and (2.3)) [38]. As shown in Table 2.1, when the gauge-origin is at the center of gravity (Z = 0.0 Å calculation), the difference between the length and velocity forms is very small. However, when the gauge-origin is 100.0 Å away from the center of gravity (Z = 100.0 Å calculation), the results of the velocity form are much different from those of the length form. For the 1, 3, and $7^1$A excited states, the sign as well as the intensity is different between the two forms for the Z = 100.0 Å calculation. However, for the velocity form, the results of the Z = 100.0 Å calculation are completely the same as those of the Z = 0.0 Å calculation. Therefore, in this review, we will give the results of only the velocity form, because the results of the length form are gauge-origin dependent.

## 2.4 Dependence of the CD Spectra on the Conformation of HPAA

As described above, the CD spectra of chiral molecules are sensitive to their conformational changes. Since the conformational change is usually a low-energy process, it is difficult to determine its conformation from only the experimental CD

**Table 2.1** Gauge-origin dependency of the rotational strength of R-methyloxirane calculated by the length and velocity forms. (Ref. [38])

| State | EE (eV) | Osc (au) | Rot.(Z = 0.0 Å) | | Rot.(Z = 100.0 Å) | | Expt.[a] | |
|---|---|---|---|---|---|---|---|---|
| | | | Length ($10^{-40}$ cgs) | Velocity ($10^{-40}$ cgs) | Length ($10^{-40}$ cgs) | Velocity ($10^{-40}$ cgs) | EE (eV) | RS ($10^{-40}$ cgs) |
| $1^1$A | 7.02 | 0.0087 | −10.2528 | −10.9735 | 3.8350 | −10.9735 | 7.08 | −12.5 |
| $2^1$A | 7.42 | 0.0100 | −0.8958 | −1.0157 | −2.3484 | −1.0157 | | |
| $3^1$A | 7.48 | 0.0158 | 5.3549 | 5.2376 | −32.0254 | 5.2376 | | |
| $4^1$A | 7.67 | 0.0112 | 1.8980 | 2.1076 | 18.0443 | 2.1076 | 7.70 | +5.9 |
| $5^1$A | 7.85 | 0.0043 | 6.6438 | 7.7420 | 0.8239 | 7.7420 | | |
| $6^1$A | 8.39 | 0.0047 | −0.8264 | −1.8091 | 2.7451 | −1.8091 | | |
| $7^1$A | 8.44 | 0.0293 | −9.9709 | −8.9756 | 56.5058 | −8.9756 | 8.35 | −4.1 |
| $8^1$A | 8.52 | 0.0020 | −4.6979 | −4.9544 | −23.3448 | −4.9544 | | |
| $9^1$A | 8.64 | 0.0060 | 1.2102 | 1.3491 | 3.6837 | 1.3491 | | |

[a]Ref. [48]

spectrum. However, when we have a reliable theory, we can calculate the CD spectra against the conformational change and we can say the conformation of the molecule from the CD spectra alone by comparing the theoretical CD spectrum with the experimental one. Of course, we can also calculate the stable conformation from the energetic point of view. In this section, we show the conformational dependence of the CD spectra of α-hydroxyphenylacetic acid (HPAA) on the rotations around the single bonds that are low-energy processes [41].

In HPAA, carboxyl (COOH), hydroxyl (OH), and phenyl ($C_6H_5$) groups are bound to the chiral carbon atom (marked with * in Fig. 2.4) by single bonds. We calculated the potential energy curves for each rotation as a function of the dihedral angles, Δ1, Δ2, or Δ3 taking the optimized angles as 0 degree and the results are shown in Fig. 2.4. At low temperature, only the phenyl rotation is important. However, since the energy barriers are small, the hydroxyl and carboxyl groups can also rotate at high temperature.

Figures 2.5, 2.6 and 2.7 show the SAC-CI CD spectra of HPAA at several conformational angles of the phenyl, COOH, and OH rotations, compared with the experimental CD spectrum (black line) observed at 27 °C [49]. For the CD spectra of HPAA, the lowest excited state is the excitation of HOMO to LUMO (π-π*) of the phenyl group corresponding to the peak at 260 nm and the second excited state is the excitation of the non-bonding orbital to the π* orbital of the COOH group corresponding to the peak at 220 nm.

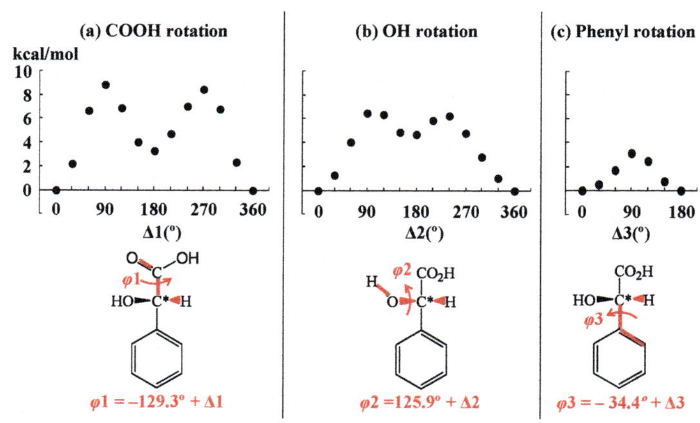


**Fig. 2.4** Potential energy curves for the **a** COOH, **b** OH, and **c** phenyl rotations of α-hydroxyphenylacetic acid, as a function of the dihedral angle change (Δ1, Δ2, Δ3) from the fully optimized dihedral angles $\varphi 1 = -129.3$, $\varphi 2 = 125.9$, and $\varphi 3 = -34.4°$. The definition of each dihedral angle is shown by the red lines in each structural formula. (Reprinted from Ref. [41] with permission from *The Journal of Physical Chemistry A* **2013**, *117*, 14065–74. Copyright 2013 American Chemical Society.)

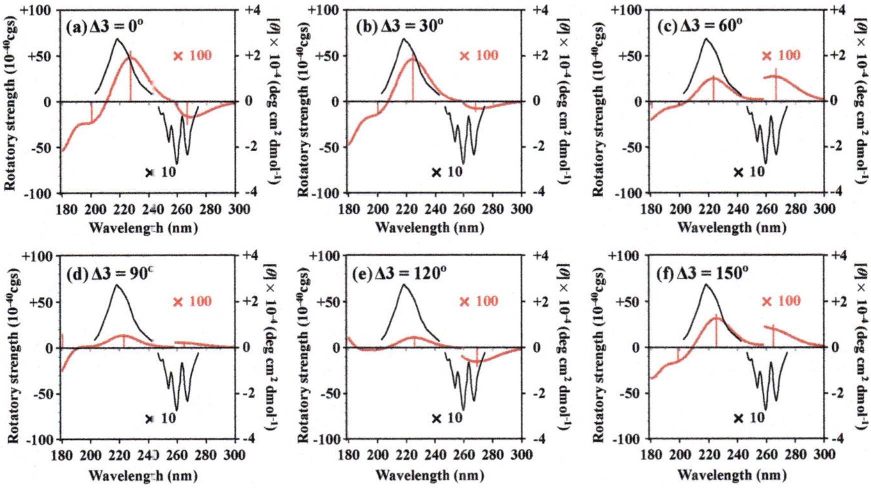

**Fig. 2.5** SAC-CI CD spectra (*red lines*) of α-hydroxyphenylacetic acid at several conformational angles of the phenyl rotation, compared with the experimental CD spectrum (*black line* [49]) (Reprinted from Ref. [41] with permission from *The Journal of Physical Chemistry A* **2013**, *117*, 14065–74. Copyright 2013 American Chemical Society.)

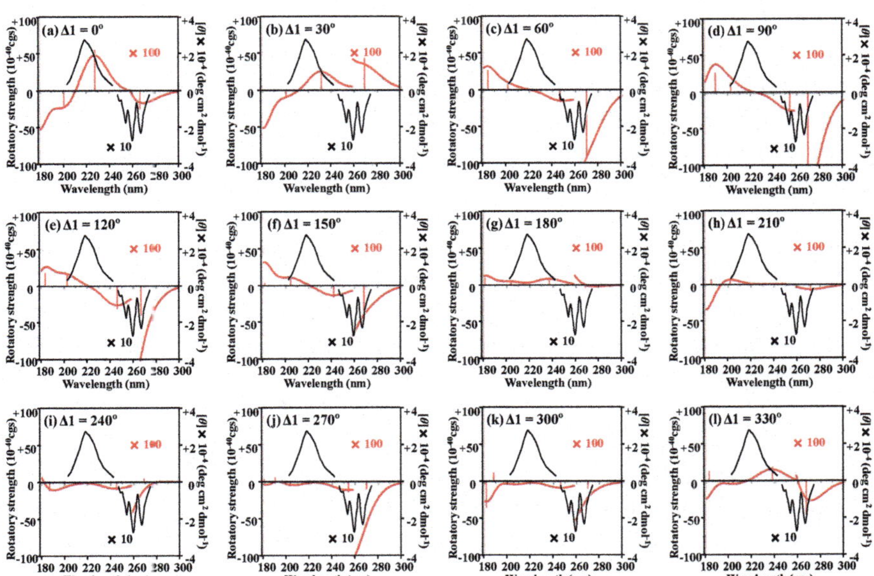

**Fig. 2.6** SAC-CI CD spectra (*red lines*) of α-hydroxyphenylacetic acid at several conformational angles of the carboxyl (COOH) rotation, compared with the experimental CD spectrum (*black line* [49]) (Reprinted from Ref. [41] with permission from *The Journal of Physical Chemistry A* **2013**, *117*, 14065–74. Copyright 2013 American Chemical Society.)

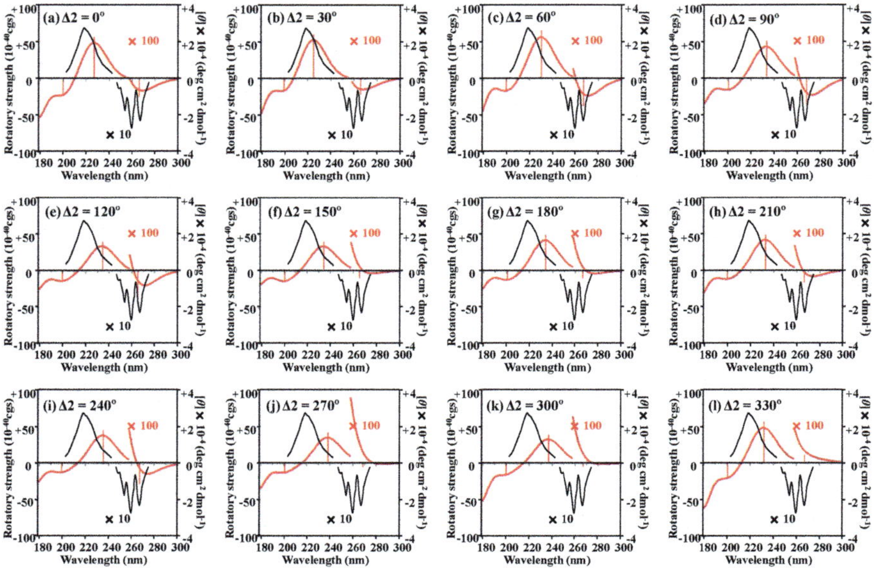

**Fig. 2.7** SAC-CI CD spectra (*red lines*) of α-hydroxyphenylacetic acid at several conformational angles of the hydroxyl (OH) rotation, compared with the experimental CD spectrum (*black line* [49]) (Reprinted from Ref. [41] with permission from *The Journal of Physical Chemistry A* **2013**, *117*, 14065–74. Copyright 2013 American Chemical Society.)

For the Phenyl rotation (Fig. 2.5), the first band (phenyl π-π*) largely depends on the phenyl rotation because of the π-π* excitation of the phenyl group. However, the second band is positive for all conformations because of the n-π* excitation of the COOH group.

For the COOH rotation (Fig. 2.6), the first band is negative except for $\Delta1 = 30°$. However, the sign and intensity of the second band (n-π* of COOH) largely change by the COOH rotation due to the n-π* excitation of the COOH group.

For the OH rotation (Fig. 2.7), the SAC-CI CD spectra are similar to each other among all conformations, because the excitation of the OH group is not included in the 200–300 nm region.

Finally, we calculated the Boltzmann averaged SAC-CI CD spectra of HPAA at several different temperatures shown in Fig. 2.8, considering that the energy barriers around the single bond rotations are relatively low. The SAC-CI CD spectrum at 27 °C (purple line) is similar to the SAC-CI CD spectrum of the most stable conformer (Figs. 2.5a, 2.6a and 2.7a). But, as the temperature increases, the intensity decreases in all regions due to the contribution of the unstable conformers.

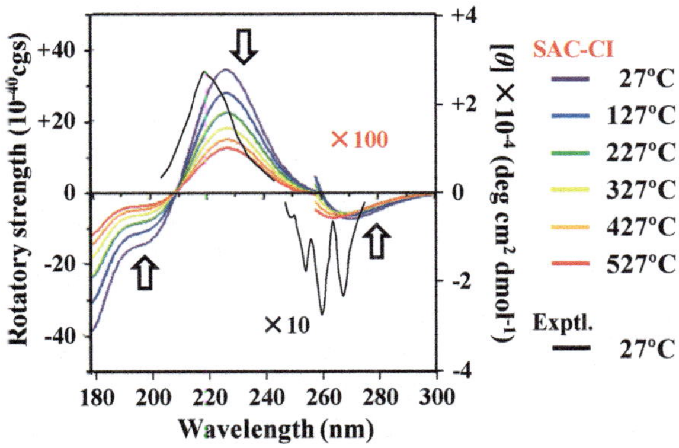


**Fig. 2.8** Boltzmann averaged SAC-CI CD spectra of α-hydroxyphenylacetic acid at several different temperatures, compared with the experimental CD spectrum [49]. *Arrows* indicates the directions of the changes caused by raising the temperature (Reprinted from Ref. [41] with permission from *The Journal of Physical Chemistry A* **2013**, *117*, 14065–74. Copyright 2013 American Chemical Society.)

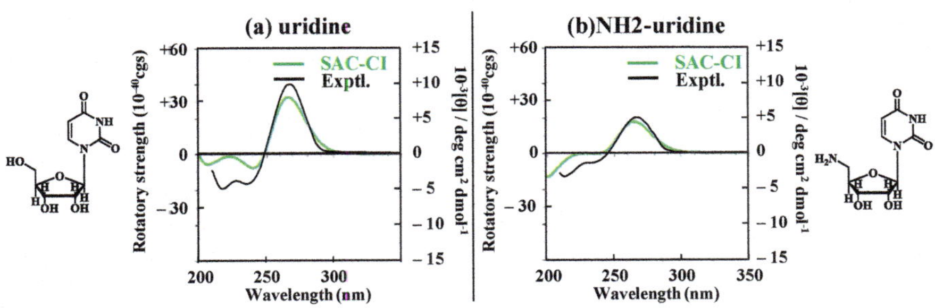


**Fig. 2.9** Experimental CD spectra (*black lines*) of uridine and NH2-uridine compared with the SAC-CI CD Boltzmann average spectra (*green lines*) (Reprinted from Ref. [42] with permission from *The Journal of Physical Chemistry A* **2014**, *118*, 2931–2941. Copyright 2014 American Chemical Society.)

## 2.5 Substituent Effect of Uridine Derivatives

Next, we study the CD spectra of uridine and NH2-uridine shown in Fig. 2.9. NH2-uridine has an NH2 group at the C5' position of the ribose. The intensity of the experimental CD spectrum is strong for uridine but weak for NH2-uridine [50, 51]. The Boltzmann averaged SAC-CI CD spectra of uridine and NH2-uridine are compared with the experimental spectra [50] in Fig. 2.9. The SAC-CI CD spectra successfully reproduced the difference between uridine and NH2-uridine.

The rotatory strength ($R_{0a}$) of the CD spectra is expressed using the angle $\theta$ between the electric transition dipole moment (ETDM, $\mu$) and the magnetic transition dipole moment (MTDM, $m$) as

$$R_{0a} = \text{Im}[|\vec{\mu}_{0a}||\vec{m}_{0a}|\cos\theta] \quad (2.4)$$

In the most stable anti-conformation, the dihedral angle between uracil and ribose was calculated to be 173.9° for uridine and 176.7° for NH2-uridine. Although the dihedral angle difference is very small, the intensity of the CD spectra is different between uridine and NH2-uridine. The first excited state corresponding to the peak at 267 nm is the n-π* excitation of uracil. However, this state also includes the π-π* excitation of uracil. The ETDM depends on the ratio of the mixing of the π-π* excitation. Since the mixing of the π-π* excitation with the n-π* excitation of uracil is large for uridine but small for NH2-uridine, the ETDM is larger for uridine (0.067) than for NH2-Uridine (0.016). This is the main reason that the intensity of the first peak is weaker for NH2-uridine than for uridine.

For the second excited state of the π-π* excitation that corresponds to the peak at 235 nm, the angle $\theta$ of NH2-uridine is closer to 90° than that of uridine as shown in Fig. 2.10. Therefore, from Eq. (2.4), the intensity of NH2-uridine is weaker than that of uridine for the second peak of the CD spectra.

The experimental CD spectra of the uridine derivatives studied here are mainly of the intramolecular excitation nature within the nucleic-acid bases. Therefore, the

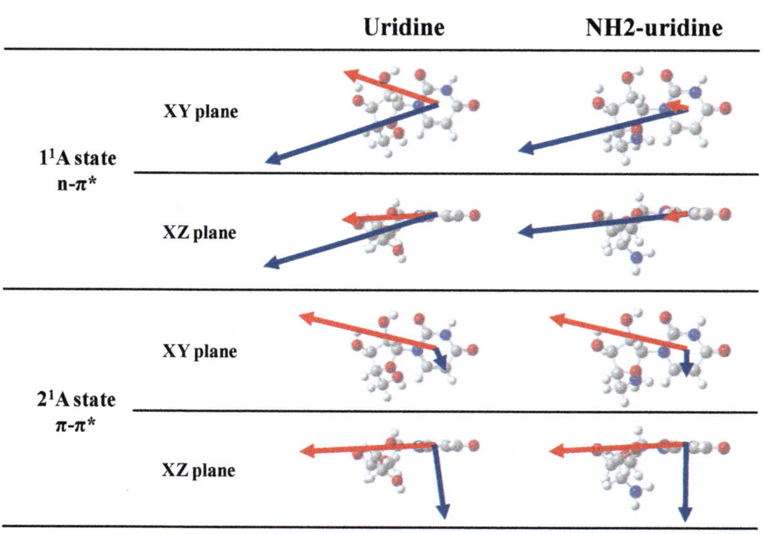


**Fig. 2.10** Direction of ETDM (*red arrows*) and MTDM (*blue arrows*) of the 1 and $2^1$A excited states of the anti-conformers of uridine and NH2-uridine. The unit of arrows of the $2^1$A states is 2.5 times larger than that of the $1^1$A states. Uracil is on the XY plane (Reprinted from Ref. [42] with permission from *The Journal of Physical Chemistry A* **2014**, *118*, 2931–2941. Copyright 2014 American Chemical Society.)

excited states of NH2-uridine are almost the same as those of uridine. However, the change of OH to NH2 leads to the difference of the dihedral angle of the optimized geometry, which weakens the intensity of the CD spectra. The differences are due to the magnitude of the ETDM for the first peak and due to the angle $\theta$ for the second peak.

## 2.6 Conformational Dependence of the CD Spectra of Deoxyguanosine

Deoxyguanosine (dG) is an important nucleic-acid base that composes DNA. The single bond between guanine and deoxyribose in dG can rotate almost freely: the energy barrier for the rotation is very low. Nevertheless, the CD spectrum is very sensitive to this rotational angle. In this section, we study the conformational dependence of the UV and CD spectra of dG. The SAC-CI spectra are given in red lines, and the experimental spectra measured in water solution are shown in black lines.

First, we studied the potential curve for the rotation around this single bond. Figure 2.11 shows the calculated potential curve for the rotation around this single bond by using the DFT method with B3LYP/6-31G(d,p). Other geometrical parameters were optimized at each rotational angle $\varphi$. The anti-dG was calculated to be more stable than the syn-dG by 0.47 kcal/mol, a very small difference. Since the energy barriers between the syn- and anti-dG are only 3–6 kcal/mol, the dG can rotate easily between the syn- and anti-conformations.

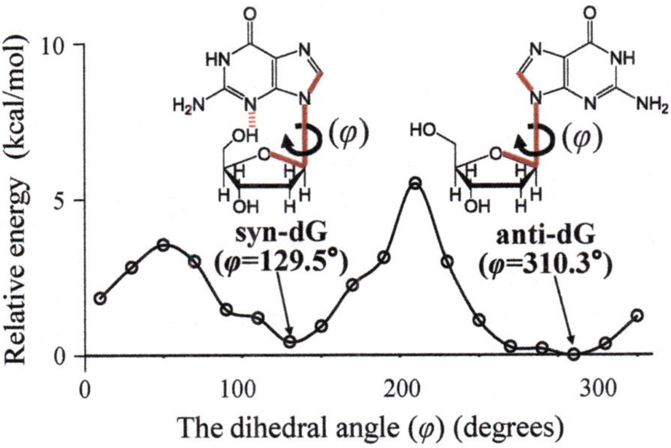


**Fig. 2.11** Potential energy curve of the ground state of dG determined as dihedral angle $\varphi$ is varied in increments of 20°. Other geometrical parameters were optimized at each $\varphi$ (Reprinted from Ref. [40] with permission from *The Journal of Physical Chemistry A* **2013**, *117*, 42–55. Copyright 2013 American Chemical Society.)

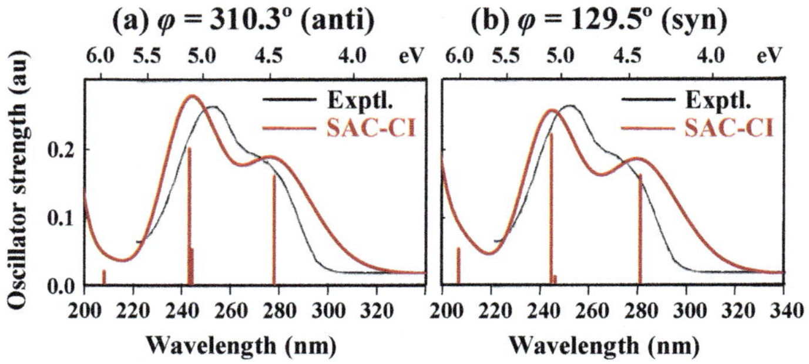


**Fig. 2.12** SAC-CI UV spectra of **a** anti- and **b** syn-dG (*red*), compared with the experimentally determined UV spectrum of dG (*black* [52]) (Reprinted from Ref. [40] with permission from *The Journal of Physical Chemistry A* **2013**, *117*, 42–55. Copyright 2013 American Chemical Society.)

We first examined the UV spectra of the syn- and anti-conformers. Figure 2.12 shows the SAC-CI UV spectra (red line) of both anti- and syn-dG compared with the experimental (black) UV spectrum of dG. The SAC-CI UV spectra of both anti- and syn-dG are similar to the experimental UV spectrum. Therefore, it is difficult to determine whether the experimental UV spectrum of dG is due to the anti or the syn conformation.

Next, we calculated the CD spectra during the rotation around the syn- and anti-conformers. Figure 2.13 shows the SAC-CI CD spectra of dG at several dihedral angles $\varphi$ between deoxyribose and guanine. For the experimental CD spectrum of dG (black line), the first band at 276 nm is very weak, the second band at 250 nm is strong negative, and the third band at 214 nm is strong positive. The experimental CD spectrum is in good agreement with the SAC-CI CD spectra of the anti-dG ($\varphi$ = 310.3°) and the dG with $\varphi$ = 330° (red lines of Fig. 2.13p and q). The Boltzmann averaged SAC-CI CD spectrum shown by the green line in Fig. 2.13s is also close to the experimental spectrum. However, the other SAC-CI CD spectra shown by red lines are different from the black experimental CD spectrum. Therefore, we concluded that the dG exists as an anti-conformation in water solution based on the comparisons with the experimental and SAC-CI CD spectra. Note that the same conclusion is also drawn from the theoretical potential energy curve of Fig. 2.11, if we can neglect the solute–solvent interactions.

Thus, the CD spectra largely depend on the rotation around the single bond though it is a very low-energy process. We can determine the conformation of the molecule in solution by comparing the SAC-CI CD spectra with the experimental CD spectrum.

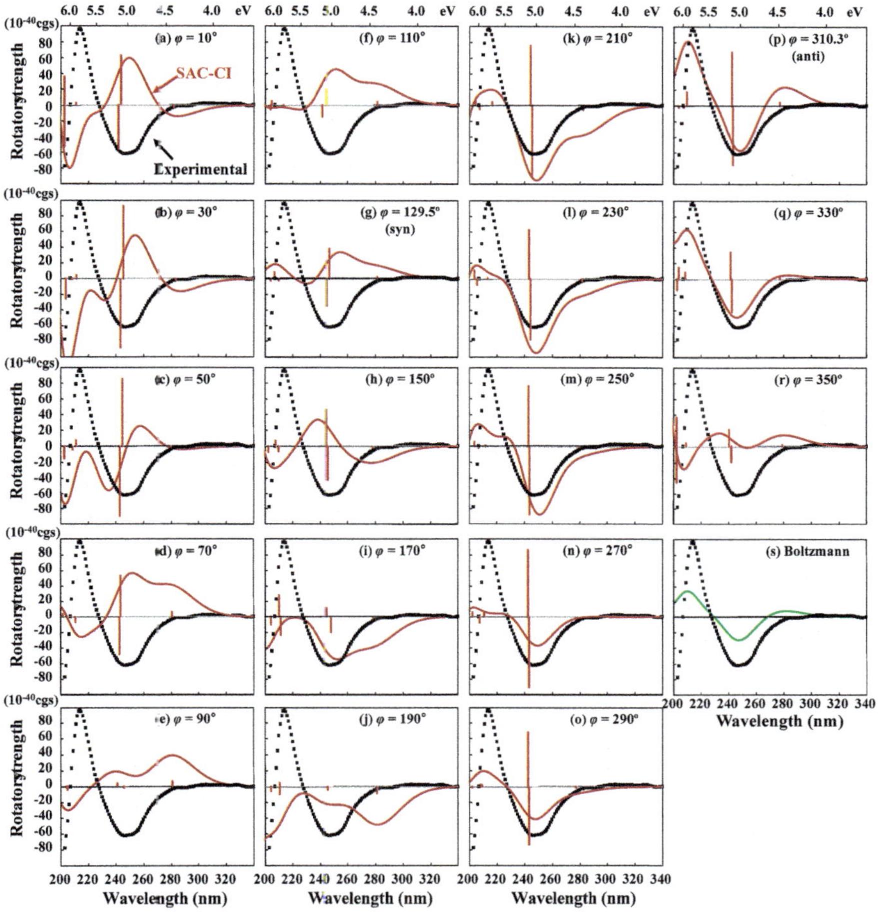


**Fig. 2.13** CD spectra of dG at several conformation angles $\varphi$. The experimental CD spectrum [52] (*black line*) of dG is compared with the SAC–CI CD spectra (*red lines*) of **a** $\varphi = 10°$, **b** $\varphi = 30°$, **c** $\varphi = 50°$, **d** $\varphi = 70°$, **e** $\varphi = 90°$, **f** $\varphi = 110°$, **g** $\varphi = 129.5°$ (syn), **h** $\varphi = 150°$, **i** $\varphi = 170°$, **j** $\varphi = 190°$, **k** $\varphi = 210°$, **l** $\varphi = 230°$, **m** $\varphi = 250°$, **n** $\varphi = 270°$, **o** $\varphi = 290°$, **p** $\varphi = 310.3°$ (anti), **q** $\varphi = 330°$ and **r** $\varphi = 350°$. The Boltzmann averaged SAC–CI CD spectrum (*green line*, (s)) of all conformers is also shown (Reprinted from Ref. [40] with permission from *The Journal of Physical Chemistry A* **2013**, *117*, 42–55. Copyright 2013 American Chemical Society.)

## 2.7 Double-Helical Structure of DNA and the Nature of Weak Interactions Involved

A well-known B-DNA has a right-handed double-helical structure, while Z-DNA has a left-handed double-helical structure. When DNA has a special sequence, the transition from B- to Z-DNA or from Z- to B-DNA is induced by the changes in salt concentration or temperature [53–58]. The left-handed Z-DNA is preferred at low temperature or at high salt concentration, but the right-handed B-DNA is preferred at the reverse conditions [53–58]. The CD spectrum of Z-DNA is much different from that of B-DNA, while the UV spectra between B- and Z-DNA are similar [54–58]. The main difference lies in the sign at around 295 nm of the CD spectra. It is positive for B-DNA but negative for Z-DNA. Therefore, the CD spectroscopy can be used to identify the helical structure of DNA or RNA in solution. However, only a few theoretical studies have investigated why the sign of the CD spectrum of DNA changes by the type of its helical structure. We studied the SAC-CI CD spectra of DNA to elucidate the origin of the negative peak at around 295 nm of the CD spectrum of Z-DNA [40].

We considered three factors to be important that affect the CD spectra of DNA. They are the conformation of monomer, the hydrogen-bonding interactions, and the stacking interactions between the nucleic-acid bases. Figure 2.14 shows the computational models taken from the X-ray crystallographic structures of the B- and Z-DNA, in which deoxyguanosine (dG) and deoxycytidine (dC) are arranged alternately. Their existence ratio is unity. We calculated the monomer, dimer, and tetramer models. The effect of the conformation of dG and dC was verified by the monomer model. The hydrogen-bonding and stacking interactions were verified by the dimer and tetramer models. The dimer models include either the hydrogen-bonding or stacking interactions, but the tetramer models include both the hydrogen-bonding and stacking interactions.

### *2.7.1 Monomer Model*

The dC has the anti-conformation in both B- and Z-DNA. However, the dG has the anti-conformation in B-DNA but syn conformation in Z-DNA. We compare the SAC-CI CD spectra of the anti- and syn-dG shown in Fig. 2.13 with the experimental CD spectra of B- and Z-DNA shown in Fig. 2.15 that was reported by Tran-Dinh et al. [54]. There, we notice that, in the second band region (near 250 nm), the SAC-CI CD spectra of anti- and syn-dG are similar to the experimental CD spectra of B- and Z-DNA, respectively.

Based on this observation, we considered the monomer model in which we examine whether the CD spectra of B- and Z-DNA could be understood from the CD spectra of monomers taking the geometries found in the B- and Z-DNA. Therefore, we calculated the SAC-CI CD spectra of the dG and dC monomers using

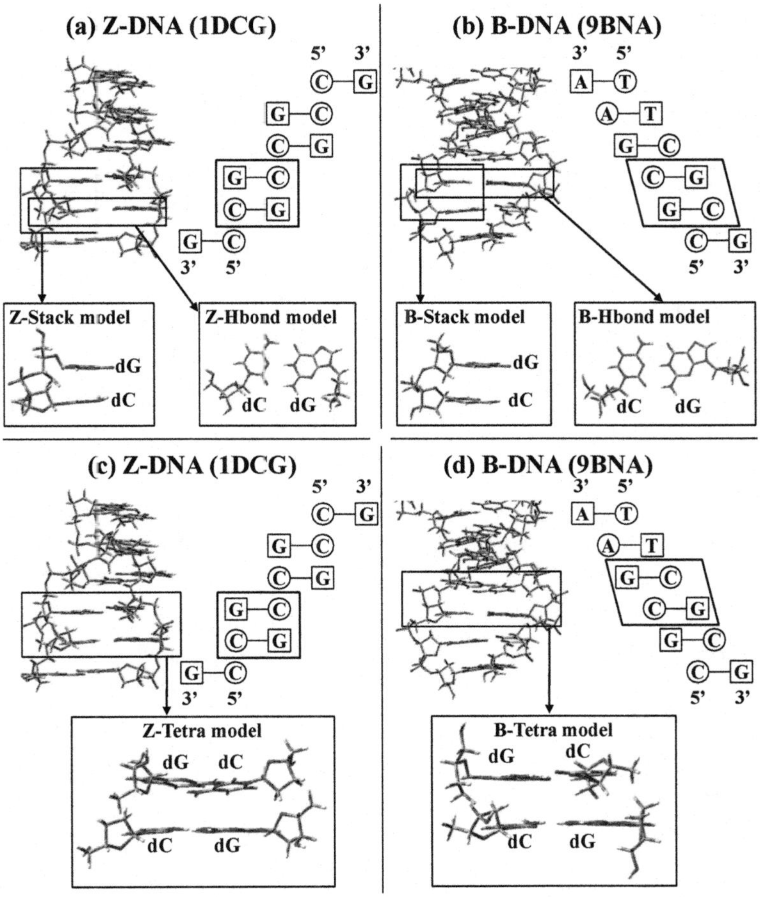


**Fig. 2.14** Computational models. **a** Hydrogen-bonding (Z-Hbond) and stacking (Z-Stack) models taken from the X-ray crystallography structure (1DCG) of Z-DNA. **b** Hydrogen-bonding (B-Hbond) and stacking (B-Stack) models taken from the X-ray crystallography structure (9BNA) of B-DNA. **c** Tetramer (Z-Tetra) model taken from the X-ray crystallography structure (1DCG) of Z-DNA. **d** Tetramer (B-Tetra) model taken from the X-ray crystallography structure (9BNA) of B-DNA (Reprinted from Ref. [40] with permission from *The Journal of Physical Chemistry A* **2013**, *117*, 42–55. Copyright 2013 American Chemical Society.)

their geometries in the X-ray crystallographic structures of the B- and Z-DNA and the results were summarized in Fig. 2.15. In this figure, the composite CD spectra are the sums of the SAC-CI CD spectra of dG and dC in Z-DNA and B-DNA.

Very roughly speaking, for B-DNA, the composite SAC-CI CD spectrum shown in a magenta color is similar to the experimental CD spectrum shown in a black color. However, for Z-DNA, the SAC-CI composite spectrum in a magenta color does not explain a large negative peak near 290 nm of the experimental spectrum

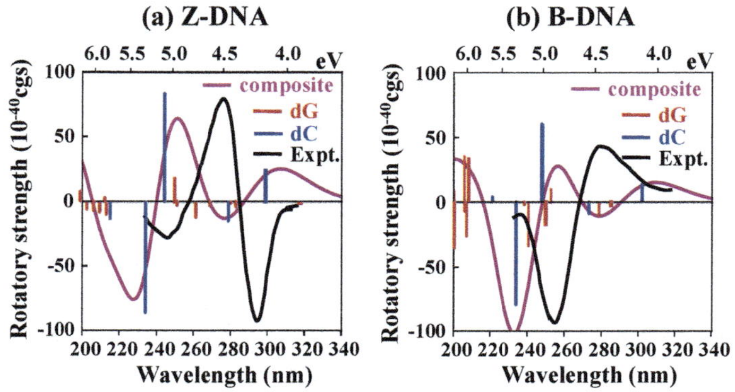


**Fig. 2.15** SAC-CI CD spectra of the composite CD spectra of **a** Z-DNA and **b** B-DNA obtained from the SAC-CI CD spectra of dG and dC, compared with the experimental CD spectra [54] (*black lines*) of Z-DNA and B-DNA. The *red* and *blue lines* represent the excited state of dG and dC, respectively. The magenta line represents the composite spectra that are the sum of the CD spectra of dG and dC (Reprinted from Ref. [40] with permission from *The Journal of Physical Chemistry A* **2013**, *117*, 42–55. Copyright 2013 American Chemical Society.)

shown in a black color. Therefore, we must say that the monomer model alone is insufficient to explain the difference in the experimental CD spectra between B- and Z-DNA. More refined models are necessary, and we explain them below.

## 2.7.2 Dimer Model

We now calculate the SAC-CI CD spectra with the dimer models that include either the hydrogen-bonding interaction or the stacking interaction between dG and dC as shown in Fig. 2.14a and b, respectively. Most importantly, the monomer model could not account for the negative peak at 295 nm. Figure 2.16 shows the SAC-CI CD spectra obtained with the dimer models, compared with the experimental CD spectra of B- and Z-DNA. In the experimental CD spectrum given with the black line, the first peak is negative at 296 nm and the second peak is positive at 275 nm for Z-DNA, while the first peak is positive at 280 nm and the second peak is negative at 255 nm for B-DNA. On the other hand, among the four SAC-CI CD spectra, the sign at 295 nm is negative with only the Z-Stack model. Since the dG strongly stacks with the dC in only the Z-Stack model as shown in Fig. 2.14a, the stacking interaction between dG and dC can account for the strong negative peak at 295 nm of the CD spectra in Z-DNA. This CD peak is due to the $1^1$A excited state from the HOMO to the LUMO in dC. The stacking interaction with dG changes the sign of the rotatory strength of the $1^1$A excited state from the positive of the monomer model to the negative of the Z-stack model. However, since the stacking interaction of B-DNA is weaker than that of Z-DNA as shown in Fig. 2.14b, such

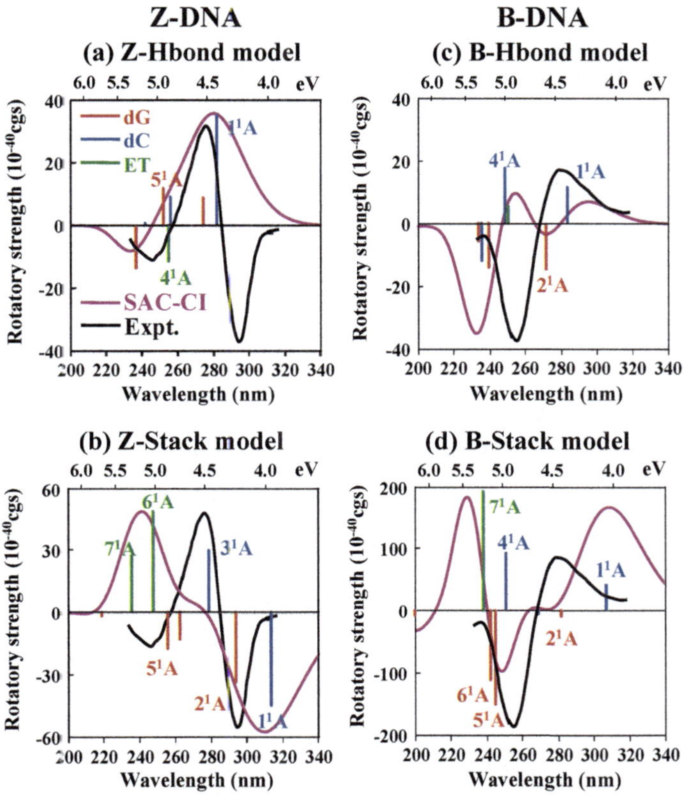


**Fig. 2.16** SAC-CI CD spectra for the dimer models. The SAC-CI spectra (*magenta line*) of **a** Z-Hbond model, **b** Z-Stack model, **c** B-Hbond model and **d** B-Stack model are compared with the experimental CD spectra [54] (*black line*) of Z-DNA (**a**, **b**) and B-DNA (**c**, **d**). The *red* and *blue lines* represent the intramolecular excited states of dG and dC, respectively. The *green lines* represent the electron transfer (ET) excited state from dG to dC (Reprinted from Ref. [40] with permission from *The Journal of Physical Chemistry A* **2013**, *117*, 42–55. Copyright 2013 American Chemical Society.)

change in the sign does not occur near 295 nm so that the negative peak does not appear at around 295 nm for the B-Stack model. Therefore, we can conclude that the negative peak at 295 nm is originated from the stacking interaction between the nucleic-acid bases.

In the hydrogen-bonding models (Z-Hbond and B-Hbond), the excitation energy of the first excited state was calculated to be at a higher energy region than those of the monomer models and the stacking dimer models. However, the rotatory strength is positive for both hydrogen-bonding models. Therefore, we concluded that the hydrogen-bonding interaction affects the excitation energies of the lowest excited states but are not the origin of the negative peak at around 295 nm of the CD spectra.

### 2.7.3 Tetramer Model

Finally, we calculated the SAC-CI CD spectra with the tetramer models that include both of the hydrogen-bonding and stacking interactions. Figure 2.17 shows the

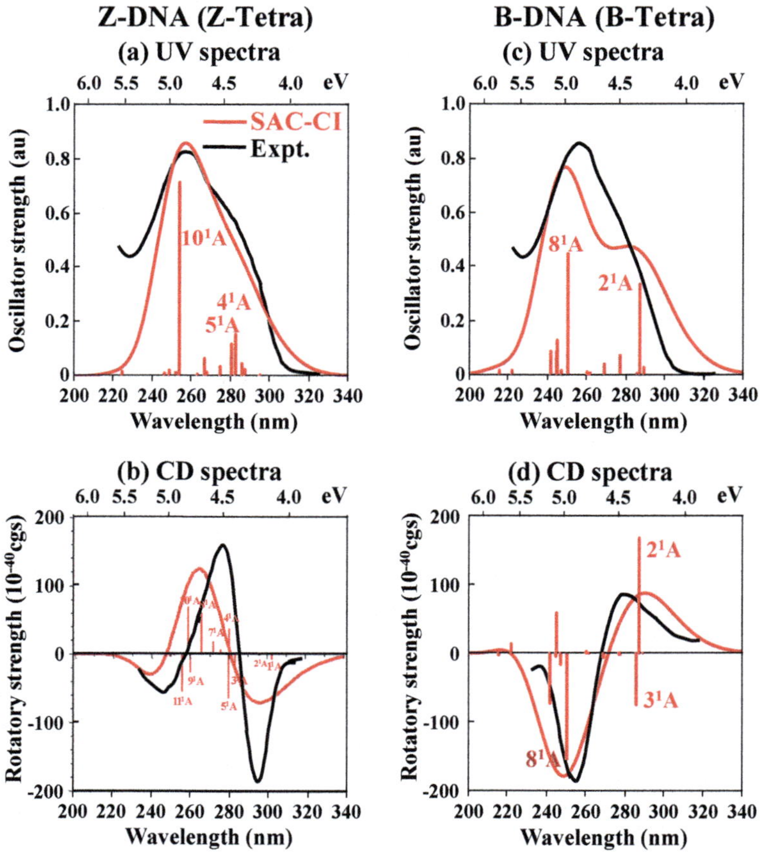


**Fig. 2.17** SAC-CI UV and CD spectra for the tetramer models. The SAC-CI spectra (*red line*) of Z-Tetra model (**a**, **b**) and B-Tetra model (**c**, **d**) compared with the experimental UV and CD spectra [54] (*black line*) of Z-DNA (**a**, **b**) and B-DNA (**c**, **d**). All excited states of SAC-CI calculations have been shifted to the lower values by 0.5 eV (Refs. [40, 43]). (Reprinted from Refs. [40] and [43] with permission from *The Journal of Physical Chemistry A* **2013**, *117*, 42–55, **2015**, *119*, 8269–8278. Copyright 2013, 2015 American Chemical Society.)

SAC-CI UV and CD spectra of the tetramer models compared with the experimental UV and CD spectra of the B- and Z-DNA. The SAC-CI UV spectrum of the Z-Tetra model reproduced the main and shoulder peaks of the experimental UV spectrum of Z-DNA. However, for the B-Tetra model, the shoulder peak is slightly different from the experiment, because the angle between the two bases in the hydrogen-bonding base pair is larger for the B-Tetra model than that of the Z-Tetra model [44].

The SAC-CI CD spectrum of the B-Tetra model was in good agreement with the experimental CD spectrum of B-DNA for the sign as well as the intensity. The SAC-CI CD spectrum of the Z-Tetra model also reproduced the general features of the experimental CD spectrum of Z-DNA.

The lowest excited state of the Z-Tetra model is the electron-transfer (ET) excited state from dG to dC through the stacking conformation, but for the B-Tetra model it is the intramolecular excitation within dC. Thus, the ET excited state is the origin of the strong negative peak at 295 nm of Z-DNA. We concluded that the hydrogen-bonding interaction changes the excitation energy of the lowest excited state and that the stacking interaction is the origin of the negative peak at 295 nm of the CD spectra.

## 2.8 CD Spectra Is an Indicator of the Stacking Interaction in Double-Helical DNA

As noted in the above section, the negative peak at 295 nm is caused by the stacking interaction between the two base pairs of dG and dC. To verify the relationship between the stacking interaction and the intensity of the peak at 295 nm of the CD spectra of DNA, we calculated the SAC-CI CD spectra for the different distances using the tetramer model of Z-DNA as shown in Fig. 2.18.

Figure 2.19 shows the SAC-CI CD spectra for the different distance R between the two base pairs using the tetramer models of Z-DNA. The experimental CD spectra of the Z- and B-DNA are shown with the solid and dotted black lines, respectively. The negative peak at 295 nm has the strongest intensity for $\Delta R = 0.0$ Å (X-ray crystallographic structure). The intensity of the negative peak at 295 nm becomes weaker and weaker as the distance between the two base pairs ($\Delta R$) increases. Note that the intensity at 295 nm is weaker for $\Delta R = -0.2$ Å than for $\Delta R = \pm 0.0$ Å.

In addition, as the distance $\Delta R$ increases, the SAC-CI CD spectra come close to the experimental CD spectrum of B-DNA (dotted black line) from that of the Z-DNA (solid black line). Although the conformations of dG and dC are much

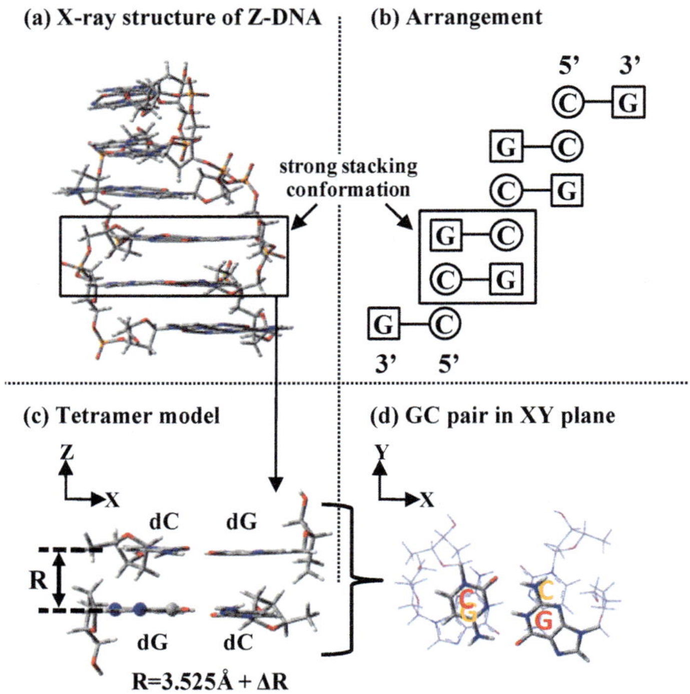


**Fig. 2.18** **a** X-ray crystallography structure of Z-DNA (PDB ID: 1DCG). **b** Arrangement of cytosine (C) and guanine (G). **c** Geometry of tetramer model. R represents the distance between the two base pairs. R is 3.525 Å from the X-ray crystallography structure. **d** Geometry of two GC (dG and dC) pairs. Three atoms indicated by "Ball" are arranged on the XY plane (Reprinted from Ref. [43] with permission from *The Journal of Physical Chemistry A* **2015**, *119*, 8269–8278. Copyright 2013 American Chemical Society.)

different from the structure of B-DNA, the SAC-CI CD spectrum of the $\Delta R = +2.0$ Å model is in good agreement with the experimental CD spectrum of B-DNA due to the weak stacking interaction.

The negative peak at 295 nm is caused by the electron-transfer (ET) excited state from dG to dC through the stacking interaction at around $\Delta R = \pm 0.0$ Å, but when the distance between dG and dC increases, the excitations at around 295 nm become intra dG and dC excitations for around $\Delta R = +2.0$ Å. The ET excited state is calculated to be at higher energy region for $\Delta R = +2.0$ Å. Since the two base pairs strongly stack for Z-DNA but weakly stack for B-DNA, the negative peak at 295 nm becomes the indicator of the stacking interaction between the two base pairs.

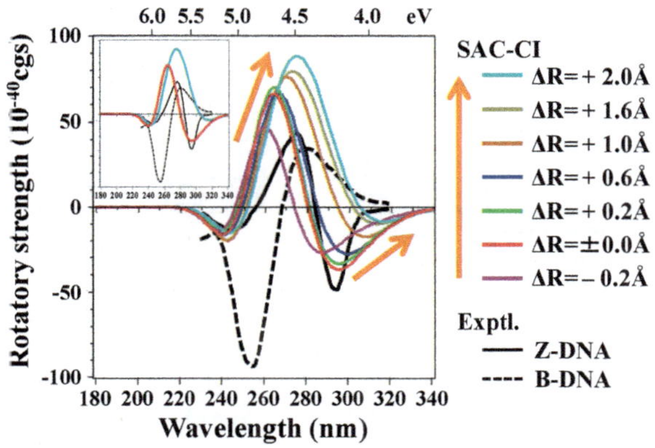


**Fig. 2.19** SAC-CI CD spectra (*colored lines*) for the different distance R between the two base pairs using the tetramer models of Z-DNA compared with the experimental CD spectra of Z- and B-DNA (*black lines*) [54]. All excited states of the SAC-CI calculation have been shifted to the lower values by 0.5 eV. The *magenta*, *red*, *green*, *blue*, *gold*, *khaki* and *cyan lines* represent the SAC-CI CD spectra with ΔR = −0.2, ± 0.0, +0.2, +0.6, +1.0, +1.6 and +2.0 Å. The black solid and dotted lines represent the experimental CD spectra of Z- and B-DNA, respectively. The orange arrows represent the direction of the SAC-CI CD spectral change in the increase of the distance R. The *inset* shows the two representative SAC-CI spectra with ΔR = ±0.0 Å and ΔR = +2.0 Å compared with the experimental CD spectra of Z-and B-DNA (Reprinted from Ref. [43] with permission from *The Journal of Physical Chemistry A* **2015**, *119*, 8269–8278. Copyright 2013 American Chemical Society.)

## 2.9 Similarities and Differences Between Double-Helical DNA and RNA

RNA (ribonucleic acid) can also form both the right- and left-handed double-helical structures. The transition between the right-handed A-RNA and the left-handed Z-RNA can be induced by the changes by salt concentration or temperature [59–61]. These phenomena are quite similar to those of DNA. However, the feature of the CD spectra of RNA is quite different from that of DNA. For the CD spectra of the right-handed A-RNA, the first band at 295 nm has a negative sign and the second band has a positive sign [59–61], which is similar to the feature of the left-handed Z-DNA. On the other hand, the features of the CD spectra of the left-handed Z-RNA are similar to those of the right-handed B-DNA. In addition, the right-handed A-RNA is preferred at low temperature [59–61], similarly to the left-handed Z-DNA. Namely, the features of the CD spectra of RNA are opposite to those of DNA.

We calculated the SAC-CI CD spectra of DNA and RNA to clarify the relationship between the helical structures and their CD spectra, and to elucidate the similarities and differences between DNA and RNA. The geometries were taken from the X-ray crystallographic structures and shown in Fig. 2.20. These structures are composed of 12 nucleic-acid bases and have the same sequence in which dG

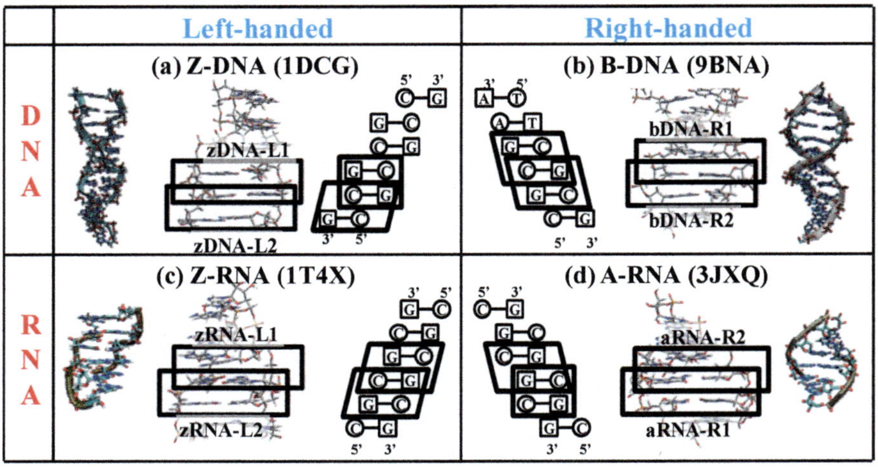


**Fig. 2.20** Tetramer models (zDNA-L1, zDNA-L2, bDNA-R1, bDNA-R2, zRNA-L1, zRNA-L2, aRNA-R1 and aRNA-R2) taken from the X-ray crystallographic structures of **a** Z-DNA (1DCG), **b** B-DNA (9BNA), **c** Z-RNA (1T4X) and **d** A-RNA (3JXQ). "G" and "C" represent guanine and cytosine, respectively (Reprinted from Ref. [44] with permission from *The Journal of Physical Chemistry A* **2016**, *120*, 9008–9018. Copyright 2016 American Chemical Society.)

and dC for DNA and guanosine (G) and cytidine (C) for RNA are arranged alternately. We calculated the CD spectra of two tetramer models for each DNA and RNA, because each DNA or RNA includes two kinds of the stacking interactions as seen from Fig. 2.20.

We show in Fig. 2.21 the SAC-CI CD spectra of the tetramer models of Z-DNA, B-DNA, Z-RNA and A-RNA, compared with their experimental CD spectra measured in water solution [58, 61]. The SAC-CI CD spectra seemed to be not in good agreement with the experimental CD spectra at first glance. However, the SAC-CI CD spectra of the zDNA-L1 and aRNA-R1 models were in good correspondence with the experimental CD spectra of Z-DNA and A-RNA, respectively. The negative peak at 295 nm corresponds to the lowest two excited states that are the electron transfer (ET) excitations from guanine to cytosine through the stacking interaction. But the SAC-CI CD spectra of the zDNA-L2 and aRNA-R2 models were opposite to those of the zDNA-L1 and aRNA-R1 models. Even though we used the tetramer models taken from the same helical structures, the negative peak did not appear at 295 nm in the CD spectra due to the weak stacking interaction (Fig. 2.20a, d). Namely, the negative peak observed at 295 nm is due to the ET excitation from guanine to cytosine through the strong stacking interaction in both DNA and RNA: The peak at 295 nm can be considered to be the indicator of the strong stacking interaction for the double-helical structures of DNA as well as RNA.

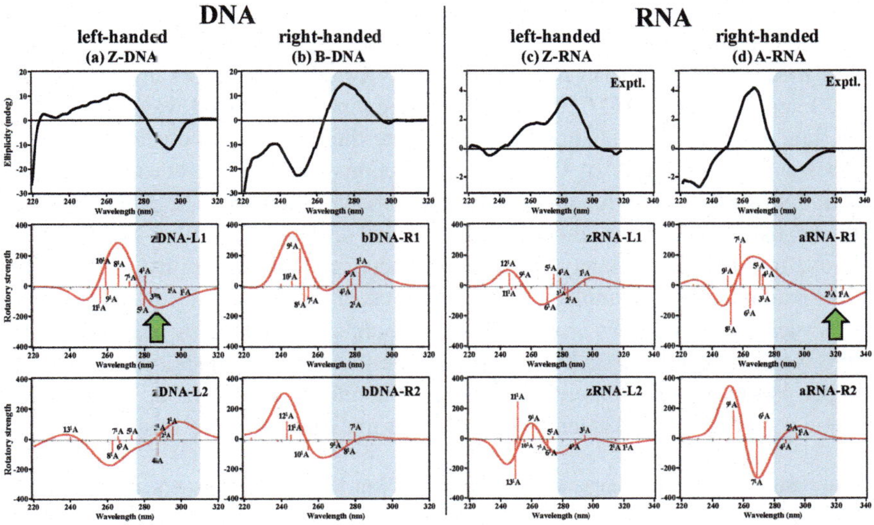


**Fig. 2.21** SAC-CI CD spectra (*red lines*) of the eight tetramer models compared with the experimental CD spectra (*black lines*) [58, 61]. All excited states of SAC-CI calculations have been shifted to the lower sides by the values of 0.5 eV for DNA and 0.7 eV for RNA, respectively. *Blue-colored squares* represent the first band region of the CD spectra of each DNA or RNA. The negative peaks at 295 nm are shown by *green arrows* (Reprinted from Ref. [44] with permission from *The Journal of Physical Chemistry A* **2016**, *120*, 9008–9018. Copyright 2016 American Chemical Society.)

## 2.10 Conclusion

The ChiraSac method is a useful tool for analyzing and clarifying the chemical information included in the CD spectra of chiral molecules and molecular systems. The CD spectra depend strongly on the conformational changes due to the rotations around the single bonds. Since this rotation is a low-energy process, it may reflect the natures of the low-energy molecular interactions with solvents and with the molecular environments. With HPAA (α-hydroxyphenylacetic acid) and dG (Deoxyguanosine), we have shown that the calculations of the CD spectra with a reliable theory like SAC-CI are useful to clarify the rotational conformation of a molecule in solution, which may be different from the stable conformation in a gas phase. Since the calculated CD spectrum is the result of the cancellations of many different peaks of positive and negative intensities, the theory must be reliable enough in both peak positions and intensities (rotatory strengths). Generally speaking, the TD-DFT calculations are not reliable enough in the qualities of these quantities, and therefore, detailed analyzes and understanding are difficult with the TD-DFT calculations. The present overview of the results of the SAC-CI calculations of the CD spectra of molecules and molecular systems supports the reliability of the SAC-CI theory. The ChiraSac method is a theoretical methodology of investigating the chemistry and physics of chiral molecules and molecular systems using the Gaussian suite of programs involving the SAC-CI method and other many useful methods. Actually, in the studies of DNA and RNA, for example, we used many features of the SAC-CI and DFT methods in systematized ways involved in Gaussian. The ChiraSac studies could elucidate the information of the low-energy processes and the nature of the weak interactions involved in the observed CD spectra by using a highly reliable excited-state theory, the SAC-CI method, together with the many efficient methodologies involved in Gaussian. With the ChiraSac method, one can study many different subjects related to chirality, such as the conformations of chiral molecules in solutions or proteins, the low-energy processes such as the rotations around the single bonds of chiral atoms and the natures of the weak interactions such as the hydrogen-bonding and stacking interactions, van der Waals forces, solvent effects, etc. Therefore, the ChiraSac method will become a useful molecular technology for the material and drug designs.

**Acknowledgements** The computations were carried out using the computers at the Research Center for Computational Science, Okazaki, Japan, whom we acknowledge sincerely. We also thank the support of Mr. Nobuo Kawakami for the researches of QCRI. This work was supported by JSPS KAKENHI, Grant Numbers 15K05408 and 16H02257.

## References

1. H. Nakatsuji, K. Hirao, Cluster expansion of the wavefunction. Symmetry-adapted-cluster expansion, its variational determination, and extension of open-shell orbital theory. J. Chem. Phys. **68**, 2053–2065 (1978)
2. H. Nakatsuji, Cluster expansion of the wavefunction. Excited states. Chem. Phys. Lett. **59**, 362–364 (1978)
3. H. Nakatsuji, Cluster expansion of the wavefunction. Electron correlations in ground and excited states by SAC (Symmetry-Adapted-Cluster) and SAC-CI theories. Chem. Phys. Lett. **67**, 329–333 (1979)
4. H. Nakatsuji, Cluster expansion of the wavefunction. Calculation of electron correlations in ground and excited states by SAC and SAC-CI theories. Chem. Phys. Lett. **67**, 334–342 (1979)
5. H. Nakatsuji, Description of two- and many-electron processes by the SAC-CI method. Chem. Phys. Lett. **177**, 331–337 (1991)
6. H. Nakatsuji, M. Ehara, Symmetry-adapted cluster-configuration interaction method applied to high-spin multiplicity. J. Chem. Phys. **98**, 7179–7184 (1993)
7. T. Nakajima, H. Nakatsuji, Analytical energy gradient of the ground, excited, ionized and electron-attached states calculated by the SAC/SAC-CI method. Chem. Phys. Lett. **280**, 79–84 (1997)
8. T. Nakajima, H. Nakatsuji, Energy gradient method for the ground, excited, ionized, and electron-attached states calculated by the SAC (symmetry-adapted cluster)/SAC-CI (configuration interaction) method. Chem. Phys. **242**, 177–193 (1999)
9. M. Ishida, K. Toyota, M. Ehara, H. Nakatsuji, Analytical energy gradients of the excited, ionized and electron-attached states calculated by the SAC-CI general-R method. Chem. Phys. Lett. **347**, 493–498 (2001)
10. M. Ishida, K. Toyota, M. Ehara, H. Nakatsuji, Analytical energy gradient of high-spin multiplet state calculated by the SAC-CI method. Chem. Phys. Lett. **350**, 351–358 (2001)

11. K. Toyota, M. Ehara, H. Nakatsuji, Elimination of singularities in molecular orbital derivatives: Minimum Orbital-Deformation (MOD) method. Chem. Phys. Lett. **356**, 1–6 (2002)
12. K. Toyota, M. Ishida, M. Ehara, M.J. Frisch, H. Nakatsuji, Singularity-free analytical energy gradients for the SAC/SAC-CI method: Coupled Perturbed Minimum Orbital-Deformation (CPMOD) approach. Chem. Phys. Lett. **367**, 730–736 (2003)
13. M. Ishida, K. Toyota, M. Ehara, M.J. Frisch, H. Nakatsuji, Analytical energy gradient of the symmetry-adapted-cluster configuration-interaction general-R method for singlet to septet ground and excited states. J. Chem. Phys. **120**, 2593–2605 (2004)
14. Gaussian 09, Revision E.01, M.J. Frisch, G.W. Trucks, H.B. Schlegel, G.E. Scuseria, M.A. Robb, J.R. Cheeseman, G. Scalmani, V. Barone, B. Mennucci, G.A. Petersson, H. Nakatsuji, M. Caricato, X. Li, H.P. Hratchian, A.F. Izmaylov, J. Bloino, G. Zheng, J.L. Sonnenberg, M. Hada, M. Ehara, K. Toyota, R. Fukuda, J. Hasegawa, M. Ishida, T. Nakajima, Y. Honda, O. Kitao, H. Nakai, T. Vreven, J.A. Montgomery Jr., J.E. Peralta, F. Ogliaro, M. Bearpark, J. J. Heyd, E. Brothers, K.N. Kudin, V.N. Staroverov, R. Kobayashi, J. Normand, K. Raghavachari, A. Rendell, J.C. Burant, S.S. Iyengar, J. Tomasi, M. Cossi, N. Rega, J.M. Millam, M. Klene, J.E. Knox, J.B. Cross, V. Bakken, C. Adamo, J. Jaramillo, R. Gomperts, R. E. Stratmann, O. Yazyev, A.J. Austin, R. Cammi, C. Pomelli, J.W. Ochterski, R.L. Martin, K. Morokuma, V.G. Zakrzewski, G.A. Voth, P. Salvador, J.J. Dannenberg, S. Dapprich, A.D. Daniels, Ö. Farkas, J.B. Foresman, J.V. Ortiz, J. Cioslowski, D.J. Fox, Gaussian, Inc., Wallingford, CT, 2009.
15. H. Nakatsuji, Electronic structures of ground, excited, ionized, and anion states studied by the SAC/SAC-CI theory. Acta Chim. Hung. Models Chem. **129**, 719–776 (1992)
16. H. Nakatsuji, SAC-CI Method: theoretical aspects and some recent topics, in *Computational Chemistry—Reviews of Current Trends*, ed. by J. Leszczynski, vol. 2 (World Scientific, Singapore, 1997), pp. 62–124
17. M. Ehara, M. Ishida, K. Toyota, H. Nakatsuji, SAC-CI general-R Method: theory and applications to the multi-electron processes, in *Reviews of Modern Quantum Chemistry (A Celebration of the Contributions of Robert G. Parr)*, ed. by K.D. Sen (World Scientific, Singapore, 2003), pp. 293–319
18. H. Nakatsuji, Deepening and extending the quantum principles in chemistry. Bull. Chem. Soc. Jpn. **78**, 1705–1724 (2005)
19. M. Ehara, J. Hasegawa, H. Nakatsuji, SAC-CI method applied to molecular spectroscopy, in *Theory and Applications of Computational Chemistry: The First Forty Years,* ed. by C.E. Dykstra, G. Frenking, K.S. Kim, G.E. Scuseria (Elsevier, Oxford, U.K., 2005), pp. 1099–1141
20. J. Hasegawa, H. Nakatsuji, Exploring photo-biology and bio-spectroscopy with the SAC-CI (Symmetry-Adapted Cluster-Configuration Interaction) method, in *Radiation Induced Molecular Phenomena in Nucleic Acid: A Comprehensive Theoretical and Experimental Analysis*, ed. by M. Shukla, J. Leszczynsk (Springer, 2008), Chapter 4, pp. 93–124
21. N. Berova, K. Nakanishi, R.W. Woody (eds.), *Circular Dichroism: Principles and Applications* (Wiley-VCH, New York, 2000)
22. N. Berova, P.L. Polavarapu, K. Nakanishi, R.W. Woody (eds.), *Comprehensive Chiroptical Spectroscopy: Applications in Stereochemical Analysis of Synthetic Compounds, Natural Products, and Biomolecules,* vol. 2 (Wiley, Hoboken, New Jersey, 2012)
23. C. Niederalt, S. Grimme, S.D. Peyerimhoff, A. Sobanski, F. Vögtle, M. Lutz, A.L. Spek, M. J. van Eis, W.H. de Wolf, F. Bickelhaupt, Chiroptical properties of 12,15-dichloro[3.0] orthometacyclophane—correlations between molecular structure and circular dichroism spectra of a biphenylophane. Tetrahedron: Asymmetry **10**, 2153–2164 (1999)
24. F. Vögtle, S. Grimme, J. Hormes, K.H. Dötz, N. Krause, Theoretical simulations of electronic circular dichroism spectra, in *Interactions in Molecules: Electronic and Steric Effects*, ed. by S.D. Peyerimhoff (Wiley-VCH, Weinheim, 2003), pp. 66–109

25. F.E. Jorge, J. Autschbach, T. Ziegler, On the origin of optical activity in tris-diamine complexes of Co(III) and Rh(III): a simple model based on time-dependent density function theory. J. Am. Chem. Soc. **127**, 975–985 (2005)
26. H. Komori, Y. Inai, Electronic CD study of a helical peptide incorporating Z-dehydrophenylalanine residues: conformation dependence of the simulated CD spectra. J. Phys. Chem. A **110**, 9099–9107 (2006)
27. T.D. Crawford, M.C. Tam, M.L. Abrams, The current state of ab initio calculations of optical rotation and electronic circular dichroism spectra. J. Phys. Chem. A **111**, 12057–12068 (2007)
28. E. Botek, B. Champagne, Circular dichroism of helical structures using semiempirical methods. J. Chem. Phys. **127**, 204101-1–9 (2007)
29. E. Giorgio, K. Tanaka, L. Verotta, K. Nakanishi, N. Berova, C. Rosini, Determination of the absolute configurations of flexible molecules: Synthesis and theoretical simulation of electronic circular dichroism/optical rotation of some pyrrolo[2,3-b]indoline alkaloids—a case study. Chirality **19**, 434–445 (2007)
30. Y. Ding, X.C. Li, D. Ferreira, Theoretical calculation of electronic circular dichroism of the rotationally restricted 3,8''-biflavonoid morelloflavone. J. Org. Chem. **72**, 9010–9017 (2007)
31. G. Bringmann, T.A.M. Gulder, M. Reichert, T. Gulder, The online assignment of the absolute configuration of natural products: HPLC-CD in combination with quantum chemical CD calculations. Chirality **20**, 628–642 (2008)
32. T. Grkovic, Y. Ding, X.C. Li, V.L. Webb, D. Ferreira, B.R. Copp, Enantiomeric discorhabdin alkaloids and establishment of their absolute configurations using theoretical calculations of electronic circular dichroism spectra. J. Org. Chem. **73**, 9133–9136 (2008)
33. A. Shimizu, T. Mori, Y. Inoue, S. Yamada, Combined experimental and quantum chemical investigation of chiroptical properties of nicotinamide derivatives with and without intramolecular cation–π interactions. J. Phys. Chem. A **113**, 8754–8764 (2009)
34. J. Fan, T. Ziegler, A theoretical study on the exciton circular dichroism of propeller-like metal complexes of bipyridine and tripodal tris(2-pyridylmethyl)amine derivatives. Chirality **23**, 155–166 (2011)
35. J. Lambert, R.N. Compton, T.D. Crawford, The optical activity of carvone: a theoretical and experimental investigation. J. Chem. Phys. **136**, 114512-1–12 (2012)
36. S. Bureekaew, J. Hasegawa, H. Nakatsuji, Electronic circular dichroism spectrum of uridine studied by the SAC-CI method. Chem. Phys. Lett. **425**, 367–371 (2006)
37. Y. Honda, A. Kurihara, M. Hada, H. Nakatsuji, Excitation and circular dichroism spectra of (-)-(3aS,7aS)-2-chalcogena-trans-hydrindans (Ch=S, Se, Te): SAC and SAC-CI calculations. J. Comput. Chem. **29**, 612–621 (2008)
38. T. Miyahara, J. Hasegawa, H. Nakatsuji, Circular dichroism and absorption spectroscopy for three-membered ring compounds using Symmetry-Adapted Cluster-Configuration Interaction (SAC-CI) method. Bull. Chem. Soc. Jpn **82**, 1215–1226 (2009)
39. Y. Honda, A. Kurihara, Y. Kenmochi, M. Hada, Excitation and circular dichroism spectra of (+)-(S, S)-bis(2-methylbutyl)chalcogenides. Molecules **15**, 2357–2373 (2010)
40. T. Miyahara, H. Nakatsuji, H. Sugiyama, Helical structure and circular dichroism spectra of DNA: a theoretical study. J. Phys. Chem. A **117**, 42–55 (2013)
41. T. Miyahara, H. Nakatsuji, Conformational dependence of the circular dichroism spectrum of α-hydroxyphenylacetic acid: a ChiraSac study. J. Phys. Chem. A **117**, 14065–14074 (2013)
42. T. Miyahara, H. Nakatsuji, T. Wada, Circular dichroism spectra of uridine derivatives: ChiraSac study. J. Phys. Chem. A **118**, 2931–2941 (2014)
43. T. Miyahara, H. Nakatsuji, Indicator of the stacking interaction in the DNA double-helical structure: ChiraSac study. J. Phys. Chem. A **119**, 8269–8278 (2015)
44. T. Miyahara, H. Nakatsuji, H. Sugiyama, Similarities and differences between RNA and DNA double-helical structures in circular dichroism spectroscopy: a SAC-CI study. J. Phys. Chem. A **120**, 9008–9018 (2016)
45. L. Rosenfeld, Quantum-mechanical theory of the natural optical activity of liquids and gases. Z. Physik **52**, 161–174 (1929)

46. M. Pericou-Cayere, M. Rerat, A. Dargelos, Theoretical treatment of the electronic circular dichroism spectrum and the optical rotatory power of $H_2S_2$. Chem. Phys. **226**, 297–306 (1998)
47. W. Moffitt, Optical rotatory dispersion of helical polymers. J. Chem. Phys. **25**, 467–478 (1956)
48. A. Breest, P. Ochmann, F. Pulm, K.H. Gödderz, M. Carnell, J. Hormes, Experimental circular dichroism and VUV spectra of substituted oxiranes and thiiranes. Mol. Phys. **82**, 539–551 (1994)
49. L. Verbit, P.J. Heffron, Optically active aromatic chromophores-IV. Circular dichroism studies of some open-chain systems. Tetrahedron **24**, 1231–1236 (1968)
50. T. Wada, N. Minamimoto, Y. Inaki, Y. Inoue, Conformational and orientational switching of uridine derivatives by borates. Chem. Lett. **10**, 1025–1026 (1998)
51. D.W. Miles, R.K. Robins, H. Eyring, Optical rotatory dispersion, circular dichroism, and absorption studies on some naturally occurring ribonucleosides and related derivatives. Proc. Natl. Acad. Sci. U.S.A. **57**, 1138–1145 (1967)
52. H. Sugiyama, unpublished
53. A. Rich, A. Nordheim, A.H.-J. Wang, The chemistry and biology of left-handed Z-DNA. Ann. Rev. Biochem. **53**, 791–846 (1984)
54. S. Tran-Dinh, J. Taboury, J.-M. Neumann, T. Huynh-Dinh, B. Genissel, B. Langlois d'Estaintot, J. Igolen, $^{1}$H NMR and circular dichroism studies of the B and Z conformations of the self-complementary deoxyhexanucleotide d(m$^{5}$C-G-C-G-m$^{5}$C-G): mechanism of the Z-B-coil transitions. Biochemistry **23**, 1362–1371 (1984)
55. H. Sugiyama, K. Kawai, A. Matsunaga, K. Fujimoto, I. Saito, H. Robinson, A.H.-J. Wang, Synthesis, structure and thermodynamic properties of 8-methylguanine-containing oligonucleotides: Z-DNA under physiological salt conditions. Nucl. Acid. Res. **24**, 1272–1278 (1996)
56. K. Kawai, I. Saito, H. Sugiyama, Conformation Dependent photochemistry of 5-halouracil-containing DNA: stereospecific 2'α-hydroxylation of deoxyribose in Z-form DNA. J. Am. Chem. Soc. **121**, 1391–1392 (1999)
57. T. Oyoshi, K. Kawai, H. Sugiyama, Efficient C2'α-hydroxylation of deoxyribose in protein-induced Z-form DNA. J. Am. Chem. Soc. **125**, 1526–1531 (2003)
58. Y. Xu, R. Ikeda, H. Sugiyama, 8-methylguanosine: a powerful Z-DNA stabilizer. J. Am. Chem. Soc. **125**, 13519–13524 (2003)
59. K. Hall, P. Cruz, I. Tinoco Jr., T.M. Jovin, J.H. van de Sande, 'Z-RNA'—a left-handed RNA double helix. Nature **311**, 584–586 (1984)
60. B.A. Brown II, K. Lowenhaupt, C.M. Wilbert, E.B. Hanlon, A. Rich, The Zα domain of the editing enzyme dsRNA adenosine deaminase binds left-handed Z-RNA as well as Z-DNA. Proc. Natl. Acad. Sci. U.S.A. **97**, 13532–13536 (2000)
61. R. Tashiro, H. Sugiyama, biomolecule-based switching devices that respond inversely to thermal stimuli. J. Am. Chem. Soc. **127**, 2094–2097 (2005)

# Chapter 3
# Frontiers of Coupled Cluster Chiroptical Response Theory

**T. Daniel Crawford**

**Abstract** We review the current state and future prospects of coupled cluster response theory for modeling chiroptical properties in both gas-phase and solvated systems. We first provide an overview of ground-state coupled cluster theory and the analytic derivative approach to computing time-independent properties, and then extend this to time-dependent perturbations. Through the well-established "quasi-energy" approach, we demonstrate the connection between the analytic derivative and response approaches, emphasizing the usefulness of the latter for frequency-dependent chiroptical properties. In addition, we review successes of response theory in the prediction of gas-phase specific rotations and electronic circular dichroism spectra for small molecules and the physical requirements of a robust computational model of such properties. We also discuss the many challenges of extending this reliability to solvated systems through several key examples for which theory and experiment diverge significantly.

**Keywords** Coupled cluster theory · Response theory · Chiroptical properties

## 3.1 Introduction

The dissymmetry of chiral molecules is the source of both their reactive uniqueness and the spectroscopic means of distinguishing their mirror images. In absorption, emission, refraction, and scattering, circularly polarized light induces opposite-sign responses between enantiomeric pairs of stereoisomers, responses that can be used to determine the absolute configuration of non-racemic samples of such compounds [1–5]. However, in order for such spectroscopic data to be of practical value, a reference must be available to tie the measured response to its chiral progenitor. Theoretical simulations of such spectra are thus central to this effort, but in order to be useful, quantum chemical response theory must be reliable, robust, accurate, and efficient [2–4, 6–41].

T. Daniel Crawford (✉)
Department of Chemistry Virginia Tech, Blacksburg, VA, USA
e-mail: crawdad@vt.edu



M.J. Wójcik et al. (eds.), *Frontiers of Quantum Chemistry*,
https://doi.org/10.1007/978-981-10-5651-2_3

A bedrock principle of molecular quantum mechanics intended to develop such reliability was established by Pople [42] in the early 1970s: the *theoretical model chemistry*. Given a form for the approximate electronic wave function, as well as the number of electrons, the spin multiplicity, and the coordinates and types nuclei in the system of interest, one can compute numerous molecular properties: structures, thermodynamic constants, potential energy surfaces and associated rate constants, spectroscopic responses, and more. Central to Pople's idea is the systematic and generic nature of a theoretical model chemistry, such that it should be applicable to any chemical system. In the decades since its introduction, this concept has motivated the development of a multitude of quantum chemical methods that are generally applicable to nearly any molecular species, though their accuracy and computational costs vary significantly.

A second pillar of modern molecular quantum mechanics is the concept of a convergent ab initio model [43]. These methods are built primarily upon wave function theories and are characterized as being formally exact, free of experimental or empirical parameters, and systematically improvable toward the limit of the solution to the Schrödinger equation. Convergence is essential to the development of reliable quantum chemical methods—indeed, it is vital to the ultimate practical applicability of Pople's theoretical model chemistry—and has led to one of the most significant accomplishments of modern quantum chemistry: its quantitative accuracy. Among such methods, coupled cluster theory [44–46] stands at the forefront of high-accuracy approaches, and in numerous examples reported in the literature, coupled cluster data have preceded, confirmed, and even overturned experimental determinations of properties [47] such as geometrical structure, vibrational and UV/Vis spectra, thermochemistry, NMR chemical shifts, nuclear spin-spin coupling constants [48–51]. This level of robustness is also why coupled cluster theory has been one of the major focal points for the development of practical response methods over the last 20 years.

In spite of its success, numerous challenges remain for coupled cluster response theory in the drive toward more complex chemical systems. Perhaps foremost among these is the high-degree polynomial scaling of coupled cluster methods [$\mathcal{O}(N^6)$ and worse] that plagues conventional formulations. This scaling precludes the routine application of coupled cluster response theory to molecular systems containing more than around a dozen heavy atoms, thus limiting its usefulness for many chemical problems, such as those involving e.g. molecular materials and other extended systems. A second, but closely related problem is that of solvation. While computational cost prevents the explicit inclusion of large numbers of (dynamic) solvent molecules in coupled cluster response calculations, implicit solvent models often fail to describe the detailed solute–solvent interactions that can dramatically impact even qualitatively correct descriptions of chiroptical response, for example.

In this chapter, we provide an overview of modern coupled cluster response theory [52–63], its applications, and future directions. The next section includes a concise presentation of the fundamental equations of both the ground-state theory and its time-dependent formulation, as well as aspects of production-level implementations. Section 3.3 focuses on recent extensions and applications of coupled cluster response theory, including both gas-phase and solvent-including calculations, and the final section offers one view of future developments. While we do not intend to provide an exhaustive summary of the coupled cluster response literature, we hope that the perspective offered here will provide newcomers to the field a starting point for further investigation.

## 3.2 Theoretical Underpinnings of Coupled Cluster Response Theory

In this section, we will briefly review the fundamental equations of coupled cluster theory and its time-/frequency-dependent formulations. We will present only the key relations and results; more complete derivations can be found in several excellent review articles and references therein [44, 47, 64, 65].

### *3.2.1 Time-Independent Coupled Cluster Theory*

Coupled cluster theory approximates the many-electron wave function using an exponential expansion of Slater determinants—anti-symmetrized products of one-electron molecular orbitals, typically obtained from a Hartree–Fock or similar mean-field calculation. With this wave function *Ansatz*, the Schrödinger equation is conveniently expressed as [64]

$$e^{-\hat{T}}\hat{H}^{(0)}e^{\hat{T}}|0\rangle = \bar{H}^{(0)}|0\rangle = E_{\mathrm{CC}}|0\rangle, \tag{3.1}$$

where $\hat{H}^{(0)}$ is the (time-independent) molecular Hamiltonian, $|0\rangle$ denotes the Hartree–Fock reference wave function, and $\hat{T}$ is a second quantized "excitation" operator whose action on $|0\rangle$ yields linear combinations of determinants in which one or more occupied orbitals have been replaced by the same number of unoccupied (or virtual) orbitals. The non-Hermitian, similarity-transformed Hamiltonian, $\bar{H}^{(0)}$, is a central quantity in modern formulations of coupled cluster theory, in part because it may be expressed as a finite commutator expansion,

$$\bar{H}^{(0)} = \hat{H}^{(0)} - \left[\hat{H}^{(0)}, \hat{T}\right] + \frac{1}{2}\left[\left[\hat{H}^{(0)}, \hat{T}\right], \hat{T}\right] + \cdots, \tag{3.2}$$

which truncates exactly at the quartic terms in $\hat{T}$ due to the two-electron nature of the electronic Hamiltonian.

Coupled cluster theory is formally exact in that inclusion of all possible excitations, and the use of a complete (infinite dimension) orbital space yields the exact solution to the electronic Schrödinger equation. In practical applications, however, a finite orbital space is employed, and $\hat{T}$ is truncated to some manageable number of terms. For example, the coupled cluster singles and doubles (CCSD) model [66] results from the choice of $\hat{T} \approx \hat{T}_1 + \hat{T}_2$, which produces all singly and doubly excited determinants from $|0\rangle$. Thus, we see that the coupled cluster wave function is systematically improvable in that incorporation of higher levels of excitation (CCSDT, CCSDTQ, etc.) and extension of the orbital space lead to better approximations to the exact wave function. These characteristics—formally exact and systematically improvable—as well as its lack of empirical parameters thus classify coupled cluster theory as a convergent ab initio method.

A useful starting point for the formulation of molecular properties is the coupled cluster energy Lagrangian [49, 67]

$$\mathscr{L}_{\rm CC} = E_{\rm CC} = \langle 0| \left(1 + \hat{\Lambda}\right) \bar{H}^{(0)} |0\rangle, \tag{3.3}$$

where the Lagrangian multipliers needed for a constrained optimization of the functional are contained within $\hat{\Lambda}$—a "de-excitation" operator analogous to its $\hat{T}$ counterpart. The wave function amplitudes within $\hat{T}$ and the multipliers within $\hat{\Lambda}$ are obtained from the stationary conditions [68]

$$\frac{\partial \mathscr{L}_{\rm CC}}{\partial \hat{\Lambda}_\phi} = \langle \phi | \bar{H}^{(0)} |0\rangle = 0 \tag{3.4}$$

and

$$\frac{\partial \mathscr{L}_{\rm CC}}{\partial \hat{T}_\phi} = \langle 0|(1 + \hat{\Lambda}) \left[\bar{H}^{(0)}, \tau_\phi\right] |0\rangle = 0, \tag{3.5}$$

where $|\phi\rangle$ represents the set of excited determinants generated by $\hat{T}$ from $|0\rangle$ through the excitation operator $\tau_\phi$. The non-Hermitian nature of the similarity-transformed Hamiltonian implies that the right- and left-hand coupled cluster wave functions are not simple adjoints of one another as would be the case in, for example, configuration interaction theory. The form of the coupled cluster Lagrangian functional suggests that while the right-hand coupled cluster wave function is

$$|\psi_{\rm CC}\rangle = e^{\hat{T}} |0\rangle, \tag{3.6}$$

its left-hand counterpart is

$$\langle \psi_{\rm CC}| = \langle 0| \left(1 + \hat{\Lambda}\right) e^{-\hat{T}}. \tag{3.7}$$

Another advantage of the Lagrangian formulation is that it provides a connection between many molecular properties and analytic derivatives of the energy through a generalization of the Hellmann–Feynman theorem, viz.

$$\frac{\partial \mathcal{L}_{\mathrm{CC}}}{\partial x} = \langle 0|\left(1+\hat{\Lambda}\right) e^{-\hat{T}} \frac{\partial \hat{H}^{(0)}}{\partial x} e^{\hat{T}} |0\rangle = \langle 0|\left(1+\hat{\Lambda}\right) \bar{H}^{(0)}_{x} |0\rangle, \tag{3.8}$$

where $x$ represents a general perturbation, and the subscript on the last term indicates differentiation with respect to $x$. (We also note that the second quantization formalism implicit in this equation implies that any dependence of the underlying orbitals on the external perturbation is contained within the derivative(s) of the Hamiltonian itself.) For example, the molecular dipole moment can be determined from the first derivative of the coupled cluster energy with respect to an external static electric field, whereas the gradient along the potential energy hypersurface requires derivatives with respect to nuclear coordinates.

In accord with the Wigner $2n+1$ rule, higher-order properties require derivatives of the right- and left-hand coupled cluster wave functions, e.g.,

$$\frac{\partial \mathcal{L}_{\mathrm{CC}}}{\partial x \partial y} = \langle 0|\left(1+\hat{\Lambda}\right) \bar{H}^{(0)}_{xy} |0\rangle + \langle 0|\left(1+\hat{\Lambda}\right) \left[\bar{H}^{(0)}_{x}, \hat{T}_{y}\right] |0\rangle + \langle 0|\hat{\Lambda}_{y} \bar{H}^{(0)}_{x} |0\rangle. \tag{3.9}$$

Properties such as quadratic force constants (leading to harmonic vibrational frequencies) are obtained if $x$ and $y$ are nuclear coordinates, whereas static dipole polarizabilities correspond to second derivatives with respect to external static electric fields. This expression may be written in a symmetric form by taking advantage of the corresponding $2n+2$ rule for the Lagrange multipliers to obtain

$$\frac{\partial \mathcal{L}_{\mathrm{CC}}}{\partial x \partial y} = P_{+}(x,y)\langle 0|\left(1+\hat{\Lambda}\right) \left(\frac{1}{2}\bar{H}^{(0)}_{xy} + \left[\bar{H}^{(0)}_{x}, \hat{T}_{y}\right] + \frac{1}{2}\left[\left[\bar{H}^{(0)}, \hat{T}_{x}\right], \hat{T}_{y}\right]\right) |0\rangle, \tag{3.10}$$

where the operator $P_{+}$ is a symmetrizer with respect to perturbations $x$ and $y$. Each of these forms offers certain computational or programming advantages [69]. The symmetric form, for example, requires only algorithms to obtain the derivatives of the $\hat{T}$ amplitudes, whereas the asymmetric form also requires effort to program the equations needed to determine the derivatives of $\hat{\Lambda}$. On the other hand, the asymmetric version offers significantly reduced persistent storage possibilities in the case where the number of perturbations in each class, $x$ and $y$, is significantly different, such as for NMR shieldings, which require second derivatives with respect to the nuclear magnetic moments (three perturbations for each nucleus) and the external magnetic field (only three perturbations total).

### 3.2.2 Time-Dependent Coupled Cluster Response Theory

Next, we consider the case in which the molecular Hamiltonian is perturbed by a time-dependent potential, as this is necessary for the desired description of chiroptical properties. We choose a periodic potential of the form [47, 60],

$$\hat{H}^{(1)}(t) = \sum_k V_k \varepsilon_k(\omega_k) e^{-i\omega_k t}, \tag{3.11}$$

where periodicity is obtained by requiring all $\omega_k$ to a multiple of a common fundamental frequency, $\omega = 2\pi/\tau$, and Hermiticity is obtained in part by requiring the summation to run over both positive and negative values of $k$ and pairing opposite-sign components. In this case, we must adopt a form of the time-dependent Schrödinger equation,

$$i\frac{\partial}{\partial t} e^{iF(T)} e^{\hat{T}(t)} = \left(\hat{H}^{(0)} + \hat{H}^{(1)}(t)\right) e^{iF(t)} e^{\hat{T}(t)} |0\rangle, \tag{3.12}$$

in which the coupled cluster right- and left-hand exponential wave functions are expressed in the phase-isolated form developed by Langhoff et al. [70],

$$|\psi_{\rm CC}(t)\rangle = e^{iF(t)} e^{\hat{T}(t)} |0\rangle \tag{3.13}$$

and

$$\langle \psi_{\rm CC}(t)| = \langle 0| \left(1 + \hat{\Lambda}(t)\right) e^{-\hat{T}(t)} e^{-iF(t)}, \tag{3.14}$$

where $F(t)$ is a real-valued function. Explicit time differentiation of the complex phase factor leads to a time-dependent Lagrangian analogous to Eq. 3.3,

$$\mathscr{L}(t) \equiv {\rm Re}\langle 0| \left(1 + \hat{\Lambda}\right) \left(\bar{H}^{(0)} + \bar{H}^{(1)} - ie^{-\hat{T}} \frac{\partial}{\partial t} e^{\hat{T}}\right) |0\rangle, \tag{3.15}$$

where $\mathscr{L}(t)$ is the time derivative of the phase factor referred to as the "quasi-energy" Lagrangian functional [59, 60, 71, 72], and we have retained only the real part of the functional in order to eliminate pure imaginary contributions from the intermediately normalized wave function [47]. (In addition, we have suppressed the explicit time dependence of the quantities on the right-hand side for notational efficiency.) This Lagrangian is subject to stationary conditions similar to those of the time-independent case in Eqs. 3.4 and 3.5,

$$\frac{\partial \mathscr{L}(t)}{\partial \hat{\Lambda}_\phi} = {\rm Re}\left[\langle \phi | \bar{H} - i\frac{\partial \hat{T}}{\partial t} |0\rangle\right] = 0, \tag{3.16}$$

and

$$\frac{\partial \mathscr{L}(t)}{\partial \hat{T}_\phi} = {\rm Re}\left[\langle 0|(1 + \hat{\Lambda}) \left[\bar{H}, \tau_\phi\right] |0\rangle + i\langle 0| \frac{\partial \hat{\Lambda}}{\partial t} |\phi\rangle\right] = 0. \tag{3.17}$$

The Lagrangian and its stationary conditions may be expanded in orders of the potential, e.g.,

$$\mathscr{L}(t) = \mathscr{L}^{(0)} + \mathscr{L}^{(1)}(t) + \mathscr{L}^{(2)} + \cdots, \tag{3.18}$$

where the zeroth-order contribution is simply the time-independent expression in Eq. 3.3 and periodicity of the potential in Eq. 3.11 carries over to the perturbed wave function amplitudes and multipliers. We may take advantage of this by introducing the time average of the Lagrangian over the common period of the perturbation components,

$$\{\mathscr{L}(t)\}_\tau = \frac{1}{\tau}\int_0^\tau \mathscr{L}(t)dt. \tag{3.19}$$

The stationary conditions above still apply to the time-averaged Lagrangian, leading to a generalized Hellmann–Feynman relation

$$\frac{\partial\{\mathscr{L}(t)\}_\tau}{\partial\varepsilon_k(\omega_k)} = \mathrm{Re}\left\{\langle 0|\left(1+\hat{\Lambda}\right)\frac{\partial\bar{H}}{\partial\varepsilon_k(\omega_k)}|0\rangle\right\}_\tau = \mathrm{Re}\left\{\langle 0|\left(1+\hat{\Lambda}\right)\bar{H}_{\varepsilon_k(\omega_k)}|0\rangle\right\}_\tau, \tag{3.20}$$

directly analogous to Eq. 3.8. When the perturbation expansion of Eq. 3.18 is inserted for the time-averaged Lagrangian above, the desired response functions appear as corresponding derivatives. Thus, again in analogy to the time-independent derivatives expressed above, e.g., for the linear response function,

$$\langle\langle V_k; V_l\rangle\rangle_{\omega_l} = \frac{\partial\left\{\mathscr{L}^{(2)}(t)\right\}_\tau}{\partial\varepsilon_k(-\omega_l)\varepsilon_l(\omega_l)} = \mathrm{Re}\left\{\langle 0|\left(1+\hat{\Lambda}\right)\left[\bar{V}_k, \hat{X}_l(\omega_l)\right] + \hat{Y}_l(\omega_l)\bar{V}_l|0\rangle\right\}_\tau, \tag{3.21}$$

we can see a clear comparison to the corresponding terms in Eq. 3.9. (The term corresponding to the second derivative of the Hamiltonian does not appear in this case, because the Hamiltonian depends only linearly on the external field.) If $V_k$ and $V_l$ are taken to be electric dipole operators, then the above equation yields the frequency-dependent dipole-polarizability, and, in the static-field limit gives results identical to that of either Eq. 3.9 or Eq. 3.10 (provided the orbital response is neglected). Other combinations of operators yield additional response functions, including those associated with chiroptical properties, such as the mixed electric dipole/magnetic dipole polarizability (the Rosenfeld tensor that yields optical rotations and circular dichroism spectra) and the electric dipole/electric quadrupole polarizability (necessary for Raman optical activity spectra, for example). In addition, transition strengths are associated with the residues of the linear response function, i.e.,

$$\langle 0|\hat{V}_k|n\rangle\langle n|\hat{V}_l|0\rangle = \lim_{\omega\to\omega_{n0}}(\omega-\omega_{n0})\langle\langle V_k; V_l\rangle\rangle_\omega. \tag{3.22}$$

To obtain oscillator strengths, $V_k$ and $V_l$ are chosen to be the electric dipole operator, while for circular dichroism rotatory strengths, the electric dipole and magnetic dipole operators are used.

A subtle, but important point for the computation of molecular chiroptical properties is that of origin invariance, i.e., if the length representation of the electric dipole operator is used in the calculation of the linear response function, then the computed property will depend on the arbitrary choice of coordinate origin used in the calculation. [1, 5, 73–75] This can be overcome in the case of optical rotation or rotatory strengths by using instead the velocity representation, which is connected to the length representation via its commutator with the Hamiltonian (in a complete basis set). While this approach is now standard for circular dichroism spectra, for optical rotations the velocity representation of the linear response function does not decay to zero at long wavelengths and is thus unphysical. To overcome this behavior approximately, the zero-frequency response function is subtracted from its finite-frequency counterpart (referred to as the "modified velocity gauge") in most practical coupled cluster response calculations [73, 75]. Note that this technique is unnecessary in methods such as DFT, where the use of gauge-including atomic orbitals corrects the origin variation automatically for the dipole-length formulation.

## 3.3 Applications of Coupled Cluster Chiroptical Response Theory

One of the earliest applications of coupled cluster response theory to chiroptical properties was the optical rotation of the $\sigma$-helicene known as (*P*)-(+)-[4]triangulane [14], a rigid tri-spiro compound comprised of four fused cyclopropane rings. Although ostensibly an alkane, de Meijere and co-workers [76] measured its surprisingly large specific rotations in $CHCl_3$ following their successful synthesis of the structure in enantiomerically pure form in 1999. At the sodium D-line (589 nm), the compound exhibited a rotation of 192.6 deg $dm^{-1}$ $(g/mL)^{-1}$, and, as the wavelength was shortened to 365 nm, the rotation increased dramatically to 648 deg $dm^{-1}$ $(g/mL)^{-1}$, in spite of the absence of a long-wavelength chromophore. Although time-dependent density functional theory (TDDFT) calculations yielded the correct sign of these rotations, they were found to overestimate the experimentally measured values by roughly 15%. Later studies revealed that this error was due to the tendency of the B3LYP functional both to underestimate electronic excitation energies and to overestimate rotatory strengths [12]. On the other hand, coupled cluster response calculations at the CCSD/aug-cc-pVDZ level of theory yielded excellent comparison with experiment, with errors less than 2% across the entire dispersion curve. However, this success is far from complete: The theoretical models did not account for factors such as solvent and vibrational effects that were found to be significant in numerous other compounds, and a more appropriate comparison between theory and experiment is offered by gas-phase measurements.

### 3.3.1 Comparison to Gas-Phase Optical Activity Data

While a number of gas-phase electronic circular dichroism spectra have been reported, to date the only such specific rotation data available have been obtained

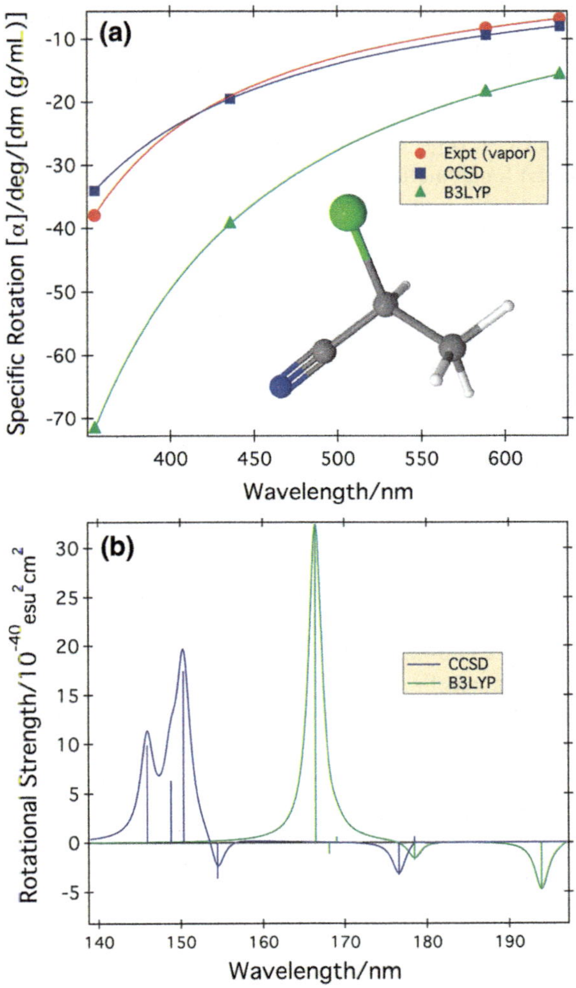


**Fig. 3.1** **a** Theoretical and experimental gas-phase optical rotatory dispersion spectra and **b** theoretical circular dichroism spectra of (*S*)-(-)-2-chloropropionitrile. Experimental data taken from Ref. [78] and theoretical data from Ref. [21]

using ultrasensitive cavity ring-down polarimetry (CRDP) techniques developed by Vaccaro and co-workers [77, 78]. In a number of studies [4, 12, 15, 21, 22, 79–84], CCSD-level specific rotations coupled with diffuse basis sets compare closely to nearly every published vapor-phase CRDP measurements at both long and short wavelengths, such as those of (S)-2-chloropropionitrile, shown in Fig. 3.1. B3LYP, on the other hand, gives rotations too large by approximately a factor of two, again due to its tendency both to underestimate the key excitation energies and to overestimate the corresponding rotational strengths, as made clear by the simulated circular dichroism spectra in Fig. 3.1b. Similarly, good agreement has been found for numerous other compounds, including rigid molecules such a (*1R*,*5R*)-$\alpha$-pinene, (*1R*,*5R*)-$\beta$-pinene, and (*1R*,*2S*,*5R*)-*c*is-pinane [15].

The importance of vibrational effects on chiroptical properties has also been examined for comparison to gas-phase measurements [11, 12, 22, 82, 83, 85–93]. For molecules such as *(1R,2S,5R)-cis*-pinane, *(1R,5R)*-$\alpha$-pinene, *(1R,5R)*-$\beta$-pinene, (*S*)-limonene, and others, excellent comparison is found between CRDP data and CC-level specific rotations with zero-point vibrational corrections derived from B3LYP calculations, though the use of high levels of electron correlation was found to be even more significant than such corrections. (*S*)-methyoxirane is a particularly thorny case, and the cumulative result of several studies is that one requires not only high levels of correlation (up to CC3, which includes triple excitations, has been utilized), but also the effects of anharmonicity on the methyl torsional motion [88], before quantitative agreement between theory and experiment is obtained. The only outlier in such analyses is *(1S,4R)*-fenchone, a large ketone for which theory and experiment differ significantly (456 deg $dm^{-1}$ $(g/mL)^{-1}$at the CCSD level versus 180 deg $dm^{-1}$ $(g/mL)^{-1}$at 355 nm). Vibrational corrections do not overcome the discrepancy in this case, and the fundamental reasons for the disagreement remain hidden.

Conformationally flexible molecules offer another level of challenge as one must consider all energetically relevant configurations thereof [94, 95]. While this is a relatively straightforward task for many spectroscopic properties, such as UV/Vis or infrared absorption spectra, CD and optical rotations are dramatically sensitive to the choice of conformer. The recent study of the chiral compound (*R*)-carvone [96] provides an excellent example of this problem: It exhibits six low-lying conformations, with the propylene group equatorial or axial relative to the hexene ring, each of which with three stable rotamers of the propylene around its C–C bond connecting it to said ring. While the equatorial structures are lower in energy, according to G3-level calculations, the axial structures still contribute roughly 20% to the overall Boltzmann distribution and thus cannot be excluded. (We note that simpler B3LYP/6—311G** calculations significantly underestimate these populations at less than 5% total, leading to serious caveats about the methods used to determine such contributions). However, whereas experimental sodium D-line measurements of the specific rotation of (*R*)-carvone in multiple (polar and nonpolar) solvents yield negative values of ca. −50 deg $dm^{-1}$ $(g/mL)^{-1}$, both DFT and CC2 calculations give *positive* rotations of ca. 57 deg $dm^{-1}$ $(g/mL)^{-1}$ and 12 deg $dm^{-1}$ $(g/mL)^{-1}$, respectively. This problem can be overcome, however, by including temperature-dependent vibrational corrections for each conformer. Using a Taylor expansion in normal coordinates, one may determine the contribution of each mode to the total specific rotation, with some modes exhibiting corrections larger than the equilibrium rotation itself. Nevertheless, once such corrections (typically computed at a lower level of theory) are taken into account, both DFT and CCSD methods give specific rotations that are semiquantitatively correct at −23 and −30 deg $dm^{-1}$ $(g/mL)^{-1}$, respectively, and quantitative accuracy at the CC2 level in conjunction with mixed aTZ/aDZ basis sets at −49 deg $dm^{-1}$ $(g/mL)^{-1}$.

This commonly used Boltzmann average is based on the assumptions of localizability of vibrational wave functions with each conformational potential well (infinitely high energetic walls separate the conformers), invariance of the optical rotation as a function of vibrational quantum number, and identical vibrational contributions to the partition functions of each conformer. While these assumptions are reasonable for most chemical systems, small barriers, in particular, between conformations may corrupt the Boltzmann average. A recent study of (*R*)-3-chloro-1-butene [93] explicitly examined these assumptions using an analytic torsional potential function for the coordinate separating the three relevant conformers. This potential was based on energy points obtained from an intrinsic reaction coordinate calculation connecting transition states to minima, followed by a fit based on a Fourier series. The resulting one-dimensional function was used in a numerical solution of the vibrational Schrödinger equation. The conclusion of the study was that the two approaches—the simple method involving separate conformers and the rigorous Boltzmann average over explicit vibrational wave functions spanning the entire range of dihedral motion—compared reasonably well over a range of temperatures, though reduction of the barrier between conformers to ca. 500 $cm^{-1}$ or less revealed larger discrepancies.

### *3.3.2 Condensed-Phase Optical Activity*

While we may safely conclude that the fundamental physical requirements of a reliable model of electronic optical activity of chiral molecules in the gas phase have been established, the same is not yet true for chiral molecules in the liquid phase, where solvent perturbations are known to impact both the magnitude and the sign of rotational strengths and specific rotation [97, 98]. Although theoretical prediction and simulation of many molecular properties in the presence of solvent has a long and distinguished history [99–102], chiroptical properties offer a much greater challenge, as can be seen from three specific examples.

#### 3.3.2.1 (*S*)-Methyloxirane

Kumata et al. demonstrated [98] more than 40 years ago that the specific rotation of methyloxirane is strongly dependent upon the choice of solvent used for the measurement. Figure 3.2 depicts experimental optical rotatory dispersion curves for (*S*)-methyloxirane in several common solvents, with water and benzene producing opposite-sign rotations at long wavelengths as well as qualitatively different shapes to the non-resonant response curves.

In 2002, Mennucci and co-workers [103] combined DFT calculations of specific rotations of methyloxirane with the implicit polarizable continuum model (PCM) [99, 102]. While this approach has yielded much fruit for a variety of spectroscopic properties, it was unable to reproduce the sodium D-line ordering of the solvents examined by Kumata and co-workers [98] and by Vaccaro and co-workers, [78] errors which they attributed to non-electrostatic effects. Pecul et al. [104] reported analogous DFT-based PCM studies of electronic circular dichroism

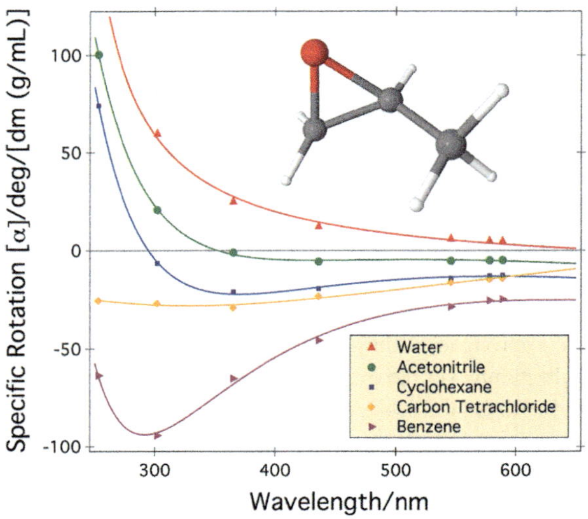


**Fig. 3.2** Experimental optical rotatory dispersion spectra of (*S*)-methyloxirane. The choice of solvent has a dramatic impact on the sign and shape of the dispersion curve, making it an ideal test case for microsolvation, QM/MM, and implicit solvation studies. Experimental data for water provided by P.H. Vaccaro (private communication); all other data from Ref. [78]

in 2005 for methyloxirane and others. They found that the reliability of their predictions depended not only on the shortcomings of available exchange-correlation functionals, but also on the type of transition in question, with Rydberg-type excitations indicated as especially problematic, perhaps due to specific solute–solvent interactions.

Kongsted et al. [86, 90] published extensive CC analyses for the optical rotation of (*S*)-methyloxirane under solvated conditions, simulated using a continuum dielectric medium surrounding a spherical solute cavity [86]. Despite the inclusion of very high levels of electron correlation (including estimates of connected triple excitations [105, 106]), this approach failed to reproduce experimental trends, yielding specific rotations at 355 nm with the wrong sign as compared to measurements in cyclohexane. More recently, Kongsted and Ruud [92] carried out a DFT-PCM/B3LYP analysis of methyloxirane specific rotations that included both solvent and zero-point vibrational effects. They found reasonable agreement in solvent shifts for cyclohexane, but not for acetonitrile and water, which they again attributed to specific molecular interactions not described by continuum-based models.

Dynamic effects were considered by Beratan and co-workers [107, 108], who carried out groundbreaking combined continuum and molecular dynamics (MD) simulations of methyloxirane in clusters of water [107] and of benzene [108], obtaining DFT specific rotations as averages over snapshots taken along the MD trajectories. While they obtained qualitative agreement with experiment in both cases and found that the chiroptical response of the solute molecule dominates in water, but the solvent cavity contributes at least as much to the optical rotation in benzene. More recently, Barone and co-workers [109] reported large-scale DFT-based MD simulations of methyloxirane in water that reproduced the shape of the corresponding curve in Fig. 3.2. However, given that DFT has been shown to fail—providing only fortuitous agreement with gas-phase experimental data in the case of methyloxirane [12, 86, 88, 92]—the general applicability of these approaches remains in doubt.

#### 3.3.2.2 (*1S,4S*)-Norbornenone

A second key example of the significance of solvent is that of norbornenone, a cage-like ketone that exhibits a particularly large specific rotation [110] at the sodium D-line of ca. 1000 deg $dm^{-1}$ $(g/mL)^{-1}$, due in part to strong exciton coupling involving its alkene and carbonyl moieties [110, 111]. Ruud and co-workers reported B3LYP specific rotations of −1216 deg $dm^{-1}$ $(g/mL)^{-1}$, in good agreement with the value of −1236 deg $dm^{-1}$ $(g/mL)^{-1}$ measured in chloroform [112], but origin-independent CCSD calculations yielded a value half that of experiment at −560 deg $dm^{-1}$ $(g/mL)^{-1}$. Over the next several years, careful examinations of potential sources of the discrepancy between theory and experiment were carried out, including basis-set effects [84], zero-point vibrations effects [113], and electron correlation [113], but none of these were substantial enough to account for the large difference between theory and experiment.

An important step forward was made with the first vapor-phase measurement of the specific rotation of norbornenone [113]. These measurements as well as corresponding computations revealed that the source of the discrepancy between theory and experiment is the liquid environment. However, while the gas- and condensed-phase specific rotations of norbornenone differ considerably (by ca. 200 deg $dm^{-1}$ $(g/mL)^{-1}$at 589 nm), the particular choice of solvent has a relatively small effects, with rotations in acetonitrile versus hexane differing by only 30 deg $dm^{-1}$ $(g/mL)^{-1}$. Thus, while there is only a small *solvent* effect, there is a surprisingly large *solvation* effect in this system. Furthermore, it has been demonstrated that polar solvents tend to yield rotations that are *closer* to the gas-phase values than nonpolar solvents, *a direct contradiction of conventional chemical wisdom.*

#### 3.3.2.3 (*P*)-Dimethylallene

Dimethylallene, a rigid structure that displays axial chirality, behaves somewhat similarly to norbornenone, exhibiting a remarkable difference between vapor- and liquid-phase specific rotations. However, in contrast to norbornenone, vapor-phase CRDP measurements yielded rotations a factor of two larger than those in the liquid-phase, though again with only small variations with the choice of solvent. In 2008, Wiberg et al. [81] reported an effort to understand these findings, combining experimental and theoretical analyses. Coupled cluster computations agreed well with vapor-phase data (Fig. 3.3), and Wiberg and co-workers speculated that torsional deformation of the solute in a condensed medium could lead to a shift in the specific rotation. However, Monte Carlo simulations of the neat liquid yielded no evidence of such a structural change. Once again, existing theoretical models were not capable of assessing the direct impact of the medium on the chiroptical response of the solute, and thus,

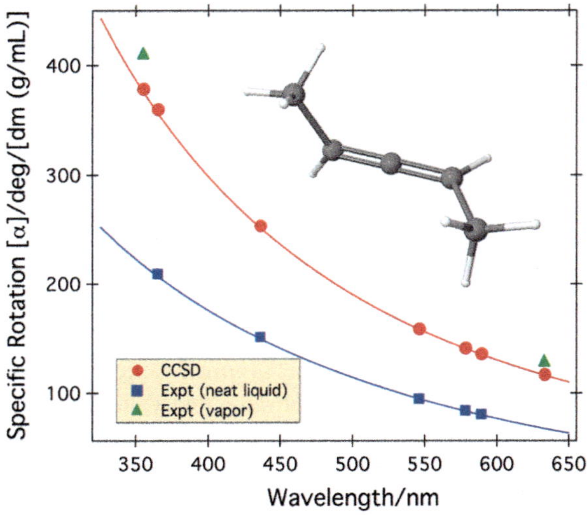


**Fig. 3.3** Experimental and theoretical optical rotatory dispersion spectra of (*P*)-(+)-2,3-pentadiene (dimethylallene). The specific rotation in the neat liquid (and numerous polar and nonpolar solvents) is roughly half that of the vapor phase, and the latter is reproduced faithfully by coupled cluster theory. All data taken from Ref. [81]

no final conclusions were possible as to the detailed reason(s) for the divergence between the vapor- and condensed-phase results.

Several questions arise from the above observations. First, what is the fundamental reason for the large difference between vapor- and liquid-phase specific rotations of compounds such as norbornenone and dimethylallene and why do such differences not occur for other compounds such as methylthiirane, [4]triangulane, $\alpha$-pinene, or 3-chloro-1-butene? Second, considering that implicit solvation models fail to reproduce the key experimental findings about methyloxirane, norbornenone, and dimethylallene, and that detailed intermolecular interactions can have a dramatic effect on molecular chiroptical response, to what extent must explicit solvation techniques in conjunction with coupled cluster linear response methods be used to obtain "the right answer for the right reason." Can one limit such calculations to the cybotactic region alone? Can the remaining bulk solvation effects be treated reliably using implicit approaches? Finally, how can such calculations be implemented with the ideal balance between accuracy and computational cost, thus serving as a viable tool for the determination of absolute stereochemical configurations.

In an effort to at least partly answer these questions, a recent collaboration between the Virginia Tech and Amsterdam groups [114] focused on a description of solvent effects that combined coupled cluster and density functional methods via the frozen-density embedding (FDE) approach. In this "wave-function-in-DFT" embedding model [115], the electron density of the solute interacting explicitly with a cluster of solvent molecules was computed using a DFT and the resulting solvent component of the density was transformed to a one-electron potential via a functional derivative. This potential was then added to the Hamiltonian for a subsequent coupled cluster linear response computation. This work employed a new interface between the PSI4 [116] and Amsterdam Density Functional (ADF) [117, 118] packages in which the FDE potential was represented on a numerical grid and then

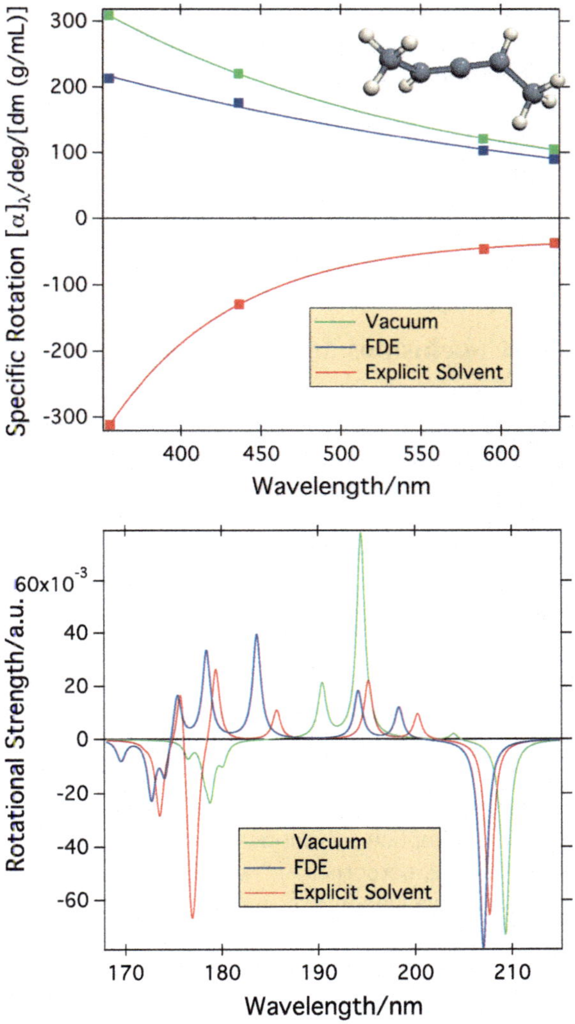


**Fig. 3.4** Coupled cluster optical rotation *(top)* and circular dichroism spectra *(bottom)* of (*P*)-dimethylallene in vacuum or interacting with two water molecules, the latter computed both with explicit water molecules and via FDE potentials. Data taken from Ref. [114]

transformed to a (Gaussian) basis set used for a CC computation of chiroptical spectra.

Unfortunately, the FDE potential does not accurately mimic the optical response of the solvated molecule. As shown in Fig. 3.4, a pair of water molecules interacting with (*P*)-dimethylallene results in a qualitative shift in the ORD curve and the ECD spectrum of the solute, including a change in sign of the specific rotation. Furthermore, low-lying Rydberg states appearing in the ECD spectrum are significantly delocalized onto the solvent molecules, increasing their contribution to both the ECD and ORD response, but these shifts are not reproduced by the FDE potential. This observation, which holds for numerous water configurations, as well as for solvents such as chloroform and other solutes such as methyloxirane and norbornenone, reveals that the chiroptical response is highly non-local in character. Neugebauer [119] has also employed FDEs in a "DFT-in-DFT" study of the optical rotation tensor of the benzaldehyde dimer and demonstrated that the failure of the FDE is a lack of response by the solvent to the external perturbation. Neugebauer clearly showed that allowing such a response essentially overcomes the error as compared to an explicit solvent calculation, though the most computationally effective way to treat this response in a "wave-function-in-DFT" FDE approach remains elusive.

## 3.4 Conclusions and Prospectus

The development and application of coupled cluster response theory for the treatment of chiroptical properties has advanced considerably in the last two decades. For properties such as specific rotation and electronic circular dichroism spectra, comparison to available gas-phase experimental measurements is excellent in nearly all cases—the fenchone molecule discussed earlier standing as a significant outlier. While some systems are straightforward, others require considerable computational effort in terms of basis-set completeness, high levels of electron correlation, inclusion of zero-point and temperature-dependent vibrational effects (including anharmonicity), and conformational averaging for a physically realistic model.

Comparison to liquid-phase measurements would be highly desirable, as the vast majority of experimental data are obtained in solution. However, the incorporation of solvation effects can dramatically increase the complexity and thus cost of the response calculation. While much effort has been expended in this area in both explicit and implicit solvent models, serious obstacles remain, as the discussion of methyloxirane, norbornenone, and dimethylallene demonstrates. Clearly, a reliable model must include some accounting for molecule-specific interactions, such that the "chiral imprint" [108] of the solute onto the solvent is correctly described. The obvious way to accomplish this goal is to include at least some solvent molecules explicitly in the linear response computation, though this also exacerbates the cost of the calculation considerably: How many solvent molecules must be included and how long should a dynamical average be carried out? If the number of required solvent species is high, then the polynomial scaling of coupled cluster theory will rapidly outstrip the available computing resources, but even if the number is low, a dynamic configurational search will still be essential for a reliable and robust model. Much progress has been made, but many questions remain and more work is to be done.

**Acknowledgements** This research was supported by a grants CHE-1465149 and ACI-1450169 from the US National Science Foundation. The author acknowledges Advanced Research Computing at Virginia Tech for providing computational resources and technical support that have contributed to the results reported within this paper.

## References


1. L.D. Barron, *Molecular Light Scattering and Optical Activity*, 2nd edn. (Cambridge University Press, Camridge, U.K., 2004)
2. M. Pecul, K. Ruud, Adv. Quantum Chem. **50**, 185 (2005)
3. T.D. Crawford, Theor. Chem. Acc. **115**, 227 (2006)
4. T.D. Crawford, M.C. Tam, M.L. Abrams, J. Phys. Chem. A **111**, 12057 (2007)
5. T.D. Crawford, in *Comprehensive Chiroptical Spectroscopy*, vol. 1, ed. by N. Berova, K. Nakanishi, R.W. Woody, P.Polavarapu (Wiley, Hoboken, New Jersey, 2012), chap. 23, pp. 675–697
6. P.L. Polavarapu, Mol. Phys. **91**(3), 551 (1997)
7. J.R. Cheeseman, M.J. Frisch, F.J. Devlin, P.J. Stephens, Chem. Phys. Lett. **252**, 211 (1996)
8. J.R. Cheeseman, M.J. Frisch, F.J. Devlin, P.J. Stephens, J. Phys. Chem. A **104**, 1039 (2000)
9. P.J. Stephens, F.J. Devlin, J.R. Cheeseman, M.J. Frisch, J. Phys. Chem. A **105**(22), 5356 (2001)
10. S. Grimme, Chem. Phys. Lett. **339**, 380 (2001)
11. K. Ruud, P.J. Stephens, F.J. Devlin, P.R. Taylor, J.R. Cheeseman, M.J. Frisch, Chem. Phys. Lett. **373**, 606 (2003)
12. M.C. Tam, N.J. Russ, T.D. Crawford, J. Chem. Phys. **121**, 3550 (2004)
13. C. Diedrich, S. Kausemann, S. Grimme, J. Comp. Meth. Sci. Eng. **4**, 1 (2004)
14. T.D. Crawford, L.S. Owens, M.C. Tam, P.R. Schreiner, H. Koch, J. Am. Chem. Soc. **127**, 1368 (2005)
15. T.D. Crawford, P.J. Stephens, J. Phys. Chem. A **112**, 1339 (2008)
16. S. Grimme, S.D. Peyerimhoff, S. Bartram, F. Vögtle, A. Breest, J. Hormes, Chem. Phys. Lett. **213**(1–2), 32 (1993)
17. F. Pulm, J. Schramm, J. Hormes, S. Grimme, S.D. Peyerimhoff, Chem. Phys. **224**, 143 (1997)
18. S. Grimme, Chem. Phys. Lett. **259**, 128 (1996)
19. S. Grimme, J. Harren, A. Sobanski, F. Vögtle, Eur. J. Org. Chem. **1998**(8), 1491 (1998)
20. M. Pecul, K. Ruud, T. Helgaker, Chem. Phys. Lett. **388**, 110 (2004)
21. T.D. Kowalczyk, M.L. Abrams, T.D. Crawford, J. Phys. Chem. A **110**, 7649 (2006)
22. T.D. Crawford, M.C. Tam, M.L. Abrams, Mol. Phys. **105**, 2607 (2007)
23. P.J. Stephens, J. Phys. Chem. **89**, 748 (1985)
24. P.J. Stephens, K.J. Jalkanene, F.J. Devlin, C.F. Chabalowski, J. Phys. Chem. **97**, 6107 (1993)
25. P.J. Stephens, F.J. Devlin, C.S. Ashvar, C.F. Chabalowski, M.J. Frisch, Faraday Discuss. **99**, 103 (1994)
26. K.L. Bak, P. Jørgensen, T. Helgaker, K. Ruud, Faraday Discuss. **99**, 121 (1994)
27. P.J. Stephens, F.J. Devlin, C.F. Chabalowski, M.J. Frisch, J. Phys. Chem. **98**(45), 11623 (1994)
28. P.J. Stephens, F.J. Devlin, Chirality **12**, 172 (2000)
29. T.B. Freedman, X. Cao, R.K. Dukor, L.A. Nafie, Chirality **15**, 743 (2003)
30. T. Helgaker, K. Ruud, K.L. Bak, P. Jørgensen, J. Olsen, Faraday Discuss. **99**, 165 (1994)
31. L.A. Nafie, Ann. Rev. Phys. Chem. **48**, 357 (1997)
32. P. Bouř, Chem. Phys. Lett. **288**, 363 (1998)
33. P. Bouř, J. Comp. Chem. **22**(4), 426 (2001)
34. W. Hug, Chem. Phys. **264**, 53 (2001)
35. W. Hug, G. Zuber, A. de Meijere, A.E. Khlebnikov, H.J. Hansen, Helv. Chim. Acta **84**, 1 (2001)
36. K. Ruud, T. Helgaker, P. Bouř, J. Phys. Chem. A **106**, 7448 (2002)
37. M. Pecul, A. Rizzo, Mol. Phys. **101**(13), 2073 (2003)
38. L.D. Barron, L. Hecht, I.H. McColl, E.W. Blanch, Mol. Phys. **102**(8), 731 (2004)
39. M. Pecul, K. Ruud, Int. J. Quantum Chem. **104**(5), 816 (2005)
40. M. Pecul, E. Larnparska, C. Cappelli, L. Frediani, K. Ruud, J. Phys. Chem. A **110**(8), 2807 (2006)
41. M. Pecul, Chirality **21**, E98 (2009)

42. J.A. Pople, in *Energy, Structure, and Reactivity*, ed. by D.W. Smith, W.B. McRae (Wiley, New York, 1973), pp. 51–61
43. H.F. Schaefer, J. Mol. Struct. (Theochem) **398**, 199 (1997)
44. R.J. Bartlett, M. Musial, Rev. Mod. Phys. **79**, 291 (2007)
45. I. Shavitt, R.J. Bartlett, *Many-Body Methods in Chemistry and Physics: MBPT and Coupled-Cluster Theory* (Cambridge University Press, Cambridge, 2009)
46. T.D. Crawford, H.F. Schaefer, in *Reviews in Computational Chemistry*, vol. 14, ed. by K.B. Lipkowitz, D.B. Boyd (VCH Publishers, New York, 2000), chap. 2, pp. 33–136
47. T. Helgaker, S. Coriani, P. Jørgensen, K. Kristensen, J. Olsen, K. Ruud, Chem. Rev. **112**, 543 (2012)
48. T.J. Lee, G.E. Scuseria, in *Quantum Mechanical Electronic Structure Calculations with Chemical Accuracy*, ed. by S.R. Langhoff (Kluwer Academic Publishers, Dordrecht, 1995), pp. 47–108
49. T. Helgaker, P. Jørgensen, J. Olsen, *Molecular Electronic Structure Theory* (Wiley, New York, 2000)
50. T. Helgaker, T.A. Ruden, P. Jørgensen, J. Olsen, W. Klopper, J. Phys. Org. Chem **17**(11), 913 (2004)
51. A. Tajti, P.G. Szalay, A.G. Császár, M. Kállay, J. Gauss, E.F. Valeev, B.A. Flowers, J. Vásquez, J.F. Stanton, J. Chem. Phys. **121**(23), 11599 (2004)
52. H.J. Monkhorst, Int. J. Quantum Chem. Symp. **11**, 421 (1977)
53. D. Mukherjee, P.K. Mukherjee, Chem. Phys. **39**, 325 (1979)
54. H. Sekino, R.J. Bartlett, Int. J. Quantum Chem. Symp. **18**, 255 (1984)
55. J. Olsen, P. Jørgensen, J. Chem. Phys. **82**(7), 3235 (1985)
56. H. Koch, P. Jørgensen, J. Chem. Phys. **93**(5), 3333 (1990)
57. O. Christiansen, C. Hättig, J. Gauss, J. Chem. Phys. **109**(12), 4745 (1998)
58. K. Hald, P. Jørgensen, C. Hättig, J. Chem. Phys. **118**(3), 1292 (2003)
59. T.B. Pedersen, H. Koch, J. Chem. Phys. **106**(19), 8059 (1997)
60. O. Christiansen, P. Jørgensen, C. Hättig, Int. J. Quantum Chem. **68**, 1 (1998)
61. T.B. Pedersen, H. Koch, J. Chem. Phys. **108**(13), 5194 (1998)
62. F. Pawlowski, J. Olsen, P. Jørgensen, J. Chem. Phys. **142**, 114109 (2015)
63. S. Coriani, F. Pawlowski, J. Olsen, P. Jørgensen, J. Chem. Phys. **144**, 024102 (2016)
64. R.J. Bartlett, Mol. Phys. **108**(21–23), 2905 (2010)
65. P. Norman, Phys. Chem. Chem. Phys. **13**, 20519 (2011)
66. G.D. Purvis, R.J. Bartlett, J. Chem. Phys. **76**, 1910 (1982)
67. P. Jørgensen, T. Helgaker, J. Chem. Phys. **89**(3), 1560 (1988)
68. P.G. Szalay, M. Nooijen, R.J. Bartlett, J. Chem. Phys. **103**(1), 281 (1995)
69. J. Gauss, J.F. Stanton, J. Chem. Phys. **103**(9), 3561 (1995)
70. P.W. Langhoff, S.T. Epstein, M. Karplus, Rev. Mod. Phys. **44**(3), 602 (1972)
71. J.E. Rice, N.C. Handy, J. Chem. Phys. **94**(7), 4959 (1991)
72. K. Sasagane, F. Aiga, R. Itoh, J. Chem. Phys. **99**(5), 3738 (1993)
73. S. Grimme, F. Furche, R. Ahlrichs, Chem. Phys. Lett. **361**, 321 (2002)
74. T.B. Pedersen, H. Koch, K. Ruud, J. Chem. Phys. **110**(6), 2883 (1999)
75. T.B. Pedersen, H. Koch, L. Boman, A.M.J.S. de Meras, Chem. Phys. Lett. **393**(4–6), 319 (2004)
76. A. de Meijere, A.F. Khlebnikov, R.R. Kostikov, S.I. Kozhushkov, P.R. Schreiner, A. Wittkopp, D.S. Yufit, Angew. Chem. Int. Ed. **38**(23), 3474 (1999)
77. T.Müller, K.B. Wiberg, P.H. Vaccaro, J. Phys. Chem. A **104**, 5959 (2000), https://doi.org/10.1021/jp000705n
78. S.M. Wilson, K.B. Wiberg, J.R. Cheeseman, M.J. Frisch, P.H. Vaccaro, J. Phys. Chem. A **109**(51), 11752 (2005), https://doi.org/10.1021/jp054283z
79. M.C. Tam, T.D. Crawford, J. Phys. Chem. A **110**, 2290 (2006), https://doi.org/10.1021/jp056093u
80. M.C. Tam, M.L. Abrams, T.D. Crawford, J. Phys. Chem. A **111**(44), 11232 (2007)

81. K.B. Wiberg, Y. Wang, S.M. Wilson, P.H. Vaccaro, W.L. Jorgensen, T.D. Crawford, M.L. Abrams, J.R. Cheeseman, M. Luderer, J. Phys. Chem. A **112**, 2415 (2008)
82. T.B. Pedersen, J. Kongsted, T.D. Crawford, Chirality **21**, S68 (2009)
83. T.B. Pedersen, J. Kongsted, T.D. Crawford, K. Ruud, J. Chem. Phys. **130**, 034310 (2009)
84. T.J. Mach, T.D. Crawford, J. Phys. Chem. A **115**, 10045 (2011)
85. K. Ruud, P.R. Taylor, P.O. Åstrand, Chem. Phys. Lett. **337**, 217 (2001)
86. J. Kongsted, T.B. Pedersen, M. Strange, A. Osted, A.E. Hansen, K.V. Mikkelsen, F. Pawlowski, P. Jørgensen, C. Hättig, Chem. Phys. Lett. **401**, 385 (2005)
87. B.C. Mort, J. Autschbach, J. Phys. Chem. A **109**(38), 8617 (2005)
88. K. Ruud, R. Zanasi, Angew. Chem. Int. Ed. Engl. **44**(23), 3594 (2005)
89. J. Neugebauer, E.J. Baerends, M. Nooijen, J. Autschbach, J. Chem. Phys. **122**, 234305 (2005)
90. J. Kongsted, T.B. Pedersen, L. Jensen, A.E. Hansen, K.V. Mikkelsen, J. Am. Chem. Soc. **128**(3), 976 (2006)
91. B.C. Mort, J. Autschbach, J. Phys. Chem. A **110**, 11381 (2006)
92. J. Kongsted, K. Ruud, Chem. Phys. Lett. **451**, 226 (2008)
93. T.D. Crawford, W.D. Allen, Mol. Phys. **107**, 1041 (2009)
94. K.B. Wiberg, P.H. Vaccaro, J.R. Cheeseman, J. Am. Chem. Soc. **125**, 1888 (2003), https://doi.org/10.1021/ja0211914
95. K.B. Wiberg, Y.G. Wang, P.H. Vaccaro, J.R. Cheeseman, M.R. Luderer, J. Phys. Chem. A **109**(15), 3405 (2005), https://doi.org/10.1021/jp040724n
96. J.M. Lambert, R.N. Compton, T.D. Crawford, J. Chem. Phys. **136**, 114512 (2012)
97. N. Berova, K. Nakanishi, R.W. Woody (eds.), *Circular Dichroism: Principles and Applications*, 2nd edn. (Wiley, New York, 2000)
98. Y. Kumata, J. Furukawa, T. Fueno, Bull. Chem. Soc. Japan **43**(12), 3920 (1970)
99. J. Tomasi, M. Persico, Chem. Rev. **94**, 2027 (1994)
100. J. Gao, Acc. Chem. Res. **29**, 298 (1996)
101. C.J. Cramer, D.G. Truhlar, Chem. Rev. **99**, 2161 (1999)
102. J. Tomasi, R. Cammi, B. Mennucci, C. Cappelli, S. Corni, Phys. Chem. Chem. Phys. **4**, 5697 (2002)
103. B. Mennucci, J. Tomasi, R. Cammi, J.R. Cheeseman, M.J. Frisch, F.J. Devlin, S. Gabriel, P.J. Stephens, J. Phys. Chem. A **106**(25), 6102 (2002)
104. M. Pecul, D. Marchesan, K. Ruud, S. Coriani, J. Chem. Phys. **122**(2), 024106 (2005)
105. H. Koch, O. Christiansen, P. Jørgensen, A.M.S. de Merás, T. Helgaker, J. Chem. Phys. **106**(5), 1808 (1997)
106. O. Christiansen, H. Koch, P. Jørgensen, J. Chem. Phys. **103**(17), 7429 (1995)
107. P. Mukhopadhyay, G. Zuber, M.R. Goldsmith, P. Wipf, D.N. Beratan, Chem. Phys. Chem. **7**(12), 2483 (2006)
108. P. Mukhopadhyay, G. Zuber, P. Wipf, D.N. Beratan, Angew. Chem. Int. Ed. **46**(34), 6450 (2007)
109. F. Lipparini, F. Egidi, C. Cappelli, V. Barone, J. Chem. Theory Comp. **9**(4), 1880 (2013)
110. M. Caricato, P.H. Vaccaro, T.D. Crawford, K.B. Wiberg, P. Lahiri, J. Phys. Chem. A **118**, 4863 (2014)
111. B. Moore, M. Srebro, J. Autschbach, J. Chem. Theory Comp. **8**, 4336 (2012)
112. D.A. Lightner, J.K. Gawroński, T.D. Bouman, J. Am. Chem. Soc. **102**, 5749 (1980)
113. P. Lahiri, K.B. Wiberg, P.H. Vaccaro, M. Caricato, T.D. Crawford, Angew. Chem. Int. Ed. Engl. **53**, 1386 (2014)
114. T.D. Crawford, A. Kumar, K. Hannon, T.J. Mach, H.R. McAlexander, S. Höfener, L. Visscher, J. Chem. Theory Comp. **11**, 5305 (2015)
115. N. Govind, Y.A. Wang, A.J.R. da Silva, E.A. Carter, Chem. Phys. Lett. **295**, 129 (1998)
116. J.M. Turney, A.C. Simmonett, R.M. Parrish, E.G. Hohenstein, F. Evangelista, J.T. Fermann, B.J. Mintz, L.A. Burns, J.J. Wilke, M.L. Abrams, N.J. Russ, M.L. Leininger, C.L. Janssen, E.T. Seidl, W.D. Allen, H.F. Schaefer, R.A. King, E.F. Valeev, C.D. Sherrill, T.D. Crawford, Wiley Interdisciplinary Reviews: Computational Molecular Science **2**, 556 (2012)

117. Amsterdam Density Functional program, Theoretical Chemistry, Vrije Universiteit, Amsterdam, http://www.scm.com
118. G. te Velde, F.M. Bickelhaupt, E.J. Baerends, C.F. Guerra, S.J.A. van Gisbergen, J.G. Snijders, T. Ziegler, J. Comp. Chem. **22**(9), 931 (2001)
119. J. Neugebauer, J. Chem. Phys. **131**, 084104 (2009)

# Chapter 4
# Response Theory and Molecular Properties

**Shane M. Parker and Filipp Furche**

**Abstract** The calculation of molecular properties, both static and dynamic, is a central goal of theoretical chemical physics. Within response theory, time-dependent properties are obtained as functional derivatives of the quantum mechanical action functional. We review how linear and nonlinear response properties may be derived from the action functional using exact electronic states, as well as within time-dependent density functional theory. Particular emphasis is given to recently discovered spurious poles in approximate nonlinear response functions.

**Keywords** Response theory · Nonlinear properties · Quantum action · Time-dependent density functional theory · Electron correlation

## 4.1 Introduction

Nonadiabatic molecular dynamics (NAMD) coupled with time-dependent density functional theory (TDDFT) has emerged as a powerful tool to elucidate mechanisms of photochemical reactions such as the photoinduced ring-opening of provitamin D [1], or the photodissociation of gas-phase acetaldehyde [2]. Each time step of a NAMD simulation requires excited-state energies and gradients as well as nonadiabatic couplings between each pair of electronic states, all of which are computed using response theory. Within response theory, transition properties between the ground state and an excited state (e.g., first-order derivative couplings) are obtained from the linear response function while transition properties between two excited states (state-to-state) are obtained from the quadratic response function. In recent years, however, several groups confirmed independently that the state-to-state derivative coupling between states $n$ and $m$ diverges unphysically when the energy

S.M. Parker · F. Furche (✉)
University of California, Irvine, 1102 Natural Sciences II, Irvine, California 92697-2025, USA
e-mail: ffurche@uci.edu

S.M. Parker
e-mail: smparker@uci.edu



M.J. Wójcik et al. (eds.), *Frontiers of Quantum Chemistry*,
https://doi.org/10.1007/978-981-10-5651-2_4

difference between states $n$ and $m$ matches any other excitation energy of the system [3–5]. These unphysical divergences are a consequence of the deficient pole structure of the approximate quadratic response function, which was first pointed out by Dalgaard in the context of time-dependent Hartree–Fock (TDHF) [6]. Furthermore, we recently showed that these unphysical divergences are not limited to TDHF or TDDFT but are found in any response method based on a nonlinear reference, including coupled cluster (CC) and multi-configurational self-consistent field (MCSCF) response theory [7]. Similar inconsistent poles have been shown to yield unphysical one-photon resonances in the TDDFT cubic response function [8]. These results call into question the validity of approximate response theory beyond linear response. The aim of this review is to provide further context for this recent development by analyzing the differences between exact and approximate response functions. We focus on TDDFT due to its popularity but emphasize that the observed divergences are not just a "TDDFT problem" but rather result from an unphysical time dependence of most effective Hamiltonians used in electronic structure theory.

The most general and widely used framework for defining time-dependent response- and excited-state properties is based on the quantum action functional [9],

$$\mathscr{S}[\boldsymbol{\Phi}] = \langle\!\langle \boldsymbol{\Phi}|\hat{H}(t) - i\frac{\partial}{\partial t}|\boldsymbol{\Phi}\rangle\!\rangle. \tag{4.1}$$

Atomic (Hartree) units are used throughout this chapter; $\boldsymbol{\Phi}$ is a normalizable time-dependent wavefunction defined on an extended Hilbert space $\mathscr{H} \oplus \mathscr{L}^2_{[0,T]}$, where $\mathscr{H}$ is the time-independent $N$-electron Hilbert space, and $\mathscr{L}^2_{[0,T]}$ is the Hilbert space of square-integrable functions on $[0, T]$ [10]. Vectors in $\mathscr{H}$ are denoted by single brackets ($|\,\rangle$) and use the usual inner product, while vectors in the extended Hilbert space are denoted by double brackets ($|\,\rangle\!\rangle$) with inner product

$$\langle\!\langle u|v\rangle\!\rangle \equiv \frac{1}{T}\int_0^T \mathrm{d}t\langle u(t)|v(t)\rangle\ . \tag{4.2}$$

The formalism applies to general time-dependent potentials [11], but we limit ourselves to time-periodic potentials here, where $T$ is chosen to be the maximum period. In this case, it is convenient to work instead in a "Floquet picture" and introduce the action Lagrangian,

$$\mathscr{A}[\boldsymbol{\Phi}, \varepsilon] = \langle\!\langle \boldsymbol{\Phi}|\hat{H}(t) - i\frac{\partial}{\partial t}|\boldsymbol{\Phi}\rangle\!\rangle - \varepsilon(\langle\!\langle \boldsymbol{\Phi}|\boldsymbol{\Phi}\rangle\!\rangle - 1), \tag{4.3}$$

where $\boldsymbol{\Phi}$ is a "Floquet picture" wavefunction that is strictly periodic with period $T$, and $\varepsilon$ is the quasi-energy, a Lagrange multiplier that enforces normalization. By construction, the action Lagrangian reduces to the quasi-energy at the stationary point (stat),

$$\mathscr{A}[\boldsymbol{\Phi}, \varepsilon]|_{\text{stat}} = \varepsilon\ , \tag{4.4}$$

while $\boldsymbol{\Phi}$ reduces to the physical Floquet wavefunction, $\boldsymbol{\Phi}|_{\text{stat}} = \Psi$. Furthermore, if the Hamiltonian is stationary, $\hat{H}(t) = \hat{H}_0$, and the reference state is an eigenstate of $\hat{H}_0$, then the quasi-energy recovers the eigenenergy of the reference state, $\mathscr{A}^{(0)}[\boldsymbol{\Phi}, \varepsilon]|_{\text{stat}} = E_n$.

By the time-dependent Hellmann–Feynman theorem, time-dependent properties of the reference state $\Psi$ are obtained as derivatives of the action Lagrangian with respect to a general external one-particle potential $v(t, x, x')$ at the stationary point. Thus, the $n$-th-order polarization propagator

$$\Pi^{(n)}(t_1, x_1, x_1', \ldots, t_n, x_n, x_n') = \left. \frac{\delta^n \mathscr{A}}{\delta v(t_1, x_1, x_1') \ldots \delta v(t_n, x_n, x_n')} \right|_{\text{stat}} \tag{4.5}$$

determines all time-dependent $n$-th-order one-particle properties of the system. For example, the time-dependent one-particle density matrix $\rho(t, x, x')$ equals $\Pi^{(1)}(t, x, x')$ and gives rise to time-dependent dipole moments, (current) densities, etc. The linear response function equals $\Pi^{(2)}(t_1, x_1, x_1', t_2, x_2, x_2')$ and gives rise to time-dependent linear polarizabilities, transition moments, excitation energies etc. If the reference state is stationary, $\Pi^{(2)}$ only depends on $t_1 - t_2$, and we will in the following consider its Fourier transform to frequency space, $\Pi^{(2)}(\omega, x_1, x_1', x_2, x_2')$.

Computing properties as action derivatives only requires access to the action as a functional of the external potential, whereas computing accurate properties from expectation values requires variational wavefunctions. Most modern electronic structure methods aim to avoid explicit representations of variational wavefunctions in favor of more compact representations of electronic degrees of freedom, such as cluster amplitudes in CC theory or Kohn–Sham orbitals in DFT. If the response functions have the correct pole structure, excited-state properties are accessible from the ground-state linear- and higher-order response functions [12]. Moreover, these properties satisfy sum rules, implying consistency between dispersive and absorptive properties such as dipole polarizabilities and transition dipole moments. While the real-time action formalism has been criticized because it does not automatically imply causality of the time-dependent response [13, 14], this appears to be less a fundamental flaw of the theory but rather a consequence of how the initial conditions are imposed on the solutions.

A crucial conceptual and computational advantage of computing properties via action derivatives is that all of the sophisticated machinery for deriving and implementing efficient analytical derivative methods for time-independent perturbation theory carries over to action derivatives in a straightforward manner. For example, commonly employed simplifications such as the $2n + 1$ rule for variational parameters (the $n$-th-order variational parameters suffice to determine the energy to order $2n + 1$) are readily applied to action derivatives [15, 16]. Furthermore, the method of Lagrange multipliers can be used for nonvariational methods in which case the $2n + 2$ rule for Lagrange multipliers (the $n$-th-order Lagrange multipliers suffice to determine the energy to order $2n + 2$) readily apply [16]. In the present formalism, the quasi-energy follows the $2n + 2$ rule.

Several other approaches are similar to the action Lagrangian approach. The first fully consistent divergence-free discussion of computing time-dependent properties from perturbation theory was Langhoff, Epstein, and Karplus [17]. Rice and Handy [18, 19] formulated response properties using what they referred to as pseudoenergy derivatives, while Sasagane et al. [20] formulated response properties in terms of a quasi-energy. The action functional used in this chapter is the time integral of the pseudoenergy of Rice and Handy or the quasi-energy of Sasagane et al.. This is directly comparable to time-averaged quasi-energy as discussed by Christiansen et al. [21], which is the integral of the quasi-energy divided by the total time. Pawłowski, Olsen, and Jørgensen introduced the quasi-energy Lagrangian in terms of an eigenvalue problem and an extended Hilbert space [22]. The response functions derived in the present chapter are equivalent to the time-averaged quasi-energy.

## 4.2 Response Functions from the Action

In this section, the general framework for deriving response functions is presented before explicit derivations for specific methods are given in Sects. 4.3 and 4.4. First, the time-dependent Hamiltonian is partitioned into

$$\hat{H}(t) = \hat{H}_0 + \hat{v}(t) \tag{4.6}$$

where $\hat{H}_0$ is the field-free time-independent Hamiltonian and $\hat{v}(t)$ is the explicitly time-dependent external operator. The external operator is chosen to be a single-particle operator parametrized using a time-dependent nonlocal potential as

$$\hat{v}(t) = \int \mathrm{d}x\mathrm{d}x' v(t, x, x')\hat{\rho}(x, x') , \tag{4.7}$$

where $\hat{\rho}(x, x')$ is the one-particle density matrix operator. The potential is expanded in a Fourier basis as

$$v(t, x, x') = \sum_{\pm\alpha} v_\alpha(x, x')\mathrm{e}^{-i\omega_\alpha t}, \tag{4.8}$$

where $\sum_{\pm\alpha}$ indicates a sum-over positive and negative indices, and $\omega_{-\alpha} = -\omega_\alpha$. Since the Hamiltonian is self-adjoint on $\mathscr{H}$ at all times, $v_{-\alpha}(x, x') = v^*_\alpha(x', x)$.

The electronic wavefunction is described by a set of nonvariational parameters, $\boldsymbol{\xi}$, and variational parameters, $\mathbf{x}(t)$. The nonvariational parameters include, for example, nuclear geometry or static external electric fields and are assumed to be time-independent. The variational parameters, on the other hand, compactly represent the time-evolving electronic wavefunction. Although any stationary state is a suitable stationary reference state, we choose the ground state as reference such that

$$\left.\frac{\delta \mathscr{A}^{(0)}}{\delta \mathbf{x}}\right|_{\text{stat}} = \frac{\delta E_0}{\delta \mathbf{x}} = 0 . \tag{4.9}$$

The frequency-dependent $n$-th-order polarization propagator is determined from the action as

$$\Pi^{(n)}(\omega_\alpha, \dots, x_1, x_1', \dots, x_n, x_n') = \left.\frac{\delta^n \mathscr{A}}{\delta v_\alpha(x_1, x_1') \dots \delta v_\gamma(x_n, x_n')}\right|_{\text{stat}} \equiv \mathscr{A}^{(\alpha \dots \gamma)} , \tag{4.10}$$

where we have introduced superscript notation for functional derivatives followed by evaluation at vanishing external field,

$$f^{(\alpha\beta\dots)} \equiv f^{(\alpha\beta\dots)}(x_1, x_1', x_2, x_2', \dots) \equiv \left.\frac{\delta^n f}{\delta v_\alpha(x_1, x_1') \delta v_\beta(x_2, x_2') \dots}\right|_{\text{stat}, v=0} , \tag{4.11}$$

and we suppress the functional arguments for brevity. A consequence of choosing the zeroth-order wavefunction to be stationary is that $\Pi^{(n)}$ contains $n-1$ unique frequencies. The first three derivatives of the action are

$$\begin{aligned}
\mathscr{A}^{(\alpha)} &= \frac{\delta \mathscr{A}}{\delta v_\alpha(x, x')} + \frac{\delta \mathscr{A}}{\delta \mathbf{x}} \mathbf{x}^{(\alpha)}, \\
\mathscr{A}^{(\alpha\beta)} &= \frac{\delta \mathscr{A}}{\delta \mathbf{x}} \mathbf{x}^{(\alpha\beta)} + \mathscr{P}^{(\alpha\beta)} \frac{\delta^2 \mathscr{A}}{\delta v_\alpha(x, x') \delta \mathbf{x}} \mathbf{x}^{(\beta)} + \frac{\delta^2 \mathscr{A}}{\delta \mathbf{x}^2} \mathbf{x}^{(\alpha)} \mathbf{x}^{(\beta)}, \\
\mathscr{A}^{(\alpha\beta\gamma)} &= \frac{\delta \mathscr{A}}{\delta \mathbf{x}} \mathbf{x}^{(\alpha\beta\gamma)} + \mathscr{P}^{(\alpha\beta\gamma)} \frac{1}{2} \left[ \frac{\delta^2 \mathscr{A}}{\delta v_\alpha(x, x') \delta \mathbf{x}} \mathbf{x}^{(\beta\gamma)} + \frac{\delta^2 \mathscr{A}}{\delta \mathbf{x}^2} \mathbf{x}^{(\alpha)} \mathbf{x}^{(\beta\gamma)} \right] \\
&\quad + \frac{\delta^3 \mathscr{A}}{\delta \mathbf{x}^3} \mathbf{x}^{(\alpha)} \mathbf{x}^{(\beta)} \mathbf{x}^{(\gamma)},
\end{aligned} \tag{4.12}$$

where it has been assumed that the action is linear in the external potential. In the previous equations, $\mathscr{P}^{(\alpha\beta\dots)}$ symmetrizes the expression to its right with respect to the labels $\alpha$, $\beta$, and so on. Extrapolating the above sequence, one sees that for $p > n/2$, the $p$-th-order variational parameter, $\mathbf{x}^{(\alpha \dots p)}$, appears linearly in the $n$-th-order action and is multiplied by a polynomial of lower-order parameters. The polynomial of lower-order parameters provides the condition that the order $n-p$ parameters must satisfy. Consequently, at the stationary point, the polynomial, and thus the contribution from the order $p$ parameter, vanishes. This result is also known as Wigner's $2n+1$ rule. For example, consider the term

$$\mathbf{x}^{(\alpha\beta)} \left( \frac{\delta^2 \mathscr{A}}{\delta v_\gamma(x, x') \delta \mathbf{x}} + \frac{\delta^2 \mathscr{A}}{\delta \mathbf{x}^2} \mathbf{x}^{(\gamma)} \right) \tag{4.13}$$

that appears in $\mathscr{A}^{(\alpha\beta\gamma)}$ above. The polynomial in parentheses is the condition that determines the first-order variational parameters, $\mathbf{x}^{(\gamma)}$, such that at the stationary point, the whole term vanishes.

## 4.3 Exact Response Functions

In this section, we use the results of the previous section to derive the response of an exact electronic state via the action formalism. We assume that absent the external potential, the electronic wavefunction is proportional to the ground eigenstate of the field-free Hamiltonian, $\hat{H}_0$, i.e.,

$$|\Psi\rangle\!\rangle|_{v=0} \equiv |\Psi^{(0)}\rangle\!\rangle = |0\rangle \tag{4.14}$$

where $|n\rangle$ labels some eigenstate of $\hat{H}_0$ with eigenvalue $E_n$. The action Lagrangian naturally produces response functions free from secular divergences [17]. In the exact case, the set of variational parameters includes both the wavefunction and its adjoint, i.e., $\mathbf{x} = \{|\Psi\rangle\!\rangle, \langle\!\langle\Psi|\}$.

### 4.3.1 Static Properties

Static (zeroth-order) properties are obtained from the first-order variation of the action with respect to external field evaluated at vanishing external field. Ignoring first-order variational parameters due to the $2n+1$ rule, the first-order action is straightforward,

$$\mathscr{A}^{(\alpha)} = \varepsilon^{(\alpha)} = \langle\!\langle\Psi^{(0)}|\hat{\rho}(x,x')\mathrm{e}^{-i\omega_\alpha t}|\Psi^{(0)}\rangle\!\rangle \,. \tag{4.15}$$

Since the time integral is nonzero only when $\omega_\alpha = 0$,

$$\Pi^{(1)}(x,x') = \rho^{00}(x,x') \,, \tag{4.16}$$

where $\rho^{nm}(x,x') = \langle n|\hat{\rho}(x,x')|m\rangle$ is the one-electron transition density matrix.

### 4.3.2 First-Order Properties

First-order properties result from the second-order variation of the action with respect to external field evaluated at vanishing external field, in analogy with the static properties of the previous subsection. Neglecting terms involving the second-order wavefunction or quasi-energy (as previously) leads to

$$\begin{aligned}\mathscr{A}^{(\alpha\beta)} &= \langle\!\langle\Psi^{(\alpha)}|\hat{H}_0 - E_0 - i\frac{\partial}{\partial t}|\Psi^{(\beta)}\rangle\!\rangle + \langle\!\langle\Psi^{(\beta)}|\hat{H}_0 - E_0 - i\frac{\partial}{\partial t}|\Psi^{(\alpha)}\rangle\!\rangle \\ &+ \langle\!\langle\Psi^{(\alpha)}|\bar{\rho}(x_2,x_2')\mathrm{e}^{-i\omega_\beta t}|\Psi^{(0)}\rangle\!\rangle + \langle\!\langle\Psi^{(0)}|\bar{\rho}(x_2,x_2')\mathrm{e}^{-i\omega_\beta t}|\Psi^{(\alpha)}\rangle\!\rangle \\ &+ \langle\!\langle\Psi^{(\beta)}|\bar{\rho}(x_1,x_1')\mathrm{e}^{-i\omega_\alpha t}|\Psi^{(0)}\rangle\!\rangle + \langle\!\langle\Psi^{(0)}|\bar{\rho}(x_1,x_1')\mathrm{e}^{-i\omega_\alpha t}|\Psi^{(\beta)}\rangle\!\rangle \,,\end{aligned} \tag{4.17}$$

where $\bar{\rho}(x,x') = \hat{\rho}(x,x') - \rho^{00}(x,x')$ is the one-particle density matrix fluctuation operator. Enforcing the stationarity of the previous equation with respect to the first-order wavefunctions gives rise to the linear response equations

$$\frac{\mathscr{A}^{(\alpha\beta)}}{\delta\langle\!\langle\Psi^{(\beta)}|} = (\hat{H}_0 - E_0 - i\frac{\partial}{\partial t})|\Psi^{(\alpha)}\rangle\!\rangle + \bar{\rho}(x_1,x_1')e^{-i\omega_\alpha t}|\Psi^{(0)}\rangle\!\rangle = 0 \tag{4.18}$$

$$\frac{\mathscr{A}^{(\alpha\beta)}}{\delta\langle\!\langle\Psi^{(\alpha)}|} = (\hat{H}_0 - E_0 - i\frac{\partial}{\partial t})|\Psi^{(\beta)}\rangle\!\rangle + \bar{\rho}(x_2,x_2')e^{-i\omega_\beta t}|\Psi^{(0)}\rangle\!\rangle = 0\,. \tag{4.19}$$

Moving the second term across the equality and moving the time-dependent phase factor to the left-hand side yields

$$e^{i\omega_\alpha t}(\hat{H}_0 - E_0 - i\frac{\partial}{\partial t})|\Psi^{(\alpha)}\rangle\!\rangle = -\bar{\rho}(x_1,x_1')|0\rangle\,. \tag{4.20}$$

The time-independence of the right-hand side implies that $|\Psi^{(\alpha)}\rangle\!\rangle = e^{-i\omega_\alpha t}|\tilde{\Psi}^{(\alpha)}\rangle$ with $|\tilde{\Psi}^{(\alpha)}\rangle$ time-independent. Thus, the first-order wavefunction is

$$|\Psi^{(\alpha)}\rangle\!\rangle = -(\hat{H}_0 - E_0 - \omega_\alpha)^{-1}\bar{\rho}(x_1,x_1')|0\rangle e^{-i\omega_\alpha t}\,. \tag{4.21}$$

Stationarizing with respect to $|\Psi^{(\alpha)}\rangle\!\rangle$ and $|\Psi^{(\beta)}\rangle\!\rangle$ and using the fact that the time derivative is an anti-Hermitian operator on the extended Hilbert space show that

$$\langle\!\langle\Psi^{(\alpha)}| = |\Psi^{(-\alpha)}\rangle\!\rangle^{\dagger} = -e^{-i\omega_\alpha t}\langle 0|\bar{\rho}(x_1,x_1')(\hat{H}_0 - E_0 + \omega_\alpha)^{-1}\,. \tag{4.22}$$

At the stationary point, by virtue of Eqs. (4.18) and (4.19), the linear response function further simplifies to

$$\mathscr{A}^{(\alpha\beta)} = \mathscr{P}^{(\alpha\beta)}\langle\!\langle\Psi^{(0)}|\bar{\rho}(x_2,x_2')e^{-i\omega_\beta t}|\Psi^{(\alpha)}\rangle\!\rangle\,. \tag{4.23}$$

Finally, this yields

$$\begin{aligned}\Pi^{(2)}(\omega,x_1,x_1',x_2,x_2') = -\Big[&\langle 0|\bar{\rho}(x_2,x_2')(\hat{H}_0 - E_0 - \omega)^{-1}\bar{\rho}(x_1,x_1')|0\rangle \\ &+ \langle 0|\bar{\rho}(x_1,x_1')(\hat{H}_0 - E_0 + \omega)^{-1}\bar{\rho}(x_2,x_2')|0\rangle\Big]\,.\end{aligned} \tag{4.24}$$

Inserting the spectral representation of the inverse operator, $(\hat{H}_0 - z)^{-1} = \sum_n \frac{|n\rangle\langle n|}{E_n - z}$, we arrive at the usual sum-over-states expression for the polarization propagator

$$\Pi^{(2)}(\omega,x_1,x_1',x_2,x_2') = -\sum_{n\neq 0}\left[\frac{\rho^{0n}(x_2,x_2')\rho^{n0}(x_1,x_1')}{\Omega_{0n} - \omega} + \frac{\rho^{0n}(x_1,x_1')\rho^{n0}(x_2,x_2')}{\Omega_{0n} + \omega}\right]\,. \tag{4.25}$$

### 4.3.3 Second-Order Properties

In this section, we derive second-order response properties and discuss extensions beyond second order. According to the $2n + 1$ rule for the variational parameters, no terms higher than first order are required, leading to

$$\mathscr{A}^{(\alpha\beta\gamma)} = \mathscr{P}^{(\alpha\beta\gamma)} \langle\!\langle \Psi^{(\alpha)} | \bar{\rho}(x_2, x_2') \mathrm{e}^{-i\omega_\beta t} | \Psi^{(\gamma)} \rangle\!\rangle \, . \tag{4.26}$$

The third-order polarization propagator is obtained by combining this with Eqs. (4.21) and (4.22) and integrating over time,

$$\begin{aligned} &\Pi^{(3)}(\omega_\alpha, \omega_\beta, x_1, x_1', x_2, x_2', x_3, x_3') = \\ &\quad \mathscr{P}^{(\alpha\beta\gamma)} \langle 0 | \bar{\rho}(x_1, x_1')(\hat{H}_0 - E_0 + \omega_\alpha)^{-1} \bar{\rho}(x_2, x_2')(\hat{H}_0 - E_0 - \omega_\gamma)^{-1} \bar{\rho}(x_3, x_3') | 0 \rangle \, , \end{aligned} \tag{4.27}$$

which has the sum-over-states representation [23]

$$\begin{aligned} &\Pi^{(3)}(\omega_\alpha, \omega_\beta, x_1, x_1', x_2, x_2', x_3, x_3') = \\ &\sum_{\substack{n\neq 0\\ m\neq 0}} \left[ \frac{\rho^{0n}(x_1, x_1')\bar{\rho}^{nm}(x_2, x_2')\rho^{m0}(x_3, x_3')}{(\Omega_{0n} - \omega_\alpha)(\Omega_{0m} + \omega_\gamma)} + \frac{\rho^{0n}(x_1, x_1')\bar{\rho}^{nm}(x_3, x_3')\rho^{m0}(x_2, x_2')}{(\Omega_{0n} + \omega_\alpha)(\Omega_{0m} - \omega_\beta)} \right. \\ &\quad + \frac{\rho^{0n}(x_2, x_2')\bar{\rho}^{nm}(x_1, x_1')\rho^{m0}(x_3, x_3')}{(\Omega_{0n} + \omega_\beta)(\Omega_{0m} - \omega_\gamma)} + \frac{\rho^{0n}(x_2, x_2')\bar{\rho}^{nm}(x_3, x_3')\rho^{m0}(x_1, x_1')}{(\Omega_{0n} + \omega_\beta)(\Omega_{0m} - \omega_\alpha)} \\ &\quad \left. + \frac{\rho^{0n}(x_3, x_3')\bar{\rho}^{nm}(x_1, x_1')\rho^{m0}(x_2, x_2')}{(\Omega_{0n} + \omega_\gamma)(\Omega_{0m} - \omega_\beta)} + \frac{\rho^{0n}(x_3, x_3')\bar{\rho}^{nm}(x_2, x_2')\rho^{m0}(x_1, x_1')}{(\Omega_{0n} + \omega_\gamma)(\Omega_{0m} - \omega_\alpha)} \right] , \end{aligned} \tag{4.28}$$

where $\bar{\rho}^{nm}(x, x') = \rho^{nm}(x, x') - \delta_{nm}\rho^{00}(x, x')$ and $\omega_\alpha + \omega_\beta + \omega_\gamma = 0$.

## 4.4 TDHF and TDDFT Response Functions

In TDDFT, the action of an interacting $N$-electron state is obtained by stationarizing the functional [24]

$$\begin{aligned} \mathscr{A}[\varphi, \varepsilon] = &\frac{1}{T} \int_0^T \mathrm{d}t \sum_j \left[ \int \mathrm{d}x \left( \frac{1}{2} |\nabla \varphi_j(t,x)|^2 - i\varphi_j^*(t,x) \frac{\partial \varphi_j(t,x)}{\partial t} \right) \right. \\ &\left. + \int \mathrm{d}x\mathrm{d}x' \varphi_j^*(t,x) v(t,x,x') \varphi_j(t,x') - \sum_k \varepsilon_{jk} \left( \int \mathrm{d}x \varphi_j^*(t,x)\varphi_k(t,x) - \delta_{jk} \right) \right] \\ &+ \mathscr{A}^{\mathrm{HXC}}[\rho] \end{aligned} \tag{4.29}$$

with respect to the time-dependent density $\rho(t,x) = \rho(t,x,x) = \sum_j |\varphi_i(t,x)|^2$ represented by the $N$ noninteracting orthonormal time-dependent Kohn–Sham (TDKS) orbitals $\varphi_j$. If the external potential is local, i.e., $v(t,x,x') = \delta(x - x')v(t,x)$, then the stationary point of $\mathscr{A}$ equals the action of the physical system by the Runge–Gross theorem [24], and $\delta\mathscr{A}/\delta v(t,x)|_{\text{stat}} = \rho(t,x)$ implies that $\rho(t,x)$ is the physical time-dependent density. Bearing this in mind, we will consider the general nonlocal case for formal and algorithmic reasons in the following. The Hartree, exchange, and correlation action $\mathscr{A}^{\text{HXC}}[\rho]$ is also a functional of the initial state, which will always be the static ground state in the following.

Enforcing the stationarity of $\mathscr{A}$ with respect to the one-particle orbital, $\varphi_j(t,x)$, shows that the TDKS orbitals evolve in time according to

$$\frac{\delta\mathscr{A}[\varphi,\varepsilon]}{\delta\varphi_j^*(t,x)} = \frac{1}{2}\nabla^2\varphi_j(t,x) + \int \mathrm{d}x' v^s(t,x,x')\varphi_j(t,x') - i\frac{\partial\varphi_j(t,x)}{\partial t} - \sum_k \varepsilon_{jk}\varphi_k(t,x) = 0\,, \quad (4.30)$$

where $v^s(t,x,x') = v(t,x,x') + v^{\text{HXC}}(t,x,x')$ and the time-dependent HXC potential is

$$v^{\text{HXC}}[\rho](t,x,x') = \left.\frac{\delta\mathscr{A}^{\text{HXC}}[\rho]}{\delta\rho(t,x,x')}\right|_{\text{ref}}, \quad (4.31)$$

where $|_{\text{ref}}$ indicates that the derivative must be evaluated at the ground-state reference density. At the stationary point, the action Lagrangian evaluates to

$$\mathscr{A}[\varphi,\varepsilon]|_{\text{stat}} = \sum_j \left(\varepsilon_{jj} - v_{jj}^{\text{HXC}}[\rho]\right) + \mathscr{A}^{\text{HXC}}[\rho]\,, \quad (4.32)$$

where $v_{jj}^{\text{HXC}}[\rho]$ is a matrix element of the HXC potential. With no external field, this simplifies to the ground-state energy,

$$\mathscr{A}^{(0)}[\varphi,\varepsilon]\Big|_{\text{stat}} = \sum_j \left(\varepsilon_{jj}^{\text{KS}} - v_{jj}^{\text{HXC}}[\rho]\right) + E^{\text{HXC}}[\rho] = \sum_j h_{jj} + E^{\text{HXC}}[\rho]\,, \quad (4.33)$$

where $h_{pq} = \int \mathrm{d}x\mathrm{d}x'\varphi_p(x)\left(\frac{1}{2}\nabla^2 + v_0(x,x')\right)\varphi_q(x')$ is a matrix element of the one-electron core Hamiltonian.

Within the widely used adiabatic approximation (AA) of TDDFT [25], the HXC action $\mathscr{A}^{\text{HXC}}$ is approximated by the time-integrated static HXC energy functional evaluated instantaneously at $\rho(t)$,

$$\mathscr{A}^{\text{HXC, AA}}[\rho] = \frac{1}{T}\int_0^T \mathrm{d}tE^{\text{HXC}}[\rho(t)]. \quad (4.34)$$

In practice, $E^{\text{HXC}}$ is further approximated by semi-local density functionals. Exact exchange is particularly important for time-dependent properties and widely used in applications. In hybrid TDDFT, the total action is usually made stationary with respect to the TDKS orbitals or density matrix instead of the density, analogous to the generalized Kohn–Sham (KS) method allowing for nonlocal exchange–correlation (XC) potentials in static DFT [26].

Just as in the previous section, we choose our initial state to be the ground state of the interacting system such that absent the external field, there is no response and the Lagrangian evaluates to the ground-state energy. For convenience, we express the TDKS orbitals in terms of the field-free (static) KS orbitals, i.e., orbitals that satisfy

$$\frac{1}{2}\nabla^2\phi_p(x) + \int \mathrm{d}x' v^s(x,x')\phi_p(x') = \left(\hat{h} + \hat{v}^{\mathrm{HXC}}\right)\phi_p(x) = \varepsilon_p^{\mathrm{KS}}\phi_p(x) \,, \tag{4.35}$$

with $\hat{h}$ being the sum of the kinetic energy and external potential terms and $\varepsilon_p^{\mathrm{KS}}$ being the Kohn–Sham eigenvalue. The time dependence of the TDKS orbitals is prescribed through a time-dependent unitary transformation $\mathbf{U}(t) = \mathrm{e}^{\kappa(t)}$ in which the unitary generator, $\kappa(t)$, is anti-Hermitian and only the occupied-virtual and virtual-occupied blocks are nonzero, i.e., $\kappa_{pq}(t) = -\kappa_{qp}^*(t)$ and $\kappa_{ij}(t) = \kappa_{ab}(t) = 0$. In the previous definitions and for the rest of this chapter, *ijkl* stand for occupied orbitals; *abcd* for virtual orbitals; and *pqrs* for general orbitals. Thus, the TDKS orbitals are expanded as

$$\varphi_j(t,x) = \phi_j(x) + \sum_a \phi_a(x)\kappa_{ja}(t) + \frac{1}{2}\sum_{ak}\phi_k(x)\kappa_{ja}(t)\kappa_{ak}(t) + \cdots \tag{4.36}$$

$$\varphi_a(t,x) = \phi_a(x) + \sum_j \phi_j(x)\kappa_{aj}(t) + \frac{1}{2}\sum_{jb}\phi_b(x)\kappa_{aj}(t)\kappa_{jb}(t) + \cdots \,. \tag{4.37}$$

### 4.4.1 Static Properties

The first derivative of the action, neglecting wavefunctions first order or higher, is

$$\begin{aligned}\mathscr{A}^{(\alpha)} = \varepsilon^{(\alpha)} &= \frac{1}{T}\int_0^T \mathrm{d}t \int \mathrm{d}x\mathrm{d}x' \sum_j \phi_j^*(x)\delta(x-x_1)\delta(x'-x_1')\phi_j(x')\mathrm{e}^{-i\omega_\alpha t} \\ &= \frac{1}{T}\int_0^T \mathrm{d}t \int \sum_j \phi_j^*(x_1)\phi_j(x_1')\mathrm{e}^{-i\omega_\alpha t} = \frac{1}{T}\int_0^T \mathrm{d}t\rho^{00}(x_1,x_1')\mathrm{e}^{-i\omega_\alpha t} \,,\end{aligned} \tag{4.38}$$

such that

$$\Pi^{(1)}(x,x') = \rho^{00}(x,x') \,. \tag{4.39}$$

While the static Kohn–Sham density matrix $\rho^{00}(x,x')$ differs from the interacting one, its diagonal $\rho^{00}(x,x)$ equals the physical ground-state density.

### 4.4.2 First-Order Properties

Following the same procedure as for exact eigenstates, we first differentiate the action with respect to the external potential twice, ignoring terms that depend on $\varphi^{(\alpha\beta)}$. Suppressing function arguments for orbitals, the differentiated action is

$$\begin{aligned}\mathscr{A}^{(\alpha\beta)} = \int_0^T \mathrm{d}t \int \mathrm{d}x \sum_j \Big[ & \varphi_j^{*(\alpha)} \left(\hat{h} - \varepsilon_j^{\mathrm{KS}} - i\frac{\partial}{\partial t}\right) \varphi_j^{(\beta)} + \varphi_j^{*(\beta)} \left(\hat{h} - \varepsilon_j^{\mathrm{KS}} - i\frac{\partial}{\partial t}\right) \varphi_j^{(\alpha)} \\ & + \varphi_j^{*(\alpha)} \bar{\rho}(x_2, x_2') \varphi_j \mathrm{e}^{-i\omega_\beta t} + \varphi_j^* \bar{\rho}(x_2, x_2') \varphi_j^{(\alpha)} \mathrm{e}^{-i\omega_\beta t} \\ & + \varphi_j^{*(\beta)} \bar{\rho}(x_1, x_1') \varphi_j \mathrm{e}^{-i\omega_\alpha t} + \varphi_j^* \bar{\rho}(x_1, x_1') \varphi_j^{(\beta)} \mathrm{e}^{-i\omega_\alpha t} \Big] \\ & + \mathscr{A}^{\mathrm{HXC}(\alpha\beta)}[\rho] \, . \end{aligned} \tag{4.40}$$

Derivatives of the HXC action are expressed using the chain rule of the functional derivative,

$$\begin{aligned}\mathscr{A}^{\mathrm{HXC}(\alpha\beta)}[\rho] = & \int \mathrm{d}\zeta v^{\mathrm{HXC}}[\rho](\tau, y, y') \rho^{(\alpha\beta)}(\tau, y, y') \\ & + \int \mathrm{d}\zeta_1 \mathrm{d}\zeta_2 f^{\mathrm{HXC}}[\rho](\tau_1 - \tau_2, y_1, y_1', y_2, y_2') \rho^{(\alpha)}(\tau_1, y_1, y_1') \rho^{(\beta)}(\tau_2, y_2, y_2') \, , \end{aligned} \tag{4.41}$$

where $\mathrm{d}\zeta_n = \mathrm{d}\tau_n \mathrm{d}y_n \mathrm{d}y_n'$ and we have defined the exchange-correlation kernel

$$f^{\mathrm{HXC}}[\rho](t_1 - t_2, x_1, x_1', x_2, x_2') \equiv \left. \frac{\delta^2 \mathscr{A}^{\mathrm{HXC}}[\rho]}{\delta \rho(t_1, x_1, x_1') \delta \rho(t_2, x_2, x_2')} \right|_{\mathrm{ref}} . \tag{4.42}$$

If the system is initially in the ground state, the frequency-dependent TDKS polarization propagator obtained from enforcing stationarity with respect to $\boldsymbol{\kappa}^{(\alpha)}(t)$ and $\boldsymbol{\kappa}^{(\beta)}(t)$ in Eq. (4.15) has a simple supermatrix form,

$$\boldsymbol{\Pi}^{(2)}_{\mathrm{TDDFT}}(\omega) = - \left[ \begin{pmatrix} \mathbf{A} & \mathbf{B} \\ \mathbf{B} & \mathbf{A} \end{pmatrix} - \omega \begin{pmatrix} \mathbf{1} & \mathbf{0} \\ \mathbf{0} & \mathbf{-1} \end{pmatrix} \right]^{-1} . \tag{4.43}$$

The orbital rotation Hessians [27]

$$\begin{aligned}(A + B)_{iajb} &= (\varepsilon_a^{\mathrm{KS}} - \varepsilon_i^{\mathrm{KS}}) \delta_{ij} \delta_{ab} + 2 f_{iajb}^{\mathrm{HXC}+}(\omega) \\ (A - B)_{iajb} &= (\varepsilon_a^{\mathrm{KS}} - \varepsilon_i^{\mathrm{KS}}) \delta_{ij} \delta_{ab} + 2 f_{iajb}^{\mathrm{HXC}-}(\omega)\end{aligned} \tag{4.44}$$

are the second derivatives of $\mathscr{A}$ with respect to the TDKS density matrix and contain the bare orbital energy differences $\varepsilon_a^{\mathrm{KS}} - \varepsilon_i^{\mathrm{KS}}$ plus corrections for Hartree, exchange, and correlation (XC) effects represented by the nondiagonal kernels $f^{\mathrm{HXC}\pm}$. Within the adiabatic approximation, the HXC kernels are frequency-independent and reduce to their static ($\omega = 0$) counterparts,

$$f_{iajb}^{\text{HXC}+}(\omega) \approx (ia\lfloor jb) - \frac{1}{2}c_x\left[(ib\lfloor ja) + (ij|ab)\right] + f_{iajb}^{\text{XC}+}$$
$$f_{iajb}^{\text{HXC}-}(\omega) \approx \frac{1}{2}c_x\left[(ib\lfloor ja) - (ij|ab)\right] + f_{iajb}^{\text{XC}-}\,, \tag{4.45}$$

with $(pq|rs)$ a matrix element of the two-electron Coulomb integral and

$$f^{\text{XC}\pm}(x_1, x_1', x_2, x_2') = \left.\frac{\delta^2 E^{\text{XC}}[\rho]}{\delta\rho^{\pm}(x_1, x_1')\delta\rho^{\pm}(x_2, x_2')}\right|_{\text{ref}}, \tag{4.46}$$

where $\rho^{\pm}$ stand for the real and imaginary parts of the TDKS density matrix [28]. The coefficient $c_x$ interpolates between the proper Kohn–Sham limit ($c_x = 0$) and TDHF ($c_x = 1$ and no correlation). The dimension of **A** and **B** is thus $N_{ph} \times N_{ph}$, where $N_{ph}$ denotes the dimension of the particle-hole space (number of occupied times number of virtual orbitals).

First-order properties are computed as

$$\Pi^{(2)}(\omega, x_1, x_1', x_2, x_2') = \begin{pmatrix}\mathbf{P}^{(\beta)}\\ \mathbf{Q}^{(\beta)}\end{pmatrix}^T \begin{pmatrix}\mathbf{X}^{(\alpha)}\\ \mathbf{Y}^{(\alpha)}\end{pmatrix}, \tag{4.47}$$

where

$$P_{ia}^{(\alpha)} = \phi_a(x_1)\phi_i(x_1')\,, \quad Q_{ia}^{(\alpha)} = \phi_i(x_1)\phi_a(x_1')\,, \tag{4.48}$$

and

$$\begin{pmatrix}\mathbf{X}^{(\alpha)}\\ \mathbf{Y}^{(\alpha)}\end{pmatrix} = \boldsymbol{\Pi}_{\text{TDDFT}}^{(2)}(\omega)\begin{pmatrix}\mathbf{P}^{(\alpha)}\\ \mathbf{Q}^{(\alpha)}\end{pmatrix}. \tag{4.49}$$

$\boldsymbol{\Pi}_{\text{TDDFT}}^{(2)}$ has poles at the physical excitation energies $\Omega_n$, which are obtained from the symplectic eigenvalue problem [27, 29–31]

$$\left[\begin{pmatrix}\mathbf{A} & \mathbf{B}\\ \mathbf{B} & \mathbf{A}\end{pmatrix} - \Omega_n\begin{pmatrix}\mathbf{1} & \mathbf{0}\\ \mathbf{0} & -\mathbf{1}\end{pmatrix}\right]\begin{pmatrix}\mathbf{X}_n\\ \mathbf{Y}_n\end{pmatrix} = \mathbf{0}, \quad \mathbf{X}_n^T\mathbf{X}_m - \mathbf{Y}_n^T\mathbf{Y}_m = \delta_{nm}. \tag{4.50}$$

The eigenvectors $\mathbf{X}_n, \mathbf{Y}_n$ yield physical transition densities between the ground state and an excited state.

### 4.4.3 Second-Order Properties

Second-order properties within TDDFT result from the third derivative of the action,

$$\mathscr{A}^{(\alpha\beta\gamma)} = \int_0^T \mathrm{d}t \int \mathrm{d}x\mathrm{d}x' \sum_j \mathscr{P}^{(\alpha\beta\gamma)}\varphi_j^{*(\alpha)}\bar{\rho}(x_3, x_3')\mathrm{e}^{-i\omega_\gamma t}\varphi_j^{(\beta)} + \mathscr{A}^{\text{HXC}(\alpha\beta\gamma)}[\rho]\,. \tag{4.51}$$

The first term in the previous equation resembles the result in the exact case and gives rise to the "unrelaxed" contribution,

$$\mathscr{A}_{\text{unrel}}^{(\alpha\beta\rho)} = \mathscr{P}^{(\alpha\beta\rho)} \int_0^T \mathrm{d}t \int \mathrm{d}x\mathrm{d}x' \sum_j \varphi_j^{*(\alpha)} \bar{\rho}(x_3, x_3')\mathrm{e}^{-i\omega_\rho t} \varphi_j^{(\beta)} . \tag{4.52}$$

The third derivative of the HXC action is

$$\begin{aligned}
&\mathscr{A}^{\text{HXC}(\alpha\beta\gamma)} = \\
&\int \mathrm{d}\zeta_1 \mathrm{d}\zeta_2 \mathrm{d}\zeta_3 g^{\text{HXC}}[\rho](\tau_1 - \tau_2, \tau_2 - \tau_3, y_1, y_1', y_2, y_2', y_3, y_3') \\
&\qquad \times \rho^{(\alpha)}(\tau_1, y_1, y_1')\rho^{(\beta)}(\tau_2, y_2, y_2')\rho^{(\gamma)}(\tau_3, y_3, y_3') \\
&+ \int \mathrm{d}\zeta_1 \mathrm{d}\zeta_2 f^{\text{HXC}}[\rho](\tau_1 - \tau_2, y_1, y_1', y_2, y_2') \frac{\mathscr{P}^{(\alpha\beta\gamma)}}{2} \rho^{(\alpha)}(\tau_1, y_1, y_1')\rho^{(\beta\gamma)}(\tau_2, y_2, y_2') \\
&+ \int \mathrm{d}\zeta v^{\text{HXC}}[\rho](\tau, y, y')\rho^{(\alpha\beta\gamma)}(\tau, y, y') ,
\end{aligned} \tag{4.53}$$

where we have defined the HXC hyperkernel,

$$g^{\text{HXC}}[\rho](t_1 - t_2, t_2 - t_3, x_1, x_1', x_2, x_2', x_3, x_3') = \left. \frac{\delta^3 \mathscr{A}^{\text{HXC}}[\rho]}{\delta\rho(t_1, x_1, x_1')\delta\rho(t_2, x_2, x_2')\delta\rho(t_3, x_3, x_3')} \right|_{\text{ref}} . \tag{4.54}$$

Note that the HXC action derivative contains a term that depends on the product of *three* first-order density matrices, while no such terms appear in the exact case. In the basis of static Kohn–Sham orbitals, the response function becomes

$$\Pi^{(3)}(\omega_\alpha, \omega_\beta) = \left[ \text{tr}(\mathbf{K}^{(\alpha\beta)}\mathbf{v}^{(\gamma)}) + \begin{pmatrix} \mathbf{P}^{(\gamma)} \\ \mathbf{Q}^{(\gamma)} \end{pmatrix}^T \Pi_{\text{TDDFT}}^{(2)}(\omega_\alpha + \omega_\beta) \begin{pmatrix} \mathbf{P}^{(\alpha\beta)} \\ \mathbf{Q}^{(\alpha\beta)} \end{pmatrix} \right] , \tag{4.55}$$

where $\mathbf{K}^{(\alpha\beta)}$ is the unrelaxed second-order density matrix,

$$\begin{aligned}
K_{ij}^{(\alpha\beta)} &= -\sum_a \mathscr{P}^{(\alpha\beta)} X_{ja}^{(\alpha)} Y_{ia}^{(\beta)} , \\
K_{ab}^{(\alpha\beta)} &= \sum_i \mathscr{P}^{(\alpha\beta)} X_{ia}^{(\alpha)} Y_{ib}^{(\beta)} , \\
K_{ia}^{(\alpha\beta)} &= K_{ai}^{(\alpha\beta)} = 0 ,
\end{aligned} \tag{4.56}$$

$\mathbf{v}^{(\gamma)}$ is the matrix representation of the perturbation,

$$v_{pq}^{(\alpha)} = \phi_p(x_1)\phi_q(x_1') , \tag{4.57}$$

and the second-order RHS is

$$
\begin{aligned}
(P+Q)_{ia}^{(\alpha\beta)} =& -\frac{1}{2}\sum_j \mathscr{P}^{(\alpha\beta)}\left[(X+Y)_{ja}^{(\beta)}U_{ji}^{+(\alpha)} - (X-Y)_{ja}^{(\beta)}U_{ji}^{-(\alpha)}\right] \\
&+\frac{1}{2}\sum_b \mathscr{P}^{(\alpha\beta)}\left[(X+Y)_{ib}^{(\beta)}U_{ab}^{+(\alpha)} - (X-Y)_{ib}^{(\beta)}U_{ab}^{-(\alpha)}\right] \\
&+H_{ai}^{+}[\mathbf{K}^{(\alpha\beta)}] + 2g_{ai}^{\mathrm{HXC}+}[\rho^{(\alpha)},\rho^{(\beta)}]\,, \\
(P-Q)_{ia}^{(\alpha\beta)} =& -\frac{1}{2}\sum_j \mathscr{P}^{(\alpha\beta)}\left[(X-Y)_{ja}^{(\beta)}U_{ji}^{+(\alpha)} - (X+Y)_{ja}^{(\beta)}U_{ji}^{-(\alpha)}\right] \\
&+\frac{1}{2}\sum_b \mathscr{P}^{(\alpha\beta)}\left[(X-Y)_{ib}^{(\beta)}U_{ab}^{+(\alpha)} - (X+Y)_{ib}^{(\beta)}U_{ab}^{-(\alpha)}\right] \\
&-H_{ia}^{-}[\mathbf{K}^{(\alpha\beta)}] + 2g_{ai}^{\mathrm{HXC}-}[\rho^{(\alpha)},\rho^{(\beta)}]\,,
\end{aligned}
\tag{4.58}
$$

where

$$
\begin{aligned}
U_{pq}^{\pm(\alpha)} &= H_{pq}^{\pm}[(X\pm Y)^{(\alpha)}] + v_{qp}^{(\alpha)} \pm v_{pq}^{(\alpha)}\,, \\
H_{pq}^{\pm}[\mathbf{M}^{(\alpha)}] &= \sum_{rs} f_{pq,rs}^{\mathrm{HXC}\pm}(\omega_\alpha)M_{rs}^{(\alpha)}\,,
\end{aligned}
\tag{4.59}
$$

and

$$
g_{pq}^{\mathrm{HXC}\pm}[\mathbf{M}^{(\alpha)},\mathbf{M}^{(\beta)}] = \sum_{rsr's'} g_{pq,rs,r's'}^{\mathrm{HXC}\pm}[\rho](\omega_\alpha,\omega_\beta)M_{rs}^{(\alpha)}M_{r's'}^{(\beta)}\,. \tag{4.60}
$$

$g^{\mathrm{HXC}\pm}$ is the double Fourier transform of the hyperkernel with $\pm$ referring to the real (+) or imaginary (−) parts of the density matrix. Within the adiabatic approximation, the HXC hyperkernel reduces to the third functional derivative of the exchange-correlation energy,

$$
g^{\mathrm{XC}\pm}[\rho](x_1,x_1',x_2,x_2',x_3,x_3') \approx \left.\frac{\delta^3 E^{\mathrm{XC}}[\rho]}{\delta\rho^{\pm}(x_1,x_1')\delta\rho(x_2,x_2')\delta\rho(x_3,x_3')}\right|_{\mathrm{ref}}\,. \tag{4.61}
$$

### *4.4.4 State-to-state Transition Properties*

The transition density matrix between two excited states $n$ and $m$ is a central quantity for excited-to-excited-state nonadiabatic couplings [3–5] and radiative transition probabilities between excited states. For the *exact* electronic ground state, the transition density matrix is obtained from a double residue of the quadratic response function,

$$
\begin{aligned}
\lim_{\substack{\omega_\alpha\to\Omega_m\\ \omega_\beta\to-\Omega_n}} &\left\{(\omega_\alpha-\Omega_m)(\omega_\beta+\Omega_n)\Pi^{(3)}(\omega_\alpha,\omega_\beta,x_1,x_1',x_2,x_2',x_3,x_3')\right\} = \\
&\rho_{0n}(x_1,x_1')\rho_{nm}(x_2,x_2')\rho_{m0}(x_3,x_3')\,.
\end{aligned}
\tag{4.62}
$$

Conventional response theory assumes that the quadratic response function obtained from an approximate action has the same pole structure as the exact one. For example, applying this argument to the TDDFT quadratic response function within the AA yields [31]

$$\rho_{nm} = \begin{pmatrix} \mathbf{K}^{nm}_{\text{occ}} & \mathbf{X}^{nm} \\ (\mathbf{Y}^{nm})^T & \mathbf{K}^{nm}_{\text{virt}} \end{pmatrix} . \tag{4.63}$$

In the last equation, the supervectors are interpreted as matrices. The unrelaxed transition density $\mathbf{K}$ is

$$K^{nm}_{ij} = -\sum_a X^m_{ja} X^n_{ia} + Y^n_{ja} Y^m_{ia}, \quad K^{nm}_{ab} = \sum_i X^m_{ia} X^n_{ib} + Y^n_{ia} Y^m_{ib} . \tag{4.64}$$

The orbital-relaxation contributions $\mathbf{X}^{nm}, \mathbf{Y}^{nm}$, are obtained from the second-order response equation

$$\begin{pmatrix} \mathbf{X}^{nm} \\ \mathbf{Y}^{nm} \end{pmatrix} = \boldsymbol{\Pi}^{(2)}_{\text{TDDFT}}(\Omega_{nm}) \begin{pmatrix} \mathbf{P}^{nm} \\ \mathbf{Q}^{nm} \end{pmatrix} . \tag{4.65}$$

The right-hand-side $\mathbf{P}_{nm}, \mathbf{Q}_{nm}$ contains third derivatives of $\mathscr{A}$ with respect to the TDKS density matrix [31] and is frequency-independent within the AA,

$$\begin{aligned}
(P+Q)^{nm}_{ia} = &-\frac{1}{2}\sum_j \left[ (X+Y)^m_{ja} H^+_{ji}[(\mathbf{X}+\mathbf{Y})^n] + (X-Y)^m_{ja} H^-_{ji}[(\mathbf{X}-\mathbf{Y})^n] + (n \leftrightarrow m) \right] \\
&+\frac{1}{2}\sum_b \left[ (X+Y)^m_{ib} H^+_{ab}[(\mathbf{X}+\mathbf{Y})^n] + (X-Y)^m_{ib} H^-_{ab}[(\mathbf{X}-\mathbf{Y})^n] + (n \leftrightarrow m) \right] \\
&+ H^+_{ai}[\mathbf{K}^{nm}] + 2g^{\text{XC}+}_{ai}(-\Omega_n, \Omega_m)[(\mathbf{X}+\mathbf{Y})^n, (\mathbf{X}+\mathbf{Y})^m] , \\
(P-Q)^{nm}_{ia} = &\ \frac{1}{2}\sum_j \left[ (X-Y)^m_{ja} H^+_{ji}[(\mathbf{X}+\mathbf{Y})^n] + (X+Y)^m_{ja} H^-_{ji}[(\mathbf{X}-\mathbf{Y})^n] - (n \leftrightarrow m) \right] \\
&-\frac{1}{2}\sum_b \left[ (X-Y)^m_{ib} H^+_{ab}[(\mathbf{X}+\mathbf{Y})^n] + (X+Y)^m_{ib} H^-_{ab}[(\mathbf{X}-\mathbf{Y})^n] - (n \leftrightarrow m) \right] \\
&- H^-_{ia}[\mathbf{K}^{nm}] + 2g^{\text{XC}-}_{ai}(-\Omega_n, \Omega_m)[(\mathbf{X}-\mathbf{Y})^n, (\mathbf{X}-\mathbf{Y})^m] ,
\end{aligned} \tag{4.66}$$

where $(n \leftrightarrow m)$ is shorthand for the previous terms of the enclosing parentheses with $n$ and $m$ interchanged. Eqs. (4.58) and (4.66) generalize our previous results [31] to current dependent and hybrid TDDFT and correct the erroneous omission of $H^+_{ai}[\mathbf{K}^{(\alpha\beta)}]$ and $H^+_{ai}[\mathbf{K}^{nm}]$.

However, within the adiabatic approximation, the linear polarization propagator in Eq. (4.65) becomes singular if the excitation energy difference $\Omega_{nm} = \Omega_{0m} - \Omega_{0n}$ approaches any ground-state excitation energy, resulting in an unphysical divergence in the orbital-relaxation contribution to the transition moment [3–5, 8]. Similarly, the orbital-relaxation contribution in the cubic response function within adiabatic TDDFT, Eq. (4.55) contains products of three poles, which do not occur in the exact case, Eq. (4.28).

### 4.4.5 One-Electron Systems

Since Hartree–Fock is exact for one-electron systems, the exact transition moment between excited states should be recovered. That is, the state-to-state transition moments between TDHF excited states should contain no poles. In this subsection, we show this explicitly by showing that the right-hand side of Eq. (4.65) vanishes for one-electron systems.

First, the eigenvalue problem for one-electron systems is simplified by noting that the off-diagonal block of the response operator, $\mathbf{B}$, vanishes. Within TDHF, the elements of $\mathbf{B}$ evaluate to $B_{ia,jb} = (ia|jb) - (ja|ib)$. However, there is only one occupied orbital in a one-electron system, i.e., only terms like $B_{ia,ib} = 0$ appear and thus $\mathbf{B} = 0$. As a consequence, $\mathbf{Y}^n = 0$ and the symplectic eigenvalue problem simplifies into the configuration interaction singles (CIS) eigenvalue problem,

$$\mathbf{A}\mathbf{X}_n = \Omega_n \mathbf{X}_n, \quad \mathbf{X}_n^T \mathbf{X}_m = \delta_{nm}. \tag{4.67}$$

Unrelaxed transition densities likewise simplify to

$$K_{ii}^{nm} = -\delta_{nm}, \quad K_{ab}^{nm} = X_{ia}^m X_{ib}^n \,. \tag{4.68}$$

Next, we split the right-hand sides into an "occupied"-like contribution,

$$\begin{aligned}(P \pm Q)_{ia}^{nm\text{-occ}} = \mp \frac{1}{2} \sum_j \left[ X_{ja}^m H_{ji}^+[\mathbf{X}^n] + X_{ja}^m H_{ji}^-[\mathbf{X}^n] \pm (n \leftrightarrow m) \right] \\ \pm H_{ai}^+[\mathbf{K}_{\text{occ}}^{nm}] \,,\end{aligned} \tag{4.69}$$

and a "virtual"-like contribution,

$$\begin{aligned}(P \pm Q)_{ia}^{nm\text{-virt}} = \pm \frac{1}{2} \sum_b \left[ X_{ib}^m H_{ab}^+[\mathbf{X}^n] + X_{ib}^m H_{ab}^-[\mathbf{X}^n] \pm (n \leftrightarrow m) \right] \\ \pm H_{ai}^+[\mathbf{K}_{\text{virt}}^{nm}] \,.\end{aligned} \tag{4.70}$$

Each term in the "occupied" contribution disappears individually because $H_{ii,ip}^{\pm} = H_{ip,ii}^{\pm} = 0$. Turning to the "virtual" contribution, we expand the linear transformations in the brackets and remove the sum-over occupied orbitals (since there is only one),

$$\begin{aligned}&\sum_b \left[ X_{ib}^m H_{ab}^+[\mathbf{X}^n] + X_{ib}^m H_{ab}^-[\mathbf{X}^n] \pm X_{ib}^n H_{ab}^+[\mathbf{X}^m] + X_{ib}^n H_{ab}^-[\mathbf{X}^m] \right] \\ &= \sum_{bc} \left[ X_{ib}^m X_{ic}^n \left( H_{ab,ic}^+ + H_{ab,ic}^- \right) \pm X_{ib}^n X_{ic}^m \left( H_{ab,ic}^+ + H_{ab,ic}^- \right) \right] \\ &= \sum_{bc} \left[ K_{bc}^{nm} \left( H_{ab,ic}^+ + H_{ab,ic}^- \pm H_{ac,ib}^+ \pm H_{ac,ib}^- \right) \right] \\ &= -2H_{ia}^{\pm}[\mathbf{K}_{\text{virt}}^{nm}] \,.\end{aligned} \tag{4.71}$$

Hence, the sum in brackets cancels the transformed unrelaxed density. Therefore, the right-hand side of Eq. (4.65) vanishes and there are no poles in state-to-state transition properties of one-electron systems.

## 4.5 Conclusions and Outlook

Conventional approximate response theories such as TDHF, adiabatic TDDFT, CC and MCSCF response theory are plagued by a qualitatively incorrect pole structure of quadratic and higher-order response functions. This deficiency results, e.g., in spurious poles in state-to-state transition moments, and incorrect frequency dependence of two-photon absorption cross sections.

The physical origin of this striking failure is the instantaneous time dependence of the effective Hamiltonian through the reference state. Consider, for example, adiabatic TDDFT: If the frequency of an external perturbation $\omega$ equals an excitation energy $\Omega_{0n}$ of the ground state, the first-order time-dependent density diverges as $1/(\omega - \Omega_{0n})$, in agreement with the exact time-dependent response, Eq. (4.25). However, this causes the $n$-th-order effective Hamiltonian to diverge as $1/(\omega - \Omega_{0n})^n$ through the instantaneous density dependence of the HXC potential. These divergences are not present in the exact many-electron Hamiltonian, and they cause the unphysical pole structure beyond linear response. These spurious divergences occur independent of whether the time or frequency domain is used and in situations where the ground state is qualitatively well described by a single determinant.

The notion of an effective potential that adjusts instantaneously to changes in the electron density appears to be unphysical for systems with more than one electron. This suggests that our current understanding of electron correlation effects in time-dependent systems is, at best, incomplete. Reliable theoretical predictions of nonlinear response properties at finite frequencies will require time-dependent electronic structure methods free of spurious divergences.

**Acknowledgements** The authors would like to acknowledge helpful discussions with Sreeganesh Balasubramani. This material is based on work supported by the US Department of Energy under Award Number DE-SC0008694. SMP is supported by an Arnold O. Beckman Postdoctoral Fellowship.

## References

1. E. Tapavicza, A.M. Meyer, F. Furche, Phys. Chem. Chem. Phys. **13**(47), 20986 (2011), https://doi.org/10.1039/c1cp21292c
2. J.C. Vincent, M. Muuronen, K.C. Pearce, L.N. Mohanam, E. Tapavicza, F. Furche, J. Phys. Chem. Lett. **7**(20), 4185 (2016), https://doi.org/10.1021/acs.jpclett.6b02037
3. Z. Li, B. Suo, W. Liu, J. Chem. Phys. **141**(24), 244105 (2014), https://doi.org/10.1063/1.4903986

4. Q. Ou, G.D. Bellchambers, F. Furche, J.E. Subotnik, J. Chem. Phys. **142**(6), 064114 (2015), https://doi.org/10.1063/1.4906941
5. X. Zhang, J.M. Herbert, J. Chem. Phys. **142**(6), 064109 (2015), https://doi.org/10.1063/1.4907376
6. E. Dalgaard, Phys. Rev. A **26**, 42 (1982), https://doi.org/10.1103/PhysRevA.26.42
7. S.M. Parker, S. Roy, F. Furche, J. Chem. Phys. **145**(13), 134105 (2016), https://doi.org/10.1063/1.4963749
8. Z. Hu, J. Autschbach, L. Jensen, J. Chem. Theory Comput. **12**(3), 1294 (2016), https://doi.org/10.1021/acs.jctc.5b01060
9. P.A.M. Dirac, Phys. Zeit. der Sowjetunion **3**, 64 (1933)
10. H. Sambe, Phys. Rev. A **7**(6), 2203 (1973), https://doi.org/10.1103/PhysRevA.7.2203
11. J.S. Howland, Math. Ann. **207**(4), 315 (1974), https://doi.org/10.1007/BF01351346
12. J. Olsen, P. Jorgensen, J. Chem. Phys. **82**, 3235 (1985), https://doi.org/10.1063/1.448223
13. R. van Leeuwen, Int. J. Mod. Phys. B **15**, 1969 (2001), https://doi.org/10.1142/S021797920100499X
14. R. van Leeuwen, Phys. Rev. Lett. **80**, 1280 (1998), https://doi.org/10.1103/PhysRevLett.80.1280
15. E. Wigner, Math. Naturwiss. Anz. Ungar. Akad. Wiss. **35**, 477 (1935)
16. T. Helgaker, P. Jørgensen, Calculation of geometrical derivatives in molecular electronic structure theory, in *Methods in Computational Molecular Physics* (Springer US, Boston, MA, 1992), pp. 353–421, https://doi.org/10.1007/978-1-4615-7419-4_15
17. P.W. Langhoff, S.T. Epstein, M. Karplus, Rev. Mod. Phys. **44**(3), 602 (1972), https://doi.org/10.1103/RevModPhys.44.602
18. J.E. Rice, N.C. Handy, J. Chem. Phys. **94**(7), 4959 (1991), https://doi.org/10.1063/1.460558
19. J.E. Rice, N.C. Handy, Int. J. Quantum Chem. **43**(1), 91 (1992), https://doi.org/10.1002/qua.560430110
20. K. Sasagane, F. Aiga, R. Itoh, J. Chem. Phys. **99**(5), 3738 (1993), https://doi.org/10.1063/1.466123
21. O. Christiansen, P. Jørgensen, C. Hättig, Int. J. Quant. Chem. **68**(1), 1 (1998), https://doi.org/10.1002/(SICI)1097-461X(1998)68:1<1::AID-QUA1>3.0.CO;2-Z
22. F. Pawłowski, J. Olsen, P. Jorgensen, J. Chem. Phys **142**(11), 114109 (2015), https://doi.org/10.1063/1.4913364
23. D.M. Bishop, J. Chem. Phys. **100**(9), 6535 (1994), https://doi.org/10.1063/1.467062
24. E. Runge, E.K.U. Gross, Phys. Rev. Lett. **52**, 997 (1984), https://doi.org/10.1103/PhysRev.140.A1133
25. E.K.U. Gross, W. Kohn, Adv. Quant. Chem. **21**, 255 (1990), https://doi.org/10.1016/S0065-3276(08)60600-0
26. R. Neumann, R.H. Nobes, N.C. Handy, Mol. Phys. **87**, 1 (1996), https://doi.org/10.1080/00268979600100011
27. R. Bauernschmitt, R. Ahlrichs, Chem. Phys. Lett. **256**, 454 (1996), https://doi.org/10.1063/1.471637
28. J.E. Bates, F. Furche, J. Chem. Phys. **137**, 164105 (2012), https://doi.org/10.1063/1.4759080
29. M.E. Casida, *Time-dependent density functional response theory for molecules, in Recent Advances in Computational Chemistry*, vol. 1 (World Scientific, Singapore, 1995), pp. 155–192
30. R.E. Stratmann, G.E. Scuseria, M.J. Frisch, J. Chem. Phys. **109**, 8218 (1998), https://doi.org/10.1021/ja00023a041
31. F. Furche, J. Chem. Phys. **114**, 5982 (2001), https://doi.org/10.1063/1.1353585

# Chapter 5
# Response Properties of Periodic Materials Subjected to External Electric and Magnetic Fields

**Bernard Kirtman, Lorenzo Maschio, Michel Rérat and Michael Springborg**

**Abstract** The ab initio treatment of external field response in crystalline solids poses particular challenges, since the operator representing the external field must be reformulated so to be consistent with periodic boundary conditions. In this chapter, we review our recent work to formulate implementable working equations for polarizability and hyperpolarizabilities in solids, in the framework of a local (gaussian) basis set. This includes the combination of electric field response with geometrical distorsion, i.e. infrared and Raman intensities, as well as piezoelectricity. In addition, we outline a pathway for including magnetic fields in a consistent way, and speculate on future possible developments.

**Keywords** Ab initio · Periodic systems · Coupled-perturbed · Density functional theory

---

B. Kirtman
Department of Chemistry and Biochemistry, University of California, Santa Barbara, California 93106, USA

L. Maschio (✉)
Dipartimento di Chimica, Università di Torino, Via P. Giuria 7, 10125 Torino, Italy
e-mail: lorenzo.maschio@unito.it

M. Rérat
Equipe de Chimie Physique, IPREM UMR5254, Université de Pau et des Pays de l'Adour, 64000 Pau, France

M. Springborg
Physical and Theoretical Chemistry, University of Saarland, 66123 Saarbrücken, Germany

M. Springborg
School of Materials Science and Engineering, Tianjin University, Tianjin 300072, China



M.J. Wójcik et al. (eds.), *Frontiers of Quantum Chemistry*,
https://doi.org/10.1007/978-981-10-5651-2_5

## 5.1 Introduction

The polarization of molecular electrons in an external electric field and the magnetization due to orbital motion in a magnetic field give rise to a large variety of physical properties that are of high interest in science and engineering. Apart from the usual spectroscopic observables, electric and magnetic fields can significantly affect chemical, mechanical and structural properties as well. For ordinary molecules the required quantum theoretical treatment and ab initio computational methods to determine these quantities are at a relatively advanced stage of development. The situation is different, however, for materials that are modeled as being infinite and periodic in 1, 2 or 3 dimensions. That is primarily because the interaction term in the Hamiltonian appropriate for ordinary molecules, is either non-periodic [6] or is periodic only for certain values of the (magnetic) field [7]. Although this difficulty was initially considered over 50 years ago, not until much more recently have suitable solutions been developed (see Refs. [2, 4, 10, 29, 30, 40, 54, 55, 58, 63]).

Despite the fact that there are unique aspects for infinite periodic systems, our general approach has been to exploit the methods that have been successfully developed in molecular quantum chemistry as much as possible. Consistent with that perspective, the implementation of computational methods has been carried out within the CRYSTAL code [9, 12, 14], which utilizes Gaussian-type orbitals (GTOs). This code has the advantage that electron exchange interactions are treated accurately and straightforwardly. It also takes maximum advantage of symmetry, and can run in parallel with a large number of cores (easily up to 4096) for large systems using a basis set containing more than $10^5$ functions. There are, of course, other periodic codes—mostly based on plane waves—that also treat some of the properties considered in this chapter.

Even though substantial progress has been made, the treatment of infinite periodic systems is far less advanced than what has been achieved for ordinary molecules. Thus, we shall indicate in our presentation some of the next steps that are on the drawing board and some that may be envisioned in the longer term. Our emphasis will be on the theoretical development although initial applications are included in the references cited.

In the next section (Sect. 5.2) we specify the translationally invariant one-electron operators that describe the interaction of static or dynamic electric and magnetic fields with periodic systems. Moreover, the important properties of these operators, with regard to their action on periodic orbitals, are discussed. This includes their hermitian character as well as the effect on the orbital phases which, surprisingly can be related to changes in the electric polarization of the material when boundary surfaces are introduced. The Hartree-Fock and Kohn-Sham (HF/KS) perturbation theory treatments of linear and nonlinear optical properties are presented in Sects. 5.3 and 5.4. This approach provides a model pathway for ultimately including magnetic properties and properties determined by the simultaneous action of electric and magnetic fields in the future. Electronic spectra are also treated in Sect. 5.4 through the frequency-dependent linear polarizability.

The next set of properties to be considered, in Sect. 5.5, concerns the effect of geometrical distortions on the electronic polarization or vice versa. This category includes, for instance, infra-red and Raman intensities, which are determined by internal unit cell deformations. Another example is the direct and converse piezo-electricity associated with (external) strain.

Finally, there is a concluding section that summarizes the current and near term developments that have been discussed in this chapter. We also speculate there about some of the advances, both theoretical and computational, that the longer term future may bring.

Let us briefly introduce the notation, adopted in the following:

- Greek indices $\mu, \nu, \ldots$ will label AO basis functions in the unit cell.
- Direct space lattice vectors will be indicated as $\mathbf{g}, \mathbf{h}, \mathbf{n} \ldots$
- Latin letters i, j, ... will indicate occupied molecular or crystalline orbitals; a, b, ... will indicate virtual orbitals; other latin letters like p, q, r, s, ... will indicate a general orbital (either occupied or virtual).
- $k$-vectors of the Brillouin Zone (BZ) sampling in reciprocal space will be indicated as $\mathbf{k}$
- A general cartesian direction (x, y, z) will be expressed by lower case letters $t$, $u$, $v$, ...
- Matrices denoted as $S(\mathbf{g}), F(\mathbf{g}), D(\mathbf{g}), \ldots$ represent direct space quantities, while reciprocal space quantities are denoted as $S(\mathbf{k}), F(\mathbf{k}), D(\mathbf{k}), \ldots$.
- Direct space and reciprocal space quantities are connected through standard Fourier and back-Fourier transforms, $\mathscr{F}$ and $\overline{\mathscr{F}}$ respectively:

$$S(\mathbf{k}) = \mathscr{F}\left[S(\mathbf{g})\right]$$
$$S(\mathbf{g}) = \overline{\mathscr{F}}\left[S(\mathbf{k})\right]$$

Atomic units will be also used all along the text, so that we set the absolute value of the electronic charge and its mass, $|e|$ and $m$, as well as $4\pi\varepsilon_0$, and $\hbar$ all equal to 1.

## 5.2 Translationally Invariant One-Electron Operators for External Fields

### *5.2.1 Basics*

In order to appreciate the challenges connected with the theoretical treatment of an infinite, periodic system exposed to electromagnetic fields, it may be useful to start with static fields (thus avoiding time dependence) for a large, finite system (thus avoiding periodic boundary conditions) that is assumed to be neutral (thus avoiding complications due to origin dependence of the dipole moment) and focus on the exact solutions to the Hartree-Fock or Kohn-Sham equations (thus avoiding complications due to a finite basis set). In that case, the complete non-relativistic electronic Hamilton operator is hermitian and equal to

$$\hat{H}_e = \sum_{\nu=1}^{N_e} \hat{h}_{\rm n}(\mathbf{r}_\nu) + \frac{1}{2} \sum_{\nu_1 \neq \nu_2 = 1}^{N_e} \hat{h}_2(\mathbf{r}_{\nu_1}, \mathbf{r}_{\nu_2}) + \sum_{\nu=1}^{N_e} \hat{h}_{\rm C+xc}(\mathbf{r}_\nu) + \sum_{\nu=1}^{N_e} \hat{h}_{\rm el}(\mathbf{r}_\nu) + \sum_{\nu=1}^{N_e} \hat{h}_{\rm k+m}(\mathbf{r}_\nu), \tag{5.1}$$

where $\hat{h}_{\rm n}(\mathbf{r}_\nu)$ is the potential due to the nuclei, $\hat{h}_2(\mathbf{r}_{\nu_1}, \mathbf{r}_{\nu_2})$ describes the electron-electron interactions, $\hat{h}_{\rm C+xc}(\mathbf{r}_\nu)$ is the Coulomb and exchange-correlation potential, and $N_e$ is the number of electrons. Within the Hartree-Fock approximation, $\hat{h}_{\rm C+xc}$ is set equal to 0, whereas $\hat{h}_2 = 0$ when using a pure Kohn-Sham density-functional approach. Various hybrid approaches may also be written in the form of Eq. (5.1) by combining the 2nd and 3rd term on the rhs appropriately.

However, our focus here is on the last two terms. In the next-to-last term we use the scalar potential to take the presence of the electrostatic field into account. Then, assuming that the field $\mathscr{E}$ is position independent

$$\hat{h}_{\rm el}(\mathbf{r}) = \mathscr{E} \cdot \mathbf{r}. \tag{5.2}$$

Any dependence of this term on the origin is compensated by a similar dependence in the equivalent nuclear term (not shown). Finally, the operator for the kinetic energy plus magnetic terms is

$$\hat{h}_{\rm k+m}(\mathbf{r}) = \frac{1}{2} \left[ -\imath \nabla + \frac{1}{2c} \mathscr{B} \times (\mathbf{r} - \mathbf{R}_G) \right]^2. \tag{5.3}$$

where $\mathscr{B}$ is the uniform external magnetic field determined by the vector potential with a gauge origin at $\mathbf{R}_G$. $\mathbf{R}_G$ is unphysical and, in the ideal case that the electronic orbitals are determined exactly, the results will not depend on its choice.

It is well-known that when approximating the exact $N_e$-electron wavefunction through a single Slater determinant $\Psi$, the requirement that the expectation value $\langle \Psi | \hat{H}_e | \Psi \rangle$ is minimal under the constraint that the orbitals are orthonormal, $\langle \psi_{j_1} | \psi_{j_2} \rangle = \delta_{j_1, j_2}$, can be cast into a single-particle Hartree-Fock or Kohn-Sham equations

$$\hat{h} \psi_j(\mathbf{r}) = \sum_{j'} \lambda_{j,j'} \psi_{j'}(\mathbf{r}). \tag{5.4}$$

The Fock or Kohn-Sham operator $\hat{h}$ is hermitian and the Lagrange multipliers $\lambda$ may be chosen so that $\lambda_{j,j'} = \varepsilon_j \delta_{j,j'}$ (the canonical choice). Then, a unitary transformation amongst the occupied orbitals will leave the calculated value of any observable unchanged. The reason we mention these well-known facts will become clear below.

For the infinite, periodic system in the absence of any electrostatic or magnetostatic field, the orbitals may be written as Bloch functions,

$$\psi_j(\mathbf{k}, \mathbf{r}) = e^{\imath\mathbf{k}\cdot\mathbf{r}} u_j(\mathbf{k}, \mathbf{r}) \tag{5.5}$$

with $u_j(\mathbf{k}, \mathbf{r})$ being lattice-periodic. Moreover, the electronic Hamilton operator for the energy per repeated unit will be identical to that of Eq. (5.1) divided by the number of units $N_u$ and evaluated in the limit $N_u \to \infty$. The number of electrons $N_e = N_u \cdot n_e$ with $n_e$ equal to the number of electrons per repeated unit.

In this case, **k** is a good quantum number, and one may calculate the orbitals for each **k** separately. Again, a unitary transformation (e.g., for each **k** separately) will leave observables unchanged.

However, the interaction with an electrostatic field as represented by Eq. (5.2) destroys translational invariance. Instead, one may use [4, 8, 23, 30, 32, 44–46]

$$\mathscr{E} \cdot \mathbf{r} \to \mathscr{E} \cdot \imath e^{\imath\mathbf{k}\cdot\mathbf{r}} \nabla_k e^{-\imath\mathbf{k}\cdot\mathbf{r}} \equiv \hat{h}_{\text{el}}(\mathbf{k}, \mathbf{r}), \tag{5.6}$$

a suggestion that goes back to the work of Blount more than 50 years ago [6]. It is easily verified that $\hat{h}_{\text{el}}(\mathbf{k}, \mathbf{r})$ in Eq. (5.6) preserves translational symmetry. Blount's operator acts on **k**. Thus, a **k**-dependent unitary transformation of the occupied orbitals, may lead to a change in the energy. Actually, even the very simple unitary transformation whereby the individual occupied are modified by **k** dependent phase factors can lead to such an effect. In fact, by making a 'pathological' choice of the phase factors it is possible to change the energy due to the electrostatic field essentially arbitrarily, which implies that there is no lowest energy and that the variational principle will fail. The solution to this dilemma is that the phase factors or, in general, the unitary transformation cannot be chosen randomly. Different choices correspond to modeling large, finite samples of the same material but with different shapes and/or surface charges [59, 60].

A second complication is that $\hat{h}_{\text{el}}(\mathbf{k}, \mathbf{r})$ is hermitian only for functions whose overlap is **k** independent, i.e., functions $\{f_n\}$ obeying $\nabla_k \langle f_p | f_q \rangle = \mathbf{0}$. This is the case for (orthonormal) electronic orbitals but not in the general case. Moreover, for static magnetic fields the corresponding condition, as shown later, is much more difficult to satisfy. Thus, we proceed here with a more general solution, which is to make the replacement

$$\hat{h}_{\text{el}}(\mathbf{k}, \mathbf{r}) \to \frac{1}{2} \left[ \hat{h}_{\text{el}}(\mathbf{k}, \mathbf{r}) + \hat{h}^{\dagger}_{\text{el}}(\mathbf{k}, \mathbf{r}) \right]. \tag{5.7}$$

If one derives the Fock or Kohn-Sham equations in the usual manner [see below Eq. (5.3)], then this substitution turns out to yield exactly the same **k**-dependent single-particle operator that was given in Eq. (5.6). Nonetheless, for the reasons mentioned above, one must still subsequently specify some criterion for how to choose the orbital phase factors in order to find a unique solution of even the canonical form of the single-particle equation,

$$\hat{h}(\mathbf{k})\psi_j(\mathbf{k},\mathbf{r}) = \varepsilon_j(\mathbf{k})\psi(\mathbf{k},\mathbf{r}) \tag{5.8}$$

The story is different, however, when a magnetostatic field is included. In that event it may be suggested to make a substitution like that of Eq. (5.6) in the expression of Eq. (5.3). Actually, the magnetic field term in Eq. (5.3) is derived from the vector potential $\mathbf{A} = \mathbf{A}_0 e^{\imath\mathbf{q}\cdot(r-R_G)}$ where the wave vector $\mathbf{q}$ for a constant field goes to zero and the magnetic field is given by $\mathscr{B} = \nabla \times \mathbf{A}$. The substitution of Eq. (5.6) can, then, be viewed as replacing $\mathscr{E} \cdot \mathbf{r}$ by $\mathscr{E} \cdot \hat{\Omega}(\mathbf{k}) = \mathscr{E} \cdot \imath e^{\imath\mathbf{k}\cdot\mathbf{r}}\nabla_k e^{-\imath\mathbf{k}\cdot\mathbf{r}} = \mathscr{E} \cdot (\mathbf{r} + \imath\nabla_k)$ [52].

Thus, one might consider replacing the linear magnetic field term in the kinetic-energy operator $\frac{1}{2}\left(\hat{\mathbf{p}} + \frac{1}{c}\mathbf{A}\right)^2$ operator by an expression that involves the angular momentum operator $\mathscr{B} \cdot \hat{L} = \frac{1}{c}\mathscr{B} \cdot (-\imath\hat{\Omega}(\mathbf{k}) \times \nabla)$. However, the resulting operator is hermitian only for special circumstances, i.e., for functions $\{f_n\}$ for which $\nabla_k \cdot \langle f_p|\hat{p}|f_q\rangle = 0$ that in most instances are irrelevant.

Hermiticity can be restored following an approach similar to that of Eq. (5.7); in this case there are two different possibilities for the single-particle operator that come readily to mind:

$$\begin{aligned} &\frac{1}{2m}\left[-\imath\nabla + \frac{1}{2c}\mathscr{B} \times (\mathbf{r} - \mathbf{R}_G)\right]^2 \\ \rightarrow &\frac{1}{2m}\left(-\imath\nabla + \frac{1}{2c}\mathscr{B} \times \imath e^{\imath\mathbf{k}\cdot(\mathbf{r}-\mathbf{R}_G)}\nabla_k e^{-\imath\mathbf{k}\cdot(\mathbf{r}-\mathbf{R}_G)}\right)^\dagger \\ &\cdot\left(-\imath\nabla + \frac{1}{2c}\mathscr{B} \times \imath e^{\imath\mathbf{k}\cdot(\mathbf{r}-\mathbf{R}_G)}\nabla_k e^{-\imath\mathbf{k}\cdot(\mathbf{r}-\mathbf{R}_G)}\right) \end{aligned} \tag{5.9}$$

or

$$\begin{aligned} &\frac{1}{2m}\left[\imath\nabla - \frac{1}{2c}\mathscr{B} \times (\mathbf{r} - \mathbf{R}_G)\right]^2 \\ \rightarrow &\frac{1}{4m}\left[\left(-\imath\nabla + \frac{1}{2c}\mathscr{B} \times \imath e^{\imath\mathbf{k}\cdot(\mathbf{r}-\mathbf{R}_G)}\nabla_k e^{-\imath\mathbf{k}\cdot(\mathbf{r}-\mathbf{R}_G)}\right)^\dagger\right]^2 \\ &+\frac{1}{4m}\left[-\imath\nabla + \frac{1}{2c}\mathscr{B} \times \imath e^{\imath\mathbf{k}\cdot(\mathbf{r}-\mathbf{R}_G)}\nabla_k e^{-\imath\mathbf{k}\cdot(\mathbf{r}-\mathbf{R}_G)}\right]^2 \\ \equiv &\frac{1}{2}\left[\hat{h}_{\mathrm{km},1}(\mathbf{k},\mathbf{r}) + \hat{h}_{\mathrm{km},2}(\mathbf{k},\mathbf{r})\right]. \end{aligned} \tag{5.10}$$

It can be shown [60] that the two expressions lead to the same expectation value for the energy due to the magnetic field. Both operators in Eqs. (5.9) and (5.10) act, again, on the **k**-dependence of the orbitals. In this case it can be demonstrated that the effect of **k**-dependent orbital phase factors and/or **k**-dependent unitary transformations among the occupied orbitals is mathematically equivalent to introducing a different magnetic field gauge origin for different orbitals. Since the gauge origins are unphysical, however, they cannot alter the magnetic energy, at least in the case that the single-particle orbitals are calculated exactly.

Again, we may derive single-particle equations by requiring that the total energy is at a minimum together with the condition that the orbitals are orthonormal. It turns out that the Fock or Kohn-Sham operator then involves either $\hat{h}_{\text{km},1}(\mathbf{k}, \mathbf{r})$ or $\hat{h}_{\text{km},2}(\mathbf{k}, \mathbf{r})$, but not both, so that it is, in general, non-hermitian. However, one can show that it is hermitian for the exact eigenfunctions.

### 5.2.2 LCAO-CO Formulation

The next step is to formulate the Fock or Kohn-Sham equation within the LCAO-CO approximation. In the following we will focus on the electric field perturbation, not the least because the magnetic field case has yet to be developed. As seen above, in either case our approach relies on a formulation of the electronic operator for interaction with the field which, in the Bloch basis, is diagonal with respect to the reciprocal space **k** vector. Provided the system has a non-zero bandgap, this operator may be written for the electric field operator as [see Eq. (5.6)]:

$$\mathcal{E} \cdot \hat{\Omega}(\mathbf{k}) = \mathcal{E} \cdot \imath e^{\imath \mathbf{k}\cdot\mathbf{r}} \nabla_k e^{-\imath \mathbf{k}\cdot\mathbf{r}} = \sum_{t=x,y,z} \mathcal{E}_t \hat{\Omega}_t(\mathbf{k}) \quad . \tag{5.11}$$

The evaluation of $\hat{\Omega}$ requires a derivative with respect to **k**. Algorithms capable of dealing with this derivative have been implemented in plane wave codes [2, 15, 16, 22] although the role of crystal orbital phase factors is often ignored, as discussed in this chapter and in Refs. [4, 35]. A rather different approach, appropriate for an atomic orbital (AO) basis, is employed herein.

The effect of the electric field interaction term in (5.11) can be treated by finite field methods or by perturbation theory. Here we focus on the latter approach.

The first-order perturbation energy, which is equal to the mean value of $\Omega_k$ determines the dipole moment. This property is ill-defined in infinite periodic systems. For instance, the polymer ...ABAB... is equivalent to ...BABA... and, in that case, the dipole moment could be either positive or negative. Mathematically, the ambiguity is associated with the phases of the Bloch orbitals [4, 29]. Physically, the possible values can be related to surface charges and/or shape of the large finite system that is being modeled [59, 61], even though the corresponding infinite periodic system does not have any surfaces. For the special case of 1D periodic systems, the dipole moment is limited to values that differ from each other by an integer multiple of the cell parameter [4, 31] Although there is an ambiguity in the value of the dipole moment, that is not the case for derivatives of the dipole moment with respect to the field(s), i.e. for (hyper)polarizabilities. This has previously been demonstrated for 1D periodic systems [4] and we will see in the next section that it can be generalized for 3D systems as well. We shall need the effect of the operator $\hat{\Omega}(\mathbf{k})$ on an arbitrary field-dependent crystal orbital (CO) expanded in an atomic orbital (AO) basis

$|\,\phi_i\rangle = \sum_\nu |\,\nu\rangle\, C_{\nu,i}$. Projecting on the left hand side with the AO basis function $\langle\mu\,|$ yields the mixed AO/CO relation [45]

$$\sum_\nu \langle\mu \mid \hat{\Omega}_t(\mathbf{k}) \mid \nu\rangle\, C_{\nu,i}(\mathbf{k}) = \sum_\nu \left[ (Z_t)_{\mu\nu}(\mathbf{k}) + \imath S^{(k_t)}_{\mu\nu}(\mathbf{k}) + \imath S_{\mu\nu}(\mathbf{k}) \frac{\partial}{\partial k_t} \right] C_{\nu,i}(\mathbf{k}) \tag{5.12}$$

Here $Z_t(\mathbf{k})$ is the AO matrix that represents the component of the electronic position coordinate along $t$ and $S^{(k_t)}(\mathbf{k})$ is the derivative of the AO overlap matrix, $S$, with respect to the wave vector $k_t$. This last relation is valid at any arbitrary field and geometry.

## 5.3 Coupled Perturbed Hartree-Fock (CPHF) and Kohn-Sham (CPKS) Static (Hyper)polarizabilities

In the previous section we have developed the proper form of the electric field interaction term to be included in the Hamiltonian for a periodic system. The wavefunction and electric dipole properties can then be obtained, for example, through a periodic coupled-pertubed Hartree-Fock (CPHF) or Kohn-Sham (CPKS) approach. Such a CPHF/CPKS treatment has been developed [18] and implemented [19, 20], by some of the present authors, in the CRYSTAL program, that uses a local Gaussian basis set [11, 13]. In this section we will further develop the formalism, present operative equations as they have been implemented, and focus on the role of the $\nabla_k$ operator in the electric field terms.

### 5.3.1 CPHF Treatment

We write the reciprocal space eigenvectors of the Fock matrix as a function of electric field using a Taylor series in the field components:

$$C(\mathbf{k}) = C^{(0)}(\mathbf{k}) + \sum_{t=x,y,z} \mathscr{E}_t\, C^{(\mathscr{E}_t)}(\mathbf{k}) + \sum_{t,u=x,y,z} \mathscr{E}_t\, \mathscr{E}_u\, C^{(\mathscr{E}_t,\mathscr{E}_u)}(\mathbf{k}) + \cdots \tag{5.13}$$

Then the perturbed first-order $k$-space density matrix $D(\mathbf{k})$ may be obtained as:

$$D^{(\mathscr{E}_t)}(\mathbf{k}) = C^{(\mathscr{E}_t)}(\mathbf{k})\, n\, C^{(0)\dagger}(\mathbf{k}) + C^{(0)}(\mathbf{k})\, n\, C^{(\mathscr{E}_t)\dagger}(\mathbf{k}) \tag{5.14}$$

where $n$ is the diagonal occupancy matrix and we will, henceforth, assume closed shells so that the diagonal elements equal 0 or 2. Similarly, the second-order expression for $D^{(\mathscr{E}_t,\mathscr{E}_u)}(\mathbf{k})$ is given by:

$$D^{(\mathscr{E}_t,\mathscr{E}_u)}(\mathbf{k}) = C^{(\mathscr{E}_t,\mathscr{E}_u)}(\mathbf{k})\, n\, C^{(0)\dagger}(\mathbf{k}) + C^{(\mathscr{E}_u)}(\mathbf{k})\, n\, C^{(\mathscr{E}_t)\dagger}(\mathbf{k})$$
$$+ C^{(\mathscr{E}_t)}(\mathbf{k})\, n\, C^{(\mathscr{E}_u)\dagger}(\mathbf{k}) + C^{(0)}(\mathbf{k})\, n\, C^{(\mathscr{E}_t,\mathscr{E}_u)\dagger}(\mathbf{k}) \quad (5.15)$$

It is common to express the perturbed coefficient vectors as a linear combination of the unperturbed vectors, i.e.

$$\frac{\partial}{\partial \mathscr{E}_t} C(\mathbf{k}) = C^{(\mathscr{E}_t)}(\mathbf{k}) = C^{(0)}(\mathbf{k})\, U^{(\mathscr{E}_t)}(\mathbf{k}) \quad (5.16)$$

$$\frac{\partial^2}{\partial \mathscr{E}_t \partial \mathscr{E}_u} C(\mathbf{k}) = C^{(\mathscr{E}_t,\mathscr{E}_u)}(\mathbf{k}) = C^{(0)}(\mathbf{k})\, U^{(\mathscr{E}_t,\mathscr{E}_u)}(\mathbf{k}) \quad (5.17)$$

where the $U$ matrices are calculated self-consistently by solving the first- and second-order non-canonical CPHF(KS) equations [18, 25]:

$$U_{ij}^{(\mathscr{E}_t)}(\mathbf{k}) = 0 \quad (5.18)$$

$$U_{ab}^{(\mathscr{E}_t)}(\mathbf{k}) = 0 \quad (5.19)$$

$$U_{ia}^{(\mathscr{E}_t)}(\mathbf{k}) = \frac{G_{ia}^{(\mathscr{E}_t)}(\mathbf{k})}{\varepsilon_a^{(0)}(\mathbf{k}) - \varepsilon_i^{(0)}(\mathbf{k})} \quad (5.20)$$

$$U_{ij}^{(\mathscr{E}_t,\mathscr{E}_u)}(\mathbf{k}) = -1/2 P_{t,u} \sum_a U_{ai}^{(\mathscr{E}_t)*}(\mathbf{k})\, U_{aj}^{(\mathscr{E}_u)}(\mathbf{k}) \quad (5.21)$$

$$U_{ab}^{(\mathscr{E}_t,\mathscr{E}_u)}(\mathbf{k}) = -1/2 P_{t,u} \sum_i U_{ia}^{(\mathscr{E}_t)*}(\mathbf{k})\, U_{ib}^{(\mathscr{E}_u)}(\mathbf{k}) \quad (5.22)$$

$$U_{ia}^{(\mathscr{E}_t,\mathscr{E}_u)}(\mathbf{k}) = \left\{ P_{t,u} \left( \sum_j G_{ij}^{(\mathscr{E}_t)}(\mathbf{k})\, U_{ja}^{(\mathscr{E}_u)}(\mathbf{k}) - \sum_b U_{ib}^{(\mathscr{E}_t)}(\mathbf{k}) G_{ba}^{(\mathscr{E}_u)}(\mathbf{k}) + \imath \frac{\partial U_{ia}^{(\mathscr{E}_t)}(\mathbf{k})}{\partial k_u} \right)\right.$$
$$\left. + G_{ia}^{(\mathscr{E}_t,\mathscr{E}_u)}(\mathbf{k}) \right\} \left[ \varepsilon_a^{(0)}(\mathbf{k}) - \varepsilon_i^{(0)}(\mathbf{k}) \right]^{-1} \quad (5.23)$$

Here $P_{t,u}$ represents a sum over the permutation of $\mathscr{E}_t$ and $\mathscr{E}_u$; also although not explicitly indicated, the unperturbed orbital energies $\varepsilon_i^{(0)}$ and $\varepsilon_a^{(0)}$ depend on $\mathbf{k}$ (for a 0D system $\mathbf{k} = \mathbf{0}$). The matrix element $G_{ia}^{(\mathscr{E}_t)}(\mathbf{k})$, for example, is obtained from the partial derivatives of the AO Fock matrix with respect to the field, that is subsequently transformed to the CO basis with $i$ an occupied and $a$ an unoccupied orbital ($\sum_{\mu\nu} C_{\mu i}^{(0)*}(\mathbf{k}) \frac{\partial F_{\mu\nu}(\mathbf{k})}{\partial \mathscr{E}_t} C_{\nu a}^{(0)}(\mathbf{k})$). This matrix element contains the bielectronic integrals multiplied by the first-order perturbed density matrix $D^{(\mathscr{E}_t)}(\mathbf{k})$ and thereby depends upon $U^{(\mathscr{E}_t)}(\mathbf{k})$. It also contains the explicit first-order electric field term of the Hamiltonian [see definition in Eq. (5.12)] which, in the CO basis, is:

$$\begin{aligned}\Omega_t(\mathbf{k}) &= C^{(0)\dagger}(\mathbf{k})\left(Z_t(\mathbf{k}) + \imath S^{(k_t)}(\mathbf{k}) + \imath\left[S(\mathbf{k})\, C^{(0)}(\mathbf{k})\, Q_t(\mathbf{k})\, C^{(0)\dagger}(\mathbf{k})\, S(\mathbf{k})\right]\right) C^{(0)}(\mathbf{k}) \\ &= C^{(0)\dagger}(\mathbf{k})\left(Z_t(\mathbf{k}) + \imath S^{(k_t)}(\mathbf{k})\right) C^{(0)}(\mathbf{k}) + \imath Q_t(\mathbf{k}) \end{aligned} \tag{5.24}$$

The CO matrix $Q_t(\mathbf{k})$ is due to the partial derivative with respect to $\mathbf{k}_t$ in $\hat{\Omega}$ acting on the unperturbed coefficient matrix. It is defined through the relation:

$$\frac{\partial}{\partial k_t} C^{(0)}(\mathbf{k}) = C^{(0)}(\mathbf{k})\, Q_t(\mathbf{k}) \ . \tag{5.25}$$

and may be calculated as (for $i \neq j$):

$$Q^{k_t}_{ij}(\mathbf{k}) = \frac{G^{k_t}_{ij}(\mathbf{k}) - \varepsilon^{(0)}_j(\mathbf{k}) S^{k_t}_{ij}(\mathbf{k})}{\varepsilon^{(0)}_j(\mathbf{k}) - \varepsilon^{(0)}_i(\mathbf{k})} \tag{5.26}$$

where $G^{k_t}_{ij}(\mathbf{k})$ and $S^{k_t}_{ij}(\mathbf{k})$ are the derivatives of the Fock and overlap matrices transformed to the CO basis, i.e. $\sum_{\mu\nu} C^{(0)*(\mathbf{k})}_{\mu i} \frac{\partial F_{\mu\nu}(\mathbf{k})}{\partial k_t} C^{(0)}_{\nu j}(\mathbf{k})$ and $\sum_{\mu\nu} C^{(0)*}_{\mu i}(\mathbf{k}) \frac{\partial S_{\mu\nu}(\mathbf{k})}{\partial k_t} C^{(0)}_{\nu j}(\mathbf{k})$, respectively. The diagonal elements of $Q^{k_t}(\mathbf{k})$ are determined by the normalization condition which leads to $-\frac{1}{2} S^{k_t}_{ii}(\mathbf{k})$ for the real part. However, there is also an undetermined imaginary part (which reflects an arbitrary phase factor for each CO coefficient vector) [4, 26, 31]. This imaginary contributionn leads to the indeterminacy in the dipole moment described earlier. It does not, however, affect the polarizability and hyperpolarizabilities as we will see later.

By applying the perturbation theory discussed above to the periodic Hartree-Fock (see later for Kohn-Sham) equations, where the Hamiltonian contains the electric field perturbation term characterized by $\Omega_t$, as defined in Eq. (5.24), one obtains the following expressions for the static dipolar electric field properties of closed-shell systems:

1. Dipole Moment

$$\mu_t = -\frac{\partial E^{TOT}}{\partial \mathcal{E}_t} = \frac{2}{n_k} \Re \sum_{\mathbf{k}}^{BZ} Tr\left[\Omega_t(\mathbf{k})\, n\right] \tag{5.27}$$

   where $n$ is the diagonal electronic occupation matrix.
   Note that the imaginary diagonal elements of the $Q$ matrix contribute to the dipole moment [cf. Eq. (5.24)]. For real finite systems this contribution will cancel when ignoring structural responses [61].
2. Polarizability

$$\alpha_{tu} = -\frac{\partial^2 E^{TOT}}{\partial \mathcal{E}_t \partial \mathcal{E}_u} = \frac{2}{n_k} \Re \left\{ P_{t,u} \sum_{\mathbf{k}}^{BZ} Tr\left[\Omega_t(\mathbf{k}) U^{(\mathcal{E}_u)}(\mathbf{k})\, n\right] \right\} \tag{5.28}$$

The diagonal blocks of $U$ vanish. Therefore, only the off-diagonal blocks of $\Omega$, and thereby $Q$, contribute to the polarizability. Thus, there is no indeterminacy in $\alpha$.

3. First hyperpolarizability

$$\beta_{tuv} = -\frac{\partial^3 E^{TOT}}{\partial \mathscr{E}_t \partial \mathscr{E}_u \partial \mathscr{E}_v} = \frac{2}{n_k} \Re \left\{ P_{t,u,v} \sum_{\mathbf{k}}^{BZ} Tr \left[ n\, U^{(\mathscr{E}_t)\dagger}(\mathbf{k}) \left( G^{(\mathscr{E}_u)}(\mathbf{k}) U^{(\mathscr{E}_v)}(\mathbf{k}) - U^{(\mathscr{E}_v)}(\mathbf{k}) G^{(\mathscr{E}_u)}(\mathbf{k}) + \imath \frac{\partial U^{(\mathscr{E}_v)}(\mathbf{k})}{\partial k_u} \right) \right] \right\} \tag{5.29}$$

Now diagonal terms of $Q$ do appear in $G^{(\mathscr{E}_u)}$, and also in $\frac{\partial U^{(\mathscr{E}_v)}}{\partial \mathbf{k}_u}$ even though the diagonal blocks of $U^{(\mathscr{E}_v)}$ itself vanish. The signs are opposite and then $\beta$, like $\alpha$, is completely determined. The 3D expression for $\frac{\partial U^{(\mathscr{E}_v)}}{\partial \mathbf{k}_u}$ and the cancellation of the diagonal terms of $Q$ is exactly analogous to the 1D treatment presented elsewhere (see Refs. [4, 30]).

4. Second hyperpolarizability

$$\gamma_{tuvw} = -\frac{\partial^4 E^{TOT}}{\partial \mathscr{E}_t \partial \mathscr{E}_u \partial \mathscr{E}_v \partial \mathscr{E}_w} =$$
$$\frac{1}{n_k} \Re \sum_{\mathbf{k}}^{BZ} P_{t,u,v,w} Tr \left[ n\, U^{(\mathscr{E}_t)\dagger}(\mathbf{k}) \left( G^{(\mathscr{E}_u)}(\mathbf{k}) U^{(\mathscr{E}_v;\mathscr{E}_w)}(\mathbf{k}) - U^{(\mathscr{E}_v;\mathscr{E}_w)}(\mathbf{k}) G^{(\mathscr{E}_u)}(\mathbf{k}) + \frac{1}{2} W^{(\mathscr{E}_u;\mathscr{E}_v)} U^{(\mathscr{E}_w)}(\mathbf{k}) - \frac{1}{2} U^{(\mathscr{E}_w)}(\mathbf{k}) \varepsilon^{(\mathscr{E}_u;\mathscr{E}_v)}(\mathbf{k}) \right) + \imath n\, U^{(\mathscr{E}_v;\mathscr{E}_w)\dagger}(\mathbf{k}) \frac{\partial U^{(\mathscr{E}_t)}(\mathbf{k})}{\partial k_u} \right] \tag{5.30}$$

In the previous equation, $W^{(\mathscr{E}_u;\mathscr{E}_v)}$ represents the second derivative of $C^{(0)\dagger}BC^{(0)}$ with respect to the field along $\mathscr{E}_u$ and $\mathscr{E}_v$ where $B$ is the sum of the Coulomb and exchange potentials in the Bloch atomic orbitals basis set.
Under the same conditions as above, it can be shown that $\gamma$ does not depend on the imaginary diagonal elements of the $Q$ matrix.

### 5.3.2 Extension to CPKS

As usual the exchange-correlation (XC) terms require special attention. In that regard we utilize the numerical approach of Pople and co-workers [50]. Thus, the electron density $\rho(r_i)$ and its derivative $\rho^t(r_i)$ with respect to the field $\mathscr{E}_t$ are expressed at each point $r_i$ of a grid as follows:

$$\rho(r_i) = \sum_{\mu,\nu} \sum_{g} D_{\mu\nu}^{g(0)} \chi_{\mu}^{0}(r_i) \chi_{\nu}^{g}(r_i) \tag{5.31}$$

$$\rho^t(r_i) = \sum_{\mu,\nu} \sum_{g} D_{\mu\nu}^{g(\mathscr{E}_t)} \chi_{\mu}^{0}(r_i) \chi_{\nu}^{g}(r_i) \tag{5.32}$$

which leads to the gradients:

$$\nabla\rho(r_i) = \sum_{\mu,\nu} \sum_{g} D_{\mu\nu}^{g(0)} \nabla(\chi_{\mu}^{0}(r) \chi_{\nu}^{g}(r))_{r_i} \tag{5.33}$$

$$\nabla\rho^t(r_i) = \sum_{\mu,\nu} \sum_{g} D_{\mu\nu}^{g(\mathscr{E}_t)} \nabla(\chi_{\mu}^{0}(r) \chi_{\nu}^{g}(r))_{r_i} \tag{5.34}$$

$D_{\mu\nu}^{g(0)}$ and $D_{\mu\nu}^{g(\mathscr{E}_t)}$ are the zeroth- and first-order perturbed density matrix elements between $\mu$ and $\nu$ atomic orbitals belonging to the origin and $g$ cells, respectively, in the direct space.

If $f^{XC}$ denotes an exchange-correlation, then the XC part of the Fock-matrix is

$$F_{DFT}^{XC} = \sum_{i} w_i \left[ \frac{\partial f^{XC}}{\partial \rho} \chi_{\mu}^{0} \chi_{\nu}^{g} + 2 \frac{\partial f^{XC}}{\partial |\nabla\rho|^2} \nabla\rho \cdot \nabla(\chi_{\mu}^{0} \chi_{\nu}^{g}) \right]_{r_i} \tag{5.35}$$

with the geometrical weights $w_i$ determined according to an atomic partition, which is field-independent. Since the atomic orbitals are also field-independent the field derivatives of $F_{DFT}^{XC}$, which are needed to obtain the first-order perturbed orbitals, as described in the previous sub-section, are given by:

$$\begin{aligned} F_{DFT}^{XC(\mathscr{E}_t)} = \sum_{i} w_i \Bigg\{ & \left[ \frac{\partial^2 f^{XC}}{\partial \rho^2} \rho^t + 2 \frac{\partial^2 f^{XC}}{\partial \rho \partial \mid \nabla\rho \mid^2} \nabla\rho \cdot \nabla\rho^t \right] \chi_{\mu}^{0} \chi_{\nu}^{g} \\ & + 2 \left[ \frac{\partial^2 f^{XC}}{\partial \rho \partial \mid \nabla\rho \mid^2} \rho^t + 2 \frac{\partial^2 f^{XC}}{(\partial \mid \nabla\rho \mid^2)^2} \nabla\rho \cdot \nabla\rho^t \right] \nabla\rho \cdot \nabla(\chi_{\mu}^{0} \chi_{\nu}^{g}) \\ & + 2 \frac{\partial f^{XC}}{\partial \mid \nabla\rho \mid^2} \nabla\rho^t \cdot \nabla(\chi_{\mu}^{0} \chi_{\nu}^{g}) \Bigg\}_{r_i} \end{aligned} \tag{5.36}$$

where the field derivatives are taken from Eqs. (5.31) to (5.34). In the expression Eq. (5.35), we have assumed that the exchange-correlation depends only on $\rho$ and $|\nabla\rho|^2$, but it can be easily generalized to include more complicated functionals. In addition, there is a direct contribution to the static $\beta$, for example, due to the third derivative of $f^{XC}$ with respect to the fields [42]:

$$\beta_{tuv}^{DFT} = \sum_i w_i P_{t,u,v} \left[ \frac{\partial^3 f^{XC}}{\partial \rho^3} \rho^t \rho^u \rho^v + 2 \frac{\partial^3 f^{XC}}{\partial \rho^2 \partial \mid \nabla \rho \mid^2} \left( \rho^t \rho^u \nabla \rho \cdot \nabla \rho^v \right) \right.$$
$$+ 4 \frac{\partial^3 f^{XC}}{\partial \rho \partial (\mid \nabla \rho \mid^2)^2} \left( \rho^t \nabla \rho \cdot \nabla \rho^u \nabla \rho \cdot \nabla \rho^v \right)$$
$$+ 8 \frac{\partial^3 f^{XC}}{\partial (\mid \nabla \rho \mid^2)^3} \nabla \rho \cdot \nabla \rho^t \nabla \rho \cdot \nabla \rho^u \nabla \rho \cdot \nabla \rho^v + 2 \frac{\partial^2 f^{XC}}{\partial \rho \partial \mid \nabla \rho \mid^2} \left( \rho^t \nabla \rho^u \cdot \nabla \rho^v \right)$$
$$\left. + 4 \frac{\partial^2 f^{XC}}{(\partial (\mid \nabla \rho \mid^2)^2} \left( \nabla \rho \cdot \nabla \rho^t \nabla \rho^u \cdot \nabla \rho^v \right) \right]_{\mathbf{r}_i} \tag{5.37}$$

A similar, but considerably more complicated, expression using also the second derivatives $\rho^{tu}(r_i)$ and $\nabla \rho^{tu}(r_i)$ with respect to $\mathscr{E}_t$ and $\mathscr{E}_u$ can be written for the second hyperpolarizability (see Ref. [41]). This expression contains functional fourth derivatives of $f^{XC}$ with respect to $\rho$ and/or its gradient. Fortunately, these derivatives can be obtained automatically (see Ref. [41]).

## 5.4 Linear and Nonlinear Optical Properties

### 5.4.1 *Dynamic Linear Polarizabilities and Optical Spectra*

If the field is frequency ($\omega$)-dependent, time-dependent perturbation theory leads to the two following $U$ matrices (see Refs. [28, 56] for the molecular or $\mathbf{k} = \mathbf{0}$ case, and Refs. [3, 17] for periodic systems):

$$U_{ia}^{(\mathscr{E}_t[\pm\omega])}(\mathbf{k}) = \lim_{\eta \to 0^+} \frac{G_{ia}^{(\mathscr{E}_t[\pm\omega])}(\mathbf{k})}{\varepsilon_a^{(0)}(\mathbf{k}) - \varepsilon_i^{(0)}(\mathbf{k}) \pm \omega + \imath\eta} = -U_{ai}^{(\mathscr{E}_t[\mp\omega])*}(\mathbf{k}) \tag{5.38}$$

where $G_{ia}^{(\mathscr{E}_t[\pm\omega])}(\mathbf{k})$ is defined in Sect. 5.3.1 except that now the electric field in Fock matrix element is $\omega$-frequency dependent. $\eta$ is a physical damping factor giving the width of the peak due to the inverse lifetime of the excited states (supposed to be the same in the UV-visible range of energy) It is not easy to evaluate the lifetime. Usually the energy resolution of the experimental equipment—typically $\eta \simeq 0.1$eV in UV-visible measurements—is used.

Solution of Eq. (5.38) for the $U^{(\mathscr{E}_t[\pm\omega])}(\mathbf{k})$ matrices can be obtained by fixing the value of $\omega$ and, then, solving iteratively. Near resonances ($\omega \simeq \varepsilon_a^{(0)} - \varepsilon_i^{(0)}$), convergence can be difficult to obtain; in that case a convergence accelerator such as direct inversion in the iterative subspace (DIIS) becomes necessary (see Refs. [21, 51]). Then an optical absorption spectrum can, in principle, be simulated through a scan of $\omega$ values in the desired UV-visible frequency range with the relative peak intensity at a given frequency being obtained from the imaginary part of the optical (high-frequency) dielectric matrix $\varepsilon^{\infty}$ for a non-zero value of the damping factor $\eta$. At low frequency ($\omega < 1$ eV or $\lambda > 1000$ nm), the vibrational contribution to the polarizability can become large depending upon the magnitude of the Born charges. This contribution will not be considered in the following. Thus, we will assume that the vibrational transition frequencies are negligible compared to $\omega$. The wavelength $\lambda$, on the other hand, is assumed to be much larger than the dimensions of the unit cell.

If we replace the static $U$ in Eq. (5.28) with its frequency-dependent counterpart from Eq. (5.38), then (after including a simple multiplicative factor to convert from polarizability to dielectric matrix) we find that the imaginary part of $\varepsilon_{tu}^{\infty}$ is:

$$\Im(\varepsilon_{tu}^{\infty}[\omega]) = \Im\left(\frac{4\pi}{V}\alpha_{t[-\omega]u[+\omega]}\right) = \frac{8\pi}{V}\frac{1}{n_k}\Im\{P_{t[-\omega],u[+\omega]}\sum_{\mathbf{k}}^{BZ}\sum_{i}^{occ}\sum_{a}^{virt}\Omega_{t,ia}(\mathbf{k})U_{ai}^{(\mathscr{E}_u[+\omega])}(\mathbf{k})\} \tag{5.39}$$

where $\alpha$ and $V$ are the unit cell polarizability and volume, respectively. Note that the dielectric matrix elements are purely real if $\eta = 0$.

In order to exhibit the pole structure of the dielectric matrix we reformulate Eq. (5.38) as:

$$U_{ia}^{(\mathscr{E}_t[\pm\omega])}(\mathbf{k}) = \lim_{\eta\to 0^+}\frac{\Omega_{t,ia}(\mathbf{k}) + \sum_{\mathbf{k}'}\sum_j\sum_b\left(U_{jb}^{(\mathscr{E}_t[\pm\omega])}(\mathbf{k}')B_{ia,jb}(\mathbf{k},\mathbf{k}') + U_{jb}^{(\mathscr{E}_t[\mp\omega])*}(\mathbf{k}')B_{ia,bj}(\mathbf{k},\mathbf{k}')\right)}{\varepsilon_a^{(0)}(\mathbf{k}) - \varepsilon_i^{(0)}(\mathbf{k}) \pm \omega + \imath\eta} \tag{5.40}$$

where $B_{ia,jb}(\mathbf{k},\mathbf{k}')$ and $B_{ia,bj}(\mathbf{k},\mathbf{k}')$ are defined as follows:

$$B_{ia,jb}(\mathbf{k},\mathbf{k}') = \left[2\langle i_{\mathbf{k}}b_{\mathbf{k}'}|a_{\mathbf{k}}j_{\mathbf{k}'}\rangle - \langle i_{\mathbf{k}}b_{\mathbf{k}'}|j_{\mathbf{k}'}a_{\mathbf{k}}\rangle\right] \tag{5.41}$$

$$B_{ia,bj}(\mathbf{k},\mathbf{k}') = \left[2\langle i_{\mathbf{k}}j_{\mathbf{k}'}|a_{\mathbf{k}}b_{\mathbf{k}'}\rangle - \langle i_{\mathbf{k}}j_{\mathbf{k}'}|b_{\mathbf{k}'}a_{\mathbf{k}}\rangle\right] \tag{5.42}$$

The notation $\langle ij|ab\rangle$ is used for the bielectronic integral $\langle i(1)j(2)|\frac{1}{r_{12}}|a(1)b(2)\rangle$. Even if the explicit calculation of such quantities poses practical challenges related to the convergence of the Coulomb Fourier series, we introduce them here since they are useful from a formal point of view to elucidate connections with the TDHF formalism.

We may, then, solve for the converged self-consistent $U(\mathbf{k})$ by re-writing Eq. (5.40):

$$
\begin{aligned}
&U_{ia}^{(\mathscr{E}_t[\pm\omega])}(\mathbf{k})\\
&= \lim_{\eta\to 0^+}\left\{\frac{\Omega_{t,ia}(\mathbf{k}) + U_{ia}^{(\mathscr{E}_t[\pm\omega])}(\mathbf{k})B_{ia,ia}(\mathbf{k},\mathbf{k}) + U_{ia}^{(\mathscr{E}_t[\mp\omega])\,*}(\mathbf{k})B_{ia,ai}(\mathbf{k},\mathbf{k})}{\varepsilon_a^{(0)}(\mathbf{k}) - \varepsilon_i^{(0)}(\mathbf{k}) \pm \omega + \imath\eta}\right.\\
&\left. + \frac{\sum_{(\mathbf{k}',j,b)\neq(\mathbf{k},i,a)}\left(U_{jb}^{(\mathscr{E}_t[\pm\omega])}(\mathbf{k}')B_{ia,jb}(\mathbf{k},\mathbf{k}') + U_{jb}^{(\mathscr{E}_t[\mp\omega])\,*}(\mathbf{k}')B_{ia,bj}(\mathbf{k},\mathbf{k}')\right)}{\varepsilon_a^{(0)}(\mathbf{k}) - \varepsilon_i^{(0)}(\mathbf{k}) \pm \omega + \imath\eta}\right\}
\end{aligned}
\tag{5.43}
$$

which yields after some manipulation[1]:

$$
\begin{aligned}
U_{ia}^{(\mathscr{E}_t[\pm\omega])}(\mathbf{k}) &= \lim_{\eta\to 0^+}\left\{\frac{\Omega_{t,ia}(\mathbf{k}) + U_{ia}^{(\mathscr{E}_t[\mp\omega])\,*}(\mathbf{k})B_{ia,ai}(\mathbf{k},\mathbf{k})}{\varepsilon_a^{(0)}(\mathbf{k}) - \varepsilon_i^{(0)}(\mathbf{k}) - B_{ia,ia}(\mathbf{k},\mathbf{k}) \pm \omega + \imath\eta}\right.\\
&\left. + \frac{\sum_{(\mathbf{k}',j,b)\neq(\mathbf{k},i,a)}\left(U_{jb}^{(\mathscr{E}_t[\pm\omega])}(\mathbf{k}')B_{ia,jb}(\mathbf{k},\mathbf{k}') + U_{jb}^{(\mathscr{E}_t[\mp\omega])\,*}(\mathbf{k}')B_{ia,bj}(\mathbf{k},\mathbf{k}')\right)}{\varepsilon_a^{(0)}(\mathbf{k}) - \varepsilon_i^{(0)}(\mathbf{k}) - B_{ia,ia}(\mathbf{k},\mathbf{k}) \pm \omega + \imath\eta}\right\}
\end{aligned}
\tag{5.44}
$$

The dielectric matrix elements become infinite when $U_{ia}^{(\mathscr{E}_t[\pm\omega])}(\mathbf{k})$ is infinite, i.e. at each $i \to a$ resonance. According to the expression of the denominator in Eq. (5.44), the pole is shifted by the real $B_{ia,ia}(\mathbf{k},\mathbf{k})$ value with respect to the transition energy $\varepsilon_a^{(0)}(\mathbf{k}) - \varepsilon_i^{(0)}(\mathbf{k})$ between unrelaxed crystalline orbitals, as in SOS, for each $\mathbf{k}$-point. This is the exciton effect for the monoexcitation $i_{\mathbf{k}} \to a_{\mathbf{k}}$, assuming there is no "overtone" due to other $j_{\mathbf{k}'} \to b_{\mathbf{k}'}$ resonances.

For computational purposes it is more convenient to obtain the transition frequencies directly using the standard TDHF format; in principle, the results are the same in either case. For this purpose we may start, again, with Eq. (5.40) but this time set $\eta$ to zero and the perturbation $\Omega_{t,ia}(\mathbf{k}) = 0$. Note that the position of the poles is unaltered by the perturbation term. After multiplying both sides of Eq. (5.44) by the denominator $(\varepsilon_a^{(0)}(\mathbf{k}) - \varepsilon_i^{(0)}(\mathbf{k}) - B_{ia,ia}(\mathbf{k},\mathbf{k}) \pm \omega)$ and some mathematical manipulations, as well as use of the relation $U_{ia}^{(\mathscr{E}_t[\pm\omega])}(\mathbf{k}) = -U_{ai}^{(\mathscr{E}_t[\mp\omega])\,*}(\mathbf{k})$ (and $B_{ai,jb}(\mathbf{k},\mathbf{k}') = B^*_{ia,bj}(\mathbf{k},\mathbf{k}')$), we obtain the following pair of equations:

$$
\begin{aligned}
&(\varepsilon_i^{(0)}(\mathbf{k}) - \varepsilon_a^{(0)}(\mathbf{k}) + B_{ia,ia}(\mathbf{k},\mathbf{k}) - \omega)U_{ia}^{(\mathscr{E}_t[+\omega])}(\mathbf{k}) - B_{ia,ai}(\mathbf{k},\mathbf{k})U_{ai}^{(\mathscr{E}_t[+\omega])}(\mathbf{k})\\
&+ \sum_{(\mathbf{k}',j,b)\neq(\mathbf{k},i,a)}\left(B_{ia,jb}(\mathbf{k},\mathbf{k}')U_{jb}^{(\mathscr{E}_t[+\omega])}(\mathbf{k}') - B_{ia,bj}(\mathbf{k},\mathbf{k}')U_{bj}^{(\mathscr{E}_t[+\omega])}(\mathbf{k}')\right) = 0
\end{aligned}
\tag{5.45}
$$

[1]In going from Eqs. (5.43) to (5.44), we use $x = \frac{a+bx}{c} \Rightarrow x(1 - b/c) = \frac{a}{c}$ which leads to $x = \frac{a}{c-b}$.

and

$$\left(\varepsilon_i^{(0)}(\mathbf{k}) - \varepsilon_a^{(0)}(\mathbf{k}) + B_{ai,ai}(\mathbf{k},\mathbf{k}) + \omega\right) U_{ai}^{(\mathcal{E}_t[+\omega])\,*}(\mathbf{k}) - B^*_{ai,ia}(\mathbf{k},\mathbf{k}) U_{ia}^{(\mathcal{E}_t[+\omega])*}(\mathbf{k})$$
$$+ \sum_{(\mathbf{k}',j,b)\neq(\mathbf{k},i,a)} \left( B^*_{ai,bj}(\mathbf{k},\mathbf{k}') U_{bj}^{(\mathcal{E}_t[+\omega])\,*}(\mathbf{k}') - B^*_{ai,jb}(\mathbf{k},\mathbf{k}') U_{jb}^{(\mathcal{E}_t[+\omega])\,*}(\mathbf{k}') \right) = 0 \tag{5.46}$$

A pair of equations is obtained for each $jb, \mathbf{k}'$ transition giving rise to the well-known non-hermitian TDHF system of $2 \times N_{occ} \times N_{virt} \times N_{\mathbf{k}}$ equations for the eigenvectors $(X, Y) = (U_{ia,\mathbf{k}}^{(\mathcal{E}_t[\omega])}, U_{ai,\mathbf{k}}^{(\mathcal{E}_t[\omega])})$ (see also Refs. [24, 27]):

$$\begin{pmatrix} A & B \\ B^* & A^* \end{pmatrix} \begin{pmatrix} X \\ Y \end{pmatrix} = \omega \begin{pmatrix} 1 & 0 \\ 0 & -1 \end{pmatrix} \begin{pmatrix} X \\ Y \end{pmatrix} \tag{5.47}$$

where

$$A = \begin{pmatrix} \varepsilon_i^{(0)}(\mathbf{k}) - \varepsilon_a^{(0)}(\mathbf{k}) + B_{ia,ia}(\mathbf{k},\mathbf{k}) & B_{ia,jb}(\mathbf{k},\mathbf{k}') \\ B_{ia,jb}(\mathbf{k},\mathbf{k}') & \varepsilon_j^{(0)}(\mathbf{k}') - \varepsilon_b^{(0)}(\mathbf{k}') + B_{jb,jb}(\mathbf{k}',\mathbf{k}') \end{pmatrix} \tag{5.48}$$

and

$$B = \begin{pmatrix} -B_{ia,ai}(\mathbf{k},\mathbf{k}) & -B_{ia,bj}(\mathbf{k},\mathbf{k}') \\ -B_{ia,bj}(\mathbf{k},\mathbf{k}') & -B_{jb,bj}(\mathbf{k}',\mathbf{k}') \end{pmatrix} \tag{5.49}$$

The matrices in Eq. (5.47) have the dimension $2 \times N_{occ} \times N_{virt} \times N_{\mathbf{k}}$ on a side with $N_{occ}$, $N_{virt}$ and $N_{\mathbf{k}}$ being, respectively, the number of occupied and virtual orbitals and the number of **k** points of the reciprocal space. Finally, the oscillator strengths are obtained from the eigenvectors and their transition moments.

Both methods of simulating the UV-visible absorption spectrum that were described here lead to the same results (poles and oscillator strengths) but via two different approaches: either by diagonalization of the unperturbed hamiltonian in the $i_{\mathbf{k}} \rightarrow a_{\mathbf{k}}$ basis set or by plotting the imaginary part of the linear response to a dynamic field.

### 5.4.2 Dynamic First Hyperpolarizabilities

In order to derive our $2n + 1$-rule working formula for the Hartree-Fock dynamic first hyperpolarizabilities of a 3D periodic system given by Eq. (5.50) below, we start with the expression for the static limit due to Ferrero et al. [see Eq. (58) in Ref. [18]]. This expression has exactly the same form as that shown below except, of course, that all frequencies are set equal to zero. The 3D static result was obtained, in fact, by generalizing the 1D periodic frequency-dependent treatment of Kirtman et al. [30] which, in turn, was based on the time-dependent Hartree-Fock formulation for molecules developed by Karna and Dupuis [28] (KD). Thus, we use KD to introduce the 3D frequency-dependence.

KD considered just four special cases (see Table VII in Ref. [28]), including static, and their treatment pertains only to the first two terms in square brackets below. Nonetheless, it is straightforward to generalize their formulation as we have done. Finally, the use of the operator $P_{t[-\omega_\sigma]u[+\omega_1]v[+\omega_2]}$ which permutes the pairs $(t[-\omega_\sigma])$, $(u[+\omega_1])$, and $(v[+\omega_2])$ has been introduced by Orr and Ward [43], and used by Bishop and Kirtman in Ref. [5].

From the approach just described the general expression for the first hyperpolarizability of closed-shell periodic systems in the presence of frequency-dependent fields may be written as (see Ref. [53]):

$$\beta_{tuv}(-\omega_\sigma;\omega_1,\omega_2) = -\frac{\partial^3 E^{TOT}}{\partial\mathcal{E}_{t[-\omega_\sigma]}\partial\mathcal{E}_{u[\omega_1]}\partial\mathcal{E}_{v[\omega_2]}}$$
$$= \frac{2}{n_k}\Re\left\{P_{t[-\omega_\sigma],u[+\omega_1],v[+\omega_2]}\sum_{\mathbf{k}}^{BZ} Tr\left[n\, U^{(\mathcal{E}_{t[-\omega_\sigma]})\dagger}(\mathbf{k})\left(G^{(\mathcal{E}_{u[+\omega_1]})}(\mathbf{k})U^{(\mathcal{E}_{v[+\omega_2]})}(\mathbf{k})\right.\right.\right.$$
$$\left.\left.\left. -U^{(\mathcal{E}_{v[+\omega_2]})}(\mathbf{k})G^{(\mathcal{E}_{u[+\omega_1]})}(\mathbf{k}) + \imath\frac{\partial U^{(\mathcal{E}_{v[+\omega_2]})}(\mathbf{k})}{\partial k_u}\right)\right]\right\} \tag{5.50}$$

where $\omega_\sigma = \omega_1 + \omega_2$. The last term of Eq. 5.50 appears only for infinite periodic systems. A working expression is:

$$\frac{\partial U_{ia}^{(\mathcal{E}_{v[+\omega_2]})}(\mathbf{k})}{\partial k_u} =$$
$$\frac{\frac{\partial}{\partial k_u}G_{ia}^{(\mathcal{E}_{v[+\omega_2]})}(\mathbf{k})}{\varepsilon_a^{(0)}(\mathbf{k}) - \varepsilon_i^{(0)}(\mathbf{k}) + \omega_2} - \frac{G_{ia}^{(\mathcal{E}_{v[+\omega_2]})}(\mathbf{k})\frac{\partial}{\partial k_u}\left(\varepsilon_a^{(0)}(\mathbf{k}) - \varepsilon_i^{(0)}(\mathbf{k})\right)}{\left(\varepsilon_a^{(0)}(\mathbf{k}) - \varepsilon_i^{(0)}(\mathbf{k}) + \omega_2\right)^2} \tag{5.51}$$

in which the development of $\frac{\partial}{\partial k_u}G_{ia}^{(\mathcal{E}_{v[+\omega_2]})}(\mathbf{k})$, the second derivative of the Fock matrix elements $F_{ia}(\mathbf{k})$ with respect to the field and $\mathbf{k}$, can be found in Ref. [30]:

$$\frac{\partial G_{ia}^{(\mathcal{E}_{v[-\omega_2]})}(\mathbf{k})}{\partial k_u} = \sum_p^{all}\left(Q_{pi}^u(\mathbf{k})\right)^* G_{pa}^{(\mathcal{E}_{v[+\omega_2]})}(\mathbf{k}) + \sum_p^{all} G_{ip}^{(\mathcal{E}_{v[+\omega_2]})}(\mathbf{k})\, Q_{pa}^u(\mathbf{k})$$
$$+\sum_{\mu,\nu} C_{\mu i}^{(0)*}(\mathbf{k})\frac{\partial F_{\mu\nu}^{(\mathcal{E}_{v[+\omega_2]})}(\mathbf{k})}{\partial k_u} C_{\nu a}^{(0)}(\mathbf{k}) \tag{5.52}$$

The evaluation of the last term on the r.h.s. of Eq. (5.52) is straightforward except for the term $\frac{\partial Q^{v}_{pq}}{\partial k_u}$ contained in it. This term, however, does not depend on the $\omega_2$ frequency, and can be thus computed as in the static case [4, 18]. As regards the calculation of $\frac{\partial \varepsilon^{(0)}_i}{\partial k_u}$ it is equal to: $\sum_{\mu,\nu} C^{(0)*}_{\mu i}(\mathbf{k}) \left( \frac{\partial F^{(0)}_{\mu\nu}}{\partial k_u} - \varepsilon^{(0)}_i \frac{\partial S_{\mu\nu}}{\partial k_u} \right) C^{(0)}_{\nu i}(\mathbf{k})$. As in the case of the static first hyperpolarizability there is an additional XC contribution in DFT to the corresponding dynamic property. Within the adiabatic approximation the generalization to frequency-dependent fields is [37]:

$$\begin{aligned}
\beta^{DFT}_{tuv}(-\omega_\sigma;\omega_1,\omega_2) = \sum_i w_i P_{t[-\omega_\sigma],u[+\omega_1],v[+\omega_2]} \Bigg[ & \frac{\partial^3 f^{XC}}{\partial \rho^3} \rho^{t}_{(-\omega_\sigma)} \rho^{u}_{(+\omega_1)} \rho^{v}_{(+\omega_2)} \\
&+ 2 \frac{\partial^3 f^{XC}}{\partial \rho^2 \partial \mid \nabla\rho \mid^2} \left( \rho^{t}_{(-\omega_\sigma)} \rho^{u}_{(+\omega_1)} \nabla\rho \cdot \nabla\rho^{v}_{(+\omega_2)} \right) \\
&+ 4 \frac{\partial^3 f^{XC}}{\partial \rho \partial (\mid \nabla\rho \mid^2)^2} \left( \rho^{t}_{(-\omega_\sigma)} \nabla\rho \cdot \nabla\rho^{u}_{(+\omega_1)} \nabla\rho \cdot \nabla\rho^{v}_{(+\omega_2)} \right) \\
&+ 8 \frac{\partial^3 f^{XC}}{\partial (\mid \nabla\rho \mid^2)^3} \nabla\rho \cdot \nabla\rho^{t}_{(-\omega_\sigma)} \nabla\rho \cdot \nabla\rho^{u}_{(+\omega_1)} \nabla\rho \cdot \nabla\rho^{v}_{(+\omega_2)} \\
&+ 2 \frac{\partial^2 f^{XC}}{\partial \rho \partial \mid \nabla\rho \mid^2} \left( \rho^{t}_{(-\omega_\sigma)} \nabla\rho^{u}_{(+\omega_1)} \cdot \nabla\rho^{v}_{(+\omega_2)} \right) \\
&+ 4 \frac{\partial^2 f^{XC}}{(\partial (\mid \nabla\rho \mid^2)^2} \left( \nabla\rho \cdot \nabla\rho^{t}_{(-\omega_\sigma)} \nabla\rho^{u}_{(+\omega_1)} \cdot \nabla\rho^{v}_{(+\omega_2)} \right) \Bigg]_{\mathbf{r}_i}
\end{aligned} \tag{5.53}$$

in which the frequency-dependent density and density gradient are:

$$\rho^{t}_{(\pm\omega)}(r_i) = \sum_{\mu,\nu} \sum_{g} D^{(\mathscr{E}_{t[\pm\omega]})}_{\mu\nu}(\mathbf{g})\, \chi^{0}_{\mu}(r_i) \chi^{g}_{\nu}(r_i) \tag{5.54}$$

and

$$\nabla\rho^{t}_{(\pm\omega)}(r_i) = \sum_{\mu,\nu} \sum_{g} D^{(\mathscr{E}_{t[\pm\omega]})}_{\mu\nu}(\mathbf{g})\, \nabla(\chi^{0}_{\mu}(r) \chi^{g}_{\nu}(r))_{r_i} \tag{5.55}$$

In principle, exactly the same approach as applied here to obtain the dynamic first hyperpolarizability can be employed for the dynamic second hyperpolarizability. That is a project, however, which remains to be done. When completed it will provide access to properties such two-photon absorption, field-induced second harmonic generation and the optical Kerr effect among others.

## 5.5 Effects of Geometric Distortions on Cell Dipole and Polarizability

In this section, we will discuss how the coupled-perturbed methods presented in previous sections can be combined with analytical gradients—with respect to either atomic positions or cell shape—to yield mixed derivatives. Some of the most interesting experimentally observable quantities arising from such treatment are:

$$\text{Infrared Intensities (Born charges)} \Longleftarrow \left. \frac{\partial}{\partial \mathscr{E}_t} \frac{\partial E^{TOT}}{\partial \mathscr{R}^A_v} \right|_{\mathscr{E}=0,\mathscr{R}_0} \tag{5.56}$$

$$\text{Raman Intensities} \Longleftarrow \left. \frac{\partial^2}{\partial \mathscr{E}_t \partial \mathscr{E}_u} \frac{\partial E^{TOT}}{\partial \mathscr{R}^A_v} \right|_{\mathscr{E}=0,\mathscr{R}_0} \tag{5.57}$$

$$\text{Piezoelectric tensor} \Longleftarrow \left. \frac{\partial}{\partial \mathscr{E}_t} \frac{\partial E^{TOT}}{\partial a_{\alpha v}} \right|_{\mathscr{E}=0,a^0} \tag{5.58}$$

$$\text{Photoelastic Tensor} \Longleftarrow \left. \frac{\partial^2}{\partial \mathscr{E}_t \mathscr{E}_u} \frac{\partial E^{TOT}}{\partial a_{\alpha v}} \right|_{\mathscr{E}=0,a^0} \tag{5.59}$$

Here, as in previous sections, $t$, $u$, $v$ are cartesian directions, $\mathscr{R}^A_v$ represents a displacement of atom $A$ along cartesian direction $v$, and $\mathscr{R}_0$ represents the equilibrium (optimized) geometry of the atoms within the unit cell. Analogously, $a_{\alpha v}$ are Cartesian components of the three (in the 3D case) lattice vectors $\mathbf{a}_\alpha$, $\alpha = 1, 2, 3$ and $a^0$ represents the equilibrium (optimized) set of lattice parameters.

When such mixed derivatives are considered, the 2n + 1 rule breaks down. In the case of Raman intensities, for example, third-order energy derivatives are needed and second-order wavefunction derivatives are required. But there is a choice. One may determine either mixed second derivatives or second-order electric field derivatives. The latter is preferred since the calculations scale more favorably with respect to the size of the unit cell due to the smaller number of perturbation equations to be solved. Details of the derivation of optimal and computable expressions for Eqs. (5.56)–(5.58) above can be found in Refs. [1, 34–36], while work for analytical implementation of photoelasticity is in progress.

Unlike the pure electric field case treated in previous sections, we now have to consider gradients of mono- and bi-electronic integrals over atomic orbitals, since now they depend on the perturbation. However, the time-consuming calculation of derivatives of orbital coefficients with respect to nuclear displacements is avoided by taking advantage of the eigenvalue-weighted density matrix. Finally, derivatives with respect to the wave-vector are carried out analytically, so that the entire treatment is analytical. In the following, we briefly review the operative equations, underlining their key aspects in connection with the discussion in previous sections.

### 5.5.1 Infrared and Raman Intensities

The mixed second derivatives of the total energy with respect to the atomic displacements and the electric field (along $b$), which determine the infrared transition intensities, are conveniently formulated in direct space as:

$$\frac{\partial}{\partial \mathscr{E}_t} \frac{\partial E^{TOT}}{\partial \mathscr{R}_v^A}\Bigg|_{\mathscr{E}=0,\mathscr{R}_0}$$
$$= \mathrm{Tr}\left( \frac{1}{2}\left(H^{(\mathscr{R}_v^A)}(\mathbf{g})+\mathfrak{F}^{(\mathscr{R}_v^A)}(\mathbf{g})\right)^{(0)} D^{(\mathscr{E}_t)}(\mathbf{g}) \right.$$
$$\left. + \frac{1}{2}\left[ \Omega_t^{(\mathscr{R}_v^A)}(\mathbf{g}) + \left(\mathfrak{F}^{(\mathscr{R}_v^A)}(\mathbf{g})\right)^{(\mathscr{E}_t)} \right] D^{(0)}(\mathbf{g}) - S^{(\mathscr{R}_v^A)}(\mathbf{g})\, D_W^{(\mathscr{E}_t)}(\mathbf{g}) \right)_{\mathscr{R}_0}$$
$$-Z_A \delta_{tv} \quad , \tag{5.60}$$

These derivatives are evaluated at zero field and equilibrium geometry. In Eq. (5.60) $Z_A$ is the nuclear charge of atom A and $H(\mathbf{g})$ is the core Hamiltonian. The notation $\mathfrak{F}^{(\mathscr{R}_v^A)}$ signifies that the derivative of the Fock matrix $F(\mathbf{g})$ is taken with respect to atomic displacements while holding the density matrix constant. Note that the definition of the "eigenvalue-weighted density matrix" $D_W(\mathbf{g})$ given by

$$D_W(\mathbf{g}) = \sum_{\mathbf{k}}^{BZ} C(\mathbf{k})\varepsilon(\mathbf{k})C^{\dagger}(\mathbf{k})e^{\imath\mathbf{k}\cdot\mathbf{g}} \tag{5.61}$$

allows the computation of its perturbed form $D_W^{(\mathscr{E}_t)}(\mathbf{g})$ as:

$$D_W^{(\mathscr{E}_t)}(\mathbf{g}) = \sum_{\mathbf{k}}^{BZ} \left( C^{(\mathscr{E}_t)}(\mathbf{k})\varepsilon(\mathbf{k})C^{\dagger}(\mathbf{k}) + C(\mathbf{k})\varepsilon^{(\mathscr{E}_t)}(\mathbf{k})C^{\dagger}(\mathbf{k}) + C(\mathbf{k})\varepsilon(\mathbf{k})C^{(\mathscr{E}_t)\,\dagger}(\mathbf{k}) \right) e^{\imath\mathbf{k}\cdot\mathbf{g}} \tag{5.62}$$

As widely discussed in previous sections, the $Q(\mathbf{k})$ matrix [Eq. (5.26)] contains undetermined imaginary diagonal elements that could, in principle, pose problems in solving Eq. 5.60. However, the derivative of these matrix elements with respect to the atomic displacements of interest vanishes because the derivatives are taken with the unit cell parameters held fixed. Thus, the dependence on the crystal boundaries vanishes and the physical quantity is correctly obtained.

The Raman intensities involve one further derivative with respect to the electric field and we obtain the following expression for the Raman tensor in Cartesian coordinates [36]:

$$
\begin{aligned}
\frac{\partial^2}{\partial\mathscr{E}_t\partial\mathscr{E}_u}\frac{\partial E^{TOT}}{\partial\mathscr{R}_v^A}\bigg|_{\mathscr{E}=0,\mathscr{R}_0} =\mathrm{Tr}\Bigg\{ & \frac{1}{2}\left(H^{(\mathscr{R}_v^A)}(\mathbf{g})+\mathfrak{F}^{(\mathscr{R}_v^A)}(\mathbf{g})\right)^{(0)} D^{(\mathscr{E}_t,\mathscr{E}_u)}(\mathbf{g}) \\
& +\frac{1}{2}P_{t,u}\left[\left(\Omega_t^{(\mathscr{R}_v^A)}(\mathbf{g})+\left(\mathfrak{F}^{(\mathscr{R}_v^A)}(\mathbf{g})\right)^{(\mathscr{E}_t)}\right)D^{(\mathscr{E}_u)}(\mathbf{g})\right] \\
& +\frac{1}{2}\left[P_{t,u}\,\Omega_t^{(\mathscr{R}_v^A,\mathscr{E}_u)}+\left(\mathfrak{F}^{(\mathscr{R}_v^A)}(\mathbf{g})\right)^{(\mathscr{E}_t,\mathscr{E}_u)}\right]D^{(0)}(\mathbf{g}) \\
& -S^{(\mathscr{R}_v^A)}(\mathbf{g})\,D_W^{(\mathscr{E}_t,\mathscr{E}_u)}(\mathbf{g})\Bigg\}_{\mathscr{R}_0}
\end{aligned}
\qquad (5.63)
$$

In Eq. (5.63) all quantities have been defined in connection to Eq. (5.60). $D_W^{(\mathscr{E}_t,\mathscr{E}_u)}(\mathbf{g})$, in particular, is obtained by differentiating Eq. (5.62) with respect to a second field $\mathscr{E}_u$). The notation $\left(\mathfrak{F}^{(\mathscr{R}_v^A)}(\mathbf{g})\right)^{(\mathscr{E}_t)}$ signifies that perturbation in the Fock matrix with respect to atomic displacements is taken at constant density, but perturbation with respect to the electric field affects the density. As in the case of infrared intensities, it can be shown [36] that the diagonal elements of $Q(\mathbf{k})$ vanish when the above expression is evaluated. This must be true since these elements do not contribute to the polarizability.

We remark that in Eqs. (5.60) and (5.63)—as well as in (5.71) below—all ingredients are obtained either by a coupled-perturbed self-consistent procedure or through analytical gradients of the AO integrals. Although not immediately obvious, no mixed derivatives of any matrix appear and there are no derivatives of the density matrix (or eigenvalue-weighted density matrix) with respect to atomic displacements.

### 5.5.2 *The Piezoelectric Tensor*

The "proper" direct piezoelectric response of a crystal is represented by a third-rank tensor that is the sum of an electronic and a nuclear term, $\mathbf{e} = \mathbf{e}^{\mathrm{ele}} + \mathbf{e}^{\mathrm{nuc}}$ [1]. The electronic term measures the variation of the current density $J$ with strain when no nuclear relaxation is allowed (clamped-nuclei condition, *clamp*) [64]:

$$
e_{tuv}^{\mathrm{ele}} = \frac{\partial J_t}{\partial \varepsilon_{uv}}\bigg|_{clamp} = \sum_{\alpha}\frac{\partial J_t}{\partial a_{\alpha u}}a_{\alpha v}\bigg|_{clamp} . \qquad (5.64)
$$

Here $\boldsymbol{\varepsilon}$ is the symmetric second-rank pure strain tensor and the $a_{\alpha v}$ are Cartesian components of the three lattice vectors $\mathbf{a}_\alpha$, $\alpha = 1, 2, 3$. The nuclear relaxation contribution can be expressed as [57]:

$$
e_{tuv}^{\mathrm{nuc}} = \frac{1}{V}\sum_{s}\frac{\partial^2 E^{TOT}}{\partial\mathscr{E}_t\partial u_s}\frac{\mathrm{d}u_s}{\mathrm{d}\varepsilon_{uv}} , \qquad (5.65)
$$

where $V = \mathbf{a}_1 \cdot (\mathbf{a}_2 \times \mathbf{a}_3)$ is the cell volume and the $u_s$ are $3N-3$ fractional coordinates ($N$ being the number of atoms per cell) obtained from the $3N$ atomic fractional coordinates by orthogonalization to the three translations along $\hat{x}$, $\hat{y}$ and $\hat{z}$.

The $\mathrm{d}u_s/\mathrm{d}\varepsilon_{jk}$ derivatives can be obtained in either of two ways: (i) optimization of the internal geometry at strained configurations or (ii) from the full Hessian matrix **H** of energy second derivatives with respect to both fractional atomic displacements and lattice deformations (with a truncation to quadratic terms in the Taylor's expansion of the energy) [65]. We have found that (ii) is preferable [1]. In that procedure the following relation is utilized:

$$\frac{\mathrm{d}u_s}{\mathrm{d}\varepsilon_{jk}} = -\sum_{s'=1}^{3N-3} H_{jk,s'} \times (H^{-1})_{s's} \,, \tag{5.66}$$

in which

$$H_{jk,s'} = \left( \frac{\partial^2 E}{\partial \varepsilon_{jk} \partial u_{s'}} \right) , \tag{5.67}$$

and $(H^{-1})_{s's}$ is an element of the inverse of the atomic Hessian:

$$H_{s's} = \left( \frac{\partial^2 E}{\partial u_{s'} \partial u_s} \right) . \tag{5.68}$$

We can now write the electronic contribution to the "proper" piezoelectric tensor in terms of the polarization **P** as [64]:

$$e^{\mathrm{ele}}_{tuv} = \left.\frac{\partial J_t}{\partial \varepsilon_{uv}}\right|_{clamp} = \left.\frac{\partial P_t}{\partial \varepsilon_{uv}}\right|_{clamp} + P_t \delta_{uv} - P_v \delta_{tu} \,, \tag{5.69}$$

with the first term on the right-hand-side being the electronic contribution to the "improper" piezoelectric tensor. By expressing the polarization in terms of a total energy derivative, and by exploiting the second equality in Eq. (5.64), one gets:

$$\left.\frac{\partial J_t}{\partial \varepsilon_{uv}}\right|_{clamp} = \frac{1}{V} \sum_{\alpha} a_{\alpha v} \left.\frac{\partial^2 E^{TOT}}{\partial \mathscr{E}_t \partial a_{\alpha u}}\right|_{clamp} - P_v \delta_{tu} \,. \tag{5.70}$$

The analogue of Eq. (5.60), where the nuclear displacements are replaced by cell deformations, can be cast (for equilibrium lattice parameters $a^0$) as:

$$\frac{\partial}{\partial \mathscr{E}_t}\frac{\partial E^{TOT}}{\partial a_{\alpha v}}\bigg|_{\mathscr{E}=0,a^0} = \text{Tr}\left(\left[H^{(a_{\alpha v})}(\mathbf{g}) + \mathfrak{F}^{(a_{\alpha v})}(\mathbf{g})\right] D^{(\mathscr{E}_t)}(\mathbf{g})\right.$$
$$\left. + \frac{1}{2}\left[\Omega_t^{(a_{\alpha v})}(\mathbf{g}) + \left(\mathfrak{F}^{(a_{\alpha v})}(\mathbf{g})\right)^{(\mathscr{E}_t)}\right] D^{(0)}(\mathbf{g}) - S^{(a_{\alpha v})}(\mathbf{g})D_W^{(\mathscr{E}_t)}(\mathbf{g})\right)_{a^0}$$
$$- \sum_A Z_A f_{A,\alpha}\delta_{tv} + e\delta_{tv}N_\alpha \,, \tag{5.71}$$

where $D_W^{(\mathscr{E}_t)}(\mathbf{g})$ is defined in Eq. (5.62). $f_{A,\alpha}$ is the fractional coordinate of atom A along lattice vector $\alpha$, that is $\mathscr{R}_t^A = \sum_\alpha f_{A,\alpha} a_{\alpha t}$, $e$ is the magnitude of the electron charge and $N_\alpha$ is an arbitrary integer, arising from the imaginary diagonal elements of $Q(\mathbf{k})$. While the latter term appears for the improper piezoelectricity [30], it vanishes for the "proper" one since it is exactly cancelled by the $-P_v\delta_{tu}$ term in Eq. (5.70).

### 5.5.3 Extension to DFT

The extension of the above expressions for the infrared, Raman and piezoelectric tensors to DFT is done along the same lines as for the polarizability and hyperpolarizability discussed in Sect. 5.3.2. We report here, as a representative case, the expression for the closed-shell Born charges (infrared) tensor:

$$\frac{\partial E_R^{XC\,(\mathscr{R}_v^A)}}{\partial \mathscr{E}_t}\Bigg|_{\mathscr{E}=0,\mathscr{R}_0}$$
$$= \sum_i w_i \left\{ \left[\frac{\partial^2 f^{XC}}{\partial \rho^2}\frac{\partial \rho}{\partial \mathscr{E}_t} + \frac{\partial^2 f^{XC}}{\partial \rho\, \partial |\nabla\rho|^2}\frac{\partial |\nabla\rho|^2}{\partial \mathscr{E}_t}\right] \rho^{(\mathscr{R}_v^A)}\right.$$
$$+ 2\left[\frac{\partial^2 f^{XC}}{\partial (|\nabla\rho|^2)^2}\frac{\partial |\nabla\rho|^2}{\partial \mathscr{E}_t}\nabla\rho + \frac{\partial^2 f^{XC}}{\partial \rho\, \partial |\nabla\rho|^2}\frac{\partial \rho}{\partial \mathscr{E}_t}\nabla\rho + \frac{\partial f^{XC}}{\partial |\nabla\rho|^2}\frac{\partial \nabla\rho}{\partial \mathscr{E}_t}\right]\cdot \nabla\rho^{(\mathscr{R}_v^A)}$$
$$\left. + \frac{\partial f^{XC}}{\partial \rho}\frac{\partial \rho^{(\mathscr{R}_v^A)}}{\partial \mathscr{E}_t} + 2\frac{\partial f^{XC}}{\partial |\nabla\rho|^2}\nabla\rho \cdot \frac{\partial \nabla\rho^{(\mathscr{R}_v^A)}}{\partial \mathscr{E}_t}\right\}_{\mathbf{r}_i}$$
$$+ \sum_i \frac{\partial w_i}{\partial \mathscr{R}_v^A}\left\{\frac{\partial f^{XC}}{\partial \rho}\frac{\partial \rho}{\partial \mathscr{E}_t} + \frac{\partial f^{XC}}{\partial |\nabla\rho|^2}\frac{\partial |\nabla\rho|^2}{\partial \mathscr{E}_t}\right\}_{\mathbf{r}_i} \tag{5.72}$$

There are two key differences due to replacing an electric field derivative by a derivative with respect to an atomic displacement. One is that the AO derivatives in the expression for the density no longer vanish. The other is that the dependence of the weights on the atomic position must be taken into account as in the last term of Eq. (5.72).

### 5.5.4 Converse Piezoelectric Effect

Piezoelectricity is a coupling between electrical and mechanical responses of a system. Whereas the piezoelectric effect describes the generation of polarization due to mechanical strain or stress, the converse piezoelectric effect describes the generation of strain or stress due to an applied electric field or voltage. Thus, in both cases structural as well as electronic responses are at the center of attention.

When structural responses to an electric field are accounted for some subtleties occur. The responses in the bulk of any large, finite sample (modeled as being infinite and periodic) may contain a finite contribution from the boundaries of the sample irrespective of its size as long as the sample is so large that the thermodynamic limit has been reached. Therefore, different piezoelectric property values may be obtained for samples of the same material that differ only in shape and/or surface charges. In addition, as a secondary effect, differences in the piezoelectric structural response, may also lead to differences in other properties.

Mathematically, when the system is described as being infinite and periodic, the information on the effects due to the shape and/or surface charges are contained in (**k** and band dependent) crystal orbital phase factors. Thus, even if the effects are due to the boundaries of the sample, they can, in principle, be calculated as a bulk property in the sense that the effects are contained in the information for the infinite, periodic system.

It is easy to verify that the crystal orbital phase factors show up in the imaginary part of the diagonal elements of the $Q$ matrices. As long as it can be assumed that the applied electric field does not change the lattice vectors, the imaginary parts of the diagonal elements of the $Q$ matrices can be ignored. This is the case for high frequency fields. However, for the converse piezoelectric effect the applied electric field has an essentially vanishing frequency and the changes in the lattice vectors due to the applied field are the quantities of interest.

If $a_{ij}$ is the $j$-th component of the $i$-th lattice vector and $F_k$ is the $k$th component of the vector describing the field (see further below), a converse piezoelectric tensor may be defined as

$$\eta_{ijk} = \left.\frac{\partial a_{ij}}{\partial F_k}\right|_{F_k=0}. \tag{5.73}$$

When $\mathbf{F} = \mathcal{E}$, it can be shown [39] that one may write

$$\eta_{ijk} = \eta_{ijk,0} + \tilde{n}_{ijk}\eta_{ijk,1}, \tag{5.74}$$

where $\eta_{ijk,0}$ is a reference value and different values of $\tilde{n}_{ijk}$ correspond to modeling samples of different shapes and/or surface charges. For chemical/physical reasons, $\tilde{n}_{ijk}$ cannot vary more than roughly $-2 \leq \tilde{n}_{ijk} \leq +2$.

At this time we do not know the precise relation between the parameters $\tilde{n}_{ijk}$ and the shape/surface charge of a given sample, which implies that when calculating the converse piezoelectric tensor, deviations between theory and experiment may be due either to theoretical and experimental inaccuracies or to differences in the shape/surface charges. A relevant question, therefore, is whether $\eta_{ijk,1}/\eta_{ijk,0}$ is so small that $\eta_{ijk,1}$ can be ignored. Exploratory calculations on some perovskite crystals [39] have shown that $\eta_{ijk,1}/\eta_{ijk,0}$ is typically about 1/10, but larger (and smaller) values can be obtained.

A special case is that of quasi-one-dimensional systems (chains). For the components parallel to the chain direction (defined as $z$), $\tilde{n}_{zzz}$ can take only integral values. Moreover, if **F** is taken to be the potential drop over one unit cell (i.e., $F_z = \mathscr{E}_z a_{zz}$), then $\eta_{zzz}$ becomes independent of $\tilde{n}_{zzz}$.

## 5.6 Summary and Prospects

Our overall goal has been to put property calculations for infinite periodic systems on the same footing as the corresponding calculations for ordinary molecules. The CRYSTAL computer program is well-suited for this task since it utilizes Gaussian-type basis functions and takes maximum advantage of symmetry. Considerable progress has been made in adapting the treatment of molecules to account for the change in boundary conditions and subsequently implementing the appropriate methodology in CRYSTAL. This is particularly so for electric field properties as well as mixed properties involving a combination of electric fields and geometric displacements. For the former, we have adopted the coupled perturbed Hartree-Fock and Kohn-Sham (CPHF/CPKS) approaches to treat linear polarizabilities and nonlinear optical properties. Although the most important second-order nonlinear optical properties, including second harmonic generation and the dc-Pockels effect, are now available the treatment of arbitrary frequencies and third-order properties remains to be accomplished in the near term. We have now also implemented the determination of optical spectra through the frequency-dependent linear polarizability. As far as the mixed derivatives are concerned, infrared and Raman spectra have been completed as well as the piezoelectric effect. Immediate next steps in this category include the hyper-Raman effect and photoelasticity.

Our treatment of the orbital electronic response to a magnetic field is still in its early stages. A new translationally invariant form for the interaction Hamiltonian, which allows for a simultaneous electric field so as to treat chiral properties, has been presented. However, the CPHF/CPKS perturbation treatment remains to be formulated and implemented.

The P (Parallel) and MPP (Massive Parallel Processing) versions of the code for the electric field properties is at an advanced stage. Right now it is possible to compute frequencies and IR intensities of systems up to about 1000 atoms in the unit cell; Raman intensities, first and second hyperpolarizabilities will become feasible for systems of this size in the near future.

Vibrational contributions to nonlinear optical properties can also be evaluated, but only in the static or high frequency limit and only in the lowest order of approximation (known as the finite field nuclear relaxation (FF-NR) treatment). FF-NR requires a geometry optimization in the presence of a finite field. At present the finite field can be applied only in a non-periodic, direction for 1D or 2D systems. This limits the tensor components that can be considered. Removal of these several restrictions can be envisioned, but certainly that will be a longer term effort.

Another longer-term perspective is the development of optical properties calculations for solids at the post-Hartree-Fock level. This could be done in connection with the local-correlation CRYSCOR code, [47–49] which presently features a very efficient implementation of the local MP2 approach for periodic systems. This implementation makes use of advanced techniques such as density fitting [38] and orbital-specific virtuals [62], as well as efficient parallel implementation [33]. The extension to local coupled-cluster is also envisaged in the future.

The treatment of properties associated with the orbital electronic response to a magnetic field is still in its infancy. An appropriate Hamiltonian for the response to a homogeneous static field was presented in Sect. 5.3. There already exist methods for the computation of magnetic shielding and magnetic susceptibilities, but this Hamiltonian opens the possibility for a new approach to these properties. Further down the line, in combination with electric fields, we should be able to obtain various optical rotatory properties, such as circular dichroism and Faraday rotation. At the moment, however, it remains to develop the CPHF/CPKS formulation along with the response to frequency-dependent magnetic fields.

## References


1. J. Baima, A. Erba, L. Maschio, C. Zicovich-Wilson, R. Dovesi, B. Kirtman, Direct piezoelectric tensor of 3D periodic systems through a coupled perturbed Hartree-Fock/Kohn-Sham method. Z. Phys. Chem. **230**, 719 (2016)
2. S. Baroni, S. de Gironcoli, A. Dal Corso, P. Giannozzi, Phonons and related crystal properties from density-functional perturbation theory. Rev. Mod. Phys. **73**, 515 (2001)
3. L. Bernasconi, S. Tomić, M. Ferrero, M. Rérat, R. Orlando, R. Dovesi, N.M. Harrison, First-principles optical response of semiconductors and oxide materials. Phys. Rev. B **83**, 195325 (2011)
4. D.M. Bishop, F.L. Gu, B. Kirtman, Coupled-perturbed Hartree-Fock theory for infinite periodic systems: calculation of static electric properties of $(LiH)_n$, $(FH)_n$, $(H_2O)_n$, $(\text{-CNH-})_n$, and $(\text{-CH=CH-})_n$. J. Chem. Phys. **114**, 7633 (2001)
5. D.M. Bishop, B. Kirtman, A perturbation method for calculating vibrational dynamic dipole polarizabilities and hyperpolarizabilities. J. Chem. Phys. **95**, 2646 (1991)
6. E.I. Blount. *Solid State Physics*, vol. 13., ed. by H. Ehrenreich, F. Seitz, D. Turnbull (Academic, New York, 1962)
7. E. Brown, Bloch electrons in a uniform magnetic field. Phys. Rev. **133**, A1038 (1964)
8. B. Champagne, J.G. Fripiat, J. André, From uncoupled to coupled HartreeFock polarizabilities of infinite polymeric chains. Pariser-Parr-Pople applications to the polyacetylene chains. J. Chem. Phys. **96**, 8330 (1992)
9. CRYSTAL14, website: http://www.crystal.unito.it/basis-sets.php
10. S. de Gironcoli, S. Baroni, R. Resta, Piezoelectric properties of III-V semiconductors from first-principles linear-response theory. Phys. Rev. Lett. **62**, 2853 (1989)
11. R. Dovesi, R. Orlando, B. Civalleri, C. Roetti, V.R. Saunders, C.M. Zicovich-Wilson, Crystal09. Z. Kristallogr. **220**, 571–573 (2005)

12. R. Dovesi, R. Orlando, C.M. Zicovich-Wilson, B. Civalleri, L. Maschio, A. Erba, S. Casassa, M. Ferrabone, M. De La Pierre, P. D'Arco, Y. Noël, M. Causà, M. Rérat, B. Kirtman, A program for the ab initio investigation of crystalline solids. Int. J. Quant. Chem. **114**, 1287 (2014)
13. R. Dovesi, V.R. Saunders, C. Roetti, R. Orlando, C.M. Zicovich Wilson et al., *CRYSTAL09 User's Manual* (Università di Torino, Torino, 2009)
14. R. Dovesi, V.R. Saunders, C. Roetti, R. Orlando, C.M. Zicovich-Wilson, F. Pascale, B. Civalleri, K. Doll, N.M. Harrison, I.J. Bush, Ph. D'Arco, M. Llunell, M. Causà, Y. Noël, Crystal14. *CRYSTAL14 User's Manual, University of Torino* (2014)
15. P. Giannozzi et al., Quantum espresso: a modular and open-source software project for quantum simulations of materials. J. Phys. Condens. Matter **21**, 395502 (2009)
16. X. Gonze et al., Abinit: first-principles approach to material and nanosystem properties. Computer Phys. Commun. **180**, 2582–2615 (2009)
17. A.M. Ferrari, R. Orlando, M. Rérat, Ab initio calculation of the ultraviolet-visible (UV-vis) absorption spectrum, electron-loss function, and reflectivity of solids. J. Chem. Theor. Comp. **11**, 3245 (2015)
18. M. Ferrero, M. Rérat, B. Kirtman, R. Dovesi, Calculation of first and second static hyperpolarizabilities of one- to three-dimensional periodic compounds. implementation in the crystal code. J. Chem. Phys. **129**, 244110 (2008)
19. M. Ferrero, M. Rérat, R. Orlando, R. Dovesi, The calculation of static polarizabilities of 1–3D periodic compounds. The implementation in the crystal code. J. Comp. Chem. **29**, 1450 (2008)
20. M. Ferrero, M. Rérat, R. Orlando, R. Dovesi, Coupled perturbed Hartree-Fock for periodic systems: The role of symmetry and related computational aspects. J. Chem. Phys. **128**, 014110 (2008)
21. A.J. Garza, G.E. Scuseria, Comparison of self-consistent field convergence acceleration techniques. J. Chem. Phys. **137**, 054110 (2012)
22. X. Gonze, First-principles responses of solids to atomic displacements and homogeneous electric fields: implementation of a conjugate-gradient algorithm. Phys. Rev. B **55**, 10337 (1997)
23. F.L. Gu, D.M. Bishop, B. Kirtman, Crystal orbital calculation of coupled Hartree-Fock dynami hyperpolarizabilities for prototype p-conjugated polymers. J. Chem. Phys. **115**, 10548 (2001)
24. S. Hirata, M. Head-Gordon, R.J. Bartlett, Configuration interaction singles, time-dependent Hartree-Fock, and time-dependent density functional theory for the electronic excited states of extended systems. J. Chem. Phys. **111**, 10774 (1999)
25. G.J.B. Hurst, M. Dupuis, E. Clementi, Ab initio analytic polarizability, first and second hyperpolarizabilities of large conjugated organic molecules: applications to polyenes $C_4H_6$ to $C_{22}H_{24}$. J. Chem. Phys. **89**, 385–395 (1988)
26. A. Izmaylov, G. Scuseria, Analytical infrared intensities for periodic systems with local basis sets. Phys. Rev. B **77**, 165131 (2008)
27. A.F. Izmaylov, E.N. Brothers, G.E. Scuseria, Linear-scaling calculation of static and dynamic polarizabilities in Hartree-Fock and density functional theory for periodic systems. J. Chem. Phys. **125**, 224105–9 (2006)
28. S.P. Karna, M. Dupuis, Frequency dependent nonlinear optical properties of molecules: formulation and implementation in the hondo program. J. Comp. Chem. **12**, 487–504 (1991)
29. R.D. King-Smith, David Vanderbilt, Theory of polarization of crystalline solids. Phys. Rev. B **47**, 1651 (1993)
30. B. Kirtman, F.L. Gu, D.M. Bishop, Extension of the Genkin and Mednis treatment for dynamic polarizabilities and hyperpolarizabilities of infinite periodic systems. I. Coupled Perturbed Hartree-Fock theory. J. Chem. Phys. **113**, 1294 (2000)
31. K. Kudin, R. Car, R. Resta, Berry phase approach to longitudinal dipole moments of infinite chains in electronic-structure methods with local basis sets. J. Chem. Phys. **126**, 234101 (2007)
32. K. Kudin, R. Car, R. Resta, Quantization of the dipole moment and of the end charges in push-pull polymers. J. Chem. Phys. **127**, 194902 (2007)
33. L. Maschio, Local MP2 with density fitting for periodic systems: a parallel implementation. J. Chem. Theory Comput. **7**(9), 2818–2830 (2011)

34. L. Maschio, B. Kirtman, M. Rérat, R. Orlando, R. Dovesi, Comment on ab initio analytical infrared intensities for periodic systems through a coupled perturbed Hartree-Fock/Kohn-Sham method [J. Chem. Phys. **137**, 204113 (2012)]. J. Chem. Phys. **139**, 167101 (2013)
35. L. Maschio, B. Kirtman, R. Orlando, M. Rérat, Ab initio analytical infrared intensities for periodic systems through a coupled perturbed Hartree-Fock/Kohn-Sham method. J. Chem. Phys. **147**, 204113 (2012)
36. L. Maschio, B. Kirtman, M. Rérat, R. Orlando, R. Dovesi, Ab initio analytical Raman intensities for periodic systems through a coupled perturbed Hartree-Fock/Kohn-Sham method in an atomic orbital basis. I. theory. J. Chem. Phys. **139**, 164101 (2013)
37. L. Maschio, M. Rérat, B. Kirtman, R. Dovesi, Calculation of the dynamic first electronic hyperpolarizability $\beta(\omega_\sigma; \omega_1, \omega_2)$ of periodic systems. theory, validation, and application to multilayer $MoS_2$. J. Chem. Phys. **143**, 244102 (2015)
38. L. Maschio, D. Usvyat, Fitting of local densities in periodic systems. Phys. Rev. B **78**, 073102 (2008)
39. M. Molayem, M. Springborg, B. Kirtman, noop2017. Phys. Chem. Chem. Phys. (in press)
40. R.W. Nunes, X. Gonze, Berry-phase treatment of the homogeneous electric field perturbation in insulators. Phys. Rev. B **63**, 155107 (2001)
41. R. Orlando, R. Bast, K. Ruud, U. Ekström, M. Ferrabone, B. Kirtman, Roberto Dovesi, The first and second static electronic hyperpolarizabilities of zigzag boron nitride nanotubes. an ab initio approach through the coupled perturbed Kohn-Sham scheme. J. Phys. Chem. A **115**, 12631 (2011)
42. R. Orlando, V. Lacivita, R. Bast, K. Ruud, Calculation of the first static hyperpolarizability tensor of three-dimensional periodic compounds with a local basis set: A comparison of LDA, PBE, PBE0, B3LYP, and HF results. J. Chem. Phys. **132**, 244106 (2010)
43. B.J. Orr, J.F. Ward, Perturbation theory of the non-linear optical polarization of an isolated system. Mol. Phys. **20**, 513 (1971)
44. P. Otto, Calculation of the polarizability and hyperpolarizabilities of periodic quasi-one-dimensional systems. Phys. Rev. B **45**, 10876 (1992)
45. P. Otto, F.L. Gu, J. Ladik, Calculation of ab initio dynamic hyperpolarizabilities of polymers. J. Chem. Phys. **110**, 2717 (1999)
46. P. Otto, A. Martinez, A. Czaja, J. Ladik, Electron correlation corrected static polarizabilities of polymers with linear and cyclic conjugated elementary cells. J. Chem. Phys. **117**, 1908 (2002)
47. C. Pisani, M. Busso, G. Capecchi, S. Casassa, R. Dovesi, L. Maschio, C. Zicovich-Wilson, M. Schütz, Local-MP2 electron correlation method for non conducting crystals. J. Chem. Phys. **122**, 094133 (2005)
48. C. Pisani, L. Maschio, S. Casassa, M. Halo, M. Schütz, D. Usvyat, Periodic local MP2 method for the study of electronic correlation in crystals: theory and preliminary applications. J. Comput. Chem. **29**, 2113 (2008)
49. C. Pisani, M. Schütz, S. Casassa, D. Usvyat, L. Maschio, M. Lorenz, A. Erba, Cryscor: a program for the post-Hartree-Fock treatment of periodic systems. Phys. Chem. Chem. Phys. **14**, 7615–7628 (2012)
50. J.A. Pople, P.M.W. Gill, B.G. Johnson, Kohn-Sham density-functional theory within a finite basis set. Chem. Phys. Lett. **199**, 557 (1992)
51. P. Pulay, Convergence acceleration of iterative sequences. The case of scf iteration. Chem. Phys. Lett. **73**, 393 (1980)
52. M. Rérat, M. Ferrero, E. Amzallag, I. Baraille, R. Dovesi, Comparison of the polarizability of periodic systems computed by using the length and velocity operators. J. Phys. Conf. Ser. **117**, 012023 (2008)
53. M. Rérat, L. Maschio, B. Kirtman, B. Civalleri, R. Dovesi, Computation of second harmonic generation for crystalline urea and KDP. An ab initio approach through the coupled perturbed Hartree-Fock/Kohn-Sham scheme. *J. Chem. Theor. Comp.*, **12**, 107 (2016)
54. R. Resta, Macroscopic polarization in crystalline dielectrics: the geometric phase approach. Rev. Mod. Phys. **66**, 899 (1994)

55. R. Resta, Quantum-mechanical position operator in extended systems. Phys. Rev. Lett. **80**, 1800 (1998)
56. J.E. Rice, R.D. Amos, S.M. Colwell, N.C. Handy, J. Sanz, Frequency dependent hyperpolarizabilities with application to formaldehyde and methyl fluoride. J. Chem. Phys. **93**, 8828–8839 (1990)
57. G. Saghi-Szabo, R.E. Cohen, H. Krakauer, First-principles study of piezoelectricity in $PbTiO_3$. Phys. Rev. Lett. **80**, 4321 (1998)
58. M. Springborg, B. Kirtman, Analysis of vector potential approach for calculating linear and nonlinear responses of infinite periodic systems to a finite static external electric field. Phys. Rev. B **77**, 045102 (2008)
59. M. Springborg, B. Kirtman, M. Molayem, The response of extended systems to electrostatic fields. *Handbook of Computational Chemistry*, p. 1 (2016)
60. M. Springborg, M. Molayem, B. Kirtman, To be published. JCP (in press)
61. M. Springborg, V. Tevekeliyska, B. Kirtman, Termination effects in electric field polarization of periodic quasi-one-dimensional systems. Phys. Rev. B **82**(16), 165442 (2010)
62. D. Usvyat, L. Maschio, M. Schütz, Periodic local MP2 method employing orbital specific virtuals. J. Chem. Phys. **143**, 102805 (2015)
63. S.J.A. van Gisbergen, F. Koostra, P.R.T. Schipper, O.V. Gritsenko, J.G. Snijders, E.J. Baerends, Density-functional-theory response-property calculations with accurate exchange-correlation potentials. Phys. Rev. A **57**, 2556 (1998)
64. D. Vanderbilt, Berry-phase theory of proper piezoelectric response. J. Phys. Chem. Solids **61**, 147 (2000)
65. X. Wu, D. Vanderbilt, D.R. Hamann, Systematic treatment of displacements, strains, and electric fields in density-functional perturbation theory. Phys. Rev. B **72**, 035105 (2005)

# Chapter 6
# Quantum Chemical Methods for Predicting and Interpreting Second-Order Nonlinear Optical Properties: From Small to Extended π-Conjugated Molecules

**Benoît Champagne, Pierre Beaujean, Marc de Wergifosse, Marcelo Hidalgo Cardenuto, Vincent Liégeois and Frédéric Castet**

**Abstract** This chapter addresses the methodological and computational aspects related to the prediction of molecular second-order nonlinear optical properties, i.e., the first hyperpolarizability (β), by using quantum chemistry methods. Both small (reference) molecules and extended push-pull π-conjugated systems are considered, highlighting contrasted effects about (i) the choice of a reliable basis set together with the convergence of β values as a function of the basis set size, (ii) the amplitude of electron correlation contributions and its estimate using wave function and density functional theory methods, (iii) the description of solvent effects using implicit and explicit solvation models, (iv) frequency dispersion effects in off-resonance conditions, and (v) numerical accuracy issues. When possible, comparisons with experiment are made. All in all, these results demonstrate that the calculations of β remain a challenge and that many issues need to be carefully addressed, pointing out difficulties toward elaborating black-box and computationally cheap protocols. Still, several strategies can be designed in order to achieve a targeted accuracy, either for reference molecules displaying small β responses or for molecules presenting large β values and a potential in optoelectronics and photonics.

---

B. Champagne (✉) · P. Beaujean · M. de Wergifosse · M.H. Cardenuto · V. Liégeois
Laboratoire de Chimie Théorique, Unité de Chimie-Physique Théorique et Structurale, Université de Namur, rue de Bruxelles 61, 5000 Namur, Belgium
e-mail: benoit.champagne@unamur.be

M.H. Cardenuto
Instituto de Física, Universidade de São Paulo, 66318, 05314-970 São Paulo, SP, Brazil

F. Castet
UMR 5255 CNRS, Institut des Sciences Moléculaires (ISM), Université de Bordeaux, Cours de la Libération, 351, 33405 Talence Cedex, France



M.J. Wójcik et al. (eds.), *Frontiers of Quantum Chemistry*,
https://doi.org/10.1007/978-981-10-5651-2_6

**Keywords** First hyperpolarizability • Wave function versus density functional theory methods • Solvation models • Frequency dispersion • Basis sets

## 6.1 Introduction

The molecular properties known as the (electric dipole) polarizability (α), first (β) and second (γ) hyperpolarizabilities, are defined by a phenomenological equation describing the change in the electric dipole moment that results from the application of external electric fields:

$$\Delta\mu_\zeta = \sum_{\eta}^{x,y,z} \alpha_{\zeta\eta}(-\omega_\sigma;\omega_1)F_\eta(\omega_1) + \frac{1}{2!}\sum_{\eta,\chi}^{x,y,z} \beta_{\zeta\eta\chi}(-\omega_\sigma;\omega_1,\omega_2)F_\eta(\omega_1)F_\chi(\omega_2) + \frac{1}{3!}\sum_{\eta,\chi,\xi}^{x,y,z} \gamma_{\zeta\eta\chi\xi}(-\omega_\sigma;\omega_1,\omega_2,\omega_3)F_\eta(\omega_1)F_\chi(\omega_2)F_\xi(\omega_3) + \ldots \quad (6.1)$$

where $F_\eta(\omega_1)$ is the amplitude of the field oscillating at pulsation $\omega_1$ and applied in the η direction and $\omega_\sigma = \sum_i \omega_i$. For more than a century in the case of the polarizability and half of that period for the higher-order responses, these properties have been the topic of intense activities among theoretical chemists and physicists. On the one hand, efforts have been devoted to the very accurate evaluation of these properties for atoms (i.e., α and γ because β is zero by symmetry) and small molecules with the aim of providing reference values for experiments, which usually rely on relative rather than on absolute measurements [1, 2]. This has stimulated extensive methods developments to account for electron correlation effects, frequency dispersion, vibrational contributions, as well as solvent effects. Small systems are also ideal to assess the reliability of new quantum chemistry methods because of reduced needs in computational resources. On the other hand, owing to their potential for achieving large nonlinear optical (NLO) responses, organic and mixed organic–inorganic chromophores were also the object of a large number of theoretical investigations, which allowed establishing structure–property relationships [3, 4]. These theoretical guidelines allowed designing molecular systems for applications in photonics and sensing devices, as well as for bio-imaging [5, 6].

This chapter discusses the use of quantum chemistry methods to calculate and interpret the NLO responses from small molecules to extended push-pull π-conjugated systems. This topic being very broad, the focus is restrained to the first hyperpolarizability rather than to both the first and second hyperpolarizabilities, and to the electronic response, leaving the pure vibrational and zero-point vibrational average counterparts for another contribution. Then, such quantum chemistry applications are illustrated in domains recently tackled by the authors. In particular, the discussion focuses on quantities that can be extracted from experimental measurements, namely the hyper-Rayleigh scattering (HRS) hyperpolarizability, $\beta_{HRS}$, and the associated depolarization ratios (DR), or the electric field-induced second harmonic generation (EFISHG) response, $\beta_{//}$. The expressions of the HRS responses involve ensemble averages over the molecular orientations:

$$\beta_{HRS}(-2\omega;\omega,\omega)=\sqrt{\{\langle\beta^2_{ZZZ}\rangle+\langle\beta^2_{ZXX}\rangle\}} \tag{6.2}$$

$$DR=\frac{\langle\beta^2_{ZZZ}\rangle}{\langle\beta^2_{ZXX}\rangle} \tag{6.3}$$

in which X and Z are axes of the laboratory frame. The expressions of $\langle\beta^2_{ZZZ}\rangle$ and $\langle\beta^2_{ZXX}\rangle$ in terms of Cartesian molecular tensor components can be found in Ref. [7]. The EFISHG response corresponds to the projection of the vector part of β on the dipole moment vector:

$$\beta_{//}(-2\omega;\omega,\omega)=\frac{1}{5}\sum_i\frac{\mu_i}{|\vec{\mu}|}\sum_j(\beta_{ijj}+\beta_{jij}+\beta_{jji})=\frac{3}{5}\sum_i\frac{\mu_i\beta_i}{|\vec{\mu}|} \tag{6.4}$$

## 6.2 Small Molecules in Gas Phase

A first element for accurately predicting the responses of small molecules is the selection of a sufficiently flexible atomic basis set, generally containing many polarization and diffuse functions. Diffuse functions are needed because a substantial part of the response originates from the outer and most diffuse part of the electron density. The need for polarization functions can easily be understood by noticing that the successive first-, second-, … order responses of a spherical (*s*) atomic orbital to an external field can be described by *p*-like, *d*-like, … functions. So, some authors have privileged adding selected diffuse and/or diffuse polarization functions to basis sets employed for geometry optimizations and thermodynamics [8–10]. Others have designed property-oriented basis sets [11, 12] or have used basis sets constructed by adding even-tempered sets of diffuse functions to the correlation consistent basis sets of Dunning and co-workers [13], showing that with multiple-augmented sets the electrical properties of small molecules converge smoothly [14]. This is illustrated in Fig. 6.1 for the main static β tensor components of the water molecule, calculated at the Hartree–Fock level. Alternatives to these atomic basis set approaches consist in employing a fully numerical approach [15, 16].

Then, electron correlation should be included at a sufficiently high level of approximation to achieve quantitative accuracy. Several levels of approximation and techniques are available. The easiest technique consists in calculating β from finite differentiation of the field-dependent energy of the molecule. This is the finite

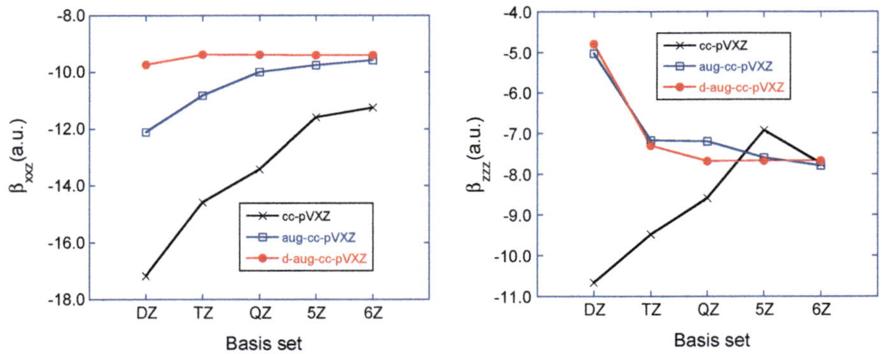


**Fig. 6.1** Evolution of the amplitude of the two dominant components of the static β tensor of the water molecule calculated at the coupled-perturbed HF level by using the cc-pVXZ, aug-cc-pVXZ, and d-aug-cc-pVXZ basis set suites

field (FF) method [17] that simply requires adding the electric dipole interaction term, $-\hat{\mu} \cdot \vec{E}$, to the Hamiltonian. Still, the field amplitudes have to be carefully chosen within a stability window to avoid two drawbacks: (i) Too small field amplitudes create a loss of accuracy on the energy values, which gets amplified during the successive numerical derivatives, and (ii) too large field amplitudes introduce high-order contaminations in the derivatives before leading to divergences as a result of the change of the ground state electron configuration [18]. Nevertheless, once the stability window is defined, the high-order contaminations can be systematically and recursively removed by using Richardson extrapolation or polynomial fitting. These aspects have been recently reviewed [19, 20]. The FF approach has been employed to calculate the static HRS response, $\beta_{HRS}$, of five reference molecules, $CCl_4$, $CHCl_3$, $CH_2Cl_2$, $CCl_3CN$, and $CH_3CN$ [14]. Results sketched in Fig. 6.2 show that electron correlation effects might be quite different as a function of the chemical nature. Considering the CCSD(T) results as the reference, general—but not systematic—trends are observed. First, the HF values generally underestimate the CCSD(T) results, up to 50% in the case of $CH_3CN$. An improvement is achieved by using the second-order Møller–Plesset (MP2) perturbation theory method so that the MP2 values underestimate or overestimate the CCSD(T) values by up to 20–25%. Finally, missing the perturbative triples, the $\beta_{HRS}$ responses are underestimated by up to 15% for $CCl_4$ and $CCl_3CN$, whereas the difference with respect to CCSD(T) is much smaller for compounds with fewer chlorine atoms (less than 1% and 6% for $CH_3CN$ and $CH_2Cl_2$, respectively).

Alternatively, β can be calculated by employing response function approaches [21, 22], which are equivalent to evaluate analytically the responses of the dipole moment to external fields oscillating at finite frequencies. At the HF level, this approach gives rise to the time-dependent HF (TDHF) method [23]. Within the density functional theory (DFT) formalism, it gives the time-dependent DFT (TDDFT) scheme [24]. A hierarchy of coupled-cluster (CC) models has also been

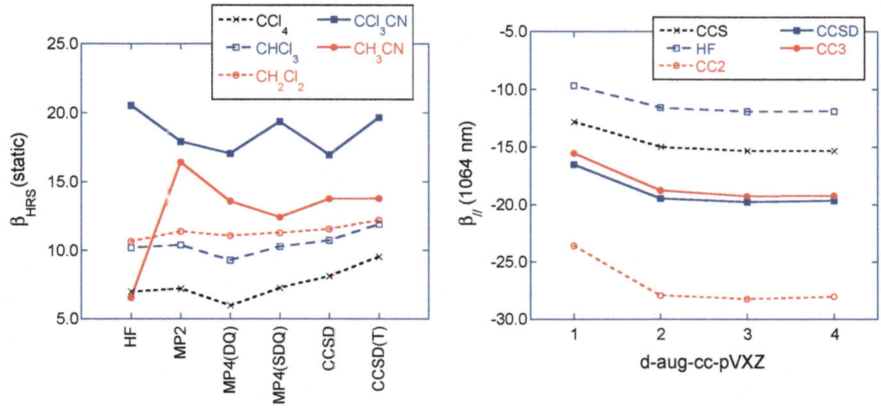


**Fig. 6.2** Basis set and electron correlation effects on β (a.u.). *Left* static $\beta_{HRS}$ of five reference molecules evaluated with the d-aug-cc-pVTZ basis set; *right* dynamic $\beta_{//}(-2\omega;\omega,\omega)$ ($\lambda$ = 1064 nm) of water

elaborated: CCS=CIS, CC2, CCSD, and CC3, which allows controlling the convergence of the responses as a function of the level of treatment of electron correlation [22]. Figure 6.2 illustrates the convergence of the EFISHG response, $\beta_{//}$, of water as a function of the basis set and of the level of electron correlation. Like in Fig. 6.1, the convergence with basis set size is smooth, though it is faster at the HF and CCS level than when using higher-order methods. Then, for a given basis set, the ordering of the $\beta_{//}$ amplitudes according to the method is:

$$\mathrm{HF} < \mathrm{CCS} < \mathrm{CC3} \approx \mathrm{CCSD} < \mathrm{CC2} \tag{6.5}$$

Again, the contribution from the triples, as estimated from the difference between the CCSD and CC3 results, is small. On the other hand, the CC2 method overestimates $\beta_{//}$ by as much as 50%, whereas the $\beta_{//}$ amplitude is strongly underestimated at the HF (40%) and CCS (20%) levels of approximation. For the CC levels, the $\beta_{//}$ amplitude ordering follows the relative values of the lowest excitation energies, dominating the β response [25] (in the case of the $^1A_1$ state, the vertical excitation energies calculated with the d-aug-cc-pVTZ basis set amount to 10.79 eV (HF), 10.81 eV (CCS), 9.82 eV (CC3), 9.81 eV (CCSD), and 9.59 eV (CC2)).

Still, frequency-dependent hyperpolarizabilities evaluated at high-order electron correlation levels are not always available, so that various approximate schemes have been proposed. They combine static correlated values with static and dynamic responses evaluated at lower levels of approximation. A multiplicative or percentage approximation (MA) has been proposed by Sekino and Bartlett [26], and an additive approximation (AA) by Rice and Handy [27]:

$$ {}^{MA}\beta_{HRS}^{CCSD}(\omega) = \beta_{HRS}^{CCSD}(0) \times \frac{\beta_{HRS}^{HF/DFT}(\omega)}{\beta_{HRS}^{HF/DFT}(0)} \tag{6.6} $$

$$ {}^{AA}\beta_{HRS}^{CCSD}(\omega) = \beta_{HRS}^{CCSD}(0) + \left[\beta_{HRS}^{HF/DFT}(\omega) - \beta_{HRS}^{HF/DFT}(0)\right] \tag{6.7} $$

Besides their widespread use, these approximations have been assessed in a limited number of studies [28 and references therein]. In the case of $CCl_4$, both MA and AA based on HF frequency dispersions underestimate $\beta_{HRS}$ ($\lambda$ = 632 nm) by about 20%, whereas for $CH_2Cl_2$, the underestimation is smaller with MA (8%) than AA (12%). More drastically, the effect of the frequency dispersion of $CCl_4$ is qualitatively wrong when adopting the HF method, since it suggests a decrease of $\beta_{HRS}$ with the photon energy (Fig. 6.3), whereas CCSD calculations predict an increase of its amplitude. A much less frequently used alternative consists in describing frequency dispersion at the TDDFT level. So, LC-BLYP and B3LYP behave better than HF for the $CCl_4$ molecule with an increase of $\beta_{HRS}$ with the frequency, though slightly slower than with CCSD. Similarly, for $CH_2Cl_2$, LC-BLYP (B3LYP) closely reproduces the CCSD frequency dispersion with small overestimations (underestimations). M06 (data not shown) and BLYP frequency dispersions are also close to the CCSD reference for $CH_2Cl_2$, whereas for $CCl_4$, M06 overestimates it substantially. These differences of frequency dispersion can be related to the relative values of the excitation energies (smaller with B3LYP than with CCSD), as well as of the transition dipoles, and excitation-induced dipole moment variations.

Comparisons of the CC3/d-aug-cc-pVQZ $\beta_{//}$ values of $H_2O$ (−19.28 and −21.77 a.u. at 1064 and 694.3 nm, respectively) with experimental values (−19.2 ± 0.9 [29] and −22/0 ± 0.9 a.u. [30]) substantiate the predictability of the method and confirm the small amplitude of the vibrational contributions (note that the pure vibrational and zero-point vibrational average contributions might also be

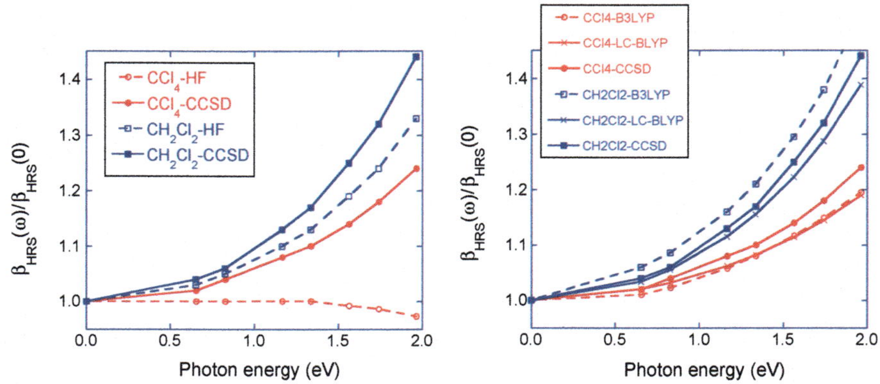


**Fig. 6.3** Frequency dispersion on $\beta_{HRS}$ described at the TDHF (*left*) or TDDFT (*right*) levels of approximation in comparison with CCSD, as determined with the d-aug-cc-pVDZ basis set

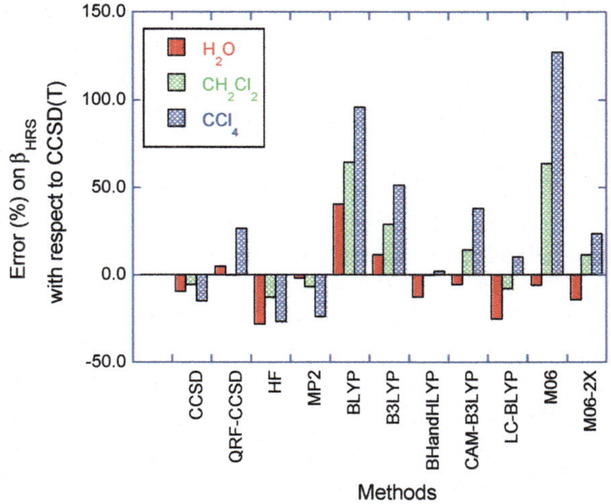


**Fig. 6.4** Error (%) on the static $\beta_{HRS}$ of $H_2O$, $CH_2Cl_2$, and $CCl_4$ for different exchange-correlation functionals and levels of approximation in comparison with CCSD(T). The calculations were performed with the d-aug-cc-pVTZ basis set. MP2, CCSD, and CCSD (T) results were obtained from the FF calculations, the QRF-CCSD calculations using the response method, and the HF and DFT calculations with the coupled-perturbed analytical differentiation procedures

non-negligible while canceling each other). In the case of $CCl_4$, Shelton [31] reported an experimental $\beta_{HRS}$ (at 1064 nm) value of 12.8 ± 1.1 a.u. in comparison with a theoretical estimate of 14.3 a.u. The latter value was obtained by adding to the electronic dynamic CCSD value (14.9 a.u.) [28], the ZPVA (−1.1 a.u.), and pure vibrational (0.5 a.u.) contributions of [32].

Though small systems such as $H_2O$, $CCl_4$, and $CH_2Cl_2$ allow using computationally demanding techniques and therefore tackling the convergence of the molecular properties as a function of electron correlation level, it is worth assessing the reliability of more approximate schemes in view of applying these to larger systems. In particular, there is an interest to assess the performance of density functional theory (DFT) with conventional as well as with most recent exchange-correlation functionals in comparison with the Hartree–Fock method and to MP2. Following Ref. [33], different levels of approximation have been assessed in comparison with FF CCSD(T) calculations (Fig. 6.4). Systematic trends are observed: (i) BLYP overestimates $\beta_{HRS}$ by as much as 100% for $CCl_4$; (ii) this overestimation decreases when adding 20% of HF exchange (B3LYP) but not for $CH_2Cl_2$ and $CCl_4$ with M06 (28% HF exchange) so that with these three functionals the accuracy is lower than with HF, which underestimates $\beta_{HRS}$ by 15–30%; (iii) then, further increasing the percentage of HF exchange (BHandHLYP (50%) and M06-2X (56%)) leads to improvements; and (iv) finally, range-separated hybrids perform also better, but the improvement is not systematic (CAM-B3LYP reproduces closely the response of $H_2O$, but the error on $CCl_4$ is as large as 40%, whereas with LC-BLYP, $\beta_{HRS}$ of $H_2O$ is underestimated by 25% and the responses of the two other compounds are close to CCSD(T)). On the other hand, Karamanis et al. [34] have shown that for doped Si clusters, hybrid GGA with 20–25% HF exchange performs better than hybrid GGAs with larger percentage of HF exchange or than long-range corrected hybrids. So, without considering push-pull π-conjugated systems for which the XC requirements are also function of their size and charge-transfer character [35–39], the reliability of DFT depends strongly on the XC functional and there is no easy clue to select a priori an accurate XC functional. Still, the best functionals are global or range-separated hybrids and include substantial amounts of HF exchange.

When using DFT approaches, an additional parameter to control is the density of the integration grid. The Gaussian09 grids are called `Fine`, `UltraFine`, and `SuperFine`, and they differ by the number of radial shells (75, 99, and 150, respectively) and of angular points per shell (302, 590, and 974). In a recent investigation [33], it was found that for TDDFT calculations of small reference molecules, the $\beta_{HRS}$ differences could attain up to 1 a.u. (up to 5%) between the UltraFine and SuperFine grids when employing meta-GGA XC functionals, whereas this difference is only of ~0.02 a.u. when using GGA or hybrid XC functionals. In fact, the situation is even more complex when considering that β tensor components can be evaluated from the FF approach as the first-, second-, and third-order derivatives. Let us consider the $\beta_{zxx}$ component:

$$\beta_{zxx} = -\left(\frac{\partial^3 E}{\partial E_z \partial E_x^2}\right)_0 = \left(\frac{\partial^2 \mu_z}{\partial E_x^2}\right)_0 = \left(\frac{\partial \alpha_{zx}}{\partial E_x}\right)_0 \tag{6.8}$$

For variational wave functions, these three quantities are expected to be identical and also identical to the fully analytical TDDFT/CPKS value. We have performed such FF calculations for $\beta_{zxx}$ of $H_2O$. A selection of Romberg's tables is provided in Table 6.1. In these calculations, the threshold on the energy has been lowered to $10^{-11}$ a.u. As discussed in the previous works [19, 20], each table presents the same structure. In particular, for n = 0, by going from k = 0 to k = 6, the β values vary randomly until a given k value where β starts increasing monotonically. At smaller k's, these variations originate from the lack of precision on the field-dependent properties (α, μ, or E). At larger k's, the monotonic behavior results from the higher-order contaminations, which can then be iteratively and systematically removed using Romberg's quadrature. This allows locating stability domains for β and therefore converged β values. Considering β as the first-order derivative of $\alpha$, a converged value of 3.37 a.u. is obtained using a `Fine` grid. An `UltraFine` grid (as well as a `SuperFine` grid, results not shown) gives a value of 3.36 a.u. Analyzing the variations of $\beta_{zxx}$ as a function of the order of Romberg's iteration *(n)* and of the field amplitude (*k*) tells that the numerical accuracy on $\beta_{zxx}$ is of the order of 0.01 a.u., consistently with an accuracy of $10^{-4}$ a.u. on the field-dependent polarizabilities. This value of $10^{-4}$ is obtained by taking the product between the accuracy on $\beta_{zxx}$ and the field amplitude for $k = 4$, which corresponds to the smallest field amplitude associated with the converged value. A similar accuracy is achieved by considering the second-order derivative of $\mu$ with non-negligible differences in the Romberg's tables when going from the `UltraFine` to the `SuperFine` grid and large differences between the `Fine` and `UltraFine` grids. This highlights a reduction of accuracy on the dipole moment values. Moreover, these data are consistent with an accuracy of about $10^{-6}$ a.u. on the field-dependent dipole moments.

**Table 6.1** Romberg's table for the evaluation of $\beta_{zxx}$ of $H_2O$ as a function of the order of the derivative and the quality of the integration grid. The LC-BLYP XC functional was employed together with the aug-cc-pVTZ basis set. $n$ represents the number of Romberg's iterations, where higher-order contaminations are removed. $k$ determines the field amplitude, $E = E_0\ 2^k$ with $E_0 = 0.0004$ a.u. Best/converged values are indicated by an arrow

| | n = 0 | n = 1 | n = 2 | n = 3 | n = 4 | n = 5 | n = 6 |
|---|---|---|---|---|---|---|---|
| *First-order derivative of α, grid =* `Fine` | | | | | | | |
| k = 0 | 3.37314 | 3.37341 | 3.37376 | 3.37387 | 3.37390 | 3.37391 | 3.37391 |
| k = 1 | 3.37234 | 3.36812 | 3.36688 | 3.36655 | 3.36647 | 3.36645 | |
| k = 2 | 3.38499 | 3.38674 | 3.38730 | 3.38744 | 3.38747 | | |
| k = 3 | 3.37975 | 3.37835 | 3.37863 | 3.37871 | | | |
| k = 4 | 3.38393 | 3.37414 | 3.37394 | | → | **3.37** | |
| k = 5 | 3.41329 | 3.37707 | | | | | |
| k = 6 | 3.52197 | | | | | | |
| *First-order derivative of α, grid =* `UltraFine` | | | | | | | |
| k = 0 | 3.36461 | 3.36776 | 3.36869 | 3.36893 | 3.36899 | 3.36901 | 3.36901 |
| k = 1 | 3.35517 | 3.35382 | 3.35355 | 3.35348 | 3.35346 | 3.35346 | |
| k = 2 | 3.35922 | 3.35792 | 3.35777 | 3.35774 | 3.35773 | | |
| k = 3 | 3.36313 | 3.36009 | 3.35993 | 3.35989 | | | |
| k = 4 | 3.37223 | 3.36257 | 3.36252 | | → | **3.36** | |
| k = 5 | 3.40122 | 3.36331 | | | | | |
| k = 6 | 3.51496 | | | | | | |
| *Second-order derivative of μ, grid =* `Fine` | | | | | | | |
| k = 0 | 4.25000 | 4.47188 | 4.52941 | 4.54396 | 4.54761 | 4.54853 | 4.54875 |
| k = 1 | 3.58457 | 3.60885 | 3.61255 | 3.61331 | 3.61349 | 3.61354 | |
| k = 2 | 3.51094 | 3.55339 | 3.56461 | 3.56745 | 3.56816 | | |
| k = 3 | 3.38359 | 3.38504 | 3.38557 | 3.38570 | | | |
| k = 4 | 3.37925 | 3.37708 | 3.37771 | | → | **3.38** | |
| k = 5 | 3.38577 | 3.36750 | | | | | |
| k = 6 | 11.49397 | | | | | | |
| *Second-order derivative of μ, grid =* `UltraFine` | | | | | | | |
| k = 0 | 3.58750 | 3.63437 | 3.64684 | 3.65002 | 3.65082 | 3.65102 | 3.65107 |
| k = 1 | 3.44688 | 3.44740 | 3.44630 | 3.44595 | 3.44585 | 3.44583 | |
| k=2 | 3.44531 | 3.46387 | 3.46843 | 3.46956 | 3.46985 | | |
| k=3 | 3.38965 | 3.39538 | 3.39714 | 3.39758 | | | |
| k=4 | 3.37246 | 3.36898 | 3.36928 | | → | **3.37** | |
| k=5 | 3.38290 | 3.36445 | | | | | |
| k=6 | 3.43824 | | | | | | |

(continued)

**Table 6.1** (continued)

| | n = 0 | n = 1 | n = 2 | n = 3 | n = 4 | n = 5 | n = 6 |
|---|---|---|---|---|---|---|---|
| *Second-order derivative of μ, grid =* `SuperFine` | | | | | | | |
| k = 0 | 3.62500 | 3.63854 | 3.63778 | 3.63731 | 3.63718 | 3.63714 | 3.63713 |
| k=1 | 3.58437 | 3.65000 | 3.66720 | 3.67156 | 3.67265 | 3.67292 | |
| k=2 | 3.38750 | 3.39193 | 3.39301 | 3.39328 | 3.39334 | | |
| k=3 | 3.37422 | 3.37567 | 3.37631 | 3.37647 | | | |
| k=4 | 3.36987 | 3.36603 | 3.36626 | | → | **3.37** | |
| k=5 | 3.38141 | 3.36261 | | | | | |
| k=6 | 3.43781 | | | | | | |
| *Third-order derivative of E, grid =* `Fine` | | | | | | | |
| k = 0 | 3.34710 | 3.34181 | 3.34038 | 3.34001 | 3.33992 | 3.33990 | 3.33989 |
| k = 1 | 3.36296 | 3.36329 | 3.36343 | 3.36347 | 3.36348 | 3.36348 | |
| k = 2 | 3.36197 | 3.36121 | 3.36120 | 3.36119 | 3.36119 | | |
| k = 3 | 3.36424 | 3.36146 | 3.36147 | 3.36147 | → | **3.361** | |
| k = 4 | 3.37256 | 3.36142 | 3.36135 | | | | |
| k = 5 | 3.40599 | 3.36251 | | | | | |
| k = 6 | 3.53643 | | | | | | |
| *Third-order derivative of E, grid =* `UltraFine` | | | | | | | |
| k = 0 | 3.35632 | 3.35569 | 3.35560 | 3.35559 | 3.35558 | 3.35558 | 3.35558 |
| k = 1 | 3.35820 | 3.35693 | 3.35664 | 3.35657 | 3.35655 | 3.35655 | |
| k = 2 | 3.36201 | 3.36129 | 3.36128 | 3.36128 | 3.36128 | | |
| k = 3 | 3.36417 | 3.36135 | 3.36135 | 3.36135 | → | **3.361** | |
| k = 4 | 3.37262 | 3.36146 | 3.36139 | | | | |
| k = 5 | 3.40608 | 3.36259 | | | | | |
| k = 6 | 3.53655 | | | | | | |

Finally, the best accuracy is achieved by considering the third-order derivative of the energy, leading to a value of 3.361 a.u. (if not 3.3613 a.u.). Using an `UltraFine` grid provides slightly more accurate results than the `Fine` grid, whereas there is almost no difference with the `SuperFine` grid. Further calculations were performed with a (very tight) threshold of $10^{-13}$ a.u. on the energy, which is not going to work for any compound and the whole set of field amplitudes, and they did not lead to improvements. So, to get highly accurate results, a `Fine` grid can be used but in combination with third-order energy derivatives, whereas the `UltraFine` grid is recommended together with first-(second-)order derivatives of $\alpha$ ($\mu$) because the number of significant digits on the energy is larger than on $\mu$ and on the CPKS $\alpha$.

## 6.3 Small Molecules in Solution

Implicit solvation models like the polarizable continuum model (PCM) [40] are usually employed to account for the effects of solvation on the first hyperpolarizability. These models describe the solvent as a structureless polarizable continuum characterized by, among other parameters, its macroscopic dielectric permittivity, which depends on the frequency of the applied field. For $CCl_4$, the dielectric constant or relative permittivity amounts to 2.23 ($\varepsilon_0$) and 2.13 ($\varepsilon_\infty$) in the zero (static) and infinite frequency limit, respectively. On the other hand, for $CH_2Cl_2$, the difference is much larger with $\varepsilon_0 = 8.93$ and $\varepsilon_\infty = 2.03$, owing to the orientation contribution related to the dipolar character of dichloromethane. This impacts directly the β responses, which are enhanced by the self-consistent reaction field. So, at the HF/d-aug-cc-pVTZ level, the $\beta_{HRS,LIQ}/\beta_{HRS,GAS}$ ratio of $CCl_4$ amounts to 1.43 and 1.39 for infinite and 1064 nm wavelengths, respectively, whereas for the dipolar $CH_2Cl_2$, it goes from 2.39 to 1.78 [14]. Then, accounting for electron correlation effects is not straightforward if the calculations are performed with the FF method, i.e., at zero frequency. Indeed, for dipolar solvents, the static dielectric constant is larger than the dynamic ones, which will lead to overestimations of the solvent effects and of the β amplitudes. A practical issue would be to use an effective static dielectric constant that only accounts for polarization contributions, neglecting the orientational contributions of the solvent. Another approach consists in correcting these static responses by including frequency dispersion with Eqs. 6.6 and 6.7:

$$^{MA}\beta_{HRS}^{CCSD,PCM}(\omega) = \beta_{HRS}^{CCSD,PCM}(0) \times \frac{\beta_{HRS}^{HF/DFT,PCM}(\omega)}{\beta_{HRS}^{HF/DFT,PCM}(0)} \tag{6.9}$$

$$^{AA}\beta_{HRS}^{CCSD,PCM}(\omega) = \beta_{HRS}^{CCSD,PCM}(0) + \left[\beta_{HRS}^{HF/DFT,PCM}(\omega) - \beta_{HRS}^{HF/DFT,PCM}(0)\right] \tag{6.10}$$

In this case, all the properties (the high-level static response as well as the static and dynamic TDHF or TDDFT responses) are performed using implicit solvation models. Consequently, two static β calculations are performed and their overestimations cancel each other, though incompletely. On the other hand, combining frequency dispersion as obtained from the two-state approximation [41] with high-level static PCM results will not be appropriate, besides in cases where an effective static dielectric constant is used.

Table 6.2 presents HRS quantities calculated for five reference molecules at the CCSD(T) level in combination with the MA scheme: $\beta_{HRS}$, the depolarization ratio (DR), which is determined by the shape of the NLOphore, the dipolar ($|\beta_{J=1}|$) and octupolar ($|\beta_{J=3}|$) components, and their ratio, the nonlinear anisotropy parameter, $\rho = |\beta_{J=3}|/|\beta_{J=1}|$. They are compared to the experimental values [14]. For all compounds, the octupolar character is overestimated although the experimental ordering of the DR (and ρ) is reproduced. Additional calculations not reported here show that this hierarchy of DR is already reproduced at the HF level, but the underestimation of the DR values is more severe than at the CCSD(T) level. Then, considering that the standard deviations on the experimental values is typically of 10%, most of the calculated $\beta_{HRS}$ values match the experimental ones, in particular for $CHCl_3$ and $CCl_3$-CN, but also for $CH_2Cl_2$ and $CH_3$-CN, with errors smaller than 25%. On the other hand, for $CCl_4$, the underestimation is substantial and attains 50%. A large part of this underestimation originates from the HF frequency dispersion that is

**Table 6.2** Dynamic ($\lambda = 1064$ nm) $\beta_{HRS}$, DR, $|\beta_{J=1}|$, $|\beta_{J=3}|$, and $\rho$ values (a.u.) as determined at the CCSD(T)/d-aug-cc-pVTZ level of approximation using the MA scheme (Eq. 6.6) and accounting for solvent effects with the PCM approach in its Integral Equation Formalism (IEF) in comparison with experiment [14]. Values in parentheses are the differences (%) with respect to the experimental values (for $\beta_{HRS}$) or the experimental values (DR and $\rho$)

| | $CCl_4$ | $CHCl_3$ | $CH_2Cl_2$ | $CCl_3$–CN | $CH_3$–CN |
|---|---|---|---|---|---|
| $\beta_{HRS}$ | 13.73 (−53) | 19.74 (+4) | 23.14 (+16) | 22.41 (−10) | 40.48 (23) |
| DR | 1.50 (1.50) | 1.90 (2.61 ± 0.02) | 3.36 (4.21 ± 0.03) | 2.16 (2.98 ± 0.03) | 6.80 (8.40 ± 0.15) |
| $\|\beta_{J=1}\|$ | 0.00 | 17.89 | 37.02 | 25.10 | 81.77 |
| $\|\beta_{J=3}\|$ | 44.49 | 57.82 | 49.24 | 61.66 | 40.07 |
| $\rho$ | ∞ (∞) | 3.22 (1.84 ± 0.02) | 1.33 (1.01 ± 0.01) | 2.46 (1.54 ± 0.02) | 0.49 (0.23 ± 0.15) |

combined with CCSD(T) static responses [14] and, when describing frequency dispersion at the CCSD level as discussed in Sect. 2, the error is reduced by a factor of 2 [28]. These successive studies demonstrate that there is an interest in performing high-level calculations with explicit solvent molecules as well as in reassessing the amplitude of the vibrational contributions, the pure vibrational contribution that is usually negligible for the second harmonic generation and the zero-point vibrational average. In particular, the non-polar but highly polarizable $CCl_4$ molecule appears to be a challenging case.

Still, for $CCl_4$, these comparisons between experiment and theory assumed that each molecule behaves like an independent light scatterer, giving rise to the so-called incoherent HRS signal, which is purely octupolar. Nevertheless, Kaatz and Shelton [42] have shown that the experimental HRS signal contains both a coherent ($\beta_{coh}$) and an incoherent ($\beta_{incoh}$) part, with $\beta_{coh}/\beta_{incoh} \sim 2/3$. This coherent response originates from the interactions between the $CCl_4$ molecules and attributes to liquid $CCl_4$ both dipolar ($|\beta_{J=1}|$) and octupolar ($|\beta_{J=3}|$) HRS responses. The description of the dual contribution to $\beta_{HRS}$ of liquid $CCl_4$ has been challenged by sequential QM/MM calculations. The method consists first in performing Monte Carlo simulations to generate uncorrelated snapshots representing the liquid structure [43, 44] and then in calculating at the QM level the first hyperpolarizability for a selection of these snapshots [45]. In these QM calculations, the solvent (surrounding molecules) is described either exclusively by point charges or by considering explicitly a few neighboring $CCl_4$ molecules, embedded in point charges of the remaining solvent molecules. It has been observed that considering explicitly a few neighboring $CCl_4$ molecules embedded in point charges enables monitoring the emergence of the dipolar contribution to $\beta_{HRS,LIQ}$, characterized by an increase of DR and a decrease of ρ. So, combining the Hartree–Fock method and the aug-cc-pVQZ basis set for feasibility purpose, with two interacting $CCl_4$ molecules, DR attains 2.34 ± 0.66 (ρ = 2.11) while when considering five interacting molecules DR = 2.78 ± 0.92 (ρ = 1.81) [46]. Figure 6.5 illustrates the convergence of $\beta_{HRS}$ as a function of the number of snapshots in the case of five interacting molecules embedded in point charges. Note that the amplitude of $\beta_{HRS}$ per $CCl_4$ molecule is little impacted when accounting for these specific intermolecular interactions. These calculations have confirmed to a large extent the experimental data and have substantiated that the dipolar contribution originates from intermolecular interactions between the $CCl_4$ molecules. Nevertheless, it remains challenging to perform high-level ab initio calculations on such $CCl_4$ clusters, which expectedly will modify the description of the intermolecular interactions and their impact on $\beta_{HRS}$, because calculating the first hyperpolarizability of $CCl_4$ requires using an extended basis set with diffuse functions.

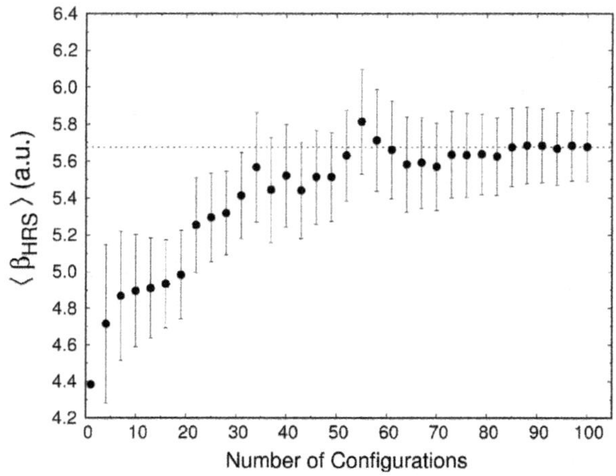


**Fig. 6.5** Statistical convergence of $\beta_{HRS}$ per $CCl_4$ molecule calculated at the HF/aug-cc-pVQZ level as a function of the number of configurations (snapshots). Each calculation was performed for a cluster of five molecules of $CCl_4$ solvated by point charge embedding. The *horizontal line* gives the averaged value and the uncertainty corresponds to the statistical error

## 6.4 Extended π-Conjugated Dyes

Basis set effects on the first hyperpolarizability of a π-conjugated dye were assessed in the case of an oxazine derivative (Scheme 6.1), which can switch between a closed and an open form upon triggering with light irradiation, as well as with pH and redox potential variations [47]. The calculations were performed at the TDHF level ($\lambda = 1064$ nm) on the closed form, which presents a rather small $\beta_{HRS}$ response, and on the zwitterionic open form, whose $\beta_{HRS}$ is enhanced due to push-pull electron delocalization effects. Solvent effects were accounted for by using the IEFPCM scheme [40]. Table 6.3 reports the $\beta_{HRS}$ values of the two forms, their contrasts, $\beta_{HRS}$(open)/$\beta_{HRS}$(closed), and the corresponding depolarization ratios. In the case of the closed form, starting from the 6-31G(d) basis set, adding *p* polarization functions on the H atoms has a negligible impact on $\beta_{HRS}$. On the other hand, adding a set of *sp* diffuse functions leads to an increase of $\beta_{HRS}$ by 20% while the addition of a second set of diffuse functions does not change significantly the $\beta_{HRS}$ values. Note that starting from the Dunning cc-pVDZ basis set, $\beta_{HRS}$ is 6% smaller than with the 6-31G(d) basis while going to cc-pVTZ, cc-pVQZ, and cc-pV5Z leads to increases of 1%, 8%, and 13% with respect to 6-31G(d), respectively. However, these basis sets contain much more contracted GTOs than 6-31G(d). The 6-31G (and, to a lower extent, 6-311G) $\beta_{HRS}$ value is also larger than the 6-31G(d) one, indicating that the inclusion of polarization and diffuse functions has counteracting effects. Finally, the values obtained with the aug-cc-pVDZ and aug-cc-pVTZ basis sets are in close agreement with those of the Pople basis set series, provided at least one set of diffuse functions is included.

The basis set effects are even smaller in the case of the oxazine open form. Indeed, going from 6-31G(d) to 6-31+G(d), $\beta_{HRS}$ increases only by 12% while from 6-31G(d) to 6-31G $\beta_{HRS}$ increases by 3%. Then, the differences with respect to 6-31G(d) amounts to –2%, 1%, 4%, and 6% for cc-pVDZ, cc-pVTZ, cc-pVQZ, and

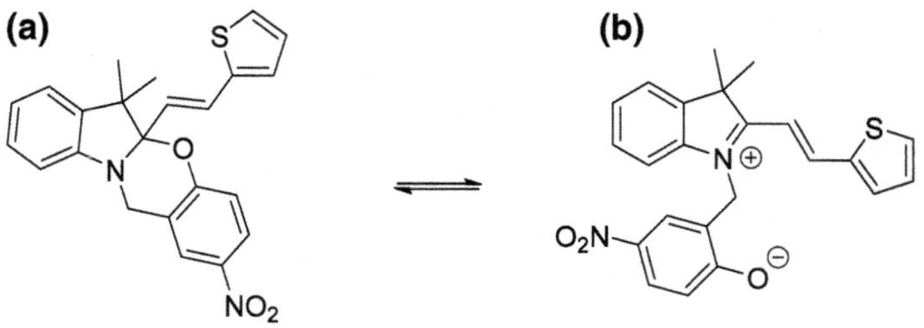


**Scheme 6.1** Closed (*left*) and open (*right*) forms of an oxazine derivative. Switching from the closed to open form can be triggered by light irradiation to form a zwitterion or by decreasing the pH to get the corresponding protonated cationic species [47]

**Table 6.3** Effect of the basis set on the HRS first hyperpolarizability ($\beta_{HRS}$) and its depolarization ratio (in parentheses) of the closed (**a**) and open (**b**) forms of an oxazine derivative as well as on the $\beta_{HRS}$ contrast ratio, $\beta_{HRS}$(open)/$\beta_{HRS}$(closed). The calculations were performed at the TDHF level with a wavelength of 1064 nm, while solvent (acetonitrile) effects were included using the IEFPCM scheme. The numbers in parentheses in the first column correspond to the number of contracted GTOs

| | $\beta_{HRS}$(closed) | $\beta_{HRS}$(open) | $\beta_{HRS}$(open)/$\beta_{HRS}$(closed) |
|---|---|---|---|
| 6-31G (305) | 841 (4.00) | 3939 (5.94) | 4.75 |
| 6-311G (445) | 816 (3.89) | 4019 (5.92) | 4.92 |
| 6-31G(d) (479) | 748 (4.11) | 3824 (5.81) | 5.12 |
| 6-31G(d,p) (539) | 749 (4.11) | 3830 (5.80) | 5.11 |
| 6-311G(d) (590) | 742 (4.00) | 3903 (5.80) | 5.26 |
| 6-31+G(d) (590) | 895 (3.93) | 4299 (5.92) | 4.80 |
| 6-31++G(d) (615) | 891 (3.90) | 4290 (5.92) | 4.81 |
| 6-311+G(d) (699) | 914 (3.94) | 4292 (5.95) | 4.70 |
| 6-311+G(d,p) (759) | 915 (3.93) | 4298 (5.94) | 4.70 |
| 6-311++G(d) (717) | 910 (3.92) | 4286 (5.95) | 4.71 |
| cc-pVDZ (510) | 701 (3.96) | 3748 (5.78) | 5.34 |
| cc-pVTZ (1154) | 755 (3.94) | 3880 (5.85) | 5.14 |
| cc-pVQZ (2199) | 809 (3.97) | 3984 (5.90) | 4.92 |
| cc-pV5Z (3743) | 849 (4.00) | 4057 (5.92) | 4.77 |
| aug-cc-pVDZ (848) | 902 (4.00) | 4194 (5.96) | 4.65 |
| aug-cc-pVTZ (1798) | 897 (4.01) | 4135 (5.95) | 4.61 |

cc-pV5Z, respectively. Similarly, the aug-cc-pVDZ $\beta_{HRS}$ value is only 2% smaller than the 6-31+G(d) value. Finally, the calculations performed with the aug-cc-pVTZ basis set, which, among those employed here, should be considered as the most flexible basis set for calculating the first hyperpolarizability, give $\beta_{HRS}$ values slightly smaller than those obtained with the aug-cc-pVDZ (–1%), 6-31+G (d) (–4%), 6-31++G(d) (–4%), 6-311+G(d) (–4%), and 6-311++G(d) (–4%) basis sets.

The combined effect of basis set extension on the first hyperpolarizability of the closed and open form results therefore in a slight decrease of the $\beta_{HRS}$(open)/$\beta_{HRS}$(closed) contrast ratio, from 5.12 with 6-31G(d) to 4.71, 4.65, and 4.61 with the 6-311++G(d), aug-cc-pVDZ, and aug-cc-pVTZ basis sets, respectively. Again, the 6-31G value is very close to the result obtained with the largest basis set. The effect of the basis set size on the depolarization ratio of both forms is also very weak, demonstrating that the dipolar versus octupolar contributions to $\beta_{HRS}$ are already estimated within ±3% with the 6-31G(d) basis set. Still, increasing the basis set size decreases DR of the closed form (i.e., leads to a relative increase of the octupolar contribution), whereas for the open form, it leads to an increase of DR, or of the dipolar contribution to $\beta_{HRS}$.

Consequently, in agreement with other studies [36], the requirements on the basis set are much less stringent for estimating $\beta_{HRS}$ and DR of a π-conjugated push-pull molecule than for a small molecule.

Electron correlation effects have then been analyzed in the case of *p*-nitroaniline, using the aug-cc-pVDZ basis set. Static β values have been obtained using the FF approach (Table 6.4). The same B3LYP/6-311G(d) geometry was used for all β calculations. Using the CCSD(T) results as references, the HF methodology strongly underestimates both $\beta_{HRS}$ and $\beta_{//}$ but most of the error is recovered by including electron correlation at the second order (MP2). Then, higher-order corrections have detrimental or canceling contributions to β. Note that the inclusion of the triples (MP4 vs. MP4SDQ and CCSD vs. CCSD(T)) amounts to a correction of

**Table 6.4** HRS and EFISHG first hyperpolarizabilities ($\beta_{HRS}$ and $\beta_{//}$) (in a.u.) and HRS depolarization ratios (DR) of *p*-nitroaniline as determined at different levels of approximation with the aug-cc-pVDZ basis set. All values have been obtained using the FF approach and are compared to the reference results obtained at the Ref = CCSD(T) level

| | $\beta_{HRS}$ (DR) | $\beta_{HRS}(X)/\beta_{HRS}(Ref)$ | $\beta_{//}$ | $\beta_{//}(X)/\beta_{//}(Ref)$ |
|---|---|---|---|---|
| HF | 397.8 (3.23) | 0.586 | 479.4 | 0.516 |
| MP2 | 668.4 (4.36) | 0.985 | 919.0 | 0.989 |
| MP3 | 579.8 (4.03) | 0.855 | 774.3 | 0.883 |
| MP4D | 621.4 (4.18) | 0.916 | 841.2 | 0.905 |
| MP4DQ | 551.1 (4.01) | 0.812 | 734.0 | 0.790 |
| MP4SDQ | 577.9 (4.09) | 0.852 | 776.3 | 0.835 |
| MP4 | 648.7 (4.28) | 0.956 | 886.2 | 0.953 |
| CCSD | 607.6 (4.14) | 0.896 | 819.6 | 0.882 |
| **CCSD(T)** | **678.4 (4.31)** | **1.000** | **929.4** | **1.000** |
| BLYP | 749.2 (4.03) | 1.104 | 1001.7 | 1.078 |
| B3LYP | 686.9 (4.00) | 1.012 | 915.8 | 0.985 |
| BHandHLYP | 567.4 (3.83) | 0.836 | 742.8 | 0.799 |
| M06 | 622.7 (3.89) | 0.918 | 820.4 | 0.883 |
| M06-2X | 551.3 (3.87) | 0.813 | 724.8 | 0.780 |
| CAM-B3LYP | 598.7 (3.95) | 0.883 | 793.6 | 0.854 |
| LC-BLYP | 527.6 (3.90) | 0.778 | 695.7 | 0.749 |

about 10%. The variations of DR are smaller but follow the same trends as those of $\beta_{HRS}$, demonstrating that the inclusion of electron correlation mostly increases the dipolar component of $\beta_{HRS}$.

Using DFT and conventional XC functionals, the results get more contrasted. BLYP overestimates the β values (but not DR) by about 10%. Then, adding larger and larger amounts of HF exchange in the functional lead to a reduction of $\beta_{HRS}$ and $\beta_{//}$ so that the agreement is excellent between B3LYP (20% HF exchange) and CCSD(T), whereas the BHandHLYP (50% HF exchange) values are underestimated. Similar effects are observed between M06 (27% HF exchange) and M06-2X (54% HF exchange), where for the later the underestimations attain 20%, similar to what is found for BHandHLYP. The use of range-separated hybrids follows also the same trend with the smallest values obtained with the LC-BLYP functional (underestimations of 22 and 25%) and then CAM-B3LYP (underestimations of 12 and 15%). Indeed, at long range ($r \rightarrow \infty$) LC-BLYP includes 100% of HF exchange, whereas CAM-B3LYP only 65%.

These results need however to be completed by those on other push-pull π-conjugated systems, with π-conjugated segments of different lengths and D/A groups of different strengths. In the case of α,ω-nitro, dimethylamino-polyenes containing 4 and 6 CH=CH units, the β(HF)/β(MP2) ratio amounts to 0.411 and 0.419, respectively [28]. Though these ratios have been obtained for the dominant longitudinal β tensor component with the 6-31+G(d) basis set, they can be compared with the $\beta_{//}$ ratio of 0.521 obtained here for *p*-nitroaniline. Similarly, the corresponding values for the α,ω-nitro, dimethylamino-polyynes with 4 and 6 C≡C units amount to 0.582 and 0.616, demonstrating that the β(HF)/β(MP2) ratio can vary by about 20% for π-conjugated segments commonly found in push-pull π-conjugated compounds. So, it is often recognized that the MP2/HF ratio for the static β values is close to 2 with a standard deviation of about 20%. Moreover, for those push-pull π-conjugated polyenes and polyynes containing 4 units, the β(MP2)/β(CCSD(T)) ratio amounts to 1.045 and 0.997 respectively, in comparison with a value of 0.989 for $\beta_{//}$ of *p*-nitroaniline, respectively. So, as substantiated by additional related investigations [48, 49], for these compounds the MP2 approach is often a good compromise between accuracy and computational needs, but it requires using approximate schemes, like Eqs. 6.9–6.10, to describe frequency dispersion [20]. Still, benchmark results (CC3, CCSD, CCSD(T) together with a decent atomic basis set) on push-pull π-conjugated systems with 20–50 C atoms would be of high interest.

Again, for systems different from (and usually larger than) *p*-nitroaniline, the selection of a reliable XC functional for evaluating β and its modifications upon chemical changes is a subtle issue, which was already addressed in depth in many studies [35–39, 48, 50]. For medium-size compounds, the range-separated LC-BLYP hybrid functional has been shown to be reliable when characterizing the changes of β upon enlarging the π-conjugated linker from 4 to 6 CH=CH (or C≡C) units or upon changing the polyyne linker into a polyene segment [36]. On the other hand, the BLYP, B3LYP, and BHandHLYP functionals—as well as functionals with similar characteristics—generally perform quantitatively better, but the above chemical/size trends are poorly described. For instance, using the

6-31G(d) basis set, the β(B3LYP)/β(CCSD(T)) ratio amounts to 0.648 and 0.714 for the substituted polyenes with N = 4 and 6 ($NMe_2$-(CH=CH)$_N$-$NO_2$) but to 0.999 and 1.372 for their polyyne analogs ($NMe_2$-(C≡C)$_N$-$NO_2$), respectively [36]. Then, going toward even larger oligomers, the unphysical delocalization inherent to conventional (LDA, GGA, meta-GGA, and global hybrids) XC functionals gives rise to overestimations of the hyperpolarizabilities by more than one order of magnitude. This can be corrected by enforcing an asymptotically correct exchange-correlation potential, i.e., 100% of Hartree–Fock exchange. Still, as shown by early works [50], global hybrids with 100% HF exchange merely reproduce the β values calculated at the Hartree–Fock level. On the other hand, range-separated or long-range corrected (LC) hybrids, where the percentage of HF exchange varies with the inter-electronic distance, bring an improvement. This is evidenced by the β results of Kamiya et al. [48] on α,ω-nitro, amino-polyenes, where the LC-BOP results are larger than the HF ones, though smaller than the MP2 values. A further refinement of the XC functional consists in adjusting the range-separated parameter (μ) such that Koopmans' theorem is obeyed as closely as possible [51]. It is found that the optimal μ values are smaller than those recommended as standard values (0.30, 0.33 or 0.47 bohr$^{-1}$) and that they increase as a function of the size of the π-conjugated segment. So, in the case of (E)-N, N-dimethyl-4-(4-nitrostyryl)aniline derivatives containing from N = 1 to 4 CH=CH units between the phenyl rings, the LC-PBE and LC-PBE0 functionals with the standard μ = 0.30 value underestimate the MP2 β values for N = 3–4 by 18% and 27%, whereas using the optimal μ value, the corresponding functionals overestimate β by 15–24% and 12–21%, respectively [38]. For these systems, using the optimal μ value does not represent a clear improvement, whereas these optimally tuned range-separated hybrids perform clearly better than the conventional ones for N = 1–2. The complexity of selecting an appropriate XC functional has been further evidenced in a recent work due to Isborn and co-workers [39], recommending a larger fraction of exact exchange when computing β than for computing excitation energies. Thus, it is not clear whether exchange hybrid functionals could be further optimized to qualitatively and quantitatively reproduce MP2–or CCSD (T)–β values of increasingly large push-pull π-conjugated systems or, in other words, whether non-local exchange is sufficient in the absence of non-local correlation. Indeed, double hybrids, which include a given percentage of MP2 correlation, can provide for medium-size push-pull π-conjugated systems β values of at least similar quality as the global hybrids. Therefore, combining the same optimally tuned range-separated strategy for both HF exchange and MP2 correlation might lead to a functional for accurate prediction of β. Note however that the inclusion of MP2 correlation should be accompanied with an improvement over MP2 because the contrary would simply substantiate the use of MP2 to approximate CCSD(T).

Finally, the selection of an appropriate XC functional can also be addressed in the case of NLO switches (Scheme 6.2), molecules characterized by their ability to alternate between two or more chemical forms displaying contrasts in one of their NLO properties (here, the second harmonic intensity) [52]. In this case, besides the absolute values, their contrast is also of interest. Scheme 6.2 gives the structure of a

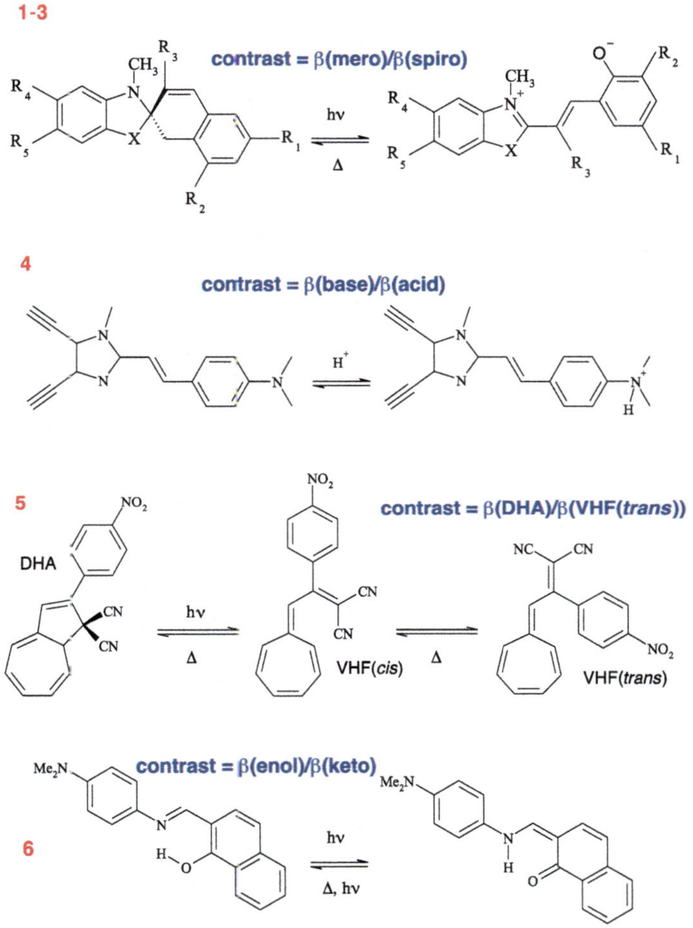


**Scheme 6.2** Second-order NLO switches; **1**–**3** merocyanine–spiropyran with different combinations of substituents [**1**: $R_1$=$NO_2$, $R_2$=OMe, $R_3$=Me, X=S, $R_4$=$R_5$=H; **2**: $R_1$=$NO_2$, $R_2$=$NMe_2$, $R_3$=$NH_2$, X=NMe, $R_4$=H, $R_5$=$NO_2$; **3**: $R_1$=$NO_2$, $R_2$=$NMe_2$, $R_3$=$NH_2$, X=NMe, $R_4$=H, $R_5$=$NO_2$] [53]; **4** neutral and protonated forms of a 4,5-dicyanoimidazole derivative [54]; **5** dihydroazulene (DHA)-vinylheptafulvene (VHF) [55]; **6** tautomeric equilibrium of the N-(2-hydroxynaphtylidene) aniline [56]

selection of NLO switches that have recently been studied [53–56] as well as the targeted contrast. The reported calculations were carried out using the 6-311+G(d) basis set. The same geometries were used for the whole set of β calculations. Some of these calculations were already reported in Refs. [52, 57]. These correspond to static $\beta_{HRS}$ values obtained in gas phase, except for compounds **6** where solvent (ethanol) effects were accounted for using the IEFPCM scheme. The results are shown in Fig. 6.6. Again, there are large variations of β amplitudes among the different methods. Considering that the MP2 method provides reference values to

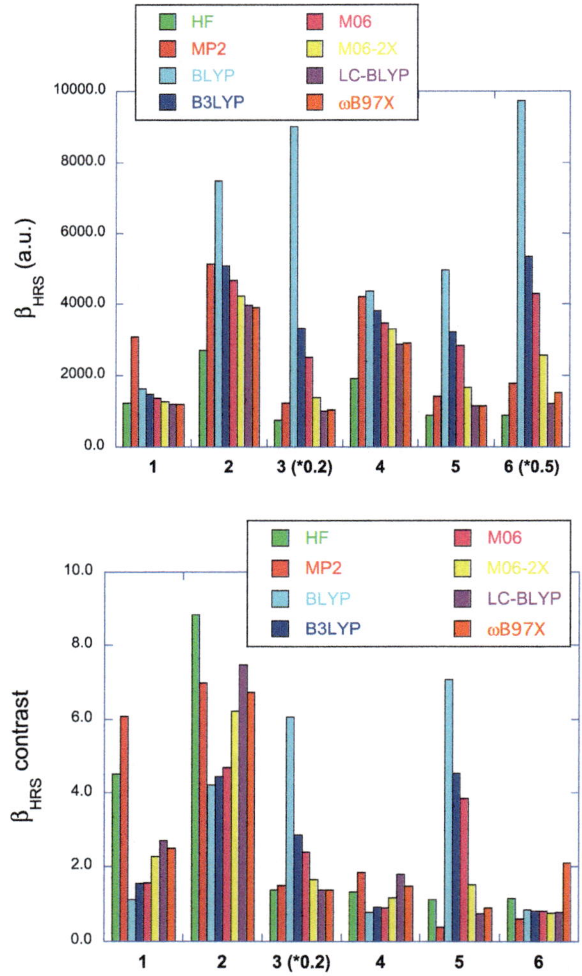


**Fig. 6.6** Comparisons between HF, MP2, and DFT with different XC functionals to evaluate the static $\beta_{HRS}$ of molecular switches. (top) $\beta_{HRS}$ of chromophores **1**–**6** in one of their forms (merocyanine form for **1**–**3**; base form for **4**, DHA form for **5**, and enol form for **6**) and (bottom) $\beta_{HRS}$ contrasts. The "*0.2" and "*0.5" labels on the x-axis mean that the corresponding $\beta_{HRS}$ or $\beta_{HRS}$ contrast values have been multiplied by the corresponding factor, for a question of readability

assess the XC functionals, the trend that the BLYP XC functional overestimates the static $\beta_{HRS}$ values of compounds **2**–**6** is confirmed (though for compound **4** the agreement is better) and is related to the lack of HF exchange. Then, adding a small percentage of HF exchange (B3LYP and M06) improves the results for compounds **2** and **4,** whereas large overestimations are still observed for **3**, **5**, and **6**. Moving from M06 to M06-2X leads to a further improvement for a third compound (**3**). Then, including long-range Hartree–Fock exchange (LC-BLYP and ωB97X) gives suitable results for compounds **2**-**6**. All methods underestimate $\beta_{HRS}$ of compound **1** in its merocyanine form, and in all cases, the HF method underestimates $\beta_{HRS}$, from 60% (**1**) to 37% (**5**).

Turning now to the contrasts, the BLYP, B3LYP, and M06 XC functionals can strongly underestimate or overestimate the $\beta_{HRS}$ contrasts. A clear improvement is achieved when using the M06-2X, LC-BLYP, and ωB97x functionals, at least for compounds **2–4**, but a similar performance is achieved at the HF level. None of the methods is suitable to describe the $\beta_{HRS}$ contrasts of compounds **5** and **6**, highlighting the difficulty in selecting a priori an XC functional to assess a broad variety of chromophores having second-order NLO responses.

**Acknowledgements** This work was supported by funds from the Belgian Government (IUAP N° P7/5 "Functional Supramolecular Systems") and the Francqui Foundation. It has also been done in the frame of the Centre of Excellence LAPHIA (Investments for the future: Programme IdEx Bordeaux–LAPHIA (ANR-10-IDEX-03-02)). V.L. thanks the Fund for Scientific Research (F.R. S.-FNRS) for his Research Associate position and M.H.C. the IUAP N° P7/5. The calculations were performed on the computing facilities of the Consortium des Équipements de Calcul Intensif (CÉCI, http://www.ceci-hpc.be), including those of the Technological Platform of High Performance Computing, for which we gratefully acknowledge the financial support of the FNRS-FNRC (Conventions 2.4.617.07.F and 2.5020.11) and of the University of Namur as well as on the "Mésocentre de Calcul Intensif Aquitain" (MCIA) of the University of Bordeaux, financed by the Conseil Régional d'Aquitaine and the French Ministry of Research and Technology.

## References

1. D.P. Shelton, J.E. Rice, Chem. Rev. **94**, 3–29 (1994)
2. T. Verbiest, K. Clays, V Rodriguez, *Second-Order Nonlinear Optical Characterization Techniques: An Introduction* (Taylor & Francis, 2009)
3. D.R. Kanis, M.A. Ratner, T.J. Marks, Chem. Rev. **94**, 195–242 (1994)
4. J.L. Brédas, C. Adant, P. Tackx, A. Persoons, B.M. Pierce, Chem. Rev. **94**, 243–278 (1994)
5. P.J. Campagnola, L.M. Loew, Nat. Biotechnol. **21**, 1356–1360 (2003); Y.C. Liang, A.S. Dvornikov, P.M. Rentzepis, Proc. Natl. Acad. Sci. USA **100**, 8109–8112 (2003)
6. P.C. Ray, Chem. Rev. **110**, 5332–5365 (2010)
7. R. Bersohn, Y.H. Pao, H.L. Frisch, J. Chem. Phys. **45**, 3184–3198 (1966)
8. S.Y. Liu, C.E. Dykstra, J. Phys. Chem. **91**, 1749–1754 (1987)
9. G.J.B. Hurst, M. Dupuis, E. Clementi, J. Chem. Phys. **89**, 385–395 (1988)
10. S. Yamada, M. Nakano, I. Shigemoto, S. Kiribayashi, K. Yamaguchi, Chem. Phys. Lett. **267**, 445–451 (1997)
11. T. Pluta, A.J. Sadlej, Chem. Phys. Lett. **297**, 391 (1998); A. Baranowska, A.J. Sadlej, J. Comput. Chem. **31**, 552–560 (2010); R. Zalesny, A. Baranowska-Laczkowska, M. Medved, J.M. Luis, J. Chem. Theor. Comput. **11**,4119-4128 (2015)
12. G. Maroulis. J. Chem. Phys. **108**, 5432–5448(1998); G. Maroulis, Theor. Chem. Acc. **129**,437–445 (2011)
13. R.A. Kendall, T.H. Dunning Jr., R.J. Harrison, J. Chem. Phys. **96**, 6796–6806(1992); D.E. Woon, T.H. Dunning Jr., J. Chem. Phys. **100**, 2975–2988 (1994)
14. F. Castet, E. Bogdan, A. Plaquet, L. Ducasse, B. Champagne, V. Rodriguez, J. Chem. Phys. **136**, 024506 (2012)
15. F.D. Vila, D.A. Strubbe, Y. Takimoto, X. Andrade, A. Rubio, S.G. Louie, J.J. Rehr, J. Chem. Phys. **133**, 034111 (2010)
16. J.D. Talman, Phys. Rev. A **86**, 022519 (2012)
17. H.D. Cohen, C.C.J. Roothaan, J. Chem. Phys. **43**, S34 (1965)
18. D.M. Bishop, S.A. Solunac, Chem. Phys. Lett. **122**, 567 (1985)
19. A.A.K. Mohammed, P.A. Limacher, B. Champagne, J. Comput. Chem. **34**, 1497–1507 (2013)

20. M. de Wergifosse, V. Liégeois, B. Champagne, Int. J. Quantum Chem. **114**, 900–910 (2014)
21. J. Linderberg, Y. Öhrn, *Propagators in Quantum chemistry* (Wiley-Interscience, Hoboken, 2004)
22. T. Helgaker, S. Coriani, P. Jørgensen, K. Kristensen, J. Olsen, K. Ruud, Chem. Rev. **112**, 543 (2012)
23. S. Karna, M. Dupuis, J. Comput. Chem. **12**, 487 (1991)
24. S.J.A. van Gisbergen, J.G. Snijders, E.J. Baerends, J. Chem. Phys. **109**, 10657 (1998)
25. P. Beaujean, B. Champagne, J. Chem. Phys. **145**, 044311 (2016)
26. H. Sekino, R.J. Bartlett, J. Chem. Phys. **94**, 3665 (1991)
27. J.E. Rice, N.C. Handy, J. Chem. Phys. **94**, 4959 (1991)
28. M. de Wergifosse, F. Castet, B. Champagne, J. Chem. Phys. **142**, 194102 (2015)
29. P. Kaatz, E.A. Donley, D.P. Shelton, J. Chem. Phys. **10**, 849 (1998)
30. J.F. Ward, C.K. Miller, Phys. Rev. A **19**, 826 (1979)
31. D.P. Shelton, J. Chem. Phys. **137**, 044312 (2012)
32. D.M. Bishop, F.L. Gu, S.M. Cybulski, J. Chem. Phys. **109**, 8407 (1998)
33. F. Castet, B. Champagne, J. Chem. Theor. Comput. **8**, 2044 (2012)
34. P. Karamanis, R. Marchal, P. Carbonnière, C. Pouchan, J. Chem. Phys. **135**, 044511 (2012)
35. K.Y. Suponitsky, S. Tafur, A.E. Masunov, J. Chem. Phys. **129**, 044109 (2008)
36. M. de Wergifosse, B. Champagne, J. Chem. Phys. **134**, 074113 (2011)
37. S.I. Lu, C.C. Chiu, Y.F. Wang, J. Chem. Phys. **135**, 134104 (2011)
38. H. Sun, J. Autschbach, ChemPhysChem **14**, 2450–2461 (2013)
39. K. Garrett, X.A. Sosa Vazquez, S.B. Egri, J. Wilmer, L.E. Johnson, B.H. Robinson, C.M. Isborn, J. Chem. Theory Comput. **10**, 3821–3831 (2014)
40. J. Tomasi, B. Mennucci, R. Cammi, Chem. Rev. **105**, 2999–3094 (2005)
41. J.L. Oudar, D.S. Chemla, J. Chem. Phys. **66**, 2664–2668 (1977)
42. P. Kaatz, D.P. Shelton, Mol. Phys. **88**, 683–691 (1996)
43. K. Coutinho, S. Canuto, Adv. Quantum Chem. **28**, 89–105 (1997)
44. K. Coutinho, R. Rivelino, H.C. Georg, S. Canuto, in *Solvation Effects in Molecules and Biomolecules: Computational Methods and Applications* (Springer, 2008), pp. 159–189
45. M. Hidalgo Cardenuto, B. Champagne, Phys. Chem. Chem. Phys. **17**, 23634–23642 (2015)
46. M. Hidalgo Cardenuto, F. Castet, B. Champagne, RSC Adv. **6**, 99558–99563 (2016)
47. E. Deniz, J. Cusido, S. Swaminathan, M. Battal, S. Impellizzeri, S. Sortino, F.M. Raymo, J. Photochem. Photobiol., A **229**, 20–28 (2012)
48. M. Kamiya, H. Sekino, T. Tsuneda, K. Hirao, J. Chem. Phys. **122**, 234111 (2005); F. Bulat, A. Toro-Labbé, B. Champagne, B. Kirtman, W. Yang, J. Chem. Phys. **123**, 014319 (2005)
49. B. Champagne, K. Kirtman, J. Chem. Phys. **125**, 024101 (2006)
50. B. Champagne, E.A. Perpète, D. Jacquemin, S.J.A. van Gisbergen, E.J. Baerends, C. Soubra-Ghaoui, K.A. Robins, B. Kirtman, J. Phys. Chem. A **104**, 4755–4763 (2000)
51. L. Kronik, T. Stein, S. Refaely-Abramson, R. Baer, J. Chem. Theory Comput. **8**, 1515–1531 (2012)
52. B.J. Coe, Chem. Eur. J. **5**, 2464–2471 (1999); J.A. Delaire, K. Nakatani K, Chem. Rev. **100**, 1817–1845 (2000); I. Asselberghs, K. Clays, A. Persoons, M.D. Ward, J. McCleverty, J. Mater. Chem. **14**, 2831–2839 (2004); F. Castet, V. Rodriguez, J.L. Pozzo, L. Ducasse, A. Plaquet, B. Champagne, Acc. Chem. Res. **46**: 2656–2665 (2013)
53. A. Plaquet, M. Guillaume, B. Champagne, F. Castet, L. Ducasse, J.L. Pozzo, V. Rodriguez, Phys. Chem. Chem. Phys. **10**, 3223–3232 (2008)
54. A. Plaquet, B. Champagne, J. Kulhanek, F. Bures, E. Bogdan, F. Castet, L. Ducasse, V. Rodriguez, ChemPhysChem **12**, 3245–3252 (2011)
55. A. Plaquet, B. Champagne, F. Castet, L. Ducasse, E. Bogdan, V. Rodriguez, J.L. Pozzo, New J. Chem. **33**, 1349–1356 (2009)
56. E. Bogdan, A. Plaquet, L. Antonov, V. Rodriguez, L. Ducasse, B. Champagne, F. Castet, J. Phys. Chem. C **114**, 12760–12768 (2010)
57. A. Plaquet, B. Champagne, L. Ducasse, E. Bogdan, F. Castet, AIP Conf. Proc. **1642**, 481–487 (2015)

# Chapter 7
# Embedding Methods in Quantum Chemistry

**Albrecht Goez and Johannes Neugebauer**

**Abstract** An overview over different embedding schemes for electronic-structure calculations is given, with the main focus on methods used for molecular systems. The in-principle exact subsystem DFT formalism is used as a reference point to classify the different approaches according to their components of the exact embedding potential. Special attention is paid to recently proposed ideas from the field of density-based embedding and density matrix reconstruction approaches.

**Keywords** Embedding methods · Subsystem methods · Fragmentation methods

## 7.1 Introduction

"Embedding" as a general term encompasses all methods where at some point in a calculation, the effect of an environment on a certain part of the total system is evaluated. The terms "embedding methods" and "fragment-based methods" are therefore closely related, since a division of the total system is unavoidable in order to formulate an embedding theory. A fragment-based method without embedding is theoretically possible, but would be restricted to the treatment of non-interacting fragments (note that we regard many-body expansions as a form of effective embedding in this context). There are two main motivations for embedding methods, which are (i) reduction of the computational cost and (ii) improved interpretability of the total system in terms of fragments and their interactions.

The first motivation can be understood by considering the scaling behavior of typical quantum-chemical methods with the system size. Even Kohn–Sham density functional theory (DFT) methods, which are among the most cost-effective ab initio methods, formally scale at least with $\mathcal{O}(N^3)$. Approaches based on wave function theory (WFT) often lead to scaling of even higher order, such as Hartree–Fock (HF) with $\mathcal{O}(N^4)$, second-order Møller–Plesset perturbation theory (MP2) with $\mathcal{O}(N^5)$,

A. Goez · J. Neugebauer (✉)
WWU Münster, Organisch-Chemisches Institut and Center for Multiscale Theory and Computation, Münster, Germany
e-mail: j.neugebauer@uni-muenster.de



M.J. Wójcik et al. (eds.), *Frontiers of Quantum Chemistry*,
https://doi.org/10.1007/978-981-10-5651-2_7

or Coupled Cluster with single/double excitations as well as perturbative triple excitations (CCSD(T), often referred to as the "gold standard" of quantum chemistry), which even has a formal scaling behavior of $\mathcal{O}(N^7)$. It is clear that the overall computational cost could be strongly reduced by splitting the total system into small parts to be evaluated sequentially or in parallel.

The second aspect is especially important from a chemist's perspective. In many cases, individual building blocks of a large system (e.g., functional groups, small molecules, cofactors in a protein) will have similar (though not identical) properties both in a complex and on their own. In fact, this is the way of thinking chemists usually apply to most problems. Quantum-chemical models, however, generally treat the system as a whole without any such classification. It is then much harder or even impossible to define and evaluate interaction terms between different components of the system. Fragment-based methods often have a decisive advantage in this respect. This is also the reason why it can be worthwhile to apply an approach which is *more* expensive than a supermolecular calculation in some cases.

The present chapter attempts to give an overview over the current state of embedding methods in quantum chemistry. However, the sheer number of individual approaches has become so vast that no single article can cover everything. Therefore, this review focuses especially on *actively developed* methods from the field of molecular quantum chemistry, whereas other approaches—though they might be historically significant—are omitted or treated in less detail. For a much broader overview, the reader is referred to a collection of reviews published in the recent Chemical Reviews issue "Calculations on Large Systems" [1], in particular Refs. [2–7], and to further reviews on density-based embedding formalisms [8–11].

This chapter is structured as follows. In Sect. 7.2, some preliminary concepts will be introduced, which are important for several of the methods described later. In addition, our classification of the presented methods will be detailed. In order to be able to relate the wealth of existing approaches, the in-principle exact subsystem DFT formalism will be treated in Sect. 7.3. The embedding potential arising from this theory has many interesting features, which can be used to compare the different approaches with respect to the incorporated interactions. The body of this chapter consists of Sects. 7.4–7.6, where a number of important embedding schemes are reviewed in three different classes. Finally, a short summary is presented in Sect. 7.7.

## 7.2 General Embedding Strategies

The interest in theoretical fragmentation or embedding methods has increased greatly over the past decades. This results in a situation where not only the number of approaches described in the literature is vast, but even the number of different classification schemes grows continually. A famous early attempt was made by Li et al. [12], who distinguished between density-based and energy-based methods. In the former case, a total density (matrix) of the system is assembled first, from which any supermolecular properties are obtained in a second step. In contrast, energy-based methods directly evaluate the total energy of the system from fragment contributions in a linear fashion. Molecular properties can then be determined by taking derivatives of the energy with respect to appropriate quantities. This classification was generalized by Gordon et al. in the framework of a comprehensive review [13]. Here, the energy-based group is renamed "one-step" group and extends the definition to quantities other than the energy, which can be determined directly from individual fragment contributions. For instance, the extension of the Molecular Fractionation with Conjugate Caps (MFCC) method (see Sect. 7.5.1) to dipole moments and the electrostatic potential [14] now falls into this category, since the total property is constructed in a strictly additive fashion from the fragment contributions. The other class of methods is accordingly termed "two-step" methods and contains all approaches where a supermolecular quantity (usually the density or density matrix) is constructed from fragment contributions and used to nonlinearly determine properties in a second step.

A different approach to classifying fragment-based methods focuses on the geometric nature of the employed fragments. Suárez et al. [15] presented an intuitive distinction between different energy-based methods, depending on whether overlapping or disjoint fragments are constructed. Mayhall and Raghavachari [16] applied the terms "top-down methods" for the former and "bottom-up methods" for the latter. To avoid double counting in top-down methods, derivative subsystems corresponding to the overlapping sections are formed according to the *inclusion–exclusion principle* (IEP) of set theory, which can be symbolically cast as

$$\begin{aligned}|A_1 \cup A_2 \cup \cdots \cup A_{N^{\mathrm{F}}}| = \sum_{K}^{N^{\mathrm{F}}} A_K - \sum_{K<L}^{N^{\mathrm{F}}} |A_K \cap A_L| + \sum_{K<L<M}^{N^{\mathrm{F}}} |A_K \cap A_L \cap A_M| + \cdots \\ + (-1)^{N^{\mathrm{F}}-1} |A_K \cap \cdots \cap A_{N^{\mathrm{F}}}| \ ,\end{aligned} \tag{7.1}$$

where $A_K$ is one of $N^{\mathrm{F}}$ fragments, $A_K \cup A_L$ represents the union of fragments $A_K$ and $A_L$, and $A_K \cap A_L$ is their intersection. Thus, overlaps of even order are subtracted, whereas overlaps of odd order enter with a positive sign. Properties of the total system (especially the energy) can be approximated by combining the fragment contributions in the same way.

In contrast, a straightforward *many-body expansion* (MBE) is applied in bottom-up methods, where no overlap exists between the individual fragments. The energy can be determined as

$$E^{\mathrm{MBE}} = \sum_{K}^{N^{\mathrm{F}}} E_K + \sum_{K<L}^{N^{\mathrm{F}}} \Delta E_{KL} + \sum_{K<L<M}^{N^{\mathrm{F}}} \Delta E_{KLM} + \cdots \tag{7.2}$$

with

$$\Delta E_{KL} = E_{KL} - (E_K + E_L) \tag{7.3}$$

$$\Delta E_{KLM} = E_{KLM} - (\Delta E_{KL} + \Delta E_{KM} + \Delta E_{LM}) - (E_K + E_L + E_M) \ . \tag{7.4}$$

Here, $E_K$ is the energy of subsystem $K$, $E_{KL}$ is the energy of a dimer formed from $K$ and $L$, and so on. While the accuracy of an MBE method is determined primarily by the order to which the expansion is carried out, the most important factor influencing the results of IEP-based approaches is the size and type of the fragments [16].

If fragments are connected by covalent bonds, many approaches apply capping groups to saturate free valences ("dangling bonds") left over after the original division of the system. It should be noted that the definition of "overlap" between fragments can be ambiguous in this situation. Many authors apply it based only on the original fragmentation; i.e., *before* capping groups are applied. This means that even though the fragments that are subjected to calculations might actually possess some spatial overlap, this is only counted if it consists of atoms originally present in the supermolecule. This is the reason why for instance the MFCC method is regarded as based on overlapping fragments (capping groups consist of the adjacent amino acids, saturated with additional hydrogen atoms), whereas the Kernel Energy Method (KEM, see Sect. 7.5.4) employs disjoint fragments (capped only with hydrogen atoms). The boundary between the two classes is blurred further if intermediate capping groups are applied (such as small amino acid termini in some variants of MFCC). The Fragment Molecular Orbital (FMO) method (see Sect. 7.5.2) is one of the few examples which can be clearly classified as employing disjoint fragments, since no capping atoms are used.

Two further classification schemes were presented by Richard and Herbert [17] and Akimov and Prezhdo [2], who both chose not to establish definitive categories, but rather define a number of attributes to determine for any given method. In this contribution, we thus divide the method space into three empirical categories: (i) QM/MM and related multilevel approaches, (ii) schemes based on an MBE, and (iii) quantum-chemical Divide-and-Conquer (DC) methods. The first class comprises methods where a certain part of the total system is treated with a more approximate description, often based on physical considerations. The second class contains all many-body expansion methods, regardless of the exact description of the interactions (e.g., supplemented by long-range Coulomb interactions or not). Methods of the third class are based on the general DC idea and often can be shown to be in principle exact. As for all classifications, there will be methods that defy categorization under the present scheme. An example is the extension of density-based embedding to the embedding of a WFT description in a DFT environment, which could be attributed to categories (i) or (iii). The fact that we file it into the latter merely reflects the development history. Likewise, Frozen Density Embedding (FDE) itself can be classified either as a QM/QM scheme, where the interaction term is evaluated with orbital-free DFT,[1] or as a DC method where the total orbital space is divided among the subsystems.

[1]This holds especially if the environmental density is obtained from model considerations instead of separate fragment calculations.

## 7.3 Exact Embedding Potential

In this section, the in-principle exact subsystem DFT framework [18–20] will be described. The embedding potential that arises from such a treatment has many interesting features that can be used to classify other embedding frameworks. The present approach is somewhat similar to the one taken by Gomes and Jacob in their review on excited-state embedding methods [9].

The central idea of subsystem DFT is to partition the electron density of the total system, $\rho^{\text{tot}}$, into several additive fragment contributions $\rho_K$,

$$\rho^{\text{tot}}(\vec{r}) = \sum_{K}^{N^{\text{F}}} \rho_K(\vec{r}) \ . \tag{7.5}$$

Here, $N^{\text{F}}$ is the total number of fragments, and $\vec{r}$ denotes a spatial coordinate. Note that in contrast to the Divide-and-Conquer approach (see Sect. 7.6.1), a fixed number of electrons are contained in each subsystem. All fragment densities can be expressed in terms of the Kohn–Sham (KS) formalism as a sum of subsystem orbital contributions,

$$\rho_K(\vec{r}) = \sum_{i}^{N_K^{\text{orb}}} |\psi_{K,i}(\vec{r})|^2 \ , \tag{7.6}$$

where $N_K^{\text{orb}}$ is the number of orbitals $\{\psi_{K,i}\}$ centered on subsystem $K$. In accordance with the regular KS formalism, the energy of an individual, non-interacting subsystem is simply given by

$$E_K^{\text{iso}}[\rho_K] = T^{\text{s}}[\{\psi_{K,i}\}] + \int v_K^{\text{ext}}(\vec{r})\rho_K(\vec{r})\text{d}\vec{r} + \iint \frac{\rho_K(\vec{r})\rho_K(\vec{r}\,')}{|\vec{r}-\vec{r}\,'|}\text{d}\vec{r}\text{d}\vec{r}\,' + E^{\text{xc}}[\rho_K] \tag{7.7}$$

with $T^{\text{s}}$ being the kinetic energy of the electrons in the fictitious reference system and

$$v_K^{\text{ext}}(\vec{r}) = -\sum_{I}^{N_K^{\text{nuc}}} \frac{Z_I}{|\vec{r}-\vec{R}_I|} \tag{7.8}$$

representing the external potential due to the $N_K^{\text{nuc}}$ nuclei with charge $Z_I$ and position $\vec{R}_I$ in the current fragment. The second term in Eq. (7.7) thus describes the electrostatic electron–nuclei interaction, while the third and fourth terms contain the Coulomb and exchange–correlation (XC) interactions, respectively, between the electrons within the isolated subsystem. For non-interacting subsystems, the total energy would simply be the sum of all the isolated contributions,

$$E^{\text{tot,non-int}}[\rho^{\text{tot}}] = \sum_{K}^{N^{\text{F}}} E_K^{\text{iso}}[\rho_K] \ , \tag{7.9}$$

but an exact expression for the real (interacting) case can be obtained in the form

$$\begin{aligned} E^{\text{sDFT}}[\rho^{\text{tot}}] = \sum_{K}^{N^{\text{F}}} \Bigg[ E_K^{\text{iso}}[\rho_K] + \overbrace{\sum_{L \neq K}^{N^{\text{F}}} \left( \int v_L^{\text{ext}}(\vec{r})\rho_K(\vec{r})\mathrm{d}\vec{r} + \frac{1}{2} \iint \frac{\rho_L(\vec{r})\rho_K(\vec{r}\,')}{|\vec{r}-\vec{r}\,'|} \mathrm{d}\vec{r}\mathrm{d}\vec{r}\,' \right)}^{E_K^{\text{emb,elstat}}} \Bigg] \\ \underbrace{+ E^{\text{xc}}[\rho^{\text{tot}}] - \sum_{K}^{N^{\text{F}}} E^{\text{xc}}[\rho_K] + T^{\text{s}}[\rho^{\text{tot}}] - \sum_{K}^{N^{\text{F}}} T^{\text{s}}[\rho_K]}_{E_K^{\text{emb,nad}}} \ . \end{aligned} \tag{7.10}$$

Although this expression seems complicated, it boils down to a sum of the electrostatic interaction energies between all possible combinations of fragments, $E_K^{\text{emb,elstat}}$, as well as the consideration of kinetic and XC effects through $E_K^{\text{emb,nad}}$. The contributions inherent to each fragment are already contained in $E_K^{\text{iso}}$. The second line of Eq. (7.10) has a more complicated form, since the described effects are of nonadditive nature. Thus, a formally exact nonadditive term is introduced in the form of $E_K^{\text{emb,nad}}$.

While the above expression gives the energy of fixed interacting subsystem densities, the energy (and total density) of the ground state can be determined by minimizing this energy with respect to the KS orbitals of the individual subsystems. This leads to one set of *Kohn–Sham Equations with Constrained Electron Density* (KSCED) [21] per fragment, which can be cast in the form

$$\left[ -\frac{\vec{\nabla}^2}{2} + v_K^{\text{eff}}[\rho_K](\vec{r}) + v_K^{\text{emb}}[\rho_K, \rho^{\text{tot}}](\vec{r}) \right] \psi_{K,i}^{\text{KSCED}}(\vec{r}) = \varepsilon_{K,i}\psi_{K,i}^{\text{KSCED}}(\vec{r}) \ , \tag{7.11}$$

yielding optimal fragment KS orbitals $\psi_{K,i}^{\text{KSCED}}$ and orbital energies $\varepsilon_{K,i}$. Here, $\vec{\nabla}$ is the Nabla operator and $v_K^{\text{eff}}$ is the effective potential of the isolated fragment, which depends on the fragment density $\rho_K$ and has the same form as in a usual KS-DFT calculation. In addition, however, the *embedding potential* $v_K^{\text{emb}}$ is contained in the Hamiltonian and incorporates the effect which the other fragments exert on $K$. This embedding potential has the form

$$\begin{aligned} v_K^{\text{emb}}[\rho_K, \rho^{\text{tot}}](\vec{r}) = \left( \sum_{L, L \neq K}^{N^{\text{F}}} v_L^{\text{ext}}(\vec{r}) \right) + v^{\text{Coul}}[\rho^{\text{tot}} - \rho_K](\vec{r}) \\ + v^{\text{kin,nad}}[\rho_K, \rho^{\text{tot}}](\vec{r}) + v^{\text{xc,nad}}[\rho_K, \rho^{\text{tot}}](\vec{r}) \ . \end{aligned} \tag{7.12}$$

The embedding potential can be related to certain physical effects by turning to intermolecular perturbation theory (see Ref. [22] for an instructive review). The first term in Eq. (7.12) represents a sum of the individual nuclear potentials of all fragments except $K$. The second term contains an analogous electronic potential. Physically speaking, these two contributions can be clearly identified with the first-order electrostatic interaction, i.e., the electrostatic potential (ESP) of the environmental subsystems acting on the active fragment. The second line of Eq. (7.12) again contains nonadditive contributions which can be expressed as

$$v^{\text{kin,nad}}[\rho_K, \rho^{\text{tot}}](\vec{r}) = \left.\frac{\delta T^{\text{s}}[\rho]}{\delta\rho(\vec{r})}\right|_{\rho=\rho^{\text{tot}}} - \left.\frac{\delta T^{\text{s}}[\rho]}{\delta\rho(\vec{r})}\right|_{\rho=\rho_K} \tag{7.13}$$

and

$$v^{\text{xc,nad}}[\rho_K, \rho^{\text{tot}}](\vec{r}) = \left.\frac{\delta E^{\text{xc}}[\rho]}{\delta\rho(\vec{r})}\right|_{\rho=\rho^{\text{tot}}} - \left.\frac{\delta E^{\text{xc}}[\rho]}{\delta\rho(\vec{r})}\right|_{\rho=\rho_K} . \tag{7.14}$$

The nonadditive XC component of the embedding potential is responsible for short-range exchange–correlation effects and can thus be related to higher-order intermolecular interactions such as dispersion. Finally, the nonadditive kinetic potential takes care of non-orthogonality effects between the different fragments and thus incorporates the Pauli repulsion. It should be stressed that the mathematical form of the embedding potential is exact and will lead to the supermolecular KS solution for the limiting case of exact nonadditive functionals. Furthermore, if an exact expression for $E^{\text{xc}}$ were known, the physically exact solution would be reached.

The total embedding potential of course depends on the densities of all subsystems. Therefore, the KSCEDs have to be either solved simultaneously or in an iterative fashion [20] (see Sect. 7.6.3) to guarantee that the mutual interactions between all fragments are treated consistently. Physically speaking, this corresponds to an incorporation of inductive effects.

Partition DFT (PDFT) [23, 24] can be considered an alternative formulation to subsystem DFT, which is also in principle exact. The real system of interacting fragments is formally replaced by a system of non-interacting fragments, which are coupled by a common "partition potential" that acts as a constraint to ensure normalization to the total number of electrons in the system. In contrast to subsystem DFT, the number of electrons per subsystem is not fixed, but used as a variational parameter to minimize the total energy. In close relation to PDFT, Huang et al. introduced an embedding theory [25] where active fragment and environment are required to share a common embedding potential, thus avoiding the general non-uniqueness of the density partitioning (see Ref. [10] for details on this problem).

## 7.4 QM/MM and Related Approaches

In this section, a variety of multilevel approaches will be presented, where a certain part of the total system is described with a more accurate method than its surroundings. In particular, Continuum Solvent Models (CSMs), QM/MM approaches, and methods based on effective potentials will be discussed.

### 7.4.1 Continuum Solvent Models

One of the oldest strategies to embed a system of interest in an environment is represented by implicit models. The general idea of these approaches is to represent the surroundings (usually a solvent) by a uniform description, often in the form of a featureless continuum. Naturally, any atomistic details about the environment are forfeited in such a representation. Specific interactions such as hydrogen bonds, $\pi$-stacking, or salt bridges should therefore be treated by including the participating molecules in the active system.

By far the most common class of implicit solvent models is of the *Apparent Surface Charge* (ASC) type. In this family of methods, a *cavity* is created around the system of interest (solute), which is then discretized and furnished with point charges located on the individual segments. The electrostatic potential due to the solvent continuum is then simply given by

$$\phi^{\mathrm{ASC}}(\vec{r}) = \sum_{\mu} \frac{q_\mu}{|\vec{r} - \vec{s}_\mu|} , \tag{7.15}$$

where $q_\mu$ and $\vec{s}_\mu$ are the magnitude and position, respectively, of point charge $\mu$. The differences between the various ASC methods lie in the way how these ASC magnitudes are determined (see below).

A crucial feature of all ASC methods is the construction of the cavity. The majority of algorithms is based on either the solvent-accessible surface (SAS) or the solvent-excluded surface (SES), which were pioneered by Connolly [26, 27]. In both cases, a preliminary surface is created by interlocking atom-centered spheres according to the individual van der Waals radii. The actual cavity is then generated by rolling a "solvent ball" of certain radius over this object and tracing either its surface (SES) or its center (SAS).

The ASC approach itself can only represent the electrostatic component of the embedding potential. However, several methods have been combined with empirical treatments to account for dispersion, and a cavitation term is also frequently included to obtain realistic free energies (see, e.g., Ref. [28]).

The polarizable continuum model (PCM) is the oldest ASC method. Its original formulation (nowadays known as D-PCM [29]) has long been superseded by the more recent integral equation formalism (IEF) [30]. In the latter, the electrostatic problem is reformulated in terms of Green's functions, which allows to derive an expression for the ASC without resorting to an evaluation of the normal component of the field on the cavity surface. The solution requires only the knowledge of the molecular ESP, which considerably simplifies the procedure (for details, the reader is referred to the original publication [30]). Besides making the evaluation cheaper, the IEF reformulation contains an inherent correction to the problem of outlying charge, i.e., the fact that part of the solute electron density will always be located outside of the cavity.

An additional variant is the so-called C-PCM method [31], which borrows the assumption of a perfectly conducting dielectric from the COSMO approach (see below). Apart from the choice of the exact screening factor, the two methods can be considered equivalent. It should be noted that the IEF-PCM contains both D-PCM and C-PCM as special cases. A recent review focusing specifically on the PCM family of methods is available [32].

In 1993, Klamt and Schüürmann presented the Conductor-Like Screening Model (COSMO) as an alternative ASC method [33]. The crucial difference between the original PCM and COSMO is the assumption that the solvent can be represented as a perfect conductor (infinite permittivity). This leads to a much simpler boundary condition which requires the total potential to vanish at the cavity surface. The corresponding surface charge distribution can be determined for a given molecular potential (due to the nuclei and electrons inside the cavity) from the equation

$$\mathbf{A}\vec{q} = -(\mathbf{B}\vec{Z} + \vec{C}) \ , \tag{7.16}$$

where

$$A_{\mu\nu} = \begin{cases} |\vec{r}_\mu - \vec{r}_\nu|^{-1} & \text{for } \mu \neq \nu \\ 1.07\sqrt{\frac{4\pi}{F_\mu}} & \text{for } \mu = \nu \end{cases} \tag{7.17}$$

$$B_{A\mu} = |\vec{r}_\mu - \vec{R}_A|^{-1} \tag{7.18}$$

$$C_\mu = \int_V V_\mu(\vec{r})\rho(\vec{r})\mathrm{d}\vec{r} \ . \tag{7.19}$$

Here, $\vec{q}$ is the vector of surface charge magnitudes, while $\vec{Z}$ contains the nuclear charges. The matrices **A** and **B** hold inverse charge–charge and charge–nuclei distances, respectively. For the special case of the diagonal elements of **A**, a "self-interaction" within the surface segment with area $F_\mu$ is described, which has been derived in the original article [33]. Finally, the vector $\vec{C}$ represents the interaction between the electron density of the system and each surface charge. Equation (7.16) can be solved directly by matrix inversion, but is more commonly subjected to an iterative procedure. To correct for the assumption of a perfect conductor, the obtained charges are scaled with a permittivity-dependent function which has the form

$$f(\varepsilon) = \frac{\varepsilon - 1}{\varepsilon + g} , \tag{7.20}$$

where $\varepsilon$ is the (real) permittivity and a value of 0.5 has been suggested for the factor $g$ by the original authors, while others set it to zero in accordance with Gauss' law [34].

The COSMO approach was soon combined with a statistical thermodynamics treatment to yield the Conductor-Like Screening Model for Real Solvents (COSMO-RS) [35], which determines solution properties on a more physical basis by evaluating the chemical potential of each species in terms of interacting surface segments. By first carrying out a regular COSMO calculation for all participating components (solute and solvent alike), a consistent description is achieved. Both COSMO and COSMO-RS are discussed and compared in a recent review [36]. An algorithm similar to COSMO-RS was presented by Lin and Sandler and termed COSMO-SAC [37].

An extension of the COSMO scheme to electronically excited states was presented by Klamt a few years after the original publication of the method [38]. A recent modification of COSMO-RS allows the calculation of response properties also in the framework of this more sophisticated approach [39]. In addition, the problem of missing correlation between different surface segments was first addressed by the COSMO-RS-DARE approach [40], which was specifically intended for dimerization reactions, and later refined to yield the COSMO-RSC method [41]. One notable recent advancement was the development of the domain decomposition COSMO (ddCOSMO) algorithm [42–44], which represents an intrinsically linear-scaling COSMO variant based on a Schwarz decomposition. Significant speedups at almost no loss of accuracy have been reported in combination with different semi-empirical and quantum-chemical methods [45]. In addition, a variant of the COSMO scheme specifically adapted to subsystem electronic-structure approaches has been published under the name "Local COSMO" (LoCOSMO) [46].

The Surface Volume and Polarization for Electrostatics (SVPE) method [47] was specifically designed to remedy the problem of outlying charge. As such, it requires a treatment not only of the cavity surface (surface polarization), but also of the adjacent continuum (volume polarization). In order to avoid the costly integration procedures resulting from this process, the related Surface and Simulation of Volume Polarization for Electrostatics (SS(V)PE) approach [48, 49] was developed. Here, an additional surface charge density distribution is introduced to account for the required boundary conditions. It has later been shown that the SS(V)PE method is equivalent to the isotropic formulation of IEF-PCM [50].

One of the best-known non-ASC continuum models is the SM$x$ family of approaches by Cramer and Truhlar [51–53], where $x$ denotes the version (currently 12). These methods are based on a Generalized Born (GB) treatment; i.e., instead of solving a non-homogeneous Poisson equation, the solute is modeled as a collection of monopoles residing on the atomic positions, which are assigned certain radii. The atomic charges are usually obtained from a suitable population analysis scheme, although a density-based scheme called SMD has been introduced recently [54]. The crucial part of setting up a GB model is the determination of suitable Coulomb radii for the individual solute monopoles. Since the original publication [51], a plethora of parameterizations and methodological extensions have been presented, which culminated in the two most recent formulations SM8 [55] and SM12 [53].

### 7.4.2 *QM/MM Approaches*

The idea of combining an active region of a system described by quantum-chemical methods with an environmental representation in terms of a force field dates back to the seminal papers by Warshel, Karplus, and Levitt [56, 57] and was eventually awarded the 2013 Nobel Prize in chemistry. Since the original formulation, an overwhelming plethora of different extensions, variants, and improvements have been developed in groups all over the world. A systematic review of the QM/MM method is clearly out of the scope of this work, which is why only the very basics and a small selection of relatively recent developments will be detailed. The reader is referred to several excellent reviews for more details on this topic [7, 58–60].

A central distinction between different QM/MM methods can be made in terms of whether they utilize a subtractive or an additive scheme. For the former, the total energy is expressed as

$$E^{\text{QM/MM}} = E^{\text{QM}}(\text{core}) + E^{\text{MM}}(\text{core} + \text{env}) - E^{\text{MM}}(\text{core}) \;, \tag{7.21}$$

where "core" and "env" represent the region of interest and the environment, respectively, while $E^{\text{QM}}$ and $E^{\text{MM}}$ denote the energy obtained with a QM or MM description in the respective region. In order to carry out a subtractive calculation, the energy of the core region thus has to be determined with both methods, in addition to an MM energy evaluation for the whole system. A clear benefit of this approach is the avoidance of a complicated coupling term between QM and MM regions. However, it relies on the assumption that the energy change of the core upon embedding in the environment is transferable between the different descriptions. Furthermore, the subtractive approach requires MM parameters for the core region, which can be problematic if nonstandard moieties are to be modeled. A prime example of this class of methods is the IMOMM scheme [61] and its better known successor, the ONIOM approach [62, 63], which will be discussed below.

In additive QM/MM schemes, the total energy is instead simply expressed as

$$E^{\text{QM/MM}} = E^{\text{QM}}(\text{core}) + E^{\text{MM}}(\text{env}) + E^{\text{QM-MM}} \;, \tag{7.22}$$

where the energies in the core and environmental regions are evaluated exclusively by either a QM or an MM description, with the additional coupling term $E^{\text{QM-MM}}$ describing the interactions between them. While the bonded terms across the boundary and the non-electrostatic contribution to the non-bonded energy are almost always described by force field terms [58], the electrostatic coupling is usually evaluated following one of three different schemes [64]:

1. Mechanical embedding. In this case, a classical representation of the core charge distribution is created and coupled electrostatically to the environmental force field charges. Only an improved total energy is obtained, while the core density/wave function stays the same.
2. Electrostatic embedding. Here, the environmental charge distribution is incorporated as an embedding potential in the QM Hamiltonian, thus directly altering the core density/wave function.
3. Polarizable embedding. Through the use of a polarizable description of the environment, a mutual polarization between QM and MM regions is modeled. While this approach best represents the underlying physics, the evaluation of environmental polarization can of course make it more costly than the other embedding schemes.

A major hurdle in all additive approaches is the modeling of a smooth boundary between the different regions of the system. This is especially problematic if cuts occur through covalent bonds, as is often the case in biological molecules. Many different solutions have been proposed, a comprehensive review of which can be found in Ref. [58].

The Integrated Molecular Orbital and Molecular Mechanics (IMOMM) method [61] is a prototypical example of a subtractive QM/MM scheme. The total energy is calculated based on Eq. (7.21), such that the energy of the QM core ("model system") has to be determined both with a QM method of choice and with MM. In addition, the total system ("real system") has to be subjected to an MM calculation. The subtraction of the MM energy of the model system eliminates any double counting. However, the latter requires the existence of force field parameters for the compounds in this region. While the bonded parameters play only a minor role (since they are mostly canceled by the subtraction), it is important to employ a good parametrization for long-range interactions [65].

A logical extension was soon presented in the form of the Integrated Molecular Orbital and Molecular Orbital (IMOMO) method [66], where two different QM methods are combined, with the less expensive method taking the role of the MM in IMOMM. Finally, a generalization of both schemes was reported under the name "Our Own $n$-Layered Integrated Molecular Orbital and Molecular Mechanics" (ONIOM) [62, 63]. Here, a general formulation for a system divided into $n$ layers was given, each of which are overlapping and described by different theoretical methods. The general notation of all derived methods is ONIOM$n$($M_1$:$M_2$:...:$M_n$), where $M_1 \ldots M_n$ denote the methods used for the different layers. For the most typical case of three layers (i.e., a partitioning of the system into model, *intermediate*, and real system), the energy expression

$$\begin{aligned} E^{\mathrm{ONIOM3}} = & E^{\mathrm{high}}(\mathrm{model}) + E^{\mathrm{medium}}(\mathrm{intermediate}) - E^{\mathrm{medium}}(\mathrm{model}) \\ & + E^{\mathrm{low}}(\mathrm{real}) - E^{\mathrm{low}}(\mathrm{intermediate}) \end{aligned} \tag{7.23}$$

is obtained, such that again no contribution is overcounted.

In the original formulation, the interaction between the two parts of the system was described purely in terms of an MM approach, corresponding to mechanical embedding. The partial charges for the model system have to be either taken from a force field (if available) or computed on the fly from the QM results [67]. By including environmental MM point charges in all calculations for the model system, a situation corresponding to electrostatic embedding can be brought about [65]. Recently, a further extension to ONIOM calculations with polarizable embedding has also been reported [68]. If all layers are described with various QM methods, a total density can be obtained from ONIOM calculations in the same spirit as the total energy [69], although it should be noted that patches of negative density can occur due to the subtraction.

If covalent bonds run through layer boundaries, this can be taken care of by including link atoms, similar to additive QM/MM calculations [69]. Expressions for the ONIOM gradient and other derivatives with and without link atoms have been derived [69]. Furthermore, the ONIOM scheme was combined with the implicit solvent model PCM [70] as well as with an adaptive atomistic description of a solvent shell [71]. In addition, there are ONIOM approaches with other fragmentation schemes applied to parts of the system, leading to methods such as the Multicentered QM:QM approach [72] and Generalized ONIOM [73]. A comprehensive review concerning the history, current developments, and applications of different ONIOM methods can be found in Ref. [5].

The Quantum-regions Interconnected by Local Descriptions (QUILD) method [74] was developed by Swart and Bickelhaupt as a generalized framework for multilevel QM/QM or QM/MM calculations. Although somewhat similar to ONIOM, the different regions can be defined in any desired way, including several overlapping parts. In this way, different interactions can be treated by different methods. For instance, a calculation for a short DNA strand was reported in the original article, where the $\pi$-stacking interactions are calculated with a different density functional than the intramolecular interactions and hydrogen bonding effects.

As mentioned above, the full mutual polarization between QM and MM regions can be described by using a polarizable force field for the classical part (polarizable embedding, according to Bakowies and Thiel [64]). Interestingly, this approach was already suggested by Warshel and Levitt in one of the first QM/MM articles in the literature [57] and further developed by Luzhkov and Warshel [75, 76]. Thompson and Schenter later presented a more consistent formulation, where the MM region is polarized by the proper QM density instead of derived classical point charges [77, 78]. This framework has since then been referred to as QM/MMpol. The approach was rederived by Curutchet et al. and applied to the description of excitation energy transfer [79]. Furthermore, it was combined with the PCM approach [80], and gradients have been derived for both ground [68, 78] and excited states [81].

### 7.4.3 *Effective Fragment Potential*

The effective fragment potential (EFP) method [82–84] is an approach where certain molecules (e.g., of a solvent) are represented by embedding potentials which are approximated on the basis of distinct physical effects. Moreover, it is a focused approach in QM/MM style, where a central molecule of interest treated with a quantum-chemical method is embedded in an approximate description of the environment.[2] Since the environmental representation is fully polarizable, EFP is often referred to as a QM-derived polarizable force field.

The original model (nowadays referred to as EFP1) [82, 83] was designed specifically for embedding a system of interest in water. Each solvent molecule ("fragment") is represented by an effective potential, which is derived from ab initio calculations. It is important to note that EFP is a rigid-body method, which means that the fragment potentials for different solvent molecules are usually chosen to be identical (except for their positions in space).

In EFP1, the total energy of the supersystem is obtained in the form

$$E^{\mathrm{EFP1}} = \left\langle \Psi \left| \hat{H}^{\mathrm{QM}} + \hat{V}^{\mathrm{Coul}} + \hat{V}^{\mathrm{ind}} + \hat{V}^{\mathrm{rep}} \right| \Psi \right\rangle + E^{\mathrm{Coul}} + E^{\mathrm{ind}} + E^{\mathrm{rep}} \;, \tag{7.24}$$

where $\Psi$ is the wave function of the central molecule and $\hat{H}^{\mathrm{QM}}$ represents its unperturbed Hamiltonian, which is modified by three different potentials stemming from the environment. $\hat{V}^{\mathrm{Coul}}$ describes the static Coulomb embedding due to the fragment potentials, whereas $\hat{V}^{\mathrm{ind}}$ and $\hat{V}^{\mathrm{rep}}$ model induction and exchange repulsion effects, respectively. In terms of intermolecular perturbation theory, these are second-order effects. The additional terms $E^{\mathrm{Coul}}$, $E^{\mathrm{ind}}$, and $E^{\mathrm{rep}}$ represent the respective interactions among the environmental fragments.

The Coulomb contribution is evaluated by placing multipoles (up to octupoles) on specific points within the solvent molecules. While all multipoles directly affect the molecular wave function (through $\hat{V}^{\mathrm{Coul}}$), the evaluation of $E^{\mathrm{Coul}}$ neglects octupole couplings except for monopole–octupole terms (all lower-order terms are coupled to each other). Since the multipole expansion is only valid for large fragment distances [85], damping terms are added to account for short-range charge penetration.

In order to model mutual polarization, the solvent fragments are furnished with anisotropic polarizability tensors at certain locations, which create induced dipoles. These interact with the mentioned (static) multipole moments and with other induced dipoles ($E^{\mathrm{ind}}$), as well as with the quantum region ($\hat{V}^{\mathrm{ind}}$). The polarization contribution is evaluated self-consistently by first iteratively converging the environmental induced dipoles for a frozen density of the QM system, whereafter the wave function or density of the latter is updated and the cycle is repeated until no significant changes occur anymore. The EFP1 exchange repulsion term is described by a highly parametrized form obtained through fitting of several dimer geometries

[2] It is also possible to describe a system on equal footing by treating all constituents as effective fragment potentials.

for the solvent at hand. By placing Gaussians (for solute–solvent interactions) or single exponentials (for solvent–solvent interactions) on the fragment atoms and centers of mass, a convenient description is obtained, albeit at the price of a rather involved fitting process, which can be considered one of the major drawbacks of the EFP1 method.

In 2001, Gordon and co-workers presented the EFP2 method [84], which addresses several problems of the original formulation. Most importantly, the repulsion term is split into exchange and charge transfer, and a dispersion term is newly introduced. All terms are evaluated on the basis of the actual structure at hand instead of a cumbersome parametrization process. The Coulomb and induction terms are determined similarly to EFP1, although much attention was paid to the construction of improved damping functions [85–87]. The exchange repulsion term is now evaluated on the fly, based on overlap and kinetic energy integrals of Localized Molecular Orbitals (LMOs) centered on the individual fragments (the LMOs and corresponding Fock matrices are determined in a preparation run before the actual calculation). Dispersion is included by modeling the interaction between induced dipoles in a typical $R^{-6}$ form. The dispersion coefficients are also determined on the fly by numerical integration over a certain set of distributed polarizability points [88]. The dispersion term is corrected for charge penetration as well. Finally, charge transfer is treated in a pairwise additive fashion on the basis of interactions between occupied valence orbitals and unoccupied orbitals on different fragments [89, 90], which are also determined before the actual run.

This leads to the favorable situation that no involved parametrization process is required to apply the EFP2 method for a specific solvent. Although several quantities are indeed computed beforehand, this is done solely on the basis of a single isolated molecule of the requested solvent. Furthermore, this process has been automatized in the MAKEFP routine implemented in GAMESS [91, 92]. Analytical gradients for both EFP1 [83, 84] and EFP2 [93, 94] are available, and several studies on EFP MD simulations have been published [95, 96]. EFP1 has been coupled to numerous QM methods (for a summary, see Ref. [13]), whereas a general interface for EFP2 is still under active development [97–99]. An extension to fragments bound covalently to the QM region has been reported [100, 101], and the EFP approach has been coupled to the Fragment Molecular Orbital (FMO) method in the framework of the EFMO [102] and the subsequent FIEFMO scheme [103].

Using a similar approach as in the EFP model, Olsen et al. developed the Polarizable Embedding (PE) method in 2010 [104] (not to be confused with the identically named QM/MM type from Ref. [64]). PE can be regarded as a special QM/MM variant, with the particular aim of obtaining various response properties of the QM system under the self-consistent influence of a structured environment. The energy of the QM core under the influence of the environment is given as

$$E^{\mathrm{PE}} = E^{\mathrm{QM}} + E^{\mathrm{es}} + E^{\mathrm{ind}} + E^{\mathrm{LJ}} \ , \tag{7.25}$$

where $E^{\mathrm{QM}}$ is the energy of the isolated QM core, $E^{\mathrm{es}}$ and $E^{\mathrm{ind}}$ give the electrostatic and induction contributions, respectively, and $E^{\mathrm{LJ}}$ covers all other interactions in the form of a classical (density-independent) Lennard-Jones potential. The electrostatic and induction contributions are evaluated as in the EFP method, by placing multipoles (up to octupoles) and anisotropic dipole–dipole polarizability tensors on certain sites of the environment (usually atomic positions and possibly bond midpoints) and self-consistently determining the respective contributions in every SCF step. The multipoles and polarizability tensors are obtained from the LoProp approach [105]. By truncating the multipole expansion at certain orders and applying isotropic or anisotropic polarizabilities, a systematic variety of QM-based embedding potentials for use in the response calculations can be constructed. The salient feature of the PE scheme is the fully self-consistent treatment of many-body environmental response up to arbitrary order (first- and second-order expressions are given in the original article [104], and a third-order form was derived later [106]), allowing for the determination of a wealth of response properties (e.g., vertical excitation energies, static and dynamic (hyper)polarizabilities, and nuclear shielding constants).

The PE formalism was originally derived for HF and DFT, but has since been extended to more sophisticated methods such as Coupled Cluster [107, 108], Multi-Configuration Self-Consistent Field (MCSCF) [109], or Density Matrix Renormalization Group (DMRG) [110] approaches. Gradients have been derived recently [111], and a short review is available [112]. In the framework of the Polarizable Density Embedding (PDE) extension [113], the environment is further partitioned into two regions, with the solvent multipoles replaced by exact densities in the "inner" environment. However, the induction term is still evaluated in the basis of distributed polarizability tensors in both environmental regions. In addition, a short-range non-electrostatic repulsion term is considered in terms of orbital contributions.

## 7.5 Many-Body and Inclusion–Exclusion-Based Methods

In this section, some approaches based on an MBE will be discussed. A common feature of all these methods is the fact that the interactions between certain fragment combinations are described in a fully quantum-mechanical fashion (by carrying out calculations for fragment oligomers or using overlapping fragments), whereas others are neglected or treated in a more approximate manner (in particular long-range electrostatics).

### *7.5.1 Molecular Fractionation with Conjugate Caps*

The Molecular Fractionation with Conjugate Caps (MFCC) method [114] was one of the earliest subsystem approaches based on overlapping subsystems. In contrast to later, more general schemes, the MFCC method was specifically intended for the description of proteins. In particular, the original scheme only aimed at reproducing protein–ligand interaction energies, but it was soon realized that an application to other quantities, such as dipole moments or electrostatic potentials, is also possible [14].

The general idea of the MFCC algorithm is to split a protein into its constituent amino acid residues (or possibly small combinations thereof) by cutting through peptide bonds. The resulting fragments of course feature dangling bonds, which is why they are patched with certain capping groups. In the original MFCC, these are chosen to be whole residues adjacent to the current fragment. In order to saturate any remaining dangling bonds (to further residues past the ones used for capping), hydrogen atoms are attached. It should be noted that the MFCC approach uses rigid structures for all calculations, such that both cap and non-cap atoms have the same position as in the original structure (except for the additional hydrogens which are not present in the actual protein).

Assuming a protein $P$ with the sequence

$$P = A_1 - A_2 - \cdots - A_{N^{\mathrm{AA}}} \ , \tag{7.26}$$

where $A_K$ are the $N^{\mathrm{AA}}$ constituent amino acids, typical capped fragments are formed as $C^{K-1*} - A_K - C^K$, where $C^{K*}$ and $C^K$ represent capping groups on different ends of residue $A_K$. The total binding energy $V^{\mathrm{MFCC}}$ to a ligand $L$ can then be calculated in the MFCC framework by evaluating the expression

$$V^{\mathrm{MFCC}}(P - L) = \sum_{K}^{N^{\mathrm{F}}} V(L - C^{K-1*} - A_K - C^K) - \sum_{M}^{N^{\mathrm{C}}} V(L - C^{M*} - C^M) \ . \tag{7.27}$$

The first sum runs over all combinations of capped protein subsystems with the ligand. Clearly, this includes artificial interactions between the ligand and the attached cap atoms. This is corrected for by the second sum, which contains binding energies of the ligand with pairs of capping groups, called “concaps” by the authors. These are small closed-shell species formed by fusing adjacent capping groups.

As mentioned above, quantities other than the binding energy can be obtained in completely analogous fashion [14]. As an important example, the total density of a protein can be expressed simply as

$$\rho^{\mathrm{MFCC}}(P) = \sum_{K}^{N^{\mathrm{F}}} \rho(L - C^{K-1*} - A_K - C^K) - \sum_{M}^{N^{\mathrm{C}}} \rho(L - C^{M*} - C^M) \ , \tag{7.28}$$

which also allows to obtain total energies by numerical integration [115], although it must be mentioned that the total kinetic energy of the electrons is approximated by simply combining the individual fragment contributions in MFCC style. The MFCC approach was put on a more general basis by Chen et al., who reformulated it in terms of density matrices [116]. In the regular MFCC spirit, each element of the total density matrix is constructed as a sum over all corresponding elements obtained for the fragments, with the cap contributions being subtracted. Naturally, many contributions vanish since most atoms are present only in a few fragments. Special care has to be taken with regard to the capping atoms not present in the supermolecule. The respective matrix elements can either be neglected (MFCC-SDM) or accounted for by employing ghost atoms (MFCC-GDM). While the former strategy is cheaper, it does not conserve the total number of electrons.

Even though not explicitly stated by the authors, the MFCC scheme works based on the IEP. From an embedding point of view, the choice of the capping groups as parts of the preceding/succeeding sequence is very important, since it represents the only influence of neighboring fragments onto each other.[3] This is also why we classify MFCC as an IEP-based approach, rather than a DC scheme. However, the fragment overlaps are created solely along the protein sequence. The original scheme thus lacks a number of effects, most importantly the following:

1. Treatment of disulfide bridges
2. Interactions between residues which are spatially close, but far from each other in the sequence
3. Long-range Coulomb interactions between remote parts of the protein
4. Mutual polarization among different fragments

All of these were addressed in different ways over the last decade. First of all, Chen et al. demonstrated that proteins with disulfide bonds can be treated essentially in the same way as regular ones, although different capping groups should be used (usually S-$CH_3$, yielding dimethyl disulfide as the concap) [119]. Non-bonded interactions between spatially close residues are incorporated in a straightforward MBE spirit in the Energy-Corrected MFCC (EC-MFCC) approach [120]. This is achieved by additionally constructing pairs of such residues (according to a user-specified cut-off radius) and determining a second-order energy correction according to Eq. (7.3). In contrast to bonded pairs, both fragments are only capped with hydrogens instead of the usual larger caps. Since proper quantum-chemical calculations are carried out for the dimers, all types of interaction are included, albeit only at the two-body level. No improved density was computed in this way (although theoretically possible), but EC-MFCC geometry optimizations are still enabled by corresponding energy gradient expressions. In the so-called MFCC-DM-PIC (Pairwise Interaction Correction) scheme [121], however, a similar correction is applied to the MFCC-DM formalism, which allows to obtain an improved density matrix for the total system.

Long-range Coulomb interactions were first treated in the context of the Electrostatic Field-Adapted MFCC (EFA-MFCC) approach by Jiang and co-workers [122]. In this method, the Coulomb interaction with charged groups in the system is included by determining the electronic structure of the individual capped fragments and concaps under the influence of point charges. These are placed on certain charge centers, such as the nitrogen atom of the ammonium group. However, only actually cationic or anionic groups are considered, whereas interactions with polar groups are still treated as in the original MFCC scheme. In the Generalized MFCC (GMFCC) scheme [123], an approach similar to EC-MFCC is pursued. In particular, quantum-chemical interaction energies between pairs of spatially close residues are calculated

[3] It should be noted that later methods based on the MFCC approach often employ an additional embedding potential, thus allowing for smaller capping groups (see, e.g., Refs. [117, 118]).

and added to the total energy. In addition, however, the interaction between distant pairs is evaluated in MM style by applying Coulomb's law to suitable partial charges as well as a Lennard-Jones potential with corresponding parameters from a force field. Again, only a correction to the total energy is obtained in this way. Some years later, Wang et al. presented the Electrostatically Embedded GMFCC (EE-GMFCC) scheme [124], where a direct Coulomb interaction is now included already on the level of the individual fragment calculations as in the EFA-MFCC scheme, but with atomistic point charges. In this way, the supermolecular density also includes polarization due to the presence of non-neighboring fragments. In addition, both the regular MFCC approach and the EE-GMFCC scheme have been combined with the C-PCM solvent model [125, 126] and the latter has been applied to geometry optimizations and the determination of vibrational frequencies [127] as well as ab initio MD simulations [128].

Finally, it should be noted that even with full electrostatic embedding, the mutual polarization among the fragments is not necessarily taken into account. Ji and co-workers addressed this problem by introducing so-called Polarized Protein-Specific Charges (PPCs) in EE-GMFCC calculations [129], which are determined self-consistently. However, only the Coulomb interaction is recovered in this way, whereas non-classical contributions are missing. An additional problem concerns the subtraction procedure present in all MFCC variants. When a total density is sought for, it is well possible that unphysical areas of negative density occur. Both of these problems are solved in the framework of the 3-FDE approach, which will be discussed in Sect. 7.6.3.

The Generalized Energy-Based Fragmentation (GEBF) formalism was presented by Li et al. in 2007 [12] and is closely related to the MFCC approach. Unlike MFCC, it is not specifically intended for the description of proteins, but of extended systems in general. In the GEBF scheme, the supersystem is partitioned in a general, albeit more complicated way. A detailed description (including an instructive example) can be found in the original article, but the most important points are as follows:

- First, the total structure is split into "fragments," which can be as small as individual functional groups or as large as small molecules (e.g., amino acid residues).
- Subsequently, overlapping "primitive subsystems" are constructed for each fragment by including all nearby fragments which are either covalently bound to the current fragment or have a hydrogen bond to it (alternatively, a distance criterion $\zeta_1$ can be used). Any remaining dangling bonds are capped with hydrogen atoms.
- Primitive subsystems which are identical to or completely contained in any other primitive subsystem are eliminated.
- Possible overcounting of interactions is avoided by using the IEP to construct "derivative subsystems" from all parts which are contained in more than one of the remaining primitive subsystems (i.e., overlap between primitive subsystems). This process is started from the largest possible fragments and repeated for ever smaller fragments until overcounting is eliminated.

The GEBF energy expression has the simple form

$$E = \sum_{K}^{N^{\mathrm{F}}} C_K E_K \ , \tag{7.29}$$

where $E_K$ are the energies of primitive or derivative subsystems and $C_K$ are the corresponding summation coefficients. While primitive subsystems have a positive coefficient (usually +1), derivative subsystems enter with a negative coefficient to cancel the respective overcounting. The magnitude of the negative coefficient depends on the number of times the current interaction would otherwise be overcounted. Note that the GEBF fragmentation would reduce to the MFCC one in case of a linear protein chain, where each primitive subsystem includes a single amino acid capped by its two neighboring residues (and additional hydrogens).

In addition to the general approach outlined above, two-body interactions between residues which are further apart than the primary cutoff radius $\zeta_1$, but less than a larger radius $\zeta_2$ are included in the same manner as in the EC-MFCC approach. Furthermore, a point charge background is used to represent the remainder of the total system already in each subsystem calculation, just like in the later proposed EE-GMFCC approach.

Geometry optimizations and the calculation of vibrational frequencies in the GEBF framework are possible through analytical derivatives [130], and the fragmentation procedure has later been improved and completely automatized by Hua et al. [131, 132]. Use of GEBF as the QM method in a QM/MM scheme has been reported under the name "Fragmentation QM/MM" [133]. Very recently, an extension to periodic boundary conditions was reported [134], and a review on the GEBF method is available [135].

### 7.5.2 *Fragment Molecular Orbital Method*

The Fragment Molecular Orbital (FMO) approach was first established by Kitaura et al. in 1999 [136], based on their earlier Pair Interaction Molecular Orbital (PIMO) method [137]. It can be regarded as a truncated MBE, where the electrostatic interactions between different fragments are taken into account self-consistently up to the full $n$th-order expression, while all other interactions are treated on a much lower order (usually two- or three-body expansion).

The FMO approach is based on disjoint fragments, without resorting to the use of capping groups. If covalent bonds have to be split, this is done in a heterolytic fashion, i.e., assigning both electrons of a bond to one of the fragments. The atom which obtains the bonding electrons is referred to as bond-attached atom (BAA), whereas the other binding partner is termed bond-detached atom (BDA). An important point concerns the proper division of basis functions among the fragments. To ensure that the BDA does not generate density contributions in the bonding region, the corresponding hybrid orbital is projected out either by a suitable projection operator [138] or Fock matrix transformation [139]. Conversely, certain orbitals of the BDA are included in the BAA-containing fragment to describe the detached bond.

The choice of fragmentation is in general left to the user, although reasonable defaults have been supplied for standard situations, such as the division of proteins [140, 141]. The workflow of an FMO calculation then proceeds as follows:

1. The electronic structures of all isolated fragments are determined.
2. An ESP contribution due to each of the fragments is extracted from the results.
3. Additional monomer calculations are set up, which include the ESP stemming from all other fragments in their Hamiltonian. Since every such calculation produces a new fragment ESP, the monomers are iterated to self-consistency.
4. Pairs of monomers are combined to dimers, which are subjected to similar calculations. However, the background field is fixed and not updated anymore (single calculation per dimer)
5. If required, higher-order combinations (trimers, tetramers, etc.) can be constructed and treated in the same way as the dimers.

The FMO total energy is then determined based on the calculated energies according to Eqs. 7.3 and 7.4 or corresponding higher-order expansions (currently, expansions up to third [142] and fourth orders [143] have been tested). Since the ESP for the monomer calculations is iterated until convergence, the electrostatic interaction is treated to the full order, while the non-electrostatic contributions are of course only contained up to the chosen order. The physical justification for this approach is the relatively short range of the latter. The incorporation of the Coulomb field into the QM calculations ensures proper coupling of the different kinds of interactions.

There are many different ways to treat the environmental ESP. Originally, the exact ESP was used by evaluating two-electron contributions for the electronic part. Since this was deemed too time-consuming, it was proposed to use either Mulliken populations or derived partial charges to approximate the ESP [144, 145]. In addition, different approximations can be used based on the inter-fragment distance, although this requires a slight reformulation of the energy expression to account for dimers where the constituents are treated on an unequal footing [144, 146]. For far-separated monomers, dimer calculations can even be forfeited and replaced solely by the electrostatic interaction [144]. In principle, an orbital-based exchange term could be included in the embedding potential for the individual fragments. However, this is discouraged on the grounds of imbalancing an otherwise favorable error compensation [147].

In order to obtain a supermolecular density and derived properties, several variants of the FMO method are available. For instance, in the FMO-MO approach [148], a supermolecular density is calculated from an MBE and used to construct a total Fock matrix, from which supermolecular orbitals can be derived. The FMO/F and FMO/FX variants proposed by Fedorov and Kitaura accomplish this without calculating the total density, directly from fragment Fock matrices [147].

Following the initial implementation for HF wave functions, many important classes of QM methods have been interfaced with the FMO scheme, e.g., DFT [149],

MP2 [150], Coupled Cluster [151], and MCSCF [152]. Recently, a combination with the semiempirical Density Functional Tight Binding (DFTB) method was also reported [153, 154]. In addition, a multilayer formalism in QM/QM style was presented [140], and based on nearly [155] or fully analytical gradients [156], several applications to geometry optimizations [157] and MD simulations [158, 159] have been reported. Furthermore, a combination with the PCM approach [160], an extension to periodic boundary conditions [161], and several excited-state treatments were published [162, 163]. For more information, the reader is referred to several exhaustive reviews [13, 164, 165] and a book about the method [141].

The FMO scheme has been combined with the EFP approach in two ways. Nagata et al. employed FMO as the QM method in an EFP1 calculation and gave a detailed description of the interactions between the two regions [166]. Steinmann et al. presented the Effective Fragment Molecular Orbital (EFMO) method [102], where the self-consistency cycle for the molecular ESP is avoided by applying the EFP polarizability model, and long-range electrostatics are described by a multipole expansion, just like in the EFP scheme. In its successor, the Fully Integrated EFMO (FIEFMO) method [103], exchange repulsion, dispersion, and charge transfer are also included in EFP style instead of merely through low-order many-body terms.

There are several methods which are related to the FMO approach. The Fast Electron Correlation (FEC) method by Hirata et al. employs the same functional principle, but approximates the electrostatic interaction solely by dipole–dipole terms [167] or even by atom-centered point charges [168]. The Electrostatically Embedded Many-Body (EE-MB) scheme [169] is also very similar to FMO, but uses a constant charge background (thus avoiding iterative optimization). In the Hybrid Many-Body Interaction (HMBI) scheme [170], higher-order terms are approximated by a polarizable force field with a special focus on applications for periodic systems. In particular, a non-periodic method for short-range interactions (e.g., within a unit cell) can be combined with a periodic treatment of all long-range interactions.

### *7.5.3 Molecular Tailoring Approach*

After the initial proposal of the Molecular Tailoring Approach (MTA) in 1994 [171], development lay silent for almost a decade until a much improved version and an automatic fragmentation scheme were presented by Babu et al. [172]. Similar to DC schemes (see Sect. 7.6.1), the total density matrix of the system is constructed ("tailored") from fragment contributions in the original approach. For each element of the total density matrix, one of many overlapping fragments is identified where the two corresponding atoms (or single atom in case of intraatomic elements) are embedded in the best representation of their molecular environment (as defined in the following). In the original implementation [172], this is evaluated by counting the minimum number of bonds to be traced until the fragment edge is encountered. A refined version was reported under the name Cardinality-Guided MTA (CG-MTA) in 2006 [173]. The sophisticated automatic fragmentation algorithm will be detailed in the following. The central quantity is the "R-goodness" parameter $R_g$, which is used on several levels. The atomic R-goodness $R_g^K(A)$ of atom $A$ in a certain fragment $K$ is defined as the maximum radius of a sphere centered on that atom, such that the sphere only includes other atoms belonging to $K$. In this way, $R_g^K(A)$ describes "how embedded" an atom is in a given fragment. Each atom will be represented by the fragment in which it has the largest R-goodness. The general R-goodness of atom $A$, $R_g(A)$, is then simply determined as the maximum of all obtained values for that atom, i.e., the value obtained for the representing fragment. Finally, the total R-goodness of a given fragmentation scheme, $R_g$, is defined as the minimum of all atomic values, giving the minimum radius of the embedding molecular environment around each atom.

The fragmentation algorithm requires two parameters, a minimum R-goodness $R_g^{\min}$ and a maximum fragment size in terms of atoms per fragment. First, certain moieties which should not be fragmented, such as aromatic rings, are identified. A large set of overlapping fragments is then created by centering a sphere of radius $R_g^{\min}$ on each heavy atom in the system. This ensures that every such atom will have at least an R-goodness of $R_g^{\min}$ in its parent fragment. A complicated set of rules is used to merge adjacent fragments, depending on their overlap. The given maximum fragment size determines the end of the merging process. Any dangling bonds are then capped with hydrogen atoms. The electronic structure now has to be determined for the resulting subsystems as well as for additional fragments created from the overlap between the basic subsystems. Instead of constructing the total density matrix as in the original approach, the energy of the system can be evaluated directly, based on the IEP (see Eq. (7.1)). Here, all overlaps of even order (e.g., between two or four fragments) are subtracted, whereas all odd-order terms (e.g., ternary intersections) are subtracted. One of the major benefits of the refined version is its treatment of bonded and non-bonded effects on the same footing, since the atom-centered spheres are bond-independent. The set-theoretical evaluation can be extended to other properties [174]. It should be noted that without any additional embedding components, the total interaction is only described through the molecular overlap between the fragments.

Gradients and other derivatives have been derived for the CG-MTA [173], and the method has been combined with several quantum-chemical methods [175, 176]. In addition, it was noticed that the MTA error in the correlation energy (as opposed to the HF energy) is approximately independent of the basis set. Thus, a "grafting correction" was introduced, where this error is determined with a smaller basis set and used to correct the MTA correlation energy in a calculation with a large basis set [177]. This concept was soon extended to the estimation of the total error at the level of the small basis set [178], and a variant which utilizes different fragment sizes to completely avoid supermolecular calculations was also reported [179]. Recently, CCSD(T) was employed in the context of grafting corrections [180].

In regular (CG-)MTA, each fragment is only embedded within directly overlapping fragments. Although the embedding potential has the correct quantum-chemical form, it usually does not represent the full environment, but merely a comparatively small region, which can lead to problems for charged or very polar systems. Ganesh et al. already suggested to create additional fragments by combining charged moieties and include them in the MTA expansion [173]. In order to account for all long-range Coulomb effects, the Electrostatically Embedded MTA (EE-MTA) approach [181] was conceived by the Truhlar group. Here, all quantum-chemical calculations are performed in an electrostatic field created by point charges representing the rest of the system (as in FMO or EE-GMFCC), which was found to clearly improve the description of polar and charged systems. The overcounting of Coulomb interactions is taken care of by a suitable energy expression. Furthermore, the use of a tuned fluorine atom was put forward in this work as an alternative to regular hydrogen caps. Many details about the MTA, especially concerning the current implementation, can be found in a recent book chapter [182], and a review is available as well [174].

### *7.5.4 Kernel Energy Method*

The Kernel Energy Method (KEM) was first reported by Huang, Massa, and Karle in 2005 [183] and specifically aims at the description of biological entities. The total system is decomposed into fragments ("kernels"), which are capped with hydrogen atoms and subjected to isolated calculations. Subsequently, pairs of these kernels are formed and additional calculations are performed. The total energy is simply obtained as

$$E^{\mathrm{KEM}} = \sum_{K}^{N^{\mathrm{F}}-1} \left( \sum_{K=1}^{N^{\mathrm{F}}-m} \sum_{L=K+m}^{N^{\mathrm{F}}-m} E_{KL} \right) - (N^{\mathrm{F}} - 2) \sum_{K=1}^{N^{\mathrm{F}}} E_K \ , \tag{7.30}$$

which has been shown to be equivalent to the regular two-body expansion in Eq. (7.3) [15]. The effect of the cap atoms is expected to cancel due to inclusion in both single and double kernels. In principle, all double kernels can be included in the calculation, although this might be very expensive for large systems. In the original publication, the double kernels were therefore restricted to bonded pairs. In applications to DNA, the important hydrogen bonds between base pairs were added by including the corresponding double kernels [184]. Straightforward extension to higher orders is possible, and tests including up to fourth-order terms have been reported [185]. In 2010, a generalized scheme based on earlier work by Deev and Collins [186, 187] was introduced, where only those combinations of kernels are taken into account which are connected in the molecular graph [188].

From an embedding point of view, the interaction between different kernels is taken into account only at the $N$-body level (where $N$ is the order of the expansion), and no additional terms are added. The KEM can thus be likened to the FMO approach without inclusion of the environmental ESP in the quantum-chemical calculations. To our knowledge, no particular automatic fragmentation algorithm has been reported for the KEM. However, for the treatment of graphene flakes, a "fissioning" process was proposed in order to cut through aromatic bonds [189]. In this approach, an aromatic bond is cut along the bond instead of across it, yielding two single bonds which are capped with hydrogen atoms.

The original tests were based on HF/STO-3G results, but the method was soon shown to be applicable in combination with a variety of methods and basis sets [190]. Gradients have not yet been derived, although a derivation according to the MBE should be straightforward. Recently, the KEM was used to determine Atoms-in-Molecules (AIM) charges for a collection of large molecules [191].

#### 7.5.4.1 Other Methods

Independent of each other, the Generalized Many-Body Expansion (GMBE) by Richard and Herbert [17] and the Many-Overlapping-Body Expansion (MOBE) by Mayhall and Raghavachari [16] were put forward in 2012. Both approaches attempt to unify methods based on the IEP (overlapping fragments) with those based on an MBE (non-overlapping fragments) in a common framework. The resulting energy expressions are very similar, although not quite identical. Furthermore, a somewhat different fragmentation scheme is used, although this is not an essential ingredient of either method [3]. A detailed comparison of the two schemes was carried out later by Richard and Herbert [192], who confirmed that the two schemes are not exactly equivalent, although very similar numerical results are obtained for many cases.

The Systematic Molecular Fragmentation (SMF) algorithm by Deev and Collins [186] is very similar to the original MFCC approach, where whole neighboring residues were used as caps. The SMF scheme is also based on creating overlapping fragments and combining their individual energies or properties according to the IEP. However, it is formulated in a more general way and employs functional groups as the smallest non-breakable units (as opposed to amino acid residues). A hierarchy of fragmentation schemes can be generated by specifying a desired "fragmentation level," which corresponds to the number of overlapping groups between adjacent fragments (and thus to the fragment size). By increasing the fragmentation level, the supermolecular result is systematically approached. Non-bonded interactions are treated by including additional two- or three-body correction terms. To avoid excessive calculations for far-separated fragments, long-range electrostatic interactions have also been modeled by a multipole expansion [187] or effective fragment potentials [193]. The fragmentation scheme has been refined in the Systematic Molecular Fragmentation by Annihilation (SMFA) approach [194], among other new features allowing for automatic ring fragmentation. The SMF and SMFA methods have been reviewed and compared to the Combined Fragmentation Method [195] in a recent article [196].

After reviewing and comparing several existing fragmentation methods, Suárez et al. presented the Fragment Energy Method (FEM) in Ref. [15], which can be regarded as a generalization of the MFCC scheme and a special case of the SMF approach. The Multilevel Fragment-Based Approach (MFBA) [197] works in a similar way, but different theoretical methods are used to determine fragment interactions based on a distance criterion. Gradients have been derived, and the method has been deemed especially suitable for fragment-based geometry optimizations. The same strategy was pursued by Mayhall and Raghavachari with their very general Molecules-in-Molecules (MIM) scheme [198]. The original formulation combines a GEBF-like fragmentation scheme with the multilevel treatment of ONIOM approaches. However, instead of treating a central fragment of interest with a more sophisticated model than its surroundings, one or more cutoff criteria are introduced to determine interaction terms in different ways. While many existing ideas were incorporated in the MIM approach, these have been combined to a conglomerate of remarkable generality, which can treat a variety of different situations (e.g., overlapping/non-overlapping fragments, different fragmentation schemes, one or several different theoretical treatments) within a common framework. Gradients for the MIM approach are available [199], and it has recently been applied to the calculation of vibrational, circular dichroism and Raman spectra [200–202].

## 7.6 Quantum-Chemical Divide-and-Conquer Methods

In this section, several methods originating from a mathematical decomposition of the supermolecular solution are presented. All approaches are at least related to the DC scheme, although the nature of the embedding can be very different. A major distinction from the methods presented so far is the fact that hardly any embedding components are included based on empirical considerations (except for those arising naturally, such as the Coulomb contribution of the subsystem DFT embedding potential).

### *7.6.1 Divide and Conquer*

The Divide-and-Conquer (DC) method was originally formulated by Yang in terms of DFT [203]. Very similar to subsystem DFT (see Sect. 7.3), the additivity of the electron density is exploited to partition the total density into fragment contributions. To this end, a positive partition function $p^K(\vec{r})$ for each subsystem is introduced under the requirement that

$$\sum_K^{N^{\mathrm{F}}} p^K(\vec{r}) = 1 \; \forall \vec{r} \;, \tag{7.31}$$

i.e., the values of the individual subsystem partition functions must add up to unity at each particular point in space. A subsystem density $\rho^K(\vec{r})$ can then be expressed in terms of the total density $\rho(\vec{r})$ as

$$\rho^K(\vec{r}) = p^K(\vec{r})\rho(\vec{r}) \tag{7.32}$$

and it thus follows that

$$\rho(\vec{r}) = \sum_{K}^{N^{\mathrm{F}}} \rho^{K}(\vec{r}) \ . \tag{7.33}$$

Clearly, the value of the partition functions will be large within and close to their corresponding subsystem, while it will be small far away from it. An example for a smooth partition function was given in the original article in terms of spherical atomic densities [203]. The crucial point of the DC formalism is to express the individual subsystem densities through localized subsystem orbitals instead of supersystem functions. This is accomplished by rewriting these densities as

$$\rho^{K}(\vec{r}) = 2p^{K}(\vec{r}) \sum_{m}^{N^{\mathrm{orb}}} f_{\beta}\left(\varepsilon_{\mathrm{F}} - \varepsilon_{m}^{K}\right) \left|\psi_{m}^{K}(\vec{r})\right|^{2} \ , \tag{7.34}$$

where $f_{\beta}(x) = [1 + \exp(-\beta x)]^{-1}$ is a smooth distribution function of width $\beta$, $\varepsilon_{\mathrm{F}}$ is the Fermi energy, and $\psi_{m}^{K}$ are *local* eigenfunctions of a subsystem Hamiltonian obtained through projection of the total KS Hamiltonian on a basis set unique to subsystem $K$ (thus, $\psi_{m}^{K}$ are localized on the latter). The Fermi energy can be obtained from the normalization condition to the total number of electrons $N^{\mathrm{el}}$,

$$N^{\mathrm{el}} = 2 \sum_{K} \sum_{m} f_{\beta}\left(\varepsilon_{\mathrm{F}} - \varepsilon_{m}^{K}\right) \left\langle \psi_{m}^{K} \left| p^{K}(\vec{r}) \right| \psi_{m}^{K} \right\rangle \ . \tag{7.35}$$

It should be noted that only the total number of electrons is fixed, but not the subsystem values. The density-based DC strategy can be summarized as follows: First, a local basis set has to be chosen for each subsystem. For instance, a straightforward choice would be the basis functions centered on a certain subset of nuclei. Furthermore, a form for the partition functions must be selected. Subsequently, the total Hamiltonian has to be projected on the local basis sets, generating a set of subsystem eigenfunctions (orbitals) and eigenvalues (orbital energies). The Fermi energy can then be determined from Eq. (7.35), which allows to determine subsystem densities from Eq. (7.34) and thus a total density. This process must be iterated to self-consistency (until identical Fermi energies for the individual subsystems are obtained), since the total Hamiltonian depends on all fragment densities and thus couples them. For a more detailed derivation, the reader is referred to the original article [203] as well as an excellent review [2].

In order to avoid the evaluation of three-dimensional integrals over the partition functions and to generalize the DC approach to wave function methods, Yang and Lee soon introduced a reformulation where the total density matrix is partitioned instead of the spatial electron density distribution [204]. The density matrix **D** is defined as (for simplicity, we consider a single-determinant closed-shell case here)

$$D_{ij} = 2 \sum_{m}^{N^{\rm orb}} C_{im} C_{jm} \ , \tag{7.36}$$

where $C_{im}$ and $C_{jm}$ are the coefficients for atomic orbitals $i$ and $j$ in the LCAO expansion for occupied molecular orbital $m$. The spatial partition functions of the original scheme are replaced by corresponding partition matrices $\mathbf{P}^K$ with the requirement that

$$\sum_K P_{ij}^K = 1 \ , \tag{7.37}$$

such that the subsystem contributions for each matrix element of $\mathbf{P}$ add up to 1. The original proposal for the form of the partition matrix was simply

$$P_{ij}^K = \begin{cases} 1 & \text{if } i \in K \text{ and } j \in K \\ 0.5 & \text{if } i \in K \text{ and } j \notin K \\ 0 & \text{if } i \notin K \text{ and } j \notin K \end{cases} \ , \tag{7.38}$$

which guarantees the above normalization condition. All further expressions correspond directly to the density-based formulation. In particular, the subsystem density matrix contribution is given as

$$D_{ij}^K = 2P_{ij}^K \sum_m f_\beta \left(\varepsilon_{\rm F} - \varepsilon_m^K\right) C_{im}^K C_{jm}^K \tag{7.39}$$

and the total density matrix can be obtained by simply summing over all subsystem contributions.

As mentioned above, a simple and chemically intuitive choice for the local basis sets would consist of the functions centered on a certain subset of nuclei. In order to improve the description, basis functions of adjacent nuclei can be included in the subsystem description, leading to the concept of "buffer regions" [205]. In its simplest form, all basis functions located within a certain cutoff distance of the current fragment are added to the subsystem basis. Buffer regions are central to the embedding aspect of DC techniques, since it is through this overlap in terms of basis functions that the individual subsystem Hamiltonians are coupled. The subsystems are usually defined by "core regions," which are mutually exclusive, and buffer regions, which are allowed to overlap with the core and/or buffer regions of adjacent fragments. In particular, the core of one fragment could play the role of buffer region for another fragment [2]. Several advanced partition matrix prescriptions utilizing multiple buffer regions have been put forward by Dixon and Merz [206, 207], where the density matrix elements corresponding to combinations of buffer functions are set to zero. This allows to improve the SCF results for a fragment, while ensuring that its density matrix contributions exclusively stem from the primary functions of that fragment. Lee et al. demonstrated that upon growing the total system but maintaining the size of the subsystems and their buffer regions, linear scaling can be achieved [208].

Some related methods, which can be regarded as non-self-consistent DC variants, were proposed by the Mezey group. In the Molecular Electron Density Lego Approach (MEDLA) [209], total densities are obtained by simply combining densities from a fragment library in an additive fashion. The fragment densities are obtained by carrying out calculations for small analogous molecules (e.g., $CH_4$ for an $sp^3$ carbon) and extracting a partial density by considering only certain density matrix contributions according to Eq. (7.38). The Adjustable Local Density Assembler Method (ALDA) [210, 211] used the same strategy, but instead of fragment densities on a grid, only the fragment density matrices are stored and evaluated on the fly, which also allows for small geometric modifications. Finally, the Adjustable Density Matrix Assembler (ADMA) approach [210, 211] is used to directly construct a supermolecular density matrix from subsystem matrices, as in other DC schemes. This approach has also been combined with a classical treatment of long-range electrostatics [212].

As mentioned above, the density matrix formalism allows to apply the DC strategy to a variety of electronic-structure methods. For instance, a combination with semiempirical methods was explored [206, 207] and Li et al. carried out first tests for correlated methods [213]. Subsequently, interfaces to HF/hybrid DFT [214], MP2 [215, 216], and Coupled Cluster [217, 218] approaches have been reported. Kobayashi et al. exploited the multilevel nature of the buffer region concept by using differently sized buffer regions (and thus basis functions) for the HF and correlation part of post-HF calculations [217, 219]. Song and co-workers separately optimized the exchange contributions for different fragments based on CCSD(T) calculations in a DC framework [220]. Guidon et al. introduced the idea of evaluating HF exchange in a smaller basis through introduction of an auxiliary density matrix, which is then corrected by a suitable density functional [221]. Recent progress on the DC family of methods includes the extension to time-dependent calculations [222, 223], a GPU implementation [224], and several developments with respect to AIMD simulations [225, 226]. Furthermore, Fornace and co-workers presented the Embedded Mean-Field Theory (EMFT) approach [227], which is also based on a division of the total density matrix into blocks, facilitating the representation of different parts of the system using different theoretical approaches. Unfortunately, no comprehensive review on DC methods seems to be available, but many useful details can be found in two perspectives [228, 229] and in Ref. [2].

### 7.6.2 *Density Matrix Embedding Theory*

Density Matrix Embedding Theory (DMET) is a very recent approach proposed by Knizia and Chan [230, 231] to treat strong embedding effects, such as between two fragments linked by a covalent bond. It is related to the earlier Dynamical Mean-Field Theory (DMFT), which employs the many-body Green's function as the central object (reviews on DMFT can be found, e.g., in Refs. [11, 232, 233]). One notable recent advancement in the context of Green's function embedding was the

Self-Energy Embedding Theory (SEET) presented by the Zgid group [234–236]. In contrast to both of these methods, DMET employs the density matrix as the partitioned object.

The central idea of DMET is to represent the total wave function of a system by a Schmidt decomposition into a linear combination of states of an embedded open quantum system (fragment) $F$ and states of the environment (bath) $B$, i.e.,

$$|\Psi\rangle = \sum_{i}^{N^{\text{fs}}} \lambda_i |F_i\rangle |B_i\rangle \ , \tag{7.40}$$

where $\lambda_i$ is a coefficient, $|\Psi\rangle$ is the total wave function, and $|F_i\rangle$ and $|B_i\rangle$ are states local to fragment or environment, respectively. It can be shown [230, 231] that the number of effective bath states $N^{\text{bs}}$ required to recover the exact solution as described above is at most the same as the number of fragment states $N^{\text{fs}}$.

This approach yields the exact wave function, but already requires it for the determination of effective bath states. In practice, solutions for individual fragments are therefore determined with a sophisticated correlation method, while the bath states are approximated by a cheap supermolecular solution (e.g., a single Slater determinant). An "embedding basis" of effective bath states that interact with the fragment and its embedding Hamiltonian is determined by several projection procedures onto the fragment states. This yields $N^{\text{fs}}$ interacting ("entangled") bath states and a number of non-interacting ("pure") bath states. Only the former need to be taken into account to express the effect of the bath on the active fragment, while the latter merely form a multiplicative "core determinant."

In order to ensure consistency between the different electronic-structure methods, the density matrix difference between the (correlated) fragment solutions and the (mean-field) supersystem solution should be minimal. This is achieved by introducing additional embedding operators $u_K$ for all fragments, which are added to the fragment Hamiltonians as well as to the supersystem Hamiltonian. As summarized in Ref. [231], the whole DMET self-consistency procedure thus takes the following form:

1. Determine the (mean-field) supersystem solution.
2. Construct an embedding basis and a fragment Hamiltonian for each fragment by projection and solve its electronic structure with a correlated method.
3. Construct embedding operators for each fragment to minimize the difference between the low- and high-level density matrices.
4. Repeat above steps with the embedding operators added to each Hamiltonian (except the fragment's own) until self-consistent operators are obtained.

DMET has been applied to physical model systems such as a 2D Hubbard model [230, 237] as well as to first chemical problems like molecular chains and rings [231, 238], solids [239], and even a chemical reaction [238]. A short review including a comparison between density-based embedding, DMFT, and DMET is available [11].

An interesting variant was recently published under the name "Bootstrap Embedding" [240]. The motivation was to improve the slow convergence of the DMET solution with respect to the fragment size, which was shown to depend on the volume-to-surface ratio. In contrast to regular DMET, the system is partitioned in multiple ways, such that each atom is close to the fragment boundary in some partitionings and buried deeper within the fragment in others. Instead of trying to match the environment density matrix of the mean-field description to the one obtained from a correlated description of the fragment (which is bound to fail due to the restriction to a single Slater determinant), the "edge" region of one fragment is matched to the "core" region of a corresponding fragment from a different partitioning scheme. In this way, the total density matrix is improved systematically. Since wave functions of the same type (e.g., Full Configuration Interaction (FCI) in the original article) are matched to each other, the condition can be fulfilled exactly, which decreases the influence of the nature of the bath. In addition, the choice of partitioning scheme clearly becomes less important, since multiple different types are constructed and made consistent with each other. Bootstrap Embedding has only been applied to lattice models so far, but could theoretically be extended to molecular systems, even though a division in terms of atomic orbitals might not be straightforward [240].

### *7.6.3 Frozen Density Embedding*

Frozen Density Embedding (FDE) [21] is one particularly interesting variant of subsystem DFT, which has been introduced in Sect. 7.3. FDE can be regarded as the special case of solving the KSCEDs for a single ("active") subsystem while keeping all environmental densities frozen. The active density is thus relaxed in the embedding potential generated by the frozen density. On the one hand, FDE could thus be used as a focused approach, with the environmental density obtained from a different method or from model considerations (see, e.g., Ref. [241]). On the other hand, the subsystem DFT solution can be recovered by successively interchanging the roles of active and frozen density (or densities) and running additional calculations. When this process is carried out until self-consistency is reached, an iterative (but equivalent) solution to the subsystem DFT problem is obtained. In the literature, these relaxation cycles are often referred to as "Freeze-and-Thaw" (FT) cycles [20]. Physically speaking, the FT cycles introduce the inductive component of the intermolecular interaction between the subsystems (see Ref. [22]), although this interpretation formally has to be taken with a grain of salt (see the discussions in, e.g., Refs. [242, 243]).

Certain approximations are usually made in practical FDE calculations. The first is to introduce approximate forms for the XC potential (fragment contributions as well as nonadditive correction) and the nonadditive kinetic energy. This approach is necessitated by the fact that exact expressions are not known. Since the total density is available in each step of an FDE calculation, all common density-dependent forms can be used to evaluate the XC contributions (e.g., the Local Density Approximation (LDA) or functionals derived from the Generalized Gradient Approximation (GGA)). Hybrid functionals, however, cannot be used directly, since the evaluation of their orbital-dependent contribution would require orbitals for the supersystem [244–246]. Optimized Effective Potential (OEP) methods [247] could offer a solution here.

For the same reason, it is difficult to obtain a suitable form for the nonadditive kinetic energy. Purely density-dependent forms can be employed for this contribution as well, but this is a much more severe approximation, as it is known that density functionals for the kinetic energy, such as the Thomas–Fermi functional [248, 249], only work for limiting cases. However, since in practice the nonadditive kinetic term is often relatively small, they can still be viable in many situations. In particular, for weakly bound complexes, good results have been obtained [20, 250, 251] with the PW91k functional [252], whereas the interactions between covalently connected subsystems are described qualitatively wrong [253, 254]. In these cases, a better description could be obtained by potential reconstruction approaches [253, 255, 256], although these require a preliminary supermolecular calculation. A comprehensive discussion of different approaches for constructing nonadditive kinetic potentials can be found in Refs. [6, 10].

It has been shown that the kinetic energy contribution vanishes if the subsystems' orbitals are orthogonal to each other [257, 258], although this is not a strict requirement to recover the exact KS solution [259]. If such an orthogonalization is desired, it can be carried out by employing the projection technique introduced by Manby et al. [260]. In this approach, localized orbitals are determined for different fragments, which allows to use a simple projection operator to orthogonalize environmental fragment orbitals, eliminating the nonadditive kinetic energy. Since a full calculation of the supersystem is required to obtain localized orbitals, the approach is mainly suitable for WFT-in-DFT embedding (see below). An alternative is the use of additional Lagrangian multipliers to enforce intersubsystem orbital orthogonality [257].

A further simplification is to employ different subsystem basis sets for the individual fragment calculations. Clearly, this reduces the computational effort tremendously, since most matrix operations during the solution of the SCF problem scale with the number of basis functions. However, one important consequence is that charge transfer between fragments becomes impossible by construction (which might be desired in some situations [261]). It should be noted that the approximation of a subsystem basis set is by no means a requirement for FDE calculations, but merely facilitates calculations for large molecules.

Early applications of the FDE method were especially concerned with the description of hydrogen-bonded and van der Waals complexes, as summarized in Ref. [262]. Soon, attention turned to molecular properties and spectra. An extension to time-dependent DFT (TDDFT) was published by Casida and Wesołowski in 2003 [263], and several studies concerning solvent/environment-induced shifts were reported [264, 265]. An extension to NMR shifts was presented by Jacob and Visscher [266]. In 2007, Neugebauer rederived the FDE-TDDFT approach and extended it to coupled excitations, delocalized over different subsystems [267]. Such a scheme is particularly suitable for the treatment of coupled chromophores, e.g., in photosynthetic light-harvesting complexes, where the subsystem picture enables a natural approach to the investigation of excitonic interactions. Several subsequent studies applied this approach for the determination of spectral properties of natural light-harvesting proteins (see Refs. [268, 269] for reviews of earlier work, Refs. [270–274] for methodological extensions and Refs. [275, 276] for more recent applications). Much effort has also been dedicated to the application of FDE for the investigation of spin densities and charge transfer processes [261, 277–279].

Already in 1998, the purely density-based FDE formalism was extended to the embedding of WFT methods in a DFT environment [280, 281]. In such a formalism, the same working equations as outlined in Sect. 7.3 can be used by determining the density of an embedded system with a WFT method and letting it enter the energy expression. In addition, the embedding potential is added to the Hamiltonian for the embedded system. The WFT-in-DFT formalism was later refined to produce self-consistent densities for all subsystems [282] and has been extended to excited states [283, 284]. Recently, thorough rederivations of the excited-state formalism for state-specific approaches [285] and response theory treatments [286] have been presented, and a corresponding implementation for coupled excitations was reported [287]. In WFT-in-DFT embedding, the bottleneck is often the high-level calculation of the embedded cluster. Thus, more sophisticated methods for estimating the nonadditive kinetic energy can be used, which require a calculation of the full system on the basis of DFT. This was first exploited by Roncero et al. [255] and later used by the Miller group to construct an accurate WFT-in-DFT embedding scheme [288].

Proteins can be considered a unique challenge for FDE, since they comprise extremely large, covalently connected units (necessitating cuts through covalent bonds), while routinely being too large even for a cheap DFT calculation (preventing the use of exact embedding schemes). An interesting variant of FDE specifically adapted for the treatment of proteins was therefore devised by Jacob and Visscher [117], based on earlier work by Casida and Wesołowski [263]. In the so-called Three-Partition FDE (3-FDE) method, the fragmentation scheme of the MFCC method (see Sect. 7.5.1) is combined with the embedding potential of FDE. Fragments are thus constructed by cutting certain peptide bonds and applying small capping groups to saturate remaining free valences. Since the fully quantum-chemical embedding potential is applied to represent the environment of each fragment, the caps can be chosen much smaller than originally intended by the MFCC developers and usually consist of neutral or methylated amino acid termini [117, 289]. An important feature of the 3-FDE method is the application of an additional cap potential, which forces the density in the region of the cap atoms to be identical to the one obtained for the concaps. This guarantees that upon subtraction of the latter, no negative density patches are obtained. The 3-FDE method has been tested on some small example systems [117, 289] and was used to obtain a fully quantum-chemical density of the Fenna–Matthews–Olson complex of green sulfur bacteria, which consists of more than 6,500 atoms [276]. Subsequently, a robust protocol to construct fully converged 3-FDE protein densities was presented [290], and a comprehensive benchmark was published [291].

Analytical gradients for regular FDE were first implemented by Wesołowski [292] and have been used for geometry optimizations [293] as well as MD simulations [294, 295]. Recently, a detailed derivation and new implementation in the Amsterdam Density Functional (ADF) program were reported [296]. In addition, approximate analytical gradients for WFT-in-DFT embedding are available [297].

Recent developments include a variant of the COSMO scheme especially suited to subsystem calculations [46], the embedding of DMRG [298] and ADC(2) [299] wave functions in a DFT environment, and the derivation of excited-state gradients [300]. Many more details about the different flavors of FDE can be found in several extensive reviews [6, 9, 10, 268, 269].

## 7.7 Summary and Conclusions

In the present chapter, we have attempted to give an overview over different classes of embedding methods, focusing especially on actively developed approaches, many of which have been put forward only recently. Three major classes of methods have been identified, which are (i) QM/MM and related approaches (Sect. 7.4), which in general require the identification of a chemically interesting part of the total system, (ii) methods based on an MBE or the IEP (Sect. 7.5), where a truncated expansion is usually supplemented by physically motivated embedding components, and (iii) DC-like approaches (Sect. 7.6), most of which are potentially able to recover the exact solution from fragment calculations. Clearly, this division is somewhat arbitrary, as the boundaries between different classes have been blurred over the last decades through the adaptation and combination of different ideas.

Especially, the approaches from the first two categories have been compared to an exact embedding potential, which emerges from the subsystem DFT approach (see Sect. 7.3). In many cases, distinct physical effects can be identified as particular terms of the interaction expressions, whereas this is more difficult for the approaches presented in Sect. 7.6. Therefore, the benefits of the first two classes of methods lie in their practical and often computationally cheap form, whereas the DC-like approaches are much more deeply rooted in formally exact theories.

Contributions from all these classes of methods will most certainly continue to be highly relevant in the field of large-scale calculations, where many systems only become accessible through fragment-based approaches, as well as with respect to the investigation of molecular interactions, where embedding facilitates interpretation and leads to chemical insight. The many recent developments presented in this chapter bear testimony to the continued interest in such approaches and their importance in both chemistry and physics.

## References


1. M.S. Gordon, L.V. Slipchenko, Chem. Rev. **115**, 5605–5606 (2015)
2. A.V. Akimov, O.V. Prezhdo, Chem. Rev. **115**, 5797–5890 (2015)
3. K. Raghavachari, A. Saha, Chem. Rev. **115**, 5643–5677 (2015)
4. M.A. Collins, R.P.A. Bettens, Chem. Rev. **115**, 5607–5642 (2015)
5. L.W. Chung, W.M.C. Sameera, R. Ramozzi, A.J. Page, M. Hatanaka, G.P. Petrova, T.V. Harris, X. Li, Z. Ke, F. Liu, H.-B. Li, L. Ding, K. Morokuma, Chem. Rev. **115**, 5678–5796 (2015)
6. T.A. Wesołowski, S. Shedge, X. Zhou, Chem. Rev. **115**, 5891–5928 (2015)
7. E. Brunk, U. Rothlisberger, Chem. Rev. **115**, 6217–6263 (2015)
8. P. Huang, E.A. Carter, Annu. Rev. Phys. Chem. **59**, 261–290 (2008)
9. A.S.P. Gomes, C.R. Jacob, Annu. Rep. Prog. Chem. C **108**, 222–277 (2012)
10. C.R. Jacob, J. Neugebauer, WIREs Comput. Mol. Sci. **4**, 325–362 (2014)
11. Q. Sun, G.K.-L. Chan, Acc. Chem. Res. (2016) (in press)
12. W. Li, S. Li, Y. Jiang, J. Phys. Chem. A **111**, 2193–2199 (2007)
13. M.S. Gordon, D.G. Fedorov, S.R. Pruitt, L.V. Slipchenko, Chem. Rev. **112**, 632–672 (2012)
14. A.M. Gao, D.W. Zhang, J.Z. Zhang, Y. Zhang, Chem. Phys. Lett. **394**, 293–297 (2004)
15. E. Suárez, N. Díaz, D. Suárez, J. Chem. Theory Comput. **5**, 1667–1679 (2009)
16. N.J. Mayhall, K. Raghavachari, J. Chem. Theory Comput. **8**, 2669–2675 (2012)
17. R.M. Richard, J.M. Herbert, J. Chem. Phys. **137**, 064113 (2012)
18. G. Senatore, K.R. Subbaswamy, Phys. Rev. B **34**, 5754–5757 (1986)
19. P. Cortona, Phys. Rev. B **44**, 8454–8458 (1991)
20. T.A. Wesołowski, J. Weber, Chem. Phys. Lett. **248**, 71–76 (1996)
21. T.A. Wesołowski, A. Warshel, J. Phys. Chem. **97**, 8050–8053 (1993)
22. B. Jeziorski, R. Moszynski, K. Szalewicz, Chem. Rev. **94**, 1887–1930 (1994)
23. P. Elliott, M.H. Cohen, A. Wasserman, K. Burke, J. Chem. Theory Comput. **5**, 827–833 (2009)
24. P. Elliott, K. Burke, M.H. Cohen, A. Wasserman, Phys. Rev. A **82**, 024501 (2010)
25. C. Huang, M. Pavone, E.A. Carter, J. Chem. Phys. **134**, 154110 (2011)
26. M.L. Connolly, Science **221**, 709–713 (1983)
27. M.L. Connolly, J. Appl. Crystallogr. **16**, 548–558 (1983)
28. J. Tomasi, B. Mennucci, R. Cammi, Chem. Rev. **105**, 2999–3094 (2005)
29. S. Miertuš, E. Scrocco, J. Tomasi, Chem. Phys. **55**, 117–129 (1981)
30. B. Mennucci, E. Cancès, J. Tomasi, J. Phys. Chem. B **101**, 10506–10517 (1997)
31. V. Barone, M. Cossi, J. Phys. Chem. A **102**, 1995–2001 (1998)
32. B. Mennucci, WIREs Comput. Mol. Sci. **2**, 386–404 (2012)
33. A. Klamt, G. Schüürmann, J. Chem. Soc., Perkin Trans. **2**, 799–805 (1993)
34. E.V. Stefanovich, T.N. Truong, Chem. Phys. Lett. **244**, 65–74 (1995)
35. A. Klamt, J. Phys. Chem. **99**, 2224–2235 (1995)
36. A. Klamt, WIREs Comput. Mol. Sci. **1**, 699–709 (2011)
37. S.-T. Lin, S.I. Sandler, Ind. Eng. Chem. Res. **41**, 899–913 (2002)
38. A. Klamt, J. Phys. Chem. **100**, 3349–3353 (1996)
39. S. Sinnecker, A. Rajendran, A. Klamt, M. Diedenhofen, F. Neese, J. Phys. Chem. A **110**, 2235–2245 (2006)
40. T. Sachsenhauser, S. Rehfeldt, A. Klamt, F. Eckert, H. Klein, Fluid Phase Equilibr. **382**, 89–99 (2014)
41. A. Klamt, J. Phys. Chem. A **120**, 2049–2056 (2016)
42. E. Cancès, Y. Maday, B. Stamm, J. Chem. Phys. **139**, 054111 (2013)
43. F. Lipparini, B. Stamm, E. Cancès, Y. Maday, B. Mennucci, J. Chem. Theory Comput. **9**, 3637–3648 (2013)
44. F. Lipparini, L. Lagardère, G. Scalmani, B. Stamm, E. Cancès, Y. Maday, J.-P. Piquemal, M.J. Frisch, B. Mennucci, J. Phys. Chem. Lett. **5**, 953–958 (2014)
45. F. Lipparini, G. Scalmani, L. Lagardère, B. Stamm, E. Cancès, Y. Maday, J.-P. Piquemal, M.J. Frisch, B. Mennucci, J. Chem. Phys. **141**, 184108 (2014)

46. A. Goez, J. Neugebauer, J. Chem. Theory Comput. **11**, 5277–5290 (2015)
47. D.M. Chipman, J. Chem. Phys. **106**, 10194–10206 (1997)
48. C.-G. Zhan, J. Bentley, D.M. Chipman, J. Chem. Phys. **108**, 177–192 (1998)
49. D.M. Chipman, J. Chem. Phys. **112**, 5558–5565 (2000)
50. E. Cancès, B. Mennucci, J. Chem. Phys. **114**, 4744–4745 (2001)
51. C.J. Cramer, D.G. Truhlar, J. Am. Chem. Soc. **113**, 8552–8554 (1991)
52. C.J. Cramer, D.G. Truhlar, Acc. Chem. Res. **41**, 760–768 (2008)
53. A.V. Marenich, C.J. Cramer, D.G. Truhlar, J. Chem. Theory Comput. **9**, 609–620 (2013)
54. A.V. Marenich, C.J. Cramer, D.G. Truhlar, J. Phys. Chem. B **113**, 6378–6396 (2009)
55. A.V. Marenich, R.M. Olson, C.P. Kelly, C.J. Cramer, D.G. Truhlar, J. Chem. Theory Comput. **3**, 2011–2033 (2007)
56. A. Warshel, M. Karplus, J. Am. Chem. Soc. **94**, 5612–5625 (1972)
57. A. Warshel, M. Levitt, J. Mol. Biol. **103**, 227–249 (1976)
58. H.M. Senn, W. Thiel, Angew. Chem. Int. Edit. **48**, 1198–1229 (2009)
59. T. Steinbrecher, M. Elstner, QM and QM/MM Simulations of Proteins, in *Biomolecular Simulations: Methods and Protocols*, ed. by L. Monticelli, E. Salonen (Humana Press, Totowa, NJ, 2013)
60. M.G. Quesne, T. Borowski, S.P. deVisser, *Chem. Eur. J.* **22**, 2562–2581 (2016)
61. F. Maseras, K. Morokuma, J. Comput. Chem. **16**, 1170–1179 (1995)
62. M. Svensson, S. Humbel, R.D.J. Froese, T. Matsubara, S. Sieber, K. Morokuma, J. Phys. Chem. **100**, 19357–19363 (1996)
63. M. Svensson, S. Humbel, K. Morokuma, J. Chem. Phys. **105**, 3654–3661 (1996)
64. D. Bakowies, W. Thiel, J. Phys. Chem. **100**, 10580–10594 (1996)
65. T. Vreven, K.S. Byun, I. Komromi, S. Dapprich, J.A. Montgomery, K. Morokuma, M.J. Frisch, J. Chem. Theory Comput. **2**, 815–826 (2006)
66. S. Humbel, S. Sieber, K. Morokuma, J. Chem. Phys. **105**, 1959–1967 (1996)
67. P. Tao, J.F. Fisher, Q. Shi, T. Vreven, S. Mobashery, H.B. Schlegel, Biochemistry **48**, 9839–9847 (2009)
68. S. Caprasecca, S. Jurinovich, L. Viani, C. Curutchet, B. Mennucci, J. Chem. Theory Comput. **10**, 1588–1598 (2014)
69. S. Dapprich, I. Komromi, K. Byun, K. Morokuma, M.J. Frisch, J. Mol. Struct. Theochem **461–462**, 1–21 (1999)
70. T. Vreven, B. Mennucci, C.O. da Silva, K. Morokuma, J. Tomasi, J. Chem. Phys. **115**, 62–72 (2001)
71. T. Kerdcharoen, K. Morokuma, Chem. Phys. Lett. **355**, 257–262 (2002)
72. B.W. Hopkins, G.S. Tschumper, J. Comput. Chem. **24**, 1563–1568 (2003)
73. L.W. Chung, H. Hirao, X. Li, K. Morokuma, WIREs Comput. Mol. Sci. **2**, 327–350 (2012)
74. M. Swart, F.M. Bickelhaupt, J. Comput. Chem. **29**, 724–734 (2008)
75. V. Luzhkov, A. Warshel, J. Am. Chem. Soc. **113**, 4491–4499 (1991)
76. V. Luzhkov, A. Warshel, J. Comput. Chem. **13**, 199–213 (1992)
77. M.A. Thompson, G.K. Schenter, J. Phys. Chem. **99**, 6374–6386 (1995)
78. M.A. Thompson, J. Phys. Chem. **100**, 14492–14507 (1996)
79. C. Curutchet, A. Muñoz-Losa, S. Monti, J. Kongsted, G.D. Scholes, B. Mennucci, J. Chem. Theory Comput. **5**, 1838–1848 (2009)
80. S. Caprasecca, C. Curutchet, B. Mennucci, J. Chem. Theory Comput. **8**, 4462–4473 (2012)
81. Q. Zeng, W. Liang, J. Chem. Phys. **143**, 134104 (2015)
82. J.H. Jensen, P.N. Day, M.S. Gordon, H. Basch, D. Cohen, D.R. Garmer, M. Kraus, W.J. Stevens, Effective fragment method for modeling intermolecular hydrogen-bonding effects on quantum mechanical calculations, in *Modeling the Hydrogen Bond*, ed. by D.A. Smith (American Chemical Society, Washington, DC, 1994), Chapter 10, pp. 139–151
83. P.N. Day, J.H. Jensen, M.S. Gordon, S.P. Webb, W.J. Stevens, M. Krauss, D. Garmer, H. Basch, D. Cohen, J. Chem. Phys. **105**, 1968–1986 (1996)
84. M.S. Gordon, M.A. Freitag, P. Bandyopadhyay, J.H. Jensen, V. Kairys, W.J. Stevens, J. Phys. Chem. A **105**, 293–307 (2001)

85. L.V. Slipchenko, M.S. Gordon, J. Comput. Chem. **28**, 276–291 (2007)
86. M.A. Freitag, M.S. Gordon, J.H. Jensen, W.J. Stevens, J. Chem. Phys. **112**, 7300–7306 (2000)
87. L.V. Slipchenko, M.S. Gordon, Mol. Phys. **107**, 999–1016 (2009)
88. I. Adamovic, M.S. Gordon, Mol. Phys. **103**, 379–387 (2005)
89. J.H. Jensen, J. Chem. Phys. **114**, 8775–8783 (2001)
90. H. Li, M.S. Gordon, J.H. Jensen, J. Chem. Phys. **124**, 214108 (2006)
91. M.W. Schmidt, K.K. Baldridge, J.A. Boatz, S.T. Elbert, M.S. Gordon, J.H. Jensen, S. Koseki, N. Matsunaga, K.A. Nguyen, S. Su, T.L. Windus, M. Dupuis, J.A. Montgomery, J. Comput. Chem. **14**, 1347–1363 (1993)
92. M.S. Gordon, M.W. Schmidt, Advances in electronic structure theory: GAMESS a decade later, in *Theory and Applications of Computational Chemistry*, ed. by C.E. Dykstra, G. Frenking, K.S. Kim, G.E. Scuseria (Elsevier, Amsterdam, 2005)
93. H. Li, H.M. Netzloff, M.S. Gordon, J. Chem. Phys. **125**, 194103 (2006)
94. H. Li, M.S. Gordon, Theor. Chem. Acc. **115**, 385–390 (2006)
95. P. Arora, L.V. Slipchenko, S.P. Webb, A. DeFusco, M.S. Gordon, J. Phys. Chem. A **114**, 6742–6750 (2010)
96. C.H. Choi, S. Re, M. Feig, Y. Sugita, Chem. Phys. Lett. **539–540**, 218–221 (2012)
97. D.D. Kemp, J.M. Rintelman, M.S. Gordon, J.H. Jensen, Theor. Chem. Acc. **125**, 481–491 (2010)
98. Q.A. Smith, K. Ruedenberg, M.S. Gordon, L.V. Slipchenko, J. Chem. Phys. **136**, 244107 (2012)
99. M.S. Gordon, Q.A. Smith, P. Xu, L.V. Slipchenko, Annu. Rev. Phys. Chem. **64**, 553–578 (2013)
100. V. Kairys, J.H. Jensen, J. Phys. Chem. A **104**, 6656–6665 (2000)
101. P.K. Gurunathan, A. Acharya, D. Ghosh, D. Kosenkov, I. Kaliman, Y. Shao, A.I. Krylov, L.V. Slipchenko, J. Phys. Chem. B **120**, 6562–6574 (2016)
102. C. Steinmann, D.G. Fedorov, J.H. Jensen, J. Phys. Chem. A **114**, 8705–8712 (2010)
103. S.R. Pruitt, C. Steinmann, J.H. Jensen, M.S. Gordon, J. Chem. Theory Comput. **9**, 2235–2249 (2013)
104. J.M. Olsen, K. Aidas, J. Kongsted, J. Chem. Theory Comput. **6**, 3721–3734 (2010)
105. L. Gagliardi, R. Lindh, G. Karlström, J. Chem. Phys. **121**, 4494–4500 (2004)
106. J.M.H. Olsen, J. Kongsted, Molecular properties through polarizable embedding, in *Advances in Quantum Chemistry*, ed. by J.R. Sabin, E. Brändas, vol. 61 (Academic Press, New York, 2011)
107. K. Sneskov, T. Schwabe, J. Kongsted, O. Christiansen, J. Chem. Phys. **134**, 104108 (2011)
108. T. Schwabe, K. Sneskov, J.M.H. Olsen, J. Kongsted, O. Christiansen, C. Hättig, J. Chem. Theory Comput. **8**, 3274–3283 (2012)
109. E.D. Hedegård, N.H. List, H.J.A. Jensen, J. Kongsted, J. Chem. Phys. **139**, 044101 (2013)
110. E.D. Hedegård, M. Reiher, J. Chem. Theory Comput. **12**, 4242–4253 (2016)
111. N.H. List, M.T.P. Beerepoot, J.M.H. Olsen, B. Gao, K. Ruud, H.J.A. Jensen, J. Kongsted, J. Chem. Phys. **142**, 034119 (2015)
112. N.H. List, J.M.H. Olsen, J. Kongsted, Phys. Chem. Chem. Phys. **18**, 20234–20250 (2016)
113. J.M.H. Olsen, C. Steinmann, K. Ruud, J. Kongsted, J. Phys. Chem. A **119**, 5344–5355 (2015)
114. D.W. Zhang, J.Z.H. Zhang, J. Chem. Phys. **119**, 3599–3605 (2003)
115. X. He, J.Z.H. Zhang, J. Chem. Phys. **122**, 031103 (2005)
116. X. Chen, Y. Zhang, J.Z.H. Zhang, J. Chem. Phys. **122**, 184105 (2005)
117. C.R. Jacob, L. Visscher, J. Chem. Phys. **128**, 155102 (2008)
118. P. Söderhjelm, U. Ryde, J. Phys. Chem. A **113**, 617–627 (2009)
119. X.H. Chen, D.W. Zhang, J.Z.H. Zhang, J. Chem. Phys. **120**, 839–844 (2004)
120. S. Li, W. Li, T. Fang, J. Am. Chem. Soc. **127**, 7215–7226 (2005)
121. X.H. Chen, J.Z.H. Zhang, J. Chem. Phys. **125**, 044903 (2006)
122. N. Jiang, J. Ma, Y. Jiang, J. Chem. Phys. **124**, 114112 (2006)
123. X. He, J.Z.H. Zhang, J. Chem. Phys. **124**, 184703 (2006)
124. X. Wang, J. Liu, J.Z.H. Zhang, X. He, J. Phys. Chem. A **117**, 7149–7161 (2013)
125. Y. Mei, C. Ji, J.Z.H. Zhang, J. Chem. Phys. **125**, 094906 (2006)
126. X. Jia, X. Wang, J. Liu, J.Z.H. Zhang, Y. Mei, X. He, J. Chem. Phys. **139**, 214104 (2013)

127. J. Liu, J.Z.H. Zhang, X. He, Phys. Chem. Chem. Phys. **18**, 1864–1875 (2016)
128. J. Liu, T. Zhu, X. Wang, X. He, J.Z.H. Zhang, J. Chem. Theory Comput. **11**, 5897–5905 (2015)
129. C. Ji, Y. Mei, J.Z.H. Zhang, Biophys. J. **95**, 1080–1088 (2008)
130. W. Hua, T. Fang, W. Li, J.-G. Yu, S. Li, J. Phys. Chem. A **112**, 10864–10872 (2008)
131. S. Hua, W. Hua, S. Li, J. Phys. Chem. A **114**, 8126–8134 (2010)
132. S. Hua, W. Li, S. Li, ChemPhysChem **14**, 108–115 (2013)
133. H. Li, W. Li, S. Li, J. Ma, J. Phys. Chem. B **112**, 7061–7070 (2008)
134. T. Fang, W. Li, F. Gu, S. Li, J. Chem. Theory Comput. **11**, 91–98 (2015)
135. S. Li, W. Li, J. Ma, Acc. Chem. Res. **47**, 2712–2720 (2014)
136. K. Kitaura, E. Ikeo, T. Asada, T. Nakano, M. Uebayasi, Chem. Phys. Lett. **313**, 701–706 (1999)
137. K. Kitaura, T. Sawai, T. Asada, T. Nakano, M. Uebayasi, Chem. Phys. Lett. **312**, 319–324 (1999)
138. T. Nakano, T. Kaminuma, T. Sato, Y. Akiyama, M. Uebayasi, K. Kitaura, Chem. Phys. Lett. **318**, 614–618 (2000)
139. D.G. Fedorov, J.H. Jensen, R.C. Deka, K. Kitaura, J. Phys. Chem. A **112**, 11808–11816 (2008)
140. D.G. Fedorov, T. Ishida, K. Kitaura, J. Phys. Chem. A **109**, 2638–2646 (2005)
141. D.G. Fedorov, K. Kitaura (eds.), *The Fragment Molecular Orbital Method: Practical Applications to Large Molecular Systems* (CRC Press, Boca Raton, FL, 2009)
142. D.G. Fedorov, K. Kitaura, J. Chem. Phys. **120**, 6832–6840 (2004)
143. T. Nakano, Y. Mochizuki, K. Yamashita, C. Watanabe, K. Fukuzawa, K. Segawa, Y. Okiyama, T. Tsukamoto, S. Tanaka, Chem. Phys. Lett. **523**, 128–133 (2012)
144. T. Nakano, T. Kaminuma, T. Sato, K. Fukuzawa, Y. Akiyama, M. Uebayasi, K. Kitaura, Chem. Phys. Lett. **351**, 475–480 (2002)
145. D.G. Fedorov, L.V. Slipchenko, K. Kitaura, J. Phys. Chem. A **114**, 8742–8753 (2010)
146. D.G. Fedorov, K. Kitaura, Chem. Phys. Lett. **433**, 182–187 (2006)
147. D.G. Fedorov, K. Kitaura, J. Chem. Phys. **131**, 171106 (2009)
148. Y. Inadomi, T. Nakano, K. Kitaura, U. Nagashima, Chem. Phys. Lett. **364**, 139–143 (2002)
149. S. Sugiki, N. Kurita, Y. Sengoku, H. Sekino, Chem. Phys. Lett. **382**, 611–617 (2003)
150. D.G. Fedorov, K. Kitaura, J. Chem. Phys. **121**, 2483–2490 (2004)
151. D.G. Fedorov, K. Kitaura, J. Chem. Phys. **123**, 134103 (2005)
152. D.G. Fedorov, K. Kitaura, J. Chem. Phys. **122**, 054108 (2005)
153. Y. Nishimoto, D.G. Fedorov, S. Irle, J. Chem. Theory Comput. **10**, 4801–4812 (2014)
154. Y. Nishimoto, D.G. Fedorov, S. Irle, Chem. Phys. Lett. **636**, 90–96 (2015)
155. K. Kitaura, S.-I. Sugiki, T. Nakano, Y. Komeiji, M. Uebayasi, Chem. Phys. Lett. **336**, 163–170 (2001)
156. T. Nagata, K. Brorsen, D.G. Fedorov, K. Kitaura, M.S. Gordon, J. Chem. Phys. **134**, 124115 (2011)
157. D.G. Fedorov, T. Ishida, M. Uebayasi, K. Kitaura, J. Phys. Chem. A **111**, 2722–2732 (2007)
158. Y. Komeiji, T. Nakano, K. Fukuzawa, Y. Ueno, Y. Inadomi, T. Nemoto, M. Uebayasi, D.G. Fedorov, K. Kitaura, Chem. Phys. Lett. **372**, 342–347 (2003)
159. Y. Komeiji, Y. Mochizuki, T. Nakano, D.G. Fedorov, J. Mol. Struct. Theochem **898**, 2–7 (2009)
160. D.G. Fedorov, K. Kitaura, H. Li, J.H. Jensen, M.S. Gordon, J. Comput. Chem. **27**, 976–985 (2006)
161. T. Fujita, T. Nakano, S. Tanaka, Chem. Phys. Lett. **506**, 112–116 (2011)
162. Y. Mochizuki, S. Koikegami, S. Amari, K. Segawa, K. Kitaura, T. Nakano, Chem. Phys. Lett. **406**, 283–288 (2005)
163. M. Chiba, D.G. Fedorov, K. Kitaura, Chem. Phys. Lett. **444**, 346–350 (2007)
164. D.G. Fedorov, K. Kitaura, J. Phys. Chem. A **111**, 6904–6914 (2007)
165. D.G. Fedorov, T. Nagata, K. Kitaura, Phys. Chem. Chem. Phys. **14**, 7562–7577 (2012)
166. T. Nagata, D.G. Fedorov, K. Kitaura, M.S. Gordon, J. Chem. Phys. **131**, 024101 (2009)

167. S. Hirata, M. Valiev, M. Dupuis, S.S. Xantheas, S. Sugiki, H. Sekino, Mol. Phys. **103**, 2255–2265 (2005)
168. M. Kamiya, S. Hirata, M. Valiev, J. Chem. Phys. **128**, 074103 (2008)
169. E.E. Dahlke, D.G. Truhlar, J. Chem. Theory Comput. **3**, 46–53 (2007)
170. G.J.O. Beran, J. Chem. Phys. **130**, 164115 (2009)
171. S.R. Gadre, R.N. Shirsat, A.C. Limaye, J. Phys. Chem. **98**, 9165–9169 (1994)
172. K. Babu, S.R. Gadre, J. Comput. Chem. **24**, 484–495 (2003)
173. V. Ganesh, R.K. Dongare, P. Balanarayan, S.R. Gadre, J. Chem. Phys. **125**, 104109 (2006)
174. N. Sahu, S.R. Gadre, Acc. Chem. Res. **47**, 2739–2747 (2014)
175. A.P. Rahalkar, V. Ganesh, S.R. Gadre, J. Chem. Phys. **129**, 234101 (2008)
176. A.P. Rahalkar, M. Katouda, S.R. Gadre, S. Nagase, J. Comput. Chem. **31**, 2405–2418 (2010)
177. J.P. Furtado, A.P. Rahalkar, S. Shanker, P. Bandyopadhyay, S.R. Gadre, J. Phys. Chem. Lett. **3**, 2253–2258 (2012)
178. N. Sahu, S.R. Gadre, J. Chem. Phys. **142**, 014107 (2015)
179. N. Sahu, S.R. Gadre, J. Chem. Phys. **144**, 114113 (2016)
180. N. Sahu, G. Singh, A. Nandi, S.R. Gadre, J. Phys. Chem. A **120**, 5706–5714 (2016)
181. M. Isegawa, B. Wang, D.G. Truhlar, J. Chem. Theory Comput. **9**, 1381–1393 (2013)
182. A.P. Rahalkar, S.D. Yeole, V. Ganesh, S.R. Gadre, Molecular tailoring: an art of the possible for ab initio treatment of large molecules and molecular clusters, in *Linear-Scaling Techniques in Computational Chemistry and Physics: Methods and Applications*, ed. by R. Zalesny, M.G. Papadopoulos, P.G. Mezey, J. Leszczynski (Springer, Netherlands, Dordrecht, 2011)
183. L. Huang, L. Massa, J. Karle, Int. J. Quantum Chem. **103**, 808–817 (2005)
184. L. Huang, L. Massa, J. Karle, Biochemistry **44**, 16747–16752 (2005)
185. L. Huang, L. Massa, J. Karle, Proc. Natl. Acad. Sci. USA **105**, 1849–1854 (2008)
186. V. Deev, M.A. Collins, J. Chem. Phys. **122**, 154102 (2005)
187. M.A. Collins, V.A. Deev, J. Chem. Phys. **125**, 104104 (2006)
188. L. Huang, L. Massa, Int. J. Quantum Chem. **111**, 2180–2186 (2011)
189. L. Huang, H.J. Bohorquez, C.F. Matta, L. Massa, Int. J. Quantum Chem. **111**, 4150–4157 (2011)
190. L. Huang, L. Massa, J. Karle, Int. J. Quantum Chem. **106**, 447–457 (2006)
191. L. Huang, C. Matta, L. Massa, Struct. Chem. **26**, 1433–1442 (2015)
192. R.M. Richard, J.M. Herbert, J. Chem. Theory Comput. **9**, 1408–1416 (2013)
193. J.M. Mullin, L.B. Roskop, S.R. Pruitt, M.A. Collins, M.S. Gordon, J. Phys. Chem. A **113**, 10040–10049 (2009)
194. M.A. Collins, Phys. Chem. Chem. Phys. **14**, 7744–7751 (2012)
195. H.-A. Le, H.-J. Tan, J.F. Ouyang, R.P.A. Bettens, J. Chem. Theory Comput. **8**, 469–478 (2012)
196. M.A. Collins, M.W. Cvitkovic, R.P.A. Bettens, Acc. Chem. Res. **47**, 2776–2785 (2014)
197. J. Řezáč, D.R. Salahub, J. Chem. Theory Comput. **6**, 91–99 (2010)
198. N.J. Mayhall, K. Raghavachari, J. Chem. Theory Comput. **7**, 1336–1343 (2011)
199. K.V.J. Jose, K. Raghavachari, J. Chem. Theory Comput. **11**, 950–961 (2015)
200. K.V.J. Jose, D. Beckett, K. Raghavachari, J. Chem. Theory Comput. **11**, 4238–4247 (2015)
201. K.V.J. Jose, K. Raghavachari, Mol. Phys. **113**, 3057–3066 (2015)
202. K.V.J. Jose, K. Raghavachari, J. Chem. Theory Comput. **12**, 585–594 (2016)
203. W. Yang, Phys. Rev. Lett. **66**, 1438–1441 (1991)
204. W. Yang, T. Lee, J. Chem. Phys. **103**, 5674–5678 (1995)
205. W. Yang, Phys. Rev. A **44**, 7823–7826 (1991)
206. S.L. Dixon, K.M. Merz, J. Chem. Phys. **104**, 6643–6649 (1996)
207. S.L. Dixon, K.M. Merz, J. Chem. Phys. **107**, 879–893 (1997)
208. T. Lee, D.M. York, W. Yang, J. Chem. Phys. **105**, 2744–2750 (1996)
209. P.D. Walker, P.G. Mezey, J. Am. Chem. Soc. **115**, 12423–12430 (1993)
210. P.G. Mezey, Local-shape analysis of macromolecular electron densities, in *Computational Chemistry: Reviews of Current Trends* (World Scientific Publishing, Singapore, 1996)
211. P.G. Mezey, J. Math. Chem. **18**, 141–168 (1995)
212. T.E. Exner, P.G. Mezey, J. Phys. Chem. A **108**, 4301–4309 (2004)

213. W. Li, S. Li, J. Chem. Phys. **121**, 6649–6657 (2004)
214. T. Akama, M. Kobayashi, H. Nakai, J. Comput. Chem. **28**, 2003–2012 (2007)
215. M. Kobayashi, T. Akama, H. Nakai, J. Chem. Phys. **125**, 204106 (2006)
216. M. Kobayashi, Y. Imamura, H. Nakai, J. Chem. Phys. **127**, 074103 (2007)
217. M. Kobayashi, H. Nakai, J. Chem. Phys. **129**, 044103 (2008)
218. M. Kobayashi, H. Nakai, J. Chem. Phys. **131**, 114108 (2009)
219. M. Kobayashi, H. Nakai, Int. J. Quantum Chem. **109**, 2227–2237 (2009)
220. G.-L. Song, Z.H. Li, Z.-P. Liu, X.-M. Cao, W. Wang, K.-N. Fan, Y. Xie, H.F. Schaefer, J. Chem. Theory Comput. **4**, 2049–2056 (2008)
221. M. Guidon, J. Hutter, J. VandeVondele, J. Chem. Theory Comput. **6**, 2348–2364 (2010)
222. T. Touma, M. Kobayashi, H. Nakai, Chem. Phys. Lett. **485**, 247–252 (2010)
223. M. Kobayashi, T. Touma, H. Nakai, J. Chem. Phys. **136**, 084108 (2012)
224. T. Yoshikawa, H. Nakai, J. Comput. Chem. **36**, 164–170 (2015)
225. F. Shimojo, S. Hattori, R.K. Kalia, M. Kunaseth, W. Mou, A. Nakano, K.-I. Nomura, S. Ohmura, P. Rajak, K. Shimamura, P. Vashishta, J. Chem. Phys. **140**, 18A529 (2014)
226. H. Nishizawa, Y. Nishimura, M. Kobayashi, S. Irle, H. Nakai, J. Comput. Chem. **37**, 1983–1992 (2016)
227. M.E. Fornace, J. Lee, K. Miyamoto, F.R. Manby, T.F. Miller, J. Chem. Theory Comput. **11**, 568–580 (2015)
228. M. Kobayashi, H. Nakai, Phys. Chem. Chem. Phys. **14**, 7629–7639 (2012)
229. K.M. Merz Jr., Acc. Chem. Res. **47**, 2804–2811 (2014)
230. G. Knizia, G.K.-L. Chan, Phys. Rev. Lett. **109**, 186404 (2012)
231. G. Knizia, G.K.-L. Chan, J. Chem. Theory Comput. **9**, 1428–1432 (2013)
232. G. Kotliar, S.Y. Savrasov, K. Haule, V.S. Oudovenko, O. Parcollet, C.A. Marianetti, Rev. Mod. Phys. **78**, 865–951 (2006)
233. D. Zgid, G.K.-L. Chan, J. Chem. Phys. **134**, 094115 (2011)
234. A.A. Kananenka, E. Gull, D. Zgid, Phys. Rev. B **91**, 121111 (2015)
235. T.N. Lan, A.A. Kananenka, D. Zgid, J. Chem. Phys. **143**, 241102 (2015)
236. T. Nguyen Lan, A.A. Kananenka, D. Zgid, J. Chem. Theory Comput. **12**, 4856–4870 (2016)
237. B.-X. Zheng, G.K.-L. Chan, Phys. Rev. B **93**, 035126 (2016)
238. S. Wouters, C.A. Jiménez-Hoyos, Q. Sun, G.K.-L. Chan, J. Chem. Theory Comput. **12**, 2706–2719 (2016)
239. I.W. Bulik, W. Chen, G.E. Scuseria, J. Chem. Phys. **141**, 054113 (2014)
240. M. Welborn, T. Tsuchimochi, T. Van Voorhis, J. Chem. Phys. **145**, 074102 (2016)
241. J.W. Kaminski, S. Gusarov, T.A. Wesołowski, A. Kovalenko, J. Phys. Chem. A **114**, 6082–6096 (2010)
242. M. Humbert-Droz, X. Zhou, S.V. Shedge, T.A. Wesołowski, Theor. Chem. Acc. **133**, 1405 (2013)
243. T. Dresselhaus, J. Neugebauer, Theor. Chem. Acc. **134**, 97 (2015)
244. S. Laricchia, E. Fabiano, F. Della Sala, J. Chem. Phys. **133**, 164111 (2010)
245. S. Laricchia, E. Fabiano, F. Della Sala, Chem. Phys. Lett. **518**, 114–118 (2011)
246. S. Laricchia, E. Fabiano, F. Della Sala, J. Chem. Phys. **137**, 14102 (2012)
247. S. Kümmel, L. Kronik, Rev. Mod. Phys. **80**, 3–60 (2008)
248. L.H. Thomas, Math. Proc. Cambridge **23**, 542–548 (1927)
249. E. Fermi, Z. Phys. **48**, 73–79 (1928)
250. S. Fux, K. Kiewisch, C.R. Jacob, J. Neugebauer, M. Reiher, Chem. Phys. Lett. **461**, 353–359 (2008)
251. D. Schlüns, K. Klahr, C. Mück-Lichtenfeld, L. Visscher, J. Neugebauer, Phys. Chem. Chem. Phys. **17**, 14323–14341 (2015)
252. A. Lembarki, H. Chermette, Phys. Rev. A **50**, 5328–5331 (1994)
253. S. Fux, C.R. Jacob, J. Neugebauer, L. Visscher, M. Reiher, J. Chem. Phys. **132**, 164101 (2010)
254. C.R. Jacob, L. Visscher, Towards the description of covalent bonds in subsystem density-functional theory, in *Recent Progress in Orbital-free Density Functional Theory*, ed. by T.A. Wesołowski, Y.A. Wang (World Scientific Publishing, Singapore, 2013)

255. O. Roncero, M.P. de Lara-Castells, P. Villarreal, F. Flores, J. Ortega, M. Paniagua, A. Aguado, J. Chem. Phys. **129**, 184104 (2008)
256. J.D. Goodpaster, N. Ananth, F.R. Manby, T.F. Miller III, J. Chem. Phys. **133**, 084103 (2010)
257. P.K. Tamukong, Y.G. Khait, M.R. Hoffmann, J. Phys. Chem. A **118**, 9182–9200 (2014)
258. D.V. Chulhai, L. Jensen, J. Chem. Theory Comput. **11**, 3080–3088 (2015)
259. J.P. Unsleber, J. Neugebauer, C.R. Jacob, Phys. Chem. Chem. Phys. **18**, 21001–21009 (2016)
260. F.R. Manby, M. Stella, J.D. Goodpaster, T.F. Miller, J. Chem. Theory Comput. **8**, 2564–2568 (2012)
261. A. Solovyeva, M. Pavanello, J. Neugebauer, J. Chem. Phys. **136**, 194104 (2012)
262. T.A. Wesołowski, One-electron equations for embedded electron density: challenge for theory and practical payoffs in multi-level modelling of complex polyatomic systems, in *Computational Chemistry: Reviews of Current Trends* (World Scientific, Singapore, 2011)
263. M.E. Casida, T.A. Wesołowski, Int. J. Quantum Chem. **96**, 577–588 (2004)
264. T.A. Wesołowski, J. Am. Chem. Soc. **126**, 11444–11445 (2004)
265. J. Neugebauer, M.J. Louwerse, E.J. Baerends, T.A. Wesołowski, J. Chem. Phys. **122**, 094115 (2005)
266. C.R. Jacob, L. Visscher, J. Chem. Phys. **125**, 194104 (2006)
267. J. Neugebauer, J. Chem. Phys. **126**, 134116 (2007)
268. J. Neugebauer, ChemPhysChem **10**, 3148–3173 (2009)
269. C. König, J. Neugebauer, ChemPhysChem **13**, 386–425 (2012)
270. A. Kovyrshin, J. Neugebauer, J. Chem. Phys. **133**, 174114 (2010)
271. C. König, N. Schlüter, J. Neugebauer, J. Chem. Phys. **138**, 034104 (2013)
272. C. Daday, C. König, O. Valsson, J. Neugebauer, C. Filippi, J. Chem. Theory Comput. **9**, 2355–2367 (2013)
273. M. Pavanello, J. Chem. Phys. **138**, 204118 (2013)
274. D.G. Artiukhin, C.R. Jacob, J. Neugebauer, J. Chem. Phys. **142**, 234101 (2015)
275. C. König, J. Neugebauer, J. Chem. Theory Comput. **9**, 1808–1820 (2013)
276. A. Goez, C.R. Jacob, J. Neugebauer, Comput. Theor. Chem. **1040–1041**, 347–359 (2014)
277. M. Pavanello, J. Neugebauer, J. Chem. Phys. **135**, 234103 (2011)
278. A. Solovyeva, M. Pavanello, J. Neugebauer, J. Chem. Phys. **140**, 164103 (2014)
279. P. Ramos, M. Pavanello, Phys. Chem. Chem. Phys. **18**, 21172–21178 (2016)
280. N. Govind, Y.A. Wang, A.J.R. da Silva, E.A. Carter, Chem. Phys. Lett. **295**, 129–134 (1998)
281. N. Govind, Y.A. Wang, E.A. Carter, J. Chem. Phys. **110**, 7677–7688 (1999)
282. P. Huang, E.A. Carter, J. Chem. Phys. **125**, 084102 (2006)
283. T. Klüner, N. Govind, Y.A. Wang, E.A. Carter, Phys. Rev. Lett. **86**, 5954–5957 (2001)
284. T. Klüner, N. Govind, Y.A. Wang, E.A. Carter, J. Chem. Phys. **116**, 42–54 (2002)
285. Y.G. Khait, M.R. Hoffmann, J. Chem. Phys. **133**, 044107 (2010)
286. S. Höfener, A.S.P. Gomes, L. Visscher, J. Chem. Phys. **136**, 044104 (2012)
287. S. Höfener, L. Visscher, J. Chem. Theory Comput. **12**, 549–557 (2016)
288. J.D. Goodpaster, T.A. Barnes, F.R. Manby, T.F. Miller, J. Chem. Phys. **137**, 224113 (2012)
289. K. Kiewisch, C.R. Jacob, L. Visscher, J. Chem. Theory Comput. **9**, 2425–2440 (2013)
290. A. Goez, J. Neugebauer, Mol. Phys. (2016) (in press)
291. A. Goez, J. Neugebauer, J. Chem. Theory Comput. **12**, 4843–4855 (2016)
292. T.A. Wesołowski, A. Goursot, J. Weber, J. Chem. Phys. **115**, 4791–4797 (2001)
293. M. Dułak, J.W. Kamiński, T.A. Wesołowski, J. Chem. Theory Comput. **3**, 735–745 (2007)
294. M. Iannuzzi, B. Kirchner, J. Hutter, Chem. Phys. Lett. **421**, 16–20 (2006)
295. S. Andermatt, J. Cha, F. Schiffmann, J. VandeVondele, J. Chem. Theory Comput. **12**, 3214–3227 (2016)
296. D. Schlüns, M. Franchini, A.W. Götz, J. Neugebauer, C.R. Jacob, L. Visscher, J. Comput. Chem. **2016** (in press)
297. J. Heuser, S. Höfener, J. Comput. Chem. **37**, 1092–1101 (2016)
298. T. Dresselhaus, J. Neugebauer, S. Knecht, S. Keller, Y. Ma, M. Reiher, J. Chem. Phys. **142**, 044111 (2015)
299. S. Prager, A. Zech, F. Aquilante, A. Dreuw, T.A. Wesołowski, J. Chem. Phys. **144**, 204103 (2016)
300. A. Kovyrshin, J. Neugebauer, Phys. Chem. Chem. Phys. **18**, 20955–20975 (2016)

# Chapter 8
# Calculation of Vibrational Spectra of Large Molecules from Their Fragments

**Shigeki Yamamoto and Petr Bouř**

**Abstract** Vibrational spectroscopies proved extremely useful to study a broad range of molecules, including their structures and interactions. Additional structural sensitivity is provided by polarized techniques measuring different absorption or scattering of left- and right circularly polarized light. The vibrational optical activity (VOA) methods include Raman optical activity (ROA) and vibrational circular dichroism (VCD). Interpretation of the spectra, however, is largely dependent on time-consuming quantum-chemical simulations limited by molecular size. To be able to handle at a reasonable precision large molecules, such as biologically relevant systems, we developed a simplified method of transfer of vibrational atomic properties from smaller fragments to the system of interest. The method found many successful applications in the past, including spectra of proteins and nucleic acids, and appears useful also in connection with the multi-level computational techniques in the foreseeable future. In this chapter, we discuss the theoretical whereabouts of the transfer method and its related computational algorithms and describe typical applications to large molecules.

**Keywords** Vibrational optical activity · Raman optical activity · Vibrational circular dichroism · Harmonic approximation · Density functional theory · Spectral simulations · Biopolymer structure

S. Yamamoto
Department of Chemistry, Graduate School of Science, Osaka University, Machikaneyama 1-1, Toyonaka, Osaka 560-0043, Japan
e-mail: aporoa@gmail.com

P. Bouř (✉)
Institute of Organic Chemistry and Biochemistry, Academy of Sciences, Flemingovo náměstí 2, 16610 Praha 6, Czech Republic
e-mail: bour@uochb.cas.cz



M.J. Wójcik et al. (eds.), *Frontiers of Quantum Chemistry*,
https://doi.org/10.1007/978-981-10-5651-2_8

## 8.1 Fragment Methodology

### *8.1.1 Introduction*

In the last decades, vibrational spectroscopy has continued to find new applications in basic and applied fields of chemistry, biology, and molecular physics. In addition to classical Raman scattering and infrared absorption (IR), new optical technologies gave rise to Raman optical activity (ROA), vibrational circular dichroism (VCD), THz-time domain (THz-TD), and multi-photon vibrational spectroscopy. However, interpretation of the spectra is not straightforward. Quite often, empirical assignment of spectral bands proved to be wrong, e.g., in ROA spectra of proteins and low-frequency vibrational spectra of condensed matter. Therefore, much more reliable quantum mechanical simulations are desirable, allowing one to interpret experimental observations more deeply and to model molecular properties in a more consistent way.

Typically, simulated spectra are compared to the experimental ones. Owing to the developments in the theory, software algorithms, their implementations, and hardware, it is nowadays possible to calculate fairly reliably vibrational spectra of small and medium molecules within a reasonable time. In general, problems start with bigger and flexible molecules, where accurate calculation is difficult because of high calculation cost in terms of processor time and computer memory.

Yet one has to acknowledge that some "direct" spectral simulations of large molecules already appeared, such as the ROA and Raman spectra of a β-domain of metallothionein composed of 31 amino acids. They have been calculated by using a density functional theory (DFT) aided by massively parallelized processors, the BP86/TZVP level of theory with the resolution of the identity density-fitting technique [1]. Still, many much larger proteins exist in nature or better precision in computation might be needed, not currently achievable by the parallelized CPU approach. Even small but flexible or solvated molecular systems, such as sugars and oligopeptides, require extensive calculations due to the need to consider many conformers and/or clusters of solvent and the solute [2–8]. In this context, we consider the fragment methodology as a useful tool to drastically cut down the computational demands and provide quantum-chemical solution with an accuracy acceptable for the purpose of vibrational spectroscopies, even though it may be one day replaced by more rigorous approaches.

One has to note that some similar fragment-like computational methodologies have been proposed by other researchers, e.g., the fragment molecular orbital method, [9] although they most often partition the electronic energy, not vibrational spectral properties, such as the force field.

### 8.1.2 Cartesian Coordinate-Based Transfer of Atomic Property Tensors

The transferability concept has been omnipresent in the history of vibrational spectroscopy. Force constants, atomic polarizabilities, etc., are supposed to be similar in similar chemical groups in different molecules. The Cartesian coordinate tensor transfer (CCT) [10] is thus only one of possible ways to overcome the difficulties of the spectral calculations of big molecules. Its advantage lies in the suitability to computer automatic or semi-automatic implementation, as the tensors are usually calculated in the Cartesian coordinates, and the transfer reduces to a rotation, eventually with a correction for origin dependence of electromagnetic tensors. The CCT method can be combined with any level of the electronic theory, although DFT calculations dominate so far, having been applied to VCD, IR, [11–14] and Raman spectra [4, 15–18].

In practical terms, an original "big" molecular system is cut into smaller fragments, which are made chemically feasible, e.g., missing bounds are capped by methyl groups (Fig. 8.1). Then their property tensors (force field, Raman, IR, ROA, and VCD intensity tensors) are calculated at the desired level of theory and transferred to the original molecule. Because the tensors comprise derivatives with respect to atomic coordinates, the transfer can proceed atom by atom for the first derivatives, or by atom pairs for the second derivatives, etc. Then, vibrational frequencies and intensities for the original molecule can be generated in the usual way.

Of course, in order to faithfully model properties of atoms in the original molecules, the fragments should be enough large to eliminate the "end effects." On the other hand, the fragment size is limited by available computer sources and

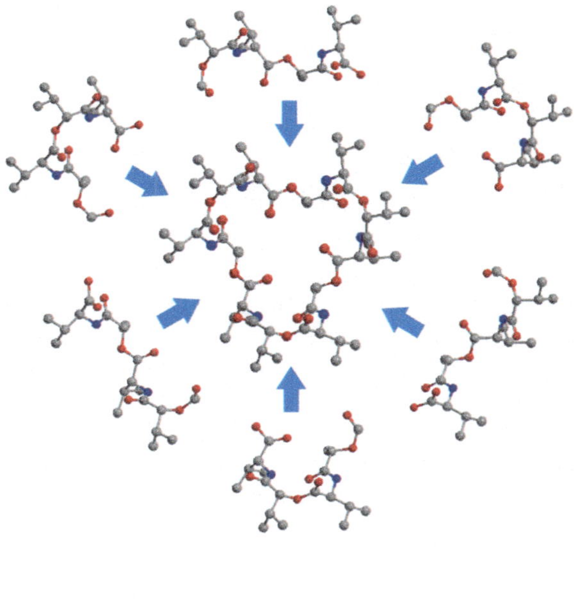

**Fig. 8.1** Example of the fragmentation of a cyclic peptide (valinomycin), used in a CCT calculation of vibrational spectra. The fragments contain four amino acids; two of them always overlapped with those in the neighboring fragments. Reproduced with permission from Springer [37]

calculation time. As an example, at this moment fragments with up to ~100 atoms give a reasonable balance between accuracy and computational cost, e.g., if the spectral calculations were done at the CAM-B3LYP/6-31++G** level with 4 parallelized processors for one fragment. The effects of fragment size and the size of their "overlapped" regions have been discussed in papers [15, 17, 19, 20].

A large overlap of the fragments eliminates the end effects, i.e., leads to more accurate simulations. Tensors for atoms in the overlapped regions thus may be taken from more fragments. They can be, for example, calculated as a weighted average, according to the position of the involved atoms with respect to the fragment mass center [10]. Typically, after a geometry optimization, terminal regions of the fragments deviate from the original structure more than the central part, which is just another reason to promote central parts of the fragments in the transfer.

The geometry optimization of the fragments is a tricky process. Ideally, we need to optimize the geometry completely, so that the atoms stay at their equilibrium positions. At the same time, we would like to keep fragment geometries similar to the original structure. The solution is obviously a *partial* optimization. This can be done with different means, for example, we can fix the torsional angles and similar coordinates not much involved in the vibrations of interest. A much more secure and in a final effect simpler method is the partial optimization in vibrational normal mode coordinates [21, 22]. Vibrational frequencies are directly addressed as the optimization criteria. For any type of optimization, one should be aware that if the fragment geometry significantly deviates from the original structure, the simulated spectra may be wrong.

Applying CCT to a particular class of compounds, e.g., peptides and proteins composed of a limited number of amino acids, it is possible to make a database of the fragments usable for molecules with similar structure [4, 15, 17]. This is also useful for coupling CCT with molecular dynamics. ROA spectra of a sugar aqueous solution [6] were calculated in this way, i.e., averaging over 10,000 conformers using a pre-calculated fragment database. For such systems, taking into account the flexibility of the molecule and its interactions with the solvent is crucial for reliable results.

### 8.1.3 Geometry Optimization in Vibrational Normal Mode Coordinates

The partial geometry optimization in the normal modes is a quite general method; within CCT it is applied to systems where we more or less know the molecular structure. For example, proteins often retain the (X-ray determined) crystal geometry also in solutions. A full structural optimization by energy minimization may often destroy it, because of the approximations used in the theoretical model. The situation is even worse for the fragments, where stabilizing contacts are lacking. As an another example, we may want to conserve the instantaneous cluster geometry obtained as an MD snapshot, such as fragments of DNA base pairs or water complexes [21]. Then, the partial optimization in normal mode coordinates can ensure minimal structural deviations while improving the geometry enough for the purpose of vibrational spectroscopy [21, 22].

The normal mode optimization (NMO) usually follows the following steps [21].

1. Estimation of the Hessian (second derivatives of the energy with respect to nuclear coordinates) so that initial vibrational normal mode coordinates ("S-matrix") can be defined. This may be done on a lower level of theory than in the other steps.
2. Calculation of the Cartesian gradient of energy.
3. Update of the Hessian and obtaining new S-matrix based on previous step.
4. Transformation of the Cartesian to normal mode gradient. If the gradient is small enough, stop the optimization.
5. Within the harmonic approximation, estimate the coordinate displacements for each normal mode (Fig. 8.2).
6. Transform the normal mode displacements to new Cartesian coordinates.
7. Go to step 2.

We implemented the normal mode optimization in a Fortran program QGRAD, which deals with energies, gradients, and second derivatives, such as those from output of the Gaussian program. QGRAD takes care about Cartesian–normal mode transformations, diagonalization of the force filed, coordinate update, etc.

Fully relaxed NMO produces the lowest-energy geometry, and its performance is roughly comparable to Cartesian coordinate-based optimization, i.e., it is relatively slow if compared to internal coordinate-based algorithms. But in a constrained NMO some vibrational frequencies can be fixed, e.g., from 300i (imaginary, often treated as negative in computer codes) to 300 $cm^{-1}$. In this way, the low-frequency vibrational modes defining the global structure (conformation, secondary/tertiary structure) are fixed. At the same time, vibrational modes important in most vibrational spectra are fully relaxed, which is the prerequisite for computation of the spectra within the harmonic approximation. The lower limit (300i $cm^{-1}$) is introduced to relax very inconvenient geometry arrangements, on the "top" of a potential hill. An important advantage also is that the constrained NMO method avoids guessing which coordinates should be frozen; unlike for internal

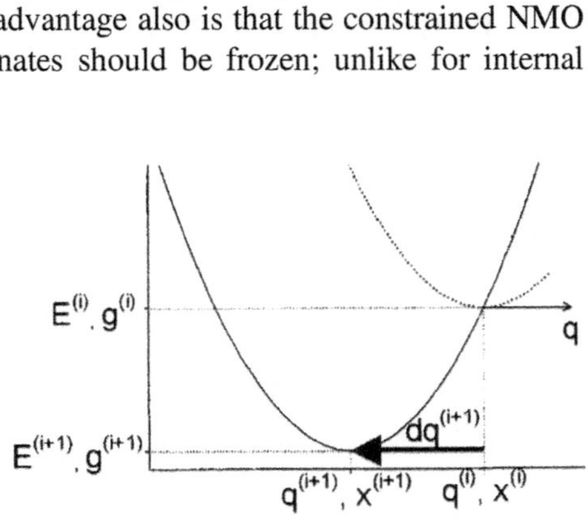


**Fig. 8.2** Normal mode optimization. At each point *i*, a quadratic energy surface is assumed so the next point *i + 1* can be estimated as its minimum. Reproduced with permission from the American Institute of Physics [21]

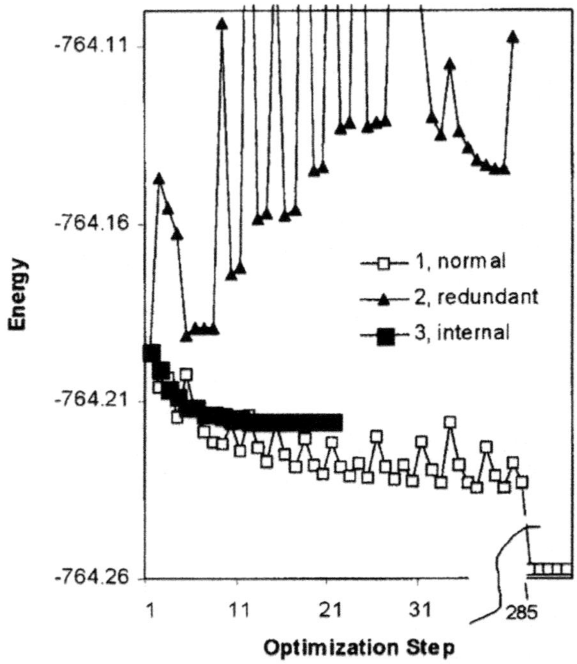


**Fig. 8.3** Convergence of three types of constrained optimizations of a water cluster consisting of 10 water molecules. The constraints were fixed normal modes below 100 $cm^{-1}$ in the normal mode method, fixed all torsion angles in the redundant internal coordinate method, and fixed torsion and bond angles in an user-defined internal coordinate method. Reproduced with permission from the American Institute of Physics [21]

coordinate-based algorithms, for example, we do not have to define thousands of torsional angles in a protein to be fixed but can use the same criterion (such as 300i–300 $cm^{-1}$) as for a small molecule. Similarly, in solvent–solute clusters definition of the internal coordinates would be very problematic.

Example of behavior of three optimization algorithms is given in Fig. 8.3, for a water cluster [21]. The NMO method exhibits some unwanted oscillation of energy but is clearly the most stable to lead the lowest-energy geometry.

## 8.2 Peptides and Proteins

### *8.2.1 Proteins*

We can exemplify application of the CCT procedure on calculations of Raman and ROA spectra of bovine insulin. The protein contains 51 amino acids and has molecular weight of ~5800 [5]. Although the measurement was performed for water solutions, the spectral calculation was based on the X-ray structure. Nevertheless, the calculations did reproduce the main features of the experimental spectra (Fig. 8.4, top and bottom) and could be extended to MD snapshots by CCT (Fig. 8.4, middle spectra).

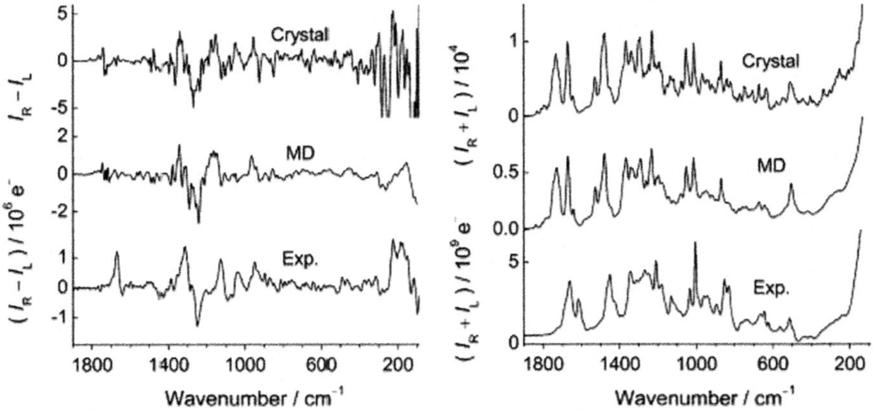


**Fig. 8.4** Calculated ROA (*left*) and Raman (*right*) spectra of bovine insulin based on the X-ray determined crystal structure (*top*), 10,000 MD snapshots (*middle*), and the experiment in water (*bottom*). The CCT calculation was based on the B3PW91/6-31++G**/CPCM (water) theory. Reproduced with permission from the American Chemical Society [5]

The computations enabled an approximate decomposition of peak intensities to contributions from molecular parts. This clarified that the amide I ROA couplet at ~1700 $cm^{-1}$ comes from amide groups with only a slight coupling with motions of the α-carbon. Another interesting observation was that while for smaller molecules there are approximately equal contributions of the electric dipole ($\alpha$), quadrupole ($A$), and magnetic dipole ($G'$) polarizabilities to ROA intensity, in the protein $\alpha$ dominates. By other words, local chirality contributions become less important in the protein and the long-range order induces a new type electric dipole-based chirality.

ROA and Raman spectra can reflect not only the equilibrium structure but also the structural fluctuations and conformer equilibria in solution phase. This was clear after averaging of ROA and Raman spectra of 10,000 MD snapshots of monomer insulin in water (Fig. 8.4, middle). The sharp Raman peaks apparent in the crystal-based spectrum vanished, in particular in the low-frequency region below 800 $cm^{-1}$. Similar smoothing effect is observed in the MD-averaged ROA spectrum, but here the spectral shapes in the higher frequency region are not conserved, which is not in favor to agreement with the experiment. This also shows that the ROA spectrum is more sensitive to protein fluctuation than the Raman, and that more accurate MD simulation, e.g., with ab initio potentials will be needed for more accurate results.

For larger globular proteins, it is important to automate the fragmentation process as much as possible; a manual definition of the fragments is not realistic anymore. This succeeded lately when Raman and ROA spectra of five globular proteins have been calculated at the B3PW91/6-31++G**/CPCM (water) level of theory, based on protein crystal structures [23]. The proteins consist of over 100 amino acids, bovine α-lactalbumin (M.W. 14178, 123 amino acids), monomer of

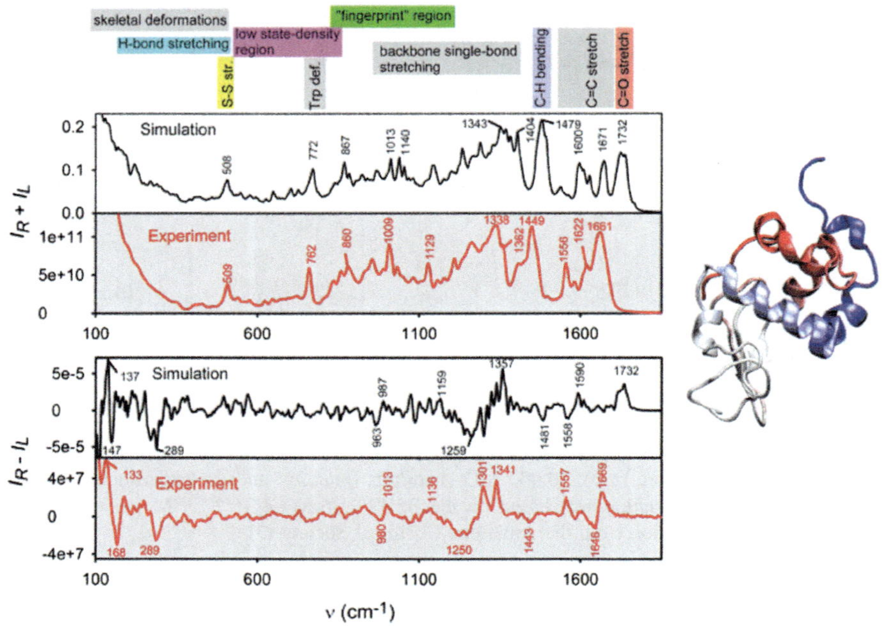


**Fig. 8.5** Comparisons of simulated (*black*) and experimental (*red*) ROA (*top*) and Raman (*bottom*) spectra of bovine α-lactalbumin. The simulation was based on CCT, the B3PW91/6-31+ +G**/CPCM (water) level of theory, and crystal structure of the protein. Reproduced with permission from the American Chemical Society [23]

concanavalin A (M.W. 26500, 235 amino acids), human serum albumin (M.W. 66437, 585 amino acids), hen egg-white lysozyme (M.W. 14307, 129 amino acids), and human lysozyme (M.W. 14700, 130 amino acids). The calculated spectral patterns were very close the experimental ones, both for Raman and ROA, which provided precious information about the relation of the spectral patterns to protein structure, cf. Figure 8.5 for bovine α-lactalbumin. Computations on dominantly α-helical human serum albumin (46% α-helix) and mostly anti-parallel β-sheet concanavalin A (46% β-sheet) reproduced the differences observed experimentally (Fig. 8.6). Even some very tiny spectral differences between hen egg- and human lysozymes having very similar structure appeared both in experiment and calculation. Analysis of the calculations showed that ROA spectra of proteins are dominated by the signal from peptide backbone, whereas Raman spectra can be roughly approximated by a sum of individual amino acid contributions. The automatic fragmentation and CCT methods provided an unprecedented precision in simulated frequencies and intensities of five globular proteins and enabled one to relate fine spectral patterns to local and higher-order structural features.

At present, it is not clear what we would get by direct DFT computations on the whole proteins; these are not possible so far. Nevertheless, the results suggest that the error of the CCT method [24] is not the limiting factor in the simulations, or at

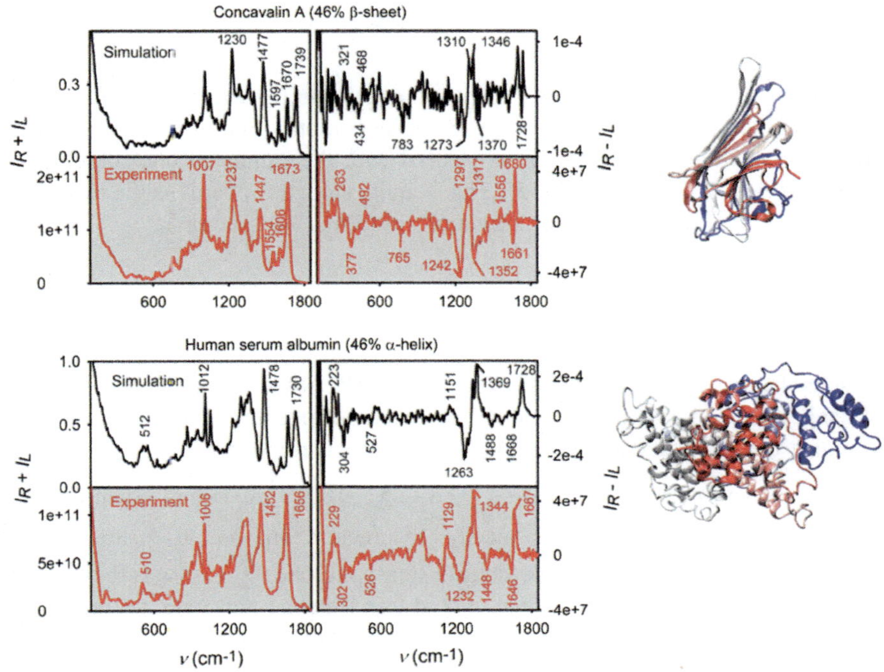


**Fig. 8.6** Comparisons of a β-sheet-dominant protein (concanavalin A) and an α-helix-dominant protein (human serum albumin). Calculated spectra with CCT-based DFT at B3PW91/6-31++G**/CPCM (water) fairly reproduced the specific spectral patterns. Reproduced with permission from the American Chemical Society [23]

least that it is comparable with the error of the underlying DFT approximation and uncertainties stemming from the geometry model including incomplete treatment of the protein dynamics. Also, one has to bear in mind the limits of the harmonic approximation; CCT to anharmonic constants can be applied as well [25] but has not been extensively explored for molecules of this size yet.

Structure and interactions of protein fibrils are important in relation to protein amyloids accompanying some neurodegenerative diseases. Layered β-sheet structures consisting of assembled five-stranded $Ala_9$ peptides were studied using CCT-based VCD, IR, and Raman spectral simulations, changing the number of layers, inter-strand and stacking arrangements [26]. CCT method was also applied to VCD and IR spectra of highly-ordered β-sheet-forming peptide, polyglutamic acid, in a combination with MD simulations [27]. It was confirmed that not only backbone amide chromophores but also the side chains (in particular the COOH groups) can produce distinct VCD spectral pattern.

### 8.2.2 Conformer Analysis of Peptides

For smaller peptides, the ROA/VCD spectroscopy coupled with computations can reveal their solution conformation. For example, conformer ratios can be estimated when the experimental spectra are compared with the simulated ones of all conformer species [1–3, 15–18, 28–35]. This is quite straightforward, as the measured spectrum ($S$) is always a sum of individual conformer subspectra $S_i$,

$$S = \sum_{i=1}^{N} c_i S_i$$

Note that, this may not be the case for NMR, where a fast conformational exchange often smears the individual signals to an average shift. Typically, we minimize the integrated square deviation between the experiments and the fit in some frequency intervals $\omega_1 \ldots \omega_2$, $\int_{\omega_1}^{\omega_2} \left( S - \sum_{i=1}^{N} c_i S_i \right)^2 d\omega \rightarrow \min$. Such decomposition is always possible algebraically. In practice, one has to ensure a good physical meaning, i.e., that the coefficients sum up to one, $\sum_{i=1}^{N} c_i = 1$, they are all positive, $N$ is not too large, false hits due to experimental noise are not produced, etc. Then, at least in some cases, the accuracy of such spectral decomposition is comparable to the more established NMR analysis [28].

The ROA spectral decomposition has also been demonstrated for conformational analysis of a cyclic twelve-membered peptide, valinomycin [15]. Valinomycin is an antibiotic ionophore and can capture selectively the potassium ion. The peptide makes a cavity for the metal, i.e., changes conformation during the binding. A spectral decomposition into calculated spectra of four conformers of free valinomycin (Fig. 8.7) revealed dominant backbone structures, which are quickly dynamically interconverting into each other in organic solvents. In 1, 4-dioxane, the "asymmetric bracelet" conformer was found to govern, and the "symmetric bracelet" was just minor one, while only two peptide-torsional angles were different from each other by ~30°. Earlier, NMR experiments could only detect the symmetric conformer, but there is no contradiction to the ROA results; NMR just

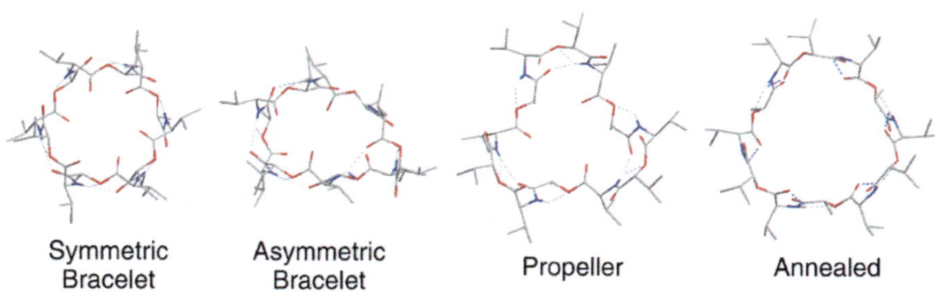


**Fig. 8.7** Most probable conformations of an antibiotic peptide, valinomycin. Reproduced with permission from Springer [37]

cannot distinguish the fast conformational equilibria. Typical reported relaxation times ranged from 20 to 2 ns; the slowest rate constant corresponds to few μs [36]. In methanol, ROA analysis combined with MD simulations mostly predicted the "propeller" conformer, with a fast bending motion of the side chains.

The ROA analysis based on the calculations can capture the side chain conformations as well. The potassium valinomycin complex can generate 6579 distinct conformers due to the rotation of the 9 isopropyl groups [17]. Two most stable conformers in methanol were determined by the combination of the ROA spectral decompositions and the energy calculations. The second stable conformer possesses only one torsional angle different from the most stable one. This high sensitivity of ROA would come from the relatively long-range interaction mediated by the ROA polarizability derivatives, as also observed in the other case [16]. The couplings of vibrations in neighboring residue strongly influence ROA intensities.

Yet further improvements of the accuracy of ROA experiment and calculations are desirable, enabling more reliable conformer analyses. To the best of our knowledge, there is no other analytical method as powerful as ROA in providing conformer populations in very fast structural equilibria.

## 8.3 Nucleic Acid

The CCT methodology has been applied to the simulation of IR and VCD spectra of large nucleic acid as well [12, 13, 38–40]. Experimental VCD and IR spectra of a model double-stranded DNA helix, poly(rA)-poly(rU) complex were computed using $(rA)_8$-$(rU)_8$ as a model (Fig. 8.8) [40]. Both the VCD/IR intensities stemming from the nucleobases, in the wavenumber region higher than 1500 $cm^{-1}$, and those from the sugar-phosphate parts, within 1070–1120 $cm^{-1}$, were reasonably well reproduced. Similarly, model helical RNA poly(rA) structures, triple-stranded DNA, and poly(rU)-poly(rA)-poly(rU) complex were also simulated with shorter model molecules and good agreements with the experimental spectra brought about better understanding of the link between the structure and the spectra and behavior of these important molecules.

Another native conformation of DNA, four-stranded DNA (quadruplex), was characterized by the CCT-based spectroscopic analysis, too [38]. Although vibrational spectroscopic methods have been used to investigate nucleic acid structure, association of particular spectral features with quadruplex structure was rather speculative. This could be done more reliably for experimental spectra of model quadruplex systems, $d(G)_8$ and deoxyguanosine-5'-monophosphate (5'-dGMP). They were compared to the calculated spectra and unambiguously assigned to the quadruplex DNA structure (Fig. 8.9). The VCD spectral pattern in a region from 1500 to 1700 $cm^{-1}$ is especially sensitive to the DNA backbone structural motif. The calculated pattern for the quadruplex very well agreed with the experiment and

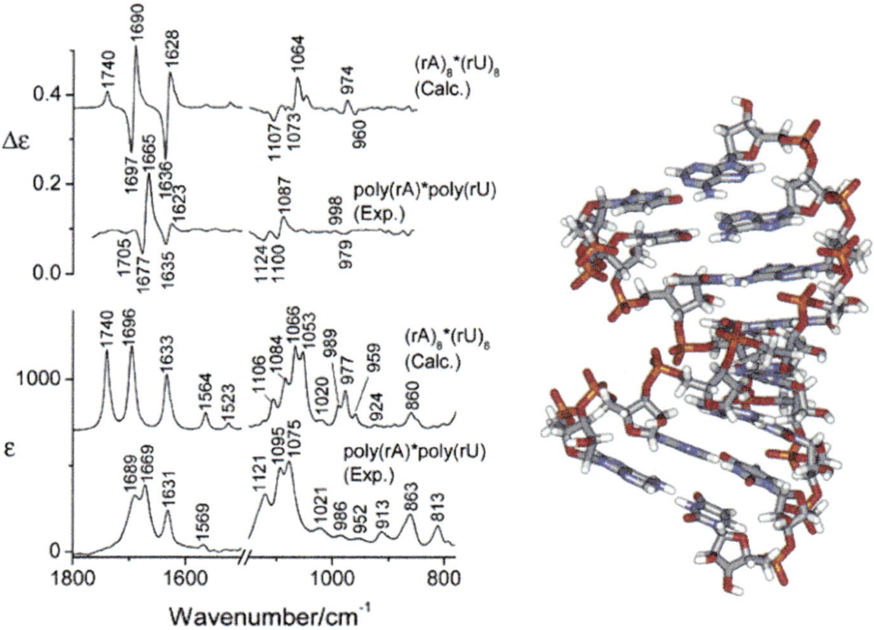


**Fig. 8.8** CCT-based VCD (*top*) and IR (*lower–middle*) spectra of the double-stranded octamer $(rA)_8$-$(rU)_8$ with the experimental spectra of poly(rA)-poly(rU). Reproduced with permission from the American Chemical Society [40]

other conformers gave much worse agreement. It was concluded that the MD simulations provided realistic geometries and conformer-averaged spectra for multiple snapshots of thermally moving DNA molecules.

## 8.4 Crystals

### *8.4.1 Polymorphism of Small Molecular Crystal*

Identification of polymorphism in molecular crystals is important for pharmaceutical industry, because the crystal state of many drugs affects physical and biological properties, such as the speed of dissolution in human body. Raman spectroscopy can be useful to monitor the polymorphism which was demonstrated for crystal forms of methacrylamide, Piracetam, and 2-thiobarbituric acid [41]. The case of methacrylamide is shown in Fig. 8.10. Spectra of both crystal forms could be well reproduced by the theory; the type of polymorphism could thus be determined by comparing the experiments and the simulations. The CCT method had to consider all important intermolecular interaction and the crystal symmetry, which was achieved by a careful selection of molecular fragments.

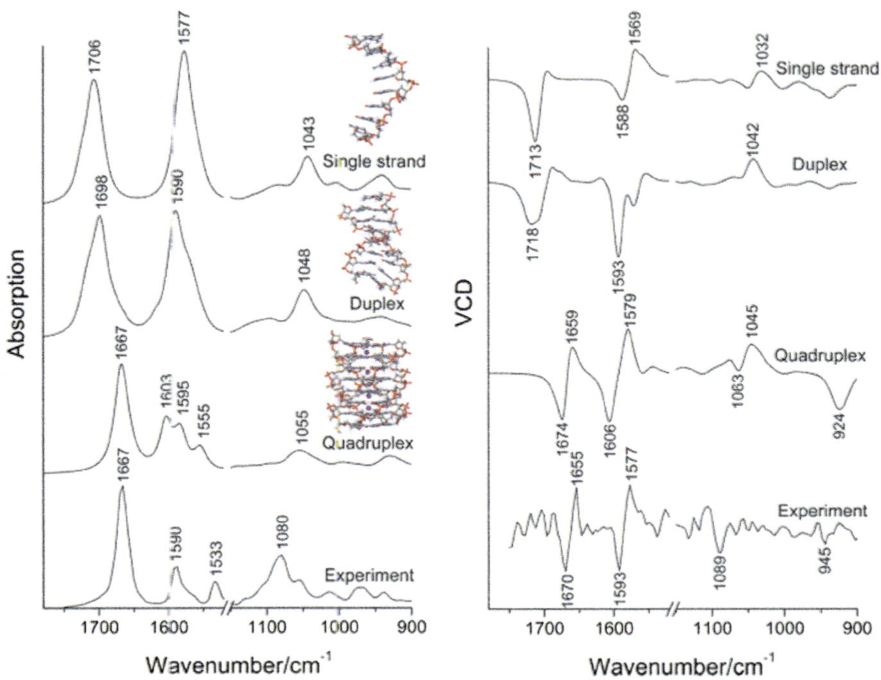


**Fig. 8.9** IR (*left*) and VCD (*right*) spectra of $d(G)_8$ in experiment (*bottom*) and CCT-based simulations of single-strand (*top*), duplex (*upper–middle*), quadruplex (*lower–middle*). The calculated spectra were MD-averaged and the lowest-energy conformers are shown for each model. Reproduced with permission from the American Chemical Society [38]

### *8.4.2 Low-Frequency Vibrational Spectra of Crystalline Polymers*

Polymeric materials continue to govern material industry for the second century. Vibrational spectra often provide useful information about intermolecular forces related to polymer stability, flexibility, biodegradability, etc. But simulations of the low-frequency polymer bands are difficult because of the size of the systems and the need to accurately account for the intermolecular interactions. On the other hand, low-frequency vibrations become experimentally easily accessible owing to the Raman technology development, such as availability of narrow filters enabling to detect light very close (~10 cm$^{-1}$) to the excitation laser line.

Also here, the CCT method has been applied. Low-frequency vibrational spectra of crystalline polyesters, poly-(*R*)-3-hydroxybutyrate (PHB) [42] and polyglycolide (PGA, to be published) were simulated and compared to the experiment. The calculations were based on the crystal structure. Unlike for the peptides and proteins, the NMO frequency limit had to be set rather low, to 20 or 40 cm$^{-1}$. This produced a reasonable agreement between the calculations and experiments, both

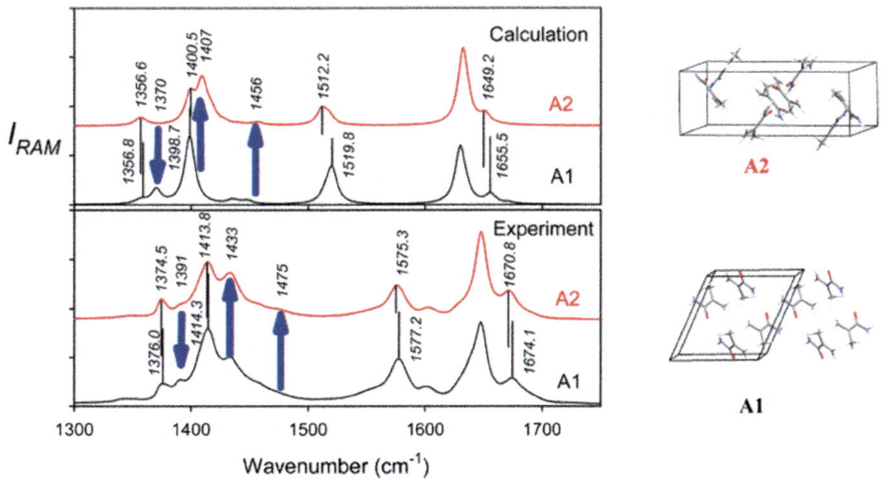


**Fig. 8.10** Calculated (*top*, plane wave) and experimental (*bottom*, NIR) Raman spectra of the two methacrylamide crystal forms. Reproduced with permission from the American Chemical Society [41]

for the Raman and FIR (THz-TD) spectra, in terms of spectral shapes, frequencies, intensities, and polarization directions.

For example, the low-frequency bands (below ~300 $cm^{-1}$) were shown to be delocalized and affected by intermolecular interactions between the polymer chains. To include the intermolecular interactions, special "intermolecular" fragments were created including atoms from two neighboring chains, which resulted to a good agreement with the observed Raman and FIR spectra (Fig. 8.11).

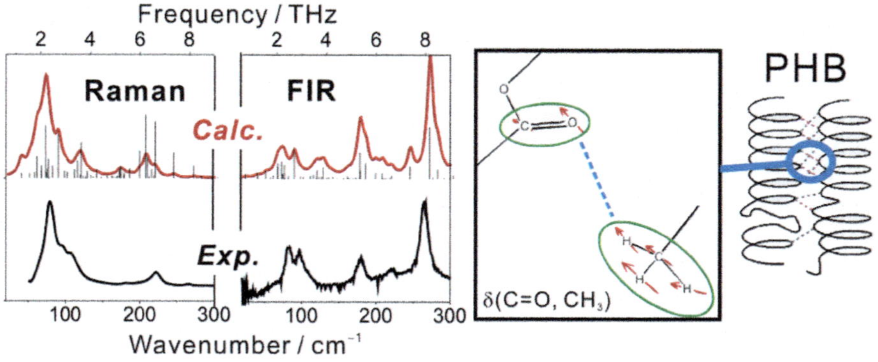


**Fig. 8.11** Calculated (*red lines*) and experimental (*black lines*) Raman and FIR spectra of crystalline PHB. Weak inter-chain H-bond interactions responsible for the 82 $cm^{-1}$ band are indicated at the *right*. Reproduced with permission from the American Chemical Society [42]

In the PHB polymer, the Raman-active band at 79 $cm^{-1}$ was assigned to an intermolecular vibrational mode of the out-of-plane $C = O + CH_3$ vibration. Polarization of a corresponding FIR band at ~82 $cm^{-1}$, observed as perpendicular to the helical direction of PHB, was reproduced as well, but only when the intermolecular fragments were included. This shows that this polarized band originates from the intermolecular interactions.

## 8.5 Conclusions

We have shown that the fragment methodology represents a relatively simple means to extend precise ab initio computation of vibrational molecular properties to fairly large molecules. The "CCT" algorithm has already found many successful applications in simulations of vibrational spectra of peptides, proteins, DNA, and crystalline polymers. Currently, such systems are not accessible by more conventional computational procedures, even with massively parallelized codes. An important advantage of the CCT approach is its flexibility and the possibility to combine computational methods, such as molecular dynamics and density functional theory. The technique can also be used to study flexible and polar molecules. Typically, it reduces computational time required during averaging of many conformers or solvent-solute clusters. We see potential of the CCT method also for the future, when improved algorithms provide more accurate simulated spectra, for example via combination of CCT with path integral or ab initio molecular dynamics, inclusion of anharmonic force field corrections, calculations accelerated by parallelization, or usage of more precise electronic theories.

**Acknowledgements** The work was supported by the Grant Agency of the Czech Republic (15-09072S and 16-05935S) to P.B. and JSPS KAKENHI (JP26708017 and JP26246037) to S.Y.

## References

1. S. Luber, M. Reiher, Theoretical Raman optical activity study of the β domain of rat metallothionein. J. Phys. Chem. B **114**, 1057–1063 (2010)
2. J. Kapitán, V Baumruk, V. Kopecký Jr., P. Bouř, Conformational flexibility of L-Alanine Zwitterion determines shapes of Raman and Raman optical activity spectral bands. J. Phys. Chem. A **110**, 4689–4696 (2006)
3. J. Kapitán, V. Baumruk, V. Kopecký Jr., R. Pohl, P. Bouř, Proline Zwitterion dynamics in solution, glass and crystalline state. J. Am. Chem. Soc. **128**(41), 13451–13462 (2006)
4. J. Kaminský, J. Kapitán, V. Baumruk, L. Bednárová, P. Bouř, Interpretation of Raman and Raman optical activity spectra of a flexible sugar derivative, the gluconic acid anion. J. Phys. Chem. A **113**, 3594–3601 (2009)
5. S. Yamamoto, J. Kaminský, P. Bouř, Structure and vibrational motion of insulin from Raman optical activity spectra. Anal. Chem. **84**, 2440–2451 (2012)

6. J. Kaminský, J. Šebek, P. Bouř, Molecular dynamics with restrictions derived from optical spectra. J. Comput. Chem. **30**, 983–991 (2009)
7. K.H. Hopmann, K. Ruud, M. Pecul, M. Dracinsky, P. Bouř, Explicit versus implicit solvent modeling of Raman optical activity spectra. J. Phys. Chem. B **115**, 4128–4137 (2011)
8. J.R. Cheeseman, M.S. Shaik, P.L.A. Popelier, E.W. Blanch, Calculation of Raman optical activity spectra of methyl-β-D-glucose incorporating a full molecular dynamics simulation of hydration effects. J. Am. Chem. Soc. **133**, 4991–4997 (2011)
9. D.G. Fedorov, T. Nagata, K. Kitaura, Exploring chemistry with the fragment molecular orbital method. Phys. Chem. Chem. Phys. **14**, 7562–7577 (2012)
10. P. Bouř, J. Sopková, L. Bednárová, P. Maloň, T.A. Keiderling, Transfer of molecular property tensors in Cartesian coordinates: a new algorithm for simulation of vibrational spectra. J. Comput. Chem. **18**, 646–659 (1997)
11. V. Andrushchenko, P. Bouř, Circular dichroism enhancement in large DNA aggregates simulated by a generalized oscillator model. J. Comput. Chem. **29**, 2693–2703 (2008)
12. V. Andrushchenko, P. Bouř, DNA oligonucleotide-cis-platin binding: ab initio interpretation of the vibrational spectra. J. Phys. Chem. A **111**, 9714–9723 (2007)
13. V. Andrushchenko, H. Wieser, P. Bouř, B-Z conformational transition of DNA monitored by vibrational circular dichroism. Ab initio interpretation of the experiment. J. Phys. Chem. B **106**, 12623–12634 (2002)
14. J. Kubelka, T.A. Keiderling, Differentiation of β-sheet forming structures: ab initio based simulations of IR absorption and vibrational CD for model peptide and protein β-sheets. J. Am. Chem. Soc. **123**, 12048–12058 (2001)
15. S. Yamamoto, H. Watarai, P. Bouř, Monitoring the backbone conformation of valinomycin by Raman optical activity. Chem. Phys. Chem. **12**, 1509–1518 (2011)
16. J. Hudecová, J. Kapitán, V. Baumruk, R.P. Hammer, T.A. Keiderling, P. Bouř, Side chain and flexibility contributions to the Raman optical activity spectra of a model cyclic hexapeptide. J. Phys. Chem. A **114**, 7642–7651 (2010)
17. S. Yamamoto, M. Straka, H. Watarai, P. Bouř, Formation and structure of the potassium complex of valinomycin in solution studied by Raman optical activity spectroscopy. Phys. Chem. Chem. Phys. **12**, 11021–11032 (2010)
18. J. Kapitán, F. Zhu, L. Hecht, J. Gardiner, D. Seebach, L.D. Barron, Solution structures of β peptides from Raman optical activity. Angew. Chem. **120**, 6492–6494 (2008)
19. S. Yamamoto, P. Bouř, On the limited precision of transfer of molecular optical activity tensors. Collect. Czech. Chem. Commun. **76**(5), 567–583 (2011)
20. N.S. Bieler, M.P. Haag, C.R. Jacob, M. Reiher, Analysis of the Cartesian tensor transfer method for calculating vibrational spectra of polypeptides. J. Chem. Theory Comput. **7**, 1867–1881 (2011)
21. P. Bouř, T.A. Keiderling, Partial optimization of molecular geometry in normal coordinates and use as a tool for simulation of vibrational spectra. J. Chem. Phys. **117**, 4126–4132 (2002)
22. P. Bouř, Convergence properties of the normal mode optimization and its combination with molecular geometry constraints. Collect. Czech. Chem. Commun. **70**, 1315–1340 (2005)
23. J. Kessler, J. Kapitán, P. Bouř, First-principles predictions of vibrational Raman optical activity of globular proteins. J. Phys. Chem. Lett. **6**, 3314–3319 (2015)
24. S. Yamamoto, X. Li, K. Ruud, P. Bouř, Transferability of various molecular property tensors in vibrational spectroscopy. J. Chem. Theory Comput. **8**, 977–985 (2012)
25. P. Bouř, V. Parchaňský, Transferability of anharmonic force fields in simulations of molecular vibrations. J. Chem. Phys. **133**, 044117 (2010)
26. W.R.W. Welch, J. Kubelka, T.A. Keiderling, Infrared, vibrational circular dichroism, and Raman spectral simulations for β-sheet structures with various isotopic labels interstrand, and stacking arrangements using density functional theory. J. Phys. Chem. B **117**, 10343–10358 (2013)
27. J. Kessler, T.A. Keiderling, P. Bouř, Arrangement of fibril side chains studied by molecular dynamics and simulated infrared and vibrational circular dichroism spectra. J. Phys. Chem. B **118**, 6937–6945 (2014)

28. M. Buděšínský, P. Daněček, L. Bednárová, J. Kapitán, V. Baumruk, P. Bouř, Comparison of quantitative conformer analyses by nuclear magnetic resonance and Raman optical activity spectra for model dipeptides. J. Phys. Chem. A **112**, 8633–8640 (2008)
29. M. Buděšínský, J. Šebestík, L. Bednárová, V. Baumruk, M. Šafařík, P. Bouř, Conformational properties of the Pro-Gly motif in the D-Ala-L-Pro-Gly-D-Ala model peptide explored by a statistical analysis of the NMR, Raman, and Raman optical activity spectra. J. Org. Chem. **73**, 1481–1489 (2008)
30. J. Kapitán, V. Baumruk, V. Kopecký Jr., P. Bouř, Demonstration of the ring conformation in polyproline by the Raman optical activity. J. Am. Chem. Soc. **128**, 2438–2443 (2006)
31. C.R. Jacob, S. Luber, M. Reiher, Calculated Raman optical activity signatures of tryptophan side chains. Chem. Phys. Chem. **9**, 2177–2188 (2008)
32. P. Bouř, J. Kapitán, V. Baumruk, Simulation of the Raman optical activity of L-Alanyl-L-Alanine. J. Phys. Chem. A **105**, 6362–6368 (2001)
33. C.R. Jacob, S. Luber, M. Reiher, Understanding the signatures of secondary-structure elements in proteins with Raman optical activity spectroscopy. Chem. Eur. J. **15**, 13491–13508 (2009)
34. F. Zhu, J. Kapitán, G.E. Tranter, P.D.A. Pudney, N.W. Isaacs, L. Hecht, L.D. Barron, Residual structure in disordered peptides and unfolded proteins from multivariate analysis and ab initio simulation of Raman optical activity data. Proteins **70**, 823–833 (2008)
35. P. Bouř, V. Sychrovský, P. Maloň, J. Hanzlíková, V. Baumruk, J. Pospíšek, M. Buděšínský, Conformation of the dipeptide cyclo (L-Pro-L-Pro) monitored by the nuclear magnetic resonance and Raman optical activity spectra. Experimental and ab initio computational study. J. Phys. Chem. A **106**, 7321–7327 (2002)
36. E. Grell, T. Funck, Dynamic properties and membrane activity of ion specific antibiotics. J. Supramol. Struct. **1**, 307–335 (1973)
37. S. Yamamoto, Conformational analyses of peptides and proteins by vibrational Raman optical activity. Anal. Bioanal. Chem. **403**, 2203–2212 (2012)
38. V. Andrushchenko, D. Tsankov, M. Krasteva, H. Wieser, P. Bouř, Spectroscopic detection of DNA quadruplexes by vibrational circular dichroism. J. Am. Chem. Soc. **133**, 15055–15064 (2011)
39. V. Andrushchenko, P. Bouř, Applications of the Cartesian coordinate tensor transfer technique in the simulations of vibrational circular dichroism spectra of oligonucleotides. Chirality **22**, E96–E114 (2010)
40. V. Andrushchenko, H. Wieser, P. Bouř, RNA structural forms studied by vibrational circular dichroism: ab initio interpretation of the spectra. J. Phys. Chem. B **108**, 3899–3911 (2004)
41. M. Dračínský, E. Procházková, J. Kessler, J. Šebestík, P. Matějka, P. Bouř, Resolution of organic polymorphic crystals by raman spectroscopy. J. Phys. Chem. B **117**, 7297–7307 (2013)
42. S. Yamamoto, Y. Morisawa, H. Sato, H. Hoshina, Y. Ozaki, Quantum mechanical interpretation of intermolecular vibrational modes of crystalline poly-(r)-3-hydroxybutyrate observed in low-frequency Raman and terahertz spectra. J. Phys. Chem. B **117**, 2180–2187 (2013)

# Chapter 9
# Describing Molecules in Motion by Quantum Many-Body Methods

**Ove Christiansen**

**Abstract** For a complete quantum description of molecular systems, it is necessary to solve Schrödinger equations for both electrons and nuclei. In this chapter, focus is given to approximate methods for solving the nuclear Schrödinger equation. Similarities and dissimilarities compared to the practice employed for the electronic case will be noted. A many-body view on potential energy surfaces will be used to motivate a many-body view on the general problem of solving the nuclear Schrödinger equation. A second quantization multimode formalism will be outlined and used to formulate many-body wave functions for nuclear motion. The vibrational self-consistent field (VSCF) method is introduced. Full vibrational configuration interaction (FVCI) is introduced as the reference, before primary attention is given to vibrational coupled cluster (VCC) theory. VCC theory is furthermore analysed from a tensor decomposition perspective and with a perspective to scaling with system size.

**Keywords** Molecules in motion · Anharmonic molecular vibrations · Vibrational coupled cluster · Many-body methods · Potential energy surfaces · Second quantization · Tensors

## 9.1 Introducing

Predicting the behaviour of molecules on the basis of quantum mechanics requires a quantum treatment of both the electrons and the atomic nuclei. In the Born-Oppenheimer approximation, we can can consider the electronic and the nuclear problem as two separate, but coupled problems. Through the years a number of computational tools have been developed for solving the electronic Schrödinger equation. Among the most successful of these methods are many-body methods such as perturbation theory and coupled cluster theory, and they are now text book methods [1, 2]. Today coupled cluster calculations are in many contexts the golden standard.

O. Christiansen (✉)
Department of Chemistry, Aarhus University, Aarhus, Denmark
e-mail: ove@chem.au.dk



M.J. Wójcik et al. (eds.), *Frontiers of Quantum Chemistry*,
https://doi.org/10.1007/978-981-10-5651-2_9

In this chapter, we consider many-body methods for the nuclear Schrödinger equation. For solving the nuclear Schrödinger equation the application of many-body methods has been considerable less common, and for nuclear motion variational treatments using a linear expansion in a basis is prevailing. Nevertheless, there is still key advantages of a many-body approach. In the next chapter, I will highlight some general differences between electronic structure theory and vibrational structure theory. Thereafter, I will introduce hierarchies of approximations for the Hamiltonian that can be denoted many-body like and subsequently discuss construction of wave functions in second quantization, including, in particular, vibrational self-consistent field (VSCF), vibrational configuration interaction (VCI) and vibrational coupled cluster (VCC) methods. I will also present a chapter with a tensor perspective on the VCC wave function.

## 9.2 Electronic and Nuclear Schrödinger Equations

Consider a molecular system described by quantum mechanics in the non-relativistic time-independent Schrödinger picture

$$(T_e + T_n + V_{ee} + V_{nn} + V_{en})\Psi_{tot}(\mathbf{r}, \mathbf{R}) = E\Psi_{tot}(\mathbf{r}, \mathbf{R}) \tag{9.1}$$

Here $\mathbf{r}$ and $\mathbf{R}$ symbolize, respectively, electronic and nuclear coordinates. We invoke from the outset the Born-Oppenheimer approximation. The electronic Schrödinger equation for a set of clamped nuclei determines the energy $E_{el}(\mathbf{R})$ as a function of nuclear coordinates

$$(T_e + V_{ee} + V_{nn} + V_{en})\Psi_{el}(\mathbf{r}; \mathbf{R}) = E_{el}(\mathbf{R})\Psi_{el}(\mathbf{r}; \mathbf{R}) \tag{9.2}$$

This electronic energy appears as the potential in the nuclear Schrödinger equation, i.e. with $V = E_{el}(\mathbf{R})$, we write

$$(T_n + V)\Psi_n(\mathbf{R}) = E_n\Psi_n(\mathbf{R}) \tag{9.3}$$

where $E_n$ is the total energy. In general, nuclear motion covers both translation, rotation and vibration.

Consider the vibrational motion of a nonlinear molecule with $N$ atoms, leading to $M = 3N - 6$ vibrational degrees of freedom denoted *modes*, separating out translational and rotational motion. The internal relative nuclear motion is thereby described by a set of $M$ coordinates denoted $q_1, q_2, \ldots, q_M$. While electronic structure theory deals with indistinguishable electrons where each is described by a set of cartesian coordinates as well as a spin coordinate, the internal motion of molecules is described by a set of distinguishable degrees of freedom. The particular set of coordinates may be a set of normal coordinates (a rectilinear coordinate set) or general curvilinear coordinates, such as bond lengths, angles and dihedrals.

The immediate consequence of this is that in the electronic case the kinetic energy operator is simple (using atomic units throughout),

$$T_e = \sum_i \frac{1}{2} \nabla_i^2 \tag{9.4}$$

In contrast, already in normal coordinates the exact kinetic operator is more involved in the exact Watson [3] form for rovibrational motion (separating out translation)

$$T = -\frac{1}{2} \sum_k \frac{\partial^2}{\partial Q_k^2} + \frac{1}{2} \sum_{\alpha\beta} (J_\alpha - \pi_\alpha)\mu_{\alpha\beta}(J_\beta - \pi_\beta) - \frac{1}{8} \sum_\alpha \mu_{\alpha\alpha} \tag{9.5}$$

Here $J_\alpha$ is the total and $\pi_\beta$ the vibrational angular momenta, respectively, $\mu_{\alpha\beta}$ is the inverse of the effective moment of inertia. Considering a non-rotating molecule through putting $J = 0$, we observed there are still some terms with the $\mu$ tensor left. These terms are often neglected for larger molecules as the inverse of the moment of inertia will decrease their importance. With their neglect a simple kinetic operator is obtained for the vibrational motion (separating out rotation)

$$T = -\frac{1}{2} \sum_k \frac{\partial^2}{\partial Q_k^2} \tag{9.6}$$

Neglecting the additional Watsonian terms can be a significant approximation for small molecules and/or vibrations with light atoms. The remaining coupling in the Watson in Eq. (9.5) derives from the interaction between rotation and vibration and the fact that the rotational set of coordinates is not a rectilinear set of coordinates. In more general curvilinear coordinates the vibrational kinetic energy operator is even more complicated. Thus, unlike electronic structure theory the kinetic energy operator may generally bring coupling between the different degrees of freedom in the quantum treatment of nuclear motion.

The primary cause of couplings between the modes is the potential. In the case of the electronic Schrödinger equation, the electron–electron interaction is of Coulomb type and thus of two-body nature. In contrast, the vibrational structure potential is the Born-Oppenheimer potential which formally couples all degrees of freedom. The Coulomb potential is known exactly and before any calculation and is similar for all molecules. In other words, the potential is universally defined. The potential of the nuclear Schrödinger equation will not be known exactly, and it is unique for each different molecule.

The above points show clearly the different outsets for electronic and nuclear theories. Since the spaces and the Hamiltonians are fundamentally different our method of solution will also differ. Nevertheless, we shall discuss how one can recover in certain ways a many-body view on describing coupling in many mode systems that is somewhat similar to many-body theory for electrons.

In passing I note that though the relevance of time-dependent electronic theory is becoming more and more clear, the overwhelming majority of quantum chemical studies up to this day are based on time-independent quantum theory, and some of the extensions to time-dependent theory are furthermore restricted to special cases, such as periodic perturbations. For nuclear motion, time-dependent approaches are generally a much more equal part of the bigger field of describing nuclear motion. For example, a general method like the multi-configurational time-dependent Hartree (MCTDH) has shown great applicability for both time-dependent and time-independent phenomena. However, in this chapter we shall stay with a time-independent description.

## 9.3 Many-Body Expansion of the Hamiltonian

The potential is decisive for accuracy and efficiency. We have already established that the potential originates from solving the electronic Schrödinger equation. The electronic Schrödinger equation cannot be solved exactly for non-trivial systems, and the first primary cause for concern is thus the accuracy of the electronic structure methodology used in a concrete case. This is the subject of electronic structure theory and will not be pursued further here. However, we shall consider other decisive factors such as how to deal with the dimensionality of the potential. How can we represent the PES in many dimensions? And how can we obtain the effect of the PES from calculating only a restricted set of points on the PES?

One classical way of obtaining a PES is through Taylor expansion. Thus, following geometry optimization and subsequent harmonic analysis, higher (than second) order derivatives are calculated up to a maximum order. The third (cubic) and fourth (quartic) order potential terms describe the lowest order anharmonicity. Their calculation and use have perfect meaning in a perturbation theoretical sense, using the normal coordinate harmonic oscillator as the zeroth order description. However, in a broader perspective Taylor expanded potentials are problematic. First of all, the radius of convergence is likely small and for larger distortions from the reference structure, the PES can be quite unphysical. It is for example common that the Taylor expanded PES have "variational holes". That is, at some distortions from the reference, the PES becomes unphysically of lower energy than at the reference structure. Applying a Taylor expanded PES in a variational calculation is obviously risky.

There are, however, also some attractive features of the Taylor representation. One attractive feature is that the corresponding Hamiltonian becomes a sum over products (at least if either the simple kinetic energy operator is used or the inverse moment of inertia in the kinetic energy terms is treated in a similar Taylor expansion). Thus, the Hamiltonian contains a sum of $q_i^{n_i} q_j^{n_j} q_k^{n_k} \cdots$ terms with prefactors. A general sum over product Hamiltonian can be written as a sum over $T$ product terms

$$\hat{H} = \sum_{t=1}^{T} c_t \prod_{m=1}^{M} \hat{h}^{m,t} \tag{9.7}$$

giving a similar simple product structure for integrals over direct product basis functions. This is in turn decisive for efficient wave function calculations, since this means that each term is separable into a product over the involved modes. This feature is in fact so important that even if we do not pursue it directly by Taylor expansion, it is often desirable to fit the potential to the sum-over-product-form of Eq. (9.7). In other words, for the PES we can make a fit of energies calculated on a grid to the particular set of fit functions we choose. Here, we may use polynomials for fitting, but it is important to distinguish the fitting case from the polynomial obtained as Taylor expansion.

The above considerations, however, does not solve the problem of dimensionality. We will now introduce a particular type of many-body expansion of the potential. The basic idea of restricted mode–mode coupling has been pursued under different names such as $n$-mode representation [4], many-body expansion [5], mode-coupling expansion, cut-HDMR (High Dimensional Model Representation) [6], cluster expansion [7] and others [8–11]. As such it has been an integrated part of much recent research. The outcome is that the full PES is obtained as a sum of many lower-dimensional functions. Let a vector of mode indices be called a mode combination (MC), and $\mathbf{m}_n$ is such a vector of $n$ indices $m_1, m_2, \ldots, m_n$. Consider an expansion point and a set of coordinates defined to be all zero for the expansion point. Assume that $V(0,0,0,0\ldots0) = 0$, noting that an additional constant term can easily be added after our treatment. We can now write the PES in terms of the particular MCs that are relevant for the representation of the potential:

$$V \approx \sum_{\mathbf{m}\in\mathbf{MCR\{V\}}} \bar{V}^{\mathbf{m}}, \tag{9.8}$$

Here MCR is a mode combination range—the set of MCs included in the potential. Equation (9.8) is the basis for introducing approximations. IA sequence of approximate potentials providing a more and more accurate description of the full coupling by including more and more mode couplings. The potentials $\bar{V}^{\mathbf{m}_n}$ are incremental potentials, defined such that they give zero if any coordinates in the set given by $\mathbf{m}_n$ are zero

$$\bar{V}^{\mathbf{m}_n}(\ldots, q_i = 0, \ldots) = 0. \tag{9.9}$$

For the two-mode example

$$\bar{V}^{m_1 m_2} = V^{m_1 m_2} - V^{m_1} - V^{m_2} \tag{9.10}$$

where

$$V^{m_1} = V(0, \ldots, 0, q_{m_1}, 0, \ldots, 0) \tag{9.11}$$

$$V^{m_1 m_2} = V(0, \ldots, 0, q_{m_1}, 0, \ldots, 0, q_{m_2}, 0, \ldots, 0) \tag{9.12}$$

etc. A two-mode representation of the PES can be written as

$$V^{(2)} = \sum_{m_1} \bar{V}^{m_1} + \sum_{m_1<m_2} \bar{V}^{m_1 m_2} = \sum_{m_1<m_2} V^{m_1 m_2} - (M-2) \sum_{m_1} V^{m_1} \tag{9.13}$$

The mode number indices $m_i$ run from mode 1 to mode $M$ and $m_i \neq m_j$. In the two-mode case, the PES of a three atomic system consists of three two-mode sub-PES and three one-mode sub-PES. Thus, the expression in Eq. (9.8) is an approximation to the full PES. Instead of computing the full PES, it is represented in terms of a number of sub-PES where each of these is restricted to a limited set of modes. This provides a much more realistic path to accurately representing the PES. Consider for example a grid-based approach for PES construction, where potential energy values are calculated on an appropriate set of points each called a single point (SP). For simplicity we assume that the number of SPs per coordinate is $N_p$ and the same for all coordinates. Including up to $n$-mode couplings for a molecule with $M$ modes gives rise to $\sum_{i=0}^{n} \binom{M}{i} (N_p)^i$ SPs as opposed to the full PES having $N_p^M$. This is a reduction from exponential to polynomial computational scaling in $M$. The leading scaling, $\binom{M}{n} (N_p)^n$ is the dominating computational effort. In the sequence of potentials converging to the full potential

$$V^{(1)}, V^{(2)}, V^{(3)}, \ldots, V^{(M)} \tag{9.14}$$

where $V^{(n)}$ include all mode couplings up to and including $n$-mode couplings, it is typically only realistic to consider low-order methods, say $n = 2, 3, 4$. Still, we have obtained a converging sequence of potentials. We also see clearly the additional complexity compared to the electronic case with its universally defined two-body Coulomb potential.

The n-mode PES approach defined above has been used and tuned by many researchers in different contexts with respect to obtaining accurate potentials with as few SPs as possible. By fitting to a basis of functions, the sub-PES can be brought into a sum-over-product form. Thus, the total Hamiltonian can be numerically represented in an n-mode expanded sum-over-product form. This is highly convenient for the subsequent calculation and use of the matrix element of the Hamiltonian for the vibrational wave function calculation. Finally, one should be aware of the fact that the PES construction and the wave function calculation are not quite independent. A given wave function spans over a particular region of configuration space, and it is important the potential is fairly accurate in this region. On the other hand, the more involved the PES really is, or its representation, the more involved the vibrational structure calculation will be. An adaptive density guided approach (ADGA) has been developed, where the PES is calculated in accordance with the convergence of the combined wave function and PES calculation [12].

The restricted mode-coupling representation of the potential discussed above is in a certain sense a many-body expansion. Such expansions can be generalized and applied to the PES in other ways [13]. In particular, the many-body view is in agreement with the so-called incremental method for calculating electronic energies. In the incremental method, the full system is considered as composed of fragments. In practice some kind of cutting and capping of the full system into subsystems is required which will not be further discussed here. The idea is now that the electronic energy of the system can be calculated in a many-body expansion similar as above. Let $A, B$ denote fragments. Then the energy of each SP can be written as

$$E_{el} = \sum_{A} E_A + \sum_{A>B} (E_{AB} - E_A - E_B) + \cdots \qquad (9.15)$$

i.e. the energy is obtained as a sum over energies calculated for different fragments and different fragment combinations. This form of the energy evaluation can now be combined with potential expansion, such that each SP in the calculation of the many small sub-PES are expanded as in Eq. (9.15). This means that we have progressed from a full dimensional PES calculated from full dimensional SPs to a PES given in terms of sums of lower-dimensional sub-PESs calculated from energies obtained as sums of lower-dimensional SPs. We denote such approaches as double incremental, as it is in a certain sense simply the simultaneous application of the same idea in two different directions.

The efficiency of the double incremental idea can be strongly boosted by using coordinates with a well-defined locality, meaning that a given coordinate is known to move only a limited set of atoms relative to each other. To understand this, note that standard normal coordinates will be formally delocalized over the full systems. This means, that in the calculation of every sub-PES the displacement of the coordinates will mean that there are slight variations in all fragment combinations. This means that all fragments must be recalculated. On the other hand, if we consider a sub-PES for coordinates that a completely localized to atoms within a few fragments, then the calculation of this sub-PES requires only calculations for the fragment combinations that actually change.

It has been shown how an algorithm for flexible adaptation of local coordinates of nuclei (FALCON) [14] can be used to provide a set of coordinates with a well-defined locality. Following the above considerations, the double incremental approach with FALCON coordinates (denoted DIF) is computationally much more efficient than the double incremental approach in normal coordinates (denoted DIN). Using the particular structure of FALCON coordinates, the efficiency can be even further improved by applying DIF with auxiliary coordinate transformations (DIFACT). With the DIFACT approach one can achieve linear scaling of the total accumulated cost of the SP calculations [13]. This should be contrasted to the original mode-coupling expansion case where the accumulated cost of the SP calculations would scale as $M^{n_{max}} \times N^s$ where $N$ is the number of orbitals of the systems and $s$ is the characteristic scaling of the applied quantum chemical method, i.e. s = 7 for CCSD(T).

The ideas in the double incremental expansion approach are likely extendable to other local coordinates than FALCON coordinates, such as various sets of internal coordinates. It does require a molecule of some size for the fragmentation procedure to make chemical sense. However, the potentially huge reductions in computational effort in the construction of the PES is highly encouraging for future applications to larger systems.

## 9.4 Second Quantization

Second quantization formulations have proven very productive in many contexts of many-body theory, for example for the electronic structure of molecules. The commutation or anti-commutation relations of the fundamental creation and annihilation operators brings to live the symmetry or antisymmetry requirements of the wave function with respect to particle permutation. The interest in developing many-body methods for nuclear motion lead to the development of SQ formulations for systems with distinguishable degrees of freedom [15, 16] and will be briefly summarized.

Assume that for mode $m$ we have a complete basis $\{\phi^m_{p^m}(q_m)\}$ indexed by $p^m = 0, 1, \ldots, N^m - 1$. Here $N^m$ is the number of one-mode basis functions for mode $m$. We may generate basis functions in the $M$-mode space as products of the one-mode basis functions

$$\Phi_{\mathbf{s}}(q_1, q_2, \ldots, q_M) = \prod_{m=1}^{M} \phi^m_{s^m}(q_m) \tag{9.16}$$

The one-mode functions shall be denoted *modals* while their $M$-mode product states are denoted Hartree products. We consider the system as composed of $M$ distinguishable degrees of freedom, and accordingly we do not consider any symmetrization

In setting up second quantization we describe the system in terms of occupation-number vectors (ONVs) where there per construction is a one-to-one correspondence between second quantization ONVs and first quantization Hartree products. The ONVs are of the form

$$|\mathbf{k}\rangle = \left| \{k^1_0, k^1_1, \ldots, k^1_{N^1-1}\}, \ldots, \{k^m_0, k^m_1, \ldots, k^m_{N^m-1}\}, \ldots, \{k^M_0, k^M_1, \ldots, k^M_{N^M-1}\} \right\rangle \tag{9.17}$$

The integer $k^m_{p^m}$ is the occupation number for modal $p^m$ of mode $m$.

The vacuum state is the state with only zero occupation $|\text{vac}\rangle = |0, 0, \ldots, 0\rangle$ and is normalized as well, i.e. $\langle\text{vac}|\text{vac}\rangle = 1$. Similar to the overlap between two Hartree products for an orthonormal basis, the ONV inner product for an orthonormal basis is

$$\langle\mathbf{k}|\mathbf{l}\rangle = \prod_{m=1}^{M} \prod_{p^m=1}^{N^m} \delta_{k^m_{p^m} l^m_{p^m}} = \delta_{\mathbf{kl}} \tag{9.18}$$

while for a general vector $|\mathbf{c}\rangle = \sum_{\mathbf{k}} c_{\mathbf{k}} |\mathbf{k}\rangle$ the inner product is defined as,

$$\langle \mathbf{c}|\mathbf{d}\rangle = \sum_{\mathbf{kl}} \langle \mathbf{k}|c^*_{\mathbf{k}} d_{\mathbf{l}}|\mathbf{l}\rangle = \sum_{\mathbf{kl}} c^*_{\mathbf{k}} d_{\mathbf{l}} \delta_{\mathbf{kl}} = \sum_{\mathbf{k}} c^*_{\mathbf{k}} d_{\mathbf{k}} \tag{9.19}$$

In the ONV space creation and annihilation operators can be introduced with commutator relations,

$$[a^{m\dagger}_{p^m}, a^{m'\dagger}_{q^{m'}}] = [a^m_{p^m}, a^{m'}_{q^{m'}}] = 0, \tag{9.20}$$

$$[a^m_{p^m}, a^{m'\dagger}_{q^{m'}}] = \delta_{mm'} \delta_{p^m q^{m'}}. \tag{9.21}$$

The action of $a^{m\dagger}_{p^m}$ is to create a vibration in mode $m$, level $p^m$ while $a^m_{q^m}$ annihilates a vibration in level $q^m$ of mode m. In particular, an annihilation operator acting on the vacuum state gives zero

$$a^m_{p^m}|\mathrm{vac}\rangle = 0 \tag{9.22}$$

The physically relevant subspace of all $|\mathbf{k}\rangle$ that correspond to the set of $M$-mode Hartree products are those where only one $k^m_{i^m}$ is equal to 1 for each mode $m$ and all the others are 0.

Defining some further terminology is useful: $\mathbf{m}$ denotes a set of modes combined in a mode combination (MC). A MC may include from 1 to $M$ modes, and we shall occasionally use a subscript to indicated the dimensionality.

The creation and annihilation operators can in turn be used to construct all SQ states and operators such that there is a one-to-one correspondence between all matrix elements and thereby all measurable quantities between calculations done in first and second quantization. The one-mode shift operators $E^m_{p^m q^m}$ that moves occupation in mode $m$ from modal $q^m$ to $p^m$ is highly useful in this regard. They are defined as

$$E^m_{p^m q^m} = a^{m\dagger}_{p^m} a^m_{q^m}. \tag{9.23}$$

The general Hartree Product above can now be expressed as

$$|\Phi_{\mathbf{s}}\rangle = \prod_{m=1}^{M} a^{m\dagger}_{s^m} |\mathrm{vac}\rangle. \tag{9.24}$$

We may pick out a reference state described by an $M$-dimensional vector $\mathbf{i} = \left(i^1, i^2, \ldots, i^M\right)$

$$|\Phi_{\mathbf{i}}\rangle = \prod_{m=1}^{M} a^{m\dagger}_{i^m} |\mathrm{vac}\rangle. \tag{9.25}$$

We can now generate other Hartree products in the $M$-mode space by applying excitation operators.

$$\tau_{\mu^{\mathbf{m}}}^{\mathbf{m}} = \prod_{m \in \mathbf{m}} a_{a^m}^{m\dagger} a_{i^m}^{m} \tag{9.26}$$

to this reference state. In doing so we "excite" all modes in the MC $\mathbf{m}$ from the occupied modals $i^m$ to the virtual modals $a^m$. The nomenclature of using indices $i, j$ to denote occupied modals and $a$, $b$, to denote modals unoccupied (virtual) in the reference product function is standard. The index $\mu$ is a compound index including the necessary information on the specific modals involved. Examples of one and two-mode excitation operators are,

$$\tau_{a^{m_1}}^{m_1} = a_{a^{m_1}}^{m_1\dagger} a_{i^{m_1}}^{m_1} = E_{a^{m_1} i^{m_1}}^{m_1} \tag{9.27}$$

$$\tau_{\mu^{\mathbf{m}_2}}^{\mathbf{m}_2} = \tau_{a^{m_1} a^{m_2}}^{m_1 m_2} = a_{a^{m_1}}^{m_1\dagger} a_{a^{m_2}}^{m_2\dagger} a_{i^{m_1}}^{m_1} a_{i^{m_2}}^{m_2} = E_{a^{m_1} i^{m_1}}^{m_1} E_{a^{m_2} i^{m_2}}^{m_2} \tag{9.28}$$

Note here that the $\mathbf{i}$ vector is often implied, meaning that it is only relevant to keep track of the unoccupied indices.

The commutator relations can be used to show that all excitation operators commute,

$$[\tau_{\mu^{\mathbf{m}}}^{\mathbf{m}}, \tau_{\nu^{\mathbf{m}'}}^{\mathbf{m}'}] = 0 \tag{9.29}$$

which is of great practical importance. The Hermitian adjoint of the excitation operator $\tau_{a^{m_1}}^{m_1}$ is a de-excitation operator

$$\tau_{a^{m_1}}^{m_1\dagger} = a_{i^{m_1}}^{m_1\dagger} a_{a^{m_1}}^{m_1} \tag{9.30}$$

From the killer condition for the annihilation operator working on the vacuum state, we have corresponding killer conditions for the excitation operators

$$(\tau_{\nu^{\mathbf{m}}}^{\mathbf{m}})^{\dagger} |\Phi_{\mathbf{i}}\rangle = 0 = \langle \Phi_{\mathbf{i}} | \tau_{\nu^{\mathbf{m}}}^{\mathbf{m}} \tag{9.31}$$

Note that $\tau_{\mu^{\mathbf{m}}}^{\mathbf{m}}$ and $(\tau_{\nu^{\mathbf{m}'}}^{\mathbf{m}'})^{\dagger}$ do not generally commute. The excited states are orthogonal to the reference ket $|\Phi_{\mathbf{i}}\rangle$ and the set of states $\{|\Phi_{\mathbf{i}}\rangle, \tau_{\mu}|\Phi_{\mathbf{i}}\rangle\}$ thus forms a basis for the full $M$-mode space when all possible $\mu$ are included.

## 9.5 Vibrational Self-consistent-field Theory and the Mode–Mode Correlation Problem

The vibrational self-consistent field (VSCF) approach [17, 18] is a mean-field theory applied to the anharmonic vibrational problem. We seek to variationally optimize the energy

$$E_{\mathbf{i}} = \langle \Phi_{\mathbf{i}} | \hat{H} | \Phi_{\mathbf{i}} \rangle \tag{9.32}$$

under variation of orthonormal one-mode functions $\phi^m_{i^m}(q_m)$ for a given Hartree Product state. The reference index vector **i** specifies which reference occupation is used for each mode as in the previously introduced notation.

The variational conditions with orthonormality constraints leads to the self-consistent-field equations

$$\hat{F}^{m,\mathbf{i}}\phi^m_{i^m}(q_m) = \varepsilon^m_{i^m}\phi^m_{i^m}(q_m) \tag{9.33}$$

where the mean-field operator $\hat{F}^{m,\mathbf{i}}\phi^m_{i^m}(q_m)$ for mode $m$ is obtained as the average of the Hamiltonian over all the other degrees of freedom (here given in first quantization)

$$\hat{F}^{m,\mathbf{i}} = \langle \prod_{m'=1,\neq m}^{M} \phi^{m'}_{i^{m'}}(q_{m'})|\hat{H}| \prod_{m''=1,\neq m}^{M} \phi^{m''}_{i^{m''}}(q_{m''})\rangle \tag{9.34}$$

After solving the coupled VSCF equations (Eq. (9.33)) self-consistently resulting in VSCF optimized modals (in analogy to Hartree-Fock orbitals for electrons). The total energy is obtained from Eq. (9.32) using the converged VSCF modals. In analogy to the electronic case, we thus have a well-defined mean-field state with a well-defined energy. This energy is based on an averaged interaction between modes as clearly seen from the averaging in Eq. (9.34). The VSCF approach thus immediately defines a correlation problem. VSCF can be implemented to give a fast and efficient evaluation of anharmonic wave functions, see Ref. [19], but will not be further discussed here. Instead we will focus on methods that goes beyond VSCF, and thereby explicitly includemany-body effects.

The basic ansatz for proceeding beyond the single Hartree Product wave function is vibrational configuration interaction (VCI). The VCI wave function can be written as

$$|\mathrm{VCI}\rangle = (1 + \mathrm{C})|\Phi_{\mathbf{i}}\rangle = |\Phi_{\mathbf{i}}\rangle + \sum_{\mu} \mathrm{C}_{\mu}\tau_{\mu}|\Phi_{\mathbf{i}}\rangle \tag{9.35}$$

using intermediate normalization $1 = \langle \Phi_{\mathbf{i}}|\mathrm{VCI}\rangle$. The intermediate normalization is convenient for later comparisons, but is not essential for the theory, and normally standard unit norm normalization is used. The **i** index vector is, as mention earlier, implied for all V excitation operators ($\tau_\mu$) and the related VCI parameters ($C_\mu$) and is not explicitly included in the notation here and in the following. The VSCF state is used for reference here, but it is trivial to use any other Hartree Product state.

We may write the operator generating the VCI wave function in more detail as

$$C = C_1 + C_2 + C_3 + \cdots C_M = \sum_{j=1}^{M}\sum_{\mu_j} c_{\mu_j}\tau_{\mu_j} \tag{9.36}$$

where the extra $j$ index denotes the excitation level.

In the case where we use all configurations possible we have a $N^M$ dimensional space, and a similar number of free parameters to determine, assuming there are $N$ modals for each of the $M$ degrees of freedom. Approximate VCI wave functions are introduced by limiting the space considered. This can be done in many ways. Here, we shall limit ourself to discuss the hierarchy in which we limit the summation over excitation levels in Eq. (9.36) to a maximum level $n$. The VCI wave function corresponding to this truncation is denoted VCI[n]. In the limit of no truncation we obtain the Full VCI solution, FVCI = VCI[M]. The linear variational approach implicit to VCI (using VSCF modals or another set of one-mode functions) has been the traditional approach to addressing the solution of the $M$-mode system [20–22]. When convergence close enough to FVCI can be obtained this is obviously fully satisfactory. The problem comes only from the increased size of the space needed for increasing systems size, which requires special treatments to be required. While treating large spaces in such linear expansions are becoming possible [23], we will here switch to the another parameterization and later show this has interesting properties for increasing system size.

## 9.6 Vibrational Coupled Cluster

The vibrational coupled cluster ansatz is given in terms of an exponential operator working on the reference wave function $|\boldsymbol{\Phi}_{\mathbf{i}}\rangle$

$$|\text{VCC}\rangle = \exp(T)|\boldsymbol{\Phi}_{\mathbf{i}}\rangle, \tag{9.37}$$

Here T is the so-called cluster operator which we in general write as

$$T = \sum_{\mu} t_\mu \tau_\mu. \tag{9.38}$$

The $t_\mu$ parameters are the cluster amplitudes and the free parameters of the methods, while $\tau_\mu$ are the corresponding excitation operators. All are indexed by $\mu$—a compound index giving all necessary information to specify the excitation. As for the VCI case the cluster operator can be written as

$$T = T_1 + T_2 + T_3 + \cdots T_M = \sum_{j=1}^{M} \sum_{\mu_j} t_{\mu_j} \tau_{\mu_j} \tag{9.39}$$

The one- and two-mode excitations can be written out as

$$T_1 = \sum_{\mu_1} t_{\mu_1} \tau_{\mu_1} = \sum_{m_1}^{M} \sum_{a^{m_1}} t_{a^{m_1}}^{m_1} \tau_{a^{m_1}}^{m_1}, \tag{9.40}$$

$$T_2 = \sum_{\mu_2} t_{\mu_2} \tau_{\mu_2} = \sum_{m_1<m_2}^{M} \sum_{a^{m_1}} \sum_{a^{m_2}} t^{m_1 m_2}_{a^{m_1} a^{m_2}} \tau^{m_1 m_2}_{a^{m_1} a^{m_2}}. \tag{9.41}$$

We may also write the cluster operator in terms of MC notation as

$$T = \sum_{\mathbf{m} \in MCR[T]} \sum_{\mu^{\mathbf{m}}} t^{\mathbf{m}}_{\mu^{\mathbf{m}}} \tau^{\mathbf{m}}_{\mu^{\mathbf{m}}} \tag{9.42}$$

where **m** is simply one of the allowed MCs. The definition of a VCC cluster operator requires thus in general a definition of the set of different MCs allowed in the cluster operator. This set (a mode combination range, MCR) is denoted $MCR[T]$ in Eq. (9.42).

Introducing now the VCC ansatz into the Schrödinger equation and premultiplying with $\exp(-T)$, we obtain

$$\exp(-T)\hat{H}\exp(T)|\Phi_{\mathrm{i}}\rangle = E|\Phi_{\mathrm{i}}\rangle \tag{9.43}$$

Projecting onto the reference state and the excited Hartree Products, we obtain respectively the VCC energy

$$E_{VCC} = \langle\Phi_{\mathrm{i}}|\exp(-T)\hat{H}\exp(T)|\Phi_{\mathrm{i}}\rangle = \langle\Phi_{\mathrm{i}}|\hat{H}\exp(T)|\Phi_{\mathrm{i}}\rangle. \tag{9.44}$$

and the VCC amplitude equations

$$e^{\mathbf{m}}_{\mu^{\mathbf{m}}} = \langle\mu^{\mathbf{m}}|\exp(-T)\hat{H}\exp(T)|\Phi_{\mathrm{i}}\rangle = 0. \tag{9.45}$$

Solving first the VCC amplitude equations in Eq. (9.45) we can calculate the energy as in Eq. (9.44).

Truncating the cluster operator at the excitation level $n$, we obtain a hierarchy of VCC methods, VCC[1], VCC[2], VCC[3], VCC[4], etc. In the limit of no truncation, we obtain the Full VCI solution, FVCC = FVCI.

The requirements for a VCC calculation is that a Hamiltonian is available in some suitable set of coordinates together with an appropriate choice of basis set such that the necessary integrals of the Hamiltonian are available. In the sum-over-product form this means that when the Hamiltonian expansion coefficients $c_t$ and the one-mode integrals of the $h^{m,t}$ operators are available (see Eq. (9.7)), we can first do a VSCF calculation defining the VCC reference state. Then we have to select the set of states to excite to. This is typically done by selecting the MCR, e.g. in the hierarchical sense of one-, two- and higher mode couplings up to a given maximum coupling level. This then defines the VCC calculation for the given reference state.

The nonlinear equations are determined by some iterative methods. From a particular guess for the solution vector **t** containing all the amplitudes, one evaluates the similarly sized error vector defined by Eq. (9.45). If this error vector is not below a given threshold in size an improved guess for the **t** is made and the process continued. Different methods of quasi-Newton nature. Sufficient here is to state that this procedure usually works relatively smoothly as long as the reference is a reasonable description of the final state. The computational most challenging part is the evaluation of the error vector. The error vector can be very complicated for high-level couplings in the Hamiltonian and high-level excitations included. For example VCC[3] with a three-mode Hamiltonian already contains thousands of terms [24]. In the current preferred approach, all these terms are automatically derived in details, analysed for computational efficiency and finally evaluated by the programme.

In principle, the above procedure is applicable to both ground and excited states. Excited states are, however, often immersed in a dense manifold of states. Solving the nonlinear VCC equations for each state in a state by state fashion can be technically difficult with respect to convergence and raises a number of theoretical concerns such as the potential lack of orthogonality between states. For that purpose, it is in most cases preferable to use response theory to access excited states or alternatively use the attractive option of directly calculating the spectrum. We shall here avoid at all entering into the long details of response theory. All our considerations, however, also apply to response theory calculations of excited states and spectra.

The above VCC theory is the one we have developed and the one we shall further investigate here. Our approach is based on the described SQ and applies to any basis and thereby any reference state, including the VSCF reference state. We note that there exist other exponential CC-like parameterizations of the vibrational wave function. The other approaches have been denoted bosonic CC and are by construction based on a harmonic oscillator like ground state [25, 26].

Let us now investigate further the wave function itself. The wave function for the case of an untruncated cluster expansion can be expanded as

$$
\begin{aligned}
|\mathrm{VCC}\rangle = |\Phi_{\mathbf{i}}\rangle + T_1|\Phi_{\mathbf{i}}\rangle + (T_2 + \frac{1}{2}T_1^2)|\Phi_{\mathbf{i}}\rangle + (T_3 + T_1T_2 + \frac{1}{6}T_1^3)|\Phi_{\mathbf{i}}\rangle \\
+ (T_4 + T_1T_3 + \frac{1}{2}T_2^2 + \frac{1}{2}T_1^2T_2 + \frac{1}{24}T_1^4)|\Phi_{\mathbf{i}}\rangle \ldots . \qquad (9.46)
\end{aligned}
$$

The linear VCI and nonlinear VCC parameterization are two alternative ways of parameterizing the exact wave functions in the untruncated limit. For an exact wave functions they are therefore related through

$$
\begin{aligned}
C_1 &= T_1 \\
C_2 &= (T_2 + \frac{1}{2}T_1^2) \\
C_3 &= (T_3 + T_1T_2 + \frac{1}{6}T_1^3) \\
C_4 &= (T_4 + T_1T_3 + \frac{1}{2}T_2^2 + \frac{1}{2}T_1^2T_2 + \frac{1}{24}T_1^4) \qquad (9.47)
\end{aligned}
$$

and so on. It is easy to turn this around and obtain

$$
\begin{aligned}
T_1 &= C_1 \\
T_2 &= C_2 - (\frac{1}{2}T_1^2) \\
T_3 &= C_3 - (T_1T_2 + \frac{1}{6}T_1^3) \\
T_4 &= C_4 - (T_1T_3 + \frac{1}{2}T_2^2 + \frac{1}{2}T_1^2T_2 + \frac{1}{24}T_1^4)
\end{aligned} \tag{9.48}
$$

and so on. We may thus transform between the linear VCI and the exponential VCC formats. This means that if we have a set of excitations in the straight linear picture, we can decompose them into a cluster representation. If such expansions has meaning it is expected that the amplitudes in the cluster representation decay faster with increasing excitation level, as compared to the linear parameterization. Thus higher-mode excitations are expected to be dominated by products of lower excitations.

## 9.7 A Tensor Perspective on VCC

A quantity with $d$ indices can be denoted a tensor, of *order d*. Hence, a vector is a first order tensor, a matrix is a second order tensor, a 3-way array is an third order tensor, etc. It is natural to represent tensors in terms of their $d$-way arrays. However, if there are $N$ possible values for each index for each dimension there in total $N^d$ numbers are required to specify the full tensor. This exponential increase in the data with increasing number of dimensions is often referred to as the curse of dimensionality. To address this, we seek in tensor decomposition to numerically represent the same tensor in terms of simple quantities. That is generally, in stead of one full $d$-way arrays we seek to represent the tensor in terms of a set of arrays of lower dimensions and/or smaller set of index values. For a general overview and descriptions of tensors and tensor decomposition we refer to the literature, see for example Ref. [27]. We will proceed by focusing specifically on the simplest variant of tensor decomposition (CP, see below) as well as on relating the idea to VCC theory.

The most obvious tensor decomposition is CANDECOMP/PARAFAC or canonical polyadic decomposition, abbreviated as CP decomposition. Here a tensor is represented as a sum of vector outer products with $\otimes$ denoting a tensor outer product

$$
\mathcal{X} \approx \sum_r^R \mathbf{a}_r^{(1)} \otimes \mathbf{a}_r^{(2)} \otimes \cdots \otimes \mathbf{a}_r^{(d)} \tag{9.49}
$$

where $R$ of Eq. (9.49) is the *rank*. The rank together with the $\mathbf{a}_r^{(n)}$ vectors which can be collected as column vectors in $d$ $\mathbf{A}^{(n)}$ factor matrices specifies decomposition. Note that the CP format is often represented with a weight factor $\lambda_r$ for each term but this prefactor is here (as in many other places) absorbed into the columns of one (or more) of the factor matrices. The smallest number of vector outer products that are needed to reproduce the tensor exactly is the tensor rank in a similar fashion to the rank of a matrix. The CP decomposition shares in this sense some similarity to the singular value decomposition of matrices. Unfortunately there is no general simple way to determine the rank of a tensor, and it becomes a numerical exercise to seek CP "fits" that reproduce the tensor to high enough accuracy.

### 9.7.1 Vibrational Coupled Cluster as a Tensor Decomposition

As is perhaps clear from the two preceding sections, a CC formatted wave function is in a certain sense a decomposition format for the FVCI wave function. We can view the equations representing the VCI excitation in terms of the VCC excitations as a kind of tensor decomposition. In other words, we can consider

$$T_n = f^1(C_n, T_1, \ldots, T_{n-1}) = f^2(C_1, C_2, \ldots, C_n) \tag{9.50}$$

Similar equations holds at the mode excitation level, where only the excitations for a given MC, **m**, is considered. The cluster excitation operator is written in the form

$$T^{\mathbf{m}} = \sum_{\mu^{\mathbf{m}}} t^{\mathbf{m}}_{\mu^{\mathbf{m}}} \tau^{\mathbf{m}}_{\mu^{\mathbf{m}}} \tag{9.51}$$

and similarly for the VCI excitations. Based upon the MC, **m** we construct the set of MCRs, SMCR[**m**], where the union of all MCs in each MCR is identical to the original MC, **m**. Summing over these MCR[s] we have

$$C^{\mathbf{m}} = \sum_{MCR[s] \in SMCR[\mathbf{m}]} \prod_{\mathbf{m}_k \in MCR[s]} T^{\mathbf{m}_k} \tag{9.52}$$

Here the sum is over all partitions of the set **m** meaning a sum over all ways the set **m** can be obtained as union of disjoint subsets. Thus a set of MCR where the union of all the non-overlapping sets in each gives the original **m**.

We may interpret Eq. (9.52) for the operators to be a tensor decomposition of the VCI correlation amplitudes

$$c^{\mathbf{m}} = \sum_{MCR[s] \in SMCR[\mathbf{m}]} \otimes_{\mathbf{m}_k \in MCR[s]} T^{\mathbf{m}_k} \tag{9.53}$$

In terms of the amplitudes, we have the following set of equations in low order

$$\begin{aligned}
c^{m_1}_{a_1} &= t^{m_1}_{a_1} \\
c^{m_1 m_2}_{a_1 a_2} &= t^{m_1 m_2}_{a_1 a_2} + t^{m_1}_{a_1} t^{m_2}_{a_2} \\
c^{m_1 m_2 m_3}_{a_1 a_2 a_3} &= t^{m_1 m_2 m_3}_{a_1 a_2 a_3} + t^{m_1}_{a_1} t^{m_2 m_3}_{a_2 a_3} + t^{m_2}_{a_2} t^{m_1 m_3}_{a_1 a_3} + t^{m_3}_{a_3} t^{m_1 m_2}_{a_1 a_2} + t^{m_1}_{a_1} t^{m_2}_{a_2} t^{m_3}_{a_3} \\
&= t^{m_1 m_2 m_3}_{a_1 a_2 a_3} + (1 + P_{12} + P_{13}) t^{m_1}_{a_1} t^{m_2 m_3}_{a_2 a_3} + t^{m_1}_{a_1} t^{m_2}_{a_2} t^{m_3}_{a_3}
\end{aligned} \tag{9.54}$$

etc., where $P_{ij}f_{ij} = f_{ji}$. Thus, recasting the FVCI wave function as a FVCC wave function corresponds implicitly to a decomposition of the correlation amplitudes. Either set of correlation amplitudes provide adequate descriptions and are permissible as departure of approximations.

An approximate VCC is not actually obtained as a decomposition of VCI parameters. Rather, approximations are introduced by restricting the excitation level/rank of the decomposition quantities as described in the above VCC sections. Thus, we formally still have a set of FVCI correlation amplitudes. But we proceed from expressing

$$c_n = F_n(t_1, \dots, t_n) \tag{9.55}$$

to

$$c_n = F_n^{\text{app}}(t_1, \dots t_{n_{max}}) \tag{9.56}$$

The cluster amplitudes are not found by "fitting of tensors" but from equations for the immersed approximation of the VCC ansatz into a projected Schrödinger equation. Still, the general decomposition perspective is enlightening and a potential source for new approximations.

### 9.7.2 Decomposed Correlation Amplitudes with CP

For the VCC wave function we may consider the full set of amplitudes as one vector, but we can also consider it as a number of sub-vectors, one vector for each MC. Each of these vectors are in turn a vectorization of a n-th order tensor, where $n$ is the number of elements of the particular MC. Thus the whole set of VCC or VCI correlation amplitudes can be considered as a stack of tensors.

We may now proceed by applying decomposition to the correlation amplitudes (on top of the inherent recasting of the VCC wave function). For example, the VCC three-mode excitations for the MC $(m_0, m_1, m_2)$ can be represented in CP format as

$$t^{m_0,m_1,m_2}_{a_0,a_1,a_2} \approx \sum_{r=1}^{R^{m_0,m_1,m_2}} t^{(m_0,m_1,m_2),0}_{r,a_0} t^{(m_0,m_1,m_2),1}_{r,a_1} t^{(m_0,m_1,m_2),2}_{r,a_2} \tag{9.57}$$

Each different MC has a different set of one-mode factor matrices, or the other way around, each factor matrix is unique to each MC, and mode for this MC. It is decisive that the rank $R^{\mathbf{m}}$ is unique to the MC $\mathbf{m}$. This is a trivial but key aspect of the approach, therein that the rank can be adjusted to the importance of each MC. Thus, it has been confirmed in numerical studies that in adaptive iterative algorithms we can solve the equations in a manner such that each individual MC achieve each their own individual rank. The advantage of this is that specific couplings are very different and have different strength. This is advantageous because by allowing the rank to be flexible this opens for handling large systems with very many MCs, where the major fraction will have low rank including often zero rank, while only a smaller fraction of particularly important MCs will have a large rank.

We note in passing that our approach is one of following the mode-coupling expansion suggested from the potential, and the hierarchical approach to VCI and VCC we have taken. A different approach is to go back to the full wave function ansatz and directly decompose the FVCI wave function. Such approaches have been undertaken by other groups in various contexts. The drawback of this is that the FVCI tensor is huge. Our hierarchial approach involves much smaller tensors (for truncated VCC) but many of them. Furthermore, there are fundamental differences in performing the decompositions for VCC and VCI.

## 9.8 Seperability and VCC and VCI Compared

To put tensor decomposition for VCI and VCC in perspective, we will here consider the separability and scalability of energies and wave functions. This is directly related to concepts such as size-extensivity and size-consistency of electronic structure theory and to the property of extensivity of thermodynamics.

Consider the case where we combine the two non-interacting subsystems A and B into a super-system AB described by the Hamiltonian

$$H_{AB} = H_A + H_B \tag{9.58}$$

The two fragments each satisfies their own Schrödinger equation $H_A|A\rangle = E_A|A\rangle$, $H_B|B\rangle = E_B|B\rangle$. Using the SQ formalism the wave functions can be written in the form

$$|A\rangle = W_A|\text{vac}\rangle, \tag{9.59}$$

where $W_A$ is a wave operator

$$W_A = \sum_{\mathbf{s}} C_{\mathbf{s}} \prod_{m=1}^{M} a^{\dagger}_{s_m} \tag{9.60}$$

The exact wave function for the compound system is required to satisfy the Schrödinger equation

$$H_{AB}|AB\rangle = E_{AB}|AB\rangle. \tag{9.61}$$

With the SQ formulation it is simple and rigorous to write up the exact wave function for the compound system in the multiplicatively separable form

$$|AB\rangle = W_{AB}|\text{vac}\rangle = W_A W_B|\text{vac}\rangle \tag{9.62}$$

Related to the product separable wave function, the exact energy is additively separable

$$E_{AB} = E_A + E_B. \tag{9.63}$$

For N non-interacting subsystems each with energy $E_1$ the total energy of the $N$-mer is N$E_1$, in complete analogy to the extensivity of properties in thermodynamics. The behaviour with increasing number of degrees of freedom is an important aspect of an approximate theory. Approximate wave functions can be classified according to whether the wave function is manifestly separable as a product and energy additively separable. This has for long been known to be an important aspect for electronic structure theory. It is also important for vibrational wave functions.

Above we considered that the Hamiltonian was additively separable, indicating that it can be separated into a part which only affects A and a part that only affects B. Actually, such a separability depends on the coordinates. Consider now the simple harmonic limit. If A and B were identical, they would have identical harmonic frequencies in separate harmonic analyses. We could use localized modes (i.e. modes either solely moving atoms in A or solely atoms in B) and have clearly a separable Hamiltonian, total energy and wave function. If we consider normal modes with the same frequencies in the case of two identical non-interacting subsystems combined into one system mixing the degenerate modes in an orthogonal transformation will produce new modes with the same frequencies but now delocalized over both A and B. Clearly, we still have an exact normal coordinate harmonic oscillator description, and as such the wave function and energy are in fact separable. It is, however, somewhat more obscured what goes on. One can then either choose to formulate the concepts to clarify the scalability in such general coordinate cases or else stick to the case of analysing for additively separable Hamiltonians above. We will follow the latter approach, as (i) it is simpler; (ii) it is the separability with distance we are interested in, and it is counterproductive to choose a coupled representation when an uncoupled one can be chosen; (iii), in continuation of the last point we prefer for real molecules also to have localized coordinates as far as possible (meaning unless their couplings become too large). We believe this is quite typical for the molecular case where we strive for low-coupling local coordinates. The case of solids may be different. Carrying out a similar analysis for solids using delocalized coordinates is possible [28], however, may to lead to considerations that are less relevant for isolated molecules where we would imagine vibrational coordinates to be somewhat localized. Our perspective here is that even with local coordinates the size-extensivity issue is non-trivial and important and there is in fact an important interplay between size-extensivity and tensor decomposition.

The VSCF wave function is trivially multiplicatively separable since the VSCF wave operator is a direct product of creation operators. Consider that we initiate the VSCF procedure with the direct product of the solution for VSCF calculations on A and B separately. In the averaging for obtaining the VSCF mean field for a mode localized to system A, we have that the averaging of $H_B$ we simply obtain $E_B^{VSCF}$.

Accordingly, the VSCF mean-field operator for each subsystem is changed only by a constant factor, which leads to the same optimized VSCF modals. Thus, with

$$W_A^{VSCF} = \prod_{m \in A}^{M_A} a_{i^m}^{m\dagger} \tag{9.64}$$

and similarly for B, we find that the compound wave operator is multiplicatively separable as the exact wave operator,

$$W_{AB}^{VSCF} = W_A^{VSCF} W_B^{VSCF} \tag{9.65}$$

while the energy evaluated as an expectation value is additively separable

$$\begin{aligned} E_{AB}^{VSCF} &= \langle AB|H_{AB}|AB\rangle \\ &= \langle vac|/W_A^{VSCF})^\dagger H_A W_A^{VSCF} (W_B^{VSCF})^\dagger W_B^{VSCF}|vac\rangle \\ &\quad + \langle vac|/W_B^{VSCF})^\dagger H_B W_B^{VSCF} (W_A^{VSCF})^\dagger W_A^{VSCF}|vac\rangle \\ &= E_A^{VSCF} + E_B^{VSCF} \end{aligned} \tag{9.66}$$

Here commutativity of creation and annihilations operators have been used for modals and modes referring to different subsystems.

Consider now a VCC calculation using the VSCF reference state with its correct separability. For a similar super-system as discussed above it is easy to show that the cluster operator (for both exact and truncated VCC treatments)

$$T_{AB} = T_A + T_B \tag{9.67}$$

is a solution to the super-system VCC equations, where $T_A$ and $T_B$ are cluster operators for the subsystems. This means that $W_A^{VCC} = \exp(T_A)W_A^{VSCF}$ generates the VCC wave functions for subsystem A and correspondingly for B. Accordingly, the VCC wave function is multiplicatively separable

$$\begin{aligned} |\text{VCC}_{AB}\rangle &= \exp(T_{AB})|\text{VSCF}_{AB}\rangle \\ &= \exp(T_A)\exp(T_B)W_A^{\text{VSCF}} W_B^{\text{VSCF}}|\text{vac}\rangle \\ &= W_A^{VCC} W_B^{VCC}|\text{vac}\rangle \end{aligned} \tag{9.68}$$

where we have used that the cluster excitation operators commute. Similarly,

$$\begin{aligned} E_{AB} &= \langle \text{VSCF}_{AB}|H_A + H_B|\text{VCC}_{AB}\rangle \\ &= E_A^{VCC} + E_B^{VCC} \end{aligned} \tag{9.69}$$

As a direct consequence of the exponential parameterization and the commutativity of the excitation operators the VCC wave function is multiplicatively separable while the energy is additively separable in accord with the exact case for both exact and truncated wave functions. We see that the separability of VCC derives from including products of $T_A$ and $T_B$ that, as we have seen before, are necessary for correct separability. Such products will be missing in truncated VCI, where the product excitations will be partly or completely outside the space included in an approximate VCI.That is a two-mode times two-mode excitations will be outside the space of VCI[3]. This is an advantage of the exponential VCC parametrization over the linear VCI parametrization for the same excitation space and the same Hamiltonian. VCI will not give the correct separability of the energy and wave function. This means that as we increase the number of non-interacting systems the VCI solution for a given excitation level decays in completeness and accuracy. Or in other words, increasing the size of the system we can in VCC theory expect a fairly constant accuracy with a constant excitation level, while in VCI constant high accuracy would require increasing the excitation space.

All though the above considerations are simplified as they are derived for non-interacting systems, we believe they are good models for real molecular systems which will only be more complicated. For larger molecular system in fairly local coordinates we can envision significant mode-coupling terms between "close" modes in the Hamiltonian while couplings between distant modes are small. The above considerations give reasons to believe that VCC methods can handle this and obtain fairly constant accuracy with increasing size of the system.

Furthermore we may relate this discussion to the tensor decomposition analysis. Consider now the case where we perform a VCC[n] calculation and choose to dynamically decompose the amplitudes to lowest rank needed for each MC. With the VCC format, we thus obtain that the cluster amplitudes belonging to subsystem A and subsystem B are significant and have some non-zero rank. In the uncoupled limit, we should actually have $T_{AB} = 0$ thus it can be represented with rank zero. On the other hand, we cannot expect VCI amplitudes to give rank zero even for uncoupled systems, and this difference between VCC and VCI has been numerically confirmed [29, 30]. For example, four-mode excitations $C_{AB,4}$ will include contributions from for example $T_{A,2}T_{B,2}$ which will not be represented as rank zero, even if $T_{AB,4}$ has in fact zero rank. Thus, if two particular cluster operators with ranks $R_A$ and $R_B$ are multiplied, this will lead to rank $R_A \times R_B$ in the corresponding VCI representation. This means in other words, that the wave function separability of VCC is important from a tensor decomposition perspective. The expected decay of amplitudes for couplings between distant localized modes will be an important aspect of the ability to apply VCC to larger systems in the future. It can be exploited both by the tensor analysis above as well as by a careful selection of the excitation space. Rather than the standard inclusion of excitations solely according to the number of modes involved, we may with reference to Eq. (9.42) choose which MCs are included or excluded by an automatic numerical screening [31]. In such a screening, unimportant distant couplings will be screened away from the outset. Both the automatic screening and the use of tensor decomposition in VCC is still in its infancy. While the promise of tensor decomposition has been illustrated, the practical use of the CP format is difficult, both due to inherent mathematical limitations of the CP format, as well as the dynamical recompression to the CP format in the context of the very complex VCC equations.

## 9.9 Conclusions

The ability to solve the nuclear Schrödinger equation is important for high accuracy calculations of spectroscopy and dynamics. Many-body methods in general and coupled cluster methods in particular have proven highly successful in solving solving Shcrödinger equations. In this chapter, I have discussed many-body expansions in the context of calculating potential energy surfaces, as well as coupled cluster methods for solving the anharmonic vibrational Schrödinger equation. It has been emphasized that these methods have attractive features in relation to scalability with system size. Here scalability refers both to maintaining a realistic low computational effort as the size of the system increases, as well as maintaining a sufficiently high accuracy. Although further work is needed for unfolding their full potential, these many-body methods are expected to be decisive for pushing the boundaries for computational methods in spectroscopy and dynamics in the coming years.

**Acknowledgements** This work has been supported by the Lundbeck Foundation and the Danish Natural Science Research Council. We acknowledge discussions with Ian Godtliebsen and Mads Bøttger Hansen. Support from the COST network Molecules in Motion is acknowledged.

## References

1. T. Helgaker, P. Jørgensen, J. Olsen, *Molecular Electronic Structure Theory* (Wiley, Chichester, 2000)
2. F. Jensen, *Introduction to Computational Chemstry*, 2nd edn. (Wiley, Chichester, 2006)
3. J.K.G. Watson, Mol. Phys. **15**, 479 (1968)
4. S. Carter, S.J. Culik, J.M. Bowman, J. Chem. Phys. **107**, 10458 (1997)
5. G.C. Schatz, Rev. Mod. Phys. **61**, 669 (1989)
6. H. Rabitz, O.F. Aliş, J. Math. Chem. **25**, 197 (1999)
7. H.-D. Meyer, WIREs: Comput. Mol. Sci. **2**, 351 (2012)
8. M. Griebel, Sparse grids and related approximation schemes for higher dimensional problems, in *Foundations of Computational Mathematics (FoCM05), Santander*, ed. by L. Pardo, A. Pinkus, E. Suli, M. Todd, (Cambridge University Press, 2006) pp. 106–161
9. J.O. Jung, R.B. Gerber, J. Chem. Phys. **105**, 10332 (1996)
10. K. Yagi, T. Taketsugu, K. Hirao, M.S. Gordon, J. Chem. Phys. **113**, 1005 (2000)
11. D. Toffoli, J. Kongsted, O. Christiansen, J. Chem. Phys. **127**, 204106 (2007)
12. M. Sparta, D. Toffoli, O. Christiansen, Theor. Chem. Acc. **123**, 413 (2009)
13. C. König, O. Christiansen, J. Chem. Phys. **145**, 064105 (2016)
14. C. König, M.B. Hansen, I.H. Godtliebsen, O. Christiansen, J. Chem. Phys. **144**, 074108 (2016)
15. O. Christiansen, J. Chem. Phys. **120**, 2140 (2004)
16. O. Christiansen, J. Chem. Phys. **120**, 2149 (2004)
17. J.M. Bowman, Acc. Chem. Res. **19**, 202 (1986)

18. R.B. Gerber, M.A. Ratner, Adv. Chem. Phys. **70**, 97 (1988)
19. M.B. Hansen, M. Sparta, P. Seidler, O. Christiansen, D. Toffoli, J. Chem. Theo. Comp. **6**, 235 (2010)
20. J.M. Bowman, T. Carrington, H. Meyer, Mol. Phys. **106**, 2145 (2008)
21. D. Oschetzki, M. Neff, P. Meier, F. Pfeiffer, G. Rauhut, Croat. Chem. Acta. **85**, 379 (2012)
22. O. Christiansen, Phys. Chem. Chem. Phys. **14**, 6672 (2012)
23. J. Brown, T. Carrington, J. Chem. Phys. **145**, 144104 (2016)
24. P. Seidler, O. Christiansen, J. Chem. Phys. **131**, 234109 (2009)
25. S. Banik, S. Pal, M.D. Prasad, J. Chem. Phys. **129**, 134111 (2008)
26. J.A. Faucheaux, S. Hirata, J. Chem. Phys. **143**, 134105 (2015)
27. T.G. Kolda, B.W. Bader, SIAM Rev. **51**(3), 455 (2009)
28. S. Hirata, M. Keçeli, K. Yagi, J. Chem. Phys. **133**, 034109 (2010)
29. I.H. Godtliebsen, B. Thomsen, O. Christiansen, J. Chem. Phys. A **117**, 7267–7279 (2013)
30. I.H. Godtliebsen, M.B. Hansen, O. Christiansen, J. Chem. Phys. **142**, 024105 (2015)
31. C. König, O. Christiansen, J. Chem. Phys. **142**, 144115 (2015)

# Chapter 10
# Relativistic Time-Dependent Density Functional Theory for Molecular Properties

**Muneaki Kamiya and Takahito Nakajima**

**Abstract** In this review article, we introduce the two-component relativistic time-dependent density functional theory (TDDFT) with spin–orbit interactions to calculate linear response properties and excitation energies. The approach is implemented in the NTChem program. Our implementation is based on a non-collinear exchange–correlation potential presented by Wang et al. In addition, various DFT functionals including the range-separated hybrid functionals have been derived and implemented with the aid of a newly developed computerized symbolic algebra system. The two-component relativistic TDDFT with spin–orbit interactions was successfully applied to the calculation of the frequency-dependent polarizabilities of $SnH_4$ and $PbH_4$ molecules containing heavy atoms and the excitation spectra of a HI molecule.

**Keywords** Relativistic time-dependent density functional theory · Spin–orbit couplings · Frequency-dependent polarizabilities · Automatic functional implementation · NTChem

## 10.1 Introduction

There is now a growing interest in obtaining electric and other response properties from first-principles theory for organic, inorganic, and organometallic compounds containing heavy elements. The inclusion of relativity is crucial for a proper description of photochemistry for systems those containing heavy elements. In particular, it is known that the spin–orbit couplings affect excited-state characters, relaxation dynamics, and radiative and nonradiative decay pathways, as well as

M. Kamiya
Faculty of Regional Studies, Gifu University, 1-1, Yanagido, Gifu 501-1193, Japan

M. Kamiya · T. Nakajima (✉)
RIKEN Advanced Institute for Computational Science, 7-1-26,
Minatojima-Minami-Machi, Chuo-ku, Kobe 650-0047, Japan
e-mail: nakajima@riken.jp



M.J. Wójcik et al. (eds.), *Frontiers of Quantum Chemistry*,
https://doi.org/10.1007/978-981-10-5651-2_10

lifetimes and reactivity [1] due to the changes induced in the splitting of the orbitals and thus potentially also in the pole structure of a molecule [2, 3].

Time-dependent density functional theory (TDDFT) [4–6] has become one of the most widely used methodologies for computing linear response properties since satisfactory accuracy can often be achieved at an acceptable computational cost. The efficient treatment of electron correlation by DFT is particularly important in calculations involving heavy atoms, since electronic correlation effects can be even more important than relativistic effects. There are several ways to calculate the excitation energy from the calculation of the linear response by TDDFT. The first is to calculate the location of the poles and the residues of the frequency-dependent polarizability through eigenvalue-type problems and is widely used in many implementations [7–9]. The other is to obtain the excitation energy by directly plotting the dynamic polarizability as a function of the frequency and analyzing the polar structure of the linear response [10, 11]. Since this approach can calculate the absorbing properties in the frequency window of interest, it can potentially be applied to excited states of large molecules with an enormous number of excitations. However, the implementation of this approach is still limited because it is necessary to compute the response of complex numbers.

Until now, in most of the TDDFT linear response calculations for heavy-element systems, only the scalar relativistic effect was taken into account while the spin–orbit effect on the excitation energies is not negligible. Scalar relativistic calculations of the first- and second-order hyperpolarizabilities using the Douglas–Kroll–Hess transformed one-component Hamiltonian [12–14], as well as using effective-core potentials [15, 16], have been reported by Norman et al. [17]. At the all-electron level, there exist four-component relativistic implementations of response theory at Hartree–Fock and Kohn–Sham levels of theory for linear [18] and quadratic response functions [19].

However, in most of these implementations, the treatment of noncollinear spin density terms, which is important due to the spin–orbit interaction, is ignored. In recent years, calculations using noncollinear kernels have been performed in excitation energy calculation by pole/residue calculation [20–23], but most are a Tamm–Dancoff approximation [24], and the full linear response calculations have hardly been performed [25].

In this review, the two-component relativistic linear-response TDDFT for molecular properties is derived and implemented on the NTChem program suite [26]. Our implementation is based on a noncollinear exchange–correlation potential presented by Wang et al. [20, 22]. In addition, since it is difficult to manually implement the latest complicated exchange–correlation functional, the computerized symbolic algebra system is developed and used for those complicated implementation in NTChem. The automatic code generator performs differentiations using the SymPy library [27], which is an open-source symbolic mathematics library for the Python programming language.

## 10.2 Theory

### 10.2.1 Polarizability

In the presence of a time-dependent external electric field, the molecular polarization is expressed as a perturbation expansion where the coefficients define the molecular properties known as polarizabilities and hyperpolarizabilities. The electric polarizability provides a measure of the distortion of an atomic or molecular charge distribution by an external field and is therefore an important property. Higher-order dynamic electric polarizabilities are extremely important in nonlinear optical materials research.

The expansion of the following frequency-dependent dipole moment to the third order introduces the linear polarizability $\alpha(-\omega;\omega)$, the first-order hyperpolarizability $\beta(-\omega_\sigma;\omega_1,\omega_2)$, and the second-order hyperpolarizability $\gamma(-\omega_\sigma;\omega_1,\omega_2,\omega_3)$:

$$\begin{aligned}\mu_\alpha(t)=\mu_\alpha^0&+\sum_{\omega}\sum_{\beta}\alpha_{\alpha\beta}(-\omega;\omega)E_\beta^\omega e^{-i\omega t}\\&+\frac{1}{2}\sum_{\omega_1,\omega_2}\sum_{\beta,\gamma}\beta_{\alpha\beta\gamma}(-\omega_\sigma;\omega_1,\omega_2)E_\beta^{\omega_1}E_\gamma^{\omega_2}e^{-i(\omega_1+\omega_2)t}\\&+\frac{1}{3!}\sum_{\omega_1,\omega_2,\omega_3}\sum_{\beta,\gamma,\delta}\gamma_{\alpha\beta\gamma\delta}(-\omega_\sigma;\omega_1,\omega_2,\omega_3)E_\beta^{\omega_1}E_\gamma^{\omega_2}E_\delta^{\omega_3}e^{-i(\omega_1+\omega_2+\omega_3)t},\end{aligned} \tag{10.1}$$

where $\mu_\alpha^0$ is the permanent electric dipole moment along the molecular axis α, $E^\omega$ are the Fourier components of the perturbing fields, and the sums are performed over both positive and negative frequency ($\omega$) components.

The real part of the linear polarizability $\alpha^R$ is connected to the refractive index, and the imaginary part $\alpha^I$ describes the absorption of light quanta. The linear absorption cross section equals

$$\sigma(\omega)=\frac{4\pi\omega}{c}\bar{\alpha}^I(-\omega;\omega), \tag{10.2}$$

where $c$ is the speed of light and the isotropic average of the polarizability has been introduced:

$$\bar{\alpha}=\frac{1}{3}\sum_{i=x,y,z}\alpha_{ii}. \tag{10.3}$$

From the standard time-dependent perturbation theory, we obtain a sum-over-states (SOS) expression for the linear polarizability involving the manifold of excited states of the unperturbed system:

$$\alpha_{\alpha\beta}(-\omega{:}\,\omega) = \frac{1}{\hbar}\sum_{n}\left\{\frac{\langle 0|\mu_\alpha|n\rangle\langle n|\mu_\beta|0\rangle}{\omega_{0n}-\omega} + \frac{\langle 0|\mu_\beta|n\rangle\langle n|\mu_\alpha|0\rangle}{\omega_{0n}+\omega}\right\}, \tag{10.4}$$

where $\mu_\alpha$ is the electric dipole moment operator along the molecular axis α, $\hbar\omega_{0n} = E_n - E_0$ are the electronic excitation energies, and the sums are performed over the manifold of states of the unperturbed system ($|n\rangle$) excluding the ground state ($|0\rangle$).

Such a formula is applicable only in the nonresonant region, since when photon energies are in the proximity of the excitation energies of the system a perturbation analysis is no longer valid. Thus, in the near-resonance case, two alternative strategies have been used: either turning to a few-states model or the inclusion of phenomenological damping terms in the SOS expressions.

The damping terms in the latter approach represent the finite lifetime of the excited states and correspond to line broadening in the absorption spectra. Having introduced damping terms, the SOS expression for the linear polarizability can be written as

$$\alpha_{\alpha\beta}(-\omega{:}\,\omega) = \frac{1}{\hbar}\sum_{n}\left\{\frac{\langle 0|\mu_\alpha|n\rangle\langle n|\mu_\beta|0\rangle}{\omega_{0n}-\omega-i\gamma} + \frac{\langle 0|\mu_\beta|n\rangle\langle n|\mu_\alpha|0\rangle}{\omega_{0n}+\omega+i\gamma}\right\}, \tag{10.5}$$

where damping factor $\gamma$ is defined using common lifetime $\tau$ of the excited states as

$$\gamma = \frac{1}{2\tau}. \tag{10.6}$$

It is immediately seen that in the nonresonant region, the imaginary part of $\alpha$ depends linearly on $\gamma$ and that it equals zero in the static limit ($\omega = 0$) regardless of $\gamma$. Furthermore, by using the identity

$$\lim_{\gamma\to 0}\left[\mathrm{Im}\left\{\frac{A}{B-i\gamma}\right\}\right] = A\pi\delta(B), \tag{10.7}$$

we see by comparing with Eqs. (10.2), (10.3), and (10.7) that

$$\lim_{\gamma\to 0}\sigma(\omega) = \frac{4\pi^2\omega}{3\hbar c}\sum_{n}\left[\delta(\omega_{0n}-\omega)\sum_{i=x,y,z}|\langle 0|\mu_i|n\rangle|^2\right], \tag{10.8}$$

and it is clear that the regular oscillator strengths are related to the infinite lifetime approximation of the absorption as described by the imaginary part of the linear polarizability.

### 10.2.2 Time-Dependent Kohn–Sham Theory

We assume that a molecule is initially in a stationary state, the electronic structure of which is suitably described by time-independent Kohn–Sham (KS) density functional theory.

In the density matrix formalism, the KS Hamiltonian and density matrices satisfy the time-independent KS equation:

$$\sum_q \left(F_{pq}^{(0)} D_{qr}^{(0)} - D_{pq}^{(0)} F_{qr}^{(0)}\right) = 0, \tag{10.9}$$

and the idempotency condition (corresponding to the orthonormality condition of orbitals):

$$\sum_q D_{pq}^{(0)} D_{qr}^{(0)} = D_{pr}^{(0)}, \tag{10.10}$$

where $F$ and $D$ are the KS Hamiltonian and density matrices, respectively, represented by Kohn–Sham orbitals for the unperturbed ground state $\{\phi_p\}$, and $p$, $q$, $r$ are spin-orbital indexes.

In two-component generalized KS equations, KS orbitals may be complex and have two spin components:

$$\phi_p = \begin{pmatrix} \phi_p^\alpha \\ \phi_p^\beta \end{pmatrix}, \tag{10.11}$$

where $\phi_p^\alpha$ and $\phi_p^\beta$ are spatial orbitals which are expanded in a linear combination of atomic orbitals $\{\chi_\mu\}$,

$$\phi_p^\omega = \sum_\mu \chi_\mu C_{\mu p}^\omega \quad (\omega = \alpha, \beta), \tag{10.12}$$

where λ, $\mu$, $\nu$, $\upsilon$, etc. are used for atomic-orbital indexes.

The KS Hamiltonian and density matrices for the ground state are simply

$$F_{pq}^{(0)} = \varepsilon_p \delta_{pq}, \tag{10.13}$$

$$D_{ij}^{(0)} = \delta_{ij}, \tag{10.14}$$

$$D_{ia}^{(0)} = D_{ai}^{(0)} = D_{ab}^{(0)} = 0, \tag{10.15}$$

where $\varepsilon_p$ is the $p$th spin-orbital energy, and we use $i, j, k, l, m, n$, etc., for occupied orbitals, $a, b, c, d, e, f$, etc., for virtual orbitals, and $p$, $q$, $r$, $s$, and $t$ for general orbitals throughout this chapter.

We then apply an oscillatory perturbation, which can be described as a single Fourier component:

$$g_{pq}^{(1)} = \frac{1}{2}\left(h_{pq}^{(1)} e^{-i\omega t} + h_{qp}^{(1)*} e^{i\omega t}\right), \tag{10.16}$$

where the matrix $h$ represents a one-electron operator describing the details of the perturbation.

The response in the density matrix $D$ to this applied perturbation consists of the first-order (linear) and higher-order terms:

$$D_{pq} = D_{pq}^{(0)} + D_{pq}^{(1)} + D_{pq}^{(2)} + D_{pq}^{(3)} + \ldots, \tag{10.17}$$

with

$$D_{pq}^{(1)} = \frac{1}{2}\left(d_{pq}^{(1)} e^{-i\omega t} + d_{pq}^{(1)*} e^{i\omega t}\right). \tag{10.18}$$

The first-order change in the KS Hamiltonian matrix arises from two sources: the direct change in the one-electron part described by Eq. (10.16) and the indirect change induced by the first- and higher-order responses in the density matrix, i.e.,

$$F_{pq} = F_{pq}^{(0)} + g_{pq}^{(1)} + \sum_{r,s} \frac{\partial F_{pq}}{\partial D_{rs}} D_{rs}^{(1)} + \ldots, \tag{10.19}$$

with

$$\frac{\partial F_{pq}}{\partial D_{rs}} = (pq|sr) - c_{\mathrm{x}}(pr|sq) - c_{\mathrm{x}}^{\mathrm{lr}}(pr|sq)^{\mathrm{lr}} + f_{pq,sr}^{\mathrm{xc}}, \tag{10.20}$$

where the prefactor $c_{\mathrm{x}}$ is the mixing ratio of Hartree–Fock (HF) exchange in the hybrid functional and $c_{\mathrm{x}}^{\mathrm{lr}}$ is the mixing ratio of long-range exchange in the range-separated hybrid functional. The regular two-electron integrals and long-range two-electron integrals are defined in the Mulliken notation as

$$(pq|sr) = \int\int \phi_p^*(\mathbf{r}_1)\phi_q(\mathbf{r}_1)\frac{1}{r_{12}}\phi_s^*(\mathbf{r}_2)\phi_r(\mathbf{r}_2)d\mathbf{r}_1 d\mathbf{r}_2, \tag{10.21}$$

and

$$(pq|sr)^{\mathrm{lr}} = \int\int \phi_p^*(\mathbf{r}_1)\phi_q(\mathbf{r}_1)\frac{\mathrm{erf}(\mu r_{12})}{r_{12}}\phi_s^*(\mathbf{r}_2)\phi_r(\mathbf{r}_2)d\mathbf{r}_1 d\mathbf{r}_2, \tag{10.22}$$

where $r_{12} = |\mathbf{r}_1 - \mathbf{r}_2|$ for coordinate vectors of electrons, $\mathbf{r}_1$ and $\mathbf{r}_2$, and $\mu$ is an adapted parameter determining the ratio of short- and long-range parts of electron repulsion operator $1/r_{12}$. The response of the exchange–correlation potential term, also called the exchange–correlation kernel, is given as:

$$f^{\mathrm{xc}}_{pq,rs} = \int\int \phi_p^*(\mathbf{r}_1)\phi_q(\mathbf{r}_2)\frac{\delta E_{\mathrm{xc}}}{\delta\rho(\mathbf{r}_1)\delta\rho(\mathbf{r}_2)}\phi_r^*(\mathbf{r}_2)\phi_s(\mathbf{r}_2)d\mathbf{r}_1 d\mathbf{r}_2. \tag{10.23}$$

We substitute the time-dependent KS and density matrices into the following time-dependent KS equation:

$$\frac{1}{i}\sum_q \left(F_{pq}D_{qr} - D_{pq}F_{qr}\right) = \frac{\partial D_{pr}}{\partial t}. \tag{10.24}$$

Collecting the terms that are linear in the perturbation with the time dependence, we obtain

$$\begin{aligned} &\sum_q F_{pq}^{(0)}d_{qr}^{(1)} - \sum_q d_{pq}^{(1)}F_{qr}^{(0)} + \sum_q h_{pq}^{(1)}D_{qr}^{(0)} + \sum_{q,s,t}\left(\frac{\partial F_{pq}}{\partial D_{st}}\right)d_{st}^{(1)}D_{qr}^{(0)} \\ &- \sum_q D_{pq}^{(0)}h_{qr}^{(1)} - \sum_{q,s,t} D_{pq}^{(0)}\left(\frac{\partial F_{qr}}{\partial D_{st}}\right)d_{st}^{(1)} = \omega d_{pr}^{(1)}. \end{aligned} \tag{10.25}$$

The terms multiplied by the $e^{i\omega t}$ factor lead to the conjugate complex of the above equation.

Because the KS (or HF) equation and energy are invariant to rotations among just occupied orbitals or among just virtual orbitals, we only need to consider the occupied virtual block of $d$, i.e., $\left\{d_{ai}^{(1)}\right\}$ and $\left\{d_{ia}^{(1)}\right\}$. Substituting Eqs. (10.13)–(10.15) into Eq. (10.25), we arrive at a pair of equations:

$$(\varepsilon_a - \varepsilon_i)d_{ai}^{(1)} + h_{ai}^{(1)} + \sum_{b,j}\left(\frac{\partial F_{ai}}{\partial D_{bj}}\right)x_{bj} + \sum_{b,j}\left(\frac{\partial F_{ai}}{\partial D_{jb}}\right)y_{bj} = \omega d_{ai}^{(1)}, \tag{10.26}$$

$$(\varepsilon_i - \varepsilon_a)d_{ia}^{(1)} - h_{ia}^{(1)} - \sum_{b,j}\left(\frac{\partial F_{ia}}{\partial D_{bj}}\right)x_{bj} - \sum_{b,j}\left(\frac{\partial F_{ia}}{\partial D_{jb}}\right)y_{bj} = \omega d_{ia}^{(1)}, \tag{10.27}$$

where $x_{ai} = d_{ai}^{(1)}$ and $y_{ai} = d_{ia}^{(1)}$. These may be cast into a compact matrix linear equation:

$$\begin{pmatrix} \mathbf{A} & \mathbf{B} \\ \mathbf{B}^* & \mathbf{A}^* \end{pmatrix}\begin{pmatrix} \mathbf{x} \\ \mathbf{y} \end{pmatrix} - \omega\begin{pmatrix} \mathbf{1} & \mathbf{0} \\ \mathbf{0} & -\mathbf{1} \end{pmatrix}\begin{pmatrix} \mathbf{x} \\ \mathbf{y} \end{pmatrix} = -\begin{pmatrix} \mathbf{h} \\ \mathbf{h}^{\dagger *} \end{pmatrix}, \tag{10.28}$$

with

$$A_{ai,bj} = \delta_{ij}\delta_{ab}(\varepsilon_a - \varepsilon_i) + (ai|jb) - c_{\mathrm{x}}(ab|ji) - c_{\mathrm{x}}^{\mathrm{lr}}(ab|ji)^{\mathrm{lr}} + f_{ai,jb}^{\mathrm{xc}}$$
$$B_{ai,bj} = (ai|bj) - c_{\mathrm{x}}(aj|bi) - c_{\mathrm{x}}^{\mathrm{lr}}(aj|bi)^{\mathrm{lr}} + f_{ai,bj}^{\mathrm{xc}}, \tag{10.29}$$

where **1** and **0** are unit matrix and zero matrix, respectively. Equation (10.28) can be solved for **x** and **y** by standard iterative techniques that use trial vectors and that work with just atomic-orbital-based integrals [28, 29]. Once the equation is solved, the frequency-dependent polarizability is readily evaluated by [30]:

$$\alpha_{\alpha}(-\omega;\omega) = -\mathrm{Tr}\left[\mathbf{h}^{(1)}\mathbf{D}^{\alpha}(\omega)\right] = -\sum_{a,i}\left\{h_{ai}^{(1)}x_{ai}^{\alpha} + h_{ia}^{(1)}y_{ia}^{\alpha}\right\}, \tag{10.30}$$

if **h** is a dipole moment matrix.

The poles of the frequency-dependent polarizability correspond to electronic excitations, occurring with an infinitesimal perturbation, i.e., $\mathbf{h}=0$. Substituting this into Eq. (10.28) leads to a nonsymmetric matrix eigenvalue problem:

$$\begin{pmatrix} \mathbf{A} & \mathbf{B} \\ \mathbf{B}^* & \mathbf{A}^* \end{pmatrix}\begin{pmatrix} \mathbf{x} \\ \mathbf{y} \end{pmatrix} = \omega\begin{pmatrix} \mathbf{1} & \mathbf{0} \\ \mathbf{0} & -\mathbf{1} \end{pmatrix}\begin{pmatrix} \mathbf{x} \\ \mathbf{y} \end{pmatrix}, \tag{10.31}$$

which can be solved for electronic excitation energies $\omega$ and corresponding **x** and **y** vectors of TDHF or RPA by standard techniques using Davidson's trial vector algorithm [31] (as adapted to a nonsymmetric problem [32]) in an atomic-orbital-based scheme [33].

In a phenomenological way, relaxation effects may be introduced into Eq. (10.24) as follows [10, 34]:

$$\frac{1}{i}\sum_q \left(F_{pq}D_{qr} - D_{pq}F_{qr}\right) - \gamma_{pr}\left(D_{pr} - D_{pr}^{(0)}\right) = \frac{\partial D_{pr}}{\partial t}, \tag{10.32}$$

where the damping terms $\gamma_{pq}$ correspond to the rate at which density matrix element $D_{pq}$ relaxes to its equilibrium value of $D_{pq}^{(0)}$. Applying the common lifetime broadening factor $\gamma$ for all excited states, Eq. (10.25) becomes

$$\sum_q F_{pq}^{(0)}d_{qr}^{(1)} - \sum_q d_{pq}^{(1)}F_{qr}^{(0)} + \sum_q h_{pq}^{(1)}D_{qr}^{(0)} + \sum_{q,s,t}\left(\frac{\partial F_{pq}}{\partial D_{st}}\right)d_{st}^{(1)}D_{qr}^{(0)}$$
$$- \sum_q D_{pq}^{(0)}h_{qr}^{(1)} - \sum_{q,s,t} D_{pq}^{(0)}\left(\frac{\partial F_{qr}}{\partial D_{st}}\right)d_{st}^{(1)} = (\omega + i\gamma)d_{pr}^{(1)}. \tag{10.33}$$

Since the difference between Eq. (10.25) and Eq. (10.33) is only in those frequency $\omega$ and $\omega + i\gamma$, substituting $\omega$ to $\omega + i\gamma$ in Eq. (10.28), we obtained the first-order damped-response equation:

$$\begin{pmatrix} \mathbf{A} & \mathbf{B} \\ \mathbf{B}^* & \mathbf{A}^* \end{pmatrix}\begin{pmatrix} \mathbf{x} \\ \mathbf{y} \end{pmatrix} - (\omega + i\gamma)\begin{pmatrix} \mathbf{1} & \mathbf{0} \\ \mathbf{0} & -\mathbf{1} \end{pmatrix}\begin{pmatrix} \mathbf{x} \\ \mathbf{y} \end{pmatrix} = -\begin{pmatrix} \mathbf{h} \\ \mathbf{h}^{\dagger *} \end{pmatrix}. \tag{10.34}$$

In the spin–orbit case, Eq. (10.34) can easily be solved since many of the intermediate quantities that are computed are already complex to begin with.

### *10.2.3 Noncollinear Formulation for Exchange–Correlation Kernel*

In relativistic density functional calculations for open-shell systems with spin–orbit couplings, the spin is no longer a good quantum number and a noncollinear formulation for the exchange–correlation potential and exchange–correlation kernels are often used [35–42]. In relativistic TDDFT, this noncollinear formula is important, since the excited state can be an open-shell electronic state by excitation even in the closed–shell system even in the ground state.

In the nonrelativistic calculation, spin electron densities $\rho_\alpha$ and $\rho_\beta$ can be written as $\rho_\alpha = 1/2(\rho + s)$ and $\rho_\beta = 1/2(\rho - s)$, using the total electron density $\rho$, the difference electron density $s$. Similarly, we can define two new quantities:

$$\begin{aligned} \rho_+ &= \frac{1}{2}(\rho + s) \\ \rho_- &= \frac{1}{2}(\rho - s), \end{aligned} \tag{10.35}$$

where the difference electron density $s$ is defined as the spin density (magnetization) vector $\mathbf{m} = \left(m_x, m_y, m_z\right)$ by

$$s = \sqrt{m_x^2 + m_y^2 + m_z^2}. \tag{10.36}$$

Here, Eq. (10.35) represents the local eigenvalues of spin density. In terms of two-component spinors, $\phi$, $\rho$, and $\mathbf{m}$ are defined as:

$$\rho(\mathbf{r}) = \sum_i \phi_i^\dagger(\mathbf{r})\phi(\mathbf{r}), \tag{10.37}$$

$$\mathbf{m}(\mathbf{r}) = \sum_i \phi_i^\dagger(\mathbf{r})\boldsymbol{\sigma}\phi(\mathbf{r}), \tag{10.38}$$

where $\boldsymbol{\sigma}$ are the Pauli spin matrix vector. It should be noted that in the generalized case, the spin density vector can change its modulus and direction at every point in space, but one would like to use only its modulus to evaluate the exchange–correlation energy.

The expression of the exchange–correlation energy is extended in a consistent way as

$$E_{\mathrm{xc}} = \int d\mathbf{r}\hat{E}_{\mathrm{xc}}(\rho_+, \rho_-, \gamma_{++}, \gamma_{--}, \gamma_{+-}), \tag{10.39}$$

with

$$\gamma_{ij} = \nabla\rho_i \cdot \nabla\rho_j \ (i,j = +, -), \tag{10.40}$$

where $\hat{E}_{xc}$ is general exchange-correlation functional.

The exchange–correlation potential and kernel can be obtained by a proper functional derivative of Eq. (10.39). In LDA case, those are expressed as

$$\begin{aligned}\int d\mathbf{r}\frac{\delta\hat{E}_{\mathrm{xc}}}{\delta\rho}\phi_p^*\phi_q &= \frac{1}{2}\int d\mathbf{r}\phi_p^*\phi_q\left(\frac{\partial\hat{E}_{\mathrm{xc}}}{\partial\rho_+} + \frac{\partial\hat{E}_{\mathrm{xc}}}{\partial\rho_-}\right) \\ &+ \frac{1}{2}\int d\mathbf{r}\left(\frac{\partial\hat{E}_{\mathrm{xc}}}{\partial\rho_+} - \frac{\partial\hat{E}_{\mathrm{xc}}}{\partial\rho_-}\right)\frac{1}{s}\sum_{i=x,y,z} m_i\left(\phi_p^*\sigma_i\phi_q\right),\end{aligned} \tag{10.41}$$

and

$$\begin{aligned}\int d\tau\phi_p^*\phi_q\frac{\delta^2\hat{E}_{\mathrm{xc}}}{\delta\rho\delta\rho'}\phi_r^*\phi_s &= \frac{1}{4}\int d\mathbf{r}\left(\frac{\partial^2\hat{E}_{\mathrm{xc}}}{\partial\rho_+^2} + \frac{\partial^2\hat{E}_{\mathrm{xc}}}{\partial\rho_-^2} + 2\frac{\partial^2\hat{E}_{\mathrm{xc}}}{\partial\rho_+\partial\rho_-}\right)\left(\phi_p^*\phi_q\right)\left(\phi_r^*\phi_s\right) \\ &+ \frac{1}{2}\int d\mathbf{r}\left(\frac{\partial\hat{E}_{\mathrm{xc}}}{\partial\rho_+} - \frac{\partial\hat{E}_{\mathrm{xc}}}{\partial\rho_-}\right)\frac{1}{s}\sum_{i=x,y,z} m_i\left(\phi_p^*\sigma_i\phi_q\right) \\ &+ \frac{1}{4}\int d\mathbf{r}\left(\frac{\partial^2\hat{E}_{\mathrm{xc}}}{\partial\rho_+^2} - \frac{\partial^2\hat{E}_{\mathrm{xc}}}{\partial\rho_-^2}\right)\frac{1}{s} \\ &\times\left[\left(\phi_p^*\phi_q\right)\sum_{i=x,y,z} m_i(\phi_r^*\sigma_i\phi_s) + \sum_{i=x,y,z} m_i\left(\phi_p^*\sigma_i\phi_q\right)\left(\phi_r^*\phi_s\right)\right] \\ &+ \frac{1}{4}\int d\mathbf{r}\left(\frac{\partial^2\hat{E}_{\mathrm{xc}}}{\partial\rho_+^2} + \frac{\partial^2\hat{E}_{\mathrm{xc}}}{\partial\rho_-^2} - 2\frac{\partial^2\hat{E}_{\mathrm{xc}}}{\partial\rho_+\partial\rho_-}\right)\frac{1}{s} \\ &\times\sum_{i=x,y,z} m_i\left(\phi_p^*\sigma_i\phi_q\right)\frac{1}{s}\sum_{i=x,y,z} m_i(\phi_r^*\sigma_i\phi_s) \\ &- \frac{1}{2}\int d\mathbf{r}\left(\frac{\partial\hat{E}_{\mathrm{xc}}}{\partial\rho_+} - \frac{\partial\hat{E}_{\mathrm{xc}}}{\partial\rho_-}\right)\frac{1}{s^3}\sum_{i=x,y,z} m_i\left(\phi_p^*\sigma_i\phi_q\right)\sum_{i=x,y,z} m_i(\phi_r^*\sigma_i\phi_s) \\ &+ \frac{1}{2}\int d\mathbf{r}\left(\frac{\partial\hat{E}_{\mathrm{xc}}}{\partial\rho_+} - \frac{\partial\hat{E}_{\mathrm{xc}}}{\partial\rho_-}\right)\frac{1}{s}\sum_{i=x,y,z}\left(\phi_p^*\sigma_i\phi_q\right)\left(\phi_r^*\sigma_i\phi_s\right).\end{aligned} \tag{10.42}$$

Note that all the derivatives are taken at the ground-state density and spin densities. Equation (10.42) can be further simplified in some special cases. For closed-shell systems, we will have the following additional conditions with $s \to 0$ [20]:

$$\frac{\delta \hat{E}_{\mathrm{xc}}}{\delta s} \to 0, \tag{10.43}$$

$$\frac{\delta^2 \hat{E}_{\mathrm{xc}}}{\delta \rho \delta s} \to 0, \tag{10.44}$$

$$\frac{1}{s}\frac{\delta \hat{E}_{\mathrm{xc}}}{\delta s} \to \frac{\delta^2 \hat{E}_{\mathrm{xc}}}{\delta s^2}. \tag{10.45}$$

Equation (10.42) is then simplified to

$$\begin{aligned}\int d\tau \phi_p^* \phi_q \frac{\delta^2 \hat{E}_{\mathrm{xc}}}{\delta \rho \delta \rho'} \phi_r^* \phi_s = \frac{1}{4} \int d\mathbf{r} \left( \frac{\partial^2 \hat{E}_{\mathrm{xc}}}{\partial \rho_+^2} + \frac{\partial^2 \hat{E}_{\mathrm{xc}}}{\partial \rho_-^2} + 2\frac{\partial^2 \hat{E}_{\mathrm{xc}}}{\partial \rho_+ \partial \rho_-} \right) \left( \phi_p^* \phi_q \right) \left( \phi_r^* \phi_s \right) \\ + \frac{1}{4} \int d\mathbf{r} \left( \frac{\partial^2 \hat{E}_{\mathrm{xc}}}{\partial \rho_+^2} + \frac{\partial^2 \hat{E}_{\mathrm{xc}}}{\partial \rho_-^2} - 2\frac{\partial^2 \hat{E}_{\mathrm{xc}}}{\partial \rho_+ \partial \rho_-} \right) \sum_{i=x,y,z} \left( \phi_p^* \sigma_i \phi_q \right) \left( \phi_r^* \sigma_i \phi_s \right).\end{aligned} \tag{10.46}$$

From Eq. (10.46), it is shown that in the nonrelativistic limit with the molecular spinors of either spin $\alpha$ or spin $\beta$, if $p$ and $q$ have the same spin and $s$ and $t$ have the same spin, the first term of Eq. (10.46) is exactly the same as the corresponding term in nonrelativistic TDDFT calculations.

When $p$ and $q$ have different spins and $s$ and $t$ have different spins, these transitions correspond to spin-flip excitations, and it has already been shown that for closed-shell systems, spin-flip transitions will result in singlet–triplet excitations and the excitation energies calculated from the second term of Eq. (10.46) are the same as those from ordinary TDDFT for singlet–triplet transitions. From this argument, we can see that the TDDFT formulation based on a noncollinear exchange–correlation potential has the correct nonrelativistic limit and can recover the threefold degeneracy of triplet excitations correctly.

## 10.3 Implementation

### *10.3.1 Trial Vector Algorithm*

In the present implementation of TDHF and TDDFT, the explicit formation of an enormous number of the two-electron integrals in Eq. (29) is avoided by invoking a trial vector algorithm, which is based on the idea of projecting the full matrices onto ones of greatly reduced dimensions. The generalization of this algorithm to a non-Hermitian eigenvalue equation has been considered by Hirao and Nakatsuji [32] and for the particular form of Eq. (10.31) by Olsen, Jensen, and Jørgensen [33], who also propose efficient algorithm for linear equation of Eq. (10.28) using trial vectors in the same paper. It should be remembered that **A** and **B** matrices are complex matrices and hence we cannot reduce Eq. (10.31) to a Hermitian eigenvalue equation of half the dimension as has usually been done in the implementations in the nonrelativistic program [7–9, 43]. In NTChem, the KAIN algorithm [44] for linear equation and Olsen's algorithm [33, 45] for non-Hermitian eigenvalue equation are implemented.

The key steps in these trial vector algorithms are to calculate matrix–trial vector products. By projecting the **A** and **B** matrices in Eq. (10.28) onto a subspace spanned by a set of trial vectors

$$\left\{ \begin{pmatrix} \mathbf{x}(1) \\ \mathbf{y}(1) \end{pmatrix}, \begin{pmatrix} \mathbf{x}(2) \\ \mathbf{y}(2) \end{pmatrix}, \ldots, \begin{pmatrix} \mathbf{x}(p) \\ \mathbf{y}(p) \end{pmatrix} \right\}, \tag{10.47}$$

the matrix–trial vector products are written as:

$$\begin{aligned} \bar{x}_{ai}^{(p)} &= \sum_{b,j} \left( A_{ai,bj} x_{bj}^{(p)} + B_{ai,bj} y_{bj}^{(p)} \right), \\ \bar{y}_{ai}^{(p)} &= \sum_{b,j} \left( B_{ai,bj}^{*} x_{bj}^{(p)} + A_{ai,bj}^{*} y_{bj}^{(p)} \right). \end{aligned} \tag{10.48}$$

Using the AO representation of the **A** and **B** matrices, Eq. (10.48) is written as:

$$\begin{aligned} \bar{x}_{ai}^{(p)} &= \sum_{\omega,\omega'} \sum_{\mu,\nu} C_{\mu a}^{\omega *} C_{\nu i}^{\omega'} G\left[\mathbf{D}^{(p)}\right]_{\mu\nu}^{\omega\omega'} \\ \bar{y}_{ai}^{(p)} &= \left( \sum_{\omega,\omega'} \sum_{\mu,\nu} C_{\mu a}^{\omega *} C_{\nu i}^{\omega'} G\left[\mathbf{D}^{(p)}\right]_{\mu\nu}^{\omega\omega'^{*}} \right)^{*}, \end{aligned} \tag{10.49}$$

where the trial density matrices $\mathbf{D}^{(p)}$ and trial Fock matrices $\mathbf{G}$ are defined as:

$$D_{\lambda\kappa}^{(p)\omega\omega'} = \sum_{b,j} C_{\lambda b}^{\omega} x_{bj}^{(p)} C_{\kappa j}^{\omega *} + \sum_{b,j} C_{\lambda j}^{\omega} y_{bj}^{(p)} C_{\kappa b}^{\omega' *}, \tag{10.50}$$

$$\begin{aligned} G\left[\mathbf{D}^{(p)}\right]_{\mu\nu}^{\omega\omega'} &= (\mu\nu|\lambda\kappa) D_{\lambda\kappa}^{(p)\omega\omega'} + f_{\mu\nu,\lambda\kappa}^{\mathrm{xc}} D_{\lambda\kappa}^{(p)\omega\omega'} \\ &- \delta_{\omega\omega'}\left[c_{\mathrm{x}}(\mu\kappa|\lambda\nu) - c_{\mathrm{x}}^{\mathrm{lr}}(\mu\kappa|\lambda\nu)^{\mathrm{lr}}\right] D_{\lambda\kappa}^{(p)\omega\omega'}, \end{aligned} \tag{10.51}$$

and

$$G\left[\mathbf{D}^{(p)}\right]_{\mu\nu}^{\omega\omega'} f_{\mu\nu,\lambda\kappa}^{\mathrm{xc}} = \int\int \chi_{\mu}(\mathbf{r}_1)\chi_{\nu}(\mathbf{r}_2) \frac{\delta^2 E_{\mathrm{xc}}}{\delta\rho(\mathbf{r}_1)\delta\rho(\mathbf{r}_2)} \chi_{\lambda}(\mathbf{r}_2)\chi_{\kappa}(\mathbf{r}_2) d\mathbf{r}_1 d\mathbf{r}_2. \tag{10.52}$$

In two-component TDDFT equations, the one-electron trial density matrix will present a nonvanishing $\alpha\beta$ block coupling the two spin components:

$$\mathbf{D}^{(p)} = \begin{pmatrix} \mathbf{D}^{(p)\alpha\alpha} & \mathbf{D}^{(p)\alpha\beta} \\ \mathbf{D}^{(p)\beta\alpha} & \mathbf{D}^{(p)\beta\beta} \end{pmatrix}. \tag{10.53}$$

Therefore, the evaluation of these matrix–trial vector products can be carried out in the direct AO-based algorithm. Since the time-consuming calculation of $\mathbf{G}$ for two trial vectors is only one time, the cost is not so different from the CIS or Tamm–Dancoff approximation using one trial vector.

The contribution to the matrix–trial vector product from the noncollinear exchange–correlation kernel (10.46) can be calculated as:

$$\sum_{\omega\omega'} f_{\mu\nu,\lambda\kappa}^{\mathrm{xc}} D_{\lambda\kappa}^{(p)\omega\omega'} = \int d\mathbf{r} \frac{\delta^2 E_{\mathrm{xc}}}{\delta\rho^2} \rho^{(p)} (\chi_{\mu}\chi_{\nu}) + \sum_{i=x,y,z} \int d\mathbf{r} \frac{\delta^2 E_{\mathrm{xc}}}{\delta m^2} m_i^{(p)} (\chi_{\mu}\sigma_i\chi_{\nu}), \tag{10.54}$$

where trial total electron density and trial spin density vector are defined by analogy of total electron density and spin density vector as:

$$\rho_0^{(p)} = \sum_{\mu\nu} D_{\mu\nu}^{(p)0} \phi_{\mu}\phi_{\nu}, \tag{10.55}$$

$$m_i^{(p)} = \sum_{\mu\nu} D_{\mu\nu}^{(p)i} \phi_{\mu}\phi_{\nu} \quad (i = x, y, z), \tag{10.56}$$

and corresponding density matrices are obtained by:

$$D_{\mu\nu}^{(p)0} = D_{\mu\nu}^{(p)\alpha\alpha} + D_{\mu\nu}^{(p)\beta\beta}, \tag{10.57}$$

$$D_{\mu\nu}^{(p)x} = D_{\mu\nu}^{(p)\beta\alpha} + D_{\mu\nu}^{(p)\alpha\beta}, \tag{10.58}$$

$$D_{\mu\nu}^{(p)y} = -i\left\{D_{\mu\nu}^{(p)\beta\alpha} - D_{\mu\nu}^{(p)\alpha\beta}\right\}, \tag{10.59}$$

$$D_{\mu\nu}^{(p)z} = D_{\mu\nu}^{(p)\alpha\alpha} - D_{\mu\nu}^{(p)\beta\beta}. \tag{10.60}$$

Here, the first term on the right-hand side of Eq. (10.54) is a term derived from singlet excitation and derived in GGA as

$$\begin{aligned}
\int d\mathbf{r}\frac{\delta^2 E_{\mathrm{xc}}}{\delta\rho^2}(\chi_\mu\chi_\nu) &= \frac{1}{4}\int d\mathbf{r}(\chi_\mu\chi_\nu)\rho_0^{(p)}\left\{\frac{\partial^2 E_{\mathrm{xc}}}{\partial\rho_\alpha\partial\rho_\alpha} + 2\frac{\partial^2 E_{\mathrm{xc}}}{\partial\rho_\alpha\partial\rho_\beta} + \frac{\partial^2 E_{\mathrm{xc}}}{\partial\rho_\beta\partial\rho_\beta}\right\} \\
&+ \frac{1}{4}\int d\mathbf{r}(\chi_\mu\chi_\nu)\left(\nabla\rho_\alpha\cdot\nabla\rho_0^{(p)}\right) \\
&\times\left\{\begin{array}{l} 2\frac{\partial^2 E_{\mathrm{xc}}}{\partial\gamma_{\alpha\alpha}\partial\rho_\alpha} + 2\frac{\partial^2 E_{\mathrm{xc}}}{\partial\gamma_{\alpha\alpha}\partial\rho_\beta} + 2\frac{\partial^2 E_{\mathrm{xc}}}{\partial\gamma_{\beta\beta}\partial\rho_\alpha} + 2\frac{\partial^2 E_{\mathrm{xc}}}{\partial\gamma_{\beta\beta}\partial\rho_\beta} \\ + 2\frac{\partial^2 E_{\mathrm{xc}}}{\partial\gamma_{\alpha\beta}\partial\rho_\alpha} + 2\frac{\partial^2 E_{\mathrm{xc}}}{\partial\gamma_{\alpha\beta}\partial\rho_\beta} \end{array}\right\} \\
&+ \frac{1}{4}\int d\mathbf{r}(\nabla(\chi_\mu\chi_\nu)\cdot\nabla\rho_\alpha)\rho_0^{(p)} \\
&\times\left\{\begin{array}{l} 2\frac{\partial^2 E_{\mathrm{xc}}}{\partial\gamma_{\alpha\alpha}\partial\rho_\alpha} + 2\frac{\partial^2 E_{\mathrm{xc}}}{\partial\gamma_{\alpha\alpha}\partial\rho_\beta} + 2\frac{\partial^2 E_{\mathrm{xc}}}{\partial\gamma_{\beta\beta}\partial\rho_\alpha} + 2\frac{\partial^2 E_{\mathrm{xc}}}{\partial\gamma_{\beta\beta}\partial\rho_\beta} \\ + 2\frac{\partial^2 E_{\mathrm{xc}}}{\partial\gamma_{\alpha\beta}\partial\rho_\alpha} + 2\frac{\partial^2 E_{\mathrm{xc}}}{\partial\gamma_{\alpha\beta}\partial\rho_\beta} \end{array}\right\} \\
&+ \frac{1}{4}\int d\mathbf{r}(\nabla(\chi_\mu\chi_\nu)\cdot\nabla\rho_\alpha)\left(\nabla\rho_\alpha\cdot\nabla\rho_0^{(p)}\right) \\
&\times\left\{\begin{array}{l} 4\frac{\partial^2 E_{\mathrm{xc}}}{\partial\gamma_{\alpha\alpha}\partial\gamma_{\alpha\alpha}} + 8\frac{\partial^2 E_{\mathrm{xc}}}{\partial\gamma_{\alpha\alpha}\partial\gamma_{\beta\beta}} + 8\frac{\partial^2 E_{\mathrm{xc}}}{\partial\gamma_{\alpha\alpha}\partial\gamma_{\alpha\beta}} + 4\frac{\partial^2 E_{\mathrm{xc}}}{\partial\gamma_{\beta\beta}\partial\gamma_{\beta\beta}} \\ + 8\frac{\partial^2 E_{\mathrm{xc}}}{\partial\gamma_{\beta\beta}\partial\gamma_{\alpha\beta}} + 4\frac{\partial^2 E_{\mathrm{xc}}}{\partial\gamma_{\alpha\beta}\partial\gamma_{\alpha\beta}} \end{array}\right\} \\
&+ \frac{1}{4}\int d\mathbf{r}\left(\nabla(\chi_\mu\chi_\nu)\cdot\nabla\rho_0^{(p)}\right)\left\{2\frac{\partial E_{\mathrm{xc}}}{\partial\gamma_{\alpha\alpha}} + 2\frac{\partial E_{\mathrm{xc}}}{\partial\gamma_{\beta\beta}} + 2\frac{\partial E_{\mathrm{xc}}}{\partial\gamma_{\alpha\beta}}\right\}
\end{aligned} \tag{10.61}$$

Similarly, the second term of Eq. (10.54) is the triplet-excitation term and written as:

$$\begin{aligned}
\int d\mathbf{r}\frac{\delta^2 E_{\mathrm{xc}}}{\delta m^2} m_i^{(p)}(\chi_\mu \sigma_i \chi_\nu) = {} & \frac{1}{4}\int d\mathbf{r}(\chi_\mu\chi_\nu) m_i^{(p)}\left\{\frac{\partial^2 E_{\mathrm{xc}}}{\partial\rho_\alpha\partial\rho_\alpha} + \frac{\partial^2 E_{\mathrm{xc}}}{\partial\rho_\beta\partial\rho_\beta} - 2\frac{\partial^2 E_{\mathrm{xc}}}{\partial\rho_\alpha\partial\rho_\beta}\right\} \\
& + \frac{1}{4}\int d\mathbf{r}(\chi_\mu\chi_\nu)\left(\nabla\rho_\alpha\cdot\nabla m_i^{(p)}\right) \\
& \times\left\{2\frac{\partial^2 E_{\mathrm{xc}}}{\partial\gamma_{\alpha\alpha}\partial\rho_\alpha} 2\frac{\partial^2 E_{\mathrm{xc}}}{\partial\gamma_{\beta\beta}\partial\rho_\beta} - 2\frac{\partial^2 E_{\mathrm{xc}}}{\partial\gamma_{\alpha\alpha}\partial\rho_\beta} - 2\frac{\partial^2 E_{\mathrm{xc}}}{\partial\gamma_{\beta\beta}\partial\rho_\alpha}\right\} \\
& + \frac{1}{4}\int d\mathbf{r}(\nabla(\chi_\mu\chi_\nu)\cdot\nabla\rho_\alpha) m_i^{(p)} \\
& \times\left\{2\frac{\partial^2 E_{\mathrm{xc}}}{\partial\gamma_{\alpha\alpha}\partial\rho_\alpha} + 2\frac{\partial^2 E_{\mathrm{xc}}}{\partial\gamma_{\beta\beta}\partial\rho_\beta} - 2\frac{\partial^2 E_{\mathrm{xc}}}{\partial\gamma_{\beta\beta}\partial\rho_\alpha} - 2\frac{\partial^2 E_{\mathrm{xc}}}{\partial\gamma_{\alpha\alpha}\partial\rho_\beta}\right\} \\
& + \frac{1}{4}\int d\mathbf{r}(\nabla(\chi_\mu\chi_\nu)\cdot\nabla\rho_\alpha)\left(\nabla\rho_\alpha\cdot\nabla m_i^{(p)}\right) \\
& \times\left\{4\frac{\partial^2 E_{\mathrm{xc}}}{\partial\gamma_{\alpha\alpha}\partial\gamma_{\alpha\alpha}} + 4\frac{\partial^2 E_{\mathrm{xc}}}{\partial\gamma_{\beta\beta}\partial\gamma_{\beta\beta}} - 8\frac{\partial^2 E_{\mathrm{xc}}}{\partial\gamma_{\alpha\alpha}\partial\gamma_{\beta\beta}}\right\} \\
& + \frac{1}{4}\int d\mathbf{r}\left(\nabla(\chi_\mu\chi_\nu)\cdot\nabla m_i^{(p)}\right)\left\{2\frac{\partial E_{\mathrm{xc}}}{\partial\gamma_{\alpha\alpha}} + 2\frac{\partial E_{\mathrm{xc}}}{\partial\gamma_{\beta\beta}} - 2\frac{\partial E_{\mathrm{xc}}}{\partial\gamma_{\alpha\beta}}\right\}
\end{aligned} \tag{10.62}$$

In NTChem, these matrix elements of the noncollinear exchange–correlation kernel and functional derivatives of exchange–correlation functional for the LDA, GGA, and meta-GGA functionals have been derived and implemented into efficient computer codes with the aid of a newly developed computerized symbolic algebra system.

### 10.3.2 Automatic Implementation

Increasing complexity of quantum chemistry methods, most of the modern exchange–correlation functionals have rather complicated forms and manual implementation of these algebraic formulas into a computer program is often impractical. In particular, the problem is more serious in implementing the *n*th order properties, which require the same order derivatives of the exchange–correlation functionals because of the nonlinear dependencies on density matrices.

Alternatively, several automatic implementations have been attempted using symbolic differentiation techniques [46, 47] or automatic differentiation techniques [48]. In NTChem, such implementation was achieved with the aid of a newly developed computerized symbolic algebra system. The automatic code generator performs differentiations using the SymPy library, which is an open-source symbolic mathematics library for the Python programming language.

The procedure for implementation of the exchange–correlation functional using our automatic implementation program is shown below.

The first thing to do is to write out the definition formula in the original paper in the input file. In this input file, the functional is described in multiple lines using the predefined input/output variable and the intermediate variable defined by the author, whereas in many programs those are defined with only one line. Figure 10.1 shows the input file of spin-polarized form of PBE exchange functional [49], which is defined as:

$$s_\sigma = \frac{|\nabla \gamma_{\sigma\sigma}|}{\rho_\sigma^{4/3}} c_s, \tag{10.63}$$

$$E_{\mathrm{X}} = \sum_{\sigma=\alpha,\beta} C_x \rho_\sigma^{\frac{4}{3}} \left(1 + \kappa - \frac{\kappa}{1 + \frac{\mu s_\sigma^2}{\kappa}}\right), \tag{10.64}$$

where $\kappa$, $\mu$, and $c_s$ are constants independent of electron density. Here, the right-hand side of each line is written in the SymPy format, which is similar to Fortran's grammar, and variables are defined in order from above in the same way as in ordinary programs. As predefined input and output variables, "Exc" indicates the exchange-correlation energy of output and "rhoa" and "gma" are the input the alpha-spin density $\rho_\alpha$ and its gradient $\gamma_{\alpha\alpha} = \nabla\rho_\alpha \cdot \nabla\rho_\alpha$, respectively (accordingly "rhob" and "gmb" denote the beta-spin quantities). In this way, those definitions can be described in the input file just like implementing into code. Importantly, at this stage, the SymPy library itself merely interprets the variables and does not optimize or deform the expressions at all, so it is possible to use the traditional programming techniques in some extent, such as avoiding zero-division and omitting digits.

After reading the input file, the autogeneration program checks the dependency on the input variables for each line. Based on the input variable dependency, the autogeneration program performs differentiation for each expression up to the desired rank. If there is a derivative of the intermediate variable that becomes zero, those terms are searched and erased in all expressions. In case of closed-shell system, duplicate calculations are erased by replacing the electron density of beta spin with alpha spin. Figure 10.2 shows the result of differentiation of the input of Fig. 10.1 for conversion to a closed-shell system. Here, "d1Exd1" indicates the first

**Fig. 10.1** Input equations of PBE exchange functional

```
sa=gma**(1/2)/(rhoa**(4/3))*Cs
sb=gmb**(1/2)/(rhob**(4/3))*Cs
gxa=1+kappa-kappa/(1+mu*sa**2/kappa)
gxb=1+kappa-kappa/(1+mu*sb**2/kappa)
Exa=Cx * rhoa**(4/3) * gxa
Exb=Cx * rhob**(4/3) * gxb
Ex=Exa + Exb
```

**Fig. 10.2** Output equations of first derivative of PBE exchange functional derived by autogeneration program

```
sa = Cs*sqrt(gma)/rhoa**(4/3)
gxa = kappa - kappa/(1 + mu*sa**2/kappa) + 1
Exa = Cx*gxa*rhoa**(4/3)
Ex = 2*Exa
d1sad1 = -4*Cs*sqrt(gma)/(3*rhoa**(7/3))
d1gxad1 = 2*d1sad1*mu*sa/(1 + mu*sa**2/kappa)**2
d1Exad1 = Cx*d1gxad1*rhoa**(4/3) + 4*Cx*gxa*rhoa**(1/3)/3
d1Exd1 = d1Exad1
d1Exd2 = d1Exad1
d1sad3 = Cs/(2*sqrt(gma)*rhoa**(4/3))
d1gxad3 = 2*d1sad3*mu*sa/(1 + mu*sa**2/kappa)**2
d1Exad3 = Cx*d1gxad3*rhoa**(4/3)
d1Exd3 = d1Exad3
d1Exd4 = d1Exad3
```

derivative of the input variables of “Ex”. At this point, we have not yet done the optimization that SymPy provides yet, so we still can find many common terms generated by chain rules.

Next is the optimization of the whole expression. This is particularly important for rather long functional expressions to reduce the number of floating-point operations by extracting common subexpressions and if nonoptimized compilers are evoked afterward. In order not to destroy the formula of the input at the input stage much here, we developed a new optimization program that considers only “break down” and “common term” instead of using SymPy’s advanced optimization. Figure 10.3 shows the result of optimizing the equations in Fig. 10.2. Despite only being optimized by extremely simple rules, the mathematical expressions generated are sufficiently optimal expressions.

The expressions are finally translated into Fortran routines with the addition of the proper skeleton common to all functionals. A conditional branch statement can also be added to this skeleton program. This capability enables us to treat classes of functionals that have stability issues, including meta-GGA and range-separated hybrid functionals, in NTChem.

**Fig. 10.3** Working equations of the first derivative of PBE exchange functional derived and optimized by autogeneration program

```
v0 = sqrt(gma)
v2 = rhoa**(1/3)
v1 = rhoa*v2
v3 = Cs/v1
sa = v0*v3
x4 = (kappa + mu*sa**2)/kappa
gxa = kappa - kappa/x4 + 1
v4 = Cx*v1
Ex = 2*gxa*v4
v5 = 2*mu*sa/x4**2
d1Exad1 = -4*Cs*v0*v4*v5/(3*rhoa*v1) + 4*Cx*gxa*v2/3
d1Exd1 = d1Exad1
d1Exd2 = d1Exad1
d1Exad3 = v3*v4*v5/(2*v0)
d1Exd3 = d1Exad3
d1Exd4 = d1Exad3
```

## 10.4 Applications

### *10.4.1 Dynamic Polarizabilities of $SnH_4$ and $PbH_4$*

As a first application, dynamic polarizabilities of $SnH_4$ and $PbH_4$ molecules were calculated by two-component TDDFT for linear response, in which noncollinear magnetism was considered for all response calculations. These calculations were performed with various exchange–correlation functional functions, where SVWN [50, 51] as LDA functional, BLYP [52, 53] as pure GGA functional, B3LYP [54] as hybrid type functional, LC-BLYP [55] as range-separated functional, and CAM-B3LYP [56] as range-separated hybrid function were used.

To compare relativistic effects, we used nonrelativistic Hamiltonian, spin-free Hamiltonian, and full relativistic Hamiltonian including spin–orbit interaction. Those Hamiltonians are obtained from spin-free and spin-dependent parts of the third- and first-order Douglas–Kroll Hamiltonians, respectively. The screened nucleus spin–orbit approximation was adopted for the spin-dependent part. Pol-DK basis set [57] was used for relativistic calculation, and Pol basis set [58] was used for nonrelativistic calculation. The molecular structures were Td symmetry, and the bond lengths $r$(Sn–H) and $r$(Pb–H) were 1.7108 Å and 1.742 Å, respectively [59].

Figures 10.4 and 10.5 show the dynamic polarizabilities of $SnH_4$ and $PbH_4$, respectively, by TDDFT, CCSD-LR, and TDHF with spin-free DK3 Hamiltonian. In both systems, LDA and BLYP overestimate the dynamical polarizability, the extent of which increases as the frequency increases. By adding long-range Hartree–Fock exchange, the overestimation is greatly improved in the results of the hybrid functional and the range-separated functional, which are in good agreement with the CCSD-LR results. However, in $PbH_4$, the correction in the hybrid functional B3LYP is not sufficient, and at a large frequency, the polarizability is overestimated. The importance of long-range exchange interaction with such frequency-dependent dynamical polarizability and its dispersion has been discussed.

In Figs. 10.6 and 10.7, dynamical polarizabilities of $SnH_4$ and $PbH_4$ with the Hamiltonian of various relativistic approximation levels obtained by RHF and LC-BLYP methods are shown. The scalar relativistic effect by DK3 increases the polarizability, and its magnitude is greater for Pb, a heavier element. The effect of electronic correlation is larger than the relativistic effect in the polarizability calculation in the comparison between RHF and LC-BLYP. From the comparison between LC-BLYP and RHF, it is shown that the contribution of the electron correlation to the polarizability is equal to or greater than those of the relativistic effect. Furthermore, by comparing the results with or without the spin–orbit interaction, it is found that the spin–orbit effect also increases the polarizability at low frequencies far from the poles, but its magnitude is small and less than 1%. However, as seen in 0.22 a.u. in Fig. 10.7, in the vicinity of the poles, the orbitals of Pb are split by spin–orbit interaction, so that a large influence of spin–orbit interaction on polarizability is observed.

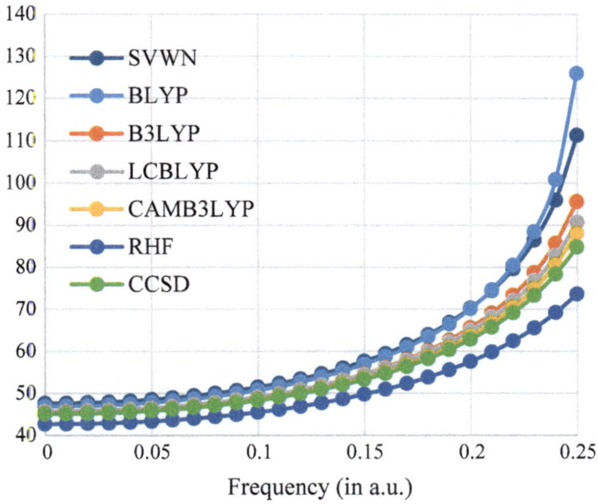


**Fig. 10.4** Dynamical polarizabilities of $SnH_4$ by TDDFT, CCSD-LR, and TDHF with spin-free DK3 Hamiltonian

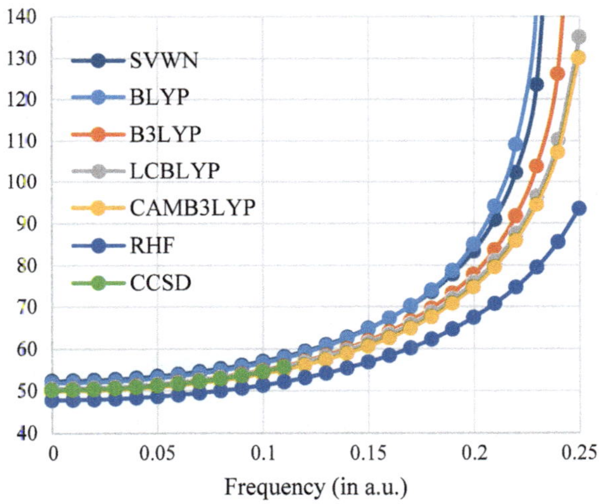


**Fig. 10.5** Dynamical polarizabilities of $PbH_4$ by TDDFT, CCSD-LR, and TDHF with spin-free DK3 Hamiltonian

### *10.4.2 Excitation Spectra of HI*

For the next test, complex dynamic polarizabilities, that is, excitation energies of HI molecule, were calculated by the damped-response calculation. Those obtained peaks were compared with excitation energy obtained from pole/residue calculation. In the pole/residue calculation, we obtained 50 roots from the bottom. The damping factor in Eq. (10.6) is set to 0.004 a.u. in the all damped-response calculations. LC-ωPBE [60–62], which is a range-separated functional, was used as the exchange–correlation functional. The Ext-Pol basis set and Ext-Pol-DK basis [63] set were used in non-relativistic and relativistic calculation, respectively. The relativistic Hamiltonians are

**Fig. 10.6** Dynamical polarizabilities of $SnH_4$ with the Hamiltonian of various relativistic approximation levels

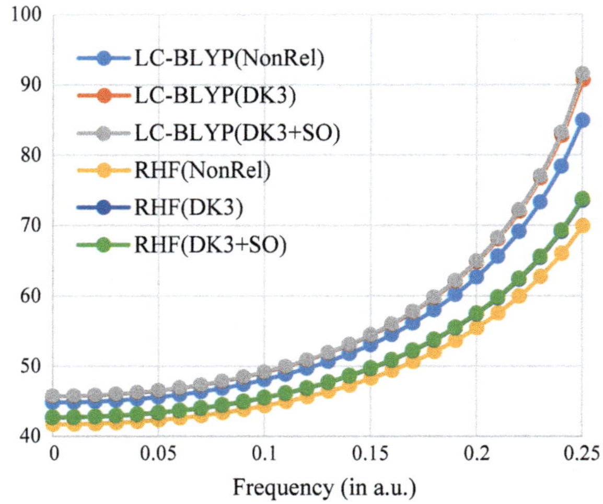


**Fig. 10.7** Dynamical polarizabilities of $PbH_4$ with the Hamiltonian of various relativistic approximation levels

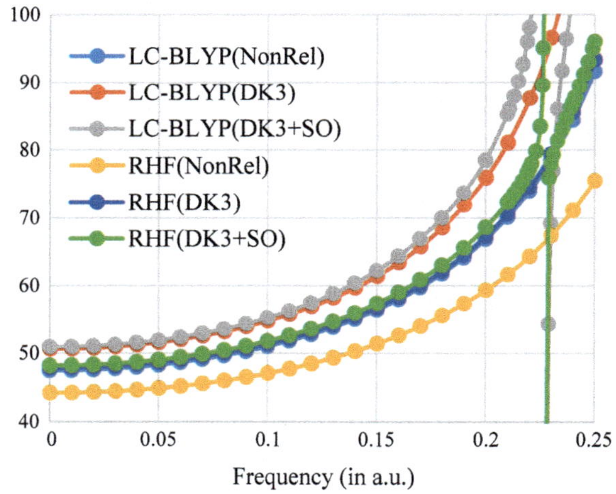


obtained from spin-free and spin-dependent parts of the third- [64] and first-order [14] Douglas–Kroll Hamiltonians, respectively. The screened nucleus spin–orbit approximation [65] was adopted for the spin-dependent part.

The calculated dynamical polarizabilities with or without damping and transition dipoles obtained by pole/residue calculation are shown in Figs. 10.8, 10.9, and 10.10. From Figs. 10.8 and 10.9, the divergence of the dynamic polarizabilities at the poles shown in Fig. 10.9 is suppressed in Fig. 10.8 by introducing the effect of the relaxation of the excitation, but instead the imaginary polarizabilities appear. The position and the relative size of the imaginary polarizability are in good agreement with the square of the magnitude of the transition dipole by the pole/residue calculation. Here, Fig. 10.10 shows the peak of the pole/residue

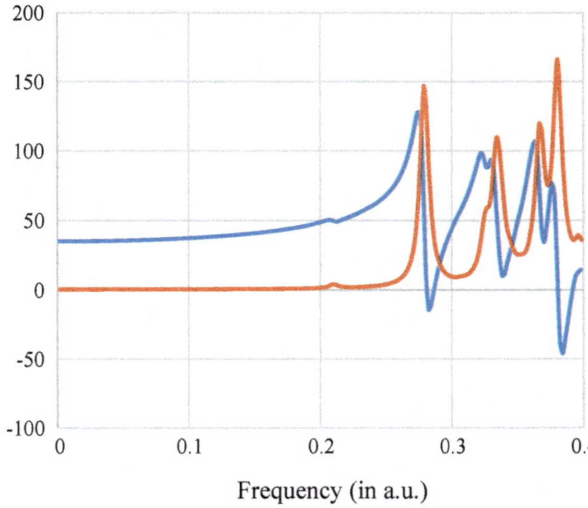


**Fig. 10.8** Dynamical polarizabilities of HI with damping

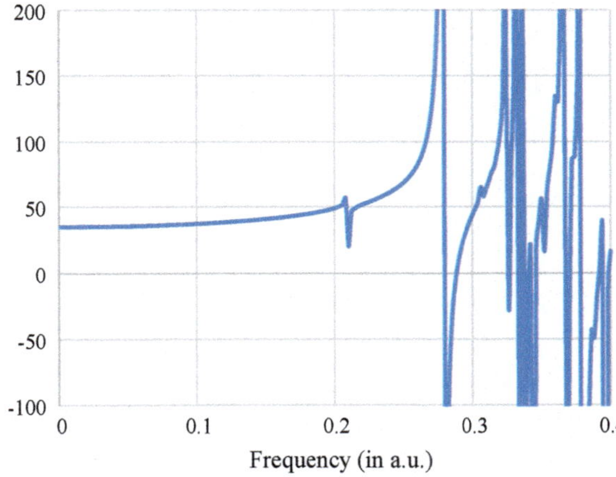


**Fig. 10.9** Dynamical polarizabilities of HI without damping

calculation is only up to 0.35 a.u. because the 50 roots obtained by the pole/residue calculation failed to cover the whole frequency range of 0.0 a.u. to 0.4 a.u. On the other hand, the imaginary polarizability obtained by the damped-response theory can be calculated with sufficient precision for all the regions of interest within similar calculation time. In the peak consisting of many excitations like the second big peak around 0.33 a.u., since the respective states are averaged by the damping factor at the calculated imaginary polarizabilities, it is difficult to analyze the characteristics of those excited states.

Figures 10.11 and 10.12 show the results of nonrelativistic and the result including scalar and so relativistic effect. As shown in Figs. 10.8 and 10.11, the real and imaginary polarizabilities are slightly shifted by the scalar relativistic effect, but the peak shape itself does not change much. Also, because the distribution of

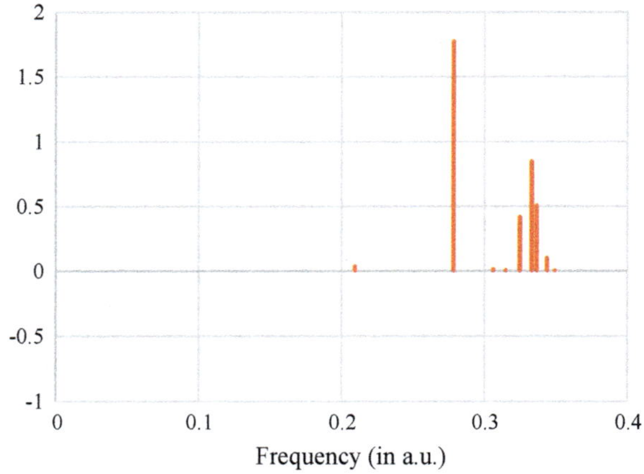


**Fig. 10.10** Transition dipoles obtained by pole/residue calculation

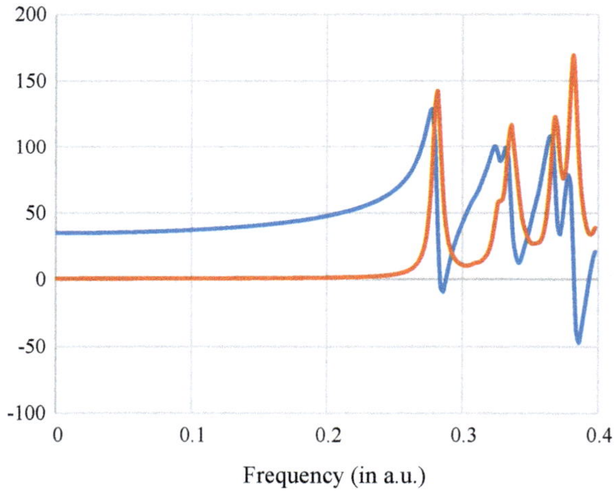


**Fig. 10.11** Dynamical polarizabilities of HI with nonrelativistic calculations

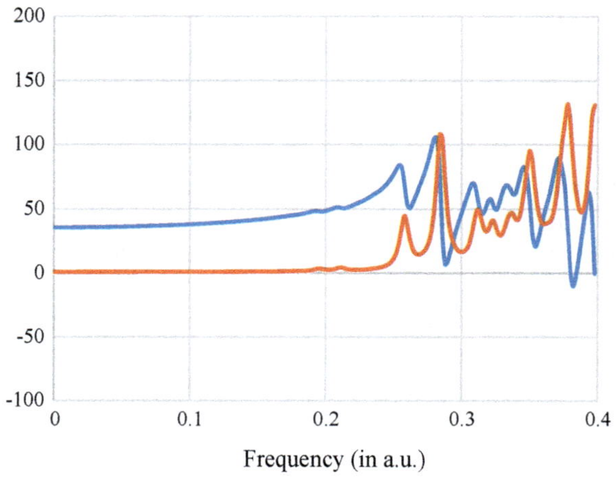


**Fig. 10.12** Dynamical polarizabilities of HI calculated with both scalar and spin–orbit relativistic effects

electron density changes due to the relativistic effect, absorption around 0.21 a.u. can be seen. However, from Figs. 10.8 and 10.12, the spin–orbit interaction greatly changes the shape of the spectrum, which can be understood from the splitting of atomic orbitals of I atom because of the large spin–orbit interaction of the I atom. From these results, we can say that the spin–orbit interaction is extremely important near the pole of the dynamical polarizabilities and excitation energies compared with other relativistic terms.

## 10.5 Conclusion

In this review article, the two-component relativistic linear-response TDDFT for molecular properties is derived and implemented into NTChem program suite. Applying an empirical damping for the electronic excited states has shown that response suitable for simulating dispersion curves can be calculated. The damping avoids singularities in the calculated real part of the polarizability around the resonance frequencies. Further, the damping technique yields both the real and imaginary parts of dipole polarizability with the latter being directly related to the radiation absorption of the system. Such an approach makes it possible to calculate absorption properties within a selected frequency window. This feature could potentially make this direct response route an alternative to common TDDFT pole-/residue-based excitation spectra calculations in the case of large molecules having a large number of excitations within and below a frequency window of interest.

Our implementation is based on a noncollinear exchange–correlation potential presented by Wang et al. In addition, since it is difficult to manually implement the latest complicated functional, the computerized symbolic algebra system is developed and used for those complicated implementation in NTChem. The automatic code generator performs differentiations using the SymPy library, which is an open-source symbolic mathematics library for the Python programming language.

By using the present approach, calculations of the frequency-dependent polarizabilities of $SnH_4$ and $PbH_4$ molecules containing heavy atoms were carried out. In the calculation of the dispersion of the polarizability, it is found that the relativistic effect and the electronic correlation effect are about the same, while the spin–orbit interaction is not so large. The two-component TDDFT was also applied to the excitation spectra of HI molecule. The calculated spectra from the imaginary part of damped dynamical polarizabilities are in excellent agreement with those obtained using poles/residues TDDFT. It is found that the spin–orbit interaction is extremely important in the dynamical polarizabilities near the pole and excitation energies.

**Acknowledgements** This work was supported by the Next-Generation Supercomputer project (the K computer project) and the FLAGSHIP2020 project within the priority study5 (Development of new fundamental technologies for high-efficiency energy creation, conversion/storage, and use) from the Ministry of Education, Culture, Sports, Science and Technology (MEXT) of Japan. This work was also supported by FOCUS Establishing Supercomputing Center of Excellence.

## References


1. C.M. Marian, Wiley Interdiscip. Rev. Comput. Molecular Sci. **2**, 187–203 (2012)
2. Y. Imamura, M. Kamiya, T. Nakajima, Chem. Phys. Lett. **635**, 152–156 (2015)
3. Y. Imamura, M. Kamiya, T. Nakajima, Chem. Phys. Lett. **648**, 60–65 (2016)
4. M.E. Casida, M. Huix-Rotllant, Ann. Rev. Phys. Chem. **63**, 287–323 (2012)
5. C. Adamo, D. Jacquemin, Chem. Soc. Rev. **42**, 845–856 (2013)
6. A.D. Laurent, D. Jacquemin, Int. J. Quantum Chem. **113**, 2019–2039 (2013)
7. H. Weiss, R. Ahlrichs, M. Häser, J. Chem. Phys. **99**, 1262–1270 (1993)
8. R. Bauernschmitt, R. Ahlrichs, Chem. Phys. Lett. **256**, 454–464 (1996)
9. R.E. Stratmann, G.E. Scuseria, M.J. Frisch, J. Chem. Phys. **109**, 8218–8224 (1998)
10. P. Norman, D.M. Bishop, H.J. Aa. Jensen, J. Oddershede, J. Chem. Phys. **115**, 10323–10334
11. K. Kristensen, J. Kauczor, T. Kjærgaard, P. Jørgensen, J. Chem. Phys. **131**, 044112 (2009)
12. M. Douglas, N.M. Kroll, Ann. Phys. (NY) **82**, 89 (1974)
13. B.A. Hess, Phys. Rev. A **32**, 756 (1985)
14. B.A. Hess, Phys. Rev. A **33**, 3742 (1986)
15. L.E. McMurchie, E.R. Davidson, J. Comput. Phys. **44**, 289–301 (1981)
16. R.M. Pitzer, N.W. Winter, Int. J. Quantum Chem. **40**, 773–780 (1991)
17. P. Norman, B. Schimmelpfennig, K. Ruud, H.J. Aa. Jensen, H. Ågren, J. Chem. Phys. **116**, 6914–6923 (2002)
18. P. Salek, T. Helgaker, T. Saue, Chem. Phys. **311**, 187–201 (2005)
19. R. Bast, T. Saue, J. Henriksson, P. Norman, J. Chem. Phys. **130**, 024109 (2009)
20. F. Wang, T. Ziegler, J. Chem. Phys. **121**, 12191–12196 (2004)
21. J. Gao, W. Zou, W. Liu, Y. Xiao, D. Peng, B. Song, C. Liu, J. Chem. Phys. **123**, 054102 (2005)
22. F. Wang, T. Ziegler, J. Chem. Phys. **122**, 074109 (2005)
23. A. Nakata, T. Tsuneda, K. Hirao, J. Chem. Phys. **135**, 224106 (2011)
24. S. Hirata, M. Head-Gordon, Chem. Phys. Lett. **314**, 291–299 (1999)
25. R. Bast, H.J.A. Jensen, T. Saue, Int. J. Quantum Chem. **109**, 2091–2112 (2009)
26. T. Nakajima, M. Katouda, M. Kamiya, Y. Nakatsuka, Int. J. Quantum Chem. **115**, 349–359 (2015)
27. SymPy: Python library for symbolic mathematics. http://sympy.org
28. J.A. Pople, R. Krishnan, H.B. Schlegel, J.S. Binkley, Int. J. Quantum Chem. **16**, 225–241 (1979)
29. P. Pulay, J. Chem. Phys. **78**, 5043–5051 (1983)
30. H. Sekino, R.J. Bartlett, J. Chem. Phys. **84**, 2726–2733 (1986)
31. E.R. Davidson, J. Comput. Phys. **17**, 87–94 (1975)
32. K. Hirao, H. Nakatsuji, J. Comput. Phys. **45**, 246–254 (1982)
33. J. Olsen, H.J. Aa. Jensen, P. Jørgensen, J. Comput. Phys. 74, 265–282 (1988)
34. R. Bast, U. Ekstrom, B. Gao, T. Helgaker, K. Ruud, A.J. Thorvaldsen, Phys. Chem. Chem. Phys. **13**, 2627–2651 (2011)
35. L. Nordström, D.J. Singh, Phys. Rev. Lett. **76**, 4420–4423 (1996)
36. T. Oda, A. Pasquarello, R. Car, Phys. Rev. Lett. **80**, 3622–3625 (1998)
37. S. Yamanaka, D. Yamaki, Y. Shigeta, H. Nagao, Y. Yoshioka, N. Suzuki, K. Yamaguchi, Int. J. Quantum Chem. **80**, 664–671 (2000)
38. P. Kurz, F. Förster, L. Nordström, G. Bihlmayer, S. Blügel, Phys. Rev. B **69**, 024415 (2004)
39. J.E. Peralta, G.E. Scuseria, J. Chem. Phys. **120**, 5875–5881 (2004)
40. J.E. Peralta, G.E. Scuseria, M.J. Frisch, Phys. Rev. B **75**, 125119 (2007)
41. M.K. Armbruster, F. Weigend, C. van Wullen, W. Klopper, Phys. Chem. Chem. Phys. **10**, 1748–1756 (2008)
42. G. Scalmani, M.J. Frisch, J. Chem. Theor. Comput. **8**, 2193–2196 (2012)
43. H. Sekino, J. Phys. Chem. A **104**, 4685–4689 (2000)
44. R.J. Harrison, J. Comput. Chem. **25**, 328–334 (2004)

45. S. Hirata, M. Head-Gordon, R.J. Bartlett, J. Chem. Phys. **111**, 10774–10786 (1999)
46. R. Strange, F.R. Manby, P.J. Knowles, Comput. Phys. Commun. **136**, 310–318 (2001)
47. P. Sałek, A. Hesselmann, J. Comput. Chem. **28**, 2569–2575 (2007)
48. U. Ekström, L. Visscher, R. Bast, A.J. Thorvaldsen, K. Ruud, J. Chem. Theor. Comput. **6**, 1971–1980 (2010)
49. J.P. Perdew, Burke K., Ernzerhof M., Phys. Rev. Lett. **77**, 3865 (1996)
50. J.C. Slater, Phys. Rev. **81**, 385 (1951)
51. S.H. Vosko, L. Wilk, M. Nusair, Can. J. Phys. **58**, 1200 (1980)
52. A.D. Becke, Phys. Rev. A **38**, 3098 (1988)
53. C. Lee, W. Yang, R.G. Parr, Phys. Rev. B **37**, 785 (1988)
54. A.D. Becke, J. Chem. Phys. **98**, 5648 (1993)
55. H. Iikura, T. Tsuneda, T. Yanai, K. Hirao, J. Chem. Phys. **115**, 3540 (2001)
56. T. Yanai, D.P. Tew, N.C. Handy, Chem. Phys. Lett. **393**, 51–57 (2004)
57. V. Kellö, A.J. Sadlej, Theor. Chim. Acta. **94**, 93–104 (1996)
58. A.J. Sadlej, Collect. Czech. Chem. Commun. **53**, 1995 (1988)
59. S. Kirpekar, J. Oddershede, Jensen HJrA. J. Chem. Phys. **103**, 2983–2990 (1995)
60. O.A. Vydrov, J. Heyd, A.V. Krukau, G.E. Scuseria, W.K, J. SL., J Chem. Phys. **125**, 074106 (2006)
61. O.A. Vydrov, G.E. Scuseria, J. Chem. Phys. **125**, 234109 (2006)
62. O.A. Vydrov, G.E. Scuseria, J.P. Perdew, J. Chem. Phys. **126**, 154109 (2007)
63. T. Pluta, A.J. Sadlej, Chem. Phys. Lett. **297**, 391–401 (1998)
64. T. Nakajima, K. Hirao, M. Douglas, N.M. Kroll, J. Chem. Phys. **113**, 7786–7789 (2000)
65. J.C. Boettger, Phys. Rev. B **62**, 7809–7815 (2000)

# Chapter 11
# Warming Up Density Functional Theory

**Justin C. Smith, Francisca Sagredo and Kieron Burke**

**Abstract** Density functional theory (DFT) has become the most popular approach to electronic structure across disciplines, especially in material and chemical sciences. In 2016, at least 30,000 papers used DFT to make useful predictions or give insight into an enormous diversity of scientific problems, ranging from battery development to solar cell efficiency and far beyond. The success of this field has been driven by usefully accurate approximations based on known exact conditions and careful testing and validation. In the last decade, applications of DFT in a new area, warm dense matter, have exploded. DFT is revolutionizing simulations of warm dense matter including applications in controlled fusion, planetary interiors, and other areas of high energy density physics. Over the past decade or so, molecular dynamics calculations driven by modern density functional theory have played a crucial role in bringing chemical realism to these applications, often (but not always) in excellent agreement with experiment. This chapter summarizes recent work from our group on density functional theory at nonzero temperatures, which we call thermal DFT. We explain the relevance of this work in the context of warm dense matter, and the importance of quantum chemistry to this regime. We illustrate many basic concepts on a simple model system, the asymmetric Hubbard dimer.

**Keywords** Thermal density functional theory · Ensemble density functional theory · Warm dense matter · Exact conditions · Thermal linear response

J.C. Smith (✉)
Department of Physics and Astronomy, University of California,
Irvine, CA 92697, USA
e-mail: justincs@uci.edu

F. Sagredo · K. Burke
Department of Chemistry, University of California, Irvine, CA 92697, USA

 
M.J. Wójcik et al. (eds.), *Frontiers of Quantum Chemistry*,
https://doi.org/10.1007/978-981-10-5651-2_11

## 11.1 Introduction

**Warm Dense Matter**: The study of warm dense matter (WDM) is a rapidly growing multidisciplinary field that spans many branches of physics, including astrophysics, geophysics, and attosecond physics [1–11]. Classical (or semiclassical) plasma physics is accurate for sufficiently high temperatures and sufficiently diffuse matter [12]. The name WDM implies too cool and too dense for such methods to be accurate, and this regime has often been referred to as the malfunction junction, because of its difficulty [2]. Many excellent schemes have been developed over the decades within plasma physics for dealing with the variety of equilibrium and non-equilibrium phenomena accessed by both people and nature under the relevant conditions [13]. These include DFT at the Thomas-Fermi level (for very high temperatures) and use of the local density approximation (LDA) within Kohn-Sham (KS) DFT at cold to moderate temperatures (at very high temperatures, sums over unoccupied orbitals fail to converge). The LDA can include thermal XC corrections based on those of the uniform gas, for which simple parametrizations have long existed [14, 15].

**Electronic Structure Theory**: On the other hand, condensed matter physicists, quantum chemists, and computational materials scientists have an enormously well-developed suite of methods for performing electronic structure calculations at temperatures at which the electrons are essentially in their ground state (GS), say, 10,000 K or less [16]. The starting point of many (but not all) such calculations is the KS method of DFT for treating the electrons [17]. Almost all such calculations are within the Born-Oppenheimer approximation, and ab initio molecular dynamics (AIMD) is a standard technique, in which KS-DFT is used for the electronic structure, while Newton's equations are solved for the ions [18].

**DFT in WDM**: In the last decade or so, standard methods from the electronic structure of materials have had an enormous impact in warm dense matter, where AIMD is often called QMD, quantum molecular dynamics [11]. Typically a standard code such as VASP is run to perform MD [8]. In WDM, the temperatures are a noticeable fraction of the Fermi energy, and thus the generalization of DFT to thermal systems must be used. Such simulations are computationally demanding but they have the crucial feature of including realistic chemical structure, which is difficult to include with any other method while remaining computationally feasible. Moreover, they are in principle exact [17, 19], if the exact temperature-dependent exchange-correlation free energy could be used because of Mermin's theorem establishing thermal DFT(thDFT). In practice, some standard ground-state approximation is usually used. (There are also quantum Monte Carlo calculations which are typically even more computationally expensive [20–26]. The beauty of the QMD approach is that it can provide chemically realistic simulations at costs that make useful applications accessible [27].) There have been many successes, such as simulation of Hugoniot curves measured by the *Z* machine [9] or a new phase diagram for high-density water which resulted in improved predictions for the structure of

Neptune [1]. Because of these successes, QMD has rapidly become a standard technique in this field.

**Missing Temperature Dependence**: However, the reliability and domain of applicability of QMD calculations are even less well understood than in GS simulations. At the equilibrium level of calculation, vital for equations of state under WDM conditions and the calculation of free energy curves, a standard generalized gradient approximation (GGA) calculation using, e.g., PBE [28], is often (but not always) deemed sufficient, just as it is for many GS materials properties. Such a calculation ignores thermal exchange-correlation (XC) corrections, i.e., the changes in XC as the temperature increases, which are related to entropic effects. We believe we know these well for a uniform gas (although see the recent string of QMC papers [25, 26] and parametrizations [15]), but such corrections will be unbalanced if applied to a GGA such as PBE. So how big a problem is the neglect of such corrections?

**(A little) Beyond Equilibrium**: On the other hand, many experimental probes of WDM extract response functions such as electrical or thermal conductivity [1]. These are always calculated from the equilibrium KS orbitals, albeit at finite temperature. Work on molecular electronics shows that such evaluations suffer both from inaccuracies in the positions of KS orbitals due to deficiencies in XC approximations, and also require further XC corrections, even if the *exact* equilibrium XC functional was used [29–31].

## 11.2 Background

**Generalities**: Everything described within uses atomic units, is non-relativistic, and does not include external magnetic fields. Unless otherwise noted, all results are for the electronic contributions within the Born-Oppenheimer approximation. While all results are stated for density functionals, in practice, they are always generalized to spin-density functionals in the usual way. In Table 11.1, we provide a summary of frequently used acronyms.

**Table 11.1** Acronyms frequently used in this chapter

| Acronym | Meaning | Acronym | Meaning |
|---|---|---|---|
| GGA | Generalized Gradient Approx. | RPA | Random Phase Approx. |
| GS | ground-state | TDDFT | Time-dependent DFT |
| HXC | Hartree XC | thDFT | thermal DFT |
| KS | Kohn-Sham | unif | uniform gas |
| LDA | Local Density Approx. | XC | exchange-correlation |
| PBE | Perdew-Burke-Ernzerhof | ZTA | Zero-Temperature Approx. |
| QMC | quantum Monte Carlo | | |

### 11.2.1 Ground-State DFT

**Hohenberg–Kohn Functional**: Just over 50 years ago, in 1964, Hohenberg and Kohn wrote down the foundations of modern DFT [32]. They start with the many-body Hamiltonian

$$\hat{H} = \hat{T} + \hat{V}_{ee} + \hat{V}, \tag{11.1}$$

where $\hat{T}$, $\hat{V}_{ee}$, and $\hat{V}$ are the kinetic, electron-electron, and potential energy operators, respectively. Assuming a non-degenerate ground state, they proved by *reductio ad absurdum* that the external potential $v(\mathbf{r})$ is a unique functional of the density $n(\mathbf{r})$, and therefore all observables are also density functionals. More directly, Levy defines the functional

$$F[n] = \min_{\Psi \to n} \langle \Psi | \hat{T} + \hat{V}_{ee} | \Psi \rangle, \tag{11.2}$$

where $\Psi$ is normalized and antisymmetric, and uses it to define the energy functional

$$E_v[n] = F[n] + \int d^3r\, v(\mathbf{r}) n(\mathbf{r}), \tag{11.3}$$

whose minimization over normalized non-negative densities with finite kinetic energy yields the ground-state energy and density [33].

**Kohn-Sham Scheme**: In 1965, Mermin generalized the Hohenberg–Kohn theorems for electrons in the grand canonical potential with fixed nonzero temperature $\tau$ and chemical potential $\mu$ [19]. Later in 1965, Kohn and Sham created an exact method to construct the universal functional (see Eq. (11.6)). The Kohn-Sham scheme imagines a system of $N$ non-interacting electrons that yield the electronic density of the original interacting $N$ electron system. These fictitious electrons sit in a new external potential called the KS potential. The KS scheme is written as a set of equations that must be solved self-consistently:

$$\left\{ -\frac{1}{2}\nabla^2 + v_s(\mathbf{r}) \right\} \phi_i(\mathbf{r}) = \epsilon_i \phi_i(\mathbf{r}), \quad n(\mathbf{r}) = \sum_i^N |\phi_i(\mathbf{r})|^2, \tag{11.4}$$

$$v_s(\mathbf{r}) = v(\mathbf{r}) + v_H(\mathbf{r}) + v_{XC}(\mathbf{r}), \quad v_{XC}(\mathbf{r}) = \frac{\delta E_{XC}[n]}{\delta n(\mathbf{r})}, \tag{11.5}$$

where $\phi_i(\mathbf{r})$ and $\epsilon_i$ are the KS orbitals and energies, $v_H(\mathbf{r})$ is the classical Hartree potential, and $v_{XC}(\mathbf{r})$ is the exchange-correlation potential defined by the unknown XC energy, $E_{XC}$, in Eq. (11.5). These must be solved self-consistently since the Hartree potential and $E_{XC}$ depend explicitly on the density. Lastly, the total energy can be found via

$$F[n] = T_s[n] + U_H[n] + E_{XC}[n] \tag{11.6}$$

where $T_s$ is the kinetic energy of the KS electrons and $U_H$ is the Hartree energy.

In practice, an approximation to $E_{\rm XC}$ must be supplied. There exists a wealth of approximations for $E_{\rm XC}$ [34]. The simplest, LDA, uses the XC per electron of the homogeneous electron gas [35]:

$$E_{\rm XC}^{\rm LDA}[n] = \int d^3r\, e_{\rm XC}^{\rm unif}(n(\mathbf{r})) \tag{11.7}$$

where $e_{\rm XC}^{\rm unif}$ is the XC energy density of a uniform gas with density $n(\mathbf{r})$. We can imagine going up a ladder by adding in more ingredients (like gradients of the density [28]) and obeying different or additional conditions to make more complicated and more accurate functionals [36]. For the exact $E_{\rm XC}$, these equations have been proven to converge [37].

### *11.2.2 Asymmetric Hubbard Dimer and Its Relevance*

Throughout this chapter we illustrate results with the simplest interesting model of an interacting system. This model is the asymmetric Hubbard dimer [38]. The Hubbard dimer crosses the divide between the weakly and strongly correlated communities. Previous work has shown that the basic concepts of ground-state KS-DFT apply to this model and here we demonstrate similar applicability to thermal DFT. The Hamiltonian is given by

$$\hat{H} = -t\sum_{\sigma}(\hat{c}_{1\sigma}^{\dagger}\hat{c}_{2\sigma} + H.c.) + \sum_{i=1}^{2}(U\hat{n}_{i\uparrow}\hat{n}_{i\downarrow} + v_i\hat{n}_i) \tag{11.8}$$

where $\hat{c}_{i\sigma}^{\dagger}(\hat{c}_{i\sigma})$ are electron creation (annihilation) operators and $\hat{n}_{i\sigma} = \hat{c}_{i\sigma}^{\dagger}\hat{c}_{i\sigma}$ are number operators. $t$ is the strength of electron hopping between sites, $U$ is the ultra-short range Coulomb repulsion for when both electrons are on the same site, and $v_i$ is the on-site external potential. We choose, without loss of generality, $v_1 + v_2 = 0$ then $\Delta v = v_2 - v_1$, and the occupation difference, the analog of density, is $\Delta n = n_2 - n_1$.

In Fig. 11.1 we plot the density $\Delta n$ versus asymmetry $\Delta v$ in the Hubbard dimer. The Hohenberg–Kohn theorem applies to this Hamiltonian and guarantees $\Delta n(\Delta v)$ is an invertible function for any value of $U$. The main physics is a competition between asymmetry and interaction strength. The weakly correlated regime is $U < \Delta v$ and the opposite is strong correlation. Increasing $\Delta v$ pushes the electrons onto a single site, thus $\Delta n$ approaches 2. Likewise, for small $\Delta v$ or large $U$ the electrons are apart and $\Delta n$ tends to 0. This is made most clear by the extreme cases, i.e.,

$$|\Delta n_{U=0}(\Delta v)| = 2\Delta v/\sqrt{(2\,t)^2 + \Delta v^2}, \qquad |\Delta n_{U\to\infty}(\Delta v)| \to 2\theta(\Delta v - U), \tag{11.9}$$

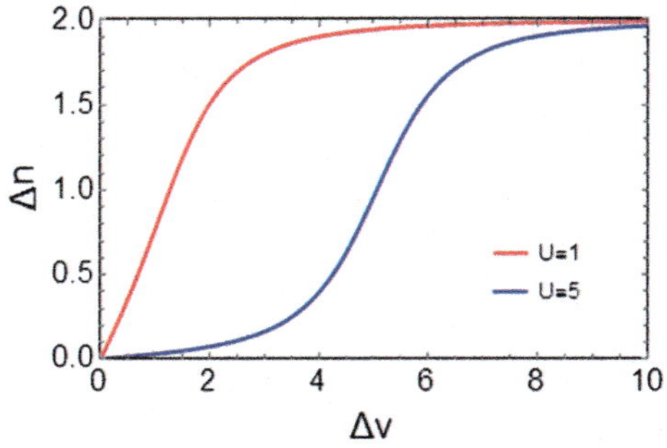


**Fig. 11.1** Difference in on-site occupations as a function of on-site potential difference for $U = 1$ and 5 in the asymmetric Hubbard dimer. The HK theorem guarantees that each function is invertible. There is a simple analytic result for $U = 0$, and for large $U$, the relation tends toward a (smoothed) step function, with the step at $\Delta v = U$

where $\theta(x)$ is the Heaviside step function. The ability to vary $U$ and move continuously from weak to strong correlation in a model that is analytically solvable makes the Hubbard dimer an excellent illustrator of how KS-DFT works [38].

### *11.2.3 Ensemble DFT as a Route to Excitation Energies*

In this section, we take a quick aside to overview ensemble DFT (eDFT), a close cousin of thermal DFT.

**Excitations in DFT**: Although time-dependent DFT (TDDFT) is the standard method used to determine the excited states of a system [39], there are still many deficiencies, due to crude approximations to the XC functional as well as being unable to approximate multiple excitations, charge transfer excitations, conical intersections, and polarizabilities of long-chain polymers; all things that can be important for photochemistry [40–44]. Ensemble DFT is a time-independent alternative to the standard TDDFT that can be a useful method for extracting excited states. Naturally, since eDFT and TDDFT are based on two different fundamental theories, it is possible to use eDFT on different systems to those of the traditional method and expect different successes and likewise different failures.

**Ensemble Variational Principle**: eDFT is based on a variational principle made up of ensembles of ground and excited states [45]. These ensembles are made of decreasing weights, with the ground state always having the highest weight.

$$E^{w} \leq \sum_{k=0}^{M-1} w_k \langle \Psi_k | \hat{H} | \Psi_k \rangle, \quad w_0 \geq w_1 \geq \ldots \geq w_k \geq 0 \tag{11.10}$$

where all $\Psi_k$ are normalized, antisymmetric, and mutually orthogonal, $\mathbf{w} = (w_0, w_1, ...w_k)$, and the sum of all weights is 1. The ensemble-weighted density is

$$n^{\mathbf{w}}(\mathbf{r}) = \sum_{k=0}^{M-1} w_k n_{\Psi_k}(\mathbf{r}). \tag{11.11}$$

Just as in the ground-state case, a one-to-one correspondence from the weighted density to the potential can be established [46], and applying this to a non-interacting system of the same weighted density can be used to construct a KS eDFT. From this KS system, it is in principle possible to extract the exact excited states of the system.

**Relation to Thermal DFT**: The connection to thermal DFT is natural and straightforward. Thermal DFT is a special case of eDFT. In thDFT, one chooses the ensemble to be the grand canonical ensemble with the usual Boltzmann factors for the weights. However, unlike eDFT, the weights themselves depend on the eigenvalues of the Hamiltonian, including the strength of the interaction. Thus, the weights in the KS system are different from those of the interacting system. In most applications of eDFT, the weights are chosen to be the same in both the physical and non-interacting systems.

**History**: Ensemble DFT was originally proposed by Gross-Olivera-Kohn in 1988 [46], but, like thermal DFT, there has been slow progress over the last 30 years due to a lack of useful approximations to the XC functional. Many of these difficulties arise from the so-called ghost interaction errors that occur in the Hartree energy [47]. More specifically, these ghost interaction errors appear when only using the ground-state definition of the Hartree energy, which causes unphysical contributions and must therefore be accounted for by using a more accurate definition of the Hartree energy for ensembles [48, 49].

**Recent Progress**: More recently, work has been done to extract the weight dependence of the KS eigenvalues, which are required in order to extract accurate transition frequencies [48, 49]. It was also found that a large cancellation of the weight-dependence occurs in the exact ensembles. Further, a new numerical method for inverting ensemble densities was derived for spherically symmetric systems, and this method was also tested for cylindrically symmetric systems. This inversion of densities to extract potentials provides a useful test of eDFT approximations.

Recent work combines linear interpolation with an extrapolation method in eDFT to extract excited states that are independent of ensemble weights [50]. Also, an exact analytical expression for the exchange energy was derived, and a generalized adiabatic connection for ensembles (GACE) was used to connect the ensemble XC functional with the ground-state functional [51]. Finally, a ghost interaction correction has been developed for range-separated eDFT [52].

**A Simple Exercise**: Here we show an example of the importance of the weight-dependence of functionals in eDFT, in a seemingly simple system. We put only one electron in the Hubbard dimer, so there is no interaction, and its a simple tight-binding Hamiltonian. We will derive the exact kinetic energy functional, which is an example of the non-interacting KS kinetic energy functional. We will also approximate it, as if we were interested in orbital-free eDFT.

There are only two levels: the ground state and a first-excited state. Thus, we can make only a bi-ensemble. The ensemble-weighted ground-state density is

$$\Delta n^w = (1-w)\Delta n_0 + w\Delta n_1 = (1-2w)\Delta n_0(x), \tag{11.12}$$

where $\Delta n_0$ and $\Delta n_1$ correspond to the occupational difference of the two sites for the ground and first-excited states, respectively, and $\Delta n_0(x)$ is half of Eq. (11.9) with $U = 0$, since we are looking at a single particle in the dimer. This last result is true only because $\Delta n_1 = -\Delta n_0$ in this simple model. The weight as previously stated is $w \leq 0.5$. Similarly, the kinetic energy for a single particle in the ground state is known, and $T_{s,1} = -T_{s,0}$, so

$$T_s^w = (1-w)T_{s,0} + wT_{s,1} = (1-2w)T_{s,0} = (1-2w)\frac{-t}{\sqrt{1+x^2}} \tag{11.13}$$

where $x = \Delta v/(2\,t)$. Note that kinetic energy is negative in the Hubbard model. Using $V^w = \Delta v \Delta n^w/2$ and adding it to $T_s^w$ yields the ensemble energy, which is exactly linear in $w$, and passes through $E_0$ at $w = 0$ and (would pass through) $E_1$ at $w = 1$.

This simple linearity with $w$ is true by construction of the ensemble, when energies are plotted against $w$ for a fixed potential. But now we show that things get complicated when we consider them as density functionals. Inverting the relation between potentials and densities we find

$$x = \frac{\Delta n^w}{\sqrt{(1-2w)^2 - (\Delta n^w)^2}}, \tag{11.14}$$

and inserting this into the kinetic energy yields

$$T_s^w[\Delta n^w] = -\sqrt{(1-2w)^2 - (\Delta n^w)^2}/2. \tag{11.15}$$

Even in this trivial case, the the $w$-dependence of the kinetic energy density functional is nonlinear.

Next, we make the most naive approximation, namely to replace $T_s^w[\Delta n]$ with its ground-state counterpart. This yields an approximate eDFT theory from which, in principle, we can estimate the energy of the first-excited state. To do this, we insert the exact $w$-dependent density of Eq. (11.12) into the ground-state functional, add the exact $V^w$, and plot the resulting energy. The exact and approximate results are shown in Fig. 11.2. The approximation is very bad, yielding an energy value that is too negative by 0.5, using the value at $w = 1/2$. But it illustrates the difficulties of capturing an accurate $w$-dependence in an ensemble density functional.

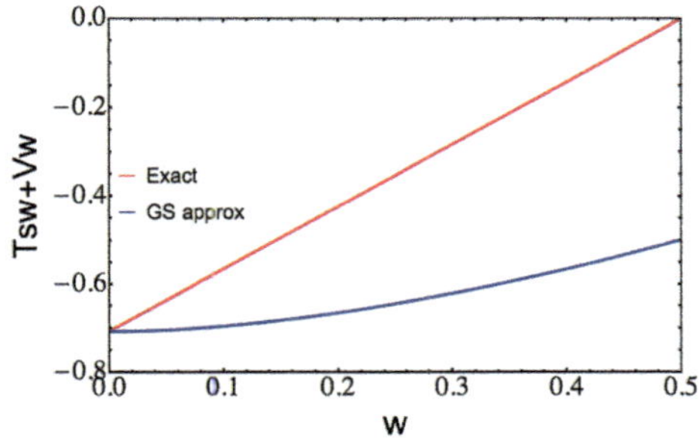


**Fig. 11.2** Exact energy curve for a single particle in a Hubbard dimer $E_{exact} = T_{\rm S}^{w}[\Delta n^{w}] + V^{w}$ (red), when $x = 1$, in units of $2\,t$. The blue line is the approximate energy when $T_{\rm S}^{w}$ is replaced by its ground state analog, $T_{\rm S}^{0}$. Notice that $w = 0$ corresponds to $E_0$, which is $1/\sqrt{2}$ in these units

### *11.2.4 Thermal DFT in a Nutshell*

**Mermin–Kohn–Sham Equations**: In a thermal system, Eqs. (11.4) and (11.5) are generalized such that the density and XC potential become $n^{\tau}(\mathbf{r})$ and $v_{\rm XC}^{\tau}(\mathbf{r})$, i.e., temperature dependent, and $E_{\rm XC}[n]$ in Eq. (11.5) becomes $A_{\rm XC}^{\tau}[n]$, the exchange-correlation *free* energy density functional. The density becomes

$$n^{\tau}(\mathbf{r}) = \sum_i f_i |\phi_i^{\tau}(\mathbf{r})|^2 \tag{11.16}$$

where the sum is now over all states and $f_i = (1 + e^{(\varepsilon_i^{\tau} - \mu)/\tau})^{-1}$, the Fermi occupation factors. One of the core difficulties in thermal DFT calculations is this sum, since a huge number of states are required once the temperature is sufficiently high. This leads to large computational demands and convergence issues. We call these the Mermin–Kohn–Sham equations.

To extract the total free energy from the MKS equations, we write

$$A^{\tau}[n] = A_{\rm S}^{\tau}[n] - U_{\rm H}[n] + A_{\rm XC}^{\tau}[n] - \int d^3r\, n(\mathbf{r}) v_{\rm XC}^{\tau}[n](\mathbf{r}), \tag{11.17}$$

where the MKS free energy is

$$A_{\rm S}^{\tau}[n] = \sum_i \varepsilon_i^{\tau}[n] - \tau S_{\rm S}^{\tau}[n], \quad S_{\rm S}^{\tau}[n] = -\sum_i \left[f_i \log(f_i) + (1 - f_i)\log(1 - f_i)\right], \tag{11.18}$$

and $S_{\rm S}$ is the MKS entropy.

**Exchange-Correlation Free Energy**: Compared to ground-state DFT, relatively few approximations have been developed for $A_{\rm XC}^{\tau}$. In active use are two approximations: thermal LDA (thLDA) and the Zero-Temperature Approximation (ZTA). The former uses the temperature-dependent XC free energy of the uniform gas instead of

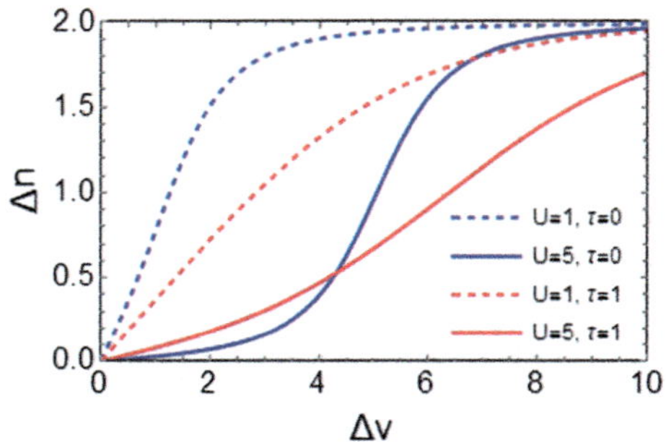


**Fig. 11.3** Effects of temperature on the difference in on-site occupations as a function of on-site potential difference for $U = 1$ and 5 in the asymmetric Hubbard dimer (see Sect. 11.3.2) with and without temperature. Increasing temperature pushes the electrons toward opposite sites and lowers $\Delta n$ while increasing $\Delta v$ pushes electrons to the same site and raises $\Delta n$

its ground-state analog in Eq. (11.7). The ZTA means simply using any ground-state XC functional instead of a temperature-dependent one. We denote use of the exact ground-state XC functional as exact ZTA (EZTA).

In Fig. 11.3, we demonstrate the effects of turning on temperature for the Hubbard dimer (see Sect. 11.3.2 for more information). Even a small increase in temperature can have a big impact on a system regardless of the strength of correlation.

## 11.3 Some Recent Developments in Thermal DFT

The formalism for thermal DFT was originally developed alongside that of ground-state DFT, but in the intervening decades ground-state DFT (and TDDFT, for that matter) has received significantly more attention and consequently more developments [16, 53]. However, In the past couple decades, thermal DFT has seen more use, and with that much more development in the past ten years. In this section, we outline some recent developments to thermal DFT from our group.

### *11.3.1 Exact Conditions and Their Relevance*

**Zero Temperature**: One of the most crucial steps in understanding and developing functionals beyond LDA is exact conditions. These conditions take many forms with some common examples being coordinate and interaction scaling conditions [54], and bounds on the XC energy [55]. There are well over a dozen conditions in ground-state DFT (a recent meta-GGA functional even uses seventeen [56]!), but the use of exact conditions is much more nascent in thermal DFT.

#### 11.3.1.1 Coordinate-Temperature Scaling and the Thermal Connection Formula

**Uniform Coordinate Scaling**: The most straightforward application of exact conditions to thermal DFT is by uniform scaling of the density [54]. The very basic conditions that this procedure generates in ground-state DFT are built into almost all modern approximations. In a sense, this is simply dimensional analysis, but while keeping the density fixed (which is the tricky bit).

Early work on exact conditions for thermal DFT [57, 58] derived basic conditions such as the signs of correlation quantities, including the separation into kinetic and potential contributions, and the adiabatic connection formula at finite temperature. More conditions come from coordinate scaling of the density, showing that is intimately related to temperature dependence. Examples of a few of these conditions are

$$F_S^{\tau'}[n] = \frac{\tau}{\tau'} F_S^{\tau}[n_{\sqrt{\tau'/\tau}}], \quad S_S^{\tau'}[n] = S_S^{\tau}[n_{\sqrt{\tau'/\tau}}], \quad A_X^{\tau'}[n] = \sqrt{\frac{\tau}{\tau'}} A_X^{\tau}[n_{\sqrt{\tau'/\tau}}], \tag{11.19}$$

where these have been rewritten in terms of temperature scaling instead of coordinate scaling. For any of these functionals, this means that, if you know the functional at *any* one finite temperature, the functional at *all* possible temperatures is available via temperature scaling as long as both temperatures are nonzero.

**New Formulas**: In recent work, many new formulas relating correlation components of the energy to one another were derived [59], such as

$$K_C^{\tau,\lambda}[n] = A_C^{\tau,\lambda}[n] - \lambda \frac{dA_C^{\tau,\lambda}[n]}{d\lambda}, \tag{11.20}$$

where $K_C^{\tau,\lambda}[n] = T_C^{\tau,\lambda}[n] - \tau S_C^{\tau,\lambda}[n]$ is the correlation kentropy. $\lambda$ is a coupling constant introduced in front of $\hat{V}_{ee}$ where superscript $\lambda$ indicates quantities defined on $\lambda \hat{V}_{ee}$. There was also a rewriting of the adiabatic connection formula [57, 60], using the relation to scaling mentioned above, yielding the XC free energy at temperature $\tau$:

$$A_{XC}^{\tau}[n] = \frac{\tau}{2} \lim_{\tau'' \to \infty} \int_{\tau}^{\tau''} \frac{d\tau'}{\tau'^2} U_{XC}^{\tau'}[n_{\sqrt{\tau'/\tau}}], \qquad n_\gamma(\mathbf{r}) = \gamma^3 n(\gamma \mathbf{r}), \tag{11.21}$$

where $U_{XC}^{\tau}[n]$ is the purely potential contribution to the XC free energy, and the scaling is the usual coordinate scaling of the density introduced by Levy and Perdew [54] for the GS problem. Note that this thermal connection formula uses only information between the desired temperature and higher ones, allowing approximations that begin from the high-temperature end instead of the low-temperature end [59]. A second set of formulas gives the many relations among the different correlation energy components (total, potential, and kentropic). These are very important in ground-state DFT [61, 62] for understanding the origins of different physical contributions to the correlation free energy and have guided the construction of many approximations.

**Entropy**: Lastly for this section, we look at a new set of exact conditions for the electronic entropy as a functional of the density [63]. The most important is that the universal functional can be written solely in terms of a temperature integral over entropy, such as

$$F^{\tau}[n] = F^{0}[n] - \int_{0}^{\tau} d\tau'\, S^{\tau'}[n], \quad A_{\rm XC}^{\tau}[n] = E_{\rm XC}[n] - \int_{0}^{\tau} d\tau'\, S_{\rm XC}^{\tau'}[n], \tag{11.22}$$

i.e., the universal contribution to the free energy functional is a simple integral over the electronic entropy, and the second shows that all thermal corrections to the XC free energy are given by an integral over the XC entropy. These formulas have no analog in ground-state DFT. They also lead to fundamental inequalities on the various thermal derivatives of both interacting and KS quantities. Such conditions have long been known for the uniform gas [64], but our results are their generalization to inhomogeneous systems. Analogs are also easily derived from statistical mechanics, but again, the tricky part is to deduce their behavior as functionals of the density rather than the external potential. This is why, for example, all derivatives are total with respect to temperature. The particle number is fixed by the density, so temperature is the sole remaining variable. Our work uses the formalism and methods of ground-state DFT, generalized to finite temperature, but the same results can also be extracted in the language of statistical mechanics [65, 66].

**Tiny Violations**: A minor illustration of the relevance of these conditions is that we found that a recent parameterization of the thermal XC free energy of the uniform gas [15] violates one of our conditions for low densities [63]. This violation is slight, and unlikely to ever influence the results of any thLDA calculation. Nonetheless, it is always better to build parameterizations that satisfy known conditions, so that the corresponding approximate calculations are guaranteed to satisfy such conditions [25, 26].

**Zero-Temperature Approximations**: This work also showed that any ZTA calculation automatically satisfies most of our conditions, whereas the inclusion of thermal XC corrections risks violating them for specific systems. For example, all approximations in Fig. 11.5 (discussed below) are guaranteed to satisfy these conditions. But practical calculations including approximate thermal XC corrections should be checked for possible violations of exact conditions in the future.

### *11.3.2 Exact Calculations on a Simple Model System*

**Importance for Ground-state DFT**: A crucial step in all DFT development is the exact solution of simple systems and the test of approximate functionals against exact quantities. There are large databases of molecular properties, based either on highly reliable and accurate experimental measurements, or on far more accurate quantum chemical calculations [67–70]. But these databases usually contain at most a few numbers per system, such as the atomization energy and bond length(s). A substantially more sophisticated test occurs when a highly accurate calculation is performed such as QMC [71, 72] or DMRG [73], and an inversion of the KS equations [74], so that essentially exact KS potentials, eigenvalues, orbitals, etc., can all be deduced. This is a much more powerful test of a DFT approximation, and usually provides detailed insight into its limitations. The QMC calculations of Umrigar and collaborators [75–77] and their high impact, testify to this fact.

**Difficulty of Exact Thermal Calculations**: Although almost all practical calculations of WDM are in a condensed phase (with hot ions), almost all high-level inversions yielding exact KS quantities are for atoms or small molecules. But even for such systems, it is difficult to imagine accurate inversions at finite temperature, as only the average particle number is fixed, and all possible particle numbers must be considered.

**Exact Calculations for Hubbard Dimer**: Exact calculations are only possible for this model because the Hilbert space is severely truncated which allows us to compute all energies analytically (see Fig. 11.4 for complete diagram of the energy spectrum). However, this means the model is not even a qualitatively realistic representation of very high temperatures (though we choose parameters such that the ceiling of the Hilbert space does not effect results). But we are able to do the inversion exactly, and so extract all the different contributions to X and C as a function of both $\tau$ and $U$. These are the first exact inversions of an interacting system at finite temperature. They show us the structure of the underlying functionals, but cannot tell us which approximations will be accurate. For example, there is no real analog of LDA for this system (although BALDA [78] somewhat plays this role).

**Paradox**: To see why such simplistic calculations are important, consider Fig. 11.5. The black line shows the exact density difference in the dimer versus temperature for moderate correlation and asymmetry. The blue curve is a Hartree-Fock calculation, while the red curve adds in the high-density limit of GS correlation. Finally, EZTA in green uses the exact GS functional (i.e., the best possible ZTA), which we had already found in Reference [38], in the MKS equations, which therefore is the best possible calculation that ignores thermal XC contributions. By construction, this becomes exact in the zero-temperature limit. But, to our surprise, we found that the relative error in the free energy and density vanishes in the *high* temperature limit. In fact, as temperature increases, the fractional errors in the energy at first increase, and then start to lessen.

**Resolution of Paradox**: How can this counterintuitive result be right? The answer is very simple. For high temperatures, the XC components of the energy remain finite (and actually shrink, see Fig. 11.6), while the KS contributions grow, at least in this simple case. Thus *any* XC approximation will produce the same effect. This is why all the approximations merge onto the exact line in Fig. 11.5 for sufficiently high temperatures. Of course, there can still be a significant absolute error in the free energy which might have important effects on quantities of interest. But the principle is clear: EZTA becomes relatively exact in both the low- and *high*-temperature limits.

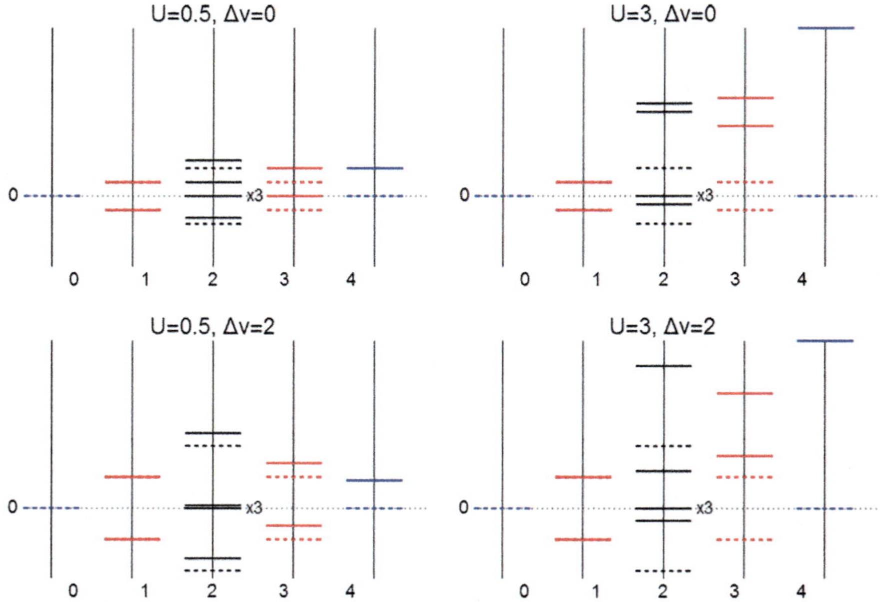


**Fig. 11.4** Energy spectrum for the Fock space of the asymmetric Hubbard dimer at various parameters ranging from weakly to strongly correlated. The x-axis is labeled by the number of particles. The *dashed lines* correspond to the non-interacting (tight-binding) case. The labels in the figures denote degeneracy. The triplet for $N = 2$ is always at $E = 0$ due to symmetry. The *top left* is symmetric and weakly correlated, so the spectral lines are close to the dashed ones. The *top right* is symmetric but strongly correlated, and the energies for $N = 2$ are substantially raised. We also see pairs of levels pushed together. The Hubbard bands of the infinite chain roughly run between these levels. In the lower panels, we turn on asymmetry and show the effects on $U$

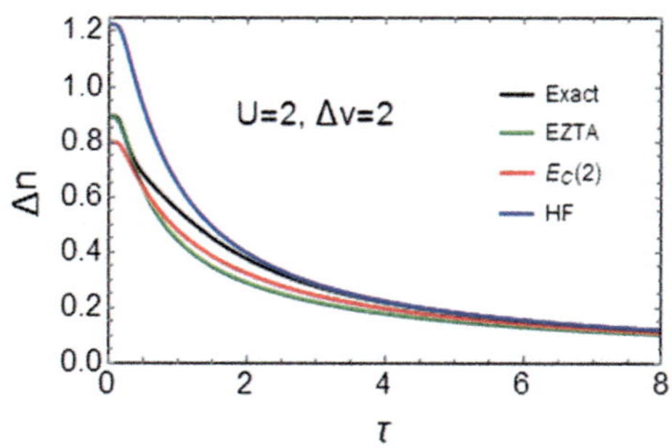


**Fig. 11.5** Difference in on-site densities as a function of temperature for an asymmetric Hubbard dimer with $U = 2$ and site-potential difference 2 (in units where the hopping parameter $t = 1/2$). The approximate calculations are all MKS-DFT-ZTA equilibrium calculations where HF denotes Hartree-Fock, $E_c(2)$ includes the leading correlation correction to HF in powers of the interaction, while EZTA denotes using the exact ground-state XC functional [38]

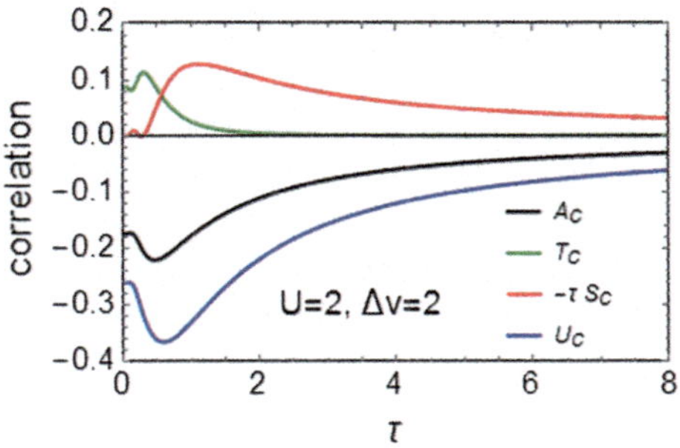


**Fig. 11.6** Correlation energy components as a function of temperature. All inequalities proven in Ref. [57] are satisfied. This figure shows that the naive assumption that $A_c^\tau$ is bounded by $E_c = A_c(\tau = 0)$ is not true. Figure 11.1 of Ref. [79] shows that the total free energy increases in magnitude as temperature increases, making the XC contribution relatively less important. Thus ZTA (or any reasonably bounded approximation) will yield relatively exact free energies, densities, and KS orbitals, in the limit of high temperatures. Additionally, Ref. [79] notes that exchange is temperature independent and has no effect on this analysis

This is also trivially true for the uniform gas, once the (infinite) Hartree energy is included, and we suspect it to be true for all systems.

**Relevance for Response Functions**: The increasing accuracy of the density with temperature has very important implications for calculations of conductivity. This means that the error in the KS orbitals, used in the construction of KS conductivities, starts to decrease beyond some temperature. This is true for *any* GS approximation for XC (within reason). All these conclusions may explain the tremendous success so far gotten by ignoring the thermal XC effects, especially for conductivities.

### *11.3.3 Beyond Equilibrium: Linear Response Thermal Time-Dependent DFT*

**Zero Temperature**: There are many applications in WDM where the system is perturbed away from equilibrium. At zero temperature, the standard approach to such problems is to apply TDDFT or many-body non-equilibrium Green's function methods [80]. TDDFT in particular [81] can handle both strong perturbations, such as atoms and molecules in intense laser fields, and weak perturbations, where the linear response formulation yields excitation energies and oscillator strengths [82].

**General Case too Difficult**: Unfortunately, the situation is very complicated if the perturbation is strong, as then a non-equilibrium treatment is needed. Theories in which the temperature is held fixed do not apply. This is the situation for example in calculations of stopping power [83]. There are many fine attempts to overcome these difficulties under a variety of practically useful conditions [84], but we have not seen a way to construct a general DFT treatment of such problems.

**Linear Response**: For a finite system (which has to be very carefully defined in the thermal case), we proved a limited theorem for the linear density response to a time-dependent external field [85]. This proof allows for (finite numbers of) degeneracies in the excited states, but not in the equilibrium state. Armed with such a theorem, all the usual XC response properties, such as the XC kernel, can be defined at finite temperature. Combined with our thermal connection formula, we have the finite-temperature generalization of the Gross-Kohn response equation [86]:

$$\chi^\tau(12) = \chi_s^\tau(12) + \int d3d4\, \chi_s^\tau(13) f_{\text{HXC}}^\tau(34)\chi^\tau(42), \tag{11.23}$$

where 1 denotes the coordinates $\mathbf{r}, t$, and 2 another pair [87], $\chi^\tau(12)$ is the density–density response function at temperature $\tau$, $\chi_s$ its KS counterpart, and $f_{\text{HXC}}^\tau(12)$ the thermal Hartree-XC kernel. This becomes the Random Phase Approximation (RPA) when $f_{\text{XC}} = 0$. Insertion of this into the thermal connection formula yields an RPA-type equation for the XC free energy [85]:

$$A_{\text{XC}}^\tau[n] = \lim_{\tau''\to\infty} \frac{\tau}{2} \int_\tau^{\tau''} \frac{d\tau'}{\tau'^2} \int_0^\infty \frac{d\omega}{2\pi} \coth\left(\frac{\omega}{2\tau}\right) \int d\mathbf{r} \int d\mathbf{r}' \frac{\Im \chi^{\tau'}[n_\gamma](\mathbf{r}, \mathbf{r}', \omega)}{|\mathbf{r} - \mathbf{r}'|} \tag{11.24}$$

where $\gamma$ is defined in Eq. (11.21). If XC contributions to the kernel are neglected, this becomes the long-known random phase approximation to the XC free energy, albeit using the KS orbitals. Since random phase approximation calculations have become standard within the GS materials world [88], there is little additional computational demand in performing them at moderate finite temperatures. Inclusion of any approximate treatment of the XC kernel yields an entirely novel approach to XC approximations for equilibrium thermal DFT. In particular, one can consider making a uniform approximation in both space and time, and also decide whether or not to include thermal corrections in an approximate kernel. All such treatments can be first tested on a uniform gas, for which the XC free energy is accurately known from QMC calculations [26].

## 11.4 Recent Applications of DFT in WDM

**Planetary Science**: The materials of interest in WDM research span the periodic table. Accurate thermal conductivities for inertial confinement fusion fuel materials such as deuterium and tritium are needed to find the calculated total neutron yield in fusion science (the simulated mixing between the fuel and coating on inertial confinement fusion capsules is very sensitive to thermal conductivities) [89]. Accurate iron thermal conductivities are used to determine whether the conventional model for how the Earth's core developed is valid because heat flux contributes to the Earth's geomagnetic field. Differential heating experiments at ALS, LCLS, Omega, and Titan facilities are all done slightly differently (heating via optical lasers, XFEL,

X-rays, or proton heating), to fit hydrodynamics models because of their high accuracy and because other approaches (SESAME, Purgatorio, and Lee-More) all differ [90–92]. Our work suggests that ignoring XC thermal corrections nevertheless yields accurate KS eigenstates and eigenvalues [79] helps explain why conductivities can be accurate in these calculations.

Much WDM research is motivated by the desire to understand planetary interiors. The Juno mission is measuring Jupiter's gravitational field extremely accurately, constraining theories of its interior [93], while Kepler has shown that many notions of planetary formation must be rethought with our new data on extra-solar planets [94]. But there is limited understanding of whether initial planetary protocores remain stable during accretion or if they dissolve into outer metallic hydrogen layers. Recent DFT-MD calculations show that MgO is surprisingly soluble in hydrogen under these conditions [95]. Similarly, the moon is thought to have formed in an enormous impact, but such a scenario depends crucially on the equation of state of MgO under extreme conditions. Recent DFT calculations and Z-machine experiments have nailed this EOS more accurately than before, and far better than unreliable extrapolations from more mundane conditions [96].

**Alternate Methods**: Path integral Monte Carlo is an excellent tool for studying WDM, has been recently extended beyond small atoms to include water and carbon, and has recently been shown to match reasonably well with DFT calculations at lower temperatures [24], validating both. Meanwhile, DFT calculations have predicted new superionic phases of $H_2O$, under conditions relevant to Uranus and Neptune interiors [97].

**DFT Failure**: A less successful application of DFT in WDM is to the liquid insulator to liquid metal transition in dense $D_2$, at about 1000 K and 300 GPa. DFT calculations with several different functionals yield very different results, none of which are in satisfactory agreement with experiment. The interpretation also depends on the accuracy of the conductivity from the DFT calculations. This system remains a challenge to WDM simulations.

**X-ray Thompson Scattering**: Some of the most exciting recent experiments have been from the LINAC at SLAC, allowing X-ray Thompson scattering (XRTS) measurements of shocked materials. These include the first highly resolved measurements of the plasmon spectrum in an ultrafast heated solid [98]. Reference [99] gives X-ray scattering results from plasmons in dynamically compressed deuterium, from which one can deduce the ionization state as a function of compression. Ionization begins at about the pressure that DFT-MD calculations show molecular dissociation. In a completely different material, X-ray diffraction showed diamond formation on nanosecond timescales, caused by shock compression to about 200 GPa [100]. This helps explain why the lonsdaleite crystal structure occurs naturally close to meteor impacts.

XRTS has been performed on a variety of materials including Be, Li, C, CH shells, and Al. Most experiments probe the electron dynamic structure factor, which is decomposed via the somewhat ad-hoc Chihara decomposition into bound, loosely bound, and free electrons [101]. But by running TDDFT at finite temperatures, one directly calculates the densities and can then test the accuracy of Chihara for determining the ionization state. The results of Ref. [85] are already being used to justify thTDDFT calculations such as Ref. [101].

## 11.5 Relation of Thermal DFT to Quantum Chemistry

At first glance, it would appear that warm dense matter has little or nothing to do with chemistry. In fact, this is not true, it is simply chemistry in an exciting new regime with which we are relatively unfamiliar.

To see this, we first note that the plasma physicists who usually study WDM think in terms of average properties of their systems, such as mean densities and numbers of electrons ionized. They are familiar with density functional methods, but traditionally only at the level of the LDA. Successes with such an approximation are often attributed to systems being somehow 'locally uniform'.

But the success of DFT methods in chemistry can be directly correlated with the arrival of the GGA and hybrids of it with Hartree-Fock. These approximations were tested on the G2 data set, and shown to yield much better energetics than LDA, because the G2 data set had already been carefully constructed and benchmarked, using both quantum chemical methods and experimental information [102–104]. This vote of confidence led to their widespread adoption in many branches of chemistry, and also led to the confidence that GGAs were better than LDA for many materials problems.

It is the same GGAs, used in MD simulations, that have led to the revolution in WDM simulations over the past decade or so [105]. The improved accuracy due to GGAs implies that the details of the electronic structure matter, and that these systems are in no way locally uniform. In fact, in many cases, there are large evanescent regions of the HOMO, just as in gas-phase molecules. The KS system is ideal for computing this, and GGAs and hybrid account for the energetic consequences. So the very success of DFT-MD for WDM implies that the detailed chemistry is vital, even if it is happening within simulations of extended systems under high temperature and pressure.

The recent work in our group is almost entirely focused on bringing GGA-level XC technology to the WDM field. Being able to distinguish among different components of the correlation energy, and switch from one to another, is a crucial part of the exact conditions that were used to construct GGAs [59]. The adiabatic connection formula is often invoked in modern DFT research to understand both exact DFT and approximations, and its recasting as a temperature integral should prove useful in the search for accurate thermal XC approximations.

On the other hand, ground-state DFT has benefited enormously from testing on benchmark data [67, 106]. But for thermal effects, even a simple H atom is difficult, as one must include sums over all possible particle numbers in the partition function. The asymmetric Hubbard dimer is the simplest imaginable exactly solvable model, and can be considered a model for $H_2$ in a minimal basis when we take the symmetric limit. While the truncated Hilbert space makes it unrealistic at higher temperatures, it also makes it practical to solve exactly. Thus our calculations on this model demonstrate the behavior of correlation at finite temperatures in one simple case. Unfortunately, due to the lack of a continuum, this cannot be used to check the performance of LDA or GGA.

Thermodynamics tells us simple relations between entropy and free energy and other quantities. But it requires very careful reasoning to deduce the corresponding relations among density functionals, as the density must be held fixed, not the external potential. Our relations between entropy and the universal part of the Mermin functional show this, and subtraction of the corresponding KS contributions yields crucial relations among correlation contributions. They also yield simple inequalities that are not automatically satisfied once thermal XC contributions are approximated.

Lastly, the recent proof of TDDFT for finite temperatures within linear response justifies the extraction of conductivities within the Kubo response formalism from KS orbitals and energies. It also shows that the random phase approximation, which is now routinely calculated for inhomogeneous systems in many codes in both quantum chemistry and materials science [107–112], might be an excellent starting point for more accurate approximations to the XC thermal corrections, using approximations to the temperature-dependent XC kernel.

Finally, our recent work explains how ignoring thermal XC effects, which is usually done in practical DFT WDM calculations, might not be as poor an approximation as it first appears. Calculations on the Hubbard dimer show that XC effects become relatively less important as the temperature increases. Thus the errors in the self-consistent density and orbitals caused by any approximation to the XC lessen with increasing temperature, so that calculations of the KS conductance should be more accurate as temperature increases (somewhat counterintuitively).

To summarize, the success of modern density functional approximations in WDM simulations strongly implies the importance of chemical phenomena in such simulations, and the need to accurately approximate the energetics.

## 11.6 Conclusion

Thermal density functional theory is an increasingly utilized tool for calculations of hot systems such as warm dense matter. These WDM systems include inertial confinement fusion, planetary interiors, and shock experiments. There have been many recent developments ranging from exact conditions, improved understanding, and extensions beyond equilibrium with more foreseeable, and exciting, improvements on the horizon. These steps forward set up the foundation for further future success of thermal DFT in the years to come.

**Acknowledgements** The authors acknowledge support from the US Department of Energy (DOE), Office of Science, Basic Energy Sciences under Award No. DE-FG02-08ER46496. J.C.S. acknowledges support through the NSF Graduate Research fellowship program under Award No. DGE-1321846.

## References


1. T.R. Mattsson, M.P. Desjarlais, Phys. Rev. Lett. **97**, 017801 (2006), https://doi.org/10.1103/PhysRevLett.97.017801
2. U.D. of Energy, Basic research needs for high energy density laboratory physics: Report of the workshop on high energy density laboratory physics research needs. Technical report, Office of Science and National Nuclear Security Administration (2009)
3. W. Lorenzen, B. Holst, R. Redmer, Phys. Rev. Lett. **102**, 115701 (2009), https://doi.org/10.1103/PhysRevLett.102.115701
4. M.D. Knudson, M.P. Desjarlais, A. Becker, R.W. Lemke, K.R. Cochrane, M.E. Savage, D.E. Bliss, T.R. Mattsson, R. Redmer, Science **348**(6242), 1455 (2015)
5. M.D. Knudson, M.P. Desjarlais, Phys. Rev. Lett. **103**, 225501 (2009), https://doi.org/10.1103/PhysRevLett.103.225501
6. M.D. Knudson, M.P. Desjarlais, A. Pribram-Jones, Phys. Rev. B **91**, 224105 (2015)
7. B. Holst, R. Redmer, M.P. Desjarlais, Phys. Rev. B **77**, 184201 (2008)
8. A. Kietzmann, R. Redmer, M.P. Desjarlais, T.R. Mattsson, Phys. Rev. Lett. **101**, 070401 (2008)
9. S. Root, R.J. Magyar, J.H. Carpenter, D.L. Hanson, T.R. Mattsson, Phys. Rev. Lett. **105**(8), 085501 (2010), https://doi.org/10.1103/PhysRevLett.105.085501
10. R.F. Smith, J.H. Eggert, R. Jeanloz, T.S. Duffy, D.G. Braun, J.R. Patterson, R.E. Rudd, J. Biener, A.E. Lazicki, A.V. Hamza, J. Wang, T. Braun, L.X. Benedict, P.M. Celliers, G.W. Collins, Nature **511**(7509), 330 (2014), https://doi.org/10.1038/nature13526
11. F. Graziani, M.P. Desjarlais, R. Redmer, S.B. Trickey (eds.), *Frontiers and Challenges in Warm Dense Matter*, *Lecture Notes in Computational Science and Engineering*, vol. 96 (Springer International Publishing, 2014)
12. S. Ichimaru, *Statistical Plasma Physics*. Frontiers in Physics (Westview, Boulder, CO, 2004), http://cds.cern.ch/record/1106845
13. C.K. Birdsall, A.B. Langdon, *Plasma physics via computer simulation* (CRC Press, 2004)
14. T. Sjostrom, J. Dufty, Phys. Rev. B **88**, 115123 (2013), https://doi.org/10.1103/PhysRevB.88.115123
15. V.V. Karasiev, T. Sjostrom, J. Dufty, S.B. Trickey, Phys. Rev. Lett. **112**, 076403 (2014), https://doi.org/10.1103/PhysRevLett.112.076403
16. K. Burke, J. Chem. Phys. **136**(15), 150901 (2012), http://link.aip.org/link/?JCP/136/150901/1
17. W. Kohn, L.J. Sham, Phys. Rev. **140**(4A), A1133 (1965), https://doi.org/10.1103/PhysRev.140.A1133
18. R. Car, M. Parrinello, Phys. Rev. Lett. **55**(22), 2471 (1985), https://doi.org/10.1103/PhysRevLett.55.2471
19. N.D. Mermin, Phys. Rev. **137**, A: 1441 (1965)
20. B. Militzer, D.M. Ceperley, Phys. Rev. Lett. **85**, 1890 (2000), https://doi.org/10.1103/PhysRevLett.85.1890
21. V.S. Filinov, M. Bonitz, W. Ebeling, V.E. Fortov, Plasma Physics and Controlled Fusion **43**(6), 743 (2001), http://stacks.iop.org/0741-3335/43/i=6/a=301
22. B. Militzer, Phys. Rev. B **79**, 155105 (2009), https://doi.org/10.1103/PhysRevB.79.155105
23. T. Schoof, M. Bonitz, A. Filinov, D. Hochstuhl, J. Dufty, Contributions to Plasma Physics **51**(8), 687 (2011), https://doi.org/10.1002/ctpp.201100012
24. K.P. Driver, B. Militzer, Phys. Rev. Lett. **108**, 115502 (2012), https://doi.org/10.1103/PhysRevLett.108.115502
25. T. Schoof, S. Groth, J. Vorberger, M. Bonitz, Phys. Rev. Lett. **115**, 130402 (2015), https://doi.org/10.1103/PhysRevLett.115.130402
26. T. Dornheim, S. Groth, T. Sjostrom, F.D. Malone, W.M.C. Foulkes, M. Bonitz, Phys. Rev. Lett. **117**, 156403 (2016), https://doi.org/10.1103/PhysRevLett.117.156403
27. J.M. McMahon, M.A. Morales, C. Pierleoni, D.M. Ceperley, Rev. Mod. Phys. **84**, 1607 (2012), https://doi.org/10.1103/RevModPhys.84.1607

28. J.P. Perdew, K. Burke, M. Ernzerhof, Phys. Rev. Lett. **77**, 3865 (1996), http://link.aps.org/doi/10.1103/PhysRevLett.77.3865
29. C. Toher, A. Filippetti, S. Sanvito, K. Burke, Phys. Rev. Lett. **95**, 146402 (2005)
30. S.Y. Quek, L. Venkataraman, H.J. Choi, S.G. Louie, M.S. Hybertsen, J.B. Neaton, Nano Letters **7**(11), 3477 (2007), https://doi.org/10.1021/nl072058i. PMID: 17900162
31. M. Koentopp, C. Chang, K. Burke, R. Car, J. Phys. Condens. Matter **20**(8), 083203 (2008), http://stacks.iop.org/0953-8984/20/i=8/a=083203
32. P. Hohenberg, W. Kohn, Phys. Rev. **136**(3B), B864 (1964), https://doi.org/10.1103/PhysRev.136.B864
33. E.H. Lieb, Rev. Mod. Phys. **53**, 603 (1981), https://doi.org/10.1103/RevModPhys.53.603
34. M.A. Marques, M.J. Oliveira, T. Burnus, Comput. Phys. Commun. **183**(10), 2272 (2012), https://doi.org/10.1016/j.cpc.2012.05.007
35. J.P. Perdew, Y. Wang, Phys. Rev. B **45**(23), 13244 (1992), https://doi.org/10.1103/PhysRevB.45.13244
36. J.P. Perdew, A. Ruzsinszky, L.A. Constantin, J. Sun, G.I. Csonka, J. Chem. Theor. Comput. **5**(4), 902 (2009), https://doi.org/10.1021/Ct800531s
37. L.O. Wagner, E.M. Stoudenmire, K. Burke, S.R. White, Phys. Rev. Lett. **111**, 093003 (2013), https://doi.org/10.1103/PhysRevLett.111.093003
38. D.J. Carrascal, J. Ferrer, J.C. Smith, K. Burke, J. Phys.C ondens. Matter **27**(39), 393001 (2015), http://stacks.iop.org/0953-8984/27/i=39/a=393001
39. N.T. Maitra, The Journal of Chemical Physics **144**(22), 220901 (2016), https://doi.org/10.1063/1.4953039
40. N.T. Maitra, F. Zhang, R.J. Cave, K. Burke, J. Chem. Phys. **120**(13), 5932 (2004), http://link.aip.org/link/?JCP/120/5932/1
41. A. Dreuw, J.L. Weisman, M. Head-Gordon, J. Chem. Phys. **119**(6) (2003)
42. D.J. Tozer, J. Chem. Phys. **119**(24) (2003)
43. B.G. Levine, C. Ko, J. Quenneville, T.J. Martnez, Mol. Phys. **104**(5–7), 1039 (2006), https://doi.org/10.1080/00268970500417762
44. M. van Faassen, P.L. de Boeij, R. van Leeuwen, J.A. Berger, J.G. Snijders, Phys. Rev. Lett. **88**, 186401 (2002)
45. A. Theophilou, J. Phys. C **12**, 5419 (1979)
46. E.K.U. Gross, L.N. Oliveira, W. Kohn, Phys. Rev. A **37**, 2809 (1988), https://doi.org/10.1103/PhysRevA.37.2809
47. E.K.U. Gross, M. Petersilka, T. Grabo, Chem. Appl. Density-Funct. Theor. **629**, 42 (1996)
48. A. Pribram-Jones, Z.H. Yang, J.R. Trail, K. Burke, R.J. Needs, C.A. Ullrich, J. Chem. Phys. **140**, 18A541 (2014)
49. Z.h. Yang, J.R. Trail, A. Pribram-Jones, K. Burke, R.J. Needs, C.A. Ullrich, Phys. Rev. A **90**, 042501 (2014), https://doi.org/10.1103/PhysRevA.90.042501
50. B. Senjean, E.D.H. rd, M.M. Alam, S. Knecht, E. Fromager. Mol. Phys. **114**(7–8), 968 (2016), https://doi.org/10.1080/00268976.2015.1119902
51. L.M.K. Deur, E. Fromager, Submitted (2016)
52. M.M. Alam, S. Knecht, E. Fromager, Phys. Rev. A **94**, 012511 (2016), https://doi.org/10.1103/PhysRevA.94.012511
53. K. Burke, J. Werschnik, E.K.U. Gross, J. Chem. Phys. **123**(6), 062206 (2005), https://doi.org/10.1063/1.1904586
54. M. Levy, J. Perdew, Phys. Rev. A **32**, 2010 (1985), https://doi.org/10.1103/PhysRevA.32.2010
55. E.H. Lieb, S. Oxford, Int. J. Quantum Chem. **19**(3), 427 (1981), http://dx.doi.org/10.1002/qua.560190306
56. J. Sun, A. Ruzsinszky, J.P. Perdew, Phys. Rev. Lett. **115**, 036402 (2015), https://doi.org/10.1103/PhysRevLett.115.036402
57. S. Pittalis, C.R. Proetto, A. Floris, A. Sanna, C. Bersier, K. Burke, E.K.U. Gross, Phys. Rev. Lett. **107**, 163001 (2011), https://doi.org/10.1103/PhysRevLett.107.163001

58. A. Pribram-Jones, S. Pittalis, E. Gross, K. Burke, in *Frontiers and Challenges in Warm Dense Matter, Lecture Notes in Computational Science and Engineering*, vol. 96, ed. by F. Graziani, M.P. Desjarlais, R. Redmer, S.B. Trickey (Springer International Publishing, 2014), pp. 25–60, https://doi.org/10.1007/978-3-319-04912-0_2
59. A. Pribram-Jones, K. Burke, Phys. Rev. B **93**, 205140 (2016), https://doi.org/10.1103/PhysRevB.93.205140
60. D. Langreth, J. Perdew, Solid State Commun. **17**, 1425 (1975)
61. W.T.D. Frydel, K. Burke, J. Chem. Phys. **112**, 5292 (2000)
62. J.P. Perdew, M. Ernzerhof, K. Burke, J. Chem. Phys. **105**(22), 9982 (1996), http://link.aip.org/link/?JCP/105/9982/1
63. K. Burke, J.C. Smith, P.E. Grabowski, A. Pribram-Jones, Phys. Rev. B **93**, 195132 (2016), https://doi.org/10.1103/PhysRevB.93.195132
64. S. Ichimaru, Rev. Mod. Phys. **54**, 1017 (1982)
65. J.W. Dufty, S.B. Trickey, Phys. Rev. B **84**, 125118 (2011), https://doi.org/10.1103/PhysRevB.84.125118
66. J.W. Dufty, S. Trickey, Mol. Phys. **114**(7–8), 988 (2016)
67. P.R.T. Schipper, O.V. Gritsenko, E.J. Baerends, Theoretical chemistry accounts: theory, computation, and modeling (Theoretica Chimica Acta) **98**, 16 (1997), https://doi.org/10.1007/s002140050273
68. P. Ziesche, S. Kurth, J.P. Perdew, Computational Materials Science **11**(2), 122 (1998), https://doi.org/10.1016/S0927-0256(97)00206-1
69. J.P.S. Kurth, P. Blaha, Int. J. Quantum Chem. **75**, 889 (1999)
70. A. Tkatchenko, M. Scheffler, Phys. Rev. Lett. **102**, 073005 (2009)
71. D. Ceperley, M. Kalos, K. Binder, *Monte Carlo Methods in Statistical Physics* (Springer, Berlin, 1979)
72. M.P. Nightingale, C.J. Umrigar, *Quantum Monte Carlo methods in physics and chemistry*. 525 (Springer Science & Business Media, 1998)
73. S.R. White, Phys. Rev. Lett. **69**(19), 2863 (1992), http://link.aps.org/doi/10.1103/PhysRevLett.69.2863
74. N. Schuch, F. Verstraete, Nat. Phys. **5**, 732 (2009), https://doi.org/10.1038/nphys1370
75. C.J. Umrigar, X. Gonze, Phys. Rev. A **50**(5), 3827 (1994)
76. C. Filippi, C.J. Umrigar, M. Taut, The Journal of Chemical Physics **100**(2), 1290 (1994), https://doi.org/10.1063/1.466658
77. C.J. Huang, C.J. Umrigar, Phys. Rev. A **56**, 290 (1997), https://doi.org/10.1103/PhysRevA.56.290
78. N. Lima, M. Silva, L. Oliveira, K. Capelle, Phys. Rev. Lett. **90**(14), 146402 (2003)
79. J.C. Smith, A. Pribram-Jones, K. Burke, Phys. Rev. B **93**, 245131 (2016), https://doi.org/10.1103/PhysRevB.93.245131
80. G. Stefanucci, R. Van Leeuwen, *Nonequilibrium Many-Body Theory of Quantum Systems: A Modern Introduction* (Cambridge University Press, 2013)
81. E. Runge, E.K.U. Gross, Phys. Rev. Lett. **52**(12), 997 (1984), https://doi.org/10.1103/PhysRevLett.52.997
82. K. Burke, J. Werschnik, E.K.U. Gross, J. Chem. Phys. **123**(6), 062206 (2005), http://link.aip.org/link/?JCP/123/062206/1
83. D. Gericke, M. Schlanges, W. Kraeft, Physics Letters A **222**(4), 241 (1996), https://doi.org/10.1016/0375-9601(96)00654-8
84. V. Rizzi, T.N. Todorov, J.J. Kohanoff, A.A. Correa, Phys. Rev. B **93**, 024306 (2016). https://doi.org/10.1103/PhysRevB.93.024306
85. d A. Pribram-Jones, P.E. Grabowski, K. Burke, Phys. Rev. Lett. **116**, 233001 (2016), https://doi.org/10.1103/PhysRevLett.116.233001
86. E. Gross, W. Kohn, Phys. Rev. Lett. **55**, 2850 (1985)
87. L. Kadanoff, G. Baym, D. Pines, *Quantum Statistical Mechanics*. Advanced Books Classics Series (Addison-Wesley, 1994)

88. L. Schimka, J. Harl, A. Stroppa, A. Grüneis, M. Marsman, F. Mittendorfer, G. Kresse, Nat. Mater. **9**(9), 741 (2010)
89. S.X. Hu, L.A. Collins, T.R. Boehly, J.D. Kress, V.N. Goncharov, S. Skupsky, Phys. Rev. E **89**, 043105 (2014), https://doi.org/10.1103/PhysRevE.89.043105
90. B. Bennett, J. Johnson, G. Kerley, G. Rood, Recent developments in the sesame equation-of-state library. Technical report, Los Alamos Scientific Lab., N. Mex. (USA) (1978)
91. Y.T. Lee, R. More, Phys. Fluids (1958–1988) **27**(5), 1273 (1984)
92. B. Wilson, V. Sonnad, P. Sterne, W. Isaacs, J. Quant. Spectrosc. Radiat. Transfer **99**(1), 658 (2006)
93. Y. Miguel, T. Guillot, L. Fayon, arXiv preprint arXiv:1609.05460 (2016)
94. D. Clery, Science **353**(6298), 438 (2016), https://doi.org/10.1126/science.353.6298.438
95. H.F. Wilson, B. Militzer, Phys. Rev. Lett. **108**, 111101 (2012), https://doi.org/10.1103/PhysRevLett.108.111101
96. S. Root, L. Shulenburger, R.W. Lemke, D.H. Dolan, T.R. Mattsson, M.P. Desjarlais, Phys. Rev. Lett. **115**, 198501 (2015), https://doi.org/10.1103/PhysRevLett.115.198501
97. H.F. Wilson, M.L. Wong, B. Militzer, Phys. Rev. Lett. **110**, 151102 (2013), https://doi.org/10.1103/PhysRevLett.110.151102
98. P. Sperling, E.J. Gamboa, H.J. Lee, H.K. Chung, E. Galtier, Y. Omarbakiyeva, H. Reinholz, G. Röpke, U. Zastrau, J. Hastings, L.B. Fletcher, S.H. Glenzer, Phys. Rev. Lett. **115**, 115001 (2015), https://doi.org/10.1103/PhysRevLett.115.115001
99. P. Davis, T. Döppner, J. Rygg, C. Fortmann, L. Divol, A. Pak, L. Fletcher, A. Becker, B. Holst, P. Sperling, et al., Nat. Commun. **7** (2016)
100. D. Kraus, A. Ravasio, M. Gauthier, D. Gericke, J. Vorberger, S. Frydrych, J. Helfrich, L. Fletcher, G. Schaumann, B. Nagler, et al., Nat. Commun. **7** (2016)
101. A.D. Baczewski, L. Shulenburger, M.P. Desjarlais, S.B. Hansen, R.J. Magyar, Phys. Rev. Lett. **116**, 115004 (2016), https://doi.org/10.1103/PhysRevLett.116.115004
102. L.A. Curtiss, K. Raghavachari, P.C. Redfern, J.A. Pople, J. Chem. Phys. **106**(3) (1997)
103. K.R.L.A. Curtiss, P.C. Redfern, J. Pople, J. Chem. Phys. **109**, 42 (1998)
104. J. Paier, R. Hirschl, M. Marsman, G. Kresse, The Journal of chemical physics **122**(23), 234102 (2005), https://doi.org/10.1063/1.1926272
105. T.J. Lenosky, S.R. Bickham, J.D. Kress, L.A. Collins, Phys. Rev. B **61**, 1 (2000), https://doi.org/10.1103/PhysRevB.61.1
106. M. Ernzerhof, G.E. Scuseria, J. Chem. Phys. **110**, 5029 (1999), http://link.aip.org/link/JCPSA6/v110/i11/p5029/s1
107. G. Kresse, J. Furthmüller, Phys. Rev. B **54**(16), 11169 (1996), https://doi.org/10.1103/PhysRevB.54.11169
108. F. Furche, Phys. Rev. B **64**, 195120 (2001)
109. F. Furche, J. Chem. Phys. **129**(11), 114105 (2008)
110. H. Eshuis, J. Yarkony, F. Furche, The Journal of Chemical Physics **132**(23), 234114 (2010), https://doi.org/10.1063/1.3442749
111. H. Eshuis, J. Bates, F. Furche, Theoretical Chemistry Accounts **131**(1), 1 (2012), https://doi.org/10.1007/s00214-011-1084-8
112. F. Furche, Ann. Rev. Phys. Chem. **68**(1) (2016)

# Chapter 12
# Quantum Chemistry at the High Pressures: The eXtreme Pressure Polarizable Continuum Model (XP-PCM)

**Roberto Cammi**

**Abstract** In this chapter, we review some recent developments in our XP-PCM method to introduce the effect of high pressure (p > 1 GPa) in the quantum chemistry study of molecular properties and processes. After a presentation of the physical basis and the computational aspects of the XP-PCM model, we give examples of its recent applications. These applications regard the study and analysis of the electron distribution, of the equilibrium geometry, and of the vibrational frequencies of molecular systems under high pressure.

**Keywords** DFT · Equilibrium geometry · Vibrational frequencies · Pressure effects · eXtreme-pressure polarizable continuum model

## 12.1 Introduction

In this chapter, we present a short review of our model in title, aimed to introduce the effects of high pressure (larger than 1 GPa) in the quantum chemical description of molecular systems. The issue of high pressure in the quantum mechanical calculation for atoms and molecules has a long history, which dates back to 1937 [1]. However, this history testimonies of models which have been restricted, so far, to the study of very simple systems as atoms or diatomic molecules, remaining excluded by the impressive developments in the modern computational quantum chemistry.

As denoted by its name, the XP-PCM model is a generalization of the well-known polarizable continuum model for the study of molecular properties and processes in solution at standard conditions of pressure and temperature [2, 3]. A first version of the XP-PCM was presented in 2008 [4]. More developed versions have been presented for applications to the study of the effect of extreme high pressures on the equilibrium geometries and vibrational frequencies of molecular systems [5–7]. The method has recently been extended to study the effect of extreme pressure on the

R. Cammi (✉)
Department of Chemical Science, Life Science and Environmental Sustainability, University of Parma, Parco Area delle Scienze, 17/A, 43100 Parma, Italy
e-mail: roberto.cammi@unipr.it



M.J. Wójcik et al. (eds.), *Frontiers of Quantum Chemistry*,
https://doi.org/10.1007/978-981-10-5651-2_12

electronic excitations of molecular systems [8], and to the study of potential energy surfaces for chemical reactions at extreme high pressure [9].

In its basic version, the XP-PCM model describes molecular systems (the solute) in a dense medium at extreme high pressure (p > 1 GPa = $10^5$ bar). The medium is represented as a continuum material distribution having a void molecular cavity in which the molecular solute is accommodated. The continuum material is characterized in terms of its dielectric permittivity and its averaged electronic charge distribution, at the given condition of pressure. The molecular solute is described at the quantum mechanical level, and hence its electron distribution spreads in the medium outside the void cavity, where it overlaps with the electron distribution of the medium. Pressure enters through the Pauli repulsive interaction originating by this overlap, and the effects of the pressure are modeled by shrinking the volume of the molecular cavity. The shrinking of the cavity increases the amount the electronic charge density of the molecular solute that lies outside of the cavity boundary, with a consequent increase of the Pauli repulsion with the solvent.

The present contribution is organized as follows: After a presentation of the essential of the computational aspects of the XP-PCM model for the study of the molecular systems at extreme pressure (Sect. 12.2), we give examples of its numerical applications to the studies of the effect of the pressure on the electron distribution (Sect. 12.3), of the effect of the pressure on the equilibrium geometry (Sect. 12.4), and of the effect of the pressure on the vibrational frequencies (Sect. 12.5). Finally, in Sect. 12.6, we draw some provisional concluding remarks on the XP-PCM model.

## 12.2 The Essential of the XP-PCM Method for Molecules Under High Pressure

In this section, we summarize the basic XP-PCM equations for the quantum mechanical description of a molecular the solute in a dense medium at extreme high pressure.

The molecular system is described, in the usual clamped nuclei approximation, by an effective electronic Hamiltonian which may be written as

$$\hat{H} = \hat{H}^o + \hat{V}_e(\Psi) + \hat{V}_r \tag{12.1}$$

where $H^o$ the electronic Hamiltonian of the isolated molecule; $\hat{V}_e(\Psi)$ represents the solute–solvent electrostatic interaction, giving origin to the solvent reaction field of the PCM model, and depending on the electronic wave-function $|\Psi>$ of the molecule [2]; The operator $\hat{V}_r$ represents the exchange–repulsion (Pauli) interaction of the molecular solute with a statistical average distribution of the surrounding molecules of the external medium. According to the theory of intermolecular interaction [10], the exchange–repulsion contribution originates from the overlap between the electronic distribution of the molecular solute and a mean electronic distribution of the solvent (i.e., the external medium). The mean electronic distribution of the solvent is approximated as a uniform distribution outside of the cavity hosting the molecular solute, and with zero density inside the cavity. As a consequence, only the portion of the electronic distribution of the molecular solute lying outside the cavity overlaps with the uniform electronic distribution of the solvent, determining the solute–solvent Pauli repulsion.

The Pauli repulsion operator $V_r$ corresponds to a repulsive step potential barrier located at the boundary of the cavity hosting the molecular solute [10][1]:

$$\hat{V}_r = \int \hat{\rho}(\mathbf{r})\mathscr{V}_0\Theta(\mathbf{r})d\mathbf{r} \tag{12.2}$$

$\hat{\rho}(\mathbf{r}) = \sum_i^N \delta(\mathbf{r} - \mathbf{r}_i)$ is the electron density operator (over the $N$ electrons of the molecular system), and $\mathscr{V}_0$ is the height of a barrier potential $\Theta(\mathbf{r})$ located at the boundary of the cavity $\mathbf{C}$:

$$\Theta(\mathbf{r}) = \begin{cases} 1 & \mathbf{r} \subseteq \mathbf{C} \\ 0 & \mathbf{r} \nsubseteq \mathbf{C} \end{cases} \tag{12.3}$$

The potential barrier height $V_0$ is given by

$$\mathscr{V}_0(s) = \frac{\mathscr{V}_0}{s^{3+\eta}} \qquad s = \left(\frac{V_c}{V_c^0}\right)^{1/3} \tag{12.4}$$

where $\mathscr{V}_O$ is the step barrier at the standard condition of pressure [2], and $V_c$ and $V_c^0$ are, respectively, the actual volume of the cavity and the volume corresponding to the standard condition of pressure. In Eq. (12.4), $\eta$ is a semi-empirical parameter that gauges the strength of the solute-solvent Pauli repulsion [13].

The effective electronic Hamiltonian (12.1) implies a quantum mechanical energy functional, $G_{e-r}$, which is defined as:

$$G_{e-r} = < \Psi|\hat{H}^o + \frac{1}{2}\hat{V}_e(|\Psi >) + \hat{V}_r|\Psi > + \tilde{V}_{nn} \tag{12.5}$$

The energy functional (12.5) has the thermodynamic status of a free energy for the whole molecule-external medium system, with respect to a reference state given by the non-interacting electrons and nuclei of the molecular solute and by the unperturbed external medium having a preformed cavity to host the solute.[2]

The time-dependent Schrödinger equation for the effective Hamiltonian (12.1) determines the electronic wave-function $|\Psi >$ of the molecular system [14]:

$$[\hat{H}^o + \hat{V}_e(|\Psi >) + \hat{V}_r]|\Psi > = E|\Psi > \tag{12.6}$$

---

[1]An alternative definition of the operator $V_r$ has been recently given by Chipman and co-workers in terms of the gradients of the electron density operator [11, 12].

[2]The factor $1/2$ in front to the operator $\hat{V}_e(|\Psi >)$ is due to the nonlinear nature of this operator, and $\tilde{V}_{nn}$ is the nuclei–nuclei interaction contribution in the presence of the external medium.

The solution of this equation, which can be approached at the available approximation levels of the Computational Quantum Chemistry (HF/DFT, Couple-Cluster, ...), gives access to the properties of the molecular systems under extreme pressure, at these various levels of theory. Within the XP-PCM model, as for the standard PCM model, the electronic response properties of the molecular solute can be expressed as derivatives of the free-energy functional $G_{e-r}$ with respect to suitable internal or external perturbations. This connection between the molecular properties and the free-energy functional $G_{e-r}$ is a consequence of the form that the Hellmann–Feynman theorem [15, 16] takes for the PCM model [3, 17].

The free-energy functional $G_{e-r}$ may also act as the Born–Oppenheimer potential energy surfaces ruling the nuclear motion of the solute, if a rigid geometry of the cavity hosting the molecular solute is assumed. This approximation can be used the case of molecules whose equilibrium geometry do not involve large displacements of the nuclei under the effect of extreme pressure. This approximation has been used in all the applications presented in this chapter. In the case of molecular processes involving large amplitude motions of the nuclei, as in the conformational changes, or along an intrinsic reaction path, the potential energy surfaces ruling the nuclear motion is obtained by adding to the electronic free-energy functional $G_{e-r}$ a free-energy contribution due to the variation of the cavity geometry within the molecular process.

The free-energy functional $G_{e-r}$ may also act as the Born–Oppenheimer potential energy surfaces ruling the nuclear motion of the solute, if a rigid geometry of the cavity hosting the molecular solute is assumed. This approximation of a rigid cavity may be used the case of molecules whose equilibrium geometry do not involve large displacements nuclei under the effect of extreme pressure. This approximation has been used in the applications presented in this chapter. On the contrary, in the case of molecular processes involving large amplitude motions of the nuclei, as in the conformational changes, or along intrinsic reaction paths, the effective potential energy surfaces ruling the nuclear motion are obtained by adding to the electronic free-energy functional $G_{e-r}$ a free-energy contribution due to the variation of the cavity geometry within the molecular process [9].

In the XP-PCM method, the pressure is not an input parameter, and it is defined in terms of the dependence of the electronic free-energy functional $G_{e-r}$ with respect to the volume of the cavity hosting the molecular solute:

$$p = -\left(\frac{\partial G_{e-r}}{\partial V_c}\right) \tag{12.7}$$

where $V_c$ is the volume of the cavity.

The cavity of the molecular solute is defined as the envelope of spheres centered on the solute atomic nuclei. The radii of these atomic spheres are given by the product of the corresponding van der Waals radii [18] times a uniform scaling factor, $f$. To increase the pressure, the volume of the cavity $V_c$ is decreased by reducing the value

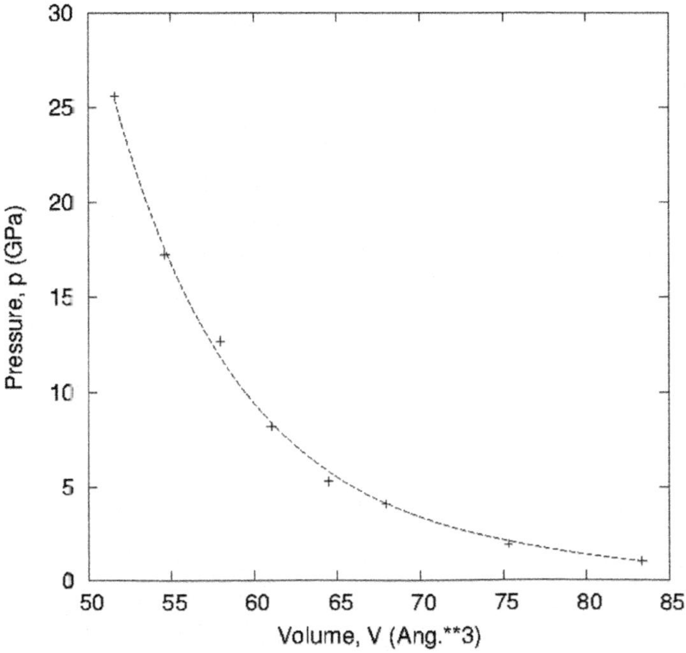


**Fig. 12.1** XP-PCM results for the pressure, $p$, as a function of the cavity volume, $V_c$, for the case of diborane as molecular solute. See Ref. [5] for computational details

of the scaling factor $f$ with respect the reference value $f_0 = 1.2$. Figure 12.1 shows the pressure as a function of the volume of the cavity for the case of the diborane molecule [5].

## 12.3 Effect of the Pressure on the Electron Density Distribution

Let us consider a molecular solute at a given fixed geometry in a dense phase at extreme pressure. Under this condition, the effect of the pressure on the molecular properties reflects the increase of the Pauli repulsion operator, $V_r$, in the effective electronic Hamiltonian (12.1). In fact, as we have described previously, XP-PCM models the effect of the pressure by reducing the volume of the cavity hosting the molecular solute, with a corresponding increase of the solute–solvent Pauli repulsion. The molecular system reacts to this perturbation rearranging its electronic wave-function, to which corresponds a change of the related electronic properties. As an example of the effect of the pressure on the electronic properties, we consider the case of the one-electron density $\rho(\mathbf{r})$.

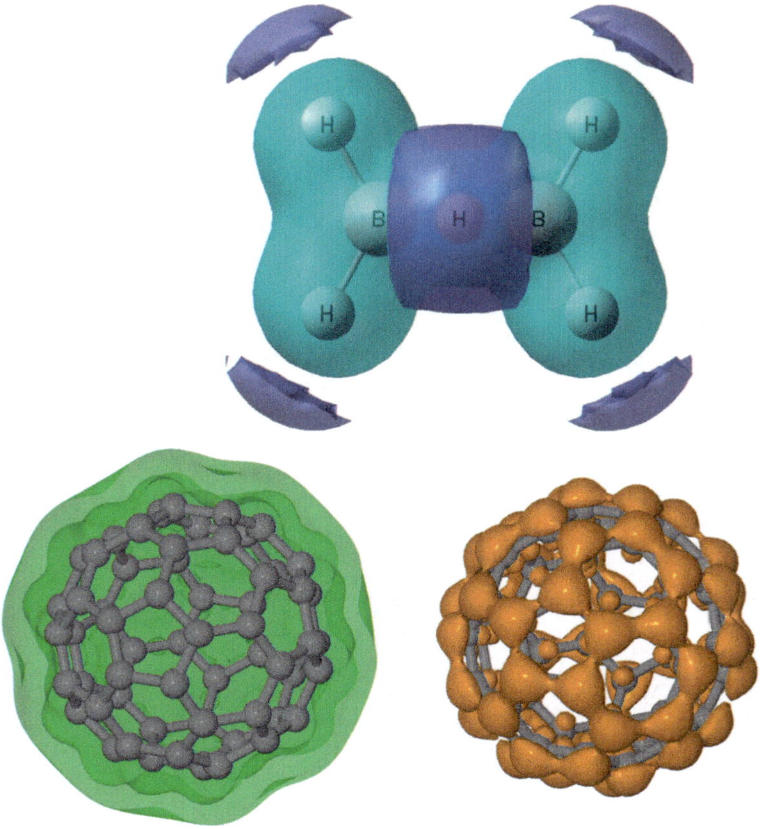


**Fig. 12.2** XP-PCM iso-surfaces of the electron density differential $\Delta\rho(p) = \rho(p) - \rho(0)$ for diborane (*top*) and Buckminsterfullerene (*bottom*) under extreme pressure ($p = 1.0$ GPa). Diborane: *blue-violet* iso-surfaces correspond to a decrease of electron density $\Delta\rho < 0$ while aquamarine iso-surface corresponds to an increase of electron density $\Delta\rho > 0$; C60: *light-green* iso-surface corresponds to a decrease of electron density $\Delta\rho < 0$ while *orange* iso-surface corresponds to an increase of electron density $\Delta\rho > 0$

Figure 12.2 shows the differential of the electron density ($\Delta\rho(p) = \rho(p) - \rho(0)$) induced by the pressure $p$ for diborane [5] and Buckminsterfullerene (C60) [6]. The effect of the pressure corresponds to a flux of electrons from outer regions of the molecular systems toward more internal regions. In particular, the increase of the $\rho(\mathbf{r})$ is concentrated in the internuclear regions of these molecular systems. The differential of the electron density $\Delta\rho(p)$ can be expressed in terms of the mixing between the molecular orbital (MO) induced by the pressure. At the first order, we can write:

$$\Delta\rho(p) = 2\sum_i^{occ} \phi_i^0 \phi_i^{(1)} \tag{12.8}$$

where the summation is over the occupied MO, $\phi_i^0$ are the unperturbed MO and $\phi_i^{(1)}$ are the first-order perturbations induced by the one-electron Pauli perturbation $v_r$.[3] In an uncoupled MO perturbation scheme,[4] $\phi_i^{(1)}$ is given by

$$\phi_i^{(1)} = \sum_a^{vac} < \phi_i^0 | v_r | \phi_a^0 > /(\varepsilon_i^0 - \varepsilon_a^0) \tag{12.9}$$

where the summation runs over the vacant MO $\phi_a^0 >$ and $(\varepsilon_i^0 - \varepsilon_a^0)$ are orbital energy differences.

The Pauli repulsion operator $v_r$ is totally symmetric with respect to the symmetry point group of the molecular system (the pressure acts isotropically), hence reasons of symmetry restrict the summation over the vacant MO that span the same symmetry (irreducible representation) of the occupied MO $\phi_i^0$. Therefore, each term of Eq. (12.8) is totally symmetric, and so the difference of the electron density $\Delta\rho(p)$, as clearly shown in Fig. 12.2.

As we will see in the next Section, the differential of the electron density ($\Delta\rho(p) = \rho(p) - \rho(0)$) plays a significant role in the discussion of the effect of the pressure on the equilibrium geometry.

## 12.4 Effect of the Pressure on Equilibrium Geometry

The XP-PCM method can study the evolution of the equilibrium geometry of molecular systems under extreme pressure. The equilibrium geometries are determined by exploiting very effective optimization processes, thanks to the available analytical form of the gradients of the electronic energy functional $G_{e-r}$ [19]. In Fig. 12.3, we show the evolution of the variations of the bond lengths distances in diborane [5] and Buckminsterfullerene [6] as functions of the pressure, up to 20 GPa. For these molecular systems, all the atomic bond lengths are shortened with the increase of the pressure, with a linear dependence with respect to the pressure.

Is there an explanation of this linear behavior of the bond lengths with pressure? What is the physical origin of this response of the molecular geometry to the pressure? Both questions find a physically coherent explanation with the XP-PCM method.

---

[3]Formally, the Pauli repulsion operator for N electrons of Eq. (12.2) may be written as $\hat{V}_r = \hat{v}_r(\mathbf{r})$ with $\hat{v}_r(\mathbf{r}) = \delta(\mathbf{r} - \mathbf{r}'_i)V_0\Theta(\mathbf{r})$.

[4]An uncoupled molecular orbital (MO) perturbation scheme is a perturbation method that evaluates the effect of the perturbation on the molecular orbital by neglecting the effects that the perturbation has on the electron–electron repulsion contribution of the Fock operator.

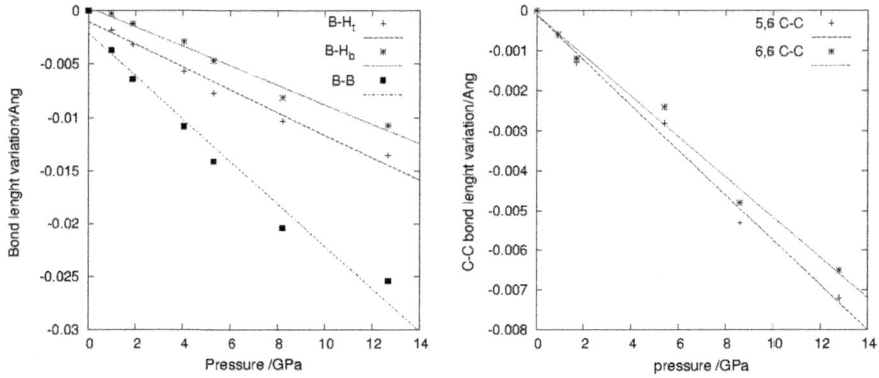


**Fig. 12.3** XP-PCM bond lengths variations in diborane (*left panel*) and Buckminsterfullerene (*right panel*) as functions of the pressure. For diborane, B–H$_t$, B–H$_b$, B–B denote, respectively boron–hydrogen (terminal), boron–hydrogen (bridge), boron–boron bonds; for fullerene C60, 5,6C–C, 6,6C–C denote, respectively, the 5,6 and 6,6 carbon–carbon bonds. See Refs. [5, 6] for the computational details

## 12.4.1 On the Linear Dependence on Pressure of the Bond Distances

As said in Sect. 12.2, the potential energy function for the nuclei motion of the molecular system under an external pressure is given by the electronic energy $G_{e-r}$. Hence, the equilibrium geometry of the molecules corresponds to a minimum of the potential energy $G_{e-r}$. The potential energy $G_{e-r}$, may be formally expanded near the equilibrium geometry in the gas phase with respect to both the nuclear coordinates $\mathbf{Q}$ and the pressure $p$. If we express the nuclear coordinates in terms of the normal vibrational coordinates, $\mathbf{Q}_i$, of the isolated molecule, $G_{e-r}(\mathbf{Q}, p)$ may be expressed up to second order as [5, 20]

$$G_{e-r}(\mathbf{Q}, p) = G_{e-r}(0, p) + \frac{1}{2}\sum_i k_i Q_i^2 + p\sum_i^{TS} \Gamma_i Q_i \tag{12.10}$$

where $G_{e-r}(0, p)$ denotes the electronic energy at the equilibrium geometry $\mathbf{Q} = 0$ at the pressure $p$, $Q_i$ and $k_i = (\partial^2 G_{e-r}/\partial^2 Q_i)_{\mathbf{Q}=0}$ are the i-th normal mode and the corresponding harmonic force constant, and $\Gamma_i$ is a pressure coupling defined as:

$$\Gamma_i = \left(\frac{\partial^2 G_{e-r}}{\partial p \partial Q_i}\right)_{\mathbf{Q=0}} = \left(\frac{\partial G_{e-r}^{Q_i}}{\partial p}\right)_{\mathbf{Q=0}} \tag{12.11}$$

The coupling parameter $\Gamma_i$ corresponds to a mixed second derivative of the basic XP-PCM energy. This coupling parameter has the physical meaning of a molecular response function which describes the response to the pressure of the component of energy gradients along the normal mode coordinates, $G_{e-r}^{Q_i} = \left(\frac{\partial G_{e-r}}{\partial Q_i}\right)_{\mathbf{Q}=\mathbf{0}}$. For symmetry reasons (as the pressure is isotropic, the symmetry point group of the molecule is conserved), the summation of Eq. (12.10) runs only over the totally symmetric (TS) normal modes of the isolated molecules.

At a given pressure $p$, the new equilibrium geometry is then determined by the condition

$$\frac{\partial G_{e-r}(\mathbf{Q}, p)}{\partial Q_i} = p\Gamma_i + k_i Q_i^{eq} = 0 \tag{12.12}$$

which gives:

$$Q_i(p)^{eq} = -\frac{\Gamma_i}{k_i} p \qquad i \subseteq TS \tag{12.13}$$

Equation (12.13) shows that *the changes of the equilibrium geometry in the molecule occur with shifts along the totally symmetric normal coordinate $Q_i$, the shifts linearly depending on the external pressure p*. This conclusion is in agreement with the linear correlation shown in Fig. 12.2.

It is useful to rewrite Eq. (12.13) in the following form

$$Q_i(p)^{eq} = -\frac{G_{e-r}^{Q_i}(p)}{k_i} \qquad i \subseteq TS \tag{12.14}$$

where $G_{e-r}^{Q_i}(p) = \Gamma_i p$ is the force induced by the pressure along the normal mode. This form is similar to the analogous equation developed within the framework of the Bell theory [21] to study the deformation of a molecular system under an external mechanical force.[5]

We now turn to the explanation of the physical origin of the force $G_{e-r}^{Q_i}(p)$ induced by the pressure on the nuclei of the molecular solute.

## 12.4.2 On the Origin of Forces Induced by the Pressure on the Nuclei

The force $G_{e-r}^{Q_i}(p)$, of Eq. (12.14), induced by the pressure on the nuclei of the molecular solute is a consequence of the primary effects that the pressure has on the electronic charge distribution that we have discussed previously in Sect. 12.3.

In Sect. 12.3, we have shown that the pressure induces, via the solute–solvent Pauli repulsion $V_0$, a difference of electron density, $\Delta\rho(\mathbf{r}; p)$, which corresponds to a flux of electrons from the outer regions of the molecular solute to its inner, inter-

[5]For reasons of space, we can no further discuss this connection between our XP-PCM theory with the Bell theory and its extensions.

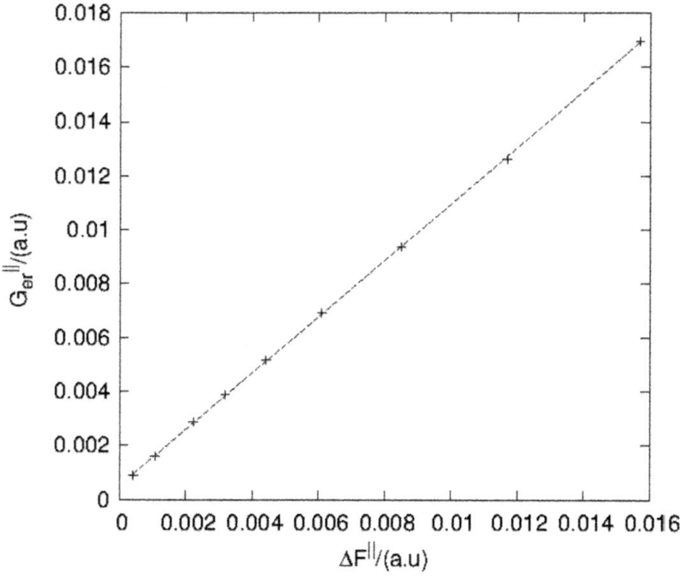


**Fig. 12.4** Correlation between the XP-PCM analytical gradients of the energy $G^{||}_{er}$ and the Hellman–Feynman electrostatic forces in diborane. The *upper-script* || denotes a component along the direction of the boron–hydrogen (terminal) direction. All quantities are expressed in the pertinent atomic units (a.u) see Ref. [5] for computational details

atomic regions. The consequences of the differential of electron density may be analyzed by using the force concept based on the Hellmann–Feynman (*H–F*) electrostatic theorem [15, 16, 22–24]. According to this theorem, the force, $G^{Q_i}_{e-r}(p)$, induced by the pressure on the nuclei of the molecular solute is given by the electric field $\Delta\mathbf{F}(p)$ originated by the differential of electron density, $\Delta\rho(\mathbf{r};p)$:

$$\Delta\mathbf{F}(\mathbf{R}_A) = -Z_A \int \Delta\rho(\mathbf{r};p)\frac{(\mathbf{r}-\mathbf{R}_A)}{|\mathbf{r}-\mathbf{R}_A|^{3/2}}d\mathbf{r} = G^{\mathbf{R}_A}_{e-r} \tag{12.15}$$

where $Z_A$ is the nuclear charge of $A$, and $\Delta\rho(\mathbf{r}\,:\,p)$.

A validation of the electrostatic origin of the force induced by the pressure on the nuclei is shown in Fig. 12.4, where the *H–F* electrostatic field of Eq. (12.15) at the terminal hydrogen atoms of diborane is compared with the corresponding energy gradient, $G^{\mathbf{R}_H}_{e-r}$, computed by analytical differentiation.

In this subsection, we have shown that the pressure has a direct effect on the electron density and that throughout this direct effect, pressure induces a further, indirect effect in molecular solute by changing its equilibrium geometry. This chain of effects triggered by the pressure does not stop here. In turn, the variation equilibrium geometry may influence, in general, many others molecular properties. In the following section, we will consider the case of the vibrational frequencies of a molecular solute under pressure.

## 12.5 Effect of the Pressure on the Vibrational Frequencies of Molecular Systems

The perspective of the theoretical modeling of the effects of the pressure on the vibrational frequencies is of considerable interest per se, as the theoretical results can be compared with available experimental data, toward a more robust validation of XP-PCM model. Furthermore, the XP-PCM model opens the way toward an understanding of the effects of the pressure on the vibrational frequencies, understanding that cannot be obtained by the experimental data alone.

Within the XP-PCM method, the vibrational frequencies (harmonic approximation) are computed at the equilibrium geometry corresponding to the given pressure (see the previous Sect. 12.4). The harmonic vibrational frequencies are determined from the Hessian matrix of the second derivatives of the electronic free-energy functional $G_{er}$ with respect to the Cartesian coordinates of the nuclei. Here, we are assuming that during the vibrational displacements of the nuclei around the equilibrium geometry, the cavity hosting the molecular solute remains fixed. As already discussed in Sect. 12.2, only under this assumption, the electronic free-energy functional $G_{e-r}$ acts as potential energy surface for the nuclei of the solute.

Specific physical arguments further motivate the approximation to held fixed cavity during the vibrational motion of the nuclei. Within the XP-PCM model, the vibrational frequencies of the molecules are evaluated taking into account the differences between the timescale of the vibrational degrees of freedom of the solvated molecule and the timescale of the various degrees of freedom (translational, rotational, vibrational, and electronic) of the molecules composing the solvent. The basic assumption is that the translational and rotational degrees of freedom of the solvent molecules remain fixed during a typical timescale of the vibrations of the molecular solute. On the other hand, the geometry of the cavity is correlated with the physical space non-accessible to the solvent molecules by the translational and rotational motion, and hence the displacements of the nuclei of the molecular solute around its equilibrium geometry are performed assuming a fixed geometry of the cavity.

### *12.5.1 On the* Curvature *and* Relaxation *Effects of the Pressure on the Vibrational Frequencies*

In Fig. 12.5, we compare the XP-PCM linear coefficients of vibrational frequencies for Buckminsterfullerene ($d\nu/dp$) [6] with the available experimental counterpart. The figure collects both the radial and tangential normal modes of Buckminsterfullerene and shows that the correlation between the theoretical and experimental data is satisfactory. Similar results have been obtained for other molecular systems [5, 7].

The XP-PCM model allows to analyze these effects of the pressure on the vibrational frequencies in terms of two different physical mechanisms [5]. The first mechanism, called the *curvature* effect, is determined by the influence of the differential

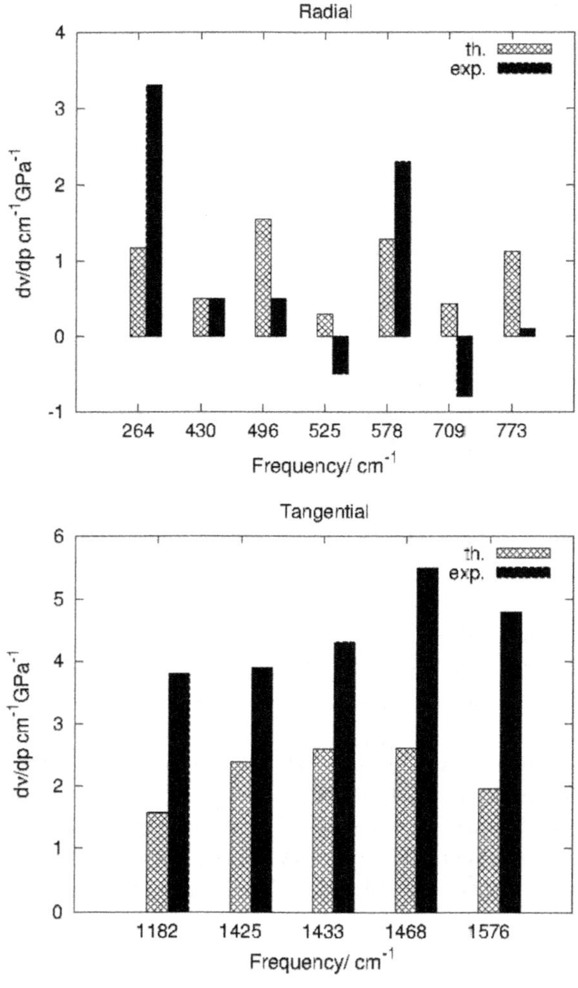


**Fig. 12.5** XP-PCM study of the effect of the pressure on the vibrational frequencies (harmonic approximation) of Buckminsterfullerene: comparison between the computed linear pressure coefficients $(d\nu/dp)_{XP-PCM}$ (*blue bars*) and the available experimental data $(d\nu/dp)_{exp}$ (*green bars*). The *upper panel* is pertinent to the normal modes with displacements of the nuclei along direction normal to the molecular surface of C60. The *lower panel* is pertinent to the normal modes with displacements of the nuclei along directions tangential to the molecular surface of C60. See Ref. [6] for computational details

electron density induced by the pressure (see Sect. 12.3) on the harmonic force constant, evaluated at the equilibrium geometry in vacuo. This effect is determined as

$$\left.\frac{dk_i}{dp}\right|_{cur} = \left(\frac{\partial k_i}{\partial p}\right)_{\mathbf{Q}(0)} \tag{12.16}$$

and the corresponding *curvature* effect on the vibrational frequencies is denoted by $\left(\frac{\partial \nu(p)}{\partial p}\right)_{cur}$.

The second mechanism, called *relaxation* effect, is related to the influence that the shift of the equilibrium geometry induced by the pressure $\mathbf{Q}_i(p)$ has on the harmonic force constant:

$$\left.\frac{\partial k_i(p)}{\partial p}\right|_{rel} = \left(\sum_j^{TS} g_{iij}(0)Q_i(p)\right) \tag{12.17}$$

where $g_{iij}(0)$ are the vibrational cubic force constants of the isolate molecule. The corresponding *relaxation* effect on the vibrational frequencies is denoted by $\left(\frac{\partial \nu(p)}{\partial p}\right)_{rel}$.

Hence, the total effect of the pressure on the vibrational frequencies is partitioned as

$$\left(\frac{\partial \nu_i(p)}{\partial p}\right)_{\mathbf{Q}(p)} = \left(\frac{\partial \nu_i(p)}{\partial p}\right)_{cur} + \left(\frac{\partial \nu_i(p)}{\partial p}\right)_{rel} \tag{12.18}$$

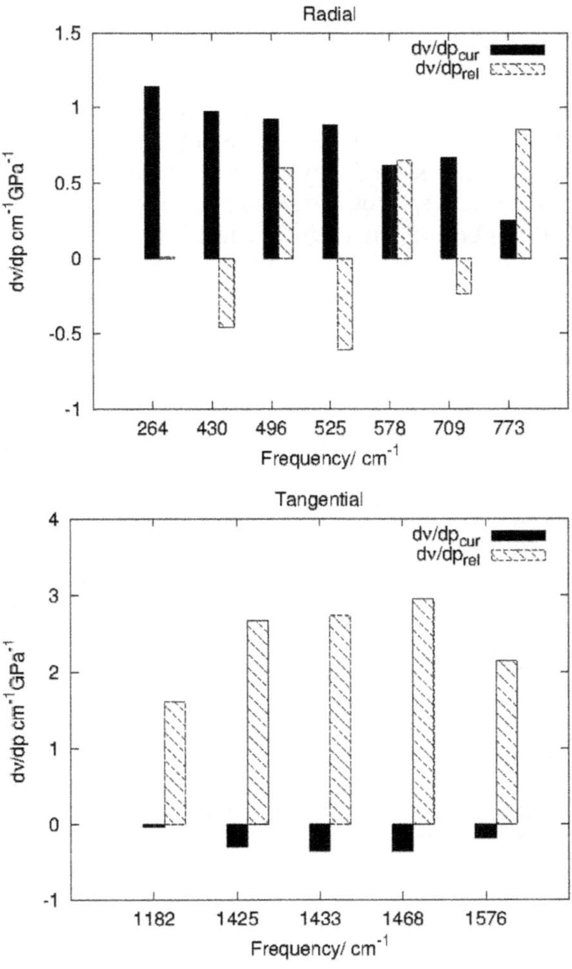


**Fig. 12.6** XP-PCM analysis of the effect of the pressure on the vibrational pressure coefficients $(d\nu/dp)_{XP-PCM}$ (harmonic approximation) of Buckminsterfullerene. *Blue bars* correspond to the direct effect $(d\nu/dp)_{cur}$ (Eq. 12.15) and the *green bars* to the indirect contribution $(d\nu/dp)_{rel}$ (Eq. 12.16). The *upper panel* is pertinent to the normal modes with displacements of the nuclei along direction normal to the molecular surface of Buckminsterfullerene. The *lower panel* is pertinent to the normal modes with displacements of the nuclei along directions tangential to the molecular surface of Buckminsterfullerene. See Ref. [6] for computational details

The results of this partition into *curvature* and *relaxation* contributions are shown in Fig. 12.6 for the vibrational normal modes of fullerene *C*60. The different nature of the effect of the pressure on the radial and tangential vibrational normal modes of *C*60 is evident. The effect of the pressure on the frequencies of the radial normal modes is dominated by the *curvature* mechanism, due to the effect of the pressure on the electron density, while the effect of the pressure on the frequencies of the tangential normal modes is dominated by the *relaxation* mechanism, due to the shift of the equilibrium geometry induced by the pressure.

## 12.6 Conclusions

In this chapter, we have reviewed some recent developments in our XP-PCM method to introduce the effect of high pressure ($p > 1$ GPa) in the quantum chemistry study of molecular properties and processes.

Since a long time, theoretical chemists [25] have been aware that chemistry under extreme condition of pressure is a field where quantum chemistry should potentially exploit its predictive power to help predicting and understanding molecular processes. In spite of this awareness, the chemistry under extreme condition of pressure has been so far a territory distant from the frontiers of modern quantum chemistry. We hope that the XP-PCM model could help extending these borders, so to include even this territory within the domain of quantum chemistry.

**Acknowledgements** The author thanks Dr. M. Frisch (Gaussian Inc.) for collaboration, Prof. B. Kirtmann (University of California, Santa Barbara) for discussions on the pressure effects on the vibrational properties, and all the Editors for the invitation to contribute to this book. The systematic application of the XP-PCM method to the study of various molecular properties and processes under high pressure has required fruitful collaborations with many people (C. Cappelli, B. Mennucci, J. Tomasi, G. Cardini, M. Pagliai, V. Schettino, M. Ehara, and R. Fukuda) that are here collectively acknowledged. Dr. M. Paglia (University of Florence, Italy) is also thanked for graphic material used in Fig. 12.2.

## References

1. A. Michels, J. De Boers, A. Bijl, Physica **4**, 981 (1937); A. Sommerfeld, H. Welker, Ann. Phys. **32**, 56 (1938); C.A. ten Seldam, S.R. de Groot, Physica **18**, 891 (1952); E.V. Ludeña, J. Chem. Phys. **69**, 1770 (1978); R. LeSar, D.R. Herschbach, J. Phys. Chem. **85**, 2798 (1981); R. LeSar, D.R. Herschbach, J. Phys. Chem. **87**, 5202 (1983); J. Gorecki, W. Byers Brown, J. Chem. Phys. **89**, 2138 (1988); J.P. Connerade, R. Smaoune, J. Phys. B **33**, 3467 (2000); K.D. Sen, B. Mayer, P.C. Schmidt, J. Garza, R. Vragas, A. Vela, Int. J. Quantum Chem. **90**, 491 (2002); S.A. Cruz, J. Soullard, Chem. Phys. Lett. **391**, 138 (2004); R.F.W. Bader, Confined atoms treated as open quantum systems, in *Advances in Quantum Chemistry*, vol. 57 (Elsevier, Amsterdam, 2009)
2. J. Tomasi, B. Mennucci, R. Cammi, Chem. Rev. **105**, 2999 (2005)
3. R. Cammi, *Molecular Response Functions for the Polarizable Continuum Model: Physical Basis and Quantum Mechanical Formalism* (Spriger, Heidelberg, 2013)

4. R. Cammi, V. Verdolino, B. Mennucci, J. Tomasi, Chem. Phys. **344**, 135 (2008)
5. R. Cammi, C. Cappelli, B. Mennucci, J. Tomasi, J. Chem. Phys. **137**, 154112 (2012)
6. M. Pagliai, G. Cardini, R. Cammi, J. Phys. Chem. A **118**, 5098 (2014)
7. M. Pagliai, G. Cardini, R. Cammi, V. Schettino, J. Phys. Chem. A **120**, 5136 (2016)
8. R. Fukuda, M. Ehara, R. Cammi, J. Chem. Theor. Comp. **11**, 2063 (2015)
9. R. Cammi, J. Comp. Chem. **36**, 2246 (2015)
10. C. Amovilli, B. Mennucci, J. Phys. Chem. B **101**, 1051 (1997)
11. A. Pomogova, D. Chipman, J. Phys. Chem. A **117**, 5212 (2014)
12. A. Pomogova, D. Chipman, J. Chem. Theo. Comp. A **10**, 211 (2014)
13. The value of parameter $\eta$ can be estimated from a comparison of the equation of state of the pressure $p$ as a function of the cavity volume $V_c$ with the macroscopic equation of state $p-V$ of several solvents [5, 6, 8]. Actual values of $\eta$ are with the range $\eta = 3 - 9$, and a higher value of the hardness parameter $\eta$ is indicative of a harder Pauli barrier potential of the medium
14. The energy eigenvalue $E$ is not the basic energy quantity of the PCM-XP model. $E$ refers to the total energy of the solute-solvent system with respect to a fictitious reference state, which depends on the wave-function $|\Psi>$. For more details on the status of the energy eigenvalue E see, for example, Ref. [2]
15. H. Hellmann, *Einführung in die Quantenchemie* (Franz Deuticke, Leipzig, 1937), p. 285
16. R. Feynman, Phys. Rev. **56**, 340 (1939)
17. R. Cammi, J. Chem. Phys. **140**, 084112 (2014)
18. A. Bondi, J. Phys. Chem. **68**, 441 (1964)
19. Here we assume that the geometry of the cavity remains fixed during the geometry optimization of the molecular solute. As pointed out in the previous section, this assumption implies that the electronic free-energy functional $G_{e-r}$ acts as potential energy surface for the nuclei of the solute
20. Y. Fujimura, S.H. Lin, H. Eyring, Proc. Natl. Acad. Sci. USA **77**, 5032 (1980)
21. S.S.M. Konda, J.N. Brantley, C.W. Bielawski, D.E. Makarov, Chemical reactions modulated by mechanical stress: extended Bell theory. J. Chem. Phys. **135**, 164103 (2011)
22. H. Nakatsuji, J. Am. Chem. Soc. **95**, 345 (1973)
23. J. Tomasi, G. Alagona, R. Bonacccorsi, C. Ghio, R. Cammi, Semiclassical interpretation of intramolecular interactions, in *Theoretical Models of Chemical Bonding*, ed. by Maksić, vol. 3 (Springer, Berlin, 1991)
24. Y. Honda, H. Nakatsuji, Chem. Phys. Lett. **293**, 230 (1998)
25. J.O. Hirschfelder, J. Chem. Edu. **43**, 457 (1966)

# Chapter 13
# Transition States of Spin-State Crossing Reactions from Organometallics to Biomolecular Excited States

**Naoki Nakatani, Akira Nakayama and Jun-ya Hasegawa**

**Abstract** Our recent studies on spin-state crossing reactions are summarized. With a constraint optimization method, a minimum-energy intersystem-crossing point (MEISCP) was located. With an energy decomposition analysis, important reaction coordinates for the intersystem-crossing were clarified. The systems investigated were twofold; one is accelerated triplet-state quenching in metal-free biomolecules such as a significantly ruffled porphyrin and a carotenoid, and the other is spin-blocked phenomena in the ligand binding of organometallic complexes. A common finding in these spin-state crossing reactions is that there is a key reaction coordinate that leads the system toward a MEISCP, a transition state of spin-crossing reaction. This coordinate is not necessary related to the primary reaction coordinate (such as C=C twisting in carotenoid, Fe–O distance in $O_2$ binding, M-H distance in metallocenes) but to the coordinate sensitive to the relative energy difference between the two spin states (such as bond-length alternation in carotenoid, porphyrin ring stretching in $O_2$ binding of heme, and cyclopentadienyl (Cp)-M-Cp angle in metallocenes). This knowledge should be useful in the catalysis design and reaction control.

**Keywords** Minimum-energy intersystem-crossing point · Triplet-state quenching · Photo-protection · Organometallics · Ligand binding

N. Nakatani · A. Nakayama · J. Hasegawa (✉)
Institute for Catalysis, Hokkaido University, N21 W10, Kita-ku, Sapporo, Hokkaido 001-0021 Japan
e-mail: hasegawa@cat.hokudai.ac.jp

*Present Address:*
N. Nakatani
Department of Chemistry, Graduate School of Science and Engineering, Tokyo Metropolitan University, Minami-Osawa 1-1, Hachioji 192-0397, Tokyo, Japan



M.J. Wójcik et al. (eds.), *Frontiers of Quantum Chemistry*,
https://doi.org/10.1007/978-981-10-5651-2_13

## 13.1 Introduction

Spin-state crossing, or intersystem-crossing (ISC), during chemical reactions is often very important for non-radiative decay processes in photochemical reactions and for catalytic reactions of transition metal complexes. As a transition between different spin states is forbidden in non-relativistic quantum mechanics, a probability of the ISC is generally small in light elements and light metals. However, ISC processes are involved as an essential part of the biological reactions although there is no heavy element. For instance, an excess energy dissipation in photosynthetic systems utilizes the ISC from triplet to singlet states [1] during a C=C double-bond twist of a carotenoid molecule. This puzzling issue has long been motivated chemists to understand the ISC in the course of the biological reactions. In some catalytic reactions, spin states of a transition metal complex exhibit completely different reactivity [2, 3]. The first-row transition metals, such as "ubiquitous" iron in heme and metalloenzymes, are included in this class of complexes and play crucial roles for their functions. These biological systems utilize the ISC very well to perform various chemical reactions under mild condition. Materials and catalysts consisting of ubiquitous metals or even metal-free ones have received much attentions in recent chemistry, since they are less expensive, abundant, and environmentally friendly. From these backgrounds, it is of considerable importance and also of interest to understand and control the spin state and to design a new chemical reaction involving the ISC.

A transition between different spin states is induced by spin-orbit coupling (SOC) interaction between $S$ and $S \pm 1$ spin states. The transition probability depends on the SOC value and the energy gap between the two spin states. Since the important part of the SOC interaction depends on nuclear charge Z due to the $Z/r^3$ term [4], large SOC values are observed in the heavier element. As we just mentioned above, because biological systems rarely involve heavy elements such as lanthanides, the mechanism of the spin-state changing reactions in biological systems must depend not only on the strength of the SOC, but also reaction coordinates which can efficiently reduce the energy gap between the different spin states.

Considering multi-dimensional reaction coordinates as a general case, the ISC occurs on a seam along the intersection between two potential energy surfaces (PESs) of different spin states. If the two PESs have energy gradient with opposite sign, the seam becomes the barrier that separates reactant and product states (see Fig. 13.1). A transition state of the ISC is defined as the minimum-energy point on this seam. Here, we call this point as minimum-energy ISC point (MEISCP). As we usually discuss mechanism of chemical reactions in terms of geometries, energies, and electronic structures of transition states, the mechanism of chemical reactions via ISC should be discussed using the energy profile including the MEISCP.

Another important feature in the spin-state crossing reactions is that the reactivity of the spin-states is sometimes completely different. If there are two reaction pathways, the reaction should proceed along the pathway with less activation energy. This reactivity concept was proposed more than two decades ago [2, 5] and

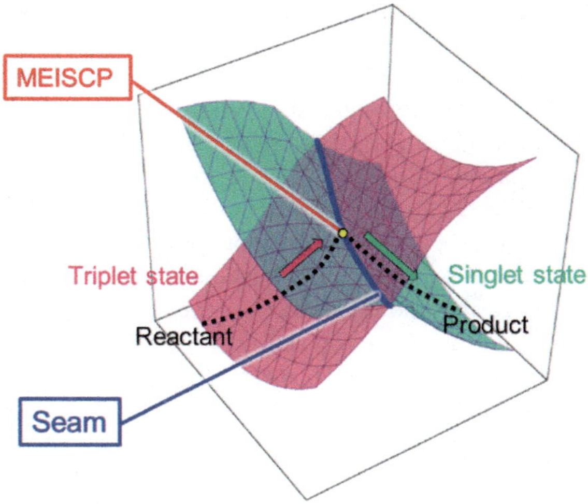


**Fig. 13.1** Schematic representation of minimum-energy intersystem-crossing point (MEISCP) in a spin-state changing reaction

called as "two-state reactivity" [2]. One of the examples is the ligand binding to transition metal complexes [6]. Slow $H_2$ binding and fast CO binding to a tungsten complex [7] was interpreted as "spin-block" or "spin-change barrier" phenomenon. Because the potential energy surface for the $H_2$ binding is repulsive in triplet state and attractive in singlet state, the reaction pathway has a maximum at the spin-state crossing point. If SOC is taken into account, the seam of crossing generates a ridge in the ground-state potential energy surface, and the lowest energy point of the seam corresponds to a transition state of a spin-state crossing reaction. This barrier has become a subject of structural optimization [8, 9]. Our effort is also moving toward this direction [10–14].

In this review, we briefly review our theoretical investigations on the reactions involving the ISC ranging from organometallics to biological systems. In the first two subsections, we discuss about the metal-free quenching processes from the triplet excited state to the singlet ground state; a significantly ruffled porphyrin [11] and a carotenoid spheroidene molecules [12]. In the latter two subsections, we discuss about the chemical reactions of transition metal complexes involving the ISC; $O_2$ binding process to a model heme compound [13], and ligand binding reactions to group 6 metallocenes [14].

## 13.2 Non-radiative Rapid Decay Mechanism of a Significantly Ruffled Porphyrin

Porphyrin and its derivatives are widely found in biological systems, such as photosystem, globin family, cytochrome P450. One of the porphyrins with a significantly non-planer structure exhibits unusual optical properties and plays an important role as a good quencher to dissipate excessive electronic energy into thermal bath [15–17]. Synthetic non-planer porphyrins are also reported to investigate their quenching mechanism as a model of biological function [18–22].

A significantly ruffled porphyrin, meso-tetra-*tert*-butylporphyrin (**$H_2T$(*t*-Bu)P**), is one of the synthetic non-planer porphyrins. Of considerable interest is that its non-radiative decay process via ISC is 200 times faster than that of free-base porphyrin (**$H_2P$**) [19]. Since both contain no heavy element, ISC depends on the potential energy and geometry of the MEISCP. It was shown that the out-of-plane vibration of porphyrin ring induces the ISC in case of **$H_2P$** [23, 24]. These results suggested that the rapid decay process of **$H_2T$(*t*-Bu)P** originates from large distortions on the porphyrin ring induced by sterically crowded substituents.

In our previous study, the relaxation processes of **$H_2T$(*t*-Bu)P** from the first singlet excited state ($S_1$) to the singlet ground state ($S_0$) via the triplet excited states ($T_1$ and $T_2$) were theoretically studied and compared with that of **$H_2P$** [11]. Particularly, a rapid decay mechanism of **$H_2T$(*t*-Bu)P** was discussed in relevance to $S_1 \rightarrow T_2$ crossing and $T_1 \rightarrow S_0$ crossing which are promoted by the distortion of porphyrin ring.

### 13.2.1 Geometrical Difference Amongst $S_0$, $S_1$, and $T_1$ States

Comparison between equilibrium geometries of the $S_0$, $S_1$, and $T_1$ states showed important structural changes in the course of relaxation pathway. First, we compared the optimized structures of **$H_2P$** and **$H_2T$(*t*-Bu)P** in the $S_0$, $S_1$, and $T_1$ at the B3LYP/6-31G(d, p) level of theory.

The $S_0$ state of **$H_2T$(*t*-Bu)P** exhibits significantly ruffled geometry to avoid steric repulsions between the bulky *tert*-butyl groups, while that of **$H_2P$** is in planer geometry (Fig. 13.2). In case of more crowded substituents, a ruffled porphyrin might have a saddle-shaped geometry [20]. The result for **$H_2T$(*t*-Bu)P** showed that the saddle-shaped geometry was a transition state which was connected to the stable ruffled conformations. The bond lengths of **$H_2P$** and **$H_2T$(*t*-Bu)P** in $S_0$, $S_1$, and $T_1$ states were compared in Fig. 13.3a, b, respectively. The optimized structure of the $S_1$ state is almost the same as that of $S_0$, though that of $T_1$ state showed a remarkable difference for either **$H_2P$** or **$H_2T$(*t*-Bu)P**.

In **$H_2T$(*t*-Bu)P**, the largest difference between the singlet and triplet states arises from C5–C6 bond and its symmetry counterpart, C15–C16 bond. Particularly in the $T_1$ state, these bonds are elongated to 1.48 Å, which is a standard C–C single bond length rather than those of double bond or conjugated bond lengths. As well as the bond lengths, dihedral angles around these bonds showed similar trend; the C4–C5–C6–N4 angles were optimized to be 18.5°, 25.5°, and 56.6° for the $S_0$, $S_1$, and $T_1$ states, respectively, (Fig. 13.3c). Considerable distortion in the $T_1$ state was clearly illustrated.

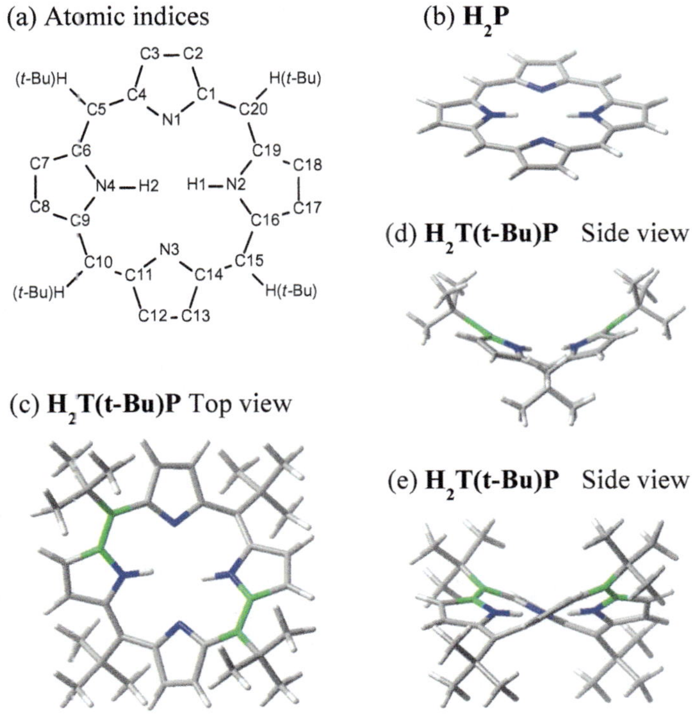


**Fig. 13.2** **a** Chemical structures of $\mathbf{H_2P}$ and $\mathbf{H_2T}$**(*t*-Bu)P**. **b** and **c** Ground-state optimized structures of $\mathbf{H_2P}$ and **H2T(t-Bu)P**, respectively. The *green area* denotes the dihedral angles C4–C5–C6–N4 and C14–C15–C16–N2 (see **a** for atomic indices). **d** and **e** The side view of **c** along C10–C20 axis and N1–N3 axis, respectively. Reprinted (adapted) with permission from (F.-Q. Bai, N. Nakatani, A. Nakayama, and J. Hasegawa, *J. Phys. Chem. A* **2014**, *118*, 4184). Copyright (2014) American Chemical Society

## *13.2.2 Excitation Energies and the Effect of Distortion*

The ring distortion effect on the photo-absorption, fluorescence, and phosphorescence energies of $\mathbf{H_2P}$ and $\mathbf{H_2T}$**(*t*-Bu)P** were investigated. We adopted DFT (B3LYP), time-dependent DFT (TD-B3LYP), and unrestricted DFT (UB3LYP) levels of theories for the $S_0$, $S_1$, and $T_1$ states, respectively. The DFT data were checked by the spin-component scaled (SCS)-second-order approximated Coupled-Cluster model with the resolution of identity approximation (RICC2) method [25].

Our results were reasonably in good agreement with the experimental data for the photo-absorption ($S_0 \rightarrow S_1$), fluorescence ($S_1 \rightarrow S_0$), phosphorescence ($T_1 \rightarrow S_0$), and adiabatic excitation energies of $S_1$ and $T_1$. To clarify the origin of a large bathochromic shifts in $\mathbf{H_2T}$**(*t*-Bu)P**, (0.21 eV from $\mathbf{H_2P}$) we also calculated the excitation energies of $\mathbf{H_2P}$ with the ruffled geometry that was taken from $\mathbf{H_2T}$**(*t*-Bu)P**;

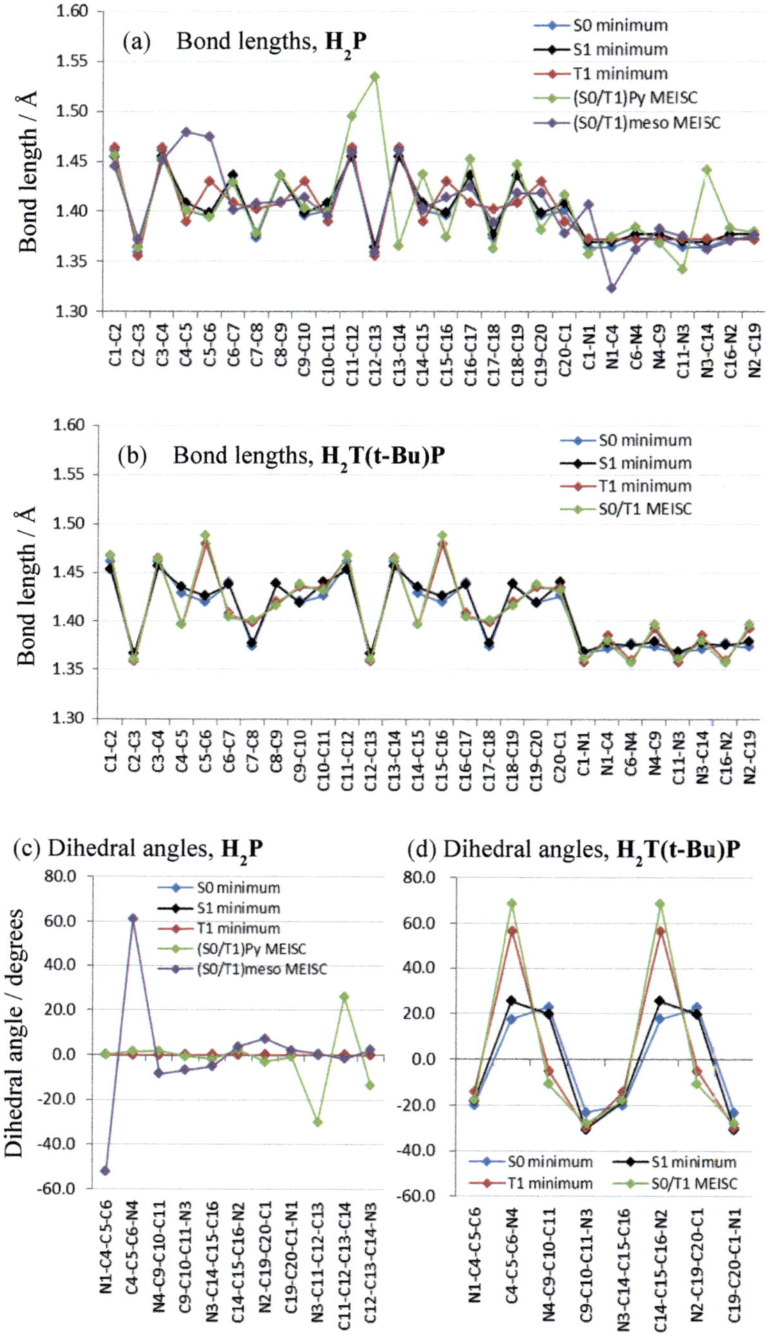


**Fig. 13.3** Optimized structural parameters for **$H_2P$** and **$H_2T(t\text{-}Bu)P$**. Bond angles of **a** **$H_2P$** and **b** **$H_2T(t\text{-}Bu)P$**. Dihedral angles of **c** **$H_2P$** and **d** **$H_2T(t\text{-}Bu)P$**. The $S_0$ and $T_1$ states were optimized at the B3LYP SCF level. For atom indices, see Fig. 2. Reprinted (adapted) with permission from (F.-Q. Bai, N. Nakatani, A. Nakayama, and J. Hasegawa, *J. Phys. Chem. A* **2014**, *118*, 4184). Copyright (2014) American Chemical Society

keeping the geometry of **$H_2$T(*t*-Bu)P**, each of *tert*-butyl group in **$H_2$T(*t*-Bu)P** was replaced by a hydrogen atom. Interestingly, the calculated photo-absorption energies of the ruffled **$H_2$P** (1.96 eV) were very close to those of **$H_2$T(*t*-Bu)P** (1.89 eV). The same results were obtained for the fluorescence, phosphorescence, and adiabatic excitation energies. Therefore, it is clearly concluded that the large bathochromic shift in **$H_2$T(*t*-Bu)P** mainly comes from the distortion of the porphyrin ring. This result also agreed with the conclusion from Parusel and co-workers that ruffled or saddled geometry induces bathochromic shift [26]. In this point, our conclusion is different from that led by DiMagno and co-workers, in which the bathochromic shift was ascribed mainly to electronic effect of substituent [27].

As summarizing these results, balance between the ruffling effect and the electronic effect highly depends on system, and therefore, it is a substantial factor for designing a good quencher. Besides, the bathochromic shift in the phosphorescence energy is equivalent to the decrease of the singlet-triplet energy gap. As we discuss in the following subsection, this is a key to explore the MEISCP and a non-radiative decay process via ISC.

### *13.2.3 Non-adiabatic Decay Mechanism*

According to Kasha's rule, any excited states decay to the lowest state in their spin multiplicity. Therefore, it is worth investigating the decay processes from $S_1$ to $T_1$ and from $T_1$ to $S_0$ both in **$H_2$P** and **$H_2$T(*t*-Bu)P**.

Because the dihedral angles around the C5–C6 and C15–C16 bonds are the most important structural changes between the $S_1$ and the $T_1$ geometry, they could be responsible for a reaction coordinate of the decay process from the $S_1$ state to the $T_1$ state. Potential energy curves along the dihedral angle C4–C5–C6–N4 and its counterpart were performed by SCS-RICC2 single-point calculations on the structures optimized by B3LYP (for $S_0$) and TD-B3LYP (for excited states) levels of theories (Fig. 13.4a). It indicated that the $S_1$ state lies very close to the $T_2$ state in a wide region of the dihedral angle. Although SOC between $S_1$ and $T_2$ is small as shown in Fig. 13.4b, the ISC could easily occur because the energy levels of these two states are near-degenerated. A non-radiative decay from $S_1$ state to $T_2$ state could occur rapidly according to the Landau–Zener formula for the non-adiabatic transition; a transition is promoted if the relative energy difference is small, and the slope of the two potential energy surfaces is similar to each other [28].

Now, the remaining issue is another ISC between $S_0$ and $T_1$ states for a relaxation process to the ground state. In case of **$H_2$P**, there are two different MEISCPs between $S_0$ and $T_1$ obtained from our computations. The first one associated with the C-H flipping motion at meso-position ($S_0/T_1^{meso}$) was located 2.52 and 1.15 eV higher in energy from the energy minimum geometry of the $S_0$ state ($S_0^{min}$) and the $T_1$ state ($T_1^{min}$), respectively, (Fig. 13.5a). The second one associated with the $C_\beta$-$C_\beta$ bond stretching of pyrrole ring ($S_0/T_1^{Py}$) was located 3.54 and 2.17 eV higher in

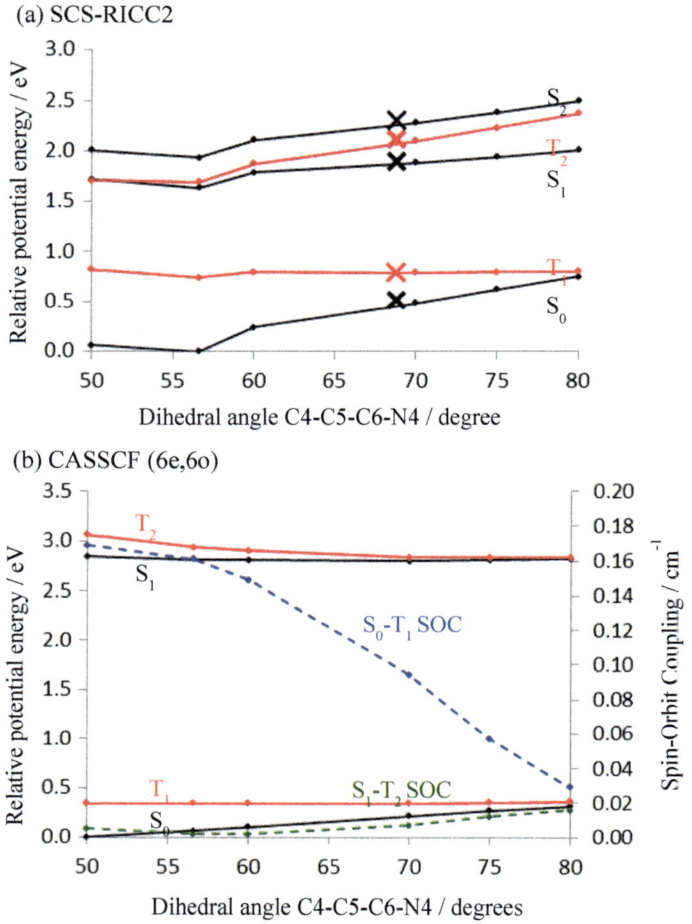


**Fig. 13.4** Potential energy curves of the singlet and triplet states of **$H_2$T(*t*-Bu)P** along the C4-C5-C6-N4 angles. Calculations were performed at the **a** SCC-RICC2/def-SV(P) and **b** CASSCF/ANO-RCC-VDZP levels. The structure was optimized for the $T_1$ state at each angle. Cross marks denote potential energies at the $S_0/T_1$ structure that were optimized at the B3LYP/6-31 + G** level. Spin-orbit coupling between the $S_0$ and $T_1$ states and between the $S_1$ and $T_2$ states was obtained with the CASSI method. Reprinted (adapted) with permission from (F.-Q. Bai, N. Nakatani, A. Nakayama, and J. Hasegawa, *J. Phys. Chem. A* **2014**, *118*, 4184). Copyright (2014) American Chemical Society

energy from the $S_0^{min}$ and the $T_1^{min}$, respectively, (Fig. 13.5a). As a result, the $T_1 \rightarrow S_0$ crossing occurs by climbing an energy barrier of 1.15 eV (26.5 kcal/mol), and this is a main bottle-neck for the non-radiative decay process in **$H_2$P**.

In contrast to **$H_2$P**, the MEISCP of **$H_2$T(*t*-Bu)P** was located very close either in energy or in geometry to the $T_1^{min}$ (Fig. 13.5b). It is only 0.04 eV (0.9 kcal/mol) higher in energy and 13.2° larger in the C4–C5–C6–N4 angle. Of another interest is

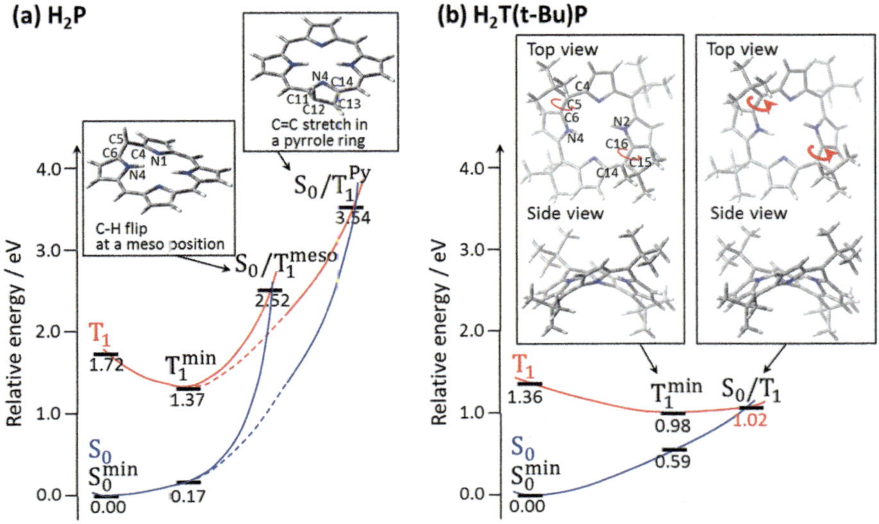


**Fig. 13.5** Potential energy profiles of the singlet and triplet excited states of **a** $\mathbf{H_2P}$ and **b** $\mathbf{H_2T}$**(*t*-Bu)P** at the equilibrium and intersystem-crossing points. Calculations were performed at the B3LYP/6-31 + G** level. Reprinted (adapted) with permission from (F.-Q. Bai, N. Nakatani, A. Nakayama, and J. Hasegawa, *J. Phys. Chem. A* **2014**, *118*, 4184). Copyright (2014) American Chemical Society

that the bond-length alternation of the MEISCP structure is very similar to that of the $T_1^{min}$ structure (Fig. 13.3). This implies that the electronic structure of the MEISCP is also very similar to that of the $T_1^{min}$. Because structural features of the MEISCP resemble with those of the $T_1^{min}$, the $S_0$ state becomes more unstable than the $T_1$ state when it approaches to the MEISCP structure, which results in decrease of the $S_0$–$T_1$ energy gap. Consequently, the MEISCP is easily accessible with almost no energy barrier from the $T_1^{min}$ and, therefore, it is clearly concluded that the $T_1 \rightarrow S_0$ crossing occurs much easier in $\mathbf{H_2T}$**(*t*-Bu)P** than in $\mathbf{H_2P}$, which is the origin of a rapid non-radiative decay process in $\mathbf{H_2T}$**(*t*-Bu)P**.

## 13.3 Energy Dissipative Photo-Protective Mechanism of Carotenoid Spheroidene

Carotenoid is conjugated hydrocarbons with 8 isoprene units and widely found in biological systems. In photosynthetic systems, carotenoid plays a dual role; one is as antenna molecule to collect photons (light-harvesting), and the other is as a quencher to dissipate excessive electronic energy (photo-protection) [29, 30]. In this subsection, we explain our previous study on the triplet energy dissipation

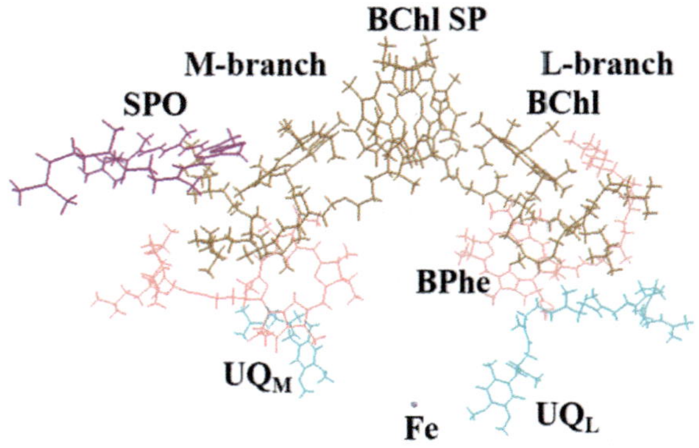


**Fig. 13.6** Reaction center of the *purple* bacteria *Rhodobacter (Rb.) sphaeroides* (PDB id 2J8C). Reproduced from Ref. [12] with permission from the PCCP Owner Societies

mechanism by carotenoids via ISC in a photo-protection mechanism of photosynthetic reaction centers (RCs) [12].

A reverse electron transfer in the reaction center produces a triplet excited state of the bacteriochlorophylls (BChls) special pair (SP), which can generate a harmful singlet oxygen [30–32]. Carotenoid spheroidene (SPO) is just next to a BChl in M-branch of the RC from *Rhodobacter* (*Rb.*) *sphaeroides* (see Fig. 13.6). This compound acts as a photo-protective quencher to protect the RC from a sensitized formation of a singlet oxygen. The SPO in this bacterial RC only takes a 15,15′-*cis* conformation nonetheless there are many possible stereoisomers. A physiological meaning of this natural selection of the 15,15′-*cis*-conformer of SPO has long been discussed both from experimental [33–46] and computational researches [47–56]. It is still not clear if the *cis*-conformer has an advantage as a triplet energy quencher against the *trans*-conformer. It has not yet been totally solved its mechanism of the triplet energy dissipation and the role of the 15,15′-*cis*-conformer.

In the below, our discussion about the triplet energy dissipation mechanism of 15,15′-*cis*-SPO via ISC is described, which provides an another example of $T_1$–$S_0$ crossing phenomena in bioorganic molecule; non-radiative decay from the $T_1$ state to the $S_0$ ground state occurs at the middle of the *cis* to all-*trans* isomerization of the C15–C15′ rotation. Similar to the ruffled porphyrin case, the bond-length alternation of conjugated carbon chain becomes an important reaction coordinate toward the MEISCP as well as the C15–C15′ rotation.

### 13.3.1 Stationary Structures of $S_0$ and $T_1$ States

We started with determination of the stationary structures of $S_0$ and $T_1$ states and to compare with each other. On the basis of the B3LYP/6-311G(d) calculations, both 15,15′-*cis*-SPO and all-*trans*-SPO take planar geometry (Fig. 13.7). According to the triplet Raman spectroscopic studies, it was suggested that the 15,15′-*cis*-SPO takes a twisted structure in the RC of *Rb. sphaeroides* [38, 40]. With these

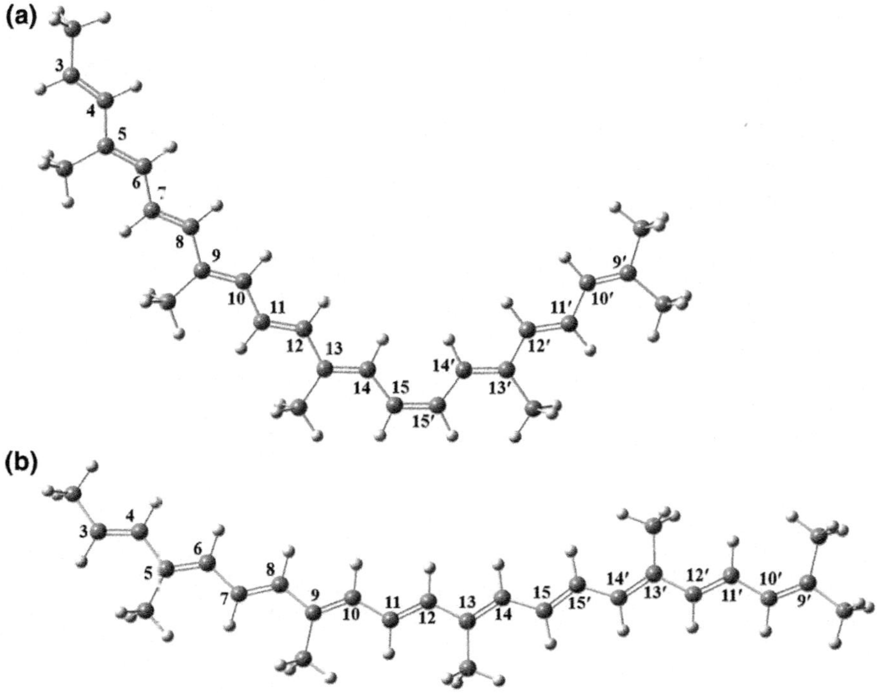


**Fig. 13.7** Structure (optimized) of the truncated **a** 15,15′-*cis*-SPO **b** *trans*-SPO. Reproduced from Ref. [12] with permission from the PCCP Owner Societies

experimental facts, it is highly expected that the RC environment in SPO-binding pocket controls the geometry of the SPO.

A main geometrical difference between the optimized structures of $S_0$ and $T_1$ states arises from the bond alternation in conjugated carbon chain (Fig. 13.8), particularly in the range from C9 to C15′. Although the C15–C15′ bond possesses a C=C double bond character in the $S_0$ ground state, it is changed to a C–C single bond character upon triplet excitation. Similar trends are also found in the C13–C14 and C11–C12 bonds. As described later, the $T_1$ minimum is shallower than the $S_0$ one. Thus, twisting of these bonds more easily occurs in the $T_1$ state, which agreed with the experimental facts from the triplet Raman spectra [38].

In the $S_0$ ground state, the *trans*-conformer was computed to be 2.5 kcal/mol more stable in energy than the *cis*-conformer. Almost the same situation was found in the $T_1$ excited state; the *trans*-conformer was computed to be 3.4 kcal/mol more stable than the *cis*-conformer. Thus, the natural selection of the *cis*-conformer is not originated from stability of the molecule itself, but from the interaction with the protein environment.

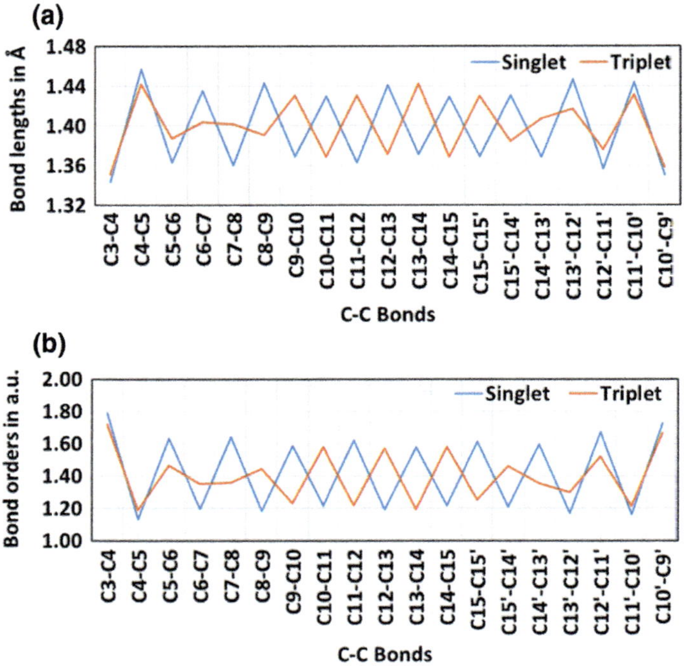


**Fig. 13.8** Conjugated **a** bond lengths and **b** bond orders in ground and excited triplet states obtained at B3LYP/6-311G(d) level. Reproduced from Ref. [12] with permission from the PCCP Owner Societies

### 13.3.2 C15–C15 Bond Rotation and the MEISCP Geometry

Non-radiative decay process of carotenoids either from the $S_1$ state via conical intersection (CI) or from the $T_1$ state via ISC is basically induced by the C15–C15′ twisting from *cis*–conformation toward all-*trans* one. Thus, the rotation around the C15–C15′ bond was adopted for the main reaction coordinate in the triplet energy dissipation. Figure 13.9 shows potential energy curves (PECs) of the $S_0$ and the $T_1$ states along the torsional angle $\phi$(C14–C15–C15′–C14′), in which the other structural parameters were optimized in the $T_1$ state. The crossing occurs at $\phi = 77°$, which is before reaching the transition state for the *cis* to all-*trans* isomerization. The PEC in the $T_1$ state is very flat compared with that in the $S_0$ state, and the crossing point was computed to be 7.8 kcal/mol (32.5 kJ/mol) higher in energy than the $T_1$ minimum on the 15,15′-*cis* geometry. This is because the nature of the C15–C15′ bond is changed to more like a C–C single bond upon the $T_1$ excitation.

Interestingly, the MEISCP is found at the smaller torsional angle $\phi = 58.4°$, indicating that another structural parameter contributes to reduce the $S_0$–$T_1$ energy gap in addition to the C15–C15′ bond rotation. The MEISCP was computed to be

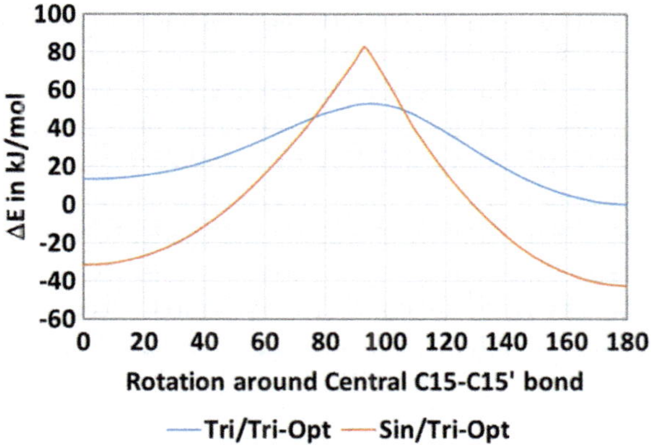


**Fig. 13.9** Potential curves for singlet and triplet states of the 15,15′-*cis*-SPO. Tri/Tri-Opt and Sin/Tri-Opt represent the triplet and singlet energies obtained at the triplet-state optimized structure. Reproduced from Ref. [12] with permission from the PCCP Owner Societies

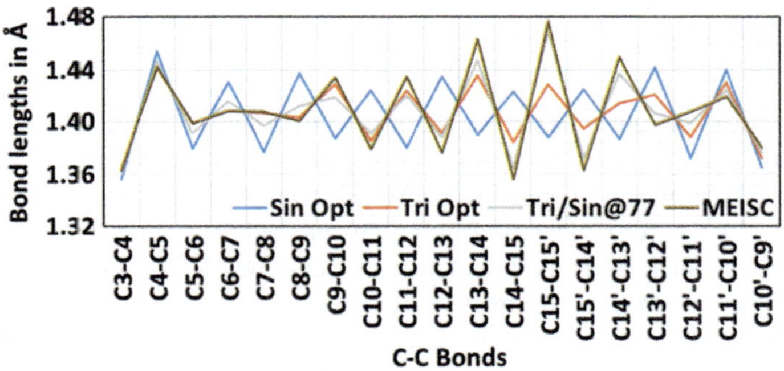


**Fig. 13.10** Conjugated C–C bond lengths in singlet, triplet, Singlet-triplet crossing ($\phi_{S\text{-}T} = 77^{\circ}$), and MEISC ($\phi = 58.4^{\circ}$) BP86/6-311G(d) optimized structures. Reproduced from Ref. [12] with permission from the PCCP Owner Societies

6.3 kcal/mol (26.5 kJ/mol) higher in energy than the $T_1$ minimum on the 15,15-*cis* geometry, and consequently, 1.5 kcal/mol (6.0 kJ/mol) lower in energy than the crossing point at $\phi = 77°$. Of further interest is that the bond-length alternation is considerably different in the MEISCP (Fig. 13.10). The magnitude of the alternation in the MEISCP is much stronger than that of the $T_1$ minimum structure, and thus, more apart from that of the $S_0$ minimum structure. As a result, the stronger bond alternation at the MEISCP geometry destabilizes the $S_0$ state more than the $T_1$ state, which helps to reach the crossing point by reducing the $S_0$–$T_1$ energy gap.

### *13.3.3 Energy Dissipative Photo-Protection Mechanism*

Koyama and co-workers suggested that the 15,15′-*cis*-conformer has an advantage in the triplet energy dissipation against the all-*trans*-conformer [34–37]. Their speculated mechanism was supported from our computational results. The all-*trans*-conformer is more stable in energy than the 15,15′-*cis*-conformer either in the $S_0$ state or in the $T_1$ state. This implies that the associated energy barrier of *cis*-to-*trans* isomerization is smaller from the 15,15′-*cis*-conformer than that from the all-*trans*-conformer, and thus, the *cis*-to-*trans* isomerization occurs more easily than the *trans*-to-*cis* isomerization. The MEISCP is located where the C15–C15′ twisting angle is around 60°, indicating that the associated energy barrier for the $T_1 \rightarrow S_0$ crossing is rather small than that for the *cis*-to-*trans* isomerization. Thus, the triplet energy dissipation can occur in the 15,15′-*cis* region. Because the *trans*-to-*cis* isomerization is more difficult in the $S_0$ state than in the $T_1$ state, it is unfavorable if the triplet energy dissipation process results in the all-*trans*-conformer in the $S_0$ state. Therefore, one of the reasons for the natural selection of the *cis*-conformer would be in the efficient triplet energy dissipation in the 15,15′-*cis* region.

## 13.4 Dioxygen Binding Pathway in a Model Heme Compound

Dioxygen ($O_2$) activation is one of the most important bioinorganic reactions for utilizing its high oxidation potential, because the ground state is inert triplet spin state ($S = 1$). Heme is a general name of iron–porphyrin compounds and is responsible for the dioxygen transportation and activation processes in various metalloproteins such as hemoglobin and cytochrome P450. An $O_2$-free form of heme (deoxyheme) takes the quintet ground state ($S = 2$). The spin state changes to the singlet spin state ($S = 0$) upon the $O_2$ binding [57–59]. Thus, the $O_2$ binding process involves an ISC.

Based on character of the potential energy surface (PES) for the $O_2$ binding process, two different mechanisms on the ISC were previously proposed. The first one is the *so-called* broad-crossing mechanism [28, 60, 61], in which potential energies for the singlet, triplet, and quintet spin states of the oxyheme are nearly degenerated in a wide range of Fe–$O_2$ distance (more than 3 Å) along the reaction coordinate. According to the Landau–Zener theory, this feature of the PES enhances the ISC. However, this degeneracy at the dissociation limit contradicts to the relative stability of the quintet spin state in the deoxyheme.

The second one is that a transition state of the $O_2$ binding process is of the ISC between the singlet and triplet spin states ($T_1/S_0^{ISC}$). Novoa and co-workers theoretically investigated the $O_2$ binding to an iron–porphyrin–imidazole system at the CASPT2 level of theory [62]. They reported that a van der Waals minimum exists around the Fe–$O_2$ distance of 2.6 Å in the triplet PES. More important point in their report was that the $T_1/S_0^{ISC}$ takes place around the Fe–$O_2$ distance of 1.9 Å. Our previous work by the B3LYP calculations also showed that the transition state of the $O_2$ binding process includes the $T_1/S_0^{ISC}$ crossing [63].

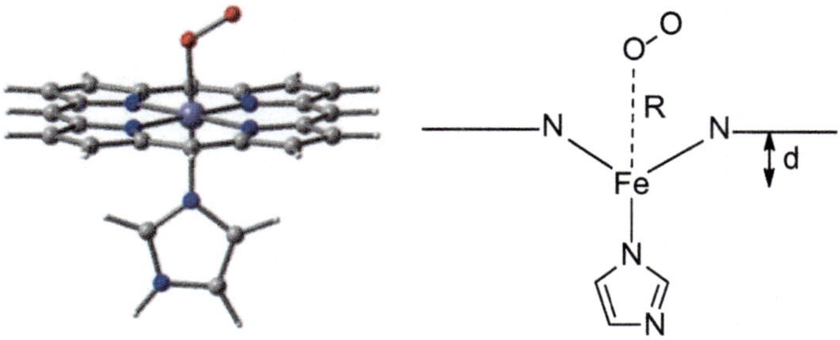


**Fig. 13.11** Computational model of oxyheme, the Fe($O_2$)PorIm complex. "R" and "d" denote the Fe–O distance and out-of-displacement of Fe atom. The latter was defined as distance between Fe and the mass center of four N atoms. Reproduced from Ref. [13] with permission from the PCCP Owner Societies

In this section, the result of our study on the $O_2$ binding reaction to a model heme, Fe(II)–Porphyrin–Imidazole [FePorIm], was reviewed. (Figure 13.11)

### *13.4.1 Dioxygen Binding Pathway via ISCP and MEISCP*

Figure 13.12 shows the PESs for the $O_2$ binding to a model heme compound in the singlet and triplet spin states at the B97D level of theory. In the long Fe–$O_2$ distance region (>2.5 Å), the ground state was computed as a triplet spin state ($T_1$) and its energy minimum was located at the Fe–$O_2$ distance of 2.92 Å. As the Fe–$O_2$ distance further decreases, the energy gap between the singlet and triplet spin states rapidly decreases to reach the ISC point at the Fe–$O_2$ distance of 1.98 Å along with the $T_1$ geometry (Fig. 13.12a). The ISC point was obtained 2.3 kcal/mol higher in energy than the $T_1$ minimum. An energy minimum of the singlet spin state ($S_0$) was located at the Fe–$O_2$ distance of 1.91 Å, which was 2.1 kcal/mol lower in energy than the $T_1$ minimum. Interestingly, the Fe–$O_2$ distance and the out-of-displacement of Fe of the ISC point and the $S_0$ minimum were very similar to each other, despite the relatively large energy difference (4.4 kcal/mol).

On the other hand, the MEISCP was computed only 1.0 kcal/mol higher in energy than the $T_1$ minimum (Fig. 13.12b) and was 1.3 kcal/mol lower in energy than the ISC point. (Figure 13.12a). It was located at the Fe–$O_2$ distance of 2.17 Å, which was remarkably longer than that of the $S_0$ minimum. From these results, the MEISCP is stabilized by geometrical parameters other than the Fe–$O_2$ distance and the out-of-displacement of Fe.

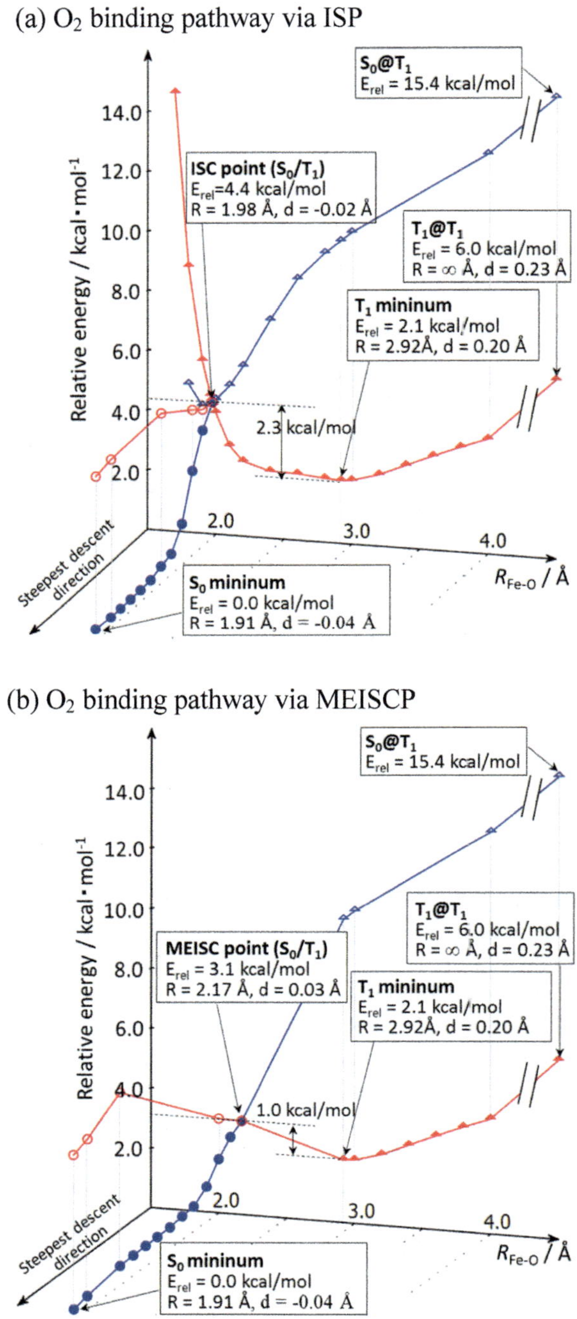


**Fig. 13.12** Two-dimensional potential energy curves for $O_2$ binding calculated at the B97D level. **a** Intersystem-crossing pathway along the Fe–O fixed coordinate and **b** that via the MEISCP. Triangle denotes that the calculations were performed with fixed Fe–O distances. *Filled triangles* denote that the structure was optimized with this triplet state, while *open triangles* denote single-point energy. *Filled circles* denote that the steepest-decent reaction pathway on the $S_0$ surface starting with the ISC or MEISC point, while *open circles* denote single-point calculations. Reproduced from Ref. [13] with permission from the PCCP Owner Societies

### 13.4.2 Reaction Coordinates to the MEISCP

In the dissociation limit (deoxyheme), the $T_1$ state was obtained as the ground state of [FePorIm], and the $S_0$ state was computed to be 9 kcal/mol higher in energy at the $T_1$ geometry. The $S_0$–$T_1$ energy gap slightly decreased to 8 kcal/mol at the $T_1$ minimum structure of [Fe($O_2$)PorIm]. This implies that the Fe–$O_2$ distance does not affect very much to reduce the $S_0$–$T_1$ energy gap, although it stabilizes either the $S_0$ state or the $T_1$ state.

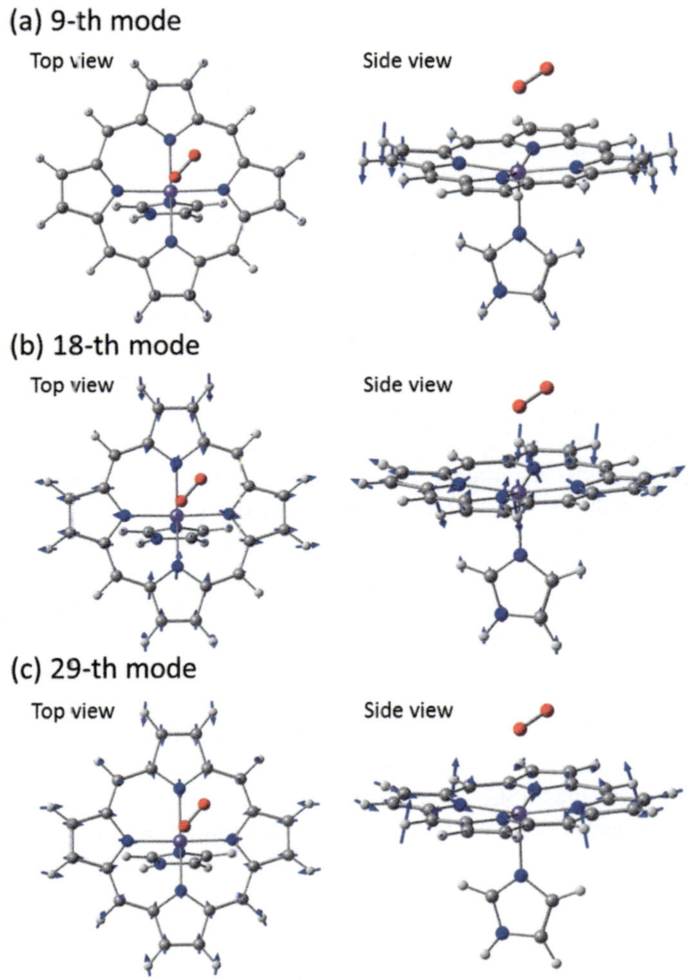


**Fig. 13.13** Ninth, eighteenth, and twenty-ninth normal mode at the $T_1$ minimum. Reproduced from Ref. [13] with permission from the PCCP Owner Societies

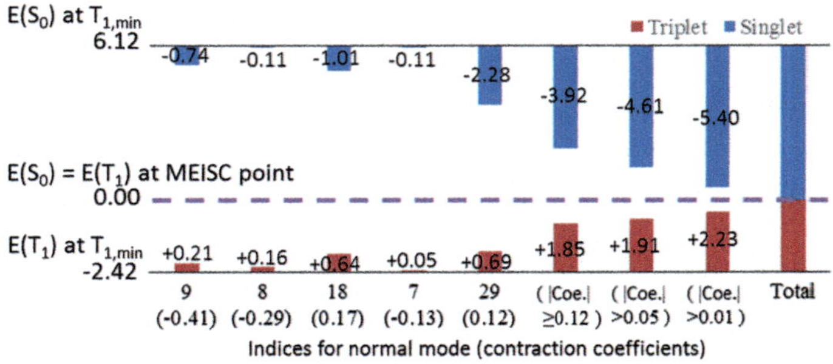


**Fig. 13.14** Energy changes of the $S_0$ and $T_1$ states introduced by structural change from the $T_1$ minimum to the MEISCP. Unit is in kcal/mol. The structural change was expressed by a linear combination of normal mode vector at the $T_1$ minimum. The *upper* and *lower lines* denote the energy levels of the triplet and singlet states at the $T_1$ minimum, respectively. Energy changes of $T_1$ and $S_0$ states caused by the structural change were represented with *red* and *blue bars*, respectively. Calculations were carried out with the B3LYP functional. Reproduced from Ref. [13] with permission from the PCCP Owner Societies

Since the $T_1$ minimum structure is initially formed during the $O_2$ binding reaction to [FePorIm], here we discuss reaction coordinates reaching to the MEISCP in terms of vibrations at the $T_1$ minimum structure. Figure 13.13 shows energy changes from $T_1$ minimum to the MEISCP in which the reaction coordinate is expanded by the normal modes at the $T_1$ minimum structure. The largest component in the reaction coordinate is of the 9th normal mode vector which is corresponding to the out-of-displacement of Fe atom (Fig. 13.14). Though the 9th mode gives the most significant change in structure (contraction coefficient is −0.41), it only reduces 1.0 kcal/mol of the $S_0$–$T_1$ energy gap.

Of considerable interest is that the 29th mode gives the largest contribution to reduce the $S_0$–$T_1$ energy gap by 3.0 kcal/mol even though its structural change is small (contraction coefficient is 0.12). The 29th mode is of symmetric shrinking of the tetrapyrrole units. This is qualitatively understood as changes in the Fe–N distances; the shorter Fe–N distance strengthens the ligand-field to stabilize the $S_0$ state.

From these results, it is clearly concluded that symmetric shrinking mode is the most important reaction coordinate to reduce the $S_0$–$T_1$ energy gap, and thus, it mainly controls the ISC. Even though the Fe–$O_2$ distance and the out-of-displacement of Fe are also very important changes in geometry during the $O_2$ binding, they are less effective to reduce the $S_0$–$T_1$ energy gap.

## 13.5 Spin-Blocking Effect in CO and $H_2$ Binding Reactions to Group 6 Metallocenes

Chemical reactions via ISC become more and more important in recent inorganic and organometallic chemistry. Particularly from a viewpoint of rare-metal alternatives, catalysts containing paramagnetic metals, such as iron and cobalt, have received attractive attentions. Therefore, it is of considerable interest and importance to explore the reaction mechanism via ISC and extract its controlling parameters for catalyst design.

Group 6 metallocene is an unsaturated 16e metal complex, which takes triplet spin state in the ground state. Because it only exists as a highly reactive intermediate, it easily reacts with various ligands to afford a stable 18e adduct in the singlet spin state. Therefore, the reaction occurs from the triplet spin state to the singlet spin state via ISC. Of particular interest is that the $H_2$ binding reaction to molybdenocene needs very high $H_2$ partial pressure (200 atm) in contract to the CO binding reaction (1 atm) [64, 65], although both $H_2$ and CO molecules easily react with most of 4d and 5d transition metal complexes. Therefore, it is considered that the difference in the reactivity of molybdenocene arises from a property of the ISC during reaction.

Pioneering theoretical works on CpCo complex by Poli and Harvey suggested that the difference in reactivity is due to spin-blocking effect; coupling the rate-determining step (RDS) with the ISC drastically lowers the reactivity in the $H_2$ binding reaction [6]. However, it was still worth further analyzing geometry and electronic interactions in this reaction, because it would give us insights into a general strategy for a catalyst design incorporating ISC. In this section, we show our previous study on the CO and $H_2$ binding reactions to group 6 metallocenes (molybdenocene [$MoCp_2$] and tungstenocene [$WCp_2$]). On the basis of our previous computation on the MEISCP, we explain the geometry and electronic structure at this particular structure [14]. We also demonstrate how to extract an important reaction coordinate which controls reactivity of spin-crossing reactions.

### *13.5.1 The CO Binding Reaction to [$MoCp_2$] and [$WCp_2$]*

Either [$MoCp_2$] or [$WCp_2$] takes triplet spin state as the ground state. The CO binding reaction first occurs in the triplet spin state with almost no activation barrier to afford a stable bound intermediate followed by the ISC to form a singlet adduct (Fig. 13.15). The binding energies in the triplet spin state were estimated to be 12.8 and 16.4 kcal/mol for [$MoCp_2CO$] ($\mathbf{T_{CO}}$) and [$WCp_2CO$] ($\mathbf{T'_{CO}}$), respectively, according to our density functional calculations using M06L functional. Interestingly, the singlet spin state is considerably stabilized upon CO binding, and it was computed to be almost the same energy level or even slightly more stable than the triplet spin state, at $\mathbf{T_{CO}}$ or $\mathbf{T'_{CO}}$ geometry. As a result, the MEISCP was also found

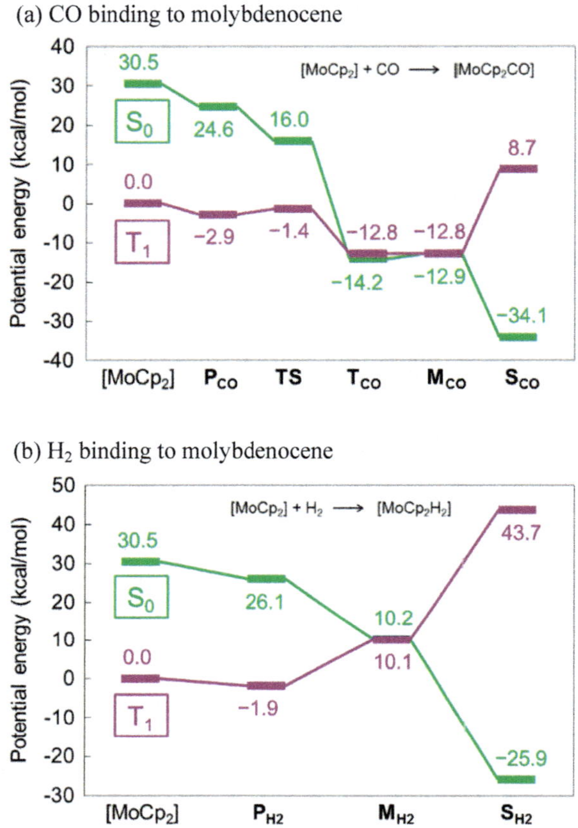


**Fig. 13.15** Potential energy profiles of **a** CO and **b** $H_2$ binding reactions to [$MoCp_2$] in the singlet spin state (*green*) and the triplet spin state (*magenta*). Reprinted (adapted) with permission from (K. Watanabe, N. Nakatani, A. Nakayama, M. Higashi, and J. Hasegawa, *Inorg. Chem.* **2016**, *55*, 8082). Copyright (2016) American Chemical Society

to be very close both in energy and in geometry to the triplet adduct; we denoted the MEISCPs of [$MoCp_2CO$] and [$WCp_2CO$] by $\mathbf{M_{CO}}$ and $\mathbf{M'_{CO}}$, respectively. Therefore, it is highly expected that the ISC easily occurs as soon as the triplet adduct is formed. After ISC occurs, the complex in singlet spin state is stabilized very much to form a CO bound complex. The overall binding energies to the singlet adduct were estimated to be 34.1and 40.8 kcal/mol for [$MoCp_2CO$] ($\mathbf{S_{CO}}$) and [$WCp_2CO$] ($\mathbf{S'_{CO}}$), respectively.

## *13.5.2 The $H_2$ Binding Reaction to [$MoCp_2$] and [$WCp_2$]*

In contract to the CO binding reaction, no bound structure was found in the triplet spin state in the $H_2$ binding reaction to either [$MoCp_2$] or [$WCp_2$]. The triplet-state potential energy monotonically increases, when $H_2$ is approaching to the metal center (Fig. 13.15). As a result, the MEISCP was located to be higher in energy by 10.1 kcal/mol for [$MoCp_2H_2$] ($\mathbf{M_{H2}}$) or by 10.6 kcal/mol for [$WCp_2H_2$] ($\mathbf{M'_{H2}}$).

After the ISC, $H_2$ is split into two hydrides with concomitant oxidation of the metal center (known as oxidative addition reaction), to result in a dihydride complex in the singlet spin state. The overall binding energies to the singlet adduct were estimated to be 25.9 and 38.3 kcal/mol for [MoCp$_2$(H)$_2$] (**S$_{H2}$**) and [WCp$_2$(H)$_2$] (**S'$_{H2}$**). Consequently, the MEISCP is to be a transition state of the $H_2$ binding reaction, from which the spin-blocking effect considerably slows the rate of the $H_2$ binding reaction.

### *13.5.3 Geometrical Parameters Controlling the MEISCP*

For a catalyst design, it is necessary to extract parameters which control reactivity and selectivity such as electronic effect and geometrical constraint from substituents. Because the MEISCP determines the reactivity in the $H_2$ binding reaction, geometry and electronic structure of the MEISCP are worth investigating. In this subsection, we discuss on the MEISCP geometry in the $H_2$ binding reaction to explore important structural parameters to reduce the $S_0$–$T_1$ energy gap.

The minimum-energy path on the $H_2$ binding reaction to [MoCp$_2$] via the MEISCP obtained by a steepest-decent reaction coordinate calculations is shown in Fig. 13.16. It is clear that the Mo-H distance and the Cp-Mo-Cp bending change simultaneously toward the MEISCP. Therefore, not only the Mo-H distance but also the Cp-Mo-Cp angle can play an important role to reach the MEISCP. Hence, change in the $S_0$–$T_1$ energy gap was decomposed into those from each geometrical parameter; (1) opening two Cp rings (2) binding of $H_2$ molecule, and (3) others to reach the MEISCP as shown in Fig. 13.17. Of considerable interest is that the opening two Cp rings reduces 78% of the $S_0$–$T_1$ energy gap from

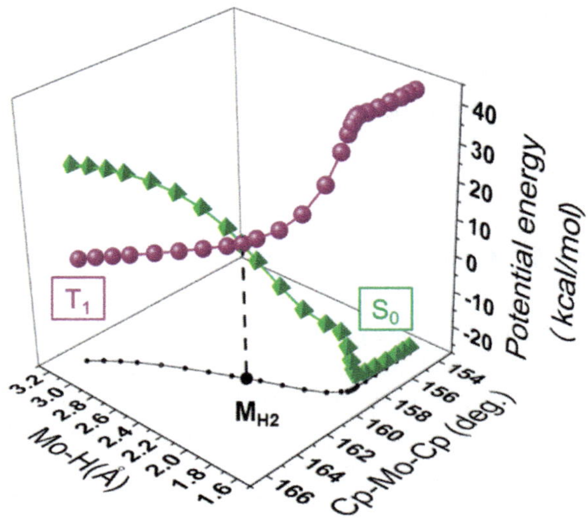


**Fig. 13.16** Minimum-energy paths from [MoCp$_2$] to **M$_{H2}$** in the triplet spin state (*magenta*) and from **M$_{H2}$** to **S$_{H2}$** in the singlet spin state (*green*). Potential energies plotted along the Mo-H distance and Cp-Mo-Cp angle. Reprinted (adapted) with permission from (K. Watanabe, N. Nakatani, A. Nakayama, M. Higashi, and J. Hasegawa, *Inorg. Chem.* **2016**, *55*, 8082). Copyright (2016) American Chemical Society

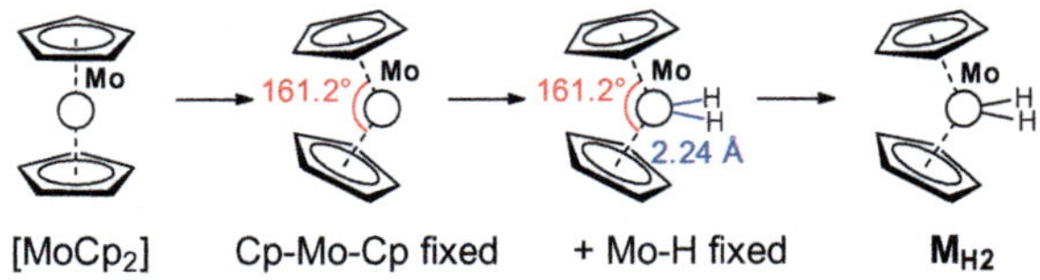


**Fig. 13.17** Reprinted (adapted) with permission from (K. Watanabe, N. Nakatani, A. Nakayama, M. Higashi, and J. Hasegawa, *Inorg. Chem.* **2016**, *55*, 8082). Copyright (2016) American Chemical Society

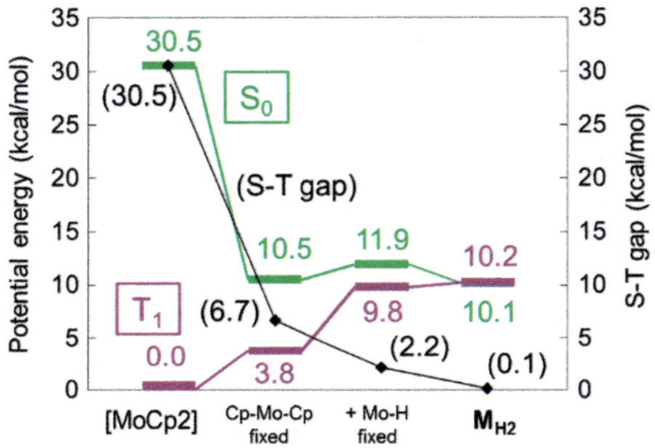


**Fig. 13.18** Energy changes in the triplet (*magenta*) and the singlet (*green*) spin states along with the decomposition scheme as defined in Scheme 1. Change in the S–T gap was also shown as a black line (values in parentheses). Reprinted (adapted) with permission from (K. Watanabe, N. Nakatani, A. Nakayama, M. Higashi, and J. Hasegawa, *Inorg. Chem.* **2016**, *55*, 8082). Copyright (2016) American Chemical Society

[$MoCp_2$], though the decrease of the Mo-H distance reduces only 15% (Fig. 13.18). It is clearly concluded that the Cp-Mo-Cp angle is the most important geometrical parameter to reach the MEISCP. This agrees with previous experimental and theoretical studies on ansa-metallocenes in which two Cp rings are tied to fix its geometry to be the ring-opened form [66–68].

## 13.6 Conclusion

Chemical reactions involving spin-state crossing are of considerable interest and importance both in scientific and industrial researches. In this chapter, we presented a general strategy to theoretically investigate such chemical reactions to explore atomistic mechanisms and to extract key structural parameters for controlling the spin-crossing reactions. Our study is based on finding the minimum-energy intersystem-crossing point (MEISCP) and also on analyzing the reaction coordinate and electronic structure, which provide significant insights into rational design of molecules and reactions.

In the first two sections, we discussed metal-free quenching reactions from triplet excited states to the singlet ground state. A significantly ruffled porphyrin exhibits a non-radiative rapid decay from triplet spin state induced by a large distortion in porphyrin ring [11]. Carotenoid spheroidene plays an important role of energy dissipative photo-protection in photosynthetic reaction centers. Though the quenching from the triplet state of carotenoid molecule is mainly induced by twisting a C=C bond, bond-length alternation of the carotenoid skeleton reduces the activation barrier of the quenching [12].

For the latter two sections, spin-state crossing in organometallic reactions was investigated. Dioxygen binding in heme complex occurs via spin-state crossing controlled not only by out-of-plane displacement of iron center, but also by a symmetric stretching vibration of four pyrroles in the porphyrin ring [13]. Group 6 metallocene is much less active in $H_2$ binding than in CO binding. This phenomenon was ascribed to the spin-blocking effect, which originates from the bonding nature of the adduct in triplet state. The butterfly motion of two cyclopentadienyl groups controls the access from the triplet minimum to the spin-state crossing transition state [14].

In the four cases from the bimolecular excited states to the ligand binding in organometallics, a salient feature which is commonly observed in the spin-state crossing reactions is that a key reaction coordinate, which controls the relative energy between two states, lead the system toward MEISCP, a transition state of spin-crossing reaction. This coordinate is not necessary related to the primary reaction coordinate (such as Fe–O in $O_2$ binding) but to the one sensitive to the relative energy level. This point should be recognized in the catalysis design and reaction control.

**Acknowledgements** This work was financially supported by JST-CREST, JSPS KAKENHI (Grant Number JP15H05805 in Precisely Designed Catalysts with Customized Scaffolding, JP15K06563, JP15K20832, JP15H03770, 26810008, and JP16H00778), and the FLAGSHIP2020 (priority study 5) program from MEXT. A part of the computations was carried out at RCCS (Okazaki, Japan) and ACCMS (Kyoto University).

## References

1. A. Telfer, S. Dhami, S.M. Bishop, D. Phillips, J. Barber, Biochemistry **33**, 14469 (1994)
2. S. Shaik, D. Danovich, A. Fiedler, D. Schröder, H. Schwarz, Helv. Chim. Acta **78**, 1393 (1995)
3. J.N. Harvey, R. Poli, K.M. Smith, Coord. Chem. Rev. **238–239**, 347 (2003)
4. C.M. Marian, WIREs Comput. Mol. Sci. **2**, 187 (2012)
5. R. Poli, Chem. Rev. **96**, 2135 (1996)
6. J.L. Carreon-Macedo, J.N. Harvey, J. Am. Chem. Soc. **126**, 5789 (2004)
7. R.R. Schrock, K.Y. Shih, D.A. Dobbs, W.M. Davis, J. Am. Chem. Soc. **117**, 6609 (1995)
8. N. Koga, K. Morokuma, Chem. Phys. Lett. **119**, 371 (1985)

9. J.N. Harvey, M. Aschi, H. Schwarz, W. Koch, Theor. Chem. Acc. **99**, 95 (1998)
10. A. Nakayama, Y. Harabuchi, S. Yamazaki, T. Taketsugu, Phys. Chem. Chem. Phys. **15**, 12322 (2013)
11. F.-Q. Bai, N. Nakatani, A. Nakayama, J. Hasegawa, J. Phys. Chem. A **118**, 4184 (2014)
12. S. Arulmozhiraja, N. Nakatani, A. Nakayama, J. Hasegawa, Phys. Chem. Chem. Phys. **17**, 23468 (2015)
13. Y. Kitagawa, Y. Chen, N. Nakatani, A. Nakayama, J. Hasegawa, Phys. Chem. Chem. Phys. **18**, 18137 (2006)
14. K. Watanabe, N. Nakatani, A. Nakayama, M. Higashi, J. Hasegawa, Inorg. Chem. **55**, 8082 (2016)
15. J.A. Shelnutt, X.Z. Song, J.G. Ma, S.L. Jia, W. Jentzen, C.J. Medforth, Chem. Soc. Rev. **27**, 31 (1998)
16. E. Sigfridsson, U. Ryde, J. Biol. Inorg. Chem. **8**, 273 (2003)
17. C. Olea, J. Kuriyan, M.A. Marletta, J. Am. Chem. Soc. **132**, 12794 (2010)
18. K.M. Barkigia, M.D. Berber, J. Fajer, C.J. Medforth, M.W. Renner, K.M. Smith, J. Am. Chem. Soc. **112**, 8851 (1990)
19. T. Ema, M.O. Senge, N.Y. Nelson, H. Ogoshi, K.M. Smith, Angew. Chem. Int. Ed. **33**, 1879 (1994)
20. S. Gentemann, C.J. Medforth, T.P. Forsyth, D.J. Nurco, K.M. Smith, J. Fajer, D. Holten, J. Am. Chem. Soc. **116**, 7363 (1994)
21. S. Gentemann, C.J. Medforth, T. Ema, N.Y. Nelson, K.M. Smith, J. Fajer, D. Holten, Chem. Phys. Lett. **245**, 441 (1995)
22. S. Gentemann, N.Y. Nelson, L. Jaquinod, D.J. Nurco, S.H. Leung, C.J. Medforth, K.M. Smith, J. Fajer, D. Holten, J. Phys. Chem. B **107**, 1247 (1997)
23. S. Perun, J. Tatchen, C.M. Marian, Chem. Phys. Chem. **9**, 282 (2008)
24. B. Minaev, H. Agren, Chem. Phys. **315**, 215 (2005)
25. A. Hellweg, S.A. Grun, C. Hattig, Phys. Chem. Chem. Phys. **10**, 4119 (2008)
26. A.B.J. Parusel, T. Wondimagegn, A. Ghosh, J. Am. Chem. Soc. **122**, 6371 (2000)
27. A.K. Wertsching, A.S. Koch, S.G. DiMagno, J. Am. Chem. Soc. **123**, 3932 (2001)
28. K.P. Kepp, Chem. Phys. Chem. **14**, 3551 (2013)
29. R.J. Cogdell, H.A. Frank, Biochim. Biophys. Acta **895**, 63 (1987)
30. H.A. Frank, R.J. Cogdell, in *Carotenoids in Photosynthesis*, ed. by A. Young, G. Britton (Chapman & Hall, London, 1993), p. 252
31. A. deWinter, S.G. Boxer, J. Phys. Chem. B 103, 8786 (1999)
32. H.A. Frank, in *The photosynthetic reaction center*, ed. by J.R. Norris, J. Deisenhofer (Academic Press, New York, 1993), p. 221
33. F. Boucher, G. Gingras, Photochem. Photobiol. **40**, 277 (1984)
34. M. Kuki, M. Naruse, T. Kakuno, Y. Koyama, Photochem. Photobiol. **62**, 502 (1995)
35. H. Hashimoto, Y. Koyama, K. Ichimura, T. Kobayashi, Chem. Phys. Lett. **162**, 517 (1989)
36. H. Hashimoto, Y. Koyama, J. Phys. Chem. **92**, 2101 (1988)
37. M. Kuki, Y. Koyama, H. Nagae, J. Phys. Chem. **95**, 7171 (1991)
38. N. Ohashi, N. Kochi, M. Kuki, T. Shimamura, R.J. Cogdell, Y. Koyama, Biospectroscopy **2**, 59 (1996)
39. J.A. Bautista, V. Chynwat, A. Cua, F.J. Jansen, J. Lugtenburg, D. Gosztola, M.R. Wasielewski, H.A. Frank, Photosynth. Res. **55**, 49 (1998)
40. Y. Mukai-Kuroda, R. Fujii, N. Ko-chi, T. Sashima, Y. Koyama, J. Phys. Chem. A **106**, 3566 (2002)
41. R. Fujii, K. Furuichi, J.P. Zhang, H. Nagae, H. Hashimoto, Y. Koyama, J. Phys. Chem. A **106**, 2410 (2002)
42. Z.D. Pendon, I. der Hoef, J. Lugtenburg, H.A. Frank, Photosynth. Res. **88**, 51 (2006)
43. Z.D. Pendon, G.N. Gibson, I. van der Hoef, J. Lugtenburg, H.A. Frank, J. Phys. Chem. B **109**, 21172 (2005)
44. Z.D. Pendon, J.O. Sullivan, I. van der Hoef, J. Lugtenburg, A. Cua, D.F. Bocian, R.R. Birge, H.A. Frank, Photosynth. Res. **86**, 5 (2005)

45. A.W. Roszak, K. McKendrick, A.T. Gardiner, I.A. Mitchell, N.W. Isaacs, R.J. Cogdell, H. Hashimoto, H.A. Frank, Structure **12**, 765 (2004)
46. Y. Kakitani, R. Fujii, Y. Koyama, H. Nagae, L. Walker, B. Salter, A. Angerhofer, Biochemistry **45**, 2053 (2006)
47. K. Schulten, M. Karplus, Chem. Phys. Lett. **14**, 305 (1972)
48. K. Schulten, I. Ohmine, M. Karplus, J. Chem. Phys. **64**, 4422 (1976)
49. D. Ghosh, J. Hachmann, T. Yanai, G.K.L. Chan, J. Chem. Phys. **128**, 144117 (2008)
50. C.M. Marian, S.C. Kock, C. Hundsdorfer, H.D. Martin, W. Stahl, E. Ostroumov, M.G. Muller, A.R. Holzwarth, Photochem. Photobiol. Sci. **8**, 270 (2009)
51. E. Ostroumov, M.G. Muller, C.M. Marian, M. Kleinschmidt, A.R. Holzwarth, Phys. Rev. Lett. **103**, 108302 (2009)
52. M. Kleinschmidt, C.M. Marian, M. Waletzke, S. Grimme, J. Chem. Phys. 130 (2009)
53. J.P. Ceron-Carrasco, A. Requena, C.M. Marian, Chem. Phys. **373**, 98 (2010)
54. S. Knecht, C.M. Marian, J. Kongsted, B. Mennucci, J. Phys. Chem. B **117**, 13808 (2013)
55. J.P. Gotze, D. Kroner, S. Banerjee, B. Karasulu, W. Thiel, Chem. Phys. Chem. **15**, 3391 (2014)
56. O. Andreussi, S. Knecht, C.M. Marian, J. Kongsted, B. Mennucci, J. Chem. Theory Comput. **11**, 655 (2015)
57. M. Momenteau, C.A. Reed, Chem. Rev. **94**, 659 (1994)
58. H. Chen, W. Lai, S. Shaik, J. Phys. Chem. B **115**, 1727 (2011)
59. L. Pauling, C.D. Coryell, Proc. Natl. Acad. Sci. U.S.A. **22**, 210 (1936)
60. K.P. Jensen, U. Ryde, J. Biol. Chem. **279**, 14561 (2004)
61. S. Franzen, Proc. Natl. Acad. Sci. U.S.A. **99**, 16754 (2002)
62. J. Ribas-Arino, J.J. Novoa, Chem. Commun. 3160 (2007)
63. H. Nakashima, J. Hasegawa, H. Nakatsuji, J. Comput. Chem. **27**, 426 (2006)
64. J.L. Thomas, B. Hh, J. Am. Chem. Soc. **94**, 1386 (1972)
65. M.L.H. Green, P.J. Knowles, J. Chem. Soc. A 1508 (1971)
66. L. Labella, A. Chernega, M.L.H. Green, J. Chem. Soc. Dalton 395 (1995)
67. J.C. Green, Chem. Soc. Rev. **27**, 263 (1998)
68. J.C. Green, J.N. Harvey, R. Poli, Dalton Trans. 1861 (2002)

# Chapter 14
# Electron Communications and Chemical Bonds

**Roman F. Nalewajski**

**Abstract** Recent developments in *Orbital Communication Theory* (OCT) of the chemical bond are summarized. Conditional probabilities defining molecular information networks are generated using the bond-projected *Superposition Principle* (SP) of quantum mechanics. The communications between atomic orbitals (AO) are proportional to squares of elements in the *Charge-and-Bond-Order* (CBO) matrix, thus being related to Wiberg's quadratic indices of the chemical bond-order. Molecular propagation of information exhibits the communication-*noise* due to electron delocalization via the system chemical bonds, the entropic bond-*covalency*, measured by the channel *conditional entropy*. Information bond-*ionicity* is reflected by the channel *mutual information* (information flow) descriptor. The *amplitude* and *probability* information systems are distinguished, with the former being capable of the communication interference. This *phase*-aspect is lost in the probabilistic treatment of classical Information Theory (IT). The OCT perspective identifies the "indirect" chemical bonds in the "cascade" propagation realized via AO intermediates. These through-*bridge* "bonds" extend range of chemical interactions and supplement the "direct" (through-*space*) bonds realized by the constructive AO interference or *direct* communications between them. The *flexible*-input extension of OCT provides a continuous description of the fragment *dissociation*, giving a fair agreement between chemical expectations, OCT bond-multiplicities, and Wiberg bond-orders. The *occupational* orbital decoupling in OCT is properly represented, when separate communication systems for each occupied MO are used and their occupation-weighted entropy/information contributions are classified as *bonding* (positive) or *antibonding* (negative), in accordance with the signs of CBO matrix elements reflecting AO phases. These developments are illustrated in *two*-orbital model and representative $\pi$-electron systems. Resultant amplitude communications in multiconfiguration theory are explored, AO communications in *Valence-Bond* (VB) structures of $H_2$ are examined, and intrinsic nonorthogonality problem of VB description is discussed. The *covalent* structure

R.F. Nalewajski (✉)
Department of Theoretical Chemistry, Jagiellonian University,
R. Ingardena 3, 30-060 Cracow, Poland
e-mail: nalewajs@chemia.uj.edu.pl



M.J. Wójcik et al. (eds.), *Frontiers of Quantum Chemistry*,
https://doi.org/10.1007/978-981-10-5651-2_14

generates noiseless channel involving solely the offdiagonal (*inter*-orbital) propagations, while the *ionic* state corresponds to deterministic channel of the diagonal (*intra*-atomic) communications.

**Keywords** Communication theory of chemical bonds • Direct/indirect bonds • Entropic bond multiplicities • Interference of amplitude communications • Molecular information channels • Orbital decoupling

The following tensor notation is adopted throughout this chapter: *A* denotes a *scalar*, ***A*** is the row (or column) *vector*, **A** represents a square or rectangular *matrix*, and the dashed symbol $\hat{\mathbf{A}}$ stands for the quantum-mechanical operator of the physical property *A*. The logarithm of the Shannon-type information measure is taken to an arbitrary but fixed base. In keeping with the custom in works on IT, the logarithm taken to base 2, log = log2, corresponds to the information content measured in *bits* (binary digits), while selecting log = ln expresses the amount of information in *nats* (natural units): 1 nat = 1.44 bits.

## 14.1 Introduction

The Information Theory (IT) [1–6] plays a unifying role in physics by facilitating derivation of its basic laws from a common principle of the *Extreme Physical Information* (EPI) [7–13]. Successful applications of IT to electronic structure phenomena have also demonstrated the theory potential in extracting chemical interpretations from calculated electron distributions, in terms of bonded atoms and chemical bonds [11–26]. The theory allows one to describe hypothetical stages in a reconstruction of the electron distribution in chemical processes. The density-based concepts and techniques of classical IT of Fisher and Shannon [1, 2] have been successfully used to probe molecular electronic structure, the system chemical bonds, and reactivity. Displacements in the Shannon entropy and entropy deficiency relative to the promolecular reference constitute effective probes into the information origins of the chemical bonds [11–13, 25–35]. They provide efficient tools for an objective definition of bonded atoms [11, 18, 36], detecting the direct chemical bonds [26, 32–35], and monitoring the promotion/hybridization changes the *Atoms-in-Molecules* (AIM) undergo in their molecular environments. The nonadditive Fisher information in Atomic-Orbital (AO) resolution [8, 11–13, 37] generates the *Contragradience* (CG) probe [11–13, 26, 32–35] for locating bonding regions in molecules, while its analog in the *Molecular Orbital* (MO) perspective [37] defines the key concept of the *Electron Localization Function* (ELF) [38–42]. These gradient information quantities stress the importance of electronic kinetic energy in an exploration of the chemical bond pattern [33, 43–45].

Diverse information probes of molecular electron distributions increase our understanding of the chemical bond from the entropic perspective of IT. The communication-*noise* (covalency) and information-*flow* (ionic) components of entropic bond multiplicities have been shown to adequately reflect the accepted chemical intuition [11–14]. These IT descriptors can be generated for both molecule as a whole and its fragments. The OCT treatment of *Diatomics-in-Molecules* (DIM) [13, 14] reproduces the Wiberg [46] bond-orders in diatomic molecules and closely approximates these quadratic indices in polyatomic molecules. The extra computational effort of such IT analysis of the molecular bonding patterns is negligible, compared to the cost of standard SCF LCAO MO calculations of the molecular electronic structure, since practically all computations in orbital approximation already determine the CBO and kinetic energy data required in the CG probe of the bond localization or the OCT exploration of bond multiplicities and their composition.

There is a wide range of problems in the theory of electronic structure and chemical reactivity which can be successfully tackled using concepts and techniques of IT. For example, the IT-optimum pieces of electron density in molecular fragments have been derived from alternative global or local variational principles of IT [11–13, 18–22, 31, 36]. This approach has been shown to lead to the "stockholder" fragments of Hirshfeld [47]. It provides an entropic justification of this popular density-partitioning scheme, generates criteria of molecular similarity, tackles the *Polarization/Promotion* (P) and *Charge-Transfer* (CT) phenomena in the reorganization of the system constituent atoms, etc. Additional diagnostic problems in the theory of molecular electronic structure deal with the shell structure and electron localization in atoms and molecules. The theory allows one to locate electrons and chemical bonds [11–13, 26, 32–35], offers a thermodynamic-like perspective on density fluctuations and flows of electrons between constituent fragments in molecules [18, 23, 48], and formulates a new class of entropic descriptors of the chemical bond multiplicity and its covalent/ionic composition [11–14, 19, 49–60].

In molecules, the information content of their electronic states is constantly scattered between bonded atoms and AO they contribute to the bond system. The spread of information is affected via the network of chemical bonds. In a sense, AIM talk to one another. This exchange of information takes the form of either the direct communication, a "dialog" between the interacting AO, or the indirect information propagation, a "gossip" spread via AO intermediates. These two mechanisms give rise to the *direct-* and *indirect*-bond components, respectively, related to the mutual and cascade communications between AO, respectively. Therefore, the chemical interactions between the specified pair of "terminal" AO have both the direct (through-*space*) and indirect (through-*bridge*) components. The "order" of the former quickly vanishes with increasing separation between atoms. The overall bond-order between more distant atomic partners can still assume appreciable values when some of the remaining atoms form an effective chemical bridge which links these terminal AIM.

The chemical bond multiplicities, reflecting the IT bond "orders" in bits, have been adequately probed using techniques and entropic quantities developed in IT of communication devices (information systems) [2, 5, 6, 11–14, 19]. This additional perspective, called the *Orbital Communication Theory* (OCT) [61–70], increases our insight and understanding of the information origins of chemical bonds. In OCT, one introduces the entropic bond multiplicities and determines their IT-*covalent* and IT-*ionic* components. The resultant IT-*covalency* and IT-*ionicity* components have been related to the *nonadditive* and *additive* information descriptors of AO *probability*-channels [78]. The *cascade* extension allows one to identify the *indirect* mechanism of chemical interactions effected through "*bridges*" (orbital intermediates) [71–76], which complements the familiar *direct* mechanism realized through "*space,*" via a constructive interference between orbitals. The (bond-projected) *Superposition Principle* (SP) of quantum mechanics [77] has been used to generate the orbital conditional probabilities determining electron communications in molecules [11–14, 68]. The *inter*-orbital conditional probabilities it generates are proportional to the quadratic bond indices formulated in MO theory, and hence strong communications correspond to strong Wiberg [46] bond-orders. The IT approach has also revealed new stages in the bond-breaking-bond-forming mechanisms of elementary chemical reactions [79–81].

In OCT, the direct bond reflects the mutual communications between bonded atoms, while the bridge bond is realized indirectly, through the cascade communications involving remaining orbitals/atoms. The latter form effective bridges for the implicit chemical coupling between more distant atoms. The most efficient bridges are the real (chemical) bridges of bonded atoms connecting the specified "terminal" atoms. The bonded status of AIM can be thus felt also at larger separations provided there exists a real chemical bridge of the direct bonds connecting them. This indirect coupling of basis functions in molecular states has been first conjectured to explain the central bond in small propellanes, lacking the charge/information accumulation between the bridgehead carbons. This bonding phenomenon has also prompted the alternative, Valence-Bond (VB) [82] inspired explanation, through the *charge-shift* mechanism [83], which attributes such an indirect "bond" to instantaneous charge fluctuations between distant atoms.

It has already been demonstrated how important the "entropic" tools of IT are for gaining a better understanding of the "chemistry" behind the calculated molecular states, e.g., [11–14]. A chemical understanding of the electronic structure of molecular systems calls for transforming the computational results into statements in terms of such chemical concepts as bonded atoms, building blocks of molecules, their collections forming functional groups, and chemical bonds representing AIM "connectivities." A collection of the constituent free atoms shifted to their actual positions in a molecule determines the so-called "promolecule" [47], which constitutes the standard reference for extracting changes in the electron distribution due to the bond formation, reflected by the *density difference* (deformation) function [11]. In Schrödinger's quantum mechanics, the molecular electronic state is determined by the system wavefunction, the (complex) amplitude of the particle *probability* distribution, which carries its resultant information content. Both the electron density or its *shape*-factor—the *probability* distribution determined by the wavefunction *modulus*—and the system *current*-distribution related to the gradient of the wavefunction *phase* ultimately contribute to the *resultant* information content of molecular states. The probability reveals the state *classical*-information contribution, while distributions of the spatial phase or its gradient (electronic current) determine the state *nonclassical* complement in the resultant information measure [14, 84–93]. Similar generalizations of the entropy-deficiency or information-distance concepts and molecular information channels have also been established. In such a combined IT description, the principle of the extreme overall information content is used to determine the molecular *phase*-equilibria in molecules [14, 84–93]. This maximum entropy rule has been shown to give rise to the *phase*-transformation of molecular electronic states into their equilibrium counterparts. Thermodynamic analogies have been discussed, and a local phenomenological description of molecular processes has been introduced [14, 88], in spirit of the ordinary irreversible thermodynamics [94]. The continuity equation for the state resultant entropy has been explored [95, 96], and a local temporal aspect of the promolecule → molecule reconstruction in the system electronic structure has been addressed [97].

This IT perspective, complementary to the MO description of quantum chemistry, brings important new insights and enhances our understanding of the complex phenomena of the chemical bond and reactivity. It generates new probes for detecting chemical bonds in molecules, uncovers the entropic and communication origins of chemical interactions, and explains the direct and indirect sources of their resultant multiplicities. In the present overview, we emphasize some additional possibilities of OCT in exploring patterns of chemical bonds, understanding the orbital-decoupling and communication interference phenomena, and probing the competition between entropic bond-covalency and bond-ionicity components.

## 14.2 Amplitude and Probability Channels

For simplicity, we focus on molecular information channels in the electronic state described by a single Slater determinant, e.g., the ground-state configuration $\Psi_0^{UHF}(N) = \det[\boldsymbol{\psi}^0] \equiv |\psi_1^0, \psi_2^0, \ldots, \psi_N^0|$, where *spin*-MO (SMO) $\boldsymbol{\psi}^0(\boldsymbol{q}) = \{\varphi_s^0(\boldsymbol{r})\ \xi_s^0(\sigma)\}$ combine the spatial (MO) components $\boldsymbol{\varphi}^0(\boldsymbol{r}) = \{\varphi_s^0(\boldsymbol{r})\}$ and spin functions $\boldsymbol{\xi}^0(\sigma) = \{\xi_s^0(\sigma)\}$ of an electron. In typical Unrestricted Hartree–Fock (UHF) SCF LCAO MO calculations or the spin-resolved Kohn–Sham (KS) Density Functional Theory (DFT) an exploration of chemical bonds calls for the (orthonormal) AO basis set $\boldsymbol{\chi} = (\chi_1, \chi_2, \ldots, \chi_m)$, which expresses the spatial (orthonormal) MO components $\boldsymbol{\varphi} = \{\varphi_s\} = \boldsymbol{\chi}\mathbf{C}$. The ground-state $\Psi_0(N)$ is then shaped by *N*-lowest, *singly*-occupied SMO $\boldsymbol{\psi}^0$ or their spatial (MO) components $\boldsymbol{\varphi}^0 = \{\varphi_s^0\}$, the configuration bonding subspace of the occupied MO, $\boldsymbol{\varphi}^0 = \boldsymbol{\chi}\mathbf{C}^0$, which defines the associated MO-projector

$$\hat{\mathrm{P}}_{\varphi}^{0} = \sum_s |\varphi_s\rangle n_s^0 \langle\varphi_s| = |\boldsymbol{\varphi}^0\rangle \mathbf{n}^0 \langle\boldsymbol{\varphi}^0| = |\chi\rangle \mathbf{C}^0\, \mathbf{C}^{0\dagger} \langle\chi| \equiv |\chi\rangle \boldsymbol{\gamma}^0 \langle\chi|; \tag{14.1}$$

here the diagonal matrix $\mathbf{n}^0 = \{n_s^0 \delta_{s,s'}\}$ combines the MO occupation numbers in the ground-state configuration: $n_s^0 = \{1, s \le N; 0, s > N\}$. This projection operator gives rise to the corresponding (idempotent) *Charge-and-Bond-Order* (CBO) matrix in AO representation,

$$\boldsymbol{\gamma}^0 = \{\gamma_{i,j}^0\} = \langle\chi|\hat{\mathrm{P}}_{\varphi}^{0}|\chi\rangle = \langle\chi|\boldsymbol{\varphi}\rangle \mathbf{n}^0 \langle\boldsymbol{\varphi}|\chi\rangle = \langle\chi|\boldsymbol{\varphi}^0\rangle\langle\boldsymbol{\varphi}^0|\chi\rangle = \mathbf{C}^0\mathbf{C}^{0\dagger},\ (\boldsymbol{\gamma}^0)^2 = \boldsymbol{\gamma}^0. \tag{14.2}$$

In accordance with the quantum-mechanical SP [77], the joint probability of the given pair of the *input* $(\chi_i)$ and *output* $(\chi_j)$ AO events in the molecular state $\Psi_0$ defined by its bond system $\boldsymbol{\varphi}^0$ is given by the square of the corresponding amplitude proportional to the CBO matrix element coupling the two basis functions [11–14]:

$$\begin{aligned} P(\chi_i, \chi_j|\Psi_0) &\equiv P^0(i, j) = |A(\chi_i, \chi_j|\Psi_0)|^2 = \gamma_{i,j}^0 \gamma_{j,i}^0/N \\ &= \langle\chi_i|\hat{\mathrm{P}}_{\varphi}^{0}|\chi_j\rangle\langle\chi_j|\hat{\mathrm{P}}_{\varphi}^{0}|\chi_j\rangle \equiv |\chi_i\rangle \equiv \langle\chi_i|\hat{\mathrm{P}}_{\varphi}^{0}\hat{\mathrm{P}}_j\hat{\mathrm{P}}_{\varphi}^{0}|\chi_i\rangle \equiv \langle\chi_i|\hat{\hat{\mathrm{P}}}_j^0|\chi_i\rangle \\ &= P^0(j, i) = \langle\chi_j|\hat{\mathrm{P}}_{\varphi}^{0}|\chi_i\rangle\langle\chi_i|\hat{\mathrm{P}}_{\varphi}^{0}|\chi_j\rangle \equiv \langle\chi_j|\hat{\mathrm{P}}_{\varphi}^{0}\hat{\mathrm{P}}_i\hat{\mathrm{P}}_{\varphi}^{0}|\chi_j\rangle \equiv \langle\chi_j|\hat{\hat{\mathrm{P}}}_i^0|\chi_j\rangle, \end{aligned}$$

$$\sum_j P^0(i, j) = N^{-1} \sum_j \gamma_{i,j}^0 \gamma_{j,i}^0 = \gamma_{i,i}^0/N = P(i|\Psi_0) \equiv p_i^0; \tag{14.3}$$

here $\hat{\hat{\mathrm{P}}}_k^0 = \hat{\mathrm{P}}_{\varphi}^{0}\hat{\mathrm{P}}_k\hat{\mathrm{P}}_{\varphi}^{0}$ denotes the resultant AO projection within the bond-subspace $\boldsymbol{\varphi}^0$. The associated *conditional* AO probabilities $\mathbf{P}^0(\chi'|\chi)$, of the output events $\chi' = \{\chi_j\}$ given the input events $\chi = \{\chi_i\}$, $\mathbf{P}^0(\chi'|\chi) = \{P^0(j|i) \equiv P_{i\to j}^0 = P^0(i, j)/p_i^0\}$, the squared moduli of the corresponding amplitudes $\{P_{i\to j}^0 \equiv |A_{i\to j}^0|^2\}$, then read:

$$P^0(j|i) \equiv P_{i\to j}^0 \equiv |A_{i\to j}^0|^2 = (\gamma_{i,i}^0)^{-1}\gamma_{i,j}^0\gamma_{j,i}^0 \equiv N_i^0\gamma_{i,j}^0\gamma_{j,i}^0,\ \sum_j P^0(j|i) = 1; \tag{14.4}$$

here $N_i^0 = (\gamma_{i,i}^0)^{-1}$ stands for the normalization constant of the scattering probabilities originating from the specified input $\chi_i$, $P^0(\chi'|i) = \{P^0(j|i), j = 1, 2, \ldots, m\}$, the $i$th row of $\mathbf{P}^0(\chi'|\chi)$. They reflect the electron delocalization in the occupied MO subspace and identify the scattering amplitudes $\mathbf{A}^0(\chi'|\chi) = \{A^0(j|i) \equiv A_{i\to j}^0\}$ related to the corresponding elements of the CBO matrix $\boldsymbol{\gamma}^0$:

$$A_{i\to j}^0 = (\gamma_{i,i}^0)^{-1/2}\gamma_{i,j}^0 \equiv \left[N_i^0\right]^{1/2}\gamma_{i,j}^0 \tag{14.5}$$

In the RHF description of the *closed*-shell state, each of the $N/2$ lowest (*doubly*-occupied) MO, $\boldsymbol{\varphi}^o = (\varphi_1, \varphi_2, \ldots, \varphi_{N/2})$, accommodates two *spin*-paired electrons: $\Psi_0^{\mathrm{RHF}}(N) = |\varphi_1\alpha, \varphi_1\beta, \ldots, \varphi_{N/2}\alpha, \varphi_{N/2}\beta|$, so that $\hat{\mathrm{P}}_{\varphi}^{0} = 2|\boldsymbol{\varphi}^o\rangle\langle\boldsymbol{\varphi}^o| \equiv 2\hat{\mathrm{P}}_{\varphi}^{o}$. The CBO idempotency then implies

$$(\hat{\mathrm{P}}^0_\varphi)^2 = [2|\varphi^o\rangle\langle\varphi^o|]^2 = 4|\varphi^o\rangle\langle\varphi^o| = 2\hat{\mathrm{P}}^o_\varphi = \hat{\mathrm{P}}^0_\varphi \tag{14.6}$$

and hence:

$$\boldsymbol{\gamma}^0 = \langle\boldsymbol{\chi}|\hat{\mathrm{P}}^0_\varphi|\boldsymbol{\chi}\rangle = 2\langle\boldsymbol{\chi}|\hat{\mathrm{P}}^o_\varphi|\boldsymbol{\chi}\rangle \equiv 2\boldsymbol{\gamma}^0,\ (\boldsymbol{\gamma}^0)^2 = \boldsymbol{\gamma}^0 = \{\gamma^o_{i,j}\},\ (\boldsymbol{\gamma}^0)^2 = 4(\boldsymbol{\gamma}^0)^2 = 4\boldsymbol{\gamma}^0 = 2\boldsymbol{\gamma}^0. \tag{14.7}$$

For such states, the representative conditional probability of the molecular AO channel reads:

$$P^0(j|i) \equiv P^0_{i\to j} = |A^0_{i\to j}|^2 = (2\gamma^0_{i,i})^{-1}\gamma^0_{i,j}\gamma^0_{j,i} \equiv N^0_i\gamma^0_{i,j}\gamma^0_{j,i} = (\gamma^o_{i,i})^{-1}\gamma^o_{i,j}\gamma^o_{j,i} \equiv N^o_i\gamma^o_{i,j}\gamma^o_{j,i},$$
$$A^0(j|i) \equiv A^0_{i\to j} = (N^0_i)^{-1/2}\gamma^0_{i,j} = (N^o_i)^{-1/2}\gamma^o_{i,j}. \tag{14.8}$$

The classical, *probability*-channel, is determined by these conditional AO probabilities $\mathbf{P}(\chi'|\chi) = \{P(j|i) = P_{i\to j}\}$,

$$\chi - \mathbf{P}(\chi'|\chi) \to \chi'. \tag{14.9}$$

It loses the memory of the AO phases in the scattering amplitudes $\mathbf{A}(\chi'|\chi) = \{A_{i\to j}\}$, i.e., phases of $\{\gamma^0_{i,j}\}$. These "coherencies" are preserved only in the associated *amplitude* channel for the *direct* electron communications in a molecule,

$$|\chi\rangle - \mathbf{A}(\chi'|\chi) \to |\chi'\rangle, \tag{14.10}$$

which is thus capable of reflecting the quantum-mechanical interference between such elementary communications.

This observation also applies to a sequential ("product") arrangements of several direct channels, called the information "cascades," for indirect ("bridge") communications between atomic orbitals in molecules. For example, the *single*-AO intermediates $\chi''$ in the sequential *three*-orbital scatterings $\chi \to \chi'' \to \chi'$ define the *probability*- and *amplitude*-cascades:

$$\chi - [\mathbf{P}(\chi''|\chi) \to \chi'' - \mathbf{P}(\chi'|\chi'')] \to \chi' \equiv \chi - \mathbf{P}[(\chi'|\chi);\chi''] \to \chi',$$
$$|\chi\rangle - [\mathbf{A}(\chi''|\chi) \to |\chi''\rangle - \mathbf{A}(\chi'|\chi'')] \to |\chi'\rangle \equiv |\chi\rangle - \mathbf{A}[(\chi'|\chi);\chi''] \to |\chi'\rangle. \tag{14.11}$$

The *indirect* conditional probabilities between AO events and their amplitudes are then given by products of the elementary *two*-orbital communications in each direct subchannel:

$$P[(j|i);\ k] \equiv P_{i\to j;k} = P_{i\to k}P_{k\to j}, \quad A[(j|i);\ k] \equiv A_{i\to j;k} = A_{i\to k}A_{k\to j}. \tag{14.12}$$

Therefore, such *bridge* probabilities and underlying amplitudes can be straightforwardly derived from the *direct* scatterings. They satisfy the relevant *bridge*-normalizations, expressed by the *sum*-rules over AO events in the intermediate and final outputs:

$$\sum_k (\sum_j P_{i\to j;k}) = \sum_k P_{i\to k} = 1. \tag{14.13}$$

This *single*-cascade development can be straightforwardly generalized to any bridge-*order* $t$. The amplitude $A_{i\to j}^{(t)}$ for the *complete* $t$-cascade, consisting of $t$ consecutive direct channels involving all AO, preserves the direct scattering probabilities [14, 73],

$$\mathbf{P}[(\chi'|\chi); t-\chi] = \{|A_{i\to j}^{(t)}|^2 = |A_{i\to j}|^2 = P_{i\to j}\} = \mathbf{P}(\chi'|\chi), \tag{14.14}$$

thus satisfying the important consistency requirement of the *stationary* character of the molecular channel. The relevant *sum*-rules for such *bridge* conditional probabilities read:

$$\begin{aligned}\sum_k \sum_l \cdots \sum_m \sum_n \left[\sum_j P_{i\to j;k,l,\ldots,m,n}\right] &= \sum_k \sum_l \cdots \sum_m \left[\sum_n P_{i\to n;k,l,\ldots,m}\right] \\ &= \sum_k \sum_l \cdots \left[\sum_m P_{i\to m;k,l,\ldots}\right] = \cdots \\ &= \sum_k \left[\sum_l P_{i\to l;k}\right] = \sum_k P_{i\to k} = 1.\end{aligned} \tag{14.15}$$

For the specified pair of "terminal" AO, say $\chi_i \in \chi$ and $\chi_j \in \chi'$, one can similarly examine the indirect scatterings via the molecular bond system in the *incomplete* cascades consisting of the *remaining* ("bridge") functions $\chi^b = \{\chi_{k\neq(i,j)}\}$, with the two terminal AO being then excluded from the set of admissible intermediate scatterers. The associated *bridge*-communications give rise to the indirect (through-*bridge*) components of the entropic bond multiplicities [71–76], which complement the familiar direct (through-*space*) chemical "bond-orders" and provide a novel IT perspective on chemical interactions between more distant AIM, alternative to the fluctuational *Charge-Shift* mechanism [83] in VB theory.

## 14.3 Entropic Multiplicities of Direct and Bridge-Bonds

We continue with a brief summary of the entropy/information descriptors of a transmission of the electron-assignment "signals" in such molecular communication systems. The *classical* orbital networks propagate the conditional probabilities of electron assignments to basis functions of SCF LCAO MO calculations. The probability networks lose memory of the *phase*-aspect of this information propagation, preserved in the amplitude systems, which becomes crucial in the *multi*-stage (cascade, bridge) propagations. The underlying conditional probabilities or the associated amplitudes, for the quantum scattering from the *emitting* (input) AO states to alternative *monitoring* (output) AO states, follow from the bond-projected SP.

In a classical communication device, the signal emitted from $n$ "inputs" $\boldsymbol{a} = (a_1, a_2, \ldots, a_n)$ of the channel *source* **A** is characterized by the *input*-probability distribution $\boldsymbol{P}(\boldsymbol{a}) \equiv \boldsymbol{p} = (p_1, p_2, \ldots, p_n)$, which describes the way the channel is exploited. It is received (monitored) at $m$ alternative "outputs" $\boldsymbol{b} = (b_1, b_2, \ldots, b_m)$ in the system *receiver* **B**. The system communication *noise* is then generated by the conditional probabilities of observing specific "outputs" given "inputs",

$$\mathbf{P}(\mathbf{B}|\mathbf{A}) \equiv \mathbf{P}(\boldsymbol{b}|\boldsymbol{a}) = \{P(b_j|a_i) = P(a_i, b_j)/P(a_i) \equiv P(j|i)\}, \tag{14.16}$$

where $\mathbf{P}(\boldsymbol{a}, \boldsymbol{b}) = \{P(a_i, b_j) \equiv P(i, j)\}$ groups probabilities of the joint occurrence of the specified pair of the *input*-and-*output* events. The distribution of the output signal among the detection events $\boldsymbol{b}$ then reads:

$$P(b) \equiv \boldsymbol{q} = (q_1, q_2, \ldots, q_m) = \boldsymbol{p}\,\mathbf{P}(\boldsymbol{b}|\boldsymbol{a}). \tag{14.17}$$

In general, the input and output probabilities are mutually dependent. One thus decomposes the joint probabilities $\mathbf{P}(\boldsymbol{a}, \boldsymbol{b}) \equiv \mathbf{P}$ of the joint events $(\boldsymbol{a}, \boldsymbol{b})$ as products of the "marginal" probabilities of events in one set, say $\boldsymbol{P}(\boldsymbol{a})$, and the corresponding conditional probabilities $\mathbf{P}(\boldsymbol{b}|\boldsymbol{a}) = \{P(j|i)\}$ of outcomes in the other set $\boldsymbol{b}$, given that events $\boldsymbol{a}$ have already occurred: $\mathbf{P} = \{P(i, j) = p_i\, P(j|i)\}$. The Shannon [2] entropy of the joint distribution $\mathbf{P}(\boldsymbol{a}, \boldsymbol{b})$ can be then expressed as the sum of the average entropy $S(\boldsymbol{p}) = -\sum_i p_i \log p_i$ in the marginal probability distribution and the average *conditional entropy* $S(\boldsymbol{q}|\boldsymbol{p})$ in $\boldsymbol{q}$ given $\boldsymbol{p}$,

$$S(\mathbf{P}) = S(\boldsymbol{p}) + S(q|p) = S(\boldsymbol{q}) + S(p|q), \tag{14.18}$$

$$S(q|p) = -\sum_i \sum_j P(i, j)\, \log[P(i, j)/p_i] = -\sum_i p_i \sum_j P(j|i)\, \log\, P(j|i) \equiv S,$$

measuring the network average communication "noise" and describing the *indeterministic* (delocalization) facet of molecular communications. The channel complementary descriptor, the *mutual information* $I(\boldsymbol{p}{:}\ \boldsymbol{q})$ in the network output and input probability distributions,

$$\begin{aligned} I(\boldsymbol{p}{:}\boldsymbol{q}) &= \sum_i \sum_j P(i, j) \log\left[P(i, j)/(p_i q_j)\right] \\ &= S(\boldsymbol{p}) + S(\boldsymbol{q}) - S(\mathbf{P}) = S(\boldsymbol{q}) - S(\boldsymbol{q}|\boldsymbol{p}) = S(\boldsymbol{p}) - S(\boldsymbol{p}|\boldsymbol{q}) \equiv I, \end{aligned} \tag{14.19}$$

then reflects its *deterministic* (localization) aspect. The input probabilities $\boldsymbol{p}$ reflect the way the channel is used or probed. The conditional entropy $S(\boldsymbol{q}|\boldsymbol{p})$ measures the average *noise* in the $\mathbf{A} \to \mathbf{B}$ transmission, i.e., the residual indeterminacy about the output signal, when the input signal is known beforehand. The conditional entropy

$S(\boldsymbol{p}|\boldsymbol{q})$ reflects a fraction of $S(\boldsymbol{p}) = S(\boldsymbol{p}|\boldsymbol{q}) + I(\boldsymbol{p} : \boldsymbol{q})$ which has been transformed into "noise" as a result of the input signal being scattered in the information channel. Accordingly, $S(\boldsymbol{q}|\boldsymbol{p})$ reflects the noise part of $S(\boldsymbol{q}) = S(\boldsymbol{q}|\boldsymbol{p}) + I(\boldsymbol{p} : \boldsymbol{q})$. An observation of the output signal provides the amount of information given by mutual information $I(\boldsymbol{p} : \boldsymbol{q})$, which reflects the channel information *flow*.

In OCT, the AO molecular channel propagates signals (probabilities or amplitudes) of the electron assignments to basis functions $\chi = (\chi_1, \chi_2, \ldots, \chi_m)$. The underlying conditional probabilities of the *output* AO events, given the input AO events, $\mathbf{P}(\chi'|\chi) = \{P(\chi_j|\chi_i) \equiv P(j|i) \equiv P_{i\to j} \equiv A(j|i)^2 \equiv (A_{i\to j})^2\}$, or the associated scattering amplitudes $\mathbf{A}(\chi'|\chi) = \{A(j|i) = A_{i\to j}\}$ of the *emitting* (input) states $|\boldsymbol{a}\rangle = |\chi\rangle = \{|\chi_i\rangle\}$ among the *receiving* (output) states $|\boldsymbol{b}\rangle = |\chi'\rangle = \{|\chi_j\rangle\}$, have been discussed in the preceding section. The entropy/information indices of the overall IT-*covalent* and IT-*ionic* components (in bits) of the system chemical bonds represent the complementary descriptors of the average communication *noise* and the amount of information *flow* in the AO-resolved molecular channel.

One observes that the *molecular* input signal $\boldsymbol{P}(\boldsymbol{a}) \equiv \boldsymbol{p}^0$ generates the same distribution in the output of this network,

$$\boldsymbol{q}^0 = \boldsymbol{p}^0\,\mathbf{P}^0(\boldsymbol{b}|\boldsymbol{a}) = \{\textstyle\sum_i p_i^0\, P^0(j|i) \equiv \sum_i P^0(i,j) = p_j^0\} = \boldsymbol{p}^0, \tag{14.20}$$

thus identifying $\boldsymbol{p}^0$ and $\boldsymbol{q}^0$ as *stationary* vectors of AO probabilities in the molecular ground state. This purely molecular channel is devoid of any reference (history) of the chemical bond formation and generates the average noise index of the IT bond-*covalency* measured by the average conditional entropy of the system AO outputs given AO inputs: $S(\boldsymbol{q}^0|\boldsymbol{p}^0) \equiv S$. The AO channel with the *promolecular* input signal $\boldsymbol{P}(\boldsymbol{a}_0) = \boldsymbol{p}_0 = \{p_{i,0}\}$, of AO in the system *free* constituent atoms, refers to the initial stage in the bond-formation process. It corresponds to the ground-state occupations of AO contributed to chemical bonds by the system constituent atoms, before their mixing into MO. These reference input probabilities give rise to the average information-*flow* index of the system IT bond-*ionicity*, given by the mutual information in the channel *promolecular* inputs and *molecular* outputs:

$$\begin{aligned} I(\boldsymbol{p}_0\colon \mathbf{q}^0) &= \textstyle\sum_i \sum_j P^0(i,j)\,\log\left[p_i^0\, P^0(i,j)/\left(p_i^0\, q_j^0\, p_{i,0}\right)\right] \\ &= \textstyle\sum_i \sum_j P^0(i,j)[-\log q_j^0 + \log\left(p_i^0/p_{i,0}\right) + \log P^0(j|i)] = S\left(\boldsymbol{q}^0\right) + \Delta S(\boldsymbol{p}^0|\boldsymbol{p}_0) - S \equiv I_0, \end{aligned} \tag{14.21}$$

where the entropy deficiency (missing-information) of Kullback and Leibler [3],

$$\Delta S(\boldsymbol{p}^0|\boldsymbol{p}_0) = \textstyle\sum_i p_i^0 \log\left(p_i^0/p_{i,0}\right), \tag{14.22}$$

measures the entropic similarity (information distance) between the compared probability vectors. The common amount of information $I(\boldsymbol{p}_0 : \boldsymbol{q}^0)$ in the promolecular and molecular probability distributions reflects the fraction of the initial information content $S(\boldsymbol{p}^0)$ which has not been dissipated as noise in the molecular communication system. In particular, for the molecular input $\boldsymbol{p}^0$ and hence: $\Delta S(\boldsymbol{p}^0|\boldsymbol{p}^0) = 0$, $I(\boldsymbol{p}^0: \boldsymbol{q}^0) = S(\boldsymbol{q}^0) - S \equiv I$.

The sum of these bond-multiplicity components, e.g.,

$$M_0(\boldsymbol{p}_0; \boldsymbol{q}^0) = S + I_0 = S(\boldsymbol{q}^0) + \Delta S(\boldsymbol{p}^0|\boldsymbol{p}_0) \equiv M_0, \tag{14.23}$$

measures the overall IT bond-multiplicity index relative to the *promolecular* reference, of all bonds in the molecular system. For the *molecular* input, this quantity preserves the Shannon entropy of the molecular probabilities:

$$M_0(\boldsymbol{p}^0; \boldsymbol{q}^0) = S(\boldsymbol{q}^0|\boldsymbol{p}^0) + I(\boldsymbol{p}^0 : \boldsymbol{q}^0) = S(\boldsymbol{q}^0) \equiv M. \tag{14.24}$$

These IT bond-multiplicity descriptors can be expressed in terms of the additive and nonadditive components of the Shannon entropy contained in the CBO matrix $\boldsymbol{\gamma}^0$:

$$\begin{aligned}
S^{total}(\boldsymbol{\gamma}^0) &\equiv N\{S[\mathbf{P}^0(\boldsymbol{a}, \boldsymbol{b})] - \log N\} = -\textstyle\sum_i \sum_j \gamma^0_{i,j}\gamma^0_{j,i} \log(\gamma^0_{i,j}\gamma^0_{j,i}) \equiv S^{add.}(\boldsymbol{\gamma}^0) + S^{nadd.}(\boldsymbol{\gamma}^0),\\
S^{add.}(\boldsymbol{\gamma}^0) &= N[S(\boldsymbol{p}^0) - \log N] = -\textstyle\sum_i \gamma^0_{i,i} \log \gamma^0_{i,i},\\
S^{nadd.}(\boldsymbol{\gamma}^0) &= S^{total}(\boldsymbol{\gamma}^0) - S^{add.}(\boldsymbol{\gamma}^0) = N\{S[\mathbf{P}^0(\boldsymbol{a}, \boldsymbol{b})] - M(\boldsymbol{p}^0; \boldsymbol{q}^0)\}.
\end{aligned} \tag{14.25}$$

The covalent descriptor $S(\boldsymbol{q}^0|\boldsymbol{p}^0) = S$ is then decomposed into the difference between the *total* and *additive* contributions,

$$\begin{aligned}
S &= -\textstyle\sum_i \sum_j P^0(i, j) \log[P^0(i, j)/p^0_i] = S[\mathbf{P}^0(\boldsymbol{a}, \boldsymbol{b})] - S(\boldsymbol{p}^0)\\
&= N^{-1}[-\textstyle\sum_i \sum_j \gamma^0_{i,j}\gamma^0_{j,i} \log(\gamma^0_{i,j}\gamma^0_{j,i}) + \sum_i \gamma^0_{i,i} \log \gamma^0_{i,i}]\\
&\equiv N^{-1}\{S^{total}(\boldsymbol{\gamma}^0) - S^{add.}(\boldsymbol{\gamma}^0)\} \equiv N^{-1} S^{nadd.}(\boldsymbol{\gamma}^0),
\end{aligned} \tag{14.26}$$

which defines the associated *nonadditive* component $S^{nadd.}(\boldsymbol{\gamma}^0)$, while the molecular information-flow quantity $I(\boldsymbol{p}^0: \boldsymbol{q}^0)$ of Eq. (14.19) is determined by the difference between the additive and nonadditive components:

$$\begin{aligned}
I(\boldsymbol{p}^0:\boldsymbol{q}^0) &= \textstyle\sum_i \sum_j P^0(i, j) \log\left[P^0(i, j)/\left(p^0_i\, q^0_j\right)\right] = S(\boldsymbol{p}^0) + S(\boldsymbol{q}^0) - S[\mathbf{P}^0(\boldsymbol{a}, \boldsymbol{b})]\\
&= N^{-1}[2S^{add.}(\boldsymbol{\gamma}^0) - S^{total}(\boldsymbol{\gamma}^0)] + \log N = N^{-1}[S^{add.}(\boldsymbol{\gamma}^0) - S^{nadd.}(\boldsymbol{\gamma}^0)] + \log N.
\end{aligned} \tag{14.27}$$

These two components generate the molecular bond-multiplicity index of Eq. (14.24) related to the additive contribution alone:

$$M(\boldsymbol{p}^0; \boldsymbol{q}^0) = S(\boldsymbol{q}^0|\boldsymbol{p}^0) + I(\boldsymbol{p}^0 : \boldsymbol{q}^0) = S(\boldsymbol{q}^0) = N^{-1} S^{add.}(\boldsymbol{\gamma}^0) + \log N. \tag{14.28}$$

To summarize, in this *single*-determinant approximation, the additive part of the Shannon entropy due to molecular communications between AO generates the overall IT bond-multiplicity index, the nonadditive contribution reflects the channel covalent (indeterministic) descriptor, while their difference measures the bond ionic (deterministic) component.

Consider the 2-AO model of the chemical bond A–B, consisting of the orthonormal basis functions, e.g., the Löwdin Orthogonalized AO (OAO) contributed by atoms A and B: $\boldsymbol{\chi} = \boldsymbol{\chi}' = (\chi_A, \chi_B)$. They give rise to two independent (spatial) MO combinations $\boldsymbol{\varphi} = (\varphi_b, \varphi_a)$,

$$\begin{aligned} &\text{bonding:} \quad \varphi_b = \chi_A C_{A,b} + \chi_B C_{B,b} \equiv \chi_A (P)^{1/2} + \chi_B (Q)^{1/2} \equiv \boldsymbol{\chi}\, C_b \quad \text{and} \\ &\text{antibonding:} \quad \varphi_a = \chi_A C_{A,a} + \chi_B C_{B,a} \equiv -\chi_A (Q)^{1/2} + \chi_B (P)^{1/2} \equiv \boldsymbol{\chi}\, C_a, \end{aligned} \tag{14.29}$$

or in the compact matrix notation

$$\boldsymbol{\varphi} = \boldsymbol{\chi}[C_b|C_a] \equiv \boldsymbol{\chi}\,\mathbf{C}, \quad \mathbf{C} = \begin{bmatrix} \sqrt{P} & -\sqrt{Q} \\ \sqrt{Q} & \sqrt{P} \end{bmatrix}, \tag{14.30}$$

which have been also expressed in terms of complementary (conditional) AO probabilities:

$$\begin{aligned} P &= P(\chi_A|\varphi_b) = (C_{A,b})^2 = P(\chi_B|\varphi_a) = [C_{B,a}]^2 \quad \text{and} \\ Q &= 1 - P = P(\chi_B|\varphi_b) = (C_{B,b})^2 = P(\chi_A|\varphi_a) = [C_{A,a}]^2. \end{aligned} \tag{14.31}$$

We further assume that each atom of the "promolecule" contributes a single electron to the molecular bond system consisting of $N = 2$ electrons.

In the model ground-state configuration, when both (spin-paired) electrons occupy the bonding MO $\varphi_b$, the relevant CBO matrix $\boldsymbol{\gamma}^0$ reads:

$$\boldsymbol{\gamma}^0 = 2C_b C_b^\dagger = 2\begin{bmatrix} P & \sqrt{PQ} \\ \sqrt{QP} & Q \end{bmatrix} \equiv 2\boldsymbol{\gamma}^o. \tag{14.32}$$

It generates the following conditional probabilities $\mathbf{P}^0(\boldsymbol{b}|\boldsymbol{a})$ and their amplitudes $\mathbf{A}^0(\boldsymbol{b}|\boldsymbol{a})$ defining communications between the input $\boldsymbol{a} = \boldsymbol{\chi}$ and output $\boldsymbol{b} = \boldsymbol{\chi}'$ AO events in the molecular *probability*-channel of Fig. 14.1:

$$\mathbf{P}^0(\boldsymbol{\chi}'|\boldsymbol{\chi}) = \{P^0(j|i)\} = \begin{bmatrix} P & Q \\ P & Q \end{bmatrix}, \quad \mathbf{A}^0(\boldsymbol{\chi}'|\boldsymbol{\chi}) = \{A^0(j|i)\} = \begin{bmatrix} \sqrt{P} & \sqrt{Q} \\ \sqrt{P} & \sqrt{Q} \end{bmatrix}. \tag{14.33}$$

In this *non*symmetric binary channel, one adopts the molecular input signal, $\boldsymbol{p}^0 = (P, Q)$, to extract the bond IT-*covalency* reflected by the conditional entropy $S(\boldsymbol{q}^0|\boldsymbol{p}^0) = S(P)$ measuring the channel average communication noise. Adopting the

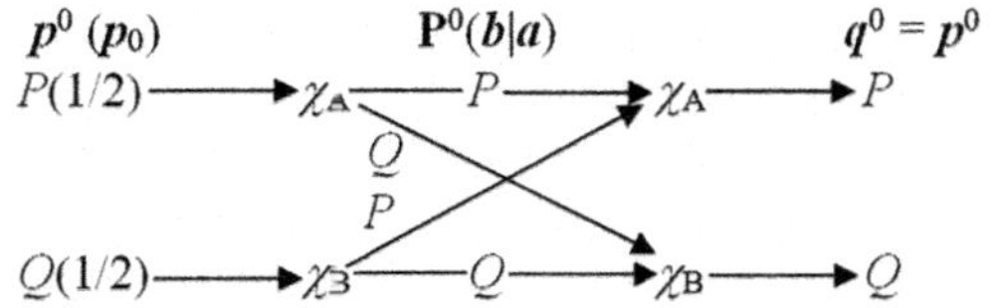


$$S(P) = -P\log_2 P - Q\log_2 Q \equiv H(P)$$
$$I_0(P) = 1 - H(P)$$
$$M_0(P) = S(P) + I_0(P) = 1 \text{ bit}$$

**Fig. 14.1** *Probability*-channel in 2-AO model of the chemical bond and its entropic descriptors (in bits) of bond multiplicity and composition

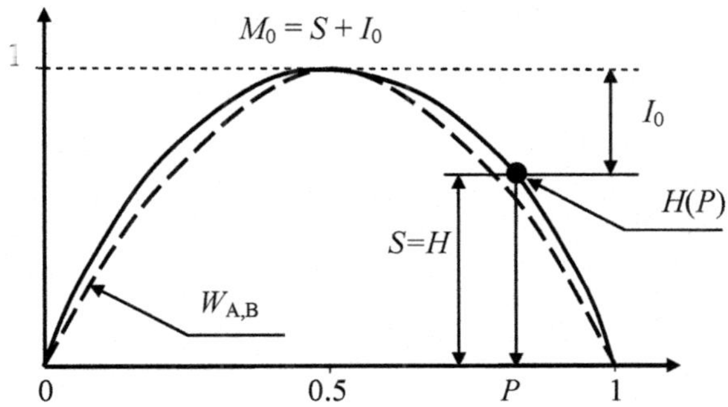


**Fig. 14.2** Conservation of the overall entropic bond multiplicity $M_0(P) = 1$ bit in the 2-AO model, combining the conditional entropy (average noise, bond IT-covalency) $S(P) = H(P)$ and the mutual information (information capacity, bond IT-ionicity) $I_0(P) = 1 - H(P)$. In MO theory, the quadratic bond-order of Wiberg is represented by the (*broken-line*) parabola $W_{A,B}(P) = 4P(1-P) \equiv 4PQ$

promolecular input signal $\boldsymbol{p}_0 = (½, ½)$, reflecting that each of the two basis functions has contributed a single electron to form the chemical bond, allows one to determine the associated index of IT-*ionicity* $I(\boldsymbol{p}_0 : \boldsymbol{q}^0) = I_0(P)$, which measures the information capacity of this model channel.

As shown in Figs. 14.1 and 14.2, these bond components preserve the overall information multiplicity of this model chemical bond: $M_0(\boldsymbol{p}_0\ ;\ \boldsymbol{q}^0) = M_0(P) = S(P) + I_0(P) = 1$ bit, for all admissible values of the bond-polarization parameter $P$: $0 \leq P \leq 1$. As shown in Fig. 14.2, the bond IT-covalency $S(P)$ is determined by Binary Entropy Function of the complementary conditional probabilities of AO in MO, $H(P) = -P\log_2 P - Q\log_2 Q$, reaching the maximum value $H(½) = 1$ bit for the symmetric bond $P = Q = ½$, e.g., the two prototype covalent bonds of chemistry: the $\sigma$ bond in $H_2$ or the $\pi$-bond in ethylene. It vanishes for the *lone*-pair configurations, when $P = (0$ or $1)$, $H(0) = H(1) = 0$, marking the alternative *ion-pair* configurations $A^+B^-$ and $A^-B^+$, respectively, relative to the initial AO occupations $N^0 = (1, 1)$ in the assumed *covalent* promolecular reference. The complementary descriptor $I_0(P) = 1 - H(P)$ of the bond IT-ionicity, which determines the channel mutual information relative to the promolecular input, is thus correctly diagnosed to reach the highest value for the two electron-transfer pairs:

$I_0(0) = I_0(1) = H(½) = 1$ bit; it is also predicted to identically vanish for the purely covalent, symmetric bond, $I_0(½) = 0$. As explicitly shown in Fig. 14.2, these two components of the chemical bond multiplicity compete with one another, yielding the conserved overall IT bond index $M_0(P) = S(P) + I_0(P) = 1$ bit, which in OCT marks the full single bond, in the whole range of admissible bond polarizations $P \in [0, 1]$. This simple model thus properly accounts for the competition between the bond covalent and ionic components, preserving the single bond-multiplicity measure of the model chemical bond.

Consider now the *quadratic* bond-multiplicity indices formulated in the MO theory [79, 98–105]. For example, for the 2-AO model, the quadratic bond-order of Wiberg [46] is given by the parabola $W_{A,B}(P) = [\gamma_{A,B}(P)]^2 = 4PQ = 4P(1 - P)$, which in Fig. 14.2 closely resembles the entropic plot $S(P) = H(P)$. Indeed, a reference to Eq. (14.4) indicates that strong diatomic Wiberg index implies a strong communication between the interacting orbitals, and hence also a high IT-covalency of the *direct* chemical bond they generate. Similar relation transpires from the *indirect*-bonds realized via orbital intermediates, delineating a chain of chemical bonds defining the bridge $\alpha$ in question [see Eq. (14.12)]. The bridge resultant conditional probability is then proportional to a product of Wiberg indices $\{W_{k,l}\}$ of each diatomic ($k \in X$, $l \in Y$), $X \neq Y$ segment of the direct {X—Y} bonds connecting the two terminal orbitals $(i, j)$ in the orbital bridge $i$—{($k$—$l$)}—$j$ under consideration:

$$W_{i,j}(\alpha) \propto \prod_{(k,l)\in\alpha} W_{k,l}.$$

Therefore, strong intermediate bonds $\{k\text{–}l\} \in \alpha$ again imply strong resultant communication between terminal orbitals, and hence high resultant entropic covalency realized via the bridge constituent *direct*-bond segments. The sum of contributions due to all relevant chemical bridges $\{\alpha\}$,

$$W_{i,j}(bridges) = \sum_{\alpha} W_{i,j}(\alpha), \tag{14.34}$$

determines the Wiberg-type bond-order of the overall indirect interactions between the specified terminal AO, which supplements the direct component $W_{i,j} = \gamma_{i,j}^2$ in the full quadratic measure of bond-multiplicity between the specified terminal orbitals $\chi_i$ and $\chi_j$ in presence of all remaining basis functions in the given molecular environment:

$$W(i, j) = W_{i,j} + W_{i,j}(bridges). \tag{14.35}$$

This indirect mechanism reflects the *implicit* dependencies between (nonorthogonal) bond-projections of AO, $|\chi^b\rangle \equiv \hat{P}_{\varphi}^{0}|\chi\rangle$. In this generalized outlook on the bond-order concept of chemistry, one thus identifies the bond multiplicity as a measure of the "*dependence*" (*non*additivity) between orbitals on different atomic centers in the whole bond system of a molecule. On one hand, this dependence between basis functions can be realized *directly*, through “space,” by a constructive interference of orbitals (probability amplitudes) on two atoms, which generally increases the electron density between them. On the other hand, it can also have an *indirect* origin, through the dependence on orbitals originating from the remaining AIM. In the nonorthogonal bond-projected AO basis $\chi^b$, these “geometrical” dependencies are embodied in the idempotent density matrix of Eq. (14.7):

$$\mathbf{D}^0 = \{D^0_{i,j} = \langle i^b | j^b \rangle\} = \mathbf{C}^o \mathbf{C}^{o\dagger}, \quad (\mathbf{D}^0)^n = \mathbf{D}^0. \tag{14.36}$$

Each pair of AO (or AIM) thus exhibits partial through-*space* and through-*bridge* components. The bond-order of the former quickly vanishes with an increasing *inter*-atomic separation and when the interacting AO are heavily engaged in forming chemical bonds with other atoms, while the latter can still assume appreciable values, when the remaining atoms form an effective bridge of the neighboring, chemically interacting atoms, which links the specified AO. The bridging atoms must be mutually bonded to generate a substantial through-bridge overlap between the bond-projections of interacting AO, so that significant bridges are in fact limited to real *chemical* bridges of atoms in the structural formula of a molecule.

As an illustration, let us summarize the indirect $\pi$-bonds between carbon atoms in benzene ring using MO from the familiar Hückel approximation. For the consecutive numbering of carbons in the $\pi$-system, the relevant CBO matrix elements read:

$$\gamma_{i,i} = 1, \quad \gamma_{i,i+1} = 2/3, \quad \gamma_{i,i+2} = 0, \quad \gamma_{i,i+3} = -1/3.$$

They generate the following (direct) $\pi$-bond multiplicities:

$$W_{i,i+1} = 0.44\,(ortho), \quad W_{i,i+2} = 0\,(meta), \quad W_{i,i+3} = 0.11\,(para).$$

These through-*space* bond-orders are complemented by the associated estimates of the resultant multiplicities of the indirect $\pi$-interactions due to all most important chemical bridges:

$$W_{i,i+1}(bridges) = 0.06\,(ortho), \quad W_{i,i+2}(bridges) = 0.30\,(meta),$$
$$W_{i,i+3}(bridges) = 0.18\,(para).$$

Together these two mechanisms give rise to the following *resultant* $\pi$-bond-orders:

$$W(para) \cong W(meta) = 0.3 < W(ortho) = 0.5.$$

The artificial distinction in Wiberg’s scheme of the $\pi$-interactions with the vanishing direct CBO matrix element as *non*bonding is thus effectively removed when the through-bridge contributions are also taken into account. One observes the difference in composition of the resultant indices for the *cross*–ring interactions: the *para* interactions exhibit comparable through-space and through-bridge components, the *meta* multiplicities are realized exclusively through bridges, while the strongest *ortho* bonds have practically direct, through-space origin.

Of interest also is a comparison of the indirect $\pi$-bond-orders in benzene realized through the ring bridges of increasing length:

$$W_{i,i+2}(i+1) = (W_{i,i+1})^2 = 0.20, \quad W_{i,i+3}(i+1, i+2) = (W_{i,i+1})^3 = 0.09,$$
$$W_{i,i+4}(i+1, i+2, i+3) = (W_{i,i+1})^4 = 0.04, \quad W_{i,i+5}(i+1, i+2, i+3, i+4) = (W_{i,i+1})^5 = 0.02.$$

Thus, the longer the bridge, the smaller the indirect bond-order it contributes. The model and HF calculations on linear polyenes and representative polymers [75, 76] indicate that the range of bridge interactions is effectively extended up to the *third*-neighbors in the chain, where the direct interactions practically disappear.

The conditional probabilities defining the molecular information channel for *direct* communications between AO generate the associated covalency (noise) and ionicity (information flow) descriptors of the through-*space* chemical bonds. They can be also used to derive the corresponding cascade communications and the associated entropy/information descriptors of the *bridge*-bonds. This allows one to generate the IT descriptors of the most important *indirect* interactions, via chemical bridges between the specified (terminal) orbitals $\chi_i$ and $\chi_j$ from descriptors of the associated AO information cascades. The resulting overall indices of such indirect IT bond multiplicities compare favorably with the bridge-generalized Wiberg-type bond-orders.

## 14.4 Orbital Decoupling

There are two challenging MO-decoupling phenomena in the bond-multiplicity theory, which have to be adequately represented in all *bona fide* approaches: the so-called *populational*-decoupling for the "frozen" shapes of MO, leading to a steady decrease of the overall bond-multiplicity descriptors accompanying an increased electron occupation of the antibonding MO, and the *shape*-decoupling accompanying the bond-breaking processes, when AO are effectively prohibited from getting involved in the chemical bond formation (orbital communications), e.g., due to the symmetry requirements or bond-elongation. The above *fixed*-input OCT using the information *probability*-channels fails to predict a steady decrease in the resultant bond-order with increasing occupation of the antibonding MO (see Fig. 14.3) and to properly represent the normalization requirements of the input signals in the *shape*-decoupling limit. A satisfactory solution [14, 64] calls for an ensemble of the separate amplitude (*phase*-dependent) communication systems, due to either the specified input events or MO, with a flexible-input signal shaped by the orbital *conditional* probabilities.

**(a)**

$$\gamma_b = n_b \begin{bmatrix} P & \sqrt{PQ} \\ \sqrt{PQ} & Q \end{bmatrix}$$

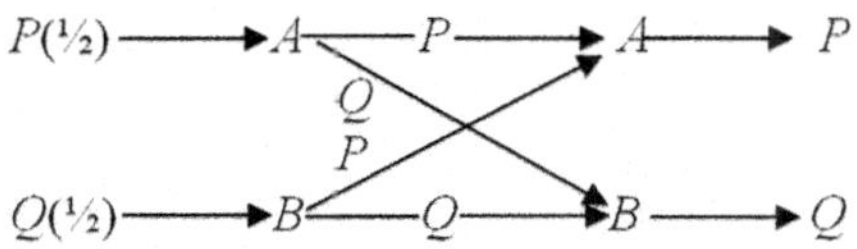


$$S(\varphi_b) = H(P)$$
$$I_0(\varphi_b) = H(½) - H(P) = 1 - H(P)$$
$$M_0(\varphi_b) = S(\varphi_b) + I_0(\varphi_b) = 1$$

**(b)**

$$\gamma_a = n_a \begin{bmatrix} Q & -\sqrt{PQ} \\ -\sqrt{PQ} & P \end{bmatrix}$$

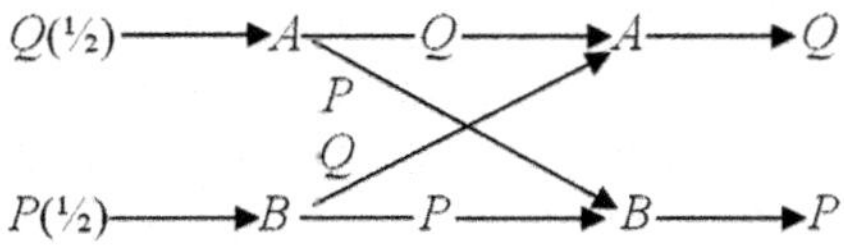


$$S(\varphi_a) = H(P)$$
$$I_0(\varphi_a) = H(½) - H(P) = 1 - H(P)$$
$$M_0(\varphi_a) = S(\varphi_a) + I_0(\varphi_a) = 1$$

**(c)**

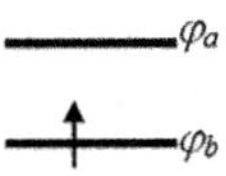


$$\gamma^0 = \begin{bmatrix} P & \sqrt{PQ} \\ \sqrt{PQ} & Q \end{bmatrix} \qquad \mathbf{P}^0(b|a) = \begin{bmatrix} P & Q \\ P & Q \end{bmatrix} \qquad p^{MO} = (1, 0)$$

$$S = ½\, S(\varphi_b) = ½ H(P) \qquad I_0 = ½\, I_0(\varphi_b) = ½[1 - H(P)] \qquad M_0 = ½ M_0(\varphi_b) = ½$$

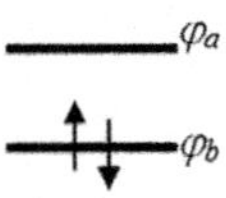


$$\gamma^0 = 2\begin{bmatrix} P & \sqrt{PQ} \\ \sqrt{PQ} & Q \end{bmatrix} \qquad \mathbf{P}^0(b|a) = \begin{bmatrix} P & Q \\ P & Q \end{bmatrix} \qquad p^{MO} = (1, 0)$$

$$S = S(\varphi_b) = H(P) \qquad I_0 = I_0(\varphi_b) = [1 - H(P)] \qquad M_0 = M_0(\varphi_b) = 1$$

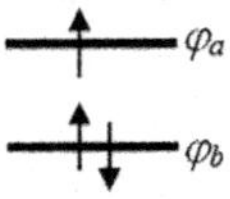


$$\gamma^0 = \begin{bmatrix} 2P+Q & \sqrt{PQ} \\ \sqrt{PQ} & 2Q+P \end{bmatrix} \qquad \mathbf{P}^0(b|a) = \begin{bmatrix} (P+1)^2/(3P+1) & PQ/(3P+1) \\ QP/(3Q+1) & (Q+1)^2/(3Q+1) \end{bmatrix} \qquad p^{MO} = (⅔, ⅓)$$

$$S = (2/3)\, S(\varphi_b) - (1/6)\, S(\varphi_a) = ½ H(P)$$
$$I_0 = (2/3)\, I_0(\varphi_b) - (1/6)\, I_0(a) = ½[1 - H(P)]$$
$$M_0 = (2/3)\, M_0(\varphi_b) - (1/6)\, M_0(\varphi_a) = ½$$

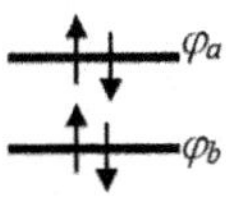


$$\gamma^0 = \begin{bmatrix} 2 & 0 \\ 0 & 2 \end{bmatrix} \qquad \mathbf{P}^0(b|a) = \begin{bmatrix} 1 & 0 \\ 0 & 1 \end{bmatrix} \qquad p^{MO} = (½, ½)$$

$$S = ½\, S(\varphi_b) - ½\, S(\varphi_a) = 0$$
$$I_0 = ½\, I_0(\varphi_b) - ½\, I_0(\varphi_a) = 0$$
$$M_0 = ½\, M_0(\varphi_b) - ½\, M_0(\varphi_a) = 0$$

**Fig. 14.3** Populational decoupling with increasing occupation of *anti*bonding MO in 2-AO model. Panels **a** and **b** summarize the bonding (**b**) and antibonding (**a**) MO probability channels, while Panel **c** reports the (occupation-weighted) entropic indices, which properly represent the populational decoupling phenomenon

Consider first the *populational*-decoupling problem in the simplest 2-AO model of the chemical bond in a diatomic molecule M = A—B. For $N = 3$, the ground-state corresponds to configuration $[\mathrm{M}(3)] = [\varphi_b^2\ \varphi_a^1]$ for which the *fixed*-input OCT predicts: $S = 0.47$, $I_0 = 0.48$ and $M_0 = 0.95$ (bits). Despite a single occupation of the antibonding MO $\varphi_a$ the predicted overall bond multiplicity $M_0$ thus remains almost the same as in the bonding configuration $[\mathrm{M}(2)]_b = [\varphi_b^2]$. Furthermore, this classical (probabilistic) approach cannot distinguish between configurations $[\mathrm{M}(1)]_b = [\varphi_b^1]$ and $[\mathrm{M}(1)]_a = [\varphi_a^1]$, or between $[\mathrm{M}(1)]_b$ and $[\mathrm{M}(2)]_b$, predicting the same bond indices reported in the first part of Fig. 14.3. For $[\mathrm{M}(4)]_n = [\varphi_b^2\ \varphi_a^2]$, one similarly predicts $S = 0$, $I_0 = M_0 = 1$ bit, which also contradicts the chemical intuition associated with this nonbonding state.

This failure of classical molecular channels is because the *probability* connections loose "memory" about relative phases of AO in MO. It is retained in elements of both the overall CBO matrix and their MO contributions, which reflect the associated *amplitudes* of AO communications, with their signs properly recognizing a chemical character of the mutual interaction between AO:

$$\begin{aligned}&bonding(\text{positive}): \gamma_{a,b}(\varphi_s) > 0; \quad nonbonding(\text{zero}): \gamma_{a,b}(\varphi_s) = 0\\ &antibonding(\text{negative}): \gamma_{a,b}(\varphi_s) < 0.\end{aligned} \tag{14.37}$$

To remedy this problem within the *probability*-channel, one thus has to examine separate MO channels $\{\mathbf{P}_s(\boldsymbol{b}|\boldsymbol{a})\}$, weight their diatomic IT bond-multiplicities in accordance with a degree of the MO occupation, and recognize in the corresponding resultant measures the character (sign) of the chemical interaction between the specified AO. Let us illustrate this procedure for the 2-AO model of the chemical bond A—B originating from the quantum–mechanical interaction between two AO: $\chi = (\chi_A \in A, \chi_B \in B)$, which define the whole AO basis of the model. The bond contributions for these two AO are determined by the conditional entropy $S_{a,b}(\varphi_s)$, mutual information $I_{a,b}(\varphi_s)$, and the overall bond-multiplicity $M_{a,b}(\varphi_s)$ descriptors of the fragment-renormalized information channel $\mathbf{P}_s(\boldsymbol{b}|\boldsymbol{a})$ for $\varphi_s$,

$$\begin{aligned}&S_{a,b}(\varphi_s) = S[\mathbf{P}_s(\boldsymbol{b}|\boldsymbol{a})], \quad I_{a,b}(\varphi_s) = S(p_s^0) - S_{a,b}(\varphi_s),\\ &M_{a,b}(\varphi_s) = S_{a,b}(\varphi_s) + I_{a,b}(\varphi_s) = S(\boldsymbol{p}_s^0),\end{aligned} \tag{14.38}$$

where $S(\boldsymbol{p}_s^0)$ stands for the Shannon entropy of the reference input probability $\boldsymbol{p}_s^0$. Alternatively, the purely molecular estimate of the mutual information descriptor $I_s[\boldsymbol{p}_{s:}\ \boldsymbol{p}_s]$ can be used to index the bond IT-iconicity.

In a larger $m > 2$ set of AO, the definition of such *two*-orbital MO indices for the specified pair $(i, j)$ of AO, $\{S_{i,j}(\varphi_s), I_{i,j}(\varphi_s), M_{i,j}(\varphi_s)\}$ requires an appropriate probability renormalization. The reference (promolecular) signal is then related to AO occupations $\boldsymbol{N}^0 = \{N_k^0\}$ in the separated (free) atoms,

$$p_s^0(ij) = \{p_k^0(ij) = N_k^0/N_{ij}^0, k \in (i,j)\}, \quad N_{ij}^0 = N_i^0 + N_j^0,$$

while its molecular analog involves the AO populations in $\varphi_s$:

$$p_s(ij) = \{p_{k,s}(ij) = N_k(\varphi_s)/N_{ij}(\varphi_s), k \in (i,j)\}, \quad N_k(\varphi_s) = \gamma_{k,k}(\varphi_s);$$
$$N_{ij}(\varphi_s) = N_i(\varphi_s) + N_j(\varphi_s).$$

One similarly extracts the appropriate *two*-orbital channel from the MO conditional-probability matrix $\mathbf{P}_s(\boldsymbol{b}|\boldsymbol{a})$:

$$\mathbf{P}_s(\boldsymbol{b}|\boldsymbol{a}); ij] = \{P_s[(l|k); ij] = P_s(l|k)/P_s(ij|k), k,\ l \in (i,j)\},$$
$$P_s(ij|k) = P_s(i|k) + P_s(j|k).$$

The bond descriptors for this pair of AO then read:

$$S_{i,j}(\varphi_s) = -\sum\nolimits_{k,l \in ij} p_{k,s}(ij)\, P_s[(l|k);\ ij] \log P_s[(l|k);\ ij],$$
$$I_{i,j}(\varphi_s) = \sum\nolimits_{k,l \in ij} p_k^0(ij)\, P_s[(l|k);\ ij] \log\{P_s[(l|k);\ ij]/p_k^0(ij)\}, \quad M_{i,j}(\varphi_s) = S_{i,j}(\varphi_s) + I_{i,j}(\varphi_s).$$

In combining such MO contributions into the corresponding resultant bond indices for the specified pair $(i, j)$ of AO, these increments should be multiplied by the MO occupation factor $\boldsymbol{f}^{MO} = \{f_s = n_s/2\}$, which recognizes that the full bonding/antibonding potential of the given MO is realized only when it is fully occupied, and by MO probability $\boldsymbol{P}^{MO} = \{P_s = n_s/N\}$. The resultant A—B descriptors are then obtained by summation of such occupation-weighted bonding or antibonding contributions from all occupied MO, which determine the bond system of a molecule:

$$S(i,\ j) = \sum_s sign\,[\gamma_{i,j}(\varphi_s)]\, P_s f_s S_{i,j}(\varphi_s), \quad I(i,\ j) = \sum_s sign\,[\gamma_{i,j}(\varphi_s)]\, P_s f_s I_{i,j}(\varphi_s),$$
$$M(i,\ j) = \sum_s sign\,[\gamma_{i,j}(\varphi_s)]\, P_s f_s M_{i,j}(\varphi_s). \tag{14.39}$$

As explicitly shown in the second part (Panel **c**) of Fig. 14.3, such weighed resultant indices adequately represent the *population*-decoupling trends in 2-AO model for $N = 1 \div 4$ electrons.

Resolving the *shape*-decoupling problem in classical probability channels calls for the *flexible*-input approach. The essence of this proposition lies in a determination of the entropy/information contributions due to each AO input $\{\chi_i\}$ in the molecular channel $\mathbf{P}(\boldsymbol{b}|\boldsymbol{a}) = \{P(j|i),\ (i,\ j) \in (1,\ 2,\ \ldots,\ m)\}$. These IT indices describe partial communications originating from $\chi_i$, i.e., the molecular subchannel determined by the $i$th row of $\mathbf{P}(\boldsymbol{b}|\boldsymbol{a})$, using separate probability distributions tailored for each AO input. The hitherto single molecular propagation of the molecular input signal $\boldsymbol{p}$ in the *fixed*-input approach is now replaced by series of $m$ molecular propagations of the separate signals $\boldsymbol{p}(i) = \{p(k; i)\}$ for each input $i = 1, 2, \ldots, m$. This partial propagation generates the associated IT-covalency descriptor,

$$S(i) = -\sum_k p(k;\, i) \sum_j P(j|k)\, \log P(j|k) \equiv \sum_k p(k;\, i)\, S(k; i) = S[\boldsymbol{p}(i)], \qquad (14.40)$$

molecular IT-ionic contribution,

$$I(i) = \sum_k p(k; i) \sum_j P(j|k)\, \log[P(j|k)/p(k; i)] \equiv \sum_k p(k; i)\, I(k; i) = I[\boldsymbol{p}(i)], \qquad (14.41)$$

and the corresponding overall bond-multiplicity index for $i$th input:

$$M(i) = I(i) + S(i) = M[\boldsymbol{p}(i)]. \qquad (14.42)$$

The bond descriptor of a molecule as a whole is then generated as the ensemble *average* of all such contributions determined in separate propagations using these separate input signals, weighted with the *molecular* AO probabilities $\boldsymbol{p} = \{p_i = \gamma_{i,i}/N\}$:

$$S_{av.} = \sum_i p_i S(i), \quad I_{av.} = \sum_i p_i I(i), \quad M_{av.} = \sum_i p_i M(i) = S_{av.} + I_{av.} \qquad (14.43)$$

There are some obvious sum-rules to be satisfied by these input-dependent probabilities. Consider first the fully coupled molecular channel, in which all orbitals are allowed to interact chemically, thus exhibiting nonvanishing direct communications with the system remainder. In this case, all molecular inputs have to be effectively probed to the full extent of the *unit* condensed probability of the molecule as a whole:

$$\sum_k p(k; i)] \equiv |\boldsymbol{p}(i)| = 1, \quad i = 1, 2, \ldots, m. \qquad (14.44)$$

This normalization requirement identifies a general category of these input-dependent probabilities as molecular *conditional* probabilities of *two*-orbital events: $p(k;\, i) \equiv P(k|i)$. Indeed, to make the AO decoupling continuous in this generalized description, the input probabilities $\{\boldsymbol{p}(i)\}$ have to reflect the actual participation of $i$th AO in chemical bonds (communications) with the remaining basis functions. One thus selects the *input*-tailored probabilities as the corresponding *rows* of the molecular conditional-probability matrix:

$$\boldsymbol{p}(i) = \boldsymbol{P}(\boldsymbol{b}|i) = \{P(k|i)\}, \quad i = 1, 2, \ldots, m. \qquad (14.45)$$

These partial input probabilities generate the following effective information system for $i$th input, called the *row*-subchannel [11],

$$\{P(k|i) \to k - P(j|k) \to j \to P(j|i)\}, \qquad (14.46)$$

where we have recognized the *complete*-cascade conservation of the direct scattering probabilities [see Eq. (14.14)]:

$$\sum_k P(k|i)P(j|k) = P(j|i). \tag{14.47}$$

In calculating the "ensemble" average bond components, the product

$$p_i \sum_k P(k|i)\, P(j|k) = p_i P(j|i) \equiv P(i,j) \tag{14.48}$$

thus represents the *joint* probability of orbitals $\chi_i$ and $\chi_j$ in the molecule.

Consider the conditional entropy contribution for $i$th input:

$$S(i) = -\sum_k P(k|i)\, [\sum_j P(j|k)\, \log_2 P(j|k)]. \tag{14.49}$$

In the *ensemble*-average quantity, this entropy covalency has to be weighted by the actual probability $p_i$ of this input in the molecule as a whole [Eq. (14.43)]. It can be directly verified that such averaging procedure indeed reproduces the molecular IT-covalency index:

$$\begin{aligned} S_{av.} &= \sum_i p_i S(i) \equiv \sum_i S_i = -\sum_i \sum_k \sum_j [p_i P(k|i)]\, P(j|k)\, \log_2 P(j|k) \\ &= -\sum_k \sum_j [\sum_i P(k,i)] P(j|k)\, \log_2 P(j|k) = -\sum_k \sum_j p_k P(j|k)\, \log_2 P(j|k) \\ &= -\sum_k \sum_j P(j,k)\, \log_2 P(j|k) = S. \end{aligned} \tag{14.50}$$

A similar averaging of the mutual information (IT-ionic) contributions,

$$\begin{aligned} I(i) &= \sum_k P(k|i)\, [\sum_j P(j|k)\, \log_2[P(j|k)/p_j] = -S(i) - \sum_j [\sum_k P(k|i)\, P(j|k)]\, \log_2 p_j \\ &= -S(i) - \sum_j P(j|i)\, \log_2 p_j, \end{aligned} \tag{14.51}$$

then gives:

$$\begin{aligned} I_{av.} &= \sum_i p_i I(i) \equiv \sum_i I_i = -S - \sum_i \sum_j p_i P(j|i)\, \log_2 p_j \\ &= -S - \sum_j [\sum_i P(i,j)]\, \log_2 p_j = -S + S(\boldsymbol{p}) = I. \end{aligned} \tag{14.52}$$

Therefore, the overall *molecular* bond index, the mean value of the input AO contributions

$$M(i) = S(i) + I(i) = -\sum_j P(j|i)\, \log_2 p_j,$$

then reproduces the Shannon entropy in molecular AO probabilities:

$$M_{av.} = \sum_i p_i M(i) \equiv \sum_i M_i = - \sum_j \left[\sum_i P(i,j)\right] \log_2 p_j = - \sum_j p_j \log_2 p_j = S(\boldsymbol{p}). \tag{14.53}$$

One also observes that the *input*-resolved quantities $\{S_i, I_i, M_i\}$ describe the following *resultant* channel for *i*th input [see Eq. (14.46)],

$$\begin{aligned} p_i \to \{P(k|i) \to k - P(j|k) \to j \to P(j|i)\} \\ \equiv \{P(i,k) \to k - P(j|k) \to j \to P(i,j)\}, \end{aligned} \tag{14.54}$$

with the *joint* probabilities of AO shaping the effective input signal of this *i*th *row*-subchannel.

## 14.5 Illustrative Applications

As an illustration, let us first consider the *flexible*-input generalization [Eq. (14.46)] of the 2-AO channel, shown in Fig. 14.4. We first observe that the *input*- and *output*-dependent distributions in this model are identical with the molecular distribution [see Eq. (14.33)]:

$$\boldsymbol{p}(\chi_A) = \boldsymbol{p}(\chi_B) = (P, Q) = \boldsymbol{p}.$$

Also reported in the diagram are the partial and average IT descriptors, including the IT-ionicity contributions relative to the reference probabilities $\boldsymbol{p}^0$ = (½, ½) of the *atomic*-promolecule, when two AO contribute a single electron each to form the chemical bond. The *flexible*-input generalization of this model channel is seen to exactly reproduce the overall IT bond-multiplicity and its components reported in Fig. 14.1.

It follows from the input (conditional) probabilities of Fig. 14.1 that in the limit of the decoupled *lone*-pair (nonbonding) MO, $\varphi_b = \chi_A$, for $P = 1$ and $Q = 0$, or $\varphi_b = \chi_B$, when $Q = 1$ and $P = 0$, the input probability of the doubly-occupied AO

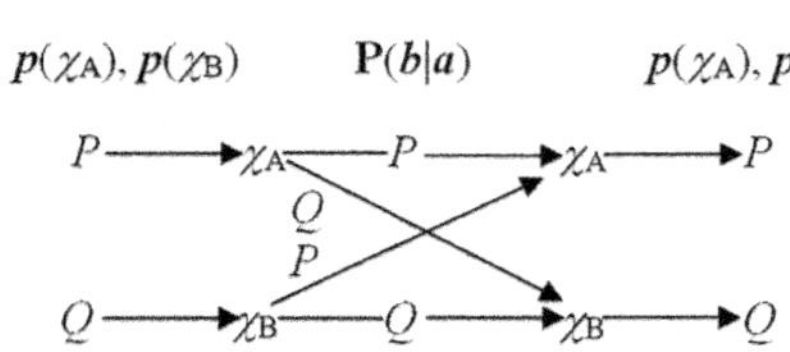


$S_A = PS(A) = PH(P)$ $\qquad$ $S_B = QS(B) = QH(P)$

$I_A = PI(A) = P[1-H(P)]$ $\qquad$ $I_B = QI(A) = Q[1-H(P)]$

$M_A = S_A + I_A = P$ $\qquad$ $M_B = S_B + I_B = Q$

$S_{av} = S_A + S_B = H(P) = S(P), \quad I_{av} = I_A + I_B = 1 - H(P) = I_0(P), \quad M_{av} = M_A + M_B = 1\text{bit} = M_0(P)$

**Fig. 14.4** *Flexible*-input generalization of the 2-AO channel. The corresponding AO partial and average entropy/information descriptors (in bits) of the chemical bond are also reported becomes 1, while that of the other, empty AO identically vanishes. The unit input probability of the doubly-occupied AO in the channel input is then deterministically transmitted to the same AO in the channel output, with the other (unoccupied) AO not participating in the channel deterministic communication, so that both orbitals do not contribute to the vanishing resultant IT-covalent index. The average ionicities, relative to *atomic*-promolecule $[\chi_A^1\,\chi_B^1]$, $I_{av.} = 1$ bit, and relative to *coordination*-promolecule $[\chi_A^2]$, $I_{av.} = 0$, then correctly predict the single bond multiplicity identifying the molecular *lone*-pair configuration $[\varphi_b^2] \equiv [\chi_A^2]$ as multiplicity of the ion-pair $[A^-B^+]$, *displaced* by an electron transfer relative to *atomic*-promolecule, and the vanishing multiplicity of the *undisplaced* configuration relative to the coordination promolecule.

In the limit $R_{AB} \to \infty$, the molecule A—B dissociates into atoms $[A] = [\chi_A^1]$ and $[B] = [\chi_B^1]$. Such decoupled AO correspond to the *equivalent* configurations $[\varphi_b^1\varphi_a^1]$ and $[\chi_A^1\,\chi_B^1]$, both producing identical Slater determinants: $|\varphi_b, \varphi_a| = |\chi_A, \chi_B|$. Indeed, using the orthogonal transformations between $\boldsymbol{\chi} = (\chi_A, \chi_B)$ and $\boldsymbol{\varphi} = (\varphi_b, \varphi_a)$,

$$\boldsymbol{\varphi} = \boldsymbol{\chi}\,\mathbf{C} \quad \text{and} \quad \boldsymbol{\chi} = \boldsymbol{\varphi}\begin{bmatrix} \sqrt{P} & \sqrt{Q} \\ -\sqrt{Q} & \sqrt{P} \end{bmatrix} \equiv \boldsymbol{\varphi}\,\mathbf{C}^{\mathrm{T}}, \quad \mathbf{C}^{\mathrm{T}}\mathbf{C} = \mathbf{C}\mathbf{C}^{\mathrm{T}} = \mathbf{I}, \qquad (14.55)$$

one can directly verify that $\boldsymbol{\gamma}[\varphi_b^1\varphi_a^1] = \mathbf{CC}^{\mathrm{T}} = \mathbf{I} = \mathbf{P}(\boldsymbol{b}|\boldsymbol{a})$, so that the decoupled AO inputs become: $\boldsymbol{p}(\boldsymbol{a}) = \boldsymbol{p}(\boldsymbol{a}^0) = (1,\ 0)$ and $\boldsymbol{p}(\boldsymbol{b}) = \boldsymbol{p}(\boldsymbol{b}^0) = (0,\ 1)$, each separately *unity*-normalized.

To summarize, while still retaining the essence of the *fixed*-input approach, the generalized proposition introduces in OCT the desired *input*-flexibility, which generates a continuity in the IT description of the fragment decoupling processes. In a common framework, the ensemble approach covers both the fully coupled AO in the molecule as well as the limiting cases of its subsystems being effectively decoupled in the molecular channel. In the former case, the resultant *input* signal corresponds to the unit norm of the probability distribution. In the case of *n*-mutually separated fragments, this flexible normalization is automatically increased to *n* by the choice of the molecular conditional probabilities as signals for each *row*-subchannel. The *flexible*-input generalization dramatically improves the agreement between predicted bond descriptors and the accepted chemical intuition [70]. It also has the conceptual and interpretative advantages, by providing a unifying description capable of tacking both the coupled and decoupled molecular fragments in a single theoretical framework. It generates a continuous description of the fragment dissociation (*shape*-decoupling) limit, when the separated subsystems exhibit vanishing mutual communications.

Consider next the $N = 3$ $\pi$-electrons in allyl, with the consecutive numbering of $2p_\pi = 2p_z \equiv z$ orbitals in the carbon chain. In Hückel's approximation, this $\pi$-system involves two MO,

$$\varphi_1 = \frac{1}{\sqrt{2}}\left[\frac{1}{\sqrt{2}}(z_1 + z_3) + z_2\right] \quad \text{(doubly occupied)},$$
$$\varphi_2 = \frac{1}{\sqrt{2}}(z_1 - z_3)\text{(singly occupied)}. \tag{14.56}$$

They generate the following MO and molecular CBO matrices:

$$\boldsymbol{\gamma}_1 = \frac{1}{2}\begin{bmatrix} 1 & \sqrt{2} & 1 \\ \sqrt{2} & 2 & \sqrt{2} \\ 1 & \sqrt{2} & 1 \end{bmatrix}, \quad \boldsymbol{\gamma}_2 = \frac{1}{2}\begin{bmatrix} 1 & 0 & -1 \\ 0 & 0 & 0 \\ -1 & 0 & 1 \end{bmatrix},$$
$$\boldsymbol{\gamma} = \boldsymbol{\gamma}_1 + \boldsymbol{\gamma}_2 = \frac{1}{2}\begin{bmatrix} 2 & \sqrt{2} & 0 \\ \sqrt{2} & 2 & \sqrt{2} \\ 0 & \sqrt{2} & 2 \end{bmatrix}. \tag{14.57}$$

The corresponding molecular and MO *probability*-channels are shown in Figs. 14.5 and 14.6, respectively [64]. The latter use the AO-input probabilities $\boldsymbol{p}_s = \{p(i|s) = \gamma_{i,i}(s)/n_s\}$. The overall channel predicts roughly 3/2 $\pi$-bonds in this molecular system, including a marginal IT-ionicity contribution, in full accord with chemical intuition.

There are no obvious combination formulas for grouping the partial MO bond indices of Fig. 14.6 into their overall analogs of Fig. 14.5. Indeed, the MO channels are determined by their own CBO structure, and a variety of their nonvanishing communication connections between AO generally differs from that of the system as a whole. Moreover, the input probabilities of Fig. 14.6 do not reflect the two MO channels being a part of the whole molecular channel. The latter requirement is satisfied only when, in spirit of the MO averaging of Fig. 14.3, the two networks are parallely coupled into the combined information system, in which the input probabilities are given by the corresponding products $\{\bar{\boldsymbol{p}}_s = P_s\, \boldsymbol{p}_s\}$, with MO probabilities $\boldsymbol{P}^{\mathrm{MO}} = \{P_s = n_s/N\} = (2/3,\ 1/3)$. In allyl, such a combination rule gives the following IT descriptors (in bits) of the two MO channels:

$$\bar{S}_1 = P_1 S(\varphi_1) = 1, \quad \bar{I}_1 = -P_1 \log_2 P_1 = 0.39 \quad \text{and}$$
$$\bar{S}_2 = P_2 S(\varphi_2) = 1/3, \quad \bar{I}_2 = -P_2 \log_2 P_2 = 0.53.$$

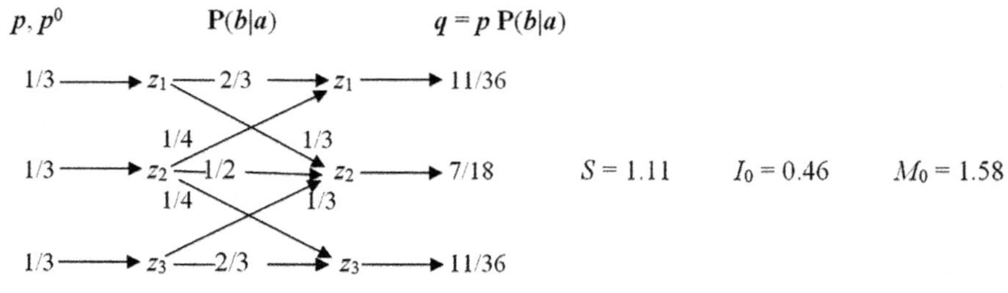


**Fig. 14.5** Overall probability channel for $\pi$-electrons in allyl and its IT bond indices in bits

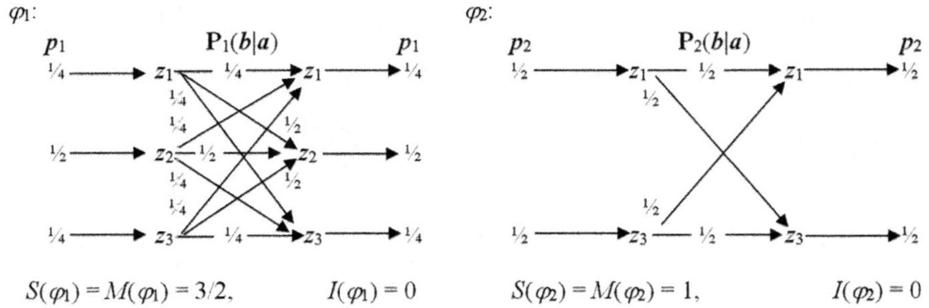


**Fig. 14.6** Molecular probability channels for two occupied $\pi$-MO in allyl. The corresponding bond contributions (in bits) for these MO communication systems are also reported

Such *molecular* inputs generate nonvanishing MO ionicities, which sum up to

$$\bar{I} = \bar{I}_1 + \bar{I}_2 = -\sum_s P_s \log_2 P_s = S(\boldsymbol{P}^{\text{MO}}) = 0.92.$$

The overall bond-multiplicity index of Fig. 14.5, $M = 1.58 = S(\boldsymbol{p}^0)$, predicting about 3/2 $\pi$-bond multiplicity in allyl, is reconstructed by adding to this ionicity measure the sum of the bonding (positive) IT-covalency $\bar{S}_1$ in first MO and the antibonding (negative) contribution $(-\bar{S}_2)$ due to the second MO:

$$\bar{S}_1 + (-\bar{S}_2) + \bar{I} = M.$$

The MO-weighting procedure of Fig. 14.3, with $f_1 = 1$ and $f_2 = ½$, predicts a roughly *single* $\pi$-bond in allyl:

$$\bar{M} = f_1(\bar{S}_1 + \bar{I}_1) - f_2(\bar{S}_2 + \bar{I}_2) \equiv f_1\bar{M}_1 - f_2\bar{M}_2 = 0.96.$$

It reflects the fact that only first (bonding) MO $\varphi_1$ is fully occupied, while the second (nonbonding) MO $\varphi_2$ groups practically separated AO on peripheral carbon atoms.

The delocalized $\pi$-bonds in butadiene are determined by two *doubly*-occupied canonical MO in Hückel approximation,

$$\varphi_1 = a(z_1 + z_4) + b(z_2 + z_3), \quad \varphi_2 = b(z_1 - z_4) + a(z_2 - z_3), \quad 2(a^2 + b^2) = 1;$$
$$a = \frac{1}{2}\sqrt{1 - 1/\sqrt{5}} = 0.372, \quad b = \frac{1}{2}\sqrt{1 + 1/\sqrt{5}} = 0.602, \tag{14.58}$$

with $\boldsymbol{P}^{\text{MO}} = (½, ½)$ and $\boldsymbol{f}^{\text{MO}} = (1, 1)$. The corresponding CBO matrices for these occupied MO and the whole $\pi$-electron system, respectively,

$$\boldsymbol{\gamma}_1 = 2\begin{bmatrix} a^2 & ab & ab & a^2 \\ ab & b^2 & b^2 & ab \\ ab & b^2 & b^2 & ab \\ a^2 & ab & ab & a^2 \end{bmatrix}, \quad \boldsymbol{\gamma}_2 = 2\begin{bmatrix} b^2 & ab & -ab & -b^2 \\ ab & a^2 & -a^2 & -ab \\ -ab & -a^2 & a^2 & ab \\ -b^2 & -ab & ab & b^2 \end{bmatrix}, \tag{14.59}$$

$$\boldsymbol{\gamma} = \boldsymbol{\gamma}_1 + \boldsymbol{\gamma}_2 = \frac{1}{\sqrt{5}}\begin{bmatrix} \sqrt{5} & 2 & 0 & -1 \\ 2 & \sqrt{5} & 1 & 0 \\ 0 & 1 & \sqrt{5} & 2 \\ -1 & 0 & 2 & \sqrt{5} \end{bmatrix}, \tag{14.60}$$

generate AO *probability*-channels shown in Figs. 14.7 and 14.8. The overall data of Fig. 14.8 correctly predict the overall double multiplicity of all $\pi$-bonds in butadiene. In this OCT treatment, they exhibit rather substantial IT-ionicity, which indicates a high degree of determinism (localization) in the orbital probability scattering. The bonding and antibonding components in $S_2$ cancel each other, when one attributes different signs to these AO contributions.

The group ionicity $\bar{I} = \bar{I}_1 + \bar{I}_2 = S(\boldsymbol{P}^{\mathrm{MO}}) = 1$, and hence, the overall MO-average bond multiplicity reads:

$$\bar{S}_1 + (\frac{1}{2}\bar{S}_2 - \frac{1}{2}\bar{S}_2) + \bar{I} = 1.925, \tag{14.61}$$

where $\bar{S}_s = P_s\, S(\varphi_s)$, thus again predicting roughly two bonds in this $\pi$-system.

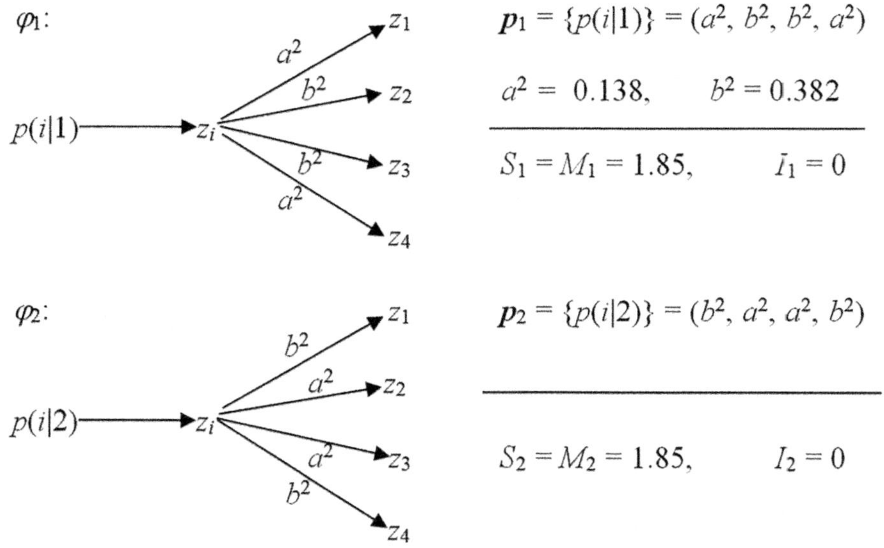


**Fig. 14.7** Hückel probability scattering in the occupied $\pi$-MO of butadiene for representative input orbital $z_i = 2p_{z,\,i}$, and the associated IT bond indices (in bits)

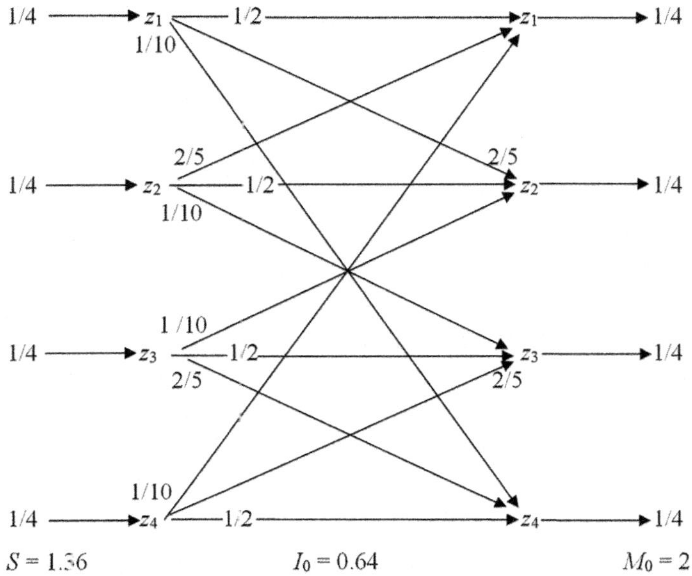


**Fig. 14.8** Overall probability channel for $\pi$-electrons in butadiene (Hückel theory) and its bond-multiplicity/composition descriptors in bits

Three occupied MO in Hückel theory, which determine $\pi$-bonds in benzene ring, $\boldsymbol{P}^{\mathrm{MO}} = (1/3)\ \boldsymbol{1}$, where $\boldsymbol{1}$ stands for the unit row matrix, read:

$$\begin{aligned}\varphi_1 &= \frac{1}{\sqrt{6}}(z_1 + z_2 + z_3 + z_4 + z_5 + z_6),\\ \varphi_2 &= \frac{1}{2}(z_1 + z_2 - z_4 - z_5),\\ \varphi_3 &= \frac{1}{\sqrt{12}}(z_1 - z_2 - 2z_3 - z_4 + z_5 + 2z_6).\end{aligned} \tag{14.62}$$

They give rise to the overall CBO matrix elements reflecting the electron population on orbital $\chi_i = z_i$, $\gamma_{i,i} = 1$, and mutual coupling between AO in relative *ortho*-, *meta*-, and *para*-positions: $\gamma_{i,i+1} = 2/3$, $\gamma_{i,i+2} = 0$, $\gamma_{i,i+3} = -1/3$. The resultant scattering of AO probabilities of $\pi$-electrons in benzene is summarized in Fig. 14.9. The predicted overall IT-multiplicity of $\pi$-bonds is somewhat lower than $M_0 = 3$ predicted for the three localized $\pi$-bonds in the alternated hexagon structure of cyclohexatriene. This is because in benzene, the $\pi$-bond alternation is prevented

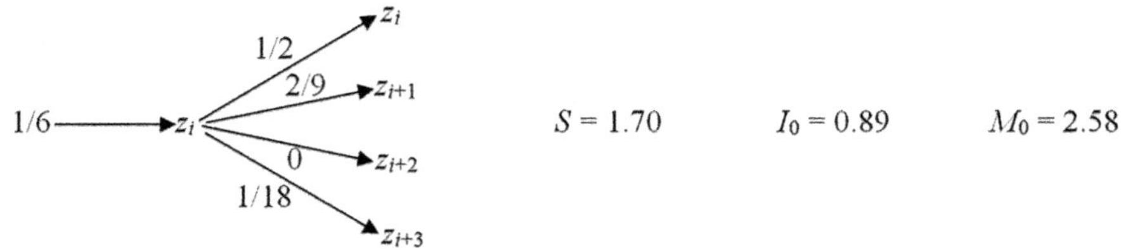


**Fig. 14.9** Probability scattering in benzene (Hückel theory) for representative input orbital $z_i = 2p_{z,\,i}$, and the molecular entropy/information descriptors of the $\pi$-electron probability channel

by stronger $\sigma$-bonds, which assume their maximum strength in the regular hexagon structure [106].

Let us explore the CBO matrices of the benzene occupied MO:

$$\boldsymbol{\gamma}_s = \langle \chi|\varphi_s\rangle n_s \langle \varphi_s|\chi\rangle = 2\langle \chi|\hat{\mathrm{P}}_s|\chi\rangle, \quad \chi = (z_1, z_2, \ldots, z_6), \quad s = 1, 2, 3.$$

In $\boldsymbol{\gamma}_1 = (1/3)\,\mathbf{1}$, where all elements in the square matrix $\mathbf{1}$ are equal to 1, all matrix elements are positive (bonding), while a half of them in $\boldsymbol{\gamma}_2$ and $\boldsymbol{\gamma}_3$ is negative, thus representing the antibonding interactions between AO. The nonvanishing elements in $\boldsymbol{\gamma}_2$ are limited to the subset $\chi' = (z_1, z_2, z_4, z_5)$:

$$\boldsymbol{\gamma}_2 = 2\langle \chi|\hat{\mathrm{P}}_2|\chi\rangle = \frac{1}{2}\begin{bmatrix} 1 & 1 & -1 & -1 \\ 1 & 1 & -1 & -1 \\ -1 & -1 & 1 & 1 \\ -1 & -1 & 1 & 1 \end{bmatrix}, \tag{14.63}$$

while $\boldsymbol{\gamma}_3$ explores the whole AO basis:

$$\boldsymbol{\gamma}_3 = 2\langle \chi|\hat{\mathrm{P}}_3|\chi\rangle = \frac{1}{6}\begin{bmatrix} 1 & -1 & -2 & -1 & 1 & 2 \\ -1 & 1 & 2 & 1 & -1 & -2 \\ -2 & 2 & 4 & 2 & -2 & -4 \\ -1 & 1 & 2 & 1 & -1 & -2 \\ 1 & -1 & -2 & -1 & 1 & 2 \\ 2 & -2 & -4 & -2 & 2 & 4 \end{bmatrix}. \tag{14.64}$$

These partial CBO matrices give rise to the following AO communications and input probabilities in the associated probability channels of the occupied MO:

$$
\mathbf{P}_1(\boldsymbol{b}|\boldsymbol{a}) = \frac{1}{6}\,\mathbf{1}, \quad \boldsymbol{p}_1 = (1/6)\,1;
$$

$$
\mathbf{P}_2(\boldsymbol{b}|\boldsymbol{a}) = \frac{1}{4}\begin{bmatrix} 1 & 1 & 0 & 1 & 1 & 0 \\ 1 & 1 & 0 & 1 & 1 & 0 \\ 0 & 0 & 0 & 0 & 0 & 0 \\ 1 & 1 & 0 & 1 & 1 & 0 \\ 1 & 1 & 0 & 1 & 1 & 0 \\ 0 & 0 & 0 & 0 & 0 & 0 \end{bmatrix}, \quad \boldsymbol{p}_2 = 1/4(1,\ 1,\ 0,\ 1,\ 1,\ 0);
$$

$$
\mathbf{P}_3(\boldsymbol{b}|\boldsymbol{a}) = \frac{1}{12}\begin{bmatrix} 1 & 1 & 4 & 1 & 1 & 4 \\ 1 & 1 & 4 & 1 & 1 & 4 \\ 1 & 1 & 4 & 1 & 1 & 4 \\ 1 & 1 & 4 & 1 & 1 & 4 \\ 1 & 1 & 4 & 1 & 1 & 4 \\ 1 & 1 & 4 & 1 & 1 & 4 \end{bmatrix}, \quad \boldsymbol{p}_3 = (1/12)(1,\ 1,\ 4,\ 1,\ 1,\ 4). \tag{14.65}
$$

Their entropy/information descriptors read:

$$
\begin{aligned}
S(\varphi_1) &= M(\varphi_1) = 2.58, I(\varphi_1) = 0; \\
S(\varphi_2) &= M(\varphi_2) = 2, I(\varphi_2) = 0; \\
S(\varphi_3) &= M(\varphi_3) = 2.25, I(\varphi_3) = 0.
\end{aligned}
$$

The resultant MO-average iconicity descriptor $\bar{I} = \bar{I}_1 + \bar{I}_2 + \bar{I}_3 = S(\boldsymbol{P}^{MO}) = 1.58$ and $\bar{S}_1 = S(\varphi_1)/3$ also give rise to roughly 2.5 bits of the IT bond-multiplicity, with the bonding (positive) and negative (antibonding) contributions in $\bar{S}_2$ and $\bar{S}_3$ approximately canceling each other.

## 14.6 Amplitude Communications in Valence-Bond Structures

In this section, we compare the probability and amplitude-averaging schemes over configurations [107], using the 2-AO model of the chemical bond A—B as an illustrative example. The original VB [82] description of the homonuclear bond in $H_2$, at equilibrium internuclear separation $R = 1.40$ a.u., uses the overlapping $1s$ orbitals contributed by both hydrogens, $\chi = (\chi_B, \chi_A)$, $S_{A,B} = \langle \chi_A | \chi_B \rangle = 0.75$, and introduces two prototype chemical structures,

$$
\Psi_{cov.}(2) = N\{[\mathrm{A-B}] + [\mathrm{B-A}]\}, \quad \Psi_{ion.}(2) = N\{[\mathrm{A^-B^+}] + [\mathrm{A^+B^-}]\}, \\ N = 0.566, \tag{14.66}
$$

expressed in terms of the four elementary AO products,

$$\begin{aligned} &covalent: \quad \{[\mathrm{A}-\mathrm{B}]=\chi_{\mathrm{A}}(1)\chi_{\mathrm{B}}(2), \quad [\mathrm{B}-\mathrm{A}]=\chi_{\mathrm{B}}(1)\chi_{\mathrm{A}}(2)\}, \quad \text{and} \\ &ionic: \quad \{[\mathrm{A}^-\mathrm{B}^+]=\chi_{\mathrm{A}}(1)\chi_{\mathrm{A}}(2), \quad [\mathrm{A}^+\mathrm{B}^-]=\chi_{\mathrm{B}}(1)\chi_{\mathrm{B}}(2)\}, \end{aligned} \tag{14.67}$$

which span the model *two*-electron Hilbert space.

One recalls that this classical description in terms of overlapping AO suffers from a severe nonorthogonality problem. For the equilibrium internuclear separation, $\langle \Psi^{ion}|\Psi^{\mathrm{cov}}\rangle = 0.96$, and this hampers a clear chemical interpretation of independent bond components in quantum chemistry and hinders an interpretation of the bond information origins and of its covalent/ionic composition in the electronic ground state:

$$\Psi^{\mathrm{VB}}(2)=0.801\,\Psi_{cov.}(2)+0.206\,\Psi_{ion.}(2). \tag{14.68}$$

One also observes that the VB structures $\Psi_{cov.}(2)$ and $\Psi_{ion.}(2)$ also represent the reference *atomic* and *ionic* dissociation limits, respectively, with the former attributing electrons to different atoms and the latter locating both electrons on a single atom. The ionic structure indeed represents an equal participation of the two admissible *ion*-pairs, relative to the assumed (atomic) promolecular reference, while the covalent structure corresponds to the equal participation of the two covalent (electron-sharing) products.

One way to extract the independent ionic component is to Schmidt-orthogonalize $\Psi_{ion.}$ (2) with respect to $\Psi_{cov.}$ (2): $\Phi_{ion.}$ (2) = 3.362 $\Psi_{ion.}$ (2)– 3.428 $\Psi_{cov.}$ (2). This gives

$$\Psi^{\mathrm{VB}}(2)=0.998\,\Psi_{cov.}(2)+0.058\,\Phi_{ion.}(2), \tag{14.69}$$

and hence predicts the bond $(0.058)^2 = 0.003$ independent ionic character, from quantum-mechanical SP, i.e., a practically pure covalent chemical bond [108]. This is in sharp contrast to 32% iconicity predicted in Shull's Natural Orbital model [108–110].

The alternative approach uses the symmetric (Löwdin) orthogonalization of AO basis into the associated orthonormal AO (OAO) set,

$$\tilde{\chi}=\chi\begin{bmatrix} 1.378 & -0.622 \\ -0.622 & 1.378 \end{bmatrix}=(\tilde{\chi}_{\mathrm{A}},\,\tilde{\chi}_{\mathrm{B}}), \tag{14.70}$$

which generates the associated orthonormal VB structures:

$$\begin{aligned} \Phi^{cov.}(2)&=2^{-1/2}[\tilde{\chi}_{\mathrm{A}}(1)\tilde{\chi}_{\mathrm{B}}(2)+\tilde{\chi}_{\mathrm{B}}(1)\tilde{\chi}_{\mathrm{A}}(2)], \\ \Phi^{ion.}(2)&=2^{-1/2}[\tilde{\chi}_{\mathrm{A}}(1)\tilde{\chi}_{\mathrm{A}}(2)+\tilde{\chi}_{\mathrm{B}}(1)\tilde{\chi}_{\mathrm{B}}(2)]. \end{aligned} \tag{14.71}$$

At the equilibrium bond-length, these two sets of VB configurations are related by transformations:

$$[\Phi^{cov.}, \Phi^{ion.}] = [\Psi_{cov.}, \Psi_{ion.}]\begin{bmatrix} 2.587 & -2.142 \\ -2.142 & 2.587 \end{bmatrix} \quad \text{and}$$
$$\Psi_{cov.}, \Psi_{ion.}] = [\Phi^{cov.}, \Phi^{ion.}]\begin{bmatrix} 0.799 & 0.599 \\ 0.599 & 0.799 \end{bmatrix}, \tag{14.72}$$

which allow one to transform Eq. (14.68) into expression in terms of orthogonal VB structures:

$$\Psi^{\mathrm{VB}}(2) = 0.764\,\Phi^{cov.}(2) + 0.645\,\Phi^{ion.}(2). \tag{14.73}$$

For $H_2$, when model probabilities equalize, $P = Q = ½$, these orthonormal structures can be expressed in terms of the spatial functions of two (orthonormal) MO configurations $\boldsymbol{\Psi} = \{\Psi_\alpha\}$ of the familiar CID expansion, "bonding," $\Psi_b(2) = \det[\varphi_b^+ \varphi_{\bar{b}}^-]$, and "antibonding," $\Psi_a(2) = \det[\varphi_a^+ \varphi_{\bar{a}}^-]$,

$$\Phi^{cov/ion} = \sum\nolimits_{\alpha = b, a} c_\alpha^{cov/ion} \Psi_\alpha, \tag{14.74}$$

determined by the two MO combinations $\varphi_b = 2^{-1/2}\ (\tilde{\chi}_{\mathrm{A}} + \tilde{\chi}_{\mathrm{B}})$ and $\varphi_a = 2^{-1/2}$ $(\tilde{\chi}_{\mathrm{B}} - \tilde{\chi}_{\mathrm{A}})$:

$$\Phi^{cov.}(2) \equiv 2^{-1/2}[\Psi_b(2) - \Psi_a(2)] \quad \text{and} \quad \Phi^{ion.}(2) \equiv 2^{-1/2}[\Psi_b(2) + \Psi_a(2)],$$
$$\Psi_b(2) \equiv 2^{-1/2}[\Phi^{cov.}(2) + \Phi^{ion.}(2)] \quad \text{and} \quad \Psi_a(2) \equiv 2^{-1/2}[\Phi^{ion.}(2) - \Phi^{cov.}(2)]. \tag{14.75}$$

The two independent "chemical" states of VB theory correspond to equal configuration probabilities $\{p_\alpha^{cov/ion} = |c_\alpha^{cov/ion}|^2\}$,

$$\boldsymbol{p}^{ion.} = \boldsymbol{p}^{cov.} = (1/2, 1/2),$$

but differ in phases of the configuration coefficients $\{c_\alpha^{cov/ion}\}$. The associated expression for the VB ground state then reads:

$$\Psi^{\mathrm{VB}}(2) = 0.996\,\Psi_b(2) - 0.084\,\Psi_a(2). \tag{14.76}$$

Let us now examine the OCT predictions from two admissible averaging schemes over configurations [107]. The classical *probability*-averaging over configurations,

$$\langle \mathbf{P}^{ion./cov.}(\tilde{\chi}'|\tilde{\chi})\rangle_{ens.} = \sum\nolimits_{\alpha = a, b} p_\alpha^{ion./cov.} \mathbf{P}^{(\alpha)}(\tilde{\chi}'|\tilde{\chi}), \tag{14.77}$$

generates identical ensemble-average communications in both VB structures,

$$\langle \mathbf{P}^{cov.}(\tilde{\chi}'|\tilde{\chi})\rangle_{ens.} = \langle \mathbf{P}^{ion.}(\tilde{\chi}'|\tilde{\chi})\rangle_{ens.} = 1/2\,[\mathbf{P}^{(b)}(\tilde{\chi}'|\tilde{\chi}) + \mathbf{P}^{(a)}(\tilde{\chi}'|\tilde{\chi})] = \frac{1}{2}\mathbf{1}, \quad (14.78)$$

where (see Fig. 14.3a, b):

$$\mathbf{P}^{(b)}(\tilde{\chi}'|\tilde{\chi}) = \begin{bmatrix} P & Q \\ P & Q \end{bmatrix} \quad \text{and} \quad \mathbf{P}^{(a)}(\tilde{\chi}'|\tilde{\chi}) = \begin{bmatrix} Q & P \\ Q & P \end{bmatrix}. \quad (14.79)$$

The *ensemble*-average channel, $\langle \mathbf{P}(\tilde{\chi}'|\tilde{\chi})\rangle_{ens.} = ½\ \mathbf{1}$, common to both VB structures, represents in OCT the purely covalent communication system, $\langle S(\tilde{\chi}'|\tilde{\chi})\rangle_{ens.} = 1$ bit and $\langle I^0(\tilde{\chi}:\tilde{\chi}')\rangle_{ens.} = 0$, of the complete dissipation of the initial (input) information into the communication "noise." It thus reflects the maximum IT-covalency and minimum IT-ionicity in the 2-AO model of the chemical bond (see Fig. 14.2).

Therefore, the classical (probability) averaging scheme does not reveal the basic chemical difference between the two VB structures, which is embodied in *phases* of coefficients multiplying the MO configurations in the two VB combinations of Eq. (14.74). It also wrongly predicts the identical *ensemble*-average bond-orders between AO in these two chemical states,

$$\langle \boldsymbol{\gamma}^{ion./cov.}\rangle_{ens.} = \sum\nolimits_{a=b,a} p_\alpha^{ion./cov.}\boldsymbol{\gamma}^{(\alpha)} = 1/2(\boldsymbol{\gamma}^{(b)} + \boldsymbol{\gamma}^{(a)}) = \begin{bmatrix} 1 & 0 \\ 0 & 1 \end{bmatrix} \equiv \mathbf{I} = \boldsymbol{\gamma}^{(nb)},$$
$$\boldsymbol{\gamma}^{(b)} = 2\begin{bmatrix} P & \sqrt{PQ} \\ \sqrt{QP} & Q \end{bmatrix} \quad \text{and} \quad \boldsymbol{\gamma}^{(a)} = 2\begin{bmatrix} Q & \sqrt{QP} \\ \sqrt{PQ} & P \end{bmatrix}, \quad (14.80)$$

which mark the nonbonded (*nb*) status of AO, for the vanishing chemical bond!

Let us next examine the *amplitude*-averaging scheme giving rise to the *resultant* covalent amplitudes:

$$\langle \mathbf{A}^{cov.}(\tilde{\chi}'|\tilde{\chi})\rangle_{res.} = \sum\nolimits_{\alpha=a,b} c_\alpha^{cov.}\mathbf{A}^{(a)}(\tilde{\chi}'|\tilde{\chi}) = \frac{1}{\sqrt{2}}\left(\begin{bmatrix} \sqrt{P} & \sqrt{Q} \\ \sqrt{P} & \sqrt{Q} \end{bmatrix} - \begin{bmatrix} \sqrt{Q} & -\sqrt{P} \\ -\sqrt{Q} & \sqrt{P} \end{bmatrix}\right)$$
$$= \frac{1}{\sqrt{2}}\begin{bmatrix} \sqrt{P}-\sqrt{Q} & \sqrt{P}+\sqrt{Q} \\ \sqrt{P}+\sqrt{Q} & \sqrt{Q}-\sqrt{P} \end{bmatrix}. \quad (14.81)$$

Their squares generate the associated *resultant* communications between AO in this covalent state (see Fig. 14.10):

$$\langle \mathbf{P}^{cov.}(\tilde{\chi}'|\tilde{\chi})\rangle_{res} = \frac{1}{2}\begin{bmatrix} 1-2\sqrt{PQ} & 1+2\sqrt{PQ} \\ 1+2\sqrt{PQ} & 1-2\sqrt{PQ} \end{bmatrix}. \quad (14.82)$$

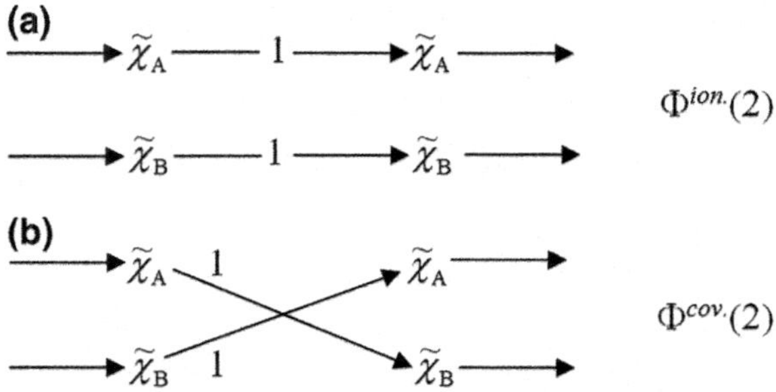


**Fig. 14.10** Diagonal (Panel **a**) and offdiagonal (Panel **b**) resultant communications, corresponding to prototype (orthonormal) ionic and covalent VB structures, respectively

In the homonuclear case, when $P = Q = ½$, this *resultant* channel represents the deterministic offdiagonal (*inter*-orbital) communications of Fig. 14.10b, between *different* AO [53], which conforms to the electron-*sharing* intuition behind the covalent bond component.

For the ionic configuration, one similarly finds the following *resultant* amplitudes,

$$\langle A^{ion.}(\tilde{\chi}'|\tilde{\chi})\rangle_{res.} = \sum_{\alpha=a,b} c_\alpha^{ion.} A^{(\alpha)}(\tilde{\chi}'|\tilde{\chi})$$
$$= \frac{1}{\sqrt{2}}\left(\begin{bmatrix}\sqrt{P} & \sqrt{Q}\\ \sqrt{P} & \sqrt{Q}\end{bmatrix} + \begin{bmatrix}\sqrt{Q} & -\sqrt{P}\\ -\sqrt{Q} & \sqrt{P}\end{bmatrix}\right) = \frac{1}{\sqrt{2}}\begin{bmatrix}\sqrt{P}+\sqrt{Q} & \sqrt{Q}-\sqrt{P}\\ \sqrt{P}-\sqrt{Q} & \sqrt{Q}+\sqrt{P}\end{bmatrix}, \tag{14.83}$$

the squares of which generate the resultant scattering probabilities:

$$\langle \mathbf{P}^{ion.}(\tilde{\chi}'|\tilde{\chi})\rangle_{res.} = \frac{1}{2}\begin{bmatrix}1+2\sqrt{PQ} & 1-2\sqrt{PQ}\\ 1-2\sqrt{PQ} & 1+2\sqrt{PQ}\end{bmatrix}. \tag{14.84}$$

In the homonuclear $H_2$ case, when $P = Q = ½$, this resultant channel represents the noiseless diagonal (*intra*-orbital) communications [53] of Fig. 14.10a.

To summarize, the amplitude-interference scheme in the CID expansion distinguishes between the chemical characters of both VB structures and gives the correct prediction of the resultant communications between AO: orbital-*mixing* in the covalent state and orbital-*localizing* in the ionic structure. These two prototype chemical states separate in the resultant channel the *additive* (ionic, diagonal) communications in $H_2$,

$$|\chi_A\rangle \rightarrow |\chi_A\rangle \quad \text{and} \quad |\chi_B\rangle \rightarrow |\chi_B\rangle,$$

from the *nonadditive* (covalent, offdiagonal) probability propagations:

$$|\chi_A\rangle \rightarrow |\chi_B\rangle \quad \text{and} \quad |\chi_B\rangle \rightarrow |\chi_A\rangle.$$

The two VB structures of the chemical bond in $H_2$ are seen to represent the independent (noiseless) 2-electron components of the effective OAO-promotion

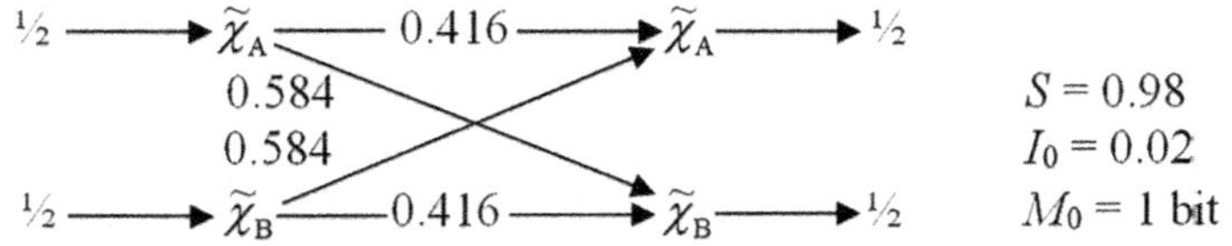


**Fig. 14.11** Resultant OAO-promotion channel in VB ground-state and its entropic bond descriptors (in bits)

channel, with the IT-covalency (noise) then being generated solely by the CI coefficients in the molecular ground state [Eq. (14.76)].

This dichotomous distinction of the electron AO communications in two chemical states of the orthogonal VB (OVB) structures allows one to generate the resultant AO-promotion amplitudes of the system ground state [see Eq. (14.73)],

$$\langle \mathbf{A}^{\mathrm{VB}}(\tilde{\chi}'|\tilde{\chi})\rangle_{res.} = 0.764\langle \mathbf{A}^{cov.}(\tilde{\chi}'|\tilde{\chi})\rangle_{res.} + 0.645\langle \mathbf{A}^{ion.}(\tilde{\chi}'|\tilde{\chi})\rangle_{res.}, \tag{14.85}$$

which generates the resultant probability-channel shown in Fig. 14.11. It is seen to generate 2% iconicity of the overall (1 bit) IT bond-multiplicity, thus again reflecting the bond practically purely covalent character. This result is qualitatively similar to that obtained from the Schmidt orthogonalization [Eq. (14.69)].

One can also directly verify that the configuration expansion of Eq. (14.76) generates the same OAO-promotion channel in VB ground state:

$$\begin{aligned} &\langle \mathbf{A}^{\mathrm{VB}}(\tilde{\chi}'|\tilde{\chi})\rangle_{res.} = 0.996\langle \mathbf{A}_b(\tilde{\chi}'|\tilde{\chi})\rangle_{res.} - 0.084\langle \mathbf{A}_a(\tilde{\chi}'|\tilde{\chi})\rangle_{res.},\\ &\langle \mathbf{A}_b(\tilde{\chi}'|\tilde{\chi})\rangle_{res.} = \frac{1}{\sqrt{2}}\begin{bmatrix} 1 & 1 \\ 1 & 1 \end{bmatrix}, \quad \langle \mathbf{A}_a(\tilde{\chi}'|\tilde{\chi})\rangle_{res.} = \frac{1}{\sqrt{2}}\begin{bmatrix} 1 & -1 \\ -1 & 1 \end{bmatrix}. \end{aligned} \tag{14.86}$$

## 14.7 Conclusion

Predictions from OCT are explicitly basis-set-dependent [64, 70], since alternative choices of these elementary atomic functions ultimately identify different discrete resolution levels ("events") of molecular communication systems. The minimum basis of AO occupied in the separated atoms or the related subset of an extended basis set transformed to exhibit the maximum resemblance to the minimum basis has been found to generate the IT interpretation in close agreement with chemical intuition [70].

The OCT extends our understanding of the complex chemical bond phenomenon from the complementary IT viewpoint. This perspective on the entropic origins of chemical bonds is thus very much in spirit of the Eugene Wigner's observation, often quoted by Walter Kohn, that the understanding in science requires insights from several different points of view. Indeed, the IT probes of molecular systems and chemical reactions generate such an additional perspective on both the genesis of chemical bonds and the elementary reaction mechanisms. It complements the familiar MO interpretations of quantum chemistry and ultimately gives rise to a deeper understanding of these complex phenomena.

## References


1. R.A. Fisher, Proc. Camb. Phil. Soc. **22**, 700 (1925)
2. C.E. Shannon, Bell Syst. Tech. J. **27**, 379, 623 (1948); C.E. Shannon, W. Weaver, *The Mathematical Theory of Communication* (University of Illinois, Urbana, 1949)
3. S. Kullback, R.A. Leibler, Ann. Math. Stat. **22**, 79 (1951)
4. S. Kullback, *Information Theory and Statistics* (Wiley, New York, 1959)
5. N. Abramson, *Information Theory and Coding* (McGraw-Hill, New York, 1963)
6. P.E. Pfeifer, *Concepts of Probability Theory*, 2nd edn. (Dover, New York, 1978)
7. B.R. Frieden, *Physics from the Fisher Information—A Unification*, 2nd edn. (Cambridge University Press, Cambridge, 2004)
8. R.F. Nalewajski, Int. J. Quantum Chem. **108**, 2230 (2008)
9. R.F. Nalewajski, Chem. Phys. Lett. **372**, 28 (2003)
10. R.F. Nalewajski, Chem. Phys. Lett. **375**, 196 (2003)
11. R.F. Nalewajski, *Information Theory of Molecular Systems* (Elsevier, Amsterdam, 2006)
12. R.F. Nalewajski, *Information Origins of the Chemical Bond* (Nova, New York, 2010)
13. R.F. Nalewajski, *Perspectives in Electronic Structure Theory* (Springer, Heidelberg, 2012)
14. R.F. Nalewajski, *Quantum Information Theory of Molecular States* (Nova, New York, 2016)
15. S.B. Sears, Ph.D. Thesis, The University of North Carolina at Chapel Hill (1980)
16. S.B. Sears, R.G. Parr, U. Dinur, Isr. J. Chem. **19**, 165 (1980)
17. S.R. Gadre, in *Reviews in Modern Quantum Chemistry: A Cellebration of the Contributions of R. G. Parr*, Vol. I, ed. by K. D. Sen (World Scientific, Singapore, 2002), p. 108
18. R.F. Nalewajski, R.G. Parr, Proc. Natl. Acad. Sci. U.S.A. **97**, 8879 (2000)
19. R.F. Nalewajski, J. Phys. Chem. A **104**, 11940 (2000)
20. R.F. Nalewajski, R.G. Parr, J. Phys. Chem. A **105**, 7391 (2001)
21. R.F. Nalewajski, Phys. Chem. Chem. Phys. **4**, 1710 (2002)
22. R.F. Nalewajski, R. Loska, Theor. Chem. Acc. **105**, 374 (2001)
23. R.F. Nalewajski, J. Phys. Chem. A **107**, 3792 (2003)
24. R.F. Nalewajski, Adv. Quantum Chem. **43**, 119 (2003)
25. R.F. Nalewajski, in *Frontiers in Modern Theoretical Chemistry: Concepts and Methods*, ed. by P.K. Chattaraj, S.K. Ghosh (Taylor & Francis/CRC, London, 2013), p. 143
26. R.F. Nalewajski, Struct. Bond. **149**, 51 (2012)
27. R.F. Nalewajski, E. Switka, A. Michalak, Int. J. Quantum Chem. **87**, 198 (2002)
28. R. F. Nalewajski and K. Jug, in *Reviews in Modern Quantum Chemistry: A Cellebration of the Contributions of R. G. Parr*, Vol. I, ed. by K.D. Sen (World Scientific, Singapore, 2002), p. 148
29. R.F. Nalewajski, E. Broniatowska, J. Phys. Chem. A **107**, 6270 (2003)
30. R.F. Nalewajski, E. Broniatowska, Chem. Phys. Lett. **376**, 33 (2003)
31. R.F. Nalewajski, E. Broniatowska, Theoret. Chem. Acc. **117**, 7 (2007)
32. R.F. Nalewajski, P. de Silva, J. Mrozek, J. Mol. Struct: THEOCHEM **954**, 57 (2010)
33. R.F. Nalewajski, J. Math. Chem. **47**, 667 (2010); R.F. Nalewajski, Found. Chem. **16**, 27 (2014)
34. R.F. Nalewajski, P. de Silva, J. Mrozek, in *Theoretical and Computational Developments in Modern Density Functional Theory*, ed. by A.K. Roy (Nova, New York, 2012), p. 561
35. R.F. Nalewajski, in *Advances in Quantum Systems Research*, ed. by Z. Ezziane (Nova Science Publishers, New York, 2014), p. 119

36. R.G. Parr, P.W. Ayers, R.F. Nalewajski, J. Phys. Chem. A **109**, 3957 (2005)
37. R.F. Nalewajski, A.M. Köster, S. Escalante, J. Phys. Chem. A **109**, 10038 (2005)
38. W.L. Luken, D.N. Beratan, Theor. Chim. Acta (Berl.) **61**, 265 (1984)
39. W.L. Luken, J.C. Culberson, Theor. Chim. Acta (Berl.) **66**, 279 (1984)
40. A.D. Becke, K.E. Edgecombe, J. Chem. Phys. **92**, 5397 (1990)
41. B. Silvi, A. Savin, Nature **371**, 683 (1994)
42. A. Savin, R. Nesper, S. Wengert, T.F. Fässler, Angew. Chem. Int. Ed. Engl. **36**, 1808 (1997)
43. M.J. Feinberg, K. Ruedenberg, E.L. Mehler, Adv. Quant. Chem. **5**, 28 (1970)
44. M.J. Feinberg, K. Ruedenberg, J. Chem. Phys. **54**, 1495 (1971); M.J. Feinberg, K. Ruedenberg, **55**, 5804 (1971)
45. W.A. Goddard, C.W. Wilson, Theoret. Chim. Acta (Berl.) **26**, 195, 211 (1972)
46. K.A. Wiberg, Tetrahedron **24**, 1083 (1968)
47. F.L. Hirshfeld, Theoret. Chim. Acta (Berl.) **44**, 129 (1977)
48. R.F. Nalewajski, Ann. Phys. (Leipzig) **13**, 201 (2004)
49. R.F. Nalewajski, Struct. Chem. **15**, 391 (2004)
50. R.F. Nalewajski, Theoret. Chem. Acc. **114**, 4 (2005)
51. R.F. Nalewajski, J. Math, Chem. **38**, 43 (2005)
52. R.F. Nalewajski, Mol. Phys.**102**, 531, 547 (2004); **103**, 451 (2005)
53. R.F. Nalewajski, Mol. Phys. **104**, 365 (2006)
54. R.F. Nalewajski, Mol. Phys. **104**, 3339 (2006)
55. R.F. Nalewajski, Mol. Phys. **104**, 493 (2006)
56. R.F. Nalewajski, J. Math. Chem. **43**, 265 (2008)
57. R.F. Nalewajski, J. Phys. Chem. A **111**, 4855 (2007)
58. R.F. Nalewajski, J. Math. Chem. **43**, 780 (2008)
59. R.F. Nalewajski, J. Math. Chem. **45**, 709 (2009)
60. R.F. Nalewajski, J. Math. Chem. **45**, 1041 (2009)
61. R.F. Nalewajski, Int. J. Quantum Chem. **109**, 425 (2009)
62. R.F. Nalewajski, Int. J. Quantum Chem. **109**, 2495 (2009)
63. R.F. Nalewajski, Adv. Quantum Chem. **56**, 217 (2009)
64. R.F. Nalewajski, D. Szczepanik, J. Mrozek, Adv. Quantum Chem. **61**, 1 (2011)
65. R.F. Nalewajski, in *Mathematical Chemistry*, ed. by W.I. Hong (Nova, New York, 2011), p. 247
66. R.F. Nalewajski, J. Math. Chem. **47**, 709 (2010)
67. R.F. Nalewajski, J. Math. Chem. **47**, 808 (2010)
68. R.F. Nalewajski, J. Math. Chem. **49**, 592 (2011)
69. R.F. Nalewajski, J. Math. Chem. **49**, 2308 (2011)
70. R.F. Nalewajski, D. Szczepanik, J. Mrozek, J. Math. Chem. **50**, 1437 (2012)
71. R.F. Nalewajski, J. Math. Chem. **49**, 371 (2011)
72. R.F. Nalewajski, J. Math. Chem. **49**, 546 (2011)
73. R.F. Nalewajski, J. Math. Chem. **49**, 806 (2011)
74. R.F. Nalewajski, P. Gurdek, J. Math. Chem. **49**, 1226 (2011)
75. R.F. Nalewajski, Int. J. Quantum Chem. **112**, 2355 (2012)
76. R.F. Nalewajski, P. Gurdek, Struct. Chem. **23**, 1383 (2012)
77. P.A.M. Dirac, *The Principles of Quantum Mechanics*, 4th edn. (Clarendon, Oxford, 1958)
78. R.F. Nalewajski, J. Math. Chem. **53**, 1 (2015)
79. R.O. Esquivel, N. Flores-Gallegos, C. Iuga, E. Carrera, J.C. Angulo, J. Antolín, Theor. Chem. Acc. **124**, 445 (2009)
80. S. López-Rosa, Information-theoretic measures of atomic and molecular systems. Ph.D Thesis, University of Granada, 2010
81. S. López-Rosa, R.O. Esquivel, J.C. Angulo, J. Antolín, J.S. Dehesa, N. Flores-Gallegos, J. Chem. Theory Comput. **6**, 145 (2010)
82. W. Heitler, F. London, Z. Physik **44**, 455 (1927)
83. S. Shaik, D. Danovich, W. Wu, P.C. Hiberty, Nat. Chem. **1**, 443 (2009)
84. R.F. Nalewajski, J. Math. Chem. **51**, 369 (2013)

85. R.F. Nalewajski, Ann. Phys. (Leipzig) **525**, 256 (2013)
86. R.F. Nalewajski, J. Math. Chem. **52**, 588 (2014)
87. R.F. Nalewajski, J. Math. Chem. **52**, 1292 (2014)
88. R.F. Nalewajski, J. Math. Chem. **52**, 1921 (2014)
89. R.F. Nalewajski, J. Math. Chem. **53**, 1549 (2015)
90. R.F. Nalewajski, J. Math. Chem. **53**, 1966 (2015)
91. R.F. Nalewajski, Mol. Phys. **112**, 2587 (2014)
92. R.F. Nalewajski, Int. J. Quantum Chem. **115**, 1274 (2015)
93. R.F. Nalewajski, Quantum Matter **4**, 12 (2015)
94. H.B. Callen, *Thermodynamics: An Introduction to the Physical Theories of the Equilibrium Thermostatics and Irreversible Thermodynamics* (Wiley, New York, 1960)
95. R.F. Nalewajski, Mol. Phys. **114**, 1225 (2016)
96. R.F. Nalewajski, J. Math. Chem. **54**, 932 (2016)
97. R.F. Nalewajski, J. Math. Chem. **53**, 1126 (2015)
98. R.F. Nalewajski, A.M. Köster, K. Jug, Theoret. Chim. Acta (Berl.) **85**, 463 (1993)
99. R.F. Nalewajski, J. Mrozek, Int. J. Quantum Chem. **51**, 187 (1994)
100. R.F. Nalewajski, S.J. Formosinho, A.J.C. Varandas, J. Mrozek, Int. J. Quantum Chem. **52**, 1153 (1994)
101. R.F. Nalewajski, J. Mrozek, G. Mazur, Can. J. Chem. **100**, 1121 (1996)
102. R.F. Nalewajski, J. Mrozek, A. Michalak, Int. J. Quantum Chem. **61**, 589 (1997)
103. J. Mrozek, R.F. Nalewajski, A. Michalak, Polish J. Chem. **72**, 1779 (1998)
104. M.S. Gopinathan, K. Jug, Theor. Chim. Acta (Berl.) **63**, 497, 511 (1983); see also: K. Jug. M.S. Gopinathan, in *Theoretical Models of Chemical Bonding*, ed. by Z.B. Maksić (Springer, Heidelberg, 1990), Vol. II, p. 77
105. I. Mayer, Chem. Phys. Lett. **97**, 270 (1983)
106. K. Jug, A.M. Köster, J. Am. Chem. Soc. **112**, 6772 (1990)
107. R.F. Nalewajski, J. Math. Chem. **52**, 42 (2014)
108. R.F. Nalewajski, J. Math. Chem. **51**, 7 (2013)
109. H. Shull, J. Am. Chem. Soc. **82**, 1287 (1960); **86**, 1469 (1964)
110. H. Shull, J. Phys. Chem. **66**, 2320 (1962)

# Chapter 15
# Molecular Dynamics Simulations of Vibrational Spectra of Hydrogen-Bonded Systems

**Mateusz Z. Brela, Marek Boczar, Łukasz Boda and Marek J. Wójcik**

**Abstract** In this chapter, we present current studies on molecular dynamics (MD) simulations of hydrogen-bonded systems with emphasis on vibrational spectra analysis. One of the most informative experimental data are spectroscopic data (infrared and Raman spectroscopy), which give information important in diverse fields, e.g. protein folding, drug design, sensors, nanotechnology, separations, etc. Spectroscopic data are very sensitive on inter- and intramolecular interactions. The processes of melting, boiling, unfolding and strand separation involve disruption of molecular interactions, that engage attractive or repulsive forces between molecules. In this chapter, we focus on calculations of IR spectra of hydrogen-bonded complexes based on linear response theory, in which the spectral density is the Fourier transform of the autocorrelation function of the dipole moment operator involved in the IR transitions. Recently, Born–Oppenheimer molecular dynamics (BOMD), Car–Parrinello molecular dynamics (CPMD), path integral molecular dynamics (PIMD), hybrid molecular dynamics (QM/MM) and other methods which use trajectories from molecular dynamics have been employed to simulate IR spectra of hydrogen-bonded systems. Each of these methods has some advantages and disadvantages which will be discussed in this chapter presenting also recent applications of these methods.

**Keywords** Molecular dynamics simulations · BOMD · CPDM · Vibrational spectra · IR · Hydrogen-bond

## 15.1 Introduction

In this review, we would like to introduce the current knowledge about molecular dynamics (MD) simulations of hydrogen-bonded systems with emphasis on vibrational spectra. Describing and understanding the interaction inside studied

M.Z. Brela · M. Boczar · Ł. Boda · M.J. Wójcik (✉)
Faculty of Chemistry, Jagiellonian University, Ingardena 3, 30-060 Krakow, Poland
e-mail: wojcik@chemia.uj.edu.pl



M.J. Wójcik et al. (eds.), *Frontiers of Quantum Chemistry*,
https://doi.org/10.1007/978-981-10-5651-2_15

systems might give us a lot of knowledge about those phenomena. One of the most important interactions is hydrogen bonds which play a crucial role in many areas of physics, chemistry and biology [1–12].

Understanding the quantum dynamics of molecular systems is one of the important steps for molecular modelling and design, and it is a big challenge for experiment as well as theory. John Pople in his Nobel lecture [13] has described the "Features of theoretical models" as prediction: "if the model has been properly validated according to some such criteria, it may be applied to chemical problems to which the answer is unknown or in dispute". That shows the main idea of cooperation and margining between experiment and theory. The good example of this concept is spectroscopy where comparison between the experimental and the calculated spectra give possibility to easily verify new model systems. On the other side, theory gives possibility to significantly improve interpretation of the experimental results.

One of the most informative experimental data is spectroscopic data [14–22], especially in the region of infrared and Raman spectroscopy, which gives information important in diverse fields, e.g. protein folding, drug design, sensors, nanotechnology, separations, etc. Spectroscopic data are very sensitive on inter- and intramolecular interactions. The processes of melting, boiling, unfolding and strand separation involve disruption of molecular interactions that involve attractive or repulsive forces between molecules.

Hydrogen bond is the most important directional intermolecular interaction. Hydrogen bonding interactions are strong, short-ranged and highly directional, and their energies lie between a dipole/dipole attraction and a covalent bond. In the last 50 years, many authors classified hydrogen bonds by many factors [23–25]. The most common is classifying hydrogen bonds in three classes: strong, moderate and weak. Unfortunately, this classification is subjective, but on the other hand as simple as possible. The main factors used for description of hydrogen bonds are as follows: stretching frequency of X–H vibration as well as geometric parameters such as distance between donor and acceptor and the H–X···Y angle. The properties of the hydrogen bond, which complicate the theoretical description of hydrogen-bonded fluids, also give rise to a number of macroscopic physical properties that are unique to such fluids. Hydrogen bonding is responsible for the remarkable properties of water, folding of proteins and is commonly exploited in the self-assembly of advanced materials.

The modelling of the molecular properties is always focused on the interactions inside model system which involve hydrogen bonds. The vibrational, especially infrared spectra of hydrogen-bonded systems have been widely studied in experimental and theoretical works [1–10]. Quantitative reconstruction and interpretation of the complex spectra of hydrogen-bonded systems require fine modelling and complete description of the interactions within and between hydrogen bonds and the molecular environment. Molecular dynamics and electronic structure calculations of infrared spectra of hydrogen-bonded systems give precise information on the nature of hydrogen bonds which are useful for understanding the molecular properties.

Hydrogen bonding generates striking changes in the infrared spectra of the X–H (D) stretching bands. These bands are shifted to lower frequencies by an amount that reflects the strength of hydrogen bond, and their widths and total intensities increase by an order of magnitude.

The correct and full description of hydrogen bonds by quantum chemistry methods involves description of the proton dynamics. Taking into account all quantum effects is very complicated. The hydrogen is the lightest atom which can easily tunnel. The dynamics of proton motion in hydrogen bonds is determined by a complex interplay of vibrational interactions [26–30]. These interactions are responsible for the complicated structure of the infrared (IR) and Raman spectra of hydrogen-bonded systems [31–33], and for the dynamics of proton tunnelling [34, 35].

In this chapter, we describe the state of research in molecular dynamics simulations of vibrational spectra of hydrogen-bonded systems. We present a theoretical quantum-mechanical model to describe the proton dynamics in the hydrogen-bonded systems. The calculations of IR spectra of hydrogen-bonded complexes are based on linear response theory, in which the spectral density is the Fourier transform of the autocorrelation function of the dipole moment operator involved in the IR transitions [36, 37]. Recently, Born–Oppenheimer molecular dynamics (BOMD), Car–Parrinello molecular dynamics (CPMD), path integral molecular dynamics (PIMD), hybrid molecular dynamics (QM/MM) and other methods which used trajectories from molecular dynamics have been used to simulate IR spectra of hydrogen-bonded systems. Each of these methods has some advantages and disadvantages which will be discussed later.

In this chapter, we also present the applications of already listed methods. The molecular dynamics methods are used for studies of the vibrational spectra of many systems from single molecule to complicated bio-systems. We consider a few kinds of such applications. The intramolecular hydrogen bonds render pronounced effects on molecular structure and properties, especially in medicinal chemistry [38]. The more complicated systems are, for example, dimer structures observed in the gas phase. Other interactions, such as lattice interactions, present in crystals, influence crystal structures with hydrogen-bond interactions.

At the end, this chapter is focused on hydrogen bonds in solution which is one of the most challenging problems. Various interactions are present in solution and have large influence on considered systems. Physical phenomena, such as diffusion or dissolution, make also a lot of difficulties for describing hydrogen bonds in solution. However, there are few recent papers, which make a big progress in understanding this complicated problem [39–42]. Hydrogen bonds are also vital interactions in biology, and all previous methods might be useful for studies of the reactions important in medicine and biology.

The last part of this chapter contains short summary and possible future direction of applications of the various methods of molecular dynamics. We discuss also the perspectives of modelling the hydrogen bonds interactions by theoretical methods.

## 15.2 Theory Background

The molecular dynamics using potentials, based on independent electronic structure calculations, is well established as a powerful tool to investigate many-body condensed matter, liquids as well as gas-phase systems. The wide and broad description of molecular dynamics technique was an aim of several monographic and reviews [43–45]. The main point of adiabatic molecular dynamics is to describe the moves of nuclei in terms of potential energy surfaces designed by electronic energy. The Born–Oppenheimer approach gives us possibility to treat the nuclei and electrons independently. Typically, the nuclei are treated as a classical particles and electrons as a quantum particles. It should be noticed that taking to account the quantization of nuclear motion should be done during MD simulation which requires fully quantum treatment. That is possible by using other approaches, e.g. [46, 47] quantum wave packet ab initio molecular dynamics (QWAIMD), but it is still computationally too expensive to be a “standard” treatment.

The calculation of the IR spectra of hydrogen-bonded complexes by MD calculations is based on linear response theory, in which the spectral density is the Fourier transform of the autocorrelation function of the dipole moment operator involved in the IR transition. Recently, molecular dynamics was performed by using several approaches which have been used to simulate the IR spectra of hydrogen-bonded systems [48, 49]. We will shortly describe few of them as the most common used in spectroscopy and the most promising. We present the study of advantages and disadvantages of these methods in terms of simulation vibrational spectra.

### *15.2.1 Born–Oppenheimer Molecular Dynamics*

The Born–Oppenheimer Molecular Dynamic approach is based on integration of the classical equation of motion on ab initio molecular potential surface [50–52]. That approach provides an information about the electronic structure from the first principles by using quantum chemistry methods, such as Hartree–Fock or density functional theory (DFT). Ab initio methods may be used to calculate potential energy surface on the fly. The potential surface calculated by ab initio methods is vital for systems with high possibility of the breaking and formation of chemical bonds. Those kinds of systems are definitely hydrogen-bonded structures, especially with strong hydrogen bonds. It should be stressed that the dynamics of proton motion in hydrogen bonds is determined by a complex interplay of vibrational interactions. These interactions are responsible for the complicated structure of the infrared (IR) and Raman spectra of hydrogen-bonded systems, and for the dynamics of proton tunnelling. For many such systems, the ab initio molecular dynamics is a matter of choice [53, 54].

The cost of MD calculations is a cost of ab initio calculations multiplied by number of trajectory steps. The long trajectories are beneficial for spectroscopic investigations, especially in the low-frequency region that represents slow motions. It should be pointed out that time step is an important parameter for a MD simulation. The decreasing time step in numerically calculated trajectory leads to the more accurate simulations. Nevertheless, computational cost increases accordingly with number of steps. In practice, the limit of the maximum time step is determined by the vibrational period of the mode with the highest energy.

### *15.2.2 Car–Parrinello Molecular Dynamics*

The Car–Parrinello Molecular Dynamics (CPMD) [55] couples the electronic degrees of freedom with the classical coordinates system by assigning electrons fictitious mass. This approximation excludes minimization of the electrons wave function at every step in the trajectory. CPMD uses fictitious dynamics to keep the electrons close to the ground state, preventing the need for a costly self-consistent iterative minimization at each time step. In order to stay in the Born–Oppenheimer surface, the time step should be comparatively small. In consequence, the computational cost of simulation steps is relatively small. The costly self-consistent iterative minimization is done only at the first step of the MD simulation.

In practice, for many investigations, the key parameter of a simulation is the simulation length. Longer trajectories provide better statistics. Kuo et al. [56] compared the Car–Parrinello and Born–Oppenheimer molecular dynamics simulations using the properties of liquid water as an example. They demonstrated that the fictitious electronic energy increased even for a relatively small fictitious electronic mass ($\mu = 100$ au.), which reflected a decrease in ionic temperature. The decrease in ionic temperature might be particularly noticeable in long trajectories and is one of the disadvantages of the Car–Parrinello approach.

### *15.2.3 Path Integral Molecular Dynamics*

The Path Integral Molecular Dynamics (PIMD) was developed by Marx and Parrinello in 1994 [57, 58]. This new method treated nuclei as quantum particles by the path integral formulation of quantum statistical mechanics. A combined extended Lagrangian for both quantum nuclei and electrons defines a dynamical system and yields molecular dynamics trajectories that can be analysed to obtain quantum statistical expectation values of time-independent operators. The nuclei are treated as quantum particles by mapping each quantum nucleus into classical system producing an effective Hamiltonian, which is derived from Feynman's path integral. The main advantage of this method is low computational cost as compared with other methods taking into account the quantum effects of nuclei.

The PIMD is one of the most promising methods for understanding the nuclear quantum effects in hydrogen-bonded systems. Further, all effects due to quantum character of nuclei, such as isotope effects, are naturally incorporated into PIMD approach. Also, the calculations concerning the kinetic isotope effects (KIE) done using path integral schemes reveal significant improvement of theoretical investigation [59, 60].

### 15.2.4 The Hybrid Molecular Dynamics

The QM/MM method was introduced by Warshel and Levit in 1976 [61]. The enormous increase of systems sizes was a trigger to develop the new method that merges accurate and exact molecular dynamics based on BO approach with the fast and efficient molecular mechanics. This method is based on selection of two regions: first, treated by quantum mechanics and second, treated by classic molecular mechanics. The main advantage of such approach is the possibility of treatment of chemical reaction in such big systems as proteins and biomolecules. However, an open question is always the selection of size of QM region. A recent study [62] shows that there is no advantage to investigate a big QM region. It should be pointed out that there is small possibility of merging QM and MM regions. Such merging is done, for example in one of the most popular ONIOM models [63–65].

The hybrid molecular dynamics is more useful for the chemical process analysis than for spectroscopic investigation. However, that method is still expanding. The new QM/MM approaches are developed recently as well as the new investigations [65]. In last decade, the QM/MM method was successfully applied to understand the vibrational spectra of biomolecules.

### 15.2.5 Post-molecular Dynamics Analysis

Post-molecular dynamics analysis is based on the trajectories obtained from MD simulations and provides information about time dependence of some system descriptors. The descriptors concern, e.g. electron densities, atomic charges, interaction energies or nuclei characterization such as proton potentials. The one of the examples is quantization of nuclear motion by snapshot methodology developed by Mavri and Stare [66–68]. That methodology is based on the analysis of the proton potentials (1D, 2D or multidimensional) along ab initio trajectories. In this method, nuclear quantum effects were studied by assuming the time-independent formalism, and this technique was successfully applied to understand hydrogen bond interactions in different phases.

One of the particular analyses is the exploration of changes in electronic structure during MD simulations. Especially, an approach based on Wannier localization function is very popular [69]. The information about distribution of Wannier centres that represent pairs of electrons (in close shell calculations) or one electron (in open shell calculations) gives us possibility to discuss changes in interaction character and its direction along MD. The Wannier localization function may be used also for the assignment of the dipole moment, which is useful for crystal structures treated by periodic boundary conditions.

The ab initio MD simulations give us very commanding descriptor that is electronic energy. Reactivity descriptors based on electronic energy, such as frontiers orbitals, may be analysed and Energy Decomposition Analysis (EDA, ETS) performed simulations [70]. That is very powerful method, especially for characterization of spontaneous processes. This methodology is still pioneering with wide perspectives.

## 15.3 Applications

In this section, we present applications of the theoretical model described in Sect. 2 to simulation of the IR spectra of selected hydrogen-bonded systems.

Hydrogen bonding is an ubiquitous interaction that is important in a diverse range of applications, including inter and intramolecular interactions, solvation, self-assembly of macromolecules, crystals and protein folding. However, despite being discovered more than a century ago, the qualifying features of hydrogen bonds require refinement and enhancement. As an example, there has been some new investigation for weak hydrogen bonds between carbonyl methyl groups in polymer structures [71]. A variety of different experimental and theoretical evidence are considered characteristic for hydrogen bonding. Let us discuss the selected applications starting from intramolecular interactions through hydrogen bonds in gas phase, crystals and solution to biologic systems.

### *15.3.1 Intramolecular Hydrogen Bonds*

The intramolecular hydrogen bonds have great role for preference of the structure with that interactions. The study of the energetics between different conformers is useful for understanding the stability of different isomers, for example, by taking into account the spatial orientation and through-space interactions of substituents. In addition, conformational analysis can be used to predict and explain product selectivity, mechanisms and rates of reactions [72–74]. The formation of an intramolecular hydrogen bond is entropically favourable over intermolecular contacts because of the formation of a pseudo-ring.

The formation of intramolecular hydrogen bonds has a very pronounced effect on molecular structure and properties. Kuhn et al. [38] derive propensities for intramolecular hydrogen-bond formation of 5–8-membered rings on the basis of exhaustive searches in crystal structure databases. A number of motifs are seen in Fig. 15.1, several of which are clearly underutilised in drug discovery, are analysed in more detail by comparing small molecule and protein-ligand X-ray structures. They performed the analysis of topologies and properties of this type of interaction. They also have provided a useful basis for its more directed and rational use.

The systems with intramolecular hydrogen bonds are very often interesting in terms of the proton delocalization between donor and acceptor atoms. Those kinds of proton transfer are very often a trigger of chemical processes such as creation of zwitterion compounds (a dipolar ion is a neutral molecule with both positive and negative electrical charges). The most typical example is an amino acid that contains both acidic (carboxylic acid fragment) and basic (amine fragment) centres [75]. The intramolecular hydrogen bond between those fragments reflects the zwitterionic nature of that compound. The studies concerning that class of compounds are mostly based on the temperature effects in vibrational spectra. Increasing or decreasing the temperature is a key of population control of compounds forms.

The investigation of molecular dynamics that includes the quantum effects of proton, and temperature dependence may give valuable information on vibrational properties of systems with intramolecular hydrogen bonds. Jezierska and Panek showed the importance of intramolecular hydrogen bonds in examples of aromatic Schiff and Mannich bases [76]. Their computational investigations showed that the Car–Parrinello molecular dynamics is able to describe faithfully the molecular properties of the studied Schiff and Mannich bases. The application of path integral molecular dynamics has elucidated the case of the bridged proton of the studied Schiff base. Taking to the account quantum effects improves significantly the description of the proton's position in the intramolecular hydrogen bond and clearly reflect the possibility of proton transfer. The vibrational properties of the two compounds were analysed on the basis of Fourier transformation of the autocorrelation function of both the atomic velocities and dipole moments. These results gave possibility to elucidate the contribution of bridged proton power spectra to the all atoms power spectra.

### *15.3.2 Hydrogen Bonds in Gas Phase*

Proton transfer and migration are among the most fundamental processes in chemistry and biology. In the last few decades, many theoretical and experimental studies on proton transfer in gas phase along hydrogen bonds have been reported and analysed. The most popular model structures are carboxylic acids cyclic dimers. Acetic acid dimer infrared spectrum has been a subject of many theoretical studies [77–79]. There are also experimental studies using new techniques, for example, ultrafast femtosecond infrared spectroscopy.

Durlak et al. showed for the first time, the ab initio CPMD results, for acetic acid and its cyclic dimer [77]. They performed calculation for acetic acid cyclic dimers with various deuterium substitution rates. Moreover, they presented results for the deuterated bridge analogue in order to examine isotopic substitution effects on the reaction mechanism. The results of these investigations have been compared with the existing experimental and static calculation data, see Fig. 15.2. That work shows the superiority of molecular dynamics simulations over static calculations in terms of Car–Parrinello molecular dynamics method proved to reproduce very well the infrared spectra of studied systems. It gave good agreement with the experimental data and reproduced well isotope effects. The CPMD IR spectra of monomer and dimer of acetic acid, especially the complicated band shape of the O–H or O–D stretching modes were in very good agreement with the experimentally observed spectra. The authors pointed out the main advantages of using molecular dynamics for IR spectra simulations. However, they also noted that MD investigation is very expensive and requires significant hardware. They discussed also the dependence of results stability on time step and energy cut-off selection.

Mackeparng et al. [80] have presented the isolated gas-phase infrared spectra of formic acid dimer and its deuterated derivative at room temperature. They obtained the formic acid dimer spectrum by spectral subtraction of a spectrum of formic acid vapour recorded at low pressure from that recorded at a higher pressure. The spectra of formic acid vapour contain features from both formic acid monomer and formic acid dimer, but at low and high pressures of formic acid, the equilibrium is pushed towards the monomer and dimer, respectively. A similar approach was used for the formic-d acid dimer. In that case, authors used classical molecular dynamics (molecular mechanics with proton transfer, MMPT) simulations for interpretation the OH region in the experimental spectrum. The combination of experimental and computational methods allows them to estimate barriers for proton transfer in the gas-phase systems. At the end, the authors demonstrated that MD simulations at the semi-empirical DFTB level of theory agree reasonably well for the band positions but underestimate the width of the OH-stretching band, see Fig. 15.3.

Latajka et al. have presented the results of theoretical studies on the molecular structure, conformational preferences, topological and vibrational analysis of allicin, Fig. 15.4 [81]. That work reported both types of simulations (static and dynamic) at the MP2, DFT and CP/PIMD levels. The authors stress the flexible character of the allicin molecule. They suggest that the weak hydrogen bond in the allicin may stabilize the structure, with relatively lower energies on the potential energy surface. However, in the course of the CPMD and PIMD simulations fast proton transfer (FPT), that includes quantum effects, was not observed in the hydrogen bond, see Fig. 15.4. The IR spectrum calculated by Car–Parrinello molecular dynamics was in good agreement with the experimental data.

Many other cyclic dimers have been studied in detail taking into the account anharmonic nature of hydrogen bonds [82–84] Systematic studies of monomers, dimers and more complicated systems with the experimental vibrational spectra reveal what kind of structures are occurring in analysed samples [82–84].

| Topology | | % hb | CSD example | PDB example |
|---|---|---|---|---|
| cC3aC3a | 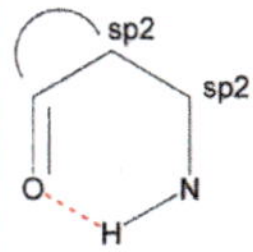 | 93.0 | 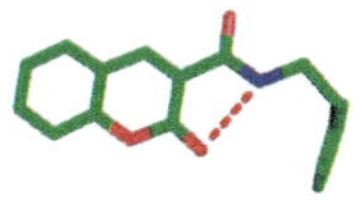 ACEMEB | 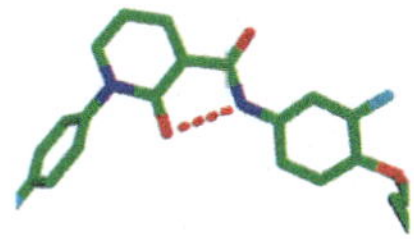 c-met kinase 3ce3 |
| aC3cC3a | 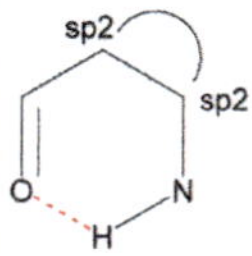 | 89.5 | 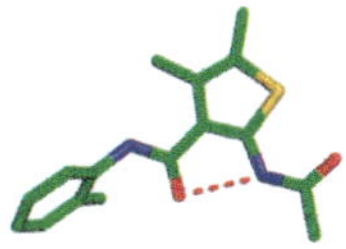 ABEKIC | 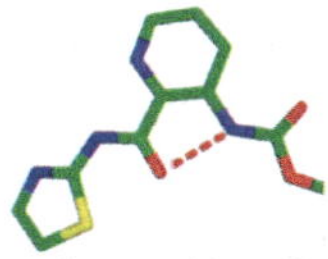 met-aminopeptidase 2nq6 |
| aNaC3c | 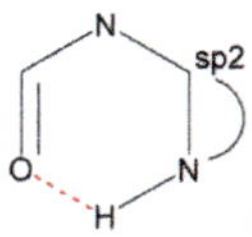 | 93.5 | 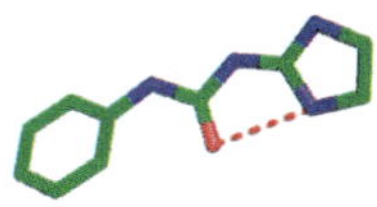 CIVRUV | 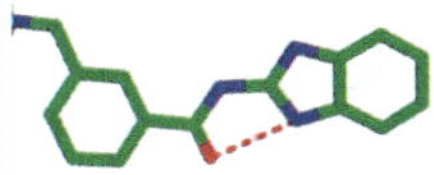 hiv protease 1bwb |
| aNaC3a | 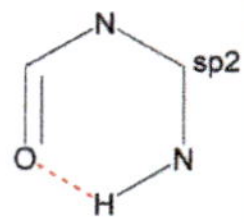 | 85.3 | 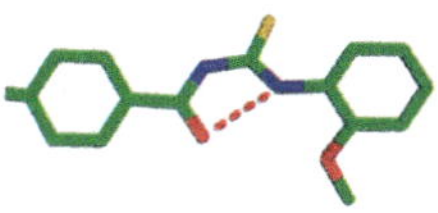 ABEFAP | 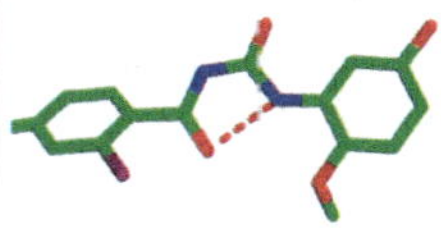 glyco-phosphorylase 2ati |
| aC4aC3a | 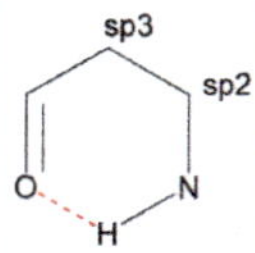 | 17.5 | 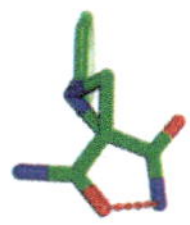 COYMOS | 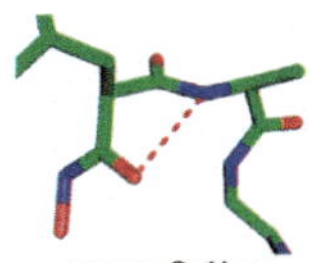 mmp-8 1jaq |
| aNaC3a | 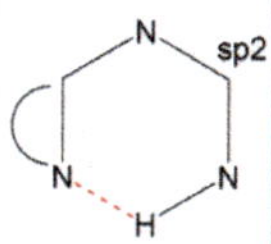 | 80.9 | 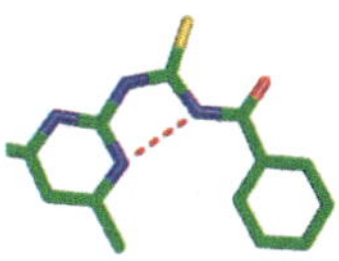 AWUBID | 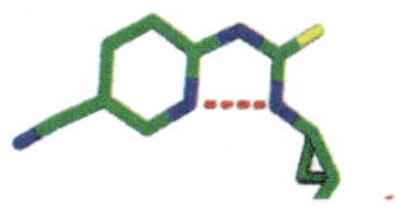 hiv rev-transcriptase 1iky |
| aC3aC3a | 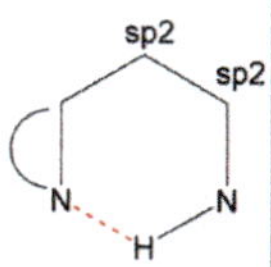 | 66.7 | 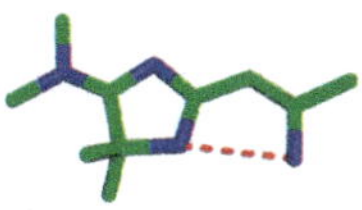 BOJYII | 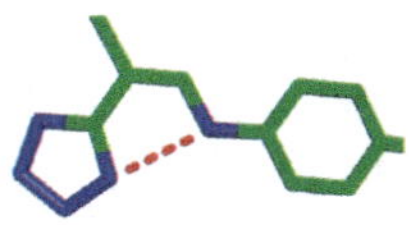 beta-lactamase 1pzp |

**Fig. 15.1** Selected topologies, Cambridge structural database hit statistics, and crystal structure examples of intramolecular NH···O=C and NH···N interactions in six-membered rings. The topology abbreviation lists the sequence of bond and atom types in the linker connecting hydrogen-bond acceptor with donor. C3 and C4 are sp2 and sp3 carbon atoms, and **a** and **c** denote acyclic and cyclic bond types, respectively. Reprinted (adapted) with permission from [38]. Copyright (2016) American Chemical Society

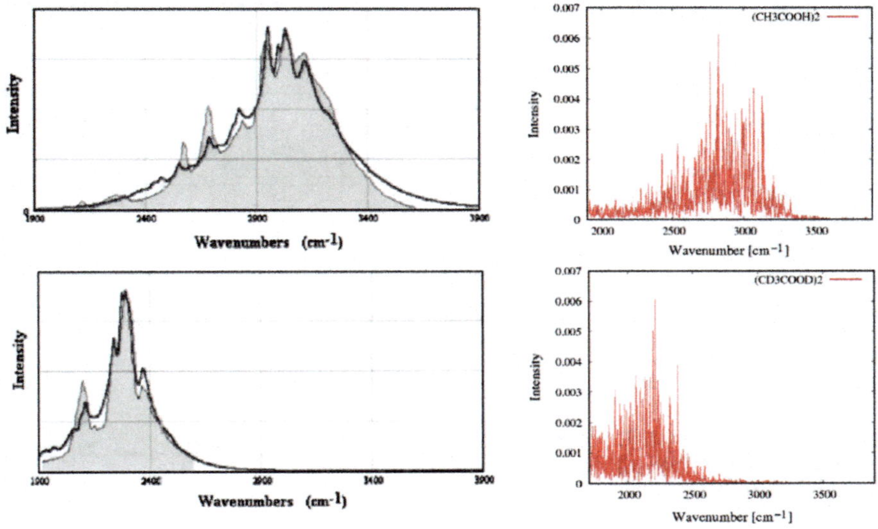


**Fig. 15.2** The comparison complicated shape spectra of –OH and –OD group in **1** $(CD_3COOH)_2$ and **2** $(CD_3COOD)_2$, *grey shape*-experimental, *black shape*-theoretical spectrum and simulated IR spectra from CPMD method **3** $(CD_3COOH)_2$ ,and **4** $(CD_3COOD)_2$. Reprinted (adapted) with permission from [77]. Copyright (2009) 2009 Elsevier B.V.A

### *15.3.3 Molecular Modelling of Hydrogen Bonding Interactions in the Crystal Field*

The interactions in crystals are in the spotlight from decades. The hydrogen bonds direction and strength may create crystal phases. One of the best examples is crystals of water that exist in many phases with different hydrogen bond network inside. The ab initio molecular dynamics allows us to describe important intra- and intermolecular interactions. In the first approximation conditions in the crystal approximate conditions present in the living cell. From that point, the MD simulations of hydrogen bonds in crystal field and analysis the vibrational spectra are very popular [85–100].

Infrared spectral studies give us information on the role of molecules in the formation of stars and planets. For example, infrared spectroscopy has shown that water, vital molecule for life, is present in space, mostly in its crystalline form–ice. Some crystal forms, such as ice XI, were observed only in Universe [101]. Ice XI is the structural variant of well-known ice Ih. The discussion about ice XI ferroelectric properties as an additional factor in planet formation processes is present in literature from years [102, 103]. Gług et al. performed the Car–Parrinello molecular dynamics for analysing the infrared spectra of ice Ih and ferroelectric ice XI [100]. The results have shown that librational region exhibits especially large differences in the simulated spectra of two considered forms of ice, see Fig. 15.5. They clearly present that theoretical IR spectra of ice forms can be used with success for

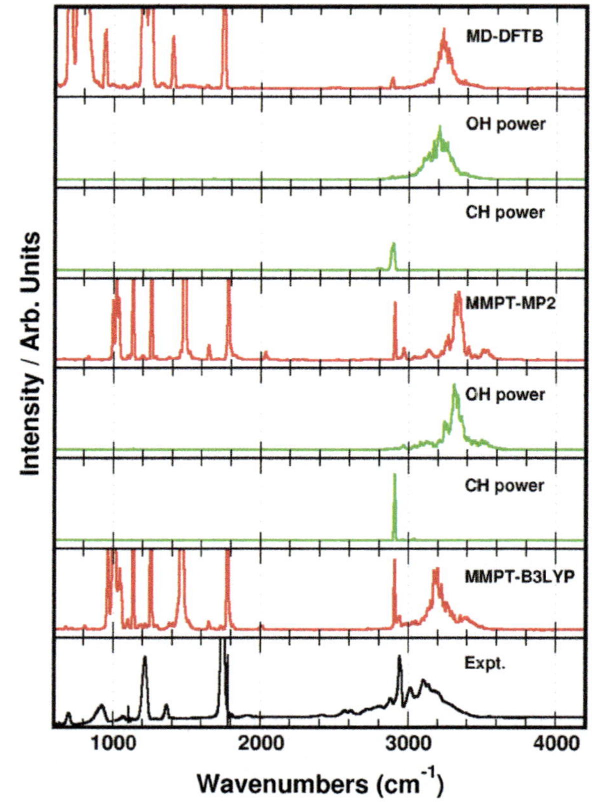


**Fig. 15.3** Spectra obtained from MD simulations with different energy functions. From *top* to the *bottom*: MD-DFTB (DPT barrier of 7.5 kcal mol1) and its OH and CH power spectra (*green*); MMPT-MP2 (DPT barrier of 8.2 kcal mol1) and the OH and CH power spectra (*green*) and MMPTB3LYP (DPT barrier of 5.4 kcal mol1). The experimentally recorded spectrum of FAD has been included for comparison (*bottom*). Reprinted (adapted) with permission from [80]. Copyright (2016) RSC

analysing experimental data obtained by IR telescopes and distinguish the forms of ice present in the Universe.

The big advantage of using molecular dynamics was also shown on example of analysis the medium-strong intermolecular hydrogen bonding in the 2-hydroxy-5-nitrobenzamide [94]. The authors compared the static calculation of dimers present in the crystal structure and performed harmonic analysis of the crystal cell. They did Car–Parrinello molecular dynamics simulation and one- and two-dimensional quantization of the proton motion in the instantaneous potentials influenced by the fluctuating environment and compared calculated spectra with the experimental ATR IR spectra. The analysis of infrared spectra associated with the O–H and N–H stretching modes was presented. That region is very sensitive to the strength of hydrogen bonding. It should be pointed out that harmonic frequencies calculated for an isolated dimer in the harmonic approximation gave poor agreement with the experiment. Inclusion of the crystal field in conjunction with the harmonic approximation slightly improved the results. Significant improvement was obtained by Fourier transform in the time course of the dipole moment function obtained from the Car–Parrinello trajectory. The best agreement with the experiment was obtained by solving the Schrödinger equation for the snapshots of 1D and

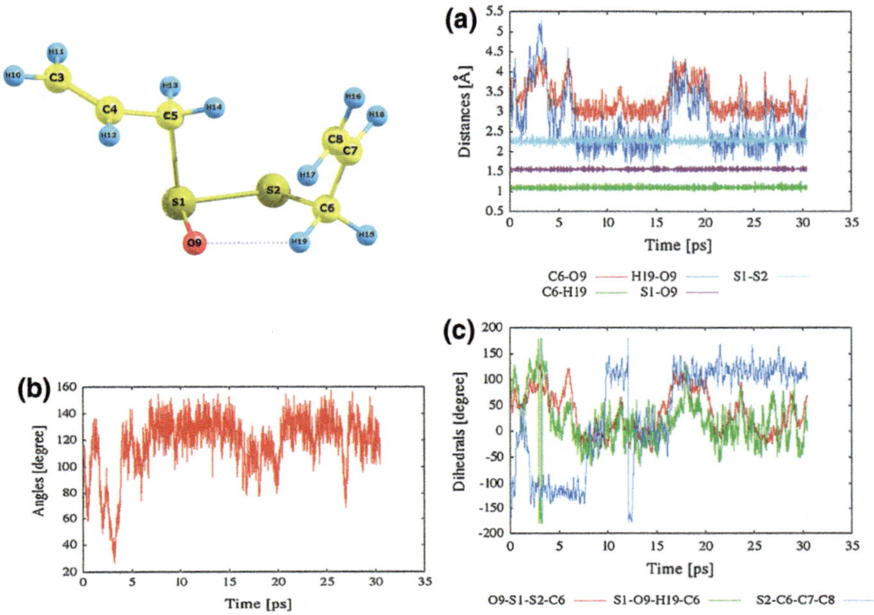


**Fig. 15.4** Structure and atoms labelling of the most stable conformation of allicin (I). Time evolutions of the bonds (**a**) angles, (**b**) and dihedrals, (**c**) involved in the hydrogen bonds according to the CPMD calculations at 300 K. Reprinted (adapted) with permission from [81]. Copyright (2015) Elsevier B.V.A

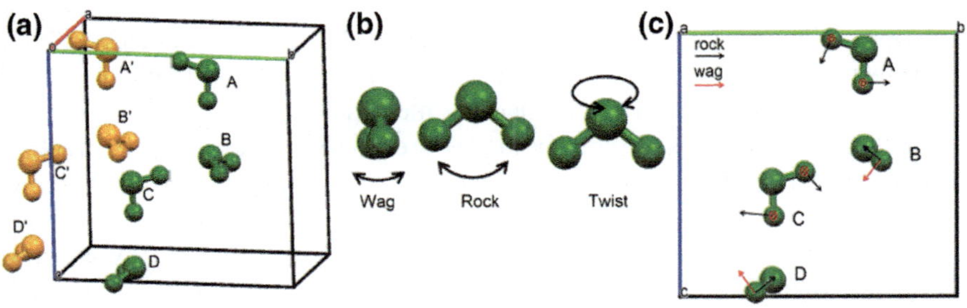


**Fig. 15.5** Subsystems ABCD and A'B'C'D' in crystal structure of ice XI, **b** three librational modes of water molecule, **c** positive directions of wagging and rocking modes of water molecules for subsystem ABCD. Reprinted (adapted) with permission from [100]. Copyright (2015) Elsevier B.V.A

2D proton potentials. In that work, authors presented the practical method for the 1D proton potential construction, see Fig. 15.6. The authors have shown also that the plain elongation of the O–H bond length gave the best agreement with the 2D treatment.

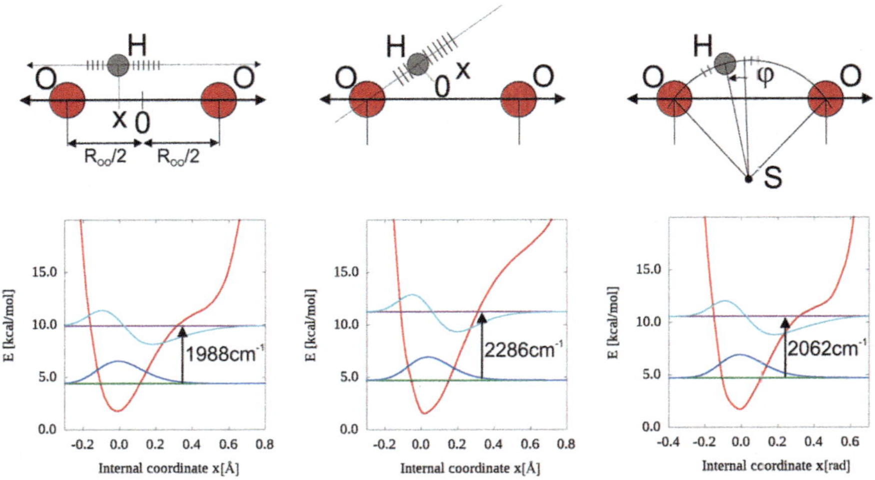


**Fig. 15.6** Definition of pathways for one-dimensional proton potentials (*upper figures*). The *lower figures* show the corresponding proton potentials with the first two eigenvalues and eigenfunctions. All three proton potentials were obtained from the same snapshot structure. Reprinted (adapted) with permission from [94]. Copyright (2012) ACS

Brela et al. presented results of the IR spectra simulation of the oxalic acid dihydrate crystal ant its isotopically substituted derivatives [104]. That investigation gave a possibility to take into account together anharmonicity effects and crystal field interactions. Infrared spectrum calculated by the CPMD method reproduced reasonably well the high-frequency bands of the experimental spectrum. Also, the deuterium shifts of vibrational bands were calculated in good accordance with the experimental data. The structure of the O–H stretching band was reproduced well, especially after investigation of the snapshot methodology, see Fig. 15.7. Authors concluded that couplings of similar modes are present in the crystal structure. That investigation proved that CPMD method is sufficient for analysis of hydrogen-bonded systems.

Several other IR spectra of crystal structures were reconstructed by ab initio molecular dynamics simulations such as the imidazole crystal [91], ascorbic acid [95] and aspirin [99]. These recent CPMD, BOMD and PIMD calculations show that the ab initio MD methods are adequate for spectroscopic investigations of complex systems with hydrogen bonds, since it takes into account most of the mechanisms determining the hydrogen bond dynamics (anharmonicity, couplings between vibrational modes and intermolecular interactions in crystals). The results reproduce both the frequencies and intensities of the experimental IR spectra of crystals reasonably well, which result from the application of dipole moment dynamics. At the end, it should be stressed that the further improvement is possible by the quantization of the nuclear motions [94, 95].

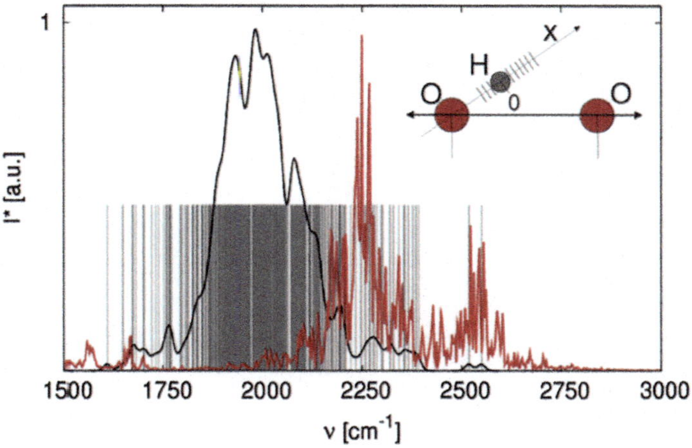


**Fig. 15.7** O–H stretching band contours of oxalic acid dihydrate: *Black line* contour was calculated from individual fundamental vibrational transitions as superposition of Gaussian functions with a half width of 10 cm$^{-1}$. Grey d functions represent fundamental vibrational transitions. The red contour represents the spectrum calculated by Fourier transform of the autocorrelation function of the atoms position obtained from Car-Parrinello trajectory. The bands were normalized. Reprinted (adapted) with permission from [104]. Copyright (2013) Elsevier B.V.A

## *15.3.4 The Hydrogen Bonds in Solution*

The investigation of the hydrogen-bond systems in the liquid phase is one of the most difficulties tasks [105–111]. Many effects should be taken into account, such as concentration, molecule mobility and diffusion. In recent years, it has been shown that ab initio molecular dynamics is a valuable tool for liquid systems. [112–119] The MD simulations in vibrational spectroscopy gave possibility to analysis the ionic liquids and process that occurs in solution. We present here selected applications.

The H-bond interactions of methyl acetate in methanol have been studied using ab initio molecular dynamics by Pagliai et al. [109]. That work represents an important development in the capability of molecular dynamics simulations to explain experimental data obtained by time-resolved spectroscopic methods. Assignment of vibrational modes and their frequencies during the simulations has allowed association of different interactions with the solvent to the vibrational properties. The H-bond effects on the spectroscopic properties of methyl acetate in methanol have been interpreted by wavelet transform analysis in conjunction with the structural and dynamic characterization of the solvation cages, see Fig. 15.8.

Eilmes et al. calculated the IR spectrum of the 1-ethyl-3-methylimidazolium bis (trifluoromethylsulfonyl) imide bulk liquid using ab initio molecular dynamics (AIMD) simulations by Fourier transform of the autocorrelation function of the dipole moment [120]. The concept at that work is presented in Fig. 15.9.

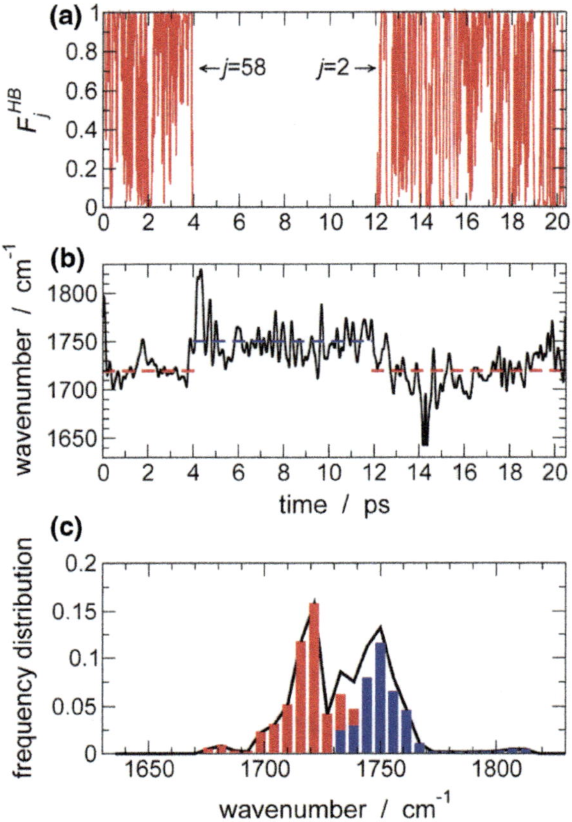


**Fig. 15.8** **a** Evolution of Fj HB for the molecule with j = 2 and 58. A molecule with j = 58 forms a H-bond in the first part of the simulation, whereas a molecule with j = 2 interacts with MA in the last part of the simulation, **b** time evolution of the maximum of the CdO stretching band during the simulation. The *dashed lines* represent the average frequency of the CdO stretching mode in the presence (*red*) or in the absence (*blue*) of the H-bond, **c** CdO stretching frequency distribution for MA during the simulation. The *red* and *blue bars* indicate the frequency distribution of MA involved and not involved in H-bond interaction, respectively. The frequencies have been uniformly scaled by a factor of 1.095. Reprinted (adapted) with permission from [109]. Copyright (2009) ACS

Description of the spectra has been improved by an explicit solvent model was compared with the gas-phase calculations or an implicit solvent model. The authors obtained good agreement between computed and experimental spectra. In that work, it has been demonstrated that the ab initio MD with Fourier transform is powerful methodology for calculating bulk IR spectra of ionic liquids.

For analysis, the ions–solvent interactions, often the QM/MM simulations have been performed [121]. The QM region is limited to the first or second solvation shells, and long range interactions with the distantly solvent molecules are approximated by the MM part. Payak et al. performed the HF/MM and B3LYP/MM molecular dynamics simulations for analysis the $CH_3COO^-$—water hydrogen bonds in dilute aqueous solution [121]. In that work, the authors demonstrated that the hydrogen bonds between $CH_3COO^-$ oxygens and first-shell waters are relatively strong compared with the water–water hydrogen bonds in the bulk [121], see Fig. 15.10. That and other works [120, 122] reveal the QM/MM method of approximation for close solvent shells in the QM part is truthful.

In that point, it is worth to notice that ab initio molecular dynamics and IR spectroscopy might be a powerful tool for the study of interface interactions. For

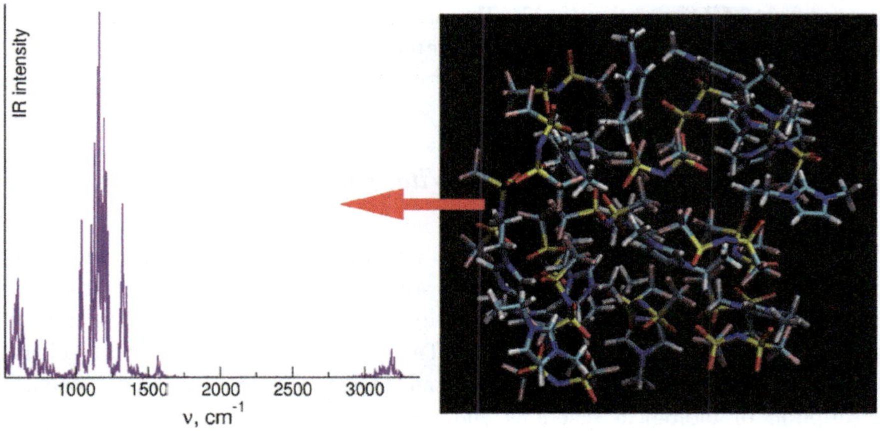


Fig. 15.9 Graphic abstract of article. Reprinted (adapted) with permission from [120]. Copyright (2016) ACS

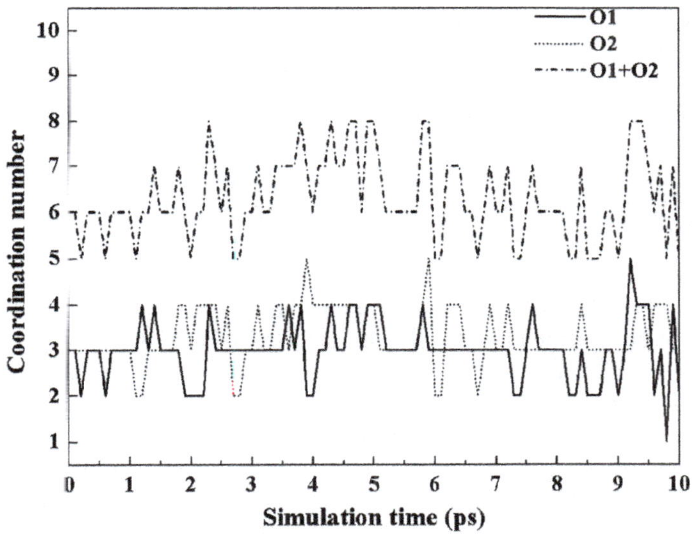


Fig. 15.10 Time dependence of the number of first-shell waters at CH3COO-oxygen atoms, selecting only the first 10 ps of the HF/MM simulation. Reprinted (adapted) with permission from [121]. Copyright (2009) Elsevier ACS

example, the study of mineral–water interfaces is of great importance to a variety of applications including oil and gas extraction, gas subsurface storage, environmental contaminant treatment and nuclear waste repositories. Understanding the fundamental properties of such interfaces is very important. Cygan et al. have presented such analysis of vibrational spectra of cation-exchanged minerals using power spectra from MD simulations [122]. This study provided molecular insights into the structure, dynamics and interactions of confined water and cations in the interlayer of clay minerals.

### *15.3.5 The Hydrogen Bonds as Vital Interactions in Biology*

The last application group concerns the biochemical systems. There are many important groups of biochemical systems with hydrogen bonds in their structures. The trigger of that field expansion was discovery of the molecular structure of nucleic acid by Watson and Crick [123]. The hydrogen bonds build up two chains in the structure of DNA. Other subjects are explanation of drug action, detailed description of biological reactions and understanding the nature of the proton mobility. In this section, we would like briefly present the direction when ab initio molecular dynamics together with vibrational spectroscopy improve analysis and interpretation of motions in biochemical structures.

The one of the most important processes of life is a replication, which on a molecular level is realized by DNA base pairing. However, the scheme with four nucleic bases of DNA does not necessarily represent the only way to achieve molecular replication. There are alternate pairing schemes and structures and interactions between the bases that can lead to mutations, for example by proton transfers that lead to different tautomers. To each of such tautomers, an IR spectrum may be assigned [124–126]. The study of interactions between individual nucleobases and the properties of isolated base pairs at the most fundamental level are very important [127–129].

Important biochemical compounds are also vitamins. They are natural organic substances which play key role in human life. There are 13 most essential vitamins, and they can be classified into two kinds: water-soluble and fat-soluble. One of the eight water-soluble vitamins is L-ascorbic acid (vitamin C). It should be pointed out that all of them have important role in enzymatic reactions in human organism. The pharmacological use of vitamins justified interest in studying properties of hydrogen-bonded systems present in their structures, such as present in the crystal of L-ascorbic acid [95]. This system allows us to describe important intermolecular interactions. In the first approximation conditions in the crystal (see Fig. 15.11)

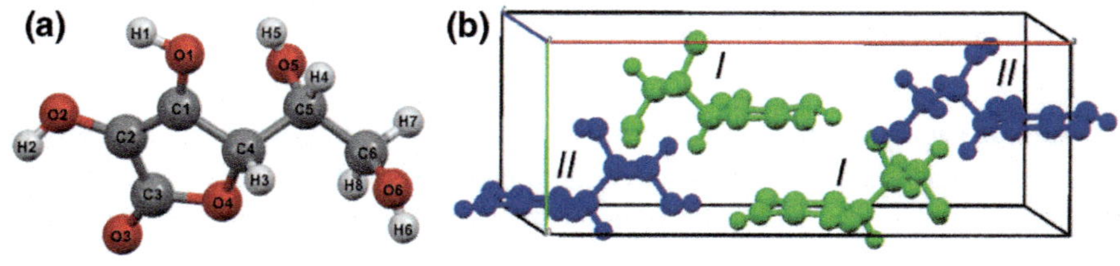


**Fig. 15.11** Ascorbic acid molecules—labelling of atoms and crystal structure of the ascorbic acid. Reprinted (adapted) with permission from [95]. Copyright (2015) ACS

model interactions in the living cell, though periodic boundary approximation is not equivalent to those present in the cell.

The isotope substitution is used as common techniques for studying the biochemical reactions. The kinetic isotope effects (KIE) analysis is one of the most essential and sensitive tools for the study of reaction mechanisms. In the isotopic effects, the nuclear quantum effects are present [130–139]. Other quantum effects, such as tunnelling have been shown, e.g. for tunnelling in protein SLO-1 (Caenorhabditis elegans) or monoamine oxidase B catalyzing decomposition of dopamine. Figure 15.12. presents structure of MAO inside residue. For studying that complex problem, the molecular dynamics was used [135–137].

It should be stressed that molecular dynamics is presently widely used in analysing the biochemical processes. The NMR spectroscopy is much more popular than vibrational spectroscopy in this kind of studies. Nevertheless, the deep analysis of the infrared spectra of proteins may be fruitful, and in that case, the assisted theoretical analysis is irreplaceable.

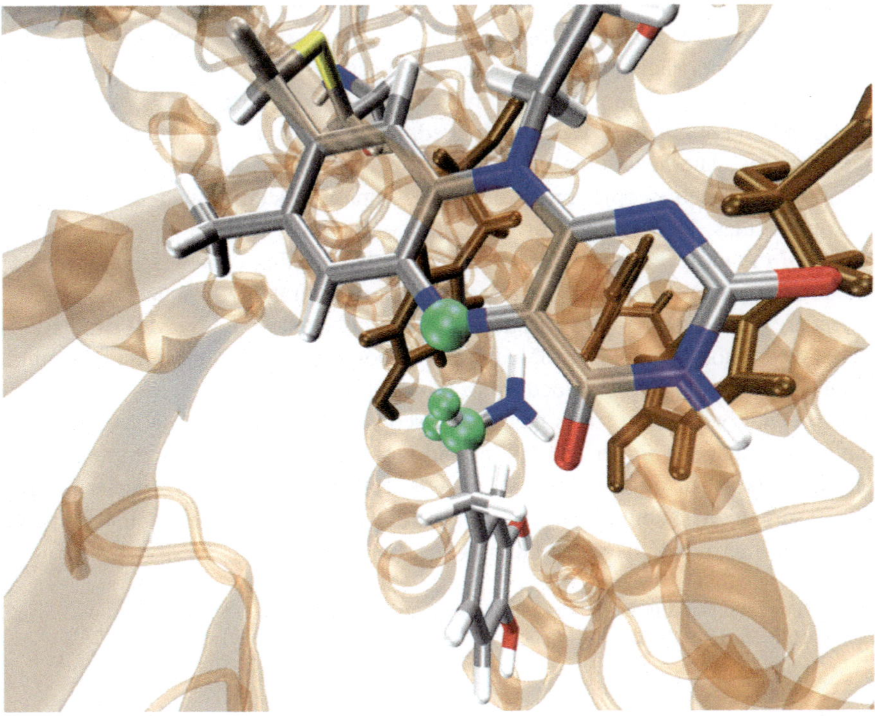

**Fig. 15.12** Structure of MAO B active site with neutral dopamine. The active site aromatic tyrosine residues are shown in *brown grey*, while for FAD prosthetic group and dopamine atom specific colouring are applied. Atom motion for the flavin N5 atom and the dopamine α-methylene group was quantized by QCP approach, and those atoms are labelled *light green*. Reprinted (adapted) with permission from [140]. Copyright (2016) ACS

## 15.4 Summary and Perspectives

This chapter is aimed to shortly introduce the ab initio molecular dynamics techniques as a powerful tool for simulations of vibrational spectra of hydrogen-bonded systems. The main advantage of molecular dynamics is taking into account the anharmonicity that plays important role for hydrogen-bonded systems. Despite enormous development of static approaches, the description of anharmonicity and coupling between motions is still challenging area of research. The molecular dynamics simulation gives us possibility to take into account thermal effects by using thermostat. That has great importance for interpreting spectroscopic data that are sensitive for temperature changes.

The molecular dynamics simulations give also a choice to include quantization of motion. The proton transfer reaction is one of the most important in enzymological and biological reactions. It was shown by Warshel [131], that preorganization and electrostatic field of neighbourhood have a pivotal role for reactions to occur. However, a calculation of accurate reaction rate constants requires consideration of many other effects, such as quantization of motion. The state of art methods based on MD simulations provides precise and truthful results.

After years, the popularity of molecular dynamics for simulation of vibronic spectra is still increasing. In this chapter, we discussed examples concerning many kinds of hydrogen-bonding systems present in different environments. All kinds of hydrogen bond interactions were considered. However, there are still places for new investigations. The near future will give deeper interpretations based on MD calculations of spectroscopic properties of complex systems such as polymers, macromolecules, biomolecules or inorganic compounds.

One of the most promising perspectives is interpretation of new spectral region. The low-frequency motions, overtone region and terahertz frequencies are the newest areas with many open questions that might be answered by MD simulations [141, 142]. Those complex problems merge couplings, resonances and quantum effects that are naturally included in ab initio MD simulations.

At the end, we should stress that spectroscopy is still evolving science. The new compounds, that are bigger and bigger and more multifaceted and complex, are in the main interest of infrared, near-IR, far-IR, Raman and terahertz spectroscopies. We are sure the MD simulations are useful, valuable and tenured tools that help and strengthen interpretations of vibrational spectra.

**Acknowledgements** This work was financially supported by National Science Centre, Poland, grant 2016/21/B/ST4/02102.

## References


1. D. Hadži, H.W. Thompson, International Union of Pure and Applied Chemistry, *Hydrogen Bonding: Papers Presented at the Symposium on Hydrogen Bonding Held at Ljubljana, 29 July–3 Aug 1957* (Pergamon Press, London, 1959), pp. xii, 571
2. G.C. Pimentel, A.L. McClellan, *The Hydrogen Bond*, ed. by W.H. Freeman (trade distributor) (Reinhold Pub. Corp., New York, San Francisco, 1960), p. 475
3. P. Schuster, G. Zundel, C. Sandorfy, *The Hydrogen Bond: Recent Developments in Theory and Experiments* (North-Holland Pub. Co. distributor, American Elsevier Pub. Co., Amsterdam, New York, 1976)
4. S. Scheiner, *Hydrogen Bonding: A Theoretical Perspective* (Oxford University Press, New York; Oxford, 1997), pp. xix, 375
5. G.A. Jeffrey, *An Introduction to Hydrogen Bonding* (Oxford University Press, New York, Oxford, 1997), pp. vii, 303
6. D. Hadži, *Theoretical Treatments of Hydrogen Bonding* (Wiley, Chichester, 1997), pp. xiii, 318
7. S. Grabowski, *Springer Link (Online service), Hydrogen Bonding New Insights* (Springer, Dordrecht, 2006)
8. Y. Maréchal, *Molecular Interactions*, vol. 1, ed. by H. Ratajczak, W.J. Orville-Thomas (Willey, Chichester, 1980)
9. C. Sandorfy, Top. Curr. Chem. **120**, 41 (1984)
10. O. Henri-Rousseau, P. Blaise, Adv. Chem. Phys. **103**, 1 (1998)
11. M.J. Wójcik, Adv. Chem. Phys. **160**, 311–346 (2016)
12. J.A. Pople, Nobel lecture (1998)
13. S. Bratož, D. Hadži **27**(5), 991 (1957)
14. Y. Maréchal, **48**(8), 3697 (1968)
15. A. Witkowski, M. Wojcik, Chem. Phys. **1**(1), 9 (1973)
16. M.J. Wojcik, Int. J. Quantum Chem. **10**(5), 747 (1976)
17. M.J. Wojcik, J. Mol. Struct. **47**, 303 (1978)
18. H.T. Flakus. Theochem. J. Mol. Struc. **104**(3), 281 (1993)
19. H. Ratajczak, A.M. Yaremko, Chem. Phys. Lett. **314**(1–2), 122 (1999)
20. O. Henri-Rousseau, D. Chamma, Chem. Phys. **229**(1), 37 (1998)
21. A.M. Yaremko, H. Ratajczak, J. Baran, A.J. Barnes, E. Mozdor, B. Silvi, Chem. Phys. **306** (1–3), 57 (2004)
22. Y. Ozaki, D.H. Pliura, P.R. Carey, A.C. Storer, Biochemistry **21**(13), 3102 (1982)
23. Y. Wu, K. Murayama, Y. Ozaki, J. Phys. Chem. B **105**(26), 6251 (2001)
24. Y. Ozaki, K. Aoyagi, K. Iriyama, H. Ogoshi, T. Kitagawa. J. Phys. Chem. **93**(9), 3842 (1989)
25. P. Blaise, M.J. Wojcik, O. Henri-Rousseau, J. Chem. Phys. **122** (6), (2005)
26. C. Sandorfy, J. Mol. Struct. **790**(1–3), 50 (2006)
27. M. Boczar, L. Boda, M.J. Wojcik, J. Chem. Phys. **124** (8), (2006)
28. M.J. Wojcik, M. Boczar, L. Boda, J. Chem. Phys. **127** (8), (2007)
29. M.J. Wojcik, J. Kwiendacz, M. Boczar, L. Boda, Y. Ozaki, Chem. Phys. **372**(1–3), 72 (2010)
30. M.J. Wójcik, V. Buch, J.P. Devlin, J. Chem. Phys. **99**, 2332 (1993)
31. M. Boczar, Ł. Boda, M.J. Wójcik, J. Chem. Phys. **125**, 084709 (2006)
32. M. Boczar, Ł. Boda, M.J. Wójcik, J. Chem. Phys. **127**, 084307 (2007)
33. M. Boczar, J. Kwiendacz, M.J. Wójcik, J. Chem. Phys. **128**, 164506 (2008)
34. M.J. Wójcik, H. Nakamura, S. Iwata, W. Tatara, J. Chem. Phys. **112**, 6322 (2000)
35. M.J. Wójcik, Ł. Boda, M. Boczar, J. Chem. Phys. **130**, 164306 (2009)
36. R. Ramírez, T. López-Ciudad, P.P. Kumar, D. Marx, J. Chem. Phys. **121**, 3973–3983 (2004)
37. M. Thomas, M. Brehm, O. Hollóczki, Z. Kelemen, L. Nyulászi, T. Pasinszki, B. Kirchner, J. Chem. Phys. **141**, 024510 (2014)

38. B. Kuhn, P. Mohr, M. Stahl, J. Med. Chem. **53**(6), 2601–2611 (2010)
39. T. Anh Pham, P. Huang, E. Schwegler, G. Galli, J. Phys. Chem. A **116**(37), 925 (2012)
40. N. Sieffert, M. Bühl, M.-P. Gaigeot, C.A. Morrison, J. Chem. Theory Comput. **9**(1), 106 (2013)
41. E. Pluhařová, M. Ončák, R. Seidel, C. Schroeder, W. Schroeder, B. Winter, S.E. Bradforth, P. Jungwirth, P. Slavíček, J. Phys. Chem. B **116**(44), 13254 (2012)
42. W. Lin, F. Paesani, J. Phys. Chem. A **119**(19), 4450 (2015)
43. T.R. Cundari, M.T. Benson, M.L. Lutz, S.O. Sommerer, *Reviews in Computational Chemistry*, vol. 8, ed. by K.B. Lipkowitz, D.B. Boyd 131 (VCH, New York, 1996), p. 145
44. D. Marx, Classical and quantum dynamics in condensed phase simulations, Chap. 15, ed. by B.J. Berne, G. Ciccotti, D.F. Coker (World Scientific, Singapore, 1998)
45. M.E. Tuckerman, A. Hughes, Classical and quantum dynamics in condensed phase simulations, Chap. 14, ed. by B.J. Berne, G. Ciccotti, D.F. Coker (World Scientific, Singapore, 1998), p. 311
46. I. Sumner, S.S. Iyengar, J. Phys. Chem. A **111**, 10313 (2007)
47. X. Li, S.S. Iyengar, J. Phys. Chem. A **115**, 6269 (2011)
48. J. VandeVondele, P. Tröster, P. Tavan, G. Mathias. J. Phys. Chem. A **116**(10), 2466 (2012)
49. E. Pluhařová, M.D. Baer, C.J. Mundy, B. Schmidt, P. Jungwirth, J. Phys. Chem. Lett. **5**(13), 2235 (2014)
50. M. Born, R. Oppenheimer, Ann. Phys. **84**, 457–484 (1927)
51. D. Marx, J. Hutter, *Ab Initio Molecular Dynamics* (Cambridge University Press, Cambridge, 2009), pp. 45–46
52. M. Krack, M. Parrinello, Forschungszentrum Jülich. NIC Series **25**, 29 (2004)
53. Y. Zhao, H. Li, X.C. Zeng, J. Am. Chem. Soc. **135**(41), 15549 (2013)
54. A. Mondal, S. Balasubramanian, J. Phys. Chem. B **119**(5), 1994 (2015)
55. R. Car, M. Parrinello, Phys. Rev. Lett. **55**(22), 2471–2474 (1985)
56. I.-F.W. Kuo, C.J. Mundy, M.J. McGrath, J.I. Siepmann, J. Chem. Theory Comput. **2**, 1274–1281 (2006)
57. D. Marx, M. Parrinello, Zeitschrift für Physik B Condensed Matter **95**(2), 143 (1994)
58. D. Marx, M. Parrinello, J. Chem. Phys. **104**, 4077 (1996)
59. J. Gao, K.-Y. Wong, D.T. Major, J. Comput. Chem. **29**, 514 (2008)
60. D.T. Major, M. Garcia-Viloca, J. Gao, J. Chem. Theory Comput. **2**(2), 236 (2006)
61. A. Warshel, M. Levitt, J. Mol. Biol. **103**(2), 227–249 (1976)
62. G. Jindal, A. Warshel, J. Phys. Chem. B **120**(37), 9913 (2016)
63. S. Dapprich, I. Komáromi, K.S. Byun, K. Morokuma, M.J. Frisch, J. Mol. Struct. (Theochem) **462**, 1 (1999)
64. T. Vreven, K. Morokuma, Ö. Farkas, H.B. Schlegel, M.J. Frisch, J. Comp. Chem. **24**, 760 (2003)
65. T. Vreven, K. Morokuma, Annual Reports Comp. Chem. **2**, 35–51 (2006)
66. Jernej Stare Aneta Jezierska Gabriela Ambrožič Iztok J. Košir Jurka Kidrič Aleksander Koll Janez Mavri Dušan Hadži. J. Am. Chem. Soc. **126**(13), pp. 4437–4443 (2004)
67. G.G. Balintkurti, R.N. Dixon, C.C. Marston, Int. Rev. Phys. Chem. **11**(2), 317–344 (1992)
68. G. Pirc, J. Mavri, J. Stare, P Vib. Spectrosc. **58**, 153–162 (2012)
69. G.H. Wannier, Phys. Rev. A **52**(3), 7 (1937)
70. M. Brela, A. Michalak, P.P. Power, T. Ziegler, Inorg. Chem. **53**(4), 2325 (2014)
71. S. Yamamoto, Y. Morisawa, H. Sato, H. Hoshina, Y. Ozaki, J. Phys. Chem. B **117**, 2180–2187 (2013)
72. P. Durlak, Z. Latajka, S. Berski, J. Chem. Phys. **131**(2), (2009)
73. P. Dopieralski, Z. Latajka, I. Olovsson, Acta Cryst. B **66**, 222–228 (2010)
74. P. Dopieralski, Z. Latajka, I. Olovsson, Chem. Phys. Lett. **476**, 223–226 (2009)
75. W.D. Price, R.A. Jockusch, E.R. Williams, J. Am. Chem. Soc. **119**(49), 11988–11989 (1997)
76. A. Jezierska, J.J. Panek, J. Chem. Theory Comput. **4**(3), 375–384 (2008)
77. P. Durlak, Z. Latajka, Chem. Phys. Lett. **477**(4–6), 249–254 (2009)

78. M. Wojcik, B Acad. Pol. Sci. Chim. **22**(2), 71–73 (1974)
79. J.J. Panek, A. Jezierska-Mazzarello, P. Lipkowski, A. Martyniak, A. Filarowski, J. Chem. Inf. Model. **54**, 86–95 (2014)
80. K. Mackeprang, Z.-H. Xu, Z. Maroun, M. Meuwly, H.G. Kjaergaard, Phys. Chem. Chem. Phys. **18**, 24654 (2016)
81. P. Durlak, S. Berski, Z. Latajka, Chem. Phys. Lett. **644**, 5–13 (2015)
82. M. Boczar, M.J. Wójcik, K. Szczeponek, D. Jamróz, S. Ikeda, Int. J. Quantum Chem. **90**, 689 (2002)
83. M. Boczar, M.J. Wójcik, K. Szczeponek, D. Jamróz, A. Zięba, B. Kawałek, Chem. Phys. **286**, 63 (2003)
84. M. Boczar, Ł. Boda, M.J. Wójcik, J. Chem. Phys. **124**, 084306 (2006)
85. Mavri, J.; Pirc, G.; Stare, J., *J Chem Phys* **2010,** *132* (22)
86. M.J. Wojcik, J. Kwiendacz, M. Boczar, Chem. Phys. Lett. **501**(4–6), 623–627 (2011)
87. J. Mavri, A. Jezierska, J.J. Panek, A. Koll, J. Chem. Phys. **126**(20), (2007)
88. J. Stare, J. Panek, J. Eckert, J. Grdadolnik, J. Mavri, D. Hadzi, J. Phys. Chem. A **112**(7), 1576–1586 (2008)
89. M.J. Wojcik, M. Boczar, L. Boda, J. Chem. Phys. **125**(8), (2006)
90. M.J. Wojcik, M. Boczar, R. Kurczab, Vib. Spectrosc. **52**(1), 39–47 (2010)
91. M.J. Wojcik, M. Boczar, J. Kwiendacz, J. Chem. Phys. **128**(16), (2008)
92. A.N. Manin, A.P. Voronin, A.V. Shishkina, M.V. Vener, A.V. Churakov, G.L. Perlovich, J. Phys. Chem. B **119**, 10466–10477 (2015)
93. V.V. Zhurov, A.A. Pinkerton, J. Phys. Chem. A **119**(52), 13092–13100 (2015)
94. M. Brela, J. Stare, G. Pirc, M. Sollner Dolenc, M. Boczar, M.J. Wójcik, J. Mavri, J. Phys. Chem. B **116**, 4510–4518 (2012)
95. M.Z. Brela, M.J. Wójcik, M. Boczar, Ł. Witek, M. Yasuda, Y. Ozaki, J. Phys. Chem. B **119** (25), 7922–7930 (2015)
96. M.J. Wójcik, M. Gług, M. Boczar, Ł. Boda, Chem. Phys. Lett. **612**, 162–166 (2014)
97. J. Stare, J. Panek, J. Eckert, J. Grdadolnik, J. Mavri, D. Hadzi, J. Phys. Chem. A **112**, 1576–1586 (2008)
98. P. Dopieralski, Z. Latajka, I. Olovsson, J. Chem. Theory Comput. **6**, 1455–1461 (2010)
99. M.Z. Brela, M.J. Wójcik, Ł.J. Witek, M. Boczar, E. Wrona, R. Hashim, Y. Ozaki, J. Phys. Chem. B **120**(16), 3854–3862 (2016)
100. M. Gług, M. Boczar, Ł. Boda, M.J. Wójcik, Chem. Phys. **459**(28), 102–111 (2015)
101. M.E. Brown, W.M. Calvin, Science **287**, 107 (2000)
102. M.J. Iedema, M.J. Dresser, D.L. Doering, J.B. Rowland, W.P. Hess, A.A. Tsekouras, J. P. Cowin, J. Phys. Chem. B **102**, 9203 (1998)
103. H. Wang, R.C. Bell, M.J. Iedema, A.A. Tsekouras, J.P. Cowin, Astrophys. J. **620**, 1027 (2005)
104. M.Z. Brela, M.J. Wójcik, M. Boczar, R. Hashim, Chem. Phys. Lett. **558**, 88–92 (2013)
105. Y. Ozaki, Y. Futami, Y. Ozaki, Y. Hamada, M.J. Wojcik, J. Phys. Chem. A **115**(7), 1194–1198 (2011)
106. J. Stare, J. Mavri, J. Grdadolnik, J. Zidar, Z.B. Maksic, R. Vianello, J. Phys. Chem. B **115** (19), 5999–6010 (2011)
107. D. Janezic, A. Jezierska, J. Panek, U. Borstnik, J. Mavri, J. Phys. Chem. B **111**(19), 5243–5248 (2007)
108. I-F.W. Kuo, C.J. Mundy, M.J. McGrath, J.I. Siepmann, J. Chem. Theory Comput. **2**(5), 1274–1281 (2006)
109. M. Pagliai, F. Muniz-Miranda, G. Cardini, R. Righini, V. Schettino, J. Phys. Chem. Lett. **1**, 2951–2955 (2010)
110. S. Park, M. Odelius, K.J. Gaffney, J. Phys. Chem. B **113**, 7825–7835 (2009)
111. P. Dopieralski, C.L. Perrin, Z.J. Latajka, Chem. Theory. Comput. **7**, 3505–3513 (2011)
112. K. Laasonen, F. Csajka, M. Parrinello, Chem. Phys. Leu. **194**, 172 (1992)
113. C. Lee, D. Vanderbilt, K. Laasonen, R. Car, M. Parrinello, Phys. Rev. Lett. **69**(1992) 462; Phys. Rev B **47**, 4863 (1993)

114. K. Laasonen, M. Parrinello, R. Car, Ch. Lee, D. Vanderbilt, Chem. Phys. Lett. **207**, 208 (1993)
115. K. Laasonen, M. Sprik, M. Parrinello, R. Car, J. Chem. Phys. **99**, 9080 (1993)
116. E.S. Fois, M. Sprik, M. Parrinello, Chem. Phys. Lett. **223**, 411 (1994)
117. K. Laasonen, M.L. Klein, J. Am. Chem. Soc. **116**, 11620 (1994)
118. M. Tuckerman, K. Laasonen, M. Sprik, M. Parrinelto, J. Phys. Condens. Matter **6** (1994) A93; J. Phys. Chem. **99**, 5749 (1995); J. Chem. Phys. **103**, 150 (1995)
119. M. Spfik, J. Hutter, M. Parrinello, J. Chem. Phys. **105**, 1142 (1996)
120. A. Eilmes, P. Kubisiak, B. Brela, J. Phys. Chem. B **120**, 11026 (2016)
121. A. Payaka, A. Tongraar, B.M. Rode, J. Phys. Chem. A **114**, 10443–10453 (2010)
122. Randall T. Cygan, Luke L. Daemen, Anastasia G. Ilgen, James L. Krumhansl, M. Tina, J. Nenoff, Phys. Chem. C **119**, 28005–28019 (2015)
123. J.D. Watson, F.H. Crick, Nature **171**, 737–738 (1953)
124. A.L. Sobolewski, W. Domcke, Phys. Chem. Chem. Phys. **6**, 2763 (2004)
125. T. Schultz, E. Samoylova, W. Radloff, I.V. Hertel, A.L. Sobolewski, W. Domcke, Science **306**, 1765 (2004)
126. S. Perun, A.L. Sobolewski, W. Domcke, J. Phys. Chem. A **110**, 9031 (2006)
127. A.L. Sobolewski, W. Domcke, Europhys. News **37**, 20 (2006)
128. Z. Lan, L.M. Frutos, A.L. Sobolewski, W. Domcke, Proc. Natl. Acad. Sci. **105**, 12707 (2008)
129. Z. Lan, W. Domcke, Chem. Phys. **350**, 125 (2008)
130. D. Shemesh, A.L. Sobolewski, W. Domcke, J. Am. Chem. Soc. **131**, 1374 (2009)
131. A. Warshel, *Computer Modeling of Chemical Reactions in Enzymes and Solutions* (Wiley, New York, 1991), pp. xiv, 236
132. Z.D. Nagel, J.P. Klinman, Nat. Chem. Biol. **5**(8), 543–550 (2009)
133. J.P. Klinman, M.J. Knapp, K. Rickert, J. Am. Chem. Soc. **124**(15), 3865–3874 (2002)
134. A. Warshel, M.H.M. Olsson, J. Mavri, Philos. T. R. Soc. B **361**(1472), 1417–1432 (2006)
135. A. Warshel, J. Mavri, H.B. Liu, M.H.M. Olsson, J. Phys. Chem. B **112**(19), 5950–5954 (2008)
136. F.H. Westheimer, Chem. Rev. **61**, 265–273 (1961)
137. L. Melander, W.H. Saunders Jr., *Reaction Rates of Isotopic Molecules* (Wiley, New York, 1980)
138. J. Bigeleisen, M. Wolfsberg, Adv. Chem. Phys. **1**, 15–76 (1958)
139. J.A. Smiley, P. Paneth, M.H. O'Leary, J.B. Bell, M.E. Jones, Biochemistry **30**, 6216–6223 (1991)
140. J. Mavri, R.A. Matute, Z.T. Chu, R. Vianello, J. Phys. Chem. B **120**, 3488–3492 (2016)
141. U. Palanivela, S. Lakshmipathia, J. Biomol. Struct. Dyn. (2015). doi:10.1080/07391102.2015.1085903
142. R. Elber, J. Chem. Phys. **144**, 060901 (2016)

# Chapter 16
# Nuclear Quantum Effect and H/D Isotope Effect on Hydrogen-Bonded Systems with Path Integral Simulation

**Kimichi Suzuki, Yukio Kawashima and Masanori Tachikawa**

**Abstract** In the past two decades, ab initio path integral (PI) simulation, in particular, ab initio path integral molecular dynamics simulation has reached its maturity and has been widely used to take account of nuclear quantum effects, such as zero-point vibrational energy and tunneling, in complex many-body systems. In particular, this method has significantly contributed to provide important insights into structures and fluctuation of the hydrogen-bonded systems as well as their isotopomers at finite temperature. In this chapter, we will review the recent advances in ab initio PI simulation. The development of an efficient algorithm for ab initio PI simulation and some applications will be featured. The efficient algorithm for path integral hybrid Monte Carlo method based on the second- and fourth-order Trotter expansion, which realizes large reduction of computational effort without loss of accuracy, will be described in detail. The applications focusing on the hydrogen-bonded systems, protonated and deprotonated water dimers ($H_5O_2{}^+$ and $H_3O_2{}^-$), $F^-(H_2O)_n$ (n = 1–3) clusters, and hydrogen maleate anion demonstrate the ability and powerfulness of PI simulation.

**Keywords** Ab initio path integral simulation · Nuclear quantum effect · Hydrogen-bonded structure · Geometrical isotope effect

K. Suzuki · M. Tachikawa (✉)
Quantum Chemistry Division, Graduate School of Science, Yokohama City University, 22-2 Seto, Kanazawa-ku, Yokohama 236-0027, Japan
e-mail: tachi@yokohama-cu.ac.jp

Y. Kawashima
RIKEN Advanced Institute for Computational Science, 7-1-26 Minatojima-Minami-Machi, Chuo-ku, Kobe 650-0047, Japan
e-mail: yukio.kawashima@riken.jp

 
M.J. Wójcik et al. (eds.), *Frontiers of Quantum Chemistry*,
https://doi.org/10.1007/978-981-10-5651-2_16

## 16.1 Introduction

Path integral (PI) simulation, which is a numerically exact method for quantum many-body systems, allows us to treat the nuclear quantum effect at finite temperature. Feynman's PI formulation has been presented as mathematically equivalent to the non-relativistic quantum theory of Schrödinger's or Heisenberg's equations [1]. The pioneering success of Fosdick and Jordan, combining PI with Monte Carlo (MC) simulation, which applied PI to nuclear quantum effect, has opened a new field in molecular simulation [2–4]. Now, PI simulations are essential for the description of chemical phenomenon, where the nuclear quantum effect plays an important role, such as the isotope effect on hydrogen-bonded systems discussed later on. The reader who is interested in the quantum statistics for boson or fermion by the PI simulations will find detailed description in some reviews [5, 6]. We summarize some important breakthroughs for PI simulation methodology and its algorithms in this section.

Sprik and co-workers have introduced the staging algorithm to relieve the sampling problem called "the stiff harmonic force problem" [7]. This problem was derived from the harmonic spring term in the effective potential. Friesner and Levy introduced an alternative method, the normal mode transformation method, to overcome the problem mentioned above [8]. Chandler and Wolynes presented the relation between quantum and classical statistics, now well known as the "classical isomorphism" [9]. Their important finding showed that the isomorphism enables us to perform simulations of the quantum system by *P* classical systems with harmonic nearest neighbor interactions, where *P* refers to number of partitions of the Boltzmann factor.

From the practical point of view, however, the number of *P* is limited to small numbers in simulations, which may lead to significant errors. To efficiently improve the accuracy with limited number of *P*, higher order correction (fourth-order Trotter expansion) algorithms have been proposed [10–12] on the basis of the Trotter decomposition formulation by Suzuki [13]. A more practical formulation has been given by Li and Broughton shortly after [14]. It has been shown that fourth-order Trotter expansion gives better energy convergence than the second-order one with respect to the Trotter number. Following developments and improvements of PI simulation techniques, a wide variety of applications emerged such as simulations of liquid $^4$He and rare gases, excess electron in alkali halide, water clusters [15]. Some applications and numerical problems of PI simulations at early stage have been summarized in reviews [15, 16]. It should be noted here that these developed techniques play a central role in PI simulation up to this day! Most of the effort of theory development afterward was based on these techniques, and their focus was on the reduction of computational cost.

In the first half of 1990s, many attempts to efficiently sample the canonical distributions with PI methods were reported. In most of the early works, MC simulations have been used for path integration. In principle, molecular dynamics (MD) method is also applicable for path integration. Parrinello and Rahman suggested the idea of PI molecular dynamics (PIMD) simulation, employing MD instead of MC, to treat the nuclear quantum effect via molecular simulation [17]. The main difference between the two sampling methods is that MD updates the nuclear configurations of the entire system at each simulation step, while they are locally updated in MC simulation. However, PIMD simulations have suffered the non-ergodic problem pointed out by Hall and Berne [18]. Tuckerman and co-workers have greatly contributed to relieve this problem by combining the staging coordinates and multiple timescale integration [19]. They also developed a new algorithm, which is a combination of PI technique and hybrid Monte Carlo (HMC) methods [19].

The first application of PI technique combined with Car-Parrinello MD (CPMD) approach (PICPMD) has been presented by Marx and Parrinello in 1994 [20]. The development and evolution of supercomputers with advanced theoretical methods enabled the use of MD simulations based on the first-principle calculations. They have achieved full quantum mechanical treatment of nuclei and electrons by PI technique and density-functional calculation with plane wave basis sets, respectively. Following the pioneering work of Marx and Parrinello, PICPMD method has been applied to various molecular systems [21–26]. The basic idea and efficient algorithm of PI combined with CPMD have been summarized in literatures [27–29]. PICPMD has been successful and useful; however, difficulties arouse from the non-trivial selection of pseudo-potentials and exchange-correlation functionals for density-functional theory calculations [30, 31].

An alternative way to achieve full quantum mechanical treatment of nuclei and electrons is to apply ab initio molecular orbital (MO) theory to electronic structure calculation with Born-Oppenheimer MD. A benefit of the use of ab initio potential is the systematic improvement of the accuracy with the selection of the MO method. The accuracy of the selected MO method with the chosen basis set is well established. Based on this idea, Shiga et al. have developed the ab initio PIMD simulation method [32]. The advantage of ab initio PIMD has been proved in various works. The combination with electronic structure calculation realized simulation of properties such as proton NMR chemical shift [33] and absorption spectra [34]. PIMD technique is also a powerful tool to investigate the temperature dependence of the above properties.

Large number of reports focused on PI simulations employing electronic structure calculation for hydrogen-bonded systems. More specifically, ionic hydrogen-bonded systems are one of the most important targets for PI simulations. Experimental works based on vibrational spectroscopy of small anionic clusters with ionic hydrogen bonds have found the protons of these hydrogen bonds located in the center of the hydrogen bonds [35–38]. This differs from typical hydrogen bonds, which have protons localized near a heavy atom found on both ends. The past works have shown that ab initio PIMD simulation is a powerful tool to tackle these ionic hydrogen bond structures, because conventional ab initio calculation fails to provide accurate description for these ionic hydrogen bonds [23, 39–49].

In biology, these ionic hydrogen bonds are categorized as the low-barrier hydrogen bonds (LBHBs). The LBHBs have especially low-barrier heights for proton transfer, and are short and strong bonds relative to typical hydrogen bonds [50–54]. It is known that in LBHBs, the hydrogen-bonded proton is located at the center between donor and acceptor atoms, as seen in ionic hydrogen bonds of small ion clusters. From recent works, it has been indicated that LBHBs play an important role in the functionality of biological molecules and enzyme catalysis [55–61]. However, some theoretical works denied the existence of LBHB [62–65], which leaves the existence of LBHB in biological systems an open question. More investigation on these hydrogen bonds is necessary to end this debate.

Most of the previous works have focused on the local structure of the hydrogen bond; however, Tuckerman and Marx have provided us a new insight of the nuclear quantum effect on hydrogen bonds and proton transfer. They studied the quantum effect on the intermolecular hydrogen bond of malonaldehyde using PICPMD simulation. They found that the quantum effect of heavy-atom skeleton enhances proton tunneling [66]. Their finding shows that nuclear quantum effect on the hydrogen bonds is not local, but delocalized to the entire molecular system. Then, a question arises: Does nuclear quantum effect on LBHB is delocalized to the entire molecule as well? Moreover, does the nuclear quantum effect on the LBHB influence the other hydrogen bonds or structures or fluctuation of the entire molecular system? This is critical for simulating biomolecular systems with LBHB, because hydrogen bonds play an important role in formation of three-dimensional structure and its functionality. To this end, we will show the nuclear quantum effect on the LBHB influences the formation of stable structure and the fluctuation of the molecular systems as well as the microscopic details of the hydrogen-bonded structure.

In this chapter, we will review some of the important contributions in the field of PI simulation. Development of new efficient algorithms and applications of PI, which provide new insights of LBHB, and assist the realization of time-consuming simulations for LBHB in biomolecules, will be introduced. An efficient algorithm for PIHMC method based on the second- and fourth-order Trotter expansion will be described in Sect. 16.2. In Sect. 16.3, we will show the ability and the powerfulness of PI simulation by applying it to ionized hydrogen-bonded systems: protonated and deprotonated water dimers ($H_5O_2^+$ and $H_3O_2^-$), $F^-(H_2O)_n$ ($n = 1-3$) clusters, and hydrogen maleate anion.

## 16.2 Development of a New Efficient Algorithm for PI Simulation

Despite the remarkable success of PICPMD and ab initio PIMD simulations, further improvements have been recently attempted to reduce the computational cost [67, 68]. In PICPMD and ab initio PIMD simulations, the main time-consuming bottleneck is the energy and gradient evaluations based on electronic structure calculations at each step. To reduce the computational cost, Nakayama and co-workers have proposed ab initio PIHMC algorithm combined with ab initio PI and HMC techniques [67]. They have used dual level of ab initio potentials, where less expensive and target (expensive) potentials are used to generate short trial trajectories and to finally judge the acceptance of the trial configurations, respectively. This method does not require energy gradient calculation of target ab initio potential.

We proposed an ab initio PIHMC based on fourth-order Trotter expansion algorithm to realize accurate simulation with small computational effort [68]. In ab initio PIMD based on fourth-order Trotter expansion, it is necessary to compute the Hessian matrix, which requires large computational cost. Our new algorithm provides the accuracy of fourth-order Trotter expansion without evaluating the cost-consuming Hessian matrix [40, 69]. In this section, we describe this algorithm in detail.

### 16.2.1 A Path Integral Simulation Based on Second-Order Trotter Expansion

In the conventional PI simulations, the partition function is approximated by the second-order Trotter expansion of $\exp[\tau(\hat{A}+\hat{B})]$ with respect to $\tau$, where $\hat{A}$ and $\hat{B}$ are kinetic and potential operators, respectively. The partition function based on second-order Trotter expansion is given by

$$Z^{2nd}(\beta) = \prod_{i=1}^{N}\left[\left(\frac{mp}{2\pi\hbar^2\beta}\right)^{3P/2}\int d\mathbf{r}_i^{(1)}\cdots d\mathbf{r}_i^{(P)}\right]\exp\left[-\beta V_{eff}^{2nd}\right], \tag{16.1}$$

where the effective potential $V_{eff}^{2nd}$ is defined as

$$V_{eff}^{2nd} = \frac{P}{2\beta^2\hbar^2}\sum_{i=1}^{N} m_i \sum_{s=1}^{P}\left(\mathbf{r}_i^{(s+1)} - \mathbf{r}_i^{(s)}\right)^2 - \frac{1}{P}\sum_{s=1}^{P} V\left(\mathbf{r}_1^{(s)},\ldots,\mathbf{r}_N^{(s)}\right), \tag{16.2}$$

and $N$ and $m_i$ are the number of nuclei and mass of the $i$th nucleus, respectively. The effective force $\mathbf{F}_i^{(s)2nd}$ of the $s$th bead on the $i$th nucleus is given by

$$\mathbf{F}_i^{(s)2nd} = \frac{m_i P}{\beta^2\hbar^2}\left(2\mathbf{r}_i^{(s)} - \mathbf{r}_i^{(s+1)} - \mathbf{r}_i^{(s-1)}\right) - \frac{1}{P}\sum_{s=1}^{P}\nabla_i^{(s)} V\left(\mathbf{r}_1^{(s)},\ldots,\mathbf{r}_N^{(s)}\right). \tag{16.3}$$

It would be interesting to compare the computational aspects for algorithms of ab initio PI simulations. The algorithm of ab initio PIMC consists of two steps. The first one is to generate a trial configuration, and the second step is to judge whether a trial configuration is accepted or rejected by Eq. (16.2). On the other hand, in the case of ab initio PIMD, one needs to evaluate not only the effective potential but also the effective force for the global update of the entire configuration. In these algorithms, the evaluation for the potential energy or its derivatives must be carried out $P$ times at each step. The total computational cost becomes expensive for PI simulations, since the number of statistical samplings from 50,000 to 100,000 steps for each bead is required. To reduce the computational efforts, Nakayama and co-workers have proposed ab initio PIHMC with dual level of potentials [67]. The HMC method has the advantages of MC and MD methods, which allow the global update of the whole system configuration by MD technique with reasonable acceptance ratio. The effective force in Eq. (16.3) with the low-level ab initio MO method is used to generate a trial configuration, while the effective potential in Eq. (16.2) with high-level potential is used to judge whether a trial configuration is accepted or rejected. This algorithm might be effective not only for large molecular systems, but also when analytic gradient with higher level of theory is unavailable such as CCSD(T)-F12.

### 16.2.2 A Path Integral Simulation Method Based on Fourth-Order Trotter Expansion

In general, the fourth-order Trotter expansion with respect to $\tau$ not only gives more accurate calculation results, but also realizes faster convergence with the increasing number of beads. In the second-order case, the number of beads should be increased as the temperature $T$ decreases, roughly by the factor of $1/T^2$. This shows that the total computation in PI simulation becomes more expensive at lower temperature. To remedy this problem, several kinds of higher order factorizations [10–12, 70] have been proposed. Among them, the fourth-order Trotter expansion suggested by Takahashi and Imada [11, 12] and reformatted by Li and Broughton [14] has been used in our algorithm. The partition function based on fourth-order Trotter expansion in PI representation is obtained by the same procedure of second-order expansion and expressed as

$$Z^{4th}(\beta) = \lim_{P\to\infty}\left(\frac{mP}{2\pi\hbar^2\beta}\right)^{3P/2}\int d\mathbf{r}^{(1)}\cdots d\mathbf{r}^{(P)}\exp\left[-\beta V_{eff}^{4th}\right], \tag{16.4}$$

where the effective potential $V_{eff}^{4th}$ is given by

$$V_{eff}^{4th} = V_{eff}^{2nd} + \sum_{i=1}^{N}\sum_{s=1}^{P}\frac{\beta^2\hbar^2}{24P^3m_i}\left[\nabla_i^{(s)}V\left(\mathbf{r}_1^{(s)},\ldots,\mathbf{r}_N^{(s)}\right)\right]. \tag{16.5}$$

The second term in Eq. (16.5) corresponds to the correction term derived from the fourth-order expansion. It should be noted that what distinguishes the fourth order from the second order is the necessity to evaluate the gradient vector. The effective force of the $s$th bead on the $i$th nucleus is evaluated by

$$\mathbf{F}_i^{(s)4th} = \mathbf{F}_i^{(s)2nd} - \frac{\beta^2\hbar^2}{12P^3}\sum_{j=1}^{N}\frac{1}{m_j}\nabla_j^{(s)}V\left(\mathbf{r}_1^{(s)},\ldots,\mathbf{r}_N^{(s)}\right)\cdot\nabla_i^{(s)}\nabla_j^{(s)}V\left(\mathbf{r}_1^{(s)},\ldots,\mathbf{r}_N^{(s)}\right). \tag{16.6}$$

As one can realize, the effective force of the fourth order requires the computation of the Hessian matrix. We will discuss the details in the next subsection.

We next compare the effective potentials of second- and fourth-order PI simulations, to understand the difference of the computational cost between these two simulations. Equations (16.2) and (16.5) show that the effective potential fourth-order expansion has to evaluate both potential energy and gradient vector, while only potential energy calculation is necessary for second-order expansion. When fourth-order PI technique combines ab initio MD and HMC methods, one has to evaluate computationally expensive Hessian matrix at each MD step. Thus, the application of ab initio PI simulation techniques might be limited to small systems. However, efficient ab initio PIHMC based on fourth-order expansion has been proposed by Suzuki et al. [68]. In this approach, the second-order effective potential in Eq (16.2) and its force in Eq. (16.3) are used to only generate trial configurations, to avoid the calculation of Hessian matrix. Then, the fourth-order effective potential in Eq. (16.5) is used to determine whether the trial configuration is accepted or rejected. Thus, this scheme keeps fourth-order accuracy without computing the Hessian matrix.

Our algorithm is summarized as follows:

1. The current positions of all the $P$ beads are stored. Momenta of all the beads are generated according to Maxwell distributions.
2. The effective force of the second order in Eq. (16.3) is evaluated.
3. The equation of motion is integrated by the Velocity Verlet algorithm with time step size $\Delta t$
4. The process consisting of steps 2 and 3 is a MD step. This process is repeated $N_{\mathrm{MD/MC}}$ times, where $N_{\mathrm{MD/MC}}$ is the number of MD steps in one MC cycle.
5. The trial configuration is accepted or rejected according to the Metropolis criterion in the fourth order using Eq. (16.5). Details are described below.
6. Go back to Step 1. The process consisting of steps 1–5 is a MC cycle. This is repeated $N_{\mathrm{MC}}$ times, where $N_{\mathrm{MC}}$ represents the total number of MC cycles.

The acceptance probability $P_{acc}$ of the Metropolis criterion in Step 5 is as follows:

$$P_{acc} = \min[1, \exp(-\beta\Delta H_{eff}^{4th})], \tag{16.7}$$

where $\Delta H_{eff}^{4th}$ is the difference of the effective energy $H_{eff}^{4th}$ before and after $N_{\mathrm{MD/MC}}$ MD steps which is described as

$$H_{eff}^{4th} = \sum_{s=1}^{P} \sum_{i=1}^{N} \frac{(\mathbf{P}_i^{(s)})^2}{2\mu_i^{(s)}} + V_{eff}^{4th}. \tag{16.8}$$

Parameters $\mathbf{P}_i^{(s)}$ and $\mu_i^{(s)}$ are the momentum conjugate of $\mathbf{r}_i^{(s)}$ and its fictitious mass, respectively.

We next consider the estimator for the fourth-order expansion, which is essential to obtain expectation values of various physical parameters. The partition function of the fourth-order Trotter expansion is different from that of the second-order one; thus, estimators for expectation values for fourth-order Trotter expansion are also different from those for the second-order one. However, the estimator for the expectation value of a physical parameter $\hat{B}$, which depends on the coordinates, is obtained by a similar procedure for the second-order case [32]. Following the same procedure in the case of the second-order expansion, we get the following expression,

$$\begin{aligned}\langle\hat{B}\rangle = &\frac{1}{Z^{4th}}\left[\prod_{i=1}^{N}\left(\frac{m_i P}{2\pi\beta\hbar^2}\right)^{3P/2}\int d\mathbf{r}_i^{(1)}\cdots d\mathbf{r}_i^{(P)}\right]\\ &\frac{1}{P}\sum_{s=1}^{P}\left(B^{(s)}+\sum_{i=1}^{N}\frac{\hbar^2}{12m_i}\left(\frac{P}{\beta}\right)^2\nabla_i^{(s)}V\left(\mathbf{r}_i^{(1)}\cdots d\mathbf{r}_i^{(P)}\right)\cdot\nabla_i^{(s)}B^{(s)}\right)\\ &\exp\left[-\beta V_{eff}^{4th}\right].\end{aligned} \tag{16.9}$$

Partial integration of Eq. (16.9) will lead to another well-known estimator given by

$$\begin{aligned}\langle\hat{B}\rangle = &\frac{1}{Z^{4th}}\left[\prod_{i=1}^{N}\left(\frac{m_i P}{2\pi\beta\hbar^2}\right)^{3P/2}\int d\mathbf{r}_i^{(1)}\cdots d\mathbf{r}_i^{(P)}\right]\\ &\frac{1}{P}\sum_{s=1}^{P}B^{(s)}\left(1-\sum_{i=1}^{N}\frac{\hbar^2}{12m_i}\left(\frac{P}{\beta}\right)^2\nabla_i^{(s)}\nabla_i^{(s)}V\left(\mathbf{r}_i^{(1)}\cdots\mathbf{r}_i^{(P)}\right)\right.\\ &\left.+\frac{1}{\exp\left(-\beta V_{eff}^{4th}\right)}\nabla_i^{(s)}\exp\left(-\beta V_{eff}^{4th}\right)\cdot\nabla_i^{(s)}V\left(\mathbf{r}_i^{(1)}\cdots\mathbf{r}_i^{(P)}\right)\right)\\ &\exp\left[-\beta V_{eff}^{4th}\right].\end{aligned} \tag{16.10}$$

We note that Eq. (16.10) includes the Hessian matrix. We need to avoid calculating the Hessian matrix to realize a computationally efficient algorithm; thus, we derive the estimators for expectation values of internuclear distance $\left(\hat{B}=\left|\hat{\mathbf{r}}_i-\hat{\mathbf{r}}_j\right|\right)$ starting from Eq. (16.9). The estimator is given by

$$\langle\hat{B}\rangle = \frac{1}{Z^{4th}}\left[\prod_{i=1}^{N}\left(\frac{m_iP}{2\pi\beta\hbar^2}\right)^{3P/2}\int d\mathbf{r}_i^{(1)}\cdots d\mathbf{r}_i^{(P)}\right]$$
$$\frac{1}{P}\sum_{s=1}^{P}\left\{\left|\mathbf{r}_i^{(s)}-\mathbf{r}_j^{(s)}\right| + \frac{\hbar^2}{12m_i}\left(\frac{P}{\beta}\right)^2\left(\frac{1}{m_i}\nabla_i^{(s)}V\left(\mathbf{r}_i^{(1)}\cdots d\mathbf{r}_i^{(P)}\right)\right.\right. \tag{16.11}$$
$$\left.\left. -\frac{1}{m_j}\nabla_j^{(s)}V\left(\mathbf{r}_i^{(1)}\cdots d\mathbf{r}_i^{(P)}\right)\right)\frac{\mathbf{r}_i^{(s)}-\mathbf{r}_j^{(s)}}{\left|\mathbf{r}_i^{(s)}-\mathbf{r}_j^{(s)}\right|}\right\}\exp\left[-\beta V_{eff}^{4th}\right].$$

As shown in Eq. (16.11), the derivative of the internuclear distance must be considered.

Next, let us derive the distribution function for fourth-order expansion. To obtain the distribution function with respect to internuclear distance ($f(r)$), we insert $\hat{B}=\delta(r-|\mathbf{r}_i-\mathbf{r}_j|)$ in Eq. (16.9), which is expressed as

$$f(r) = \frac{1}{Z^{4th}}\left[\prod_{i=1}^{N}\left(\frac{m_iP}{2\pi\beta\hbar^2}\right)^{3P/2}\int d\mathbf{r}_i^{(1)}\cdots d\mathbf{r}_i^{(P)}\right]$$
$$\frac{1}{P}\sum_{s=1}^{P}\left\{\left(1+\sum_{i=1}^{N}\frac{\hbar^2}{12m_i}\left(\frac{P}{\beta}\right)^2\nabla_i^{(s)}V\left(\mathbf{r}_i^{(1)}\cdots d\mathbf{r}_i^{(P)}\right)\cdot\nabla_i^{(s)}\right)\right. \tag{16.12}$$
$$\left.\delta(r-|\mathbf{r}_i-\mathbf{r}_j|)\right\}\exp\left[-\beta V_{eff}^{4th}\right].$$

Equation (16.12) cannot be directly used as a distribution function, since the derivative of the delta function is included. We therefore describe the delta function as the limit of Gaussian distribution as follows:

$$\delta(r-|\mathbf{r}_i-\mathbf{r}_j|) = \lim_{\alpha\to\infty}\sqrt{\frac{\alpha}{\pi}}\exp\left[-\alpha(r-|\mathbf{r}_i-\mathbf{r}_j|)^2\right], \tag{16.13}$$

where α is a real parameter. Equation (16.12) can be rewritten by inserting Eq. (16.13) as

$$f(r) = \frac{1}{Z^{4th}}\left[\prod_{i=1}^{N}\left(\frac{m_iP}{2\pi\beta\hbar^2}\right)^{3P/2}\int d\mathbf{r}_i^{(1)}\cdots d\mathbf{r}_i^{(P)}\right]$$
$$\frac{1}{P}\sum_{s=1}^{P}\left[\sqrt{\frac{\alpha}{\pi}}\exp\left(-\alpha(r-|\mathbf{r}_i-\mathbf{r}_j|)^2\right)\left\{1-2\alpha(r-|\mathbf{r}_i-\mathbf{r}_j|)\left(\frac{P}{\beta}\right)^2\right.\right. \tag{16.14}$$
$$\left(\frac{\hbar^2}{12m_i}\nabla_i^{(s)}V\left(\mathbf{r}_i^{(1)}\cdots d\mathbf{r}_i^{(P)}\right)\cdot\nabla_i^{(s)}\right.$$
$$\left.\left.\left. +\frac{\hbar^2}{12m_j}\nabla_j^{(s)}V\left(\mathbf{r}_i^{(1)}\cdots d\mathbf{r}_i^{(P)}\right)\cdot\nabla_j^{(s)}\right)|\mathbf{r}_i-\mathbf{r}_j|\right\}\right]\exp\left[-\beta V_{eff}^{4th}\right].$$

Equation (16.14) can be approximated by employing the Taylor series, and we finally arrive at the distribution function of the internuclear distance excluding the Hessian matrix as follows:

$$f(r) \approx \frac{1}{Z^{4th}} \left[ \prod_{i=1}^{N} \left( \frac{m_i P}{2\pi\beta\hbar^2} \right)^{3P/2} \int d\mathbf{r}_i^{(1)} \cdots d\mathbf{r}_i^{(P)} \right] \frac{1}{P} \sum_{s=1}^{P} \delta\Bigg( r - \Bigg[ \left| \mathbf{r}_i^{(s)} - \mathbf{r}_j^{(s)} \right| + \left( \frac{P}{\beta} \right)^2 \left\{ \frac{\hbar^2}{12m_i} \nabla_i^{(s)} V\left( \mathbf{r}_i^{(1)} \cdots d\mathbf{r}_i^{(P)} \right) - \frac{\hbar^2}{12m_j} \nabla_j^{(s)} V\left( \mathbf{r}_i^{(1)} \cdots d\mathbf{r}_i^{(P)} \right) \right\} \frac{\mathbf{r}_i^{(s)} - \mathbf{r}_j^{(s)}}{\left| \mathbf{r}_i^{(s)} - \mathbf{r}_j^{(s)} \right|} \Bigg] \Bigg) \exp\left[ -\beta V_{eff}^{4th} \right]. \quad (16.15)$$

This approximation is generalized to other distribution functions such as bond or dihedral angles.

### 16.2.3 Numerical Calculations of Hydrogen Molecule

We performed numerical calculations of the hydrogen molecular system to demonstrate the efficiency of our algorithm. Table 16.1 shows the average energy and bond length of hydrogen molecule by second- and fourth-order PIMD, conventional second order and our fourth PIHMC scheme [68]. It is clearly seen that our fourth-order PIHMC scheme gives the same average values of fourth-order PIMD for all $P$ values. This shows that our algorithm maintains the fourth-order accuracy without evaluating the Hessian matrix elements. Our fourth-order PIHMC algorithm significantly improves the average values of the second order. For instance, the average bond length is mostly converged at $P = 64$ in second-order PIHMC, while it is converged at $P = 16$ in our algorithm.

**Table 16.1** Convergence of total energy (kcal/mol) and average bond length (angstrom) of $H_2$ molecule at 300 K against the number of classical beads for second- and fourth-order PIMD and PIHMC simulations

| No. of Beads | PIMD | | | | PIHMC | | | |
|---|---|---|---|---|---|---|---|---|
| | Second | | Fourth | | Second | | Fourth | |
| | $\langle E \rangle$ | $\langle R_{HH} \rangle$ | $\langle E \rangle$ | $\langle R_{HH} \rangle$ | $\langle E \rangle$ | $\langle R_{HH} \rangle$ | $\langle E \rangle$ | $\langle R_{HH} \rangle$ |
| 1 | 2.08 | 0.7468 | 2.08 | 0.7468 | 2.08 | 0.7463 | 2.08 | 0.7463 |
| 4 | 3.79 | 0.7532 | 5.55 | 0.7596 | 3.69 | 0.7526 | 5.62 | 0.7603 |
| 8 | 5.34 | 0.7588 | 6.72 | 0.7654 | 5.25 | 0.7594 | 7.06 | 0.7663 |
| 16 | 6.52 | 0.7645 | 7.44 | 0.7687 | 6.49 | 0.7649 | 7.44 | 0.7683 |
| 32 | 7.20 | 0.7683 | 7.67 | 0.7693 | 7.43 | 0.7676 | 7.67 | 0.7684 |
| 64 | 7.55 | 0.7689 | 7.69 | 0.7684 | 7.83 | 0.7680 | 7.77 | 0.7683 |

## 16.3 Applications of PI Simulation: Hydrogen-Bonded Systems

### 16.3.1 Protonated and Deprotonated Water Dimers

The protonated and deprotonated water dimers ($H_5O_2^+$ and $H_3O_2^-$) have received long-standing attention, and their molecular structures and vibrations have been focused on experimental and theoretical works [23, 24, 35, 39, 40, 69, 71–100], since they are the minimum unit of acidic and basic aqueous solutions. From ab initio MO calculations, it is known that most stable structure of $H_3O_2^-$ becomes asymmetric due to the existence of a potential barrier along the proton transfer coordinate, while that of $H_5O_2^+$ has $C_2$ symmetry, where the two $H_2O$ molecules are twisted and the hydrogen-bonded proton (H*) is located at the center of two oxygen atoms.

Tuckerman and co-workers have carried out the pioneering work of these clusters including the nuclear quantum effect by PICPMD simulation at ambient temperature [23]. They found that H* treated as quantum nuclei in $H_3O_2^-$ is distributed around the center of two oxygen atoms due to the zero-point vibration with respect to proton transfer motion, while classical H* is delocalized around the two oxygen atoms on both ends of the hydrogen bond. Following the work by Tuckerman et al., Tachikawa and Shiga have performed ab initio PIMD simulations for these dimers as well as the deuterated systems to study the GIE at ambient temperature [75]. They found two important GIEs. The first GIE is the difference of OO distance: It is longer in $D_3O_2^-$ than in $H_3O_2^-$. The second GIE is the location of the shared proton/deuteron in the hydrogen bond: Deuteron (D*) is localized around the two oxygen atoms of the end of the hydrogen bond, while shared proton (H*) is delocalized between two oxygen atoms. McCoy and co-workers have also reported the GIE for these systems using a diffusion Monte Carlo approach (DMC) at 0 K [78]. However, they have found contrary GIE results to those of PIMD simulation results. We will discuss the temperature dependence on the GIE of the hydrogen-bonded structures employing PIMD simulation based on fourth-order expansion in this subsection.

We focus on the hydrogen-bonded structure to investigate the temperature dependence of GIE for $H_3O_2^-$ and $H_5O_2^+$. The average values of $\delta_{OH*}$ and OO distance ($R_{OO}$) for $H_3O_2^-$ are shown in Table 16.2, where $|\delta_{OH*}|$ is defined as the deviation of H* from the center of $R_{OO}$. The average values of $\delta_{OH*}$ and OO distance gradually enlarge as the temperature rises. It is found that the average OO distance and $|\delta_{OH*}|$ become smaller as the isotopomer is heavier at low temperature, while these orderings reverse for both $H_3O_2^-$ and $H_5O_2^+$ at high temperature. For example, $|\delta_{OH*}|$ and $R_{OO}$ of H-species in $H_3O_2^-$ are the longest among all isotopomers at low temperature, while those are the smallest at high temperature. How can these results be explained? The analysis of distribution function is helpful to interpret the obtained results. This is important to see both $\delta_{OH*}$ and $R_{OO}$ coordinates, since these two coordinates strongly correlate with each other. The two-dimensional distributions with respect to $\delta_{OH*}$ and $R_{OO}$ coordinates of $H_3O_2^-$

**Table 16.2** Temperature (K) dependence of the average values of $|\delta_{OH^*}|$ ($|\delta_{OD^*}|$ and $|\delta_{OT^*}|$ for the corresponding isomers and $R_{OO}$ distance in the quantum (PIMD) and classical simulations. Units are in Å

| Temperature (K) | $\delta_{OH^*}$ | | | $R_{OO}$ | | |
|---|---|---|---|---|---|---|
| | $H_3O_2^-$ | $D_3O_2^-$ | $T_3O_2^-$ | $H_3O_2^-$ | $D_3O_2^-$ | $T_3O_2^-$ |
| 50 | 0.237 | 0.219 | 0.215 | 2.484 | 2.479 | 2.477 |
| 100 | 0.238 | 0.222 | 0.215 | 2.487 | 2.480 | 2.478 |
| 200 | 0.247 | 0.244 | 0.237 | 2.492 | 2.488 | 2.485 |
| 400 | 0.286 | 0.297 | 0.311 | 2.509 | 2.508 | 2.513 |
| 600 | 0.337 | 0.370 | 0.380 | 2.529 | 2.543 | 2.551 |

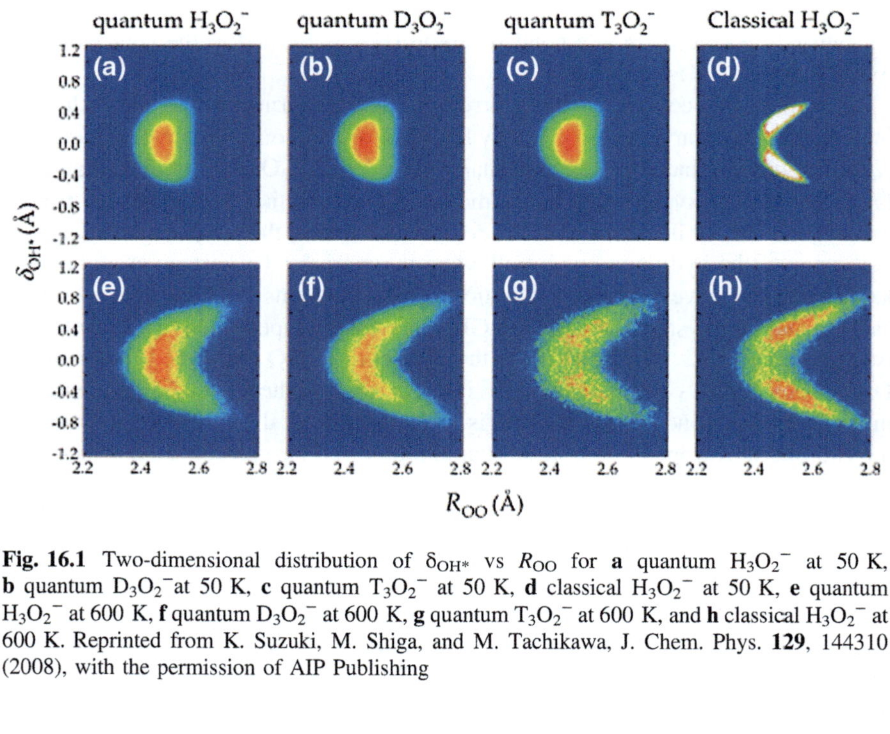


**Fig. 16.1** Two-dimensional distribution of $\delta_{OH^*}$ vs $R_{OO}$ for **a** quantum $H_3O_2^-$ at 50 K, **b** quantum $D_3O_2^-$at 50 K, **c** quantum $T_3O_2^-$ at 50 K, **d** classical $H_3O_2^-$ at 50 K, **e** quantum $H_3O_2^-$ at 600 K, **f** quantum $D_3O_2^-$ at 600 K, **g** quantum $T_3O_2^-$ at 600 K, and **h** classical $H_3O_2^-$ at 600 K. Reprinted from K. Suzuki, M. Shiga, and M. Tachikawa, J. Chem. Phys. **129**, 144310 (2008), with the permission of AIP Publishing

are shown in Fig. 16.1. The distributions for the quantum simulations show single peak structures. This shows that H*, D*, and T* isotopomers have delocalized structures (O…X…O). The distribution with respect to $\delta_{OH^*}$ coordinate of H species is more broadened than that of D- or T-species due to the difference of zero-point energy. In addition, average OO distance of H-species is the longest among all species at low temperature, since OO distance is elongated by the zero-point energy of two OH* stretching motion under the anharmonicity of the potential. We found the same tendency for $H_5O_2^+$ as well.

At high temperature, on the other hand, OO distance and $\delta_{OH^*}$ coordinates are strongly correlated with each other. It can be seen that the distributions of D- and

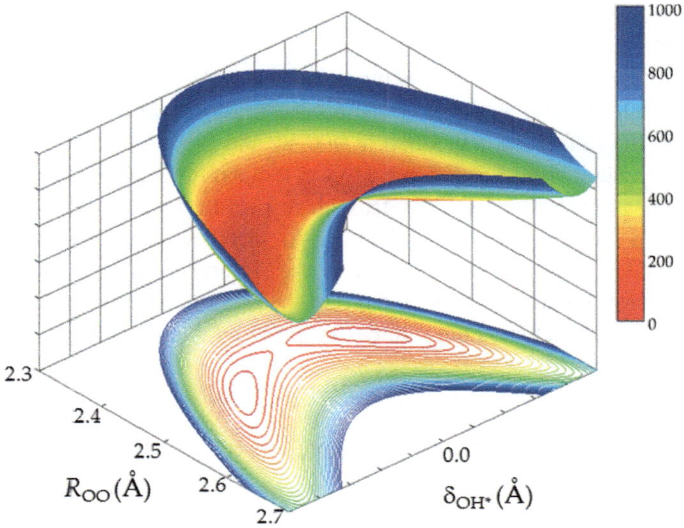


**Fig. 16.2** Two-dimensional potential energy surface of $\delta_{OH*}$ vs $R_{OO}$ in $H_3O_2^-$. Unit in $cm^{-1}$

T-species have two peaks, while that of H-species shows a single peak structure. This shows that D* and T* isotopomers have localized structures (O–X…H), while H* is still delocalized between two oxygen atoms. The results of $H_5O_2^+$ are similar to the case of $H_3O_2^-$.

We next investigated the potential surface to understand why D* and T* are likely to be localized on the oxygen atoms at high temperature. Two-dimensional potential surface with respect to $\delta_{OH*}$ and $R_{OO}$ coordinates for $H_3O_2^-$ is shown in Fig. 16.2. The potential surface clearly shows that the potential barrier height in the center of two oxygen atoms appears and becomes higher as OO bond distance lengthens. Thus, D* or T* is likely to be more localized either two oxygen atoms than H* due to less quantum nature when OO bond distance elongates by thermal activation. The potential surface of $H_5O_2^+$ also shows the same tendency. At this point, the order relation of average OO bond length at high temperature is explained as the Ubbelohde effect. This idea is based on the assumption that H/D mediates the attractive interaction of two oxygen atoms. Chemically, the resonance of O–H…O and O…H–O is likely to occur, and therefore, the two oxygen atoms are more attractive by proton rather than deuteron because the delocalization of the proton is enhanced compared to deuteron or triton. In the case of $T_3O_2^-$at high temperature, the tritiation makes T* more localized to the oxygen atoms on both ends of the hydrogen bond, then loosens the hydrogen bond, and therefore lengthens the OO distance. The situation of $H_5O_2^+$ at high temperatures is similar, but somewhat different from that of $H_3O_2^-$. The order relation of average values for $H_3O_2^-$ alters above 400 K, while that for $H_5O_2^+$ does not change. The harmonic frequencies of OO stretching mode for $H_3O_2^-$ (323 $cm^{-1}$ = 464 K), $D_3O_2^-$ (279 $cm^{-1}$ = 401 K), and $T_3O_2^-$ (245 $cm^{-1}$ = 343 K) are systematically smaller than those of $H_5O_2^+$

(628 cm$^{-1}$ = 898 K), $D_5O_2^+$ (593 cm$^{-1}$ = 847 K), and $T_5O_2^+$ (534 cm$^{-1}$ = 763 K). Thus, the OO stretching mode in $H_3O_2^-$ is activated at temperatures lower than that in $H_5O_2^+$. According to the above explanation, the order relation of average values should reverse for $H_5O_2^+$ at higher temperature. However, we could not observe this clearly since the difference was small at 900 K, and the system becomes unstable due to (pre-)dissociation of $H_3O^+$ and $H_2O$ at temperatures 1200 K or higher.

The temperature dependence of GIE with respect to hydrogen-bonded structure was observed for $H_3O_2^-$ and $H_5O_2^+$. At low temperature, GIE on the hydrogen-bonded structure can be seen as the difference of zero-point vibration under the anharmonicity of the potential. On the other hand, the coupling between proton transfer and OO stretching modes is the key for GIE at high temperature.

### *16.3.2 Fluoride Ion-Water Clusters*

Fluoride ion-water clusters $F^-(H_2O)_n$ have attracted considerable attention due to the strong interaction between fluoride ion and water compared to typical hydrogen bonds [23, 39, 50, 68, 76, 101–111]. Some reports have suggested that the ion-water interaction energy is over 25 kcal/mol [103, 105, 107–109]. There have been many experimental and theoretical reports discussing the hydrogen-bonded structure with respect to the coordination number of water molecules in fluoride ion-water clusters. Experiments employing IR spectra found that the frequency of ionic hydrogen-bonded stretching mode blue-shifts as the coordination number of water molecules increases, and the second solvation shell emerges at the coordination number of five [102, 103]. From ab initio calculations, their reports have shown that binding energies of fluoride ion-water clusters are stronger than those of other halogen ion-water clusters. Consequently, the elongation of distance for OH* (H* represents hydrogen atoms in fluoride ion-water hydrogen bonds) of $F^-$...H*O hydrogen bond in water molecule originates from the strong ion-water interaction. They have also reported that OH* distance becomes shorter as the coordination number of water molecules increases, while the distance of $F^-$...H*O (between fluoride and oxygen atoms) elongates. It has been qualitatively shown that the interaction between fluoride ion and hydrogen-bonded hydrogen weakens as the coordination number of water molecules increases. In this subsection, we discuss the dependence of the coordination number of water molecules on the hydrogen bond structures of fluoride ion-water clusters including the nuclear quantum and thermal fluctuation effects with the aid of our developed ab initio PIHMC algorithm based on fourth-order Trotter expansion.

The values of OH* bond length $R_{\mathrm{OH}*}$, $F^-$...H* length $R_{\mathrm{FH}*}$, $F^-$...O length $R_{\mathrm{FO}}$, and H*FO angle $\theta_{\mathrm{H*FO}}$ for $F^-(H_2O)_n$ (n = 1–3) are shown in Table 16.3. The average values in classical and quantum simulations change systematically as the coordination of number of water molecules increases. The hydrogen bond parameters $R_{\mathrm{FH}*}$, $R_{\mathrm{FO}}$, and $\theta_{\mathrm{H*FO}}$ increase, while $R_{\mathrm{OH}*}$ decreases. We next compare these parameters among equilibrium and averages of classical and quantum simulations.

**Table 16.3** Average structural parameters of $F^-(H_2O)_n$ ($R_{OH*}$, $R_{FH*}$, $R_{FO}$, $\theta_{H*FO}$) for equilibrium structure, classical and quantum simulations. The units of bond length and angles are angstroms and degrees, respectively

| | | Equilibrium | Classical | Quantum |
|---|---|---|---|---|
| $\langle R_{OH^*} \rangle$ | $F^-(H_2O)$ | 1.056 | 1.063 | 1.114 |
| | $F^-(H_2O)_2$ | 1.019 | 1.022 | 1.048 |
| | $F^-(H_2O)_3$ | 1.000 | 1.005 | 1.026 |
| $\langle R_{FH^*} \rangle$ | $F^-(H_2O)$ | 1.414 | 1.431 | 1.377 |
| | $F^-(H_2O)_2$ | 1.529 | 1.566 | 1.552 |
| | $F^-(H_2O)_3$ | 1.632 | 1.654 | 1.671 |
| $\langle R_{FO} \rangle$ | $F^-(H_2O)$ | 2.469 | 2.482 | 2.465 |
| | $F^-(H_2O)_2$ | 2.544 | 2.569 | 2.566 |
| | $F^-(H_2O)_3$ | 2.613 | 2.634 | 2.654 |
| $\langle \theta_{H^*FO} \rangle$ | $F^-(H_2O)$ | 1.3 | 4.2 | 6.5 |
| | $F^-(H_2O)_2$ | 2.5 | 4.8 | 6.6 |
| | $F^-(H_2O)_3$ | 5.3 | 5.2 | 7.4 |

We found that the average values of the bond distance in the classical simulations were larger than the values of equilibrium structures due to the thermal fluctuation effects. However, the dependence of the coordinate number of water molecules on the average values for quantum simulations showed an interesting behavior. For parameters $R_{OH*}$ and $\theta_{H*FO}$, the equilibrium structure had the smallest values and the quantum simulation had the largest value for all calculated systems. On the other hand, the relation for the other two parameters altered with respect to the coordination number of water molecules. The average value of $R_{FH*}$ and $R_{FO}$ for the quantum simulation was the smallest for $F^-(H_2O)$, but the largest for $F^-(H_2O)_3$. The nuclear quantum effects on these parameters seemed to be inconsistent with the number of water molecules. The origin of this could not be found just by investigating the hydrogen bond between the fluoride ion and water molecule.

We next studied the nuclear quantum effects on the interaction between the water molecules to understand the origin of the peculiar behavior found in the ion-water hydrogen bonds. One-dimensional distributions of OO distance ($R_{OO}$) between the water molecules for $F^-(H_2O)_3$ are shown in Fig. 16.3. The distribution of the classical simulation is highly delocalized within the range of 3.0 Å to 5.3 Å, while that in the quantum simulation is observed in the localized range between 2.8 Å and 3.7 Å. This shows that a hydrogen bond forms between the two water molecules in the quantum simulation. This indicates that water–water hydrogen bond formation is enhanced by the nuclear quantum effects.

We show some representative snapshots in Fig. 16.4 to illustrate the formation of the hydrogen bonds. Hydrogen atoms of water, which bind with $F^-$, point toward the ion in both simulations. On the other hand, other hydrogen atoms do not interact with other water for classical simulation. In the quantum simulation, however, some hydrogen atoms point toward an oxygen atom of water, as shown in Fig. 16.4c, d. Water molecules in classical simulation have negligible interaction with each other and fluctuate largely, while water molecules with water–water interaction in quantum simulation develop certain structures.

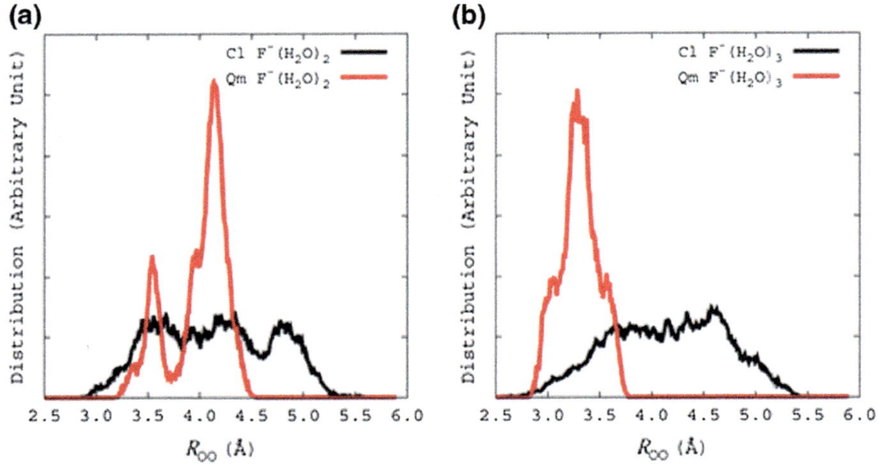


**Fig. 16.3** One-dimensional distributions of O…O distance for **a** $F^{-}(H_2O)_2$ and **b** $F^{-}(H_2O)_3$ with classical (*black*) and quantum (*red*) simulations. Reprinted with permission from Y. Kawashima, K. Suzuki, M. Tachikawa, J. Phys. Chem. A **117**, 5205 (2013). Copyright 2013 American Chemical Society

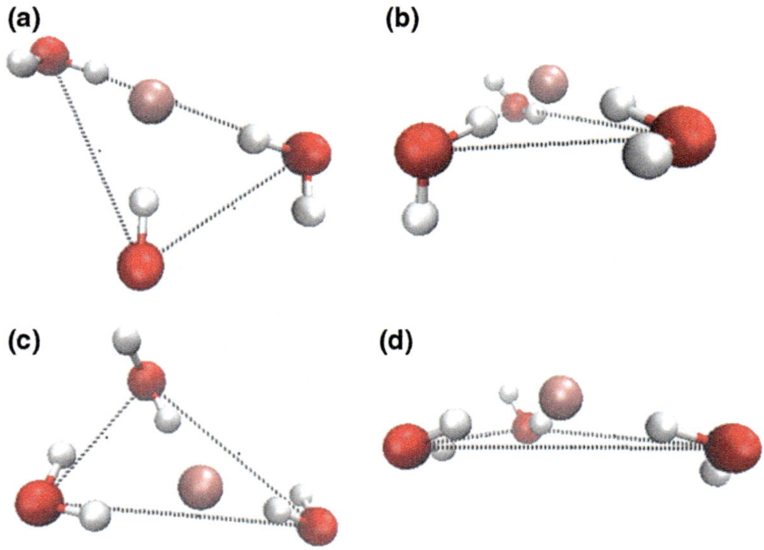


**Fig. 16.4** Snapshot structures of $F^{-}(H_2O)_3$ obtained from classical and quantum simulations: **a** *top view* and **b** *side view* of snapshot structure from classical simulation, and **c** *top view* and **d** *side view* of snapshot structure of the centroids for each atom from quantum simulation. Reprinted with permission from Y. Kawashima, K. Suzuki, M. Tachikawa, J. Phys. Chem. A **117**, 5205 (2013). Copyright 2013 American Chemical Society

The emergence of such nuclear quantum effects on water–water hydrogen bonds leads to the decrease of nuclear quantum effects on $F^-$-water hydrogen bonds. The nuclear quantum effects on the different hydrogen bonds, $F^-$-water and water–water hydrogen bonds, compete against each other. Our results suggest that the competition of the two different hydrogen bonds is the origin of the inversion of the relation for $R_{FH^*}$ and $R_{FO}$ as the coordination number of water molecules increases from $F^-(H_2O)$ to $F^-(H_2O)_3$. We found that the nuclear quantum effect on hydrogen bonds influences the stable structure formation.

The competition of nuclear quantum effect on the different kinds of hydrogen bonds should be indispensable for the quantitative analysis for systems including different hydrogen bonds such as ion liquid-water systems, or simulating biomolecules functioning in water such as proton relay in bacteriorhodopsin. The nuclear quantum effect on hydrogen bonds influences the stable structure formation and thus may influence the function of biomolecules.

### *16.3.3 Hydrogen Maleate Anion*

Hydrogen maleate anion (HM, $H(OOC–CH=CH–COO)^-$) is one of the simplest symmetric intramolecular hydrogen-bonded systems with symmetric and short hydrogen bond. The structures of HM are depicted in Fig. 16.5. The hydrogen bond structure and dynamics of HM have been widely studied by both experimental [37, 112–126] and theoretical [48, 127–141] works. HM continues to receive certain attention, because it is considered as a simple model for studying low-barrier hydrogen bonds (LBHB) in biomolecular systems [55–61]. Fillaux et al. found a

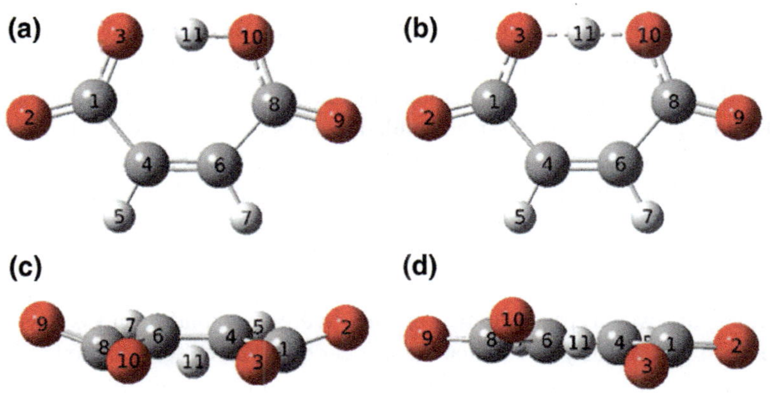


**Fig. 16.5** Structures of hydrogen maleate anion molecule with atom numbering for **a** equilibrium $C_s$ structure, **b** transition state $C_{2v}$ structure, **c** representative snapshot with plane symmetric distorted structure from quantum simulation, **d** representative snapshot with centrosymmetric distorted structure from quantum simulation. Reprinted with permission from Y. Kawashima, M. Tachikawa, J. Chem. Theory Comput. **10**, 153 (2014). Copyright 2014 American Chemical Society

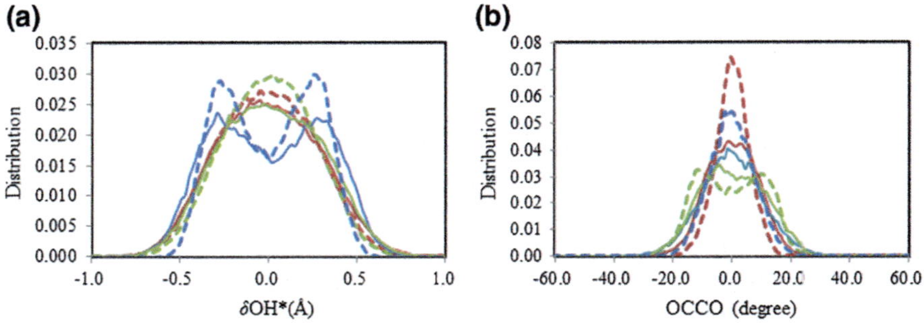


**Fig. 16.6** Distributions of **a** $\delta_{OH*}$, and **b** ∠OCCO for classical simulation (*blue line*), quantum simulation for DM (*green line*), and quantum simulation for HM (*red line*). The *solid* and *dashed lines* represent simulations at 300 K and 150 K, respectively. Reprinted with permission from Y. Kawashima, M. Tachikawa, J. Chem. Theory Comput. **10**, 153 (2014). Copyright 2014 American Chemical Society

unique feature of the structure of HM [120]. They studied the dynamics of potassium HM crystal using inelastic neutron scattering. From the viewpoint that the crystal of potassium HM is symmetric, they suggested a proton potential function with three non-equivalent minima along the out-of-plane ring-bending mode [119]. The potential indicates the existence of a stable structure with out-of-plane ring deformation. However, the three non-equivalent minima have not been confirmed by other experimental work and further investigation was necessary. No theoretical work has explored the out-of-plane ring deformation of HM to confront or support the triple-minima potential surface proposed by Fillaux et al. [119]. We studied HM employing ab initio PIMD simulations, to investigate Fillaux's suggestion and to understand how the nuclear quantum effects influence the fluctuation of the entire molecular system.

We analyzed the proton transfer coordinate $\delta_{OH*}$ for understanding the location of H* in the intramolecular hydrogen bond of HM. The deuterated HM (DM) is studied as well. The distribution of $\delta_{OH*}$ found in quantum and classical simulations is depicted in Fig. 16.6a. The classical simulation shows two peaks with large absolute $\delta_{OH*}$ values. The dominant hydrogen-bonded structure found in classical simulation for both 150 K and 300 K has $C_s$ symmetry (Fig. 16.5a), or H* is localized on one oxygen atom. The localization of the proton on oxygen atom at both ends of the hydrogen bond is stronger at 150 K than at 300 K. The distributions of quantum simulations show one dominant peak at $\delta_{OH*} = 0$, which show that H* is located at the center of two oxygen atoms, or the hydrogen bond has $C_{2v}$ symmetry (Fig. 16.5b). This clearly shows that the nuclear quantum effects lead to barrier-less proton transfer. No significant isotope effect is found in $\delta_{OH*}$ distribution.

We next focused on an additional geometric parameter of the hydrogen bond: dihedral angle ($O_3C_1C_8O_{10}$). The dihedral angle is related to the hydrogen bond fluctuation and is a good index to seek the out-of-plane ring deformation. The distributions of the dihedral angle are illustrated in Fig. 16.6b. These distributions

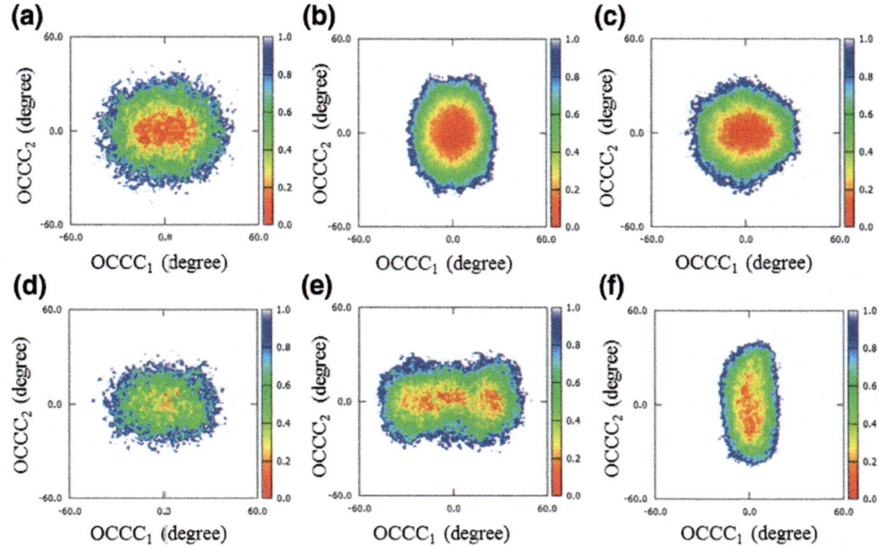


**Fig. 16.7** Two-dimensional free-energy landscape of ∠$OCCC_1$ and ∠$OCCC_2$ for **a** classical simulation at 150 K, **b** quantum simulation for DM at 150 K, **c** quantum simulation for HM at 150 K, **d** classical simulation at 300 K, **e** quantum simulation for DM at 300 K, **f** quantum simulation for HM at 300 K. The unit of free energy (potential of mean force) is kcal/mol. Reprinted with permission from Y. Kawashima, M. Tachikawa, J. Chem. Theory Comput. **10**, 153 (2014). Copyright 2014 American Chemical Society

show large thermal effects: Distributions for simulations at 150 K are localized compared to those at 300 K. While distributions of quantum simulation for HM have a sharp peak at the center, distributions of classical simulations are slightly more broadened. Distributions of DM show broad distribution compared to other simulations, which indicates that the ring deformation is large. We found large isotope effect for this dihedral angle.

We constructed a free-energy landscape from two largest modes describing the fluctuation of the system obtained from principal component analysis to focus on the out-of-plane ring deformation. We constructed a two-dimensional free-energy landscape, which mimics the two modes with largest fluctuation. We later denote these parameters as ∠$OCCC_1$ and ∠$OCCC_2$. Large absolute ∠$OCCC_1$ values correspond to structures shown in Fig. 16.5c, and large absolute ∠$OCCC_2$ values correspond to structures shown in Fig. 16.5d. The free-energy landscapes are illustrated in Fig. 16.7. Simulations at 150 K have dense distribution at the center indicating that the molecular structure is fluctuating near the planar structure, while the structure is more widely distributed with less area of small free-energy region at 300 K. For quantum simulation of HM, the shape of the distribution differs among simulations at 150 K and 300 K. For quantum simulation of DM, the difference between the distribution at 150 K and 300 K is significant. At 150 K, the distribution is described as a single peak at the center with wider distribution in the $\angle OCCC_2$ axis. At 300 K, we found three minima in the $\angle OCCC_1$ axis. This shows that structures with ring deformation are stable as the planar structure at 300 K for DM. We succeeded in finding a stable structure with out-of-plane ring deformation in addition to the planar symmetric structure. We found large isotope effects on the free-energy landscape.

We also studied the potential energy surface and found that the proton transfer plays a key role in the stabilization of the structures with out-of-plane ring deformation [49]. We proposed that the "proton transfer-driven" mechanism is the origin of stabilization for the structure with out-of-plane ring deformation. In this study, we have shown the capability of PIMD simulation for studying the nuclear quantum effects on the entire molecule. We found that the nuclear quantum effects influence the fluctuation of the entire molecule. Large isotope effect was found in the stable structure of HM.

## 16.4 Summary

In this review, we have introduced some recent developments of PI simulations. We developed an efficient algorithm to realize PI simulations of large-scale molecular systems under various temperatures with the accuracy of the fourth-order Trotter expansion. We confirmed that our newly developed PIHMC scheme provides similar accuracy compared to the conventional fourth-order PIMD without evaluating the Hessian matrix, which requires large amount of computational cost. We also showed some applications of PI simulation to low-barrier hydrogen-bonded systems. We first showed the study of the temperature and isotope effects on hydrogen-bonded structures of $H_5O_2^+$ and $H_3O_2^-$. GIE of the two systems was well described. We next investigated the dependence of the coordination number of water molecules on the hydrogen bond structures of fluoride ion-water clusters. We found the nuclear quantum effect on the different kind of hydrogen bonds competes with each other. We finally employed PIMD simulation to hydrogen maleate anion and found a stable structure with out-of-plane ring deformation. We have shown the ability and powerfulness of PI simulations to tackle molecular systems in which nuclear quantum effects play critical roles.

**Acknowledgements** This work is partly supported by Grants-in-Aid for Scientific Research (KAKENHI) of the Ministry of Education, Culture, Sports, Science and Technology (MEXT), Grant Numbers of 16K05676 (YK), 15KT0067, and 16H00780 (MT). Theoretical calculations were partly performed at the Research Center for Computational Science, Institute for Molecular Science, Japan, and the K computer and other computers provided by the RIKEN Advanced Institute for Computational Science through the HPCI System Research project.

## References


1. R.P. Feynman, Rev. Mod. Phys. **20**, 367 (1948)
2. L.D. Fosdick, J. Math. Phys. **3**, 1251 (1962)
3. L.D. Fosdick, H.F. Jordan, Phys. Rev. **143**, 58 (1966)
4. H.F. Jordan, L.D. Fosdick, Phy. Rev. **171**, 128 (1968)
5. C. Chakravarty, Int. Rev. Phys. Chem. **16**, 421 (1997)
6. D.M. Ceperley, Rev. Mod. Phys. **67**, 279 (1995)
7. M. Sprik, M L. Klein, D. Chandler, Phys. Rev. B **31**, 4234 (1985)
8. R.A. Friesner, R.M. Levy, J. Chem. Phys. **80**, 4488 (1984)
9. D. Chandler, P. Wolynes, J. Chem. Phys. **74**, 4078 (1981)
10. H.D. Raedt, B.D. Raedt, Phys. Rev. A **28**, 3576 (1983)
11. M. Takahashi, M. Imada, J. Phys. Soc. Jpn. **53**, 964 (1984)
12. M. Takahashi, M. Imada, J. Phys. Soc. Jpn. **11**, 3675 (1984)
13. M. Suzuki, Comm. Math. Phys. **51**, 183 (1976)
14. X.-P. Li, J.Q. Broughton, J. Chem. Phys. **86**, 5094 (1987)
15. B.J. Berne, D. Thirumalai, Ann. Rev. Phys. Chem. **37**, 401 (1986)
16. M.J. Gillan, in *Computer Modelling of Fluids Polyers and Solids*, ed. by C.R.A. Catlow, S. C. Parker, M.P. Allen (Kluwer, Dordrecht, The Netherlands, 1990), p. 155
17. M. Parrinello, A. Rahman, J. Chem. Phys. **80**, 860 (1984)
18. R.W. Hall, B.J. Berne, J. Chem. Phys. **81**, 3641 (1984)
19. M.E. Tuckerman, B.J. Berne, G.J. Martyna, M.L. Klein, J. Chem. Phys. **99**, 2796 (1993)
20. D. Marx, M. Parrinello, Z. Phys. B, **95**, 143 (1994)
21. D. Marx, M. Parrinello, Nature **375**, 216 (1995)
22. D. Marx, M. Parrinello, Science, **271**, 179 (1996)
23. M.E. Tuckerman, D. Marx, M.L. Klein, M. Parrinello, Science **275**, 817 (1997)
24. D. Marx, M.E. Tuckerman, J. Hutter, M. Parrinelo, Nature **397**, 601 (1999)
25. T. Miyake, T. Ogitsu, S. Tsuneyuki, Phys. Rev. Lett. **81**, 1873 (1998)
26. H. Kitamura, S. Tsuneyuki, T. Ogitsu, T. Miyake, Nature **404**, 259 (2000)
27. D. Marx, M. Parrinello, J. Chem. Phys. **104**, 4077 (1996)
28. M.E. Tuckerman, D. Marx, M.L. Klein, M. Parrinello, J. Chem. Phys. **104**, 5579 (1996)
29. D. Marx, J. Hutter, Cambridge Univ. Press, Ab initio Molecular Mechanics: Basic Theory and Advanced Methods (2009)
30. B.J. Johnson, P.M.W. Gill, J.A. Pople, J. Chem. Phys. **98**, 5612 (1993)
31. J. Florián, B.J. Johnson, J. Phys. Chem. **99**, 5899 (1995)
32. M. Shiga, M. Tachikawa, S. Miura, J. Chem. Phys. **115**, 9149 (2001)
33. M. Shiga, K. Suzuki, M. Tachikawa, J. Chem. Phys. **132**, 114104 (2010)
34. A. Kaczmarek, M. Shiga, D. Marx, J. Phys. Chem. A **113**, 1985 (2009)
35. E.G. Diken, J.H. Headrick, J.R. Roscioli, J.C. Bopp, M.A. Johnson, J. Phys. Chem. A **109**, 1487 (2005)
36. J.M. Price, M.W. Crofton, Y.T. Lee, J. Phys. Chem. **95**, 2182 (1991)
37. H.M.E. Cardwell, J.D. Dunitz, L.E. Orgel, J. Chem. Soc., 3740 (1953)
38. C.L. Perrin, Pure Appl. Chem. **81**, 571 (2009)
39. M. Tachikawa, M. Shiga, J. Am. Chem. Soc. **127**, 11908 (2005)
40. K. Suzuki, M. Shiga, M. Tachikawa, J. Chem. Phys. **129**, 144310 (2008)
41. J.A. Platts, K.E. Laidig, J. Phys. Chem. **99**, 6487 (1995)
42. T. Asada, H. Haraguchi, K. Kitaura, J. Phys. Chem. A **105**, 7423 (2001)
43. H. Ishibashi, A. Hayashi, M. Shiga, M. Tachikawa, Chem. Phys. Chem. **9**, 383 (2008)
44. P. George, C.W. Bock, M. Trachtman, J. Phys. Chem. **87**, 1839 (1983)
45. C.C. Wilson, L.H. Thormas, C.A. Morrison, Chem. Phys. Lett. **381**, 102 (2003)
46. Y. Kawashima, M. Tachikawa, Chem. Phys. Lett. **571**, 23 (2013)
47. Y. Ogata, M. Daido, Y. Kawashima, M. Tachikawa, RSC Adv. **3**, 25252 (2013)
48. Y. Kawashima, K. Suzuki, M. Tachikawa, J. Phys. Chem. A **117**, 5205 (2013)

49. Y. Kawashima, M. Tachikawa, J. Chem. Theor. Comput. **10**, 153 (2014)
50. G.A. Jeffrey, *An Introduction to Hydrogen Bonding* (Oxford University Press, New York, 1997)
51. A. Kohen, H.H. Limbach (eds.), *Isotope Effects in Chemistry and Biology* (CRC Press, Boca Raton, 2005)
52. T. Steiner, Angew. Chem. Int. Ed. **41**, 48 (2002)
53. W.H. Robertson, M.A. Johnson, Annu. Rev. Phys. Chem. **54**, 173 (2003)
54. M. Kabeláč, P. Hobza, Chem. Eur. J. **7**, 2067 (2001)
55. J.A. Gerlt, P.G. Gassman, Biochemistry **32**, 11943 (1993)
56. W.W. Cleland, M.M. Kreevoy, Science **264**, 1887 (1994)
57. P.A. Frey, S.A. Whitt, J.B. Tobin, Science **264**, 1927 (1994)
58. S. Yamaguchi, H. Kamikubo, K. Kurihara, R. Kuroki, N. Niimura, N. Shimizu, Y. Yamazaki, M. Kataoka, Proc. Natl. Acad. Sci. U.S.A. **106**, 440 (2009)
59. S.O. Shan, S. Loh, D. Herschlag, Science **272**, 97 (1996)
60. D.B. Northrop, Acc. Chem. Res. **34**, 790 (2001)
61. T. Ishida, Biochemistry **45**, 5413 (2006)
62. S. Scheiner, T. Kar, J. Am. Chem. Soc. **117**, 6970 (1995)
63. J.P. Guthrie, Chem. Biol. **3**, 163 (1996)
64. A. Warshel, A. Papazyan, Proc. Natl. Acad. Sci. U.S.A. **93**, 13665 (1996)
65. K. Saito, H. Ishikita, Biochemistry **51**, 1171 (2012)
66. M.E. Tuckerman, D. Marx, Phys. Rev. Lett. **86**, 4946 (2001)
67. A. Nakayama, T. Taketsugu, M. Shiga, Chem. Lett. **38**, 976 (2009)
68. K. Suzuki, M. Tachikawa, M. Shiga, J. Chem. Phys. **132**, 144108 (2010)
69. K. Suzuki, M. Tachikawa, M. Shiga, J. Chem. Phys. **138**, 184307 (2013)
70. S.A. Chin, Phys. Rev. E **69**, 046118 (2004)
71. J.M. Headrick, J.C. Bopp, M.A. Johnson, J. Chem. Phys. **121**, 11523 (2004)
72. N.I. Hammer, E.G. Diken, J.R. Roscioli, M.A. Johnson, E.M. Myshakin, K.D. Jordan, A.B. McCoy, X. Huang, J.M. Bowman, S. Carter, J. Chem. Phys. **122**, 224301 (2005)
73. L.R. McCunn, J.R. Roscioli, M.A. Johnson, A.B. McCoy, J. Phys. Chem. B **112**, 321 (2008)
74. D.J. Wales, J. Chem. Phys. **110**, 10403 (1999)
75. M.E. Tuckerman, D. Marx, M.L. Klein, M. Parrinello, Nature **417**, 925 (2002)
76. M. Meot-Ner, Chem. Rev. **105**, 213 (2005)
77. X. Huang, B, J. Braams, J. M. Bowman. J. Chem. Phys. **122**, 044308 (2005)
78. A.B. McCoy, X. Huang, S. Carter, M.Y. Landeweer, J.M. Bowman, J. Chem. Phys. **122**, 061101 (2005)
79. D. Marx, Chem. Phys. Chem. **7**, 1848 (2006)
80. O. Vendrell, F. Gatti, D. Lauvergnat, H.-D. Meyer, J. Chem. Phys. **127**, 184302 (2007)
81. O. Vendrell, F. Gatti, H.-D. Meyer, J. Chem. Phys. **127**, 184303 (2007)
82. O. Vendrell, F. Gatti, H.-D. Meyer, Angew. Chem. Int. Ed. **46**, 6918 (2007)
83. Y. Yang, O. Kühn, Z. Phys, Chem. **222**, 1375 (2008)
84. O. Vendrell, M. Brill, F. Gatti, D. Lauvergnat, H.-D. Meyer, J. Chem. Phys. **130**, 234305 (2009)
85. O. Vendrell, F. Gatti, H.-D. Meyer, J. Chem. Phys. **131**, 034308 (2009)
86. M. Kaledin, A.L. Kaledin, J.M. Bowman, J. Ding, J.M. Bowman, J. Phys. Chem. A **113**, 7671 (2009)
87. D. Marx, A. Chandra, M.E. Tuckerman, Chem. Rev. **110**, 2174 (2010)
88. X.-Z. Li, B. Walker, A. Michaelides, Proc. Natl. Acad. Sci. U.S.A. **108**, 6369 (2011)
89. T.L. Guasco, M.A. Johnson, A.B. McCoy, J. Phys. Chem. A **115**, 5847 (2011)
90. T. Ishimoto, M. Koyama, Chem. Phys. **392**, 166 (2012)
91. N. Agmon, Chem. Phys. Lett. **319**, 247 (2000)
92. L. Turi, W.S. Sheu, P. Rossky, Science **309**, 914 (2005)
93. M. Tuckerman, K. Laasonen, M. Spirk, M. Parrinello, J. Chem. Phys. **103**, 150 (1995)
94. J. Lobaugh, G.A. Voth, J. Chem. Phys. **104**, 2056 (1996)
95. H. Lapid, N. Agmon, M.K. Petersen, G.A. Voth, J. Chem. Phys. **122**, 014506 (2005)

96. W.H. Robertson, E.G. Diken, E.A. Price, J.W. Shin, M.A. Johnson, Science **299**, 1367 (2003)
97. E.A. Price, W.H. Robertson, E.G. Diken, G.H. Weddle, M.A. Johnson, Chem. Phys. Lett. **366**, 412 (2002)
98. C.C.M. Samson, W. Klopper, J. Mol. Struct. (THEOCHEM) **586**, 201 (2002)
99. A.B. McCoy, X. Huang, S. Carter, J.M. Bowman, J. Chem. Phys. **123**, 064317 (2005)
100. X. Huang, B.J. Braams, S. Carter, J.M. Bowman, J. Am. Chem. Soc. **126**, 5042 (2004)
101. W.H. Robertson, M.A. Johnson, Ann. Rev. Phys. Chem. **54**, 173 (2003)
102. W.H. Robertson, E.G. Diken, E.A. Price, J.-W. Shin, M.A. Johnson, Science **299**, 1367 (2003)
103. S.S. Xantheas, T.H. Dunning, J. Phys. Chem. **98**, 13489 (1994)
104. J. Kim, H.M. Lee, S.B. Suh, D. Majumdar, K.S. Kim, J. Chem. Phys. **113**, 5259 (2000)
105. G.M. Chaban, S.S. Xantheas, R.B. Gerber, J. Phys. Chem. A **107**, 4952 (2003)
106. S. Horvath, A.B. McCoy, J.R. Roscioli, M.A. Johnson, J. Phys. Chem. A **112**, 12337 (2008)
107. S.S. Xantheas, L. Dang, J. Phys. Chem. **100**, 3989 (1996)
108. S.S. Xantheas, J. Chem. Phys. **104**, 8821 (1996)
109. P. Weis, P.R. Kemper, T.M. Bowers, S.S. Xantheas, J. Am. Chem. Soc. **121**, 3531 (1999)
110. S.S. Xantheas, J. Phys. Chem. **100**, 9703 (1996)
111. O.M. Cabarcos, C.J. Weinheimer, J.M. Lisy, S.S. Xantheas, J. Chem. Phys. **110**, 5 (1999)
112. S.W. Peterson, H.A. Levy, J. Chem. Phys. **29**, 948 (1958)
113. S. Darlow, W. Cochran, Acta Crystallogr. **14**, 1250 (1961)
114. K. Nakamoto, Y.A. Sarma, G.T. Behnke, J. Chem. Phys. **42**, 1662 (1965)
115. H.R. Zelsmann, Z. Mielke, M.M. Ilczyszyn, Spectrochim. Acta **44A**, 705 (1988)
116. M.M. Ilczynszyn, J. Baran, H. Ratajczak, A.J. Barnes, J. Mol. Struct. **270**, 499 (1992)
117. J. Tomkinson, I.J. Braid, J. Howard, T.C. Waddington, Chem. Phys. **64**, 151 (1982)
118. J. Howard, J. Tomkinson, J. Eckert, J.A. Goldstone, A.D. Taylor, J. Chem. Phys. **78**, 3150 (1983)
119. F. Fillaux, N. Leygue, J. Tomkinson, A. Cousson, W. Paulus, Chem. Phys. **120**, 387 (1999)
120. S. Forsen, J. Chem. Phys. **31**, 852 (1959)
121. A. Wong, K.J. Pike, R. Jenkins, G.J. Clarkson, T. Anupõld, A.P. Howes, D.H.G. Crout, A. Samoson, R. Dupree, M.E. Smith, J. Phys. Chem. A **110**, 1824 (2006)
122. M. Fukai, T. Matsuo, H. Suga, J. Chem. Thermodyn. **20**, 1337 (1988)
123. G. Olovsson, I. Olovsson, M.S. Lehmann, Acta Crystallogr. C **40**, 1521 (1984)
124. C.L. Perrin, Science **266**, 1665 (1994)
125. G.E. Hall, J. Am. Chem. Soc. **100**, 8264–8266 (1978)
126. P. Schah-Mohammedi, I.G. Shenderovich, C. Deterring, H.-H. Limbach, P.M. Tolstoy, S.N. Smirnov, G.S. Denisov, N.S. Golubev, J. Am. Chem. Soc. **122**, 12878–12879 (2000)
127. Z. Latajka, H. Ratajczak, Chem. Phys. Lett. **49**, 407 (1977)
128. P. George, C.W. Bock, M. Trachtman, J. Phys. Chem. **87**, 1839 (1983)
129. M. Hodoscek, D. Hadzi, J. Mol. Struct. (Theochem) **209**, 411 (1990)
130. M.A. Rios, J. Rodriguez, Can. J. Chem. **71**, 303 (1993)
131. M. Garcia-Viloca, A. Gonzalez-Lafont, J.M. Lluch, J. Am. Chem. Soc. **119**, 1081 (1997)
132. M. Garcia-Viloca, A. Gonzalez-Lafont, J.M. Lluch, J. Am. Chem. Soc. **121**, 9198 (1999)
133. R.D. Bach, O. Dmitrenko, M.N. Glukhovtsev, J. Am. Chem. Soc. **123**, 7134 (2001)
134. H.-K. Woo, X.-B. Wang, L.-S. Wang, K.-C. Lau, J. Phys. Chem. A **109**, 10633 (2005)
135. S.X. Tian, H.-B. Li, J. Phys. Chem. A **111**, 4404 (2007)
136. F. Avbelj, M. Hodoscek, D. Hadzi, Spectrochim. Acta **41A**, 89 (1985)
137. H. Ratajczak, A. Barnes, J. Baran, A. Yaremko, Z. Latajka, P. Dopieralski, J. Mol. Struct. (Theochem) **887**, 9 (2008)
138. M.V. Vener, A.V. Manaev, V.G. Tsirelson, J. Phys. Chem. A **112**, 13628 (2008)
139. C.C. Wilson, L.H. Thomas, C.A. Morrison, Chem. Phys. Lett. **381**, 102 (2003)
140. C.C. Wilson, L.H. Thomas, C.A. Morrison, Chem. Phys. Lett. **399**, 292 (2004)
141. P. Dopieralski, C.L. Perrin, Z. Latajka, J. Chem. Theory Comput. **7**, 3505 (2011)

# Chapter 17
# Vibrational Linear and Nonlinear Optical Properties: Theory, Methods, and Application

**Heribert Reis, David Lauvergnat, Josep M. Luis and Robert Zaleśny**

**Abstract** Methods for the calculation of vibrational contributions to nonresonant and resonant linear and nonlinear optical properties of molecules are reviewed. Both the Bishop–Kirtman perturbation theory and the finite-field/nuclear relaxation–curvature approach are treated. Emerging methods using variational approaches instead of perturbation theory and the methodology required to employ curvilinear coordinates instead of the more usual normal coordinates are briefly described, too. Finally, methods using a time-dependent approach are also treated.

**Keywords** Vibrational (hyper)polarizabilities · Bishop–Kirtman double perturbation theory · Finite-field/nuclear relaxation–curvature approach · Variational vibrational (hyper)polarizabilities · Vibrational effects to one- and two-photon absorption

H. Reis (✉)
Institute of Biology, Medicinal Chemistry and Biotechnology, National Hellenic Research Foundation, 48 Vas. Constantinou Avenue, 11635 Athens, Greece
e-mail: hreis@eie.gr

D. Lauvergnat
Laboratoire de Chimie Physique, CNRS,
UMR 8000, Bâtiment 349, 91405 Orsay, France
e-mail: david.lauvergnat@u-psud.fr

D. Lauvergnat
University of Paris-Sud, 91405 Orsay, France

J.M. Luis
Institute of Computational Chemistry and Catalysis and Department of Chemistry,
University of Girona, Campus de Montilivi, 17071 Girona, Catalonia, Spain
e-mail: josepm.luis@udg.edu

R. Zaleśny
Department of Physical and Quantum Chemistry, Faculty of Chemistry,
Wroclaw University of Science and Technology, Wyb. Wyspiańskiego 27,
50370 Wrocław, Poland
e-mail: robert.zalesny@pwr.edu.pl



M.J. Wójcik et al. (eds.), *Frontiers of Quantum Chemistry*,
https://doi.org/10.1007/978-981-10-5651-2_17

## 17.1 Introduction

This chapter summarizes some of the recent and still ongoing efforts to compute the vibrational contributions to the linear and nonlinear polarizabilities caused by externally applied electric fields. It is well recognized by now that these contributions have to be added to the pure electronic contribution in order to obtain a full description of the electric field effect on the molecule. Although much progress has been made in the last few decades, there is still no general treatment available for the full range of energies covered by nuclear motions in polyatomic molecules. The way to go at the moment is therefore to choose the appropriate method according to the kind of dominant vibrations exhibited by the molecule under investigation. Some general considerations concerning this point are thus mentioned in Sect. 17.2. Two main approaches available for actual computations of vibrational contributions to electric properties are then reviewed. The first and older one rests on perturbation theory and is treated in Sect. 17.3, while the second one starts from a consideration of the electric field effect on the position and shape of the molecular potential surface and is summarized in Sect. 17.4. The connection of this so-called nuclear relaxation/curvature approach with the perturbational expressions is brought out. In Sect. 17.5, a combination of the nuclear relaxation/curvature approach with variational methods for the solution of the vibrational Schrödinger equation is reviewed, which shows the promise of overcoming many of the restrictions inherent to the perturbational treatment, albeit at a higher computational cost. Methods based on the vibrational SCF (VSCF) treatment as well as methods employing curvilinear coordinates are treated. The solution of the vibrational Schrödinger equation in curvilinear coordinates requires special care, in particular in connection with the application of external electric fields, and is therefore subject of a separate subsection (Sect. 17.5.3).

Most of the text mentioned above is concerned with nonresonant properties. As vibrational effects on resonant effects have also been studied and are still under active investigation, we have included a Subsect. 17.3.3, where several methods available for the treatment of vibrational effects on two-photon absorption are summarized.

Finally, Sect. 17.6 treats approaches based explicitly on the time domain, instead of the frequency domain, and thus allow the study of vibrational effects on time-dependent, periodic, and nonperiodic processes.

The area of nonlinear optics is plagued by a plethora of different conventions used to define the nonlinear polarizabilities [1, 2]. We follow here the conventions most often adopted in the modern literature of computational chemistry and define the Fourier component of the total dipole moment $\mu_i(\omega_\sigma)$ of a molecule subject to an electric field described by $\underline{F}(t) = \sum_n \underline{F}^{\omega_n} \cos \omega_n t$, which may include a static field, $\omega_0 = 0$, as:

$$\mu_i(\omega_\sigma) = \mu_i^0 \delta_{\omega_\sigma 0} + \sum_j \alpha_{ij}(-\omega_\sigma;\omega_\sigma)F_j^{\omega_\sigma} + \sum_{jk}\sum_{\omega_1,\omega_2} \frac{K_{\sigma,1,2}}{2}\beta_{ijk}(-\omega_\sigma;\omega_1,\omega_2)F_j^{\omega_1}F_k^{\omega_2}$$
$$+ \sum_{jkl}\sum_{\omega_1\omega_2\omega_3} \frac{K_{\sigma,1,2,3}}{6}\gamma_{ijkl}(-\omega_\sigma;\omega_1,\omega_2,\omega_3)F_j^{\omega_1}F_k^{\omega_2}F_l^{\omega_3} \quad (17.1)$$

where $\sum_i \omega_i = \omega_\sigma$, $K_{\sigma,1,2...}$ are numerical factors depending on the number of nonzero frequencies $\omega_i$ (see, e.g., [3]).

## 17.2 How to Choose the Right Approach: Harmonic, Anharmonic, and Nonharmonic Molecules

The calculation of reliable vibrational (hyper)polarizabilities is not that straightforward as the calculation of their electronic counterparts, which is essentially a question of selecting the right basis set and the appropriate level of theory. Assuming the Born–Oppenheimer approximation to be valid, one could conceive of a black-box method for vibrational properties, which would require the calculation of the potential surfaces of the energy and several properties over all accessible vibrational modes. Currently, this is clearly unfeasible for most molecules with more than, say, seven to ten atoms. Thus, depending on the form of potential energy surface (PES), as well as the corresponding electric property surfaces, of (usually) the nuclear motions of lowest frequency, other methods have to be employed. In the theory of vibrational–rotational spectroscopy, one commonly distinguishes between quasi-rigid (also called semirigid or rigid) molecules, which are characterized by one clearly defined equilibrium structure, around which atoms undergo small amplitude motions at $T = 0$, and, on the other hand, nonrigid (also called flexible or floppy) molecules, characterized by one or more large amplitude motions (LAMs), while the other vibrations are still described by small amplitude motions [4]. Similarly, in the theory of vibrational optical properties, one may distinguish between (an)harmonic molecules, for which the relevant potential surfaces can be well approximated by Taylor series in, for example, rectilinear normal coordinates, and nonharmonic molecules, where this is not possible for one or more vibrational modes. A harmonic molecule would be a limiting case of the former class of molecules, where it is sufficient to include terms through second order in the Taylor series of the PES, and through first order for the property surfaces. This leads, within Bishop–Kirtman perturbation theory (BKPT), to the so-called double harmonic approximation (DHA), which is often quite easy to calculate (see Sect. 17.3.1). Methods to compute vibrational (hyper)polarizabilities of (an)harmonic molecules are quite well established (although not necessarily easy to apply), including, *inter alia*, BKPT, the computation of field-induced coordinates (FICs), the Bishop–Hasan–Kirtman approach, which is based on geometry optimizations in an electric field, or methods based on the vibrational self-consistent field approach. Some of these methods may also be applicable to nonharmonic molecules, if the barriers between the minima visited during the LAMs remain sufficiently high and tunneling is negligible. Once this is not the case anymore, however, more sophisticated methods have to be employed, which usually require the explicit calculation of the different relevant potential surfaces along one or several LAMs, as described, for example, in Sect. 17.5.3 below. These latter methods have just been emerging recently, and we can only provide a snapshot of the state of the art. Much progress can be anticipated in this area.

As mentioned briefly above, the Born–Oppenheimer approximation (BOA) is generally assumed to be valid in all of the approaches mentioned in this chapter. There is no guarantee, however, that this is a valid assumption for all the points along the path of a LAM (see, e.g., Ref. [5]). To our knowledge, no treatments for the computation of nonlinear optical effects beyond the BOA have been devised so far, with the exception of a calculation of the $H_2^+$ ion and some of its isotopes [6].

## 17.3 Perturbation Theory

### *17.3.1 Bishop–Kirtman Perturbation Theory*

One of the oldest and still widely applied approaches to compute the vibrational contributions to linear and nonlinear optical properties starts from the nuclear Schrödinger equation in standard *normal coordinates* and applies *double perturbation theory* to take into account both mechanical and electrical anharmonicities.

This approach starts from very general sum-over-states expressions, which were derived by Orr and Ward [7], applying perturbation theory to compute the effect of an electric field on a general state $|N\rangle$. Assuming the Born–Oppenheimer approximation to be valid, the wavefunction factors into an electronic and a nuclear wavefunction. Separating off the translational motion, and treating the rotational motion by a classical average, the nuclear wavefunction reduces to a pure vibrational wavefunction $|v\rangle$. Thus, we can write $|N\rangle \rightarrow |n\rangle|v\rangle \equiv |nv\rangle$, where $|n\rangle$ denotes the electronic state. Then, the coefficients of the Taylor expansion of the induced dipole moment in Eq. 17.1 are given by:

$$\alpha_{ij}(-\omega_\sigma;\omega_\sigma) = \frac{1}{\hbar}\sum \mathscr{P}_{-\sigma,\sigma} \sum_{nv}^{\infty}{}' \frac{\langle 0\breve{0}|\hat{\mu}_i|nv\rangle\langle nv|\hat{\mu}_j|0\breve{0}\rangle}{\omega_{nv,0\breve{0}} - \omega_\sigma} \tag{17.2}$$

$$\beta_{ijk}(-\omega_\sigma;\omega_1,\omega_2) = \frac{1}{\hbar^2}\sum \mathscr{P}_{-\sigma,1,2} \sum_{nv,pv'}{}' \frac{\langle 0\breve{0}|\hat{\mu}_i|nv\rangle\langle nv|\hat{\bar{\mu}}_j|pv'\rangle\langle pv'|\hat{\mu}_k|0\breve{0}\rangle}{(\omega_{nv,0\breve{0}} - \omega_\sigma)(\omega_{pv',0\breve{0}} - \omega_2)} \tag{17.3}$$

$$\gamma_{ijkl}(-\omega_\sigma;\omega_1,\omega_2,\omega_3) = \frac{1}{\hbar^3}\sum \mathscr{P}_{-\sigma,1,2,3} \left[\sum_{nv,mv',pv''}{}' \frac{\langle 0\check{0}|\hat{\mu}_i|nv\rangle\langle nv|\bar{\hat{\mu}}_j|mv'\rangle\langle mv'|\bar{\hat{\mu}}_k|pv''\rangle\langle pv''|\hat{\mu}_l|0\check{0}\rangle}{(\omega_{nv,0\check{0}}-\omega_\sigma)(\omega_{mv',0\check{0}}-\omega_2-\omega_3)(\omega_{pv'',0\check{0}}-\omega_3)} - \sum_{nv,mv'}{}' \frac{\langle 0\check{0}|\hat{\mu}_i|nv\rangle\langle nv|\hat{\mu}_j|0\check{0}\rangle\langle 0\check{0}|\hat{\mu}_k|mv'\rangle\langle mv'|\hat{\mu}_l|0\check{0}\rangle}{(\omega_{nv,0\check{0}}-\omega_\sigma)(\omega_{mv',0\check{0}}-\omega_3)(\omega_{mv',0\check{0}}+\omega_2)}\right], \tag{17.4}$$

where $\hat{\mu}$ is the dipole moment operator, $\bar{\hat{\mu}} = \hat{\mu} - \langle 0|\hat{\mu}|0\rangle$, $\omega_{nv,0\check{0}} = (E_{nv} - E_{0\check{0}})/\hbar$. The permutation operator $\mathscr{P}_{-\sigma,1...}$ creates a sum of all distinguishable permutations of the quantities in brackets $(-\omega_\sigma, \hat{\mu}_i), (\omega_1, \hat{\mu}_j), \ldots$. The sums run over all electronic states $n$ and their associated vibrational states $|v\rangle$, $\check{0}$ denote the vibrational ground state, and the prime indicates that the state $|nv> = |0\check{0}>$ is excluded.

These expressions are valid for *nonresonant* processes, which means that all frequencies of externally applied fields $\omega_i$ are far from any molecular resonances. If this is not the case (e.g., in two-photon absorption), imaginary damping terms need to be introduced in the denominators containing the frequency terms, which complicate the general expressions (see Ref. [7] and further below). Replacing the vibronic states $|nv\rangle$ by electronic states $|n\rangle$ in Eqs. 17.2–17.4, the electronic (hyper)polarizabilities are obtained.

In BKPT, the additional approximation $\omega_{nv,0\check{0}} \approx \omega_{n,0}$ for $n \neq 0$ is introduced, under the assumption that the vibrational energy differences are negligible compared to the electronic ones. At this point, it is natural to separate the resulting terms into a contribution with all intermediate electronic states $n \neq 0$, which is the *zero-point vibrational average* (ZPVA). The remainder, terms with at least one intermediate electronic state $n = 0$, form the so-called *pure vibrational* (PV) part. The pure vibrational nonlinear optical properties (NLOP) can be written in terms of *square bracket* contributions:

$$\alpha_{ij}^{pv} = \left[\mu^2\right] \tag{17.5}$$

$$\beta_{ijk}^{pv} = \left[\mu\alpha\right] + \left[\mu^3\right] \tag{17.6}$$

$$\gamma_{ijkl}^{pv} = \left[\mu\beta\right] + \left[\alpha^2\right] + \left[\mu^2\alpha\right] + \left[\mu^4\right] \tag{17.7}$$

where for convenience, all frequency arguments and, on the right-hand side of the equations, Cartesian indices have been suppressed. The quantities $\mu$, $\alpha$, and $\beta$ inside the square brackets indicate the corresponding electronic property expressed as functions of vibrational coordinates. In BKPT, these electronic property functions are assumed to be static, or frequency-independent. However, the vibrational properties of Eqs. 17.5–17.7 are still frequency-dependent due to the optical frequency arguments occurring in the denominators of Eqs. 17.2–17.4, which are not part of the electronic property functions. As an example for the form of the quantities occurring in Eqs. 17.5–17.7, we show here the equation for $\left[\mu^4\right]$:

$$\left[\mu^4\right] = \frac{1}{\hbar^3}\sum \mathscr{P}_{-\sigma,1,2,3}$$
$$\sum_{vv'v''}{}' \frac{\langle\check{0}|\mu_i^{el}(\underline{R}_k)|v\rangle\langle v|\bar{\mu}_j^{el}(\underline{R}_k)|v'\rangle\langle v'|\bar{\mu}_k^{el}(\underline{R}_k)|v''\rangle\langle v''|\mu_l^{el}(\underline{R}_k)|\check{0}\rangle}{(\omega_{0v,0\check{0}}-\omega_\sigma)(\omega_{0v',0\check{0}}-\omega_2-\omega_3)(\omega_{0v'',0\check{0}}-\omega_3)}$$
$$-\sum_{vv'}{}' \frac{\langle\check{0}|\mu_i^{el}(\underline{R}_k)|v\rangle\langle v|\mu_j^{el}(\underline{R}_k)|0\rangle\langle 0|\mu_k^{el}(\underline{R}_k)|v'\rangle\langle v'|\mu_l^{el}(\underline{R}_k)|\check{0}\rangle}{(\omega_{0v,0\check{0}}-\omega_\sigma)(\omega_{0v',0\check{0}}-\omega_3)(\omega_{0v',0\check{0}}+\omega_2)}, \tag{17.8}$$

where the electronic dipole moment functions of the electronic ground state $\mu_i^{el}(\underline{R}_k) = \langle 0|\hat{\mu}_i|0\rangle$ have been introduced.

In addition, one may write the full ZPVA contribution as the electronic part, evaluated usually at the optimized geometry, plus a correction due to the average over the vibrational wavefunction. Thus, for any property $P_{\{i_n\}}(\{\omega\}) = \alpha_{ij}(-\omega_\sigma;\omega_\sigma), \beta_{ijk}(-\omega_\sigma;\omega_1,\omega_2), \gamma_{ijkl}(-\omega_\sigma;\omega_1,\omega_2,\omega_3)$ one obtains:

$$P_{\{i_n\}}(\{\omega\}) = P^{pv}_{\{i_n\}}(\{\omega\};\underline{R}_0) + P^{el}_{\{i_n\}}(\{\omega\};\underline{R}_0) + \Delta P^{zpva}_{\{i_n\}}(\{\omega\};\underline{R}_0) \tag{17.9}$$

The frequency dependence of the two vibrational parts is quite different: The ZPVA contributions behave qualitatively similar to their corresponding electronic properties, which means they are smooth functions of the frequency in the nonresonant region, without large differences between different nonlinear processes. The pure vibrational contributions, on the other hand, behave very differently according to the nonlinear process under consideration, and there are no smooth connections between them. For example, for the third harmonic generation process, the pure vibrational part is essentially zero, while the static third-order contribution may be several order of magnitudes larger than the pure electronic part.

In order to obtain working expressions for the vibrational NLO properties, a second perturbation approach is adopted. The potential of the zero-order Hamiltonian is taken to be the harmonic part only. Thus, in the usual mass-weighted normal coordinates $Q_v$, with $v = 1, ..., 3N-6(3N-5)$ for nonlinear (linear) molecules, the vibrational Schrödinger equation is:

$$\hat{H}\Phi(Q_v) = [\hat{H}^{(0)} + \hat{H}^{(1)} + \hat{H}^{(2)} + \ldots]\Phi(Q_v) \tag{17.10}$$

$$\hat{H}^{(0)} = \frac{1}{2}\sum_v(\hat{p}_v^2 + \omega_v^2 Q_v^2)\Phi(Q_v), \tag{17.11}$$

$$\hat{H}^{(1)} = \frac{1}{6}\sum_{vv'v''} K_{vv'v''}Q_vQ_{v'}Q_{v''};\ K_{vv'v''} = \left(\frac{\partial^3 E_0^{el}(\underline{R}_k)}{\partial Q_v\partial Q_{v'}\partial Q_{v''}}\right)_{Q_i=0} \tag{17.12}$$

$$\hat{H}^{(2)} = \frac{1}{24}\sum_{vv'v''v'''} K_{vv'v''v'''}Q_vQ_{v'}Q_{v''}Q_{v'''};\ K_{vv'v''v'''} = \left(\frac{\partial^4 E_0^{el}(\underline{R}_k)}{\partial Q_v\partial Q_{v'}\partial Q_{v''}\partial Q_{v'''}}\right)_{Q_i=0} \tag{17.13}$$

where $\hat{p}_v = -i\hbar\frac{\partial}{\partial Q_v}$, $\omega_v^2 = \left.\frac{\partial^2 E_0^{el}(\underline{R}_k)}{\partial Q_v^2}\right|_{Q_v=0}$ are the harmonic (or quadratic) force constants, while $K_{v'v'v''}$ ($K_{vv'v''v'''}$) are cubic (quartic) force constants, which constitute the first(second)-order *mechanical anharmonicity*. The $\omega_v \equiv \omega_{0v,0\breve{0}}$ are vibrational frequencies corresponding to the electronic *ground state*. In addition, the electronic dipole moment, polarizability, and first hyperpolarizability are also expanded as Taylor series of $Q_k$, e.g.,

$$\mu_i^{el}(\underline{R}_k) = \mu_i^{el}(\underline{R}_{k0}) + \sum_v \left(\frac{\partial \mu_i^{el}}{\partial Q_v}\right)_0 Q_v + \frac{1}{2}\sum_{vv'}\left(\frac{\partial^2 \mu_i^{el}}{\partial Q_v \partial Q_{v'}}\right)_0 Q_v Q_{v'} \dots \quad (17.14)$$

where $\frac{\partial^2 \mu_i^{el}}{\partial Q_v \partial Q_{v'}}$ and higher terms account for *electrical dipole anharmonicities*.

With all these approximations, and including all terms containing up to second-order mechanical and electrical anharmonicities, the pure vibrational (hyper)polarizabilities can be finally written in the following form:

$$\alpha_{ij}^{pv} = \left[\mu^2\right]^{(0)} + \left[\mu^2\right]^{(2)} + \dots \quad (17.15)$$

$$\beta_{ijk}^{pv} = \left[\mu\alpha\right]^{(0)} + \left[\mu^3\right]^{(1)} + \left[\mu\alpha\right]^{(2)} + \dots$$

$$\gamma_{ijkl}^{pv} = \left[\mu\beta\right]^{(0)} + \left[\alpha^2\right]^{(0)} + \left[\mu^2\alpha\right]^{(1)} + \left[\mu\beta\right]^{(2)} + \left[\alpha^2\right]^{(2)} + \left[\mu^4\right]^{(2)} + \dots \quad (17.16)$$

The superscript in brackets refers to the total level of anharmonicity (mechanical + electrical). These may also be written explicitly, which in general would lead to expressions $[ab]^{(n)} = [ab]^{(n,0)} + [ab]^{(n-1,1)} + \dots[ab]^{(1,n-1)} + [ab]^{(0,n)}$, where the usual convention is that the first superscript refers to the order of electrical anharmonicity, the second one to the order of mechanical anharmonicity. In terms of matrix elements, the square bracket $\left[\mu^4\right]^{1,1}$, for example, is given by:

$$\left[\mu^4\right]^{1,1} = -\frac{1}{2}\sum \mathscr{P}_{-\sigma,1,2,3} \sum_{vv'v''v'''} K_{vv'v''}$$
$$\frac{\left(\partial\mu_i^{el}/\partial Q_v\right)_0 \left(\partial\mu_j^{el}/\partial Q_{v'}\right)_0 \left(\partial^2\mu_k^{el}/\partial Q_{v''}\partial Q_{v'''}\right)_0 \left(\partial\mu_l^{el}/\partial Q_{v'''}\right)_0}{(\omega_v^2 - \omega_\sigma^2)(\omega_{v'}^2 - \omega_1^2)(\omega_{v''}^2 - (\omega_2 + \omega_3)^2)(\omega_{v'''}^2 - \omega_3^2)} \quad (17.17)$$

Detailed expressions of the pure vibrational contributions up to total order of (2) are given in [8].

The ZPVA correction is defined as the difference between the property averaged over the zero-point vibration and the electronic property computed at the equilibrium geometry, see Eq. 17.9, i.e., $\Delta P^{zpva}(\omega) = \langle 0|P(\omega)|0\rangle - P^{el}(\omega, \underline{R}_0)$. Usually, only the *zero vibrational* level is considered, although higher levels may also be relevant, depending on temperature and vibrational spacings. In BKPT, perturbation theory is again applied to compute the vibrational matrix elements. Up to the same maximum order of anharmonicities as considered above for the pure vibrational contributions, the ZPVA correction can be written as a sum of two terms:

$$\Delta P^{zpva} = [P]^{0,1} + [P]^{1,0} = -\frac{\hbar}{4}\sum_{v}\left[\sum_{v'}\frac{K_{vv'v'}}{\omega_{v'}\omega_v^2}\frac{\partial P^{el}}{\partial Q_v} - \frac{1}{\omega_v}\frac{\partial^2 P^{el}}{\partial Q_v^2}\right] \tag{17.18}$$

where the first term is first order in mechanical anharmonicity and the second one is first order in electrical anharmonicity. The next nonvanishing higher-order terms are already of third order in total anharmonicity; in fact, all even order terms vanish [9].

Finally, we note here a further useful approximation concerning the frequency arguments. In BKPT, the frequency differences between the electronic levels are assumed to be larger than the frequencies of the occurring light waves and much larger than the frequencies between the vibrational levels of one electronic state, that is, $\omega_{0\check{0},ev} > \omega \gg \omega_{ev,ev'}$. Thus, one may assume the so-called *infinite optical frequency approximation* (IOFA) in the equations for the *vibrational contributions*, that is, perform the limit $\lim_{\omega_i\to\infty}$ $(i = 1, 2, 3, \sigma)$ for quantities occurring in the square brackets, such as Eq. 17.17. This leads to much simpler expressions, if there is at least one optical frequency involved in the process considered. For example, for the diagonal terms $\gamma_{iiii}$ in the DHA, the following equations for different third-order processes hold [10]:

$$\text{THG:} \quad \lim_{\omega\to\infty}\gamma_{iiii}^{\mathrm{DHA}}(-3\omega;\omega,\omega,\omega) = 0 \tag{17.19}$$

$$\text{EFISH:} \quad \lim_{\omega\to\infty}\gamma_{iiii}^{\mathrm{DHA}}(-2\omega;\omega,\omega,0) = \frac{1}{4}[\mu\beta]_{\omega=0}^{(0,0)} \tag{17.20}$$

$$\text{Kerr:} \quad \lim_{\omega\to\infty}\gamma_{iiii}^{\mathrm{DHA}}(-\omega;\omega,0,0) = \frac{1}{3}[\alpha^2]_{\omega=0}^{(0,0)} + \frac{1}{2}[\mu\beta]_{\omega=0}^{(0,0)} + \frac{1}{6}[\mu^2\alpha]_{\omega=0}^{(0,0)} \tag{17.21}$$

$$\text{IDRI:} \quad \lim_{\omega\to\infty}\gamma_{iiii}^{\mathrm{DHA}}(-\omega;\omega,-\omega,\omega) = \frac{2}{3}[\alpha^2]_{\omega=0}^{(0,0)} \tag{17.22}$$

### *17.3.2 Convergence Behavior*

Applications of BKPT beyond the zeroth-order or double harmonic approximation level require the computation of many high-order derivatives, which may be a very cumbersome task, especially if the calculation involves numerical derivatives, which has been nearly invariably the case for the higher-order derivatives (but see below for recently developed automated differentiation techniques). Thus, it is not an uncommon practice to stop already at the DHA. This is relatively easy to compute, but obviously does not provide any indication about the magnitude of higher-order terms and thus about the reliability of the values computed at the DHA.

The last two decades have seen a significant number of studies devoted to the assessment of the reliability of the DHA applied to medium-sized organic molecules. Most of these studies, however, remained at the Hartree–Fock (HF) level of theory. In the few cases where electron correlation effects were included [11, 12], these were found to be as important for the vibrational (hyper)polarizabilities as for the electronic ones. Thus, conclusions based solely on calculations at the HF level should be considered with care. Nevertheless, in general, the DHA was found to be a poor approximation for the vibrational contributions computed at a higher level of perturbation theory. The contributions at the DHA level were often found to be of the same order of magnitude as the higher-order contributions. In some cases, this leads to a spectacular failure of the DHA. For example, the DHA contribution to $\gamma$ of 1,1-diamino–6,6-diphosphinohexa-1,3,5-triene was computed as $[\mu\alpha]^{(0,0)} = 1.55 \times 10^4$ au, while the first-order anharmonic contribution is $[\mu^3]^{(1)} = -1.41 \times 10^4$, leading to a sum of just $0.14 \times 10^4$ au. In general, thus, the convergence of the BKPT at low order is by no means guaranteed, especially for larger molecules with possibly several low-frequency internal motions. For such systems, it is usually much more efficient to apply one of the more recent methods described below.

However, a newly developed method to compute the pure vibrational contributions in the framework of DFT and using response theory, without the need to use numerical differentiation [13], should also be noted. This method employs a general recursive and open-ended scheme to create response functions to different orders, together with an automatic differentiation approach to generate the corresponding higher-order exchange–correlation kernels, and has been applied to compute the pure vibrational contribution to the first hyperpolarizability of retinal and similar molecules [13]. This approach has been extended to the computation of several high-order derivatives (see, e.g., Ref. [14, 15]), and may, eventually, become a viable alternative to the methods described in Sect. 17.4, at least in the field of DFT.

### *17.3.3 Treatments for Two-photon Absorption: BKPT and Beyond*

The treatment in the previous sections is applicable only to nonresonant processes. As mentioned above, for the treatment of resonant or nearly resonant processes, additional phenomenological imaginary damping terms $\Gamma_{ve}$ have to be included in the denominators in Eqs. 17.2–17.4. Then, the electric properties $\alpha$–$\gamma$ acquire an imaginary part, which is associated with absorption processes, e.g., with linear absorption in the case of $\alpha$ and with two-photon absorption (TPA) in the case of $\gamma$. The introduction of damping terms complicates the equations for $\gamma$ considerably, creating eight terms instead of two in the nonresonant case, but it is reasonable to assume that only terms which include the resonant or near resonant frequency in the denominator will contribute significantly.

For TPA, two photons with the same frequency induce an excitation from the ground state to an excited state $e$. TPA is described by the imaginary part of $\gamma_{ijkl}(-\omega_\sigma;\omega_1,\omega_2,\omega_3)$ with $\omega_1=-\omega_2=\omega_3=\omega_\sigma\equiv\omega$ and $2\omega\approx\omega_{ve,0\check{0}}$, where $v$ is the vibrational state index in the excited state $e$. Using in addition the assumption that all damping terms can be neglected except for the transitions $|0\check{0}\rangle\rightarrow|ev'\rangle$ and assuming the same broadening for each vibrational state (i.e., $\Gamma_{ev'}\equiv\Gamma$), a comparatively simple equation results for the TPA process [16]:

$$\mathrm{Im}\gamma_{ijkl}(-\omega;\omega,-\omega,\omega)=\frac{\Gamma}{2\hbar^3}\sum\mathscr{P}_{ik}\mathscr{P}_{jl}\sum_{nv,mv',v''}{}' \frac{\langle 0\check{0}|\hat{\mu}_i|nv\rangle\langle nv|\hat{\bar{\mu}}_k|ev''\rangle\langle ev''|\hat{\bar{\mu}}_j|mv'\rangle\langle mv'|\hat{\mu}_l|0\check{0}\rangle}{((\omega_{ev'',0\check{0}}-2\omega)^2+\Gamma^2/4)(\omega_{nv,0\check{0}}-\omega)(\omega_{mv',0\check{0}}-\omega)}, \tag{17.23}$$

where the prime on the summation sign excludes the state $|0\check{0}\rangle$ from the summations over $nv'$ and $mv''$. Proceeding in the same vein as for the nonresonant case, we may neglect the vibrational component of all intermediate states not involving the electronic ground state, and introduce electronic transition dipole functions $\mu_i^{0e}(\underline{R}_k)=\langle 0|\hat{\mu}_i|e\rangle$ and transition polarizability functions

$$\alpha_{ik}^{0e}(\underline{R}_k,\omega)=\sum\mathscr{P}_{ik}\frac{1}{\hbar}\sum_m{}'\frac{\langle 0|\hat{\bar{\mu}}_i|m\rangle\langle m|\hat{\bar{\mu}}_k|e\rangle}{\omega_{0m}-\omega}, \tag{17.24}$$

Then, the expression for $\mathrm{Im}\gamma_{ijkl}(-\omega;\omega,-\omega,\omega)$ can be written in a similar "bracket notation" as the nonresonant cases:

$$\mathrm{Im}\gamma_{ijkl}(-\omega;\omega,-\omega,\omega)=\left\{\alpha^2\right\}+\left\{\mu^2\alpha\right\}+\left\{\mu^4\right\}. \tag{17.25}$$

This "curly bracket" notation is similar to the square bracket notation used in Eqs. 17.5–17.7. However, we use curly brackets to indicate that the terms in the two set of equations are different. Thus, for example, the term $\left\{\mu^4\right\}$ is given by:

$$\left\{\mu^4\right\}=\frac{\Gamma}{2\hbar^3}\sum\mathscr{P}_{ik}\sum\mathscr{P}_{jl}\sum_v\sum_{v'v''}{}'\frac{\langle\check{0}^0|\mu_i^{00}(\underline{R}_k)|v'^0\rangle\langle v'^0|\mu_k^{e0}(\underline{R}_k)|v^e\rangle\langle v^e|\mu_j^{e0}(\underline{R}_k)|v''^m\rangle\langle v''^m|\mu_l^{00}(\underline{R}_k)|\check{0}^0\rangle}{((\omega_{ev,0\check{0}}-2\omega)^2+\Gamma^2/4)(\omega_{0v',0\check{0}}-\omega)(\omega_{0v'',0\check{0}}-\omega)}, \tag{17.26}$$

where the notation $v^l$ denotes the $v$-th vibrational state of the electronic state $l$, and $\mu_i^{00}\equiv\mu_i^{el}$ in the notation used previously. This expression may be compared with the corresponding square bracket in Eq. 17.8. For explicit expressions of all terms on the right-hand side of Eq. 17.25, see Ref. [16]. From this point, there are at least two ways forward: either following the usual BKPT treatment and assuming Taylor expansions in normal coordinates for the transition dipoles and polarizabilities and computing the arising matrix elements of normal coordinates analytically, as has been worked out for diatomics in Ref. [16], or computing explicitly the transition polarizabilities and dipole moments as a function of the displacements $\underline{R}_k$ and calculating the matrix elements of the electronic moments over vibrational wavefunctions with the help of dedicated codes, such as ElVibRot [17]. The latter procedure is clearly more general than the former, as it does not assume that the molecule is semirigid both in the electronic ground and in the electronic excited state, but it requires software which is able to compute matrix elements over vibrational wavefunctions belonging to two different electronic states, such as required for example for the calculation of Franck–Condon factors. The first procedure mentioned in the previous paragraph has been applied to compute the TPA for the CO molecule [16], while work is in progress to implement the second method.

Finally, we mention some more approximate treatments for vibrational effects on TPA. These are based on the expansion of the transition polarizability in powers of normal coordinates (Eq. 17.24), where only linear terms are taken into account. One specific linear coupling model (LCM) has gained some popularity [18]. In this model, normal mode displacements between shifted harmonic oscillators are computed using the energy gradient in the electronic excited state. Normal mode displacements determined in such a way introduce some anharmonicity effects in the region of the vertical transition [18]. The LCM is computationally very efficient, and it has been used to study the vibrational fine structure of TPA bands of molecules with several dozens of atoms.

An even more approximate model was proposed by Drobizhev et al. [19], who used the so-called two-state approximation, which assumes that two states, the ground and the first dipole-allowed excited state, are sufficient for the major part of a property described by a sum-over-states expression. With this assumption, Drobizhev et al. reduced the transition polarizability to the dipole moment difference between electronic ground and excited state, $\Delta\mu$. As in the LCM, only linear terms with respect to normal coordinates were kept in the subsequent expansion of $\Delta\mu$. This model was used to compare spectral profiles of one- and two-photon absorption profiles of a fluorescent protein chromophore.

## 17.4 Finite-Field Methods: The Nuclear Relaxation/Curvature Approach

There is an alternative approach to BKPT for deriving the expressions of the vibrational contributions to the nonlinear optical properties. This method is based on considering the effect of a static electric field on the two contributions to the total energy of a field-free molecule, i.e., the electronic energy and the zero-point energy (ZPE) [20, 21]. A static field will change the equilibrium geometry, and thus will lead to a different electronic contribution to the NLOP compared to the field-free counterpart. Taking the latter as the electronic contribution $P^{el}$, the remainder may be called the *nuclear relaxation* contribution $P^{nr}$.

An electric field will also change the shape of the potential energy surface, and thus alter the ZPE. The total contribution of this effect is usually called the *curvature* contribution. Separating off the *zero-point vibrational average* contribution of the field-free structure as the ZPVA contribution $P^{zpva}$ to any property $P$, the remainder is then called the *curvature–ZPVA* contribution $P^{c-zpva}$.

Thus, in the nuclear relaxation/curvature approach, the total value of the NLO properties is given by:

$$P = P^{el} + P^{zpva} + P^{nr} + P^{c-zpva} \tag{17.27}$$

In order to obtain expressions for the $P^{nr}$ contributions in terms of computable quantities, one starts from a double power expansion of the potential energy of a molecule subject to an external electric field $V(Q_n, F_i)$ in some vibrational coordinates and electric field components $F_i$. The theory has first been developed with normal coordinate $Q_n$ as vibrational coordinates, and these will be assumed here, but see below and Sect. 17.5.3 for different choices. Using the stationarity condition $\partial V(Q_n, F_i)/\partial Q_n = 0$ at a *finite (static) field* $\underline{F}$, an expression for the normal coordinates in the field $\underline{F}$ as a series in $Q_n$ and $F_i$ can be derived, which, when inserted back into the expression for $V(Q_n, F_i)$, allows to identify both the $P^{el}$ and $P^{nr}$ contributions as functions of the double power expansion terms, which in turn allows a direct comparison with the terms occurring in the BKPT expressions (see Sect. 17.4.1). Analytic expressions for the $P^{nr}$ contributions in terms of derivatives of the electronic energy with respect to normal coordinates and electric field components can be derived and used in computations, but it is also possible to obtain the $P^{nr}$ contributions by computing finite-field derivatives of the electronic energy or dipole moment at field-dependent equilibrium geometries (see Sect. 17.4.3). This *finite field/nuclear relaxation* (FF/NR) approach may be the only feasible method to compute the anharmonic terms of $P^{nr}$ of larger molecules at correlated levels of theory, where the computation of the corresponding terms of the BKPT approach would be too costly in terms of computational resources.

If the zero-point energy or, more conveniently, the ZPVA dipole moment, defined as the first derivative of the ZPE with respect to the electric field, is expanded as a double power series in normal coordinates $Q_n$ and field components $F_i$, expressions for the curvature contribution to NLOP ($P^c = P^{zpva} + P^{c-zpva}$) can be obtained. Analogously to the $P^{nr}$ contributions, the $P^{c-zpva}$ contributions can also be obtained using the FF/NR approach. In this case, finite-field calculations are required to perform the numerical derivatives of $\mu^{zpva}$.

In theory, the $P^{c-zpva}$ contributions contain all the higher-order anharmonic terms not included in the $P^{nr}$ contribution. In practice, however, the order of anharmonicity included in any actual calculation depends on the order of anharmonicity included in the quantity $\mu^{zpva}$, from which the $P^{c-zpva}$ contributions are obtained. If $\mu^{zpva}$ is computed with a variational approach using an "exact" numerical PES, as described in

Sect. 17.5.3, the obtained $P^{c-zvpa}$ includes all high-order anharmonic contributions. Such a calculation, however, is only feasible for small molecules. On the other hand, if $\mu^{zpva}$ is calculated by a perturbational approach of some order $n$, this order also determines the order of anharmonicity included in $P^{c}$.

The expressions for the infinite optical frequency approximation of $P^{nr}$ and $P^{c-zpva}$ contributions can also be obtained using the nuclear relaxation/curvature approach. $\beta^{nr}(-\omega;\omega,0)_{\omega\to\infty}$ ($\beta^{zpva}(-\omega;\omega,0)_{\omega\to\infty}$) and $\gamma^{nr}(-\omega;\omega,0,0)_{\omega\to\infty}$ ($\gamma^{zpva}(-\omega;\omega,0,0)_{\omega\to\infty}$), which are the vibrational hyperpolarizabilities corresponding to Pockels and Kerr nonlinear effects, can be obtained from the double expansion of $\alpha^{el}(0,0)$ ($\alpha^{zpva}(0,0)$) with respect to the electric field and normal mode displacements. The IOFA vibrational contributions of the field-induced second harmonic generation hyperpolarizabilities $\gamma^{nr}(-2\omega;\omega,\omega,0)_{\omega\to\infty}$ and $\gamma^{zpva}(-2\omega;\omega,\omega,0)_{\omega\to\infty}$ are obtained from the double expansion of $\beta^{el}(0;0,0)$ and $\beta^{zpva}(0;0,0)$. Again, all these infinite optical frequency approximation vibrational hyperpolarizabilities can be obtained using the FF/NR approach.

## 17.4.1 Connection Between FF and PT Methods

The nuclear relaxation/curvature formulas have a direct correspondence with the expressions obtained with BKPT [22]. $P^{nr}$ contributions contain the lowest-order terms of each square bracket contribution of perturbation theory $P^{pv}$. The next higher-order nonvanishing square bracket contributions to $P^{pv}$ are given by $[P^{(c-zpva)}]^{(1)}$, which is obtained from first-order perturbation theory contribution of $[\mu^{zpva}]^{(1)}$. And the higher-order contributions (i.e., $[P^{(c-zpva)}]^{(3)}$, $[P^{(c-zpva)}]^{(5)}$, ...) are obtained from $[\mu^{zpva}]^{(3)}$, $[\mu^{zpva}]^{(5)}$, etc. For instance, the $P^{nr}$, $[P^{(c-zpva)}]^{(1)}$ and $[P^{(c-zpva)}]^{(3)}$ contributions to the second hyperpolarizability corresponding to the Kerr process are given by:

$$\gamma^{nr}_{ijkl}(-\omega;\omega,0,0)_{\omega\to\infty} = \left[\mu\beta\right]^{(0)}_{\omega\to\infty} + \left[\alpha^2\right]^{(0)}_{\omega\to\infty} + \left[\mu^2\alpha\right]^{(1)}_{\omega\to\infty} \tag{17.28}$$

$$\gamma^{(c-zpva)(1)}_{ijkl}(-\omega;\omega,0,0)_{\omega\to\infty} = \left[\mu\beta\right]^{(2)}_{\omega\to\infty} + \left[\alpha^2\right]^{(2)}_{\omega\to\infty} + \left[\mu^2\alpha\right]^{(3)}_{\omega\to\infty} \tag{17.29}$$

$$\gamma^{(c-zpva)(3)}_{ijkl}(-\omega;\omega,0,0)_{\omega\to\infty} = \left[\mu\beta\right]^{(4)}_{\omega\to\infty} + \left[\alpha^2\right]^{(4)}_{\omega\to\infty} + \left[\mu^2\alpha\right]^{(5)}_{\omega\to\infty} \tag{17.30}$$

## 17.4.2 Field-Induced Coordinates, FICs

The bottleneck of the evaluation of expressions of the $P^{pv}$, or its equivalents $P^{nr}$ and $P^{c-zpva}$, in the BKPT approach is the computational cost of the electronic energy and property derivatives. As pointed out above, one way to avoid this problem is to compute the vibrational hyperpolarizabilities with the FF/NR approach. Another possibility to reduce the computational cost of the calculation of $P^{nr}$ and $P^{c-zpva}$ is to use the so-called field-induced coordinates (FICs), instead of the usual normal mode coordinates. While the number of normal coordinates (3$N$-6) increases linearly with the size of the molecule, the number of FICs required to determine a component of a vibrational contribution to any property is quite small, and independent of the size of the molecule. For example, the calculation of $\gamma^{nr}_{iiii}(-\omega;\omega,0,0)$ using normal coordinates requires the determination of (3$N$-6) first derivatives of $\mu^{el}$, $\alpha^{el}$, and $\beta^{el}$, about $(3N-6)^2$ force constants, second derivatives of $\mu^{el}$ and $\alpha^{el}$, and in the order of $(3N-6)^3$ cubic force constants. Using FICs, in contrast, only one single derivative of each kind is required. This method thus reduces the computational cost for medium and large molecules by several orders of magnitude.

In the context of the nuclear relaxation approach, the definition of the FICs is quite straightforward. They are defined as the change of the equilibrium geometry caused by a static external electric field, which is also the first step in the determination of $P^{nr}$ and $P^{c-zpva}$ through an FF/NR calculation. This field-dependent equilibrium geometry shift can be expanded as a power series in the field, leading to analytic expressions of the first-order FIC, second-order FIC, etc. It turns out that there are only three first-order FICs, which depend only on harmonic terms, i.e.,

$$\chi_1^i = \sum_v \frac{(\partial\mu_i^{el}/\partial Q_v)_0}{(\partial^2 E^{el}/\partial Q_v^2)_0} Q_v \qquad \forall i \tag{17.31}$$

With the three first-order FICs, we can compute $\alpha^{nr}(0;0)$, $\beta^{nr}(0;0,0)$, $\lim_{\omega\to\infty}\beta^{nr}(-\omega;\omega,0)$, and $\lim_{\omega\to\infty}\gamma^{nr}(-2\omega;\omega,\omega,0)$. The second-order FICs contain both harmonic and anharmonic terms. Removing the anharmonic terms, one may define the harmonic part of the second-order FICs, which together with the $\chi_1^i$ are sufficient to compute many vibrational contributions such as the Kerr effect, $\lim_{\omega\to\infty}\gamma^{nr}(-\omega;\omega,0,0)$, and the intensity-dependent refraction index, $\lim_{\omega\to\infty}\gamma^{nr}(-\omega;\omega,-\omega,\omega)$, and are given by:

$$\chi_{2,har}^{ij} = \sum_v \frac{(\partial\alpha_{ij}^{el}/\partial Q_v)_0}{(\partial^2 E^{el}/\partial Q_v^2)_0} Q_v \qquad \forall i,j \tag{17.32}$$

Only for the static $\gamma^{nr}_{ijkl}(0;0,0,0)$, the full anharmonic second-order FICs are required.

As examples, we give below two expressions for vibrational nuclear relaxation properties in terms of FICs:

$$\lim_{\omega\to\infty}\beta^{nr}_{iii}(-\omega;\omega,0) = \left(\frac{\partial\alpha_{ii}^{el}}{\partial\chi_1^i}\right)_0 \left(\frac{\partial\mu_i^{el}}{\partial\chi_1^i}\right)_0 \frac{1}{\omega^2_{\chi_1^i}} \tag{17.33}$$

$$\lim_{\omega\to\infty}\gamma^{nr}_{iiii}(-2\omega;\omega,\omega,0) = \left(\frac{\partial\beta_{iii}^{el}}{\partial\chi_1^i}\right)_0 \left(\frac{\partial\mu_i^{el}}{\partial\chi_1^i}\right)_0 \frac{1}{\omega^2_{\chi_1^i}} \tag{17.34}$$

where $\omega_{\chi_1^i}$ is the vibrational circular frequency obtained from the diagonal of the Hessian defined by the FIC $\chi_1^i$. For a full account of all properties in terms of FICs, we refer to the original literature [23].

As expressed above, the FICs are suitable to compute vibrational electric properties either in the static limit or in the limit of infinite optical frequencies. However, general frequency-dependent properties can also by expressed in FICs, if these are defined in a slightly different manner [24]:

$$\chi_1^{i,|\omega|} = \sum_v \frac{(\partial \mu_i^{el}/\partial Q_v)_0}{(\partial^2 E^{el}/\partial Q_v^2)_0 - \omega^2} Q_v \qquad \forall i \tag{17.35}$$

$$\chi_{2,har}^{ij,|\omega|} = \sum_v \frac{(\partial \alpha_{ij}^{el}/\partial Q_v)_0}{(\partial^2 E^{el}/\partial Q_v^2)_0 - \omega^2} Q_v \qquad \forall i,j \tag{17.36}$$

General expressions for the frequency-dependent hyperpolarizabilities $\beta^{nr}(-\omega_\sigma;\omega_1,\omega_2)$ and $\gamma^{nr}(-\omega_\sigma;\omega_1,\omega_2,\omega_3)$ in terms of these dynamic FICs can be found in [24].

**FICs in the calculation of the curvature contribution**

As mentioned above, the curvature contributions $P^c$ can be decomposed into the ZPVA contributions $P^{zpva}$ and the remainder $P^{c-zpva}$ contributions. The BKPT expression for the first-order ZPVA correction $P^{zpva}$ consists of two terms (see Eq. 17.18). One of these terms, $[P]^{1,0}$, contains second derivatives of the property $P$ with respect to normal coordinates, and does not benefit from using FICs. The other term, however, may be simplified by employing FICs [25] to:

$$[P]^{0,1} = -\frac{\hbar}{4}\left(\sum_v \frac{1}{\omega_i}\frac{\partial \omega_i^2}{\partial \chi_P}\right)\frac{\partial P^{el}/\partial \chi_P}{\omega_{\chi_P}^2} \tag{17.37}$$

The kind of FICs $\chi_P$ depends on the property $P^{el}$: For $\mu^{el}$, $\alpha^{el}$, $\beta^{el}$, and $\gamma^{el}$, the corresponding FICs $\chi_P$ are $\chi_1^i$, $\chi_{2,har}^{ij}$, $\chi_{3,har}^{ijk}$, and $\chi_{4,har}^{ijkl}$, respectively, where the harmonic third- and fourth-order FICs are defined in a similar way as the harmonic second-order FIC in Eq. 17.32.

Although the remaining contributions $P^{c-zpva}$ may also be expressed using FICs, they contain derivatives of rather high order, and it is thus generally easier to evaluate these properties by the finite-field geometry optimization method described in the next subsection.

### *17.4.3 Finite-Field Geometry Optimization Approach*

As pointed out in the previous section, the nuclear relaxation contributions can be calculated from finite-field derivatives of electronic properties computed at the field-relaxed equilibrium geometries. The change in the electronic property $P^{el}$

($P^{el} = \mu^{el}, \alpha^{el}, \beta^{el}, \ldots$), evaluated at field-relaxed ($\underline{R}_F$) and field-free ($\underline{R}_0$) equilibrium geometry:

$$(\Delta P^{el})_{\underline{R}_F} = P^{el}(F, \underline{R}_F) - P^{el}(0, \underline{R}_0) \tag{17.38}$$

can be expanded as:

$$(\Delta \mu_i^{el})_{\underline{R}_F} = a_1^{el} F_j + \frac{1}{2} b_1^{el} F_j F_k + \frac{1}{6} g_1^{el} F_j F_k F_l + \ldots \tag{17.39}$$

$$(\Delta \alpha_{ij}^{el})_{\underline{R}_F} = b_2^{el} F_k + \frac{1}{2} g_2^{el} F_k F_l + \ldots \tag{17.40}$$

$$(\Delta \beta_{ijk}^{el})_{\underline{R}_F} = g_3^{el} F_l + \ldots \tag{17.41}$$

The expansion coefficients ($a_1^{el}$, $b_1^{el}$, …) can be directly linked to the electronic and nuclear relaxation (hyper)polarizabilities [26]:

$$a_1^{el} = \alpha_{ij}^{el}(0;0) + \alpha_{ij}^{nr}(0;0) \tag{17.42}$$

$$b_1^{el} = \beta_{ijk}^{el}(0;0,0) + \beta_{ijk}^{nr}(0;0,0) \tag{17.43}$$

$$g_1^{el} = \gamma_{ijkl}^{el}(0;0,0,0) + \gamma_{ijkl}^{nr}(0;0,0,0) \tag{17.44}$$

$$b_2^{el} = \beta_{ijk}^{el}(0;0,0) + \beta_{ijk}^{nr}(-\omega;\omega,0)_{\omega\to\infty} \tag{17.45}$$

$$g_2^{el} = \gamma_{ijkl}^{el}(0;0,0,0) + \gamma_{ijkl}^{nr}(-\omega;\omega,0,0)_{\omega\to\infty} \tag{17.46}$$

$$g_3^{el} = \gamma_{ijkl}^{el}(0;0,0,0) + \gamma_{ijkl}^{nr}(-2\omega;\omega,\omega,0)_{\omega\to\infty} \tag{17.47}$$

One of the advantages of the FF-NR method over the traditional BKPT approach is that field-dependent geometry optimizations can be performed using standard quantum chemistry software, as long as they allow the enforcement of the Eckart conditions during the optimization [27]. This is necessary because a geometry relaxation in an external electric field may also cause a rotation of the molecule, which needs to be avoided. Unfortunately, many standard quantum chemistry program packages do not allow to maintain the Eckart frame during a field-dependent optimization, and custom implementations may have to be used [27].

The curvature contributions $P^c$ may also be obtained with the FF geometry optimization method, if in Eq. 17.38 the electronic properties $P^{el}$ are replaced by their ZPVA counterparts $P^{zpva}$ [21]. This leads to very similar equations as in Eqs. 17.39–17.47, where just the superscripts "*el*" and "*nr*" are replaced by "*zpva*" and "*c-zpva*", respectively. As an example, we give here the ZPVA equivalent to Eq. 17.39:

$$(\Delta \mu_i^{zpva})_{\underline{R}_F} = a_1^{zpva} F_j + \frac{1}{2} b_1^{zpva} F_j F_k + \frac{1}{6} g_1^{zpva} F_j F_k F_l + \ldots \tag{17.48}$$

with

$$a_1^{zpva} \equiv \alpha_{ij}^{c}(0;0) = \alpha_{ij}^{zpva}(0;0) + \alpha_{ij}^{c-zpva}(0;0) \tag{17.49}$$

$$b_1^{zpva} \equiv \beta_{ijk}^{c}(0;0,0) = \beta_{ijk}^{zpva}(0;0,0) + \beta_{ijk}^{c-zpva}(0;0,0) \tag{17.50}$$

$$g_1^{zpva} \equiv \gamma_{ijkl}^{c}(0;0,0,0) = \gamma_{ijkl}^{zpva}(0;0,0,0) + \gamma_{ijkl}^{c-zpva}(0;0,0,0) \tag{17.51}$$

The quantities $(\Delta\mu_i^{zpva})_{\underline{R}_F}$ etc. need to be computed by BKPT (with or without using FICs), or by a variational approach (see next section). Depending on the level of electronic structure theory and basis set employed, it may be computationally less expensive to use the field-dependent change of zero-point energies, which is given by:

$$(\Delta E^{zpe})_{\underline{R}_F} = E^{zpe} - \mu_i^{zpva} F_i - \frac{1}{2} a_1^{zpva} F_j F_k - \frac{1}{6} b_1^{zpva} F_j F_k F_l - \frac{1}{24} g_1^{zpva} F_j F_k F_l F_m + \ldots \tag{17.52}$$

## 17.5 Variational Methods

### 17.5.1 *Introduction*

BKPT may be successfully applied to compute the vibrational (hyper)polarizabilities of harmonic and anharmonic molecules, in the sense defined in Sect. 17.2. However, for nonharmonic or highly anharmonic large amplitude motions, BKPT may diverge or converge very slowly. The convergence of BKPT vibrational contributions can be evaluated using the following two series [28, 29]:

A: $P^{el}, [P^{zpva}]^{(1)}, [P^{zpva}]^{(3)}, \ldots$
B: $P^{nr}, [P^{(c-zpva)}]^{(1)}, [P^{(c-zpva)}]^{(3)}, \ldots$

The two series are convergent, if the terms in each series become increasingly smaller. However, in general, only the first two terms in each series can be computed. Thus, if both ratios $[P^{zpva}]^{(1)}/P^{el}$ and $[P^{(c-zpva)}]^{(1)}/P^{nr}$ are substantially smaller than one, then both series may be considered initially convergent and the vibrational contribution $P^{nr} + [P^{zpva}]^{(1)}$ using either BKPT or the finite-field NR/curvature approach is hopefully a good approximation to the total vibrational NLOPs. If this is not the case, however, perturbation theory cannot be used and another approach has to be applied. Thus, for nonharmonic and strongly anharmonic molecules, variational methods are the obvious alternative to BKPT. For instance, variational methods may be the only reliable option to calculate the vibrational NLOPs of molecules with a LAM with a double-minimum PES and a low-energy barrier between the wells (e.g., $CH_3^-$, see Sect. 17.5.3). Other examples for which the BKPT series is divergent include the dimers of HF and $H_2O$ [30].

There are two alternatives to compute the vibrational NLOPs using a variational approach. The first one is based on applying a variational approach to calculate the

$P^{zpva}$ contribution. Then, the $P^{c-zpva}$ contributions are computed from $P^{zpva}$ using the FF methods treated in Sect. 17.4. Any variational method for the computation of the vibrational energy may be used to calculate the variational NLOP. In the second alternative, the vibrational (hyper)polarizabilities are computed analytically using response theory applied to the vibrational Hamiltonian.

### 17.5.2 Approaches Using Normal Modes

The combination of the FF/NR approach with a variational method to compute the field-dependent vibrational energies is the most straightforward variational method to determine the vibrational NLOPs. The $P^{zpva}$ and $P^{c-zpva}$ contributions obtained with this approach contain in principle all anharmonic terms not included in $P^{nr}$. In practice, however, their accuracy depends on the quality of the field-dependent vibrational wavefunction and the PES used. The reliability of the PES is mainly determined by the level of the electronic structure method employed and the degree of anharmonicity included in the calculation. The accuracy of the vibrational wavefunction depends on the vibrational coordinates used and the level of vibrational correlation included. Normal mode coordinates are the coordinates most often applied for chemical systems with a moderately anharmonic PES, such as $H_2O$, or even HSSH. On the other hand, for nonharmonic and strongly anharmonic molecules, curvilinear coordinates may be the better option, and are treated in the next subsection.

The most straightforward variational approach for the treatment of anharmonicity is the vibrational self-consistent field (VSCF) method (for review of the development and applications of VSCF, see Ref. [31]). In the VSCF, each normal mode vibrates in the average field generated by all the other modes. The VSCF ground-state vibrational wavefunction is approximated as a direct product of one-mode functions:

$$\psi_0^{\mathrm{VSCF}}(Q_1, Q_2, \ldots, Q_M) = \prod_{n=1}^{M} \phi_i^0(Q_i) \tag{17.53}$$

where $M$ is the number of vibrational modes, $Q_i$ are the normal coordinates, $\psi_0^{\mathrm{VSCF}}$ $(Q_1, Q_2, \ldots, Q_M)$ is the molecular vibrational wavefunction, and $\phi_i^0(Q_i)$ are the one-mode functions (modals). The optimal modals are obtained by solving the following Hartree-type equations:

$$\left(-\frac{\hbar^2}{2}\frac{\partial^2}{\partial Q_i^2} + \overline{V_i^0}(Q_i)\right)\varepsilon_i^0 = \varepsilon_i^0 \phi_i^0(Q_i) \qquad \forall i \tag{17.54}$$

The one-mode effective potential, i.e.,

$$\overline{V_i^0}(Q_i) = \left\langle \prod_{j \neq i}^{M} \phi_j^0(Q_j) \left| V(Q_1, Q_2, ..., Q_M) \right| \prod_{j \neq i}^{M} \phi_j^0(Q_j) \right\rangle \tag{17.55}$$

can be computed from the Taylor expansion of the potential energy in the $M$ normal modes, $V(Q_1, Q_2, ..., Q_M)$. The VSCF vibrational energy, $E_{VSCF}$, and the sum of the electronic term and the contribution due to the average over the zero-point vibrations to an electric property, $P^{el} + P^{zpva}_{VSCF}$, are given by the expectation values of the vibrational Hamiltonian and the Taylor expansion of $P^{el}$ in terms of normal modes, respectively. Then, using the FF method, we can obtain the $P^{c-zpva}_{VSCF}$ contributions from field-dependent $P^{zpva}_{VSCF}$ values [32]. $P^{c-zpva}_{VSCF}$ is roughly comparable to the sum $P^{(c-zpva)(1)} + P^{(c-zpva)(3)}$ of BKPT in the case of weak anharmonicity, such as $H_2O$. In the case of stronger anharmonic systems, such as HSSH and even HOOH, for which BKPT clearly diverges, $P^{c-zpva}_{VSCF}$ gives a reasonable first approximation. However, the inclusion of the vibrational correlation is required to obtain accurate vibrational contributions for such strongly anharmonic or nonharmonic systems.

Vibrational correlation corrections, which are due to anharmonic mode–mode interactions, may be added to the VSCF wavefunction by the vibrational analog of Møller–Plesset perturbation theory of second order (VMP2), configuration interaction (VCI), or coupled cluster methods (VCC). Any of these approaches can be used to compute the vibrational NLOPs in the same way as explained above for VSCF. Judging from the few cases investigated so far, the following conclusions can be drawn: For weakly anharmonic systems, VMP2 leads to vibrational hyperpolarizabilities, which are in good agreement with those obtained from the vibrational full CI method. For stronger anharmonic or nonharmonic systems, such as HOOH, this is not the case anymore, and higher correlated methods, such as VCI or VCC, are required.

The calculation of $P^{zpva}$ requires both an expansion of the PES, to compute the vibrational wavefunction, and an expansion of electric property surfaces, in order to compute the expectation value of the corresponding property. The order at which the Taylor expansion can be truncated to obtain accurate values of $P^{zpva}$ is much higher for the PES than for the electric property surfaces. For instance, at least a quartic PES is required to obtain accurate $P^{c-zpva}$ values for water, while a second-order expansion of $\mu^{el}$ is already sufficient to obtain a similar accuracy. A dipole surface expanded to second order would lead to accurate $P^{c-zpva}$ values even for the rather anharmonic molecule HOOH. However, to compute the vibrational wavefunction of HOOH, it is necessary to employ a very flexible expansion of the PES, which is easier to achieve if curvilinear coordinates are used.

As mentioned in Sect. 17.2, at least for small molecules, there are often only one or two modes which are responsible for the nonharmonic or strongly anharmonic character of the vibrational manifold, which also cause the main contribution to the values of $P^{c-zpva}$. This is the case, for example, in $NH_3$, where the umbrella mode together with its couplings to the other modes contributes nearly 100% to the value of $\gamma^{c-zpva}(0; 0, 0, 0)$, as computed at the full VCI level with a quartic PES [28]. Thus, taking into account only the umbrella mode and its couplings with the other modes already leads to quite accurate values for the $P^{c-zpva}$ contributions of $NH_3$. For larger molecules, of course, the number of LAMs may be quite substantial and other criteria have to be used to reduce the number of motions which need to be taken into account explicitly.

Response theory provides an interesting analytical alternative for the computation of frequency-dependent molecular properties. Christiansen et al. derived the analytical response functions to evaluate the static and dynamic vibrational linear polarizability, first and second hyperpolarizability at VSCF and VCI levels [33–35]. The vibrational response theory is the variational counterpart to BKPT, but in contrast to the latter approach, in vibrational response theory, it is not assumed that the electronic (hyper)polarizability functions in Eq. 17.14 and analogues are static. Therefore, for instance, using sum-over-states formalism, the response function for the pure vibrational linear polarizability is given by:

$$\alpha_{ij}^{pv}(-\omega_\sigma;\omega_\sigma) = \frac{1}{\hbar}\sum \mathcal{P}_{-\sigma,\sigma} {\sum_{v}^{\infty}}' \frac{\langle 0\langle 0|\hat{\mu}_i|0\rangle v\rangle\langle v\langle 0|\hat{\mu}_j|0\rangle 0\rangle}{\omega_v - \omega_\sigma} = \langle\langle\langle\hat{\mu}_i\rangle^{el}, \langle\hat{\mu}_j\rangle^{el}\rangle\rangle_{\omega_\sigma}^{pv} \quad (17.56)$$

where $\langle\hat{\mu}_i\rangle^{el} = \langle 0|\hat{\mu}_i|0\rangle$. Whereas the sum-over-states approach is in general a rather slowly converging procedure with respect to the number of states included, the linear response functions can be obtained by solving only three sets of linear equations. The feasibility of the analytical calculations of $\alpha_{ij}^{pv}(-\omega_\sigma;\omega_\sigma)$ and $\beta_{ijk}(-\omega_\sigma;\omega_1,\omega_2)$ was shown with sample calculations for hydrogen fluoride, water, formaldehyde, and pyrrole molecules.

### 17.5.3 Approaches Using Curvilinear Coordinates

A molecule with large amplitude motions may have several minima which are separated by energetically low barriers, so that tunneling becomes important. In that event, it is often advantageous to set up and solve the Schrödinger equation in curvilinear, or internal coordinates. The deformational part of the kinetic energy operator (KEO) for vanishing total angular momentum ($J$=0) and the potential energy surfaces, which are required for vibrational or vibronic calculations, is independent of the orientation of the molecular frame. However, in the context of the vibrational linear and nonlinear optical properties, which involve the application of external fields, it is essential to apply the Eckart conditions [36], both in the determination of the field-dependent potential energy surface [27] and in the calculation of the electric property surfaces [37].

Several programs are currently available which allow to compute the (ro)vibrational states of floppy molecules, such as TROVE [38], GENIUSH [39], ElVibRot [17, 40], PVSCF [41]. The programs MCTDH [42] and QUANTICS [43] are also able to deal with curvilinear coordinates, but the KEO has to be defined explicitly in the input. These programs share essentially the same structure:

(i) The selection of the kind of curvilinear coordinates is an essential step. The selection should be done in such a way as to find a well-balanced set of coordinates leading to a compact multidimensional basis, while at the same time avoiding a KEO, which is too complex to handle efficiently. For a floppy molecule consisting of $N$ atoms, the $3N-6$ internal coordinates can be split into $n$ active ones, which are treated explicitly, and $m$ inactive ones ($n+m=3N-6$), which are treated in a more approximate manner. One may choose, for example, to treat the inactive coordinates in a harmonic approximation, leading to *adiabatic separation models* [40, 44]. In *constrained models* [45–49], the inactive coordinates are either held fixed (rigid constraints) or relaxed at each point of the PES along the active coordinates (flexible constraints). To derive the KEO [50, 51], $\hat{T}$, in a numerical form [39, 46–49, 52], it is necessary to calculate the values of the Cartesian coordinates, $\underline{X}^{\lambda}_{MF}$, of the atom $\lambda$ as a function of the $n$ internal coordinates, $\underline{q}$, in the molecular frame, MF. The deformation part of the KEO reads [49]:

$$\hat{T}_{def} = -\frac{\hbar^2}{2}\sum_{i,j}^{n,n}\frac{1}{\rho(\underline{q})}\frac{\partial}{\partial q^i}\rho(\underline{q})G^{i,j}(\underline{q})\frac{\partial}{\partial q^j} + V_{\text{extra}}(\underline{q}) \tag{17.57}$$

$$d\tau_{\text{def}} = \rho(\underline{q})dq^1 \ldots dq^n \tag{17.58}$$

where the $G^{i,j}$ are the contravariant components of the mass-weighted metric tensor, $\underline{\underline{G}}$, and $\rho(\underline{q})$ is a weight function dependent on the choice of the volume element, $d\tau_{\text{def}}$. Finally, $V_{\text{extra}}(\underline{q})$ is the scalar part of the KEO, which appears if a non-Euclidean volume element is chosen. It is important to note that Eq. 17.57 can be used even when a reduced dimensionality model is chosen ($n < 3N-6$). For the numerical approach, the $\underline{\underline{G}}$ matrix is obtained by the inversion of the $\underline{\underline{g}}$ matrix:

$$\underline{\underline{G}}(\underline{q}) = [G^{i,j}] = \underline{\underline{g}}(\underline{q})^{-1} \tag{17.59}$$

$$\underline{\underline{g}}(\underline{q}) = \begin{bmatrix} \underline{\underline{S}}(\underline{q}) & \underline{\underline{C}}(\underline{q})^T \\ \underline{\underline{C}}(\underline{q}) & \underline{\underline{I}}(\underline{q}) \end{bmatrix} \tag{17.60}$$

$\underline{\underline{S}}(\underline{q})$, $\underline{\underline{C}}(\underline{q})$ are, respectively, the deformation and Coriolis parts of the metric tensor. $\underline{\underline{I}}(\underline{q})$ is the usual inertia tensor. All these matrices can be easily expressed as functions of the molecular frame Cartesian coordinates, $\underline{X}^{\lambda}_{MF}$, and their derivatives with respect to the internal coordinates. For instance, $S_{i,j}$ reads:

$$S_{i,j} = \sum_{\lambda=1}^{N} m_\lambda \left( \frac{\partial X^{\lambda,x}_{MF}}{\partial q^i}\frac{\partial X^{\lambda,x}_{MF}}{\partial q^j} + \frac{\partial X^{\lambda,y}_{MF}}{\partial q^i}\frac{\partial X^{\lambda,y}_{MF}}{\partial q^j} + \frac{\partial X^{\lambda,z}_{MF}}{\partial q^i}\frac{\partial X^{\lambda,z}_{MF}}{\partial q^j} \right), \tag{17.61}$$

where $[X_{MF}^{\lambda,x}, X_{MF}^{\lambda,y}, X_{MF}^{\lambda,z}]$ are the three MF components of the atomic Cartesian coordinates, $\underline{X}_{MF}^{\lambda}$, of atom $\lambda$ with mass $m_{\lambda}$. A substantial advantage of the numerical approach is the possibility to deal with complex internal coordinates, thus providing a large flexibility. In particular, one can apply successive transformations from the coordinate, $\underline{q}$, used for the dynamics up to the Cartesian coordinates in the MF. Therefore, all kinds of coordinates (Z-matrix coordinates, curvilinear normal modes [50], coordinates used in the Reaction Path Hamiltonian [53], etc.) may be used, without requiring a special treatment to set up the Hamiltonian matrix or the action of the Hamiltonian of a wavefunction.

An additional advantage of the numerical approach is that the transformation between different molecular frames, for instance, in order to transform into the Eckart frame [36, 54], is completely transparent. It only requires the rotation of the Cartesian coordinates from the non-Eckart frame into the Eckart frame using a rotation matrix calculated with the help of a well-defined procedure [55, 56].

(ii) In all models, the $n$ active coordinates are treated explicitly, which leads to the potential energy and property surfaces (such as the dipole moment surface), which are specific to the molecular system under study.

(iii) The wavefunction can be expanded on an $n$D-basis and/or on an $n$D-grid. The $n$D-basis functions are usually expressed as products of primitive basis functions, and each primitive basis set is associated with a coordinate or a group of coordinates (in the case of spherical harmonics as basis sets, for instance). The grids associated with the basis sets are usually based on Gauss quadrature schemes.

(iv) The final diagonalization of the Hamiltonian either can be performed directly, if the number of basis functions is not very large, or can be achieved using iterative schemes (such as the Lanczos [57, 58] or the Davidson approach [59, 60]), which avoid building the Hamiltonian matrix explicitly.

**Examples**

Variational computations of nonlinear optical properties using curvilinear coordinates are still scarce. The umbrella motion in $NH_3$ was investigated in Ref. [61, 62], and a more general treatment for a slightly asymmetric two-well potential was published in Ref. [5]. We restrict ourselves here to the symmetric double well, appropriate, for example, for the umbrella motion in pyramidal $AX_3$ molecules. Generalized van Vleck perturbation theory was applied to obtain the field-dependent energies, including tunneling through the barrier. Two quantities, the average $E^{(+)}$ of the energies of the two lowest perturbed states $a, b$ and the square of their difference, $\Delta$, can be defined as power series in the field. Exploiting the occurrence of the two equivalent potential wells due to the inversion symmetry, one obtains:

$$E^{+}(F_z) = E^{+}(0) - \frac{1}{2}\alpha_{zz}^{+}F_z^2 - \frac{1}{24}\gamma_{zzzz}^{+}F_z^4 \tag{17.62}$$

$$\Delta(F_z) = \frac{[E^{(a)}(F_z) - E^{(b)}(F_z)]^2}{4} = C^{(0)} + C_{zz}^{(2)}F_z^2 + C_{zzzz}^{(4)}F_z^4 \tag{17.63}$$

with

$$C^{(0)} = (E^-)^2 \tag{17.64}$$

$$C^{(2)} = -E^-\alpha_{zz}^- + (\mu_z^{(01)})^2 \tag{17.65}$$

$$C^{(4)} = -\frac{E^-\gamma_{zzzz}^-}{12} + \frac{(\alpha_{zz}^-)^2}{2} - \frac{\mu_z^{(01)}\beta_{zzz}^{(01)}}{3} \tag{17.66}$$

Here, $P^{\pm} = \frac{1}{2}(P^{(00)} \pm P^{(11)})$, $P = E, \mu_z, \alpha_{zz}, \beta_{zzz}, \gamma_{zzzz}$, and the superscripts 0, 1 refer to field-free reference states. The field is supposed to be aligned along the symmetry axis of the $AX_3$ molecule, here taken to be the $z$-axis. In that case, the Eckart–Sayvetz conditions are fulfilled automatically. We mention that $\Delta$ is proportional to the square of the frequency between the two vibrational states, and may thus be accessible experimentally by vibrational Stark spectroscopy.

For special cases, it is possible to write only linear combinations of the state energies as usual Taylor series in the field. One of these cases, in Ref. [5] referred to as the high-field/medium barrier case, requires that the field-induced transition moment between the two lowest states is large compared to the field-dependent splitting $E^-(F_z)$. In that event, one may use instead of $\Delta(F_z) \sim \omega_{ab}^2$ the quantity

$$|\hbar\omega_{ab}(F_z)| = |\mu_z^{(01)}F_z + \frac{1}{6}\beta_{zzz}^{(01)}F_z^3| \tag{17.67}$$

This case is applicable to the inversion of $NH_3$, where the comparatively large barrier leads to a small tunneling splitting of 0.79 $cm^{-1}$ while the same motion in $CH_3^-$ with a much lower barrier, and consequently a tunneling splitting about 50 times larger than that of $NH_3$, requires the more general treatment [5].

Using the treatment in the high-field/medium barrier limit, the umbrella motion in $NH_3$ was treated in Ref. [61, 62] and compared with the one-well VCI treatment of the same motion in Ref. [63, 64]. Several nuclear models were used for the coupling between the *active* coordinate, i.e., the coordinate describing the inversion motion, and the rest of the vibrational modes, the *inactive* coordinates. The NLO properties were found to be quite sensitive to the different models. Only the more accurate ones, using the so-called harmonic adiabatic channels, where the active coordinate $\tau$ is treated exactly at several points along the inversion motion, while the other coordinates are approximated by a harmonic potential at each point of $\tau$, provided results, which were in agreement with the VCI approach [62]. Results obtained at a lower level of accuracy originally led to erroneous conclusions with respect to the VCI treatment [61].

Efforts to extend the approach described above to more complicated large amplitude motions, such as the torsion of the methyl group in methanol, are currently underway.

## 17.6 Time-Dependent Approaches to L&NLO Processes

All the methods described up to now in this chapter are defined in the frequency domain, which is a convenient description for systems subject to continuous periodic electromagnetic radiation, and is obtained from the time domain by a Fourier transformation. Instead, it is also possible to set up the computation directly in the time domain. This generally implies a molecular simulation of the molecular system subject to a time-dependent electric field (see, e.g., Ref. [65] for a real-time calculation of the electronic NLO contribution). Computations of (non)linear spectra, such as vibrational sum-frequency generation or one- and two-photon absorption, in the time domain may be performed more efficiently than in the frequency domain [66].

To illustrate the connection between the two descriptions, we will consider the one-photon absorption (OPA) process, where the photon energy $\hbar\omega$ is comparable to the energy difference between two electronic states. In Eq. 17.2, we defined the real part of polarizability (Re[$\alpha(\omega)$]). The OPA spectrum corresponds to the imaginary part of the complex polarizability function. In order to deal with the latter, we closely follow the route paved in Sect. 17.3. In particular, we adopt the Born–Oppenheimer approximation and we keep only the first term in the expansion of dipole moment operator (see Eq. 17.14). The latter step is known as the Franck–Condon approximation. The final expression for the one-photon absorption cross section is then given by [67]:

$$\sigma_{\mathrm{OPA}}^{\mathrm{FC}}(\omega) = \frac{4\pi^2\omega}{3c} \sum_{nv'} \sum_{i} \langle 0|\hat{\mu}_i|n\rangle\langle n|\hat{\mu}_i|0\rangle \frac{\langle v^0|v'^n\rangle\langle v'^n|v^0\rangle}{(\omega_{nv'} - \omega_{0v} - \omega - i\Gamma_n)} \tag{17.68}$$

The matrix element $\mu_i^{0n} = \langle 0|\hat{\mu}_i|n\rangle$ involving electronic wavefunctions $|0\rangle$ and $|n\rangle$ is evaluated at the ground-state geometry; $|v^0\rangle$ ($|v'^n\rangle$) stands for the initial (final) vibrational state in the electronic ground (excited) state. $\Gamma_n$ is the broadening parameter. The transition into the time domain is obtained by a Fourier transformation. The final result for OPA cross section is:

$$\sigma_{\mathrm{OPA}}^{\mathrm{FC}}(\omega) = \frac{4\pi\hbar\omega}{3c} \sum_{n} |\underline{\mu}^{0n}|^2 \times \mathrm{Re} \int_0^{\infty} \langle v^0|v^n(t)\rangle e^{i\hbar(\omega_{0v}+\omega)t - g_n(t)} \mathrm{d}t \tag{17.69}$$

where $|v^n(t)\rangle$ is the wavepacket corresponding to the time-dependent nuclear wavefunction of electronic state $n$. Note that the broadening parameter $\Gamma_n$ is now replaced by a more general lifetime function $g_n(t)$. The theory of time-dependent spectroscopy yields a more physical and detailed description of the lifetime function than the simple introduction of a broadening parameter $\Gamma$ as an empirical quantity [67, 68].

The use of Eq. 17.68 for systems with a large density of eigenstates involves an excessively high computational costs. On the other hand, Eq. 17.69 allows the efficient determination of converged spectra. This is one of the reasons behind the increasing popularity of time-dependent approaches in the domain of nonlinear spectroscopies probing electronic and vibrational motions. Computer codes are available to simulate, e.g., vibrational fine structure of absorption bands in one- and two-photon absorption spectra [67, 68] or double-resonance sum-frequency generation spectra [69].

## 17.7 Conclusions and Outlook

In this chapter, we summarized most of the currently available methods for the calculation of linear and nonlinear vibrational optical properties. Although much progress has been made in the last two decades or so, there is still much work to do, as there is no general procedure available as yet for these properties. Variational methods based on VSCF are developing fast and show great promise to be able to handle larger systems, which would be necessary to investigate molecules of interest for practical applications. However, these methods still have problems with certain large amplitude motions with low barriers and multiple wells. For such motions, methods based on curvilinear coordinates and dedicated codes to solve the vibrational Schrödinger equation numerically will probably be the more adequate solution. However, these approaches currently require a lot of user intervention and are not yet able to treat large systems, although this area is also developing fast. There are still other problems, which have received only scant attention up to now, such as the effect of rotation–vibration coupling on NLOP, as well as the treatment of systems where rotation and vibration cannot be separated easily, as may happen in certain very weakly bound complexes.

Despite recent progress [70–72], still rather little is known about another very important topic, which is the effect of the molecular interactions and, in extension, of the environment in bulk systems on the vibrational NLOPs and which has not been treated at all in this review. An approach to treat vibrational NLOPs in periodic systems is currently being developed by M.Springborg, B. Kirtman, and the group of R. Dovesi, but there is much more left to be done in this area for other kinds of bulk systems, such as liquids, glasses. This would also open the way for comparison of the computed values with experimental results, and which would the also allow for more realistic assessment of the computational approaches.

## References

1. A. Willetts, J.E. Rice, D.M. Burland, D. Shelton, Problems in the comparison of theoretical and experimental hyperpolarizabilities. J. Chem. Phys. **97**, 7590–7599 (1992)
2. H. Reis, Problems in the comparison of theoretical and experimental hyperpolarizabilities revisited. J. Chem. Phys. **125**, 014506 (2006)
3. P.N. Butcher, D. Cotter, *The Elements of Nonlinear Optics* (Cambridge University Press, Cambridge, 1990)
4. G.S. Ezra, *Symmetry Properties of Molecules*. Lecture Notes in Chemistry Vol. 28, (Springer, Berlin 1982)

5. H. Reis, J.M. Luis, M. Garcia Borràs, B. Kirtman, Computation of nonlinear optical properties of molecules with large amplitude anharmonic motions. III. Arbitrary Double-Well Potentials, J. Chem. Theo. Comp. **10**, 236–242 (2015)
6. D.M. Bishop, S.A. Solunac, Breakdown of the born-oppenheimer approximation in the calculation of electric hyperpolarizabilities. Phys. Rev. Lett. **55**, 2627 (1985). erratum: Phys. Rev. Lett. 55 2627
7. B.J. Orr, J.F. Ward, Perturbation theory of the non-linear optical polarization of an isolated system. Mol. Phys. **20**, 513–526 (1971)
8. D.M. Bishop, J.M. Luis, B. Kirtman, Additional compact formulas for vibrational dynamic dipole polarizabilities and hyperpolarizabilities. J. Chem. Phys. **108**, 10013–10017 (1998)
9. B. Kirtman, B. Champagne, J.M. Luis, Efficient treatment of the effect of vibrations on electrical, magnetic, and spectroscopic properties. J. Comp. Chem. **21**, 1572–1588 (2000)
10. D.M. Bishop, Molecular vibration and nonlinear optics. Adv. Chem. Phys. **104**, 1–40 (1998)
11. M. Torrent-Sucarrat, M. Sola, M. Duran, J.M. Luis, B. Kirtman, Initial convergence of the perturbation series expansion for vibrational nonlinear optical properties. J. Chem. Phys. **116**, 5363–5373 (2002)
12. R. Zaleśny, Anharmonicity contributions to the vibrational first and second hyperpolarizability of para-disubstituted benzenes. Chem. Phys. Lett. **595–596**, 109–112 (2014)
13. B. Gao, M. Ringholm, R. Bast, K. Ruud, A.J. Thorvaldsen, M. Jaszunski, Analytic density-functional theory calculations of pure vibrational Hyperpolarizabilities: The first dipole Hyperpolarizability of retinal and related molecules. J. Phys. Chem. A **118**, 748–756 (2014). Erratum: ibid, p. 2835 (2014)
14. M. Ringholm, D. Jonsson, R. Bast, B. Gao, A.J. Thorvaldsen, U. Ekström, T. Helgaker, K. Ruud, Analytic cubic and quartic force fields using density-functional theory. J. Chem. Phys. **140**, 034103 (2014)
15. Y. Cornaton, M. Ringholm, K. Ruud, Complete analytic anharmonic hyper-Raman scattering spectra. Phys. Chem. Chem. Phys. **18**, 22331–22342 (2016)
16. D.M. Bishop, J.M. Luis, B. Kirtman, Vibration and two-photon absorption. J. Chem. Phys. **116**, 9729–9739 (2002)
17. Lauvergnat, D.: Elvibrot-Tnum, http://pagesperso.lcp.u-psud.fr/lauvergnat/ElVibRot/ElVibRot.html
18. P. Macak, Y. Luo, H. Ågren, Simulations of vibronic profiles in two-photon absorption. Chem. Phys. Lett. **330**, 447–456 (2000)
19. M. Drobizhev, N.S. Makarov, S.E. Tillo, T.E. Hughes, A. Rebane, Describing two-photon absorptivity of fluorescent proteins with a new vibronic coupling mechanism. J. Phys. Chem. B **116**, 1736–1744 (2012)
20. J.M. Luis, M. Duran, J.L. Andres, A systematic and feasible method for computing nuclear contributions to electrical properties of polyatomic molecules, J. Chem. Phys. **107**, 1501 (1997)
21. B. Kirtman, J.M. Luis, D.M. Bishop, Simple finite field method for calculation of static and dynamic vibrational hyperpolarizabilities: Curvature contributions. J. Chem. Phys. **108**, 10008–10012 (1998)
22. J.M. Luis, J. Martí, M. Duran, J.L. Andrés, B. Kirtman, Nuclear relaxation contribution to static and dynamic (infinite frequency approximation) nonlinear optical properties by means of electrical property expansions: Application to HF, CH4, CF4, and SF6. J. Chem. Phys. **108**, 4123 (1998)
23. J.M. Luis, M. Duran, B. Champagne, B. Kirtman, Determination of vibrational polarizabilities and hyperpolarizabilities using field–induced coordinated. J. Chem. Phys. **113** 5203–5213 (2000); see also the correction in ref. 23 of [21]
24. J.M. Luis, M. Duran, B. Champagne, B. Kirtman, Field-induced coordinates for the determination of dynamic vibrational nonlinear optical properties. J. Chem. Phys. **115**, 4473–4483 (2001)
25. J.M. Luis, B. Champagne, B. Kirtman, Calculation of static zero-point vibrational averaging corrections and other vibrational curvature contributions to polarizabilities and hyperpolarizabilities using field-induced coordinates. Int. J. Quant. Chem. **80**, 471–479 (2000)

26. D.M. Bishop, M. Hasan, B. Kirtman, A simple method for determining approximate static and dynamic vibrational hyperpolarizabilities. J. Chem. Phys. **103**, 4157–4159 (1995)
27. J.M. Luis, M. Duran, J.L. Andrés, B. Champagne, B. Kirtman, Finite field treatment of vibrational polarizabilities and hyperpolarizabilities: On the role of the Eckart conditions, their implementation, and their use in characterizing key vibrations. J. Chem. Phys. **111**, 875–884 (1999)
28. M. Torrent-Sucarrat, M. Solà, M. Duran, J.M. Luis, B. Kirtman, Initial convergence of the perturbation series expansion for vibrational nonlinear optical properties. J. Chem. Phys. **116**, 5363–5373 (2002)
29. M. Torrent-Sucarrat, M. Solà, M. Duran, J.M. Luis, B. Kirtman, Basis set and electron correlation effects on initial convergence for vibrational nonlinear optical properties of conjugated organic molecules. J. Chem. Phys. **120**, 6346–6355 (2004)
30. U. Eckart, A. Sadlej, Vibrational corrections to electric properties of weakly bound systems. Mol. Phys. **99**, 735–743 (2001)
31. R.B. Gerber, J.O. Jung, The vibrational self-consistent field approach and extensions: Method and applications to spectroscopy of large molecules and clusters. In: Jensen, P., Bunker, P. R.: *Computational Molecular Spectroscopy*. (Wiley and Sons, Chichester, 2000)
32. M. Torrent-Sucarrat, J.M. Luis, B. Kirtman, Variational calculation of vibrational linear and nonlinear optical properties. J. Chem. Phys. **122**, 204108 (2005)
33. O. Christiansen, J. Kongsted, M.J. Paterson, J.M. Luis, Linear response functions for a vibrational configuration interaction state. J. Chem. Phys. **125**, 124309 (2006)
34. M.B. Hansen, O. Christiansen, C. Hättig, Automated calculation of anharmonic vibrational contributions to first hyperpolarizabilities: quadratic response functions from vibrational configuration interaction wave functions. J. Chem. Phys. **131**, 154101 (2009)
35. M.B. Hansen, O. Christiansen, Vibrational contributions to cubic response functions from vibrational configuration interaction response theory. J. Chem. Phys. **135**, 154107 (2011)
36. C. Eckart, Some studies concerning rotating axes and polyatomic molecules. Phys. Rev. **47**, 552–558 (1935)
37. C.R. Le Sueur, S. Miller, J. Tennyson, B.T. Sutcliffe, On the use of variational wavefunctions in calculating vibrational band intensities. Mol. Phys. **76**, 1147–1156 (1992)
38. S.N. Yurchenko, W. Thiel, P. Jensen, Theoretical ROVibrational Energies (TROVE): A robust numerical approach to the calculation of rovibrational energies for polyatomic molecules. J. Mol. Spectrosc. **245**, 126 (2007)
39. E. Mátyus, G. Czakó, A.G. Császár, Toward black-box-type full- and reduced-dimensional variational (ro)vibrational computations. J. Chem. Phys. **130**, 28 (2009)
40. D. Lauvergnat, A. Nauts, Torsional energy levels of nitric acid in reduced and full dimensionality with ElVibRot and Tnum. Phys. Chem. Chem. Phys. **12**, 8405–12 (2010)
41. Y. Scribano, D. Lauvergnat, D.M. Benoit, Fast vibrational configuration interaction using generalized curvilinear coordinates and self-consistent basis. J. Chem. Phys. **133**, 94103 (2010)
42. M.H. Beck, A. Jäckle, G.A. Worth, H.-D. Meyer, The multiconfiguration time-dependent Hartree (MCTDH) method: A highly efficient algorithm for propagating wavepackets. Phys. Rep. **324**, 1–105 (2000)
43. G.A. Worth, K. Giri, G.W. Richings, M.H. Beck, A. Jäckle, H.-D. Meyer, QUANTICS Package, Version 1.1 (2015)
44. J.C.C. Light, Z. Bačić, Adiabatic approximation and nonadiabatic corrections in the discrete variable representation: Highly excited vibrational states of triatomic molecules. J. Chem. Phys. **87**, 8007 (1987)
45. A. Nauts, X. Chapuisat, Hamiltonians for constrained N-particle systems. Chem. Phys. Lett. **136**, 164 (1987)
46. R. Meyer, Flexible models for intramolecular motion, a versatile treatment and its application to glyoxal. J. Mol. Spectrosc. **76**, 266 (1979)
47. M.A. Harthcock, J. Laane, Calculation of kinetic energy terms for the vibrational Hamiltonian: Application to large-amplitude vibrations using one-, two-, and three-dimensional models. J. Mol. Spectrosc. **91**, 300–324 (1982)

48. Y. Smeyers, F. Melendez, M. Senent, Ab initio theoretical study of the methyl and phosphine torsion modes in ethylphosphine. J. Chem. Phys. **106**, 1709–1717 (1997)
49. D. Lauvergnat, A. Nauts, Exact numerical computation of a kinetic energy operator in curvilinear coordinates. J. Chem. Phys. **116**, 8560–8570 (2002)
50. E.B. Wilson, J.C. Decius, P.C. Cross, *Molecular Vibrations: The Theory of Infrared and Raman Vibrational Spectra*. McGraw-Hill (New York, 1955)
51. A. Nauts, X. Chapuisat, Momentum, quasi-momentum and hamiltonian operators in terms of arbitrary curvilinear coordinates, with special emphasis on molecular hamiltonians. Mol. Phys. **55**, 1287–1318 (1985)
52. D. Strobusch, C. Scheurer, Hierarchical expansion of the kinetic energy operator in curvilinear coordinates for the vibrational self-consistent field method. J. Chem. Phys. **135**, 124102 (2011)
53. W. Miller, N. Handy, J. Adams, Reaction path Hamiltonian for polyatomic molecules. J. Chem. Phys. **72**, 99 (1980)
54. P.R. Bunker, P. Jensen, *Molecular Symmetry and Spectroscopy*, NRC Press (Ottawa, 1998)
55. A.Y. Dymarsky, K.N. Kudin, Computation of the pseudorotation matrix to satisfy the Eckart axis conditions. J. Chem. Phys. **122**, 124103 (2005)
56. D. Lauvergnat, J.M. Luis, B. Kirtman, H. Reis, A. Nauts, Numerical and exact kinetic energy operator using Eckart conditions with one or several reference geometries: Application to HONO. J. Chem. Phys. **144**, 084116 (2016)
57. C. Lanczos, An iteration method for the solution of the eigenvalue problem of linear differential and integral operators. J. Res. Natl. Bur. Stand. **45**, 255 (1950)
58. J.K. Cullum, R.A. Willoughby, *Lanczos Algorithms for Large Symmetric Eigenvalue Computations*. Birkhäuser (Boston, 2002)
59. E.R. Davidson, The iterative calculation of a few of the lowest eigenvalues and corresponding eigenvectors of large real-symmetric matrices. J. Comput. Phys. **17**, 87 (1975)
60. F. Ribeiro, C. Iung, C. Leforestier, Calculation of highly excited vibrational levels: A prediagonalized Davidson scheme. Chem. Phys. Lett. **362**, 199 (2002)
61. J.M. Luis, H. Reis, M.G. Papadopoulos, B. Kirtman, Treatment of nonlinear optical properties due to large amplitude vibrational motions: Umbrella motion in $NH_3$. J. Chem. Phys. **131**, 034116 (2009)
62. S. Garcia-Borràs, D. Lauvergnat, H. Reis, J.M. Luis, B. Kirtman, A full dimensionality approach to evaluate the nonlinear optical properties of molecules with large amplitude anharmonic tunneling motions. J. Chem. Theory Comput. **9**, 520–532 (2013)
63. M. Torrent-Sucarrat, J.M. Luis, B. Kirtman, Variational calculation of vibrational linear and nonlinear optical properties. J. Chem. Phys. **122**, 204108 (2005)
64. J.M. Luis, M. Torrent-Sucarrat, O. Christiansen, B. Kirtman, Variational calculation of static and dynamic vibrational nonlinear optical properties. J. Chem. Phys. **127**, 084118 (2007)
65. Y. Takimoto, F.D. Vila, J.J. Rehr, Real-time time-dependent density functional theory approach for frequency-dependent nonlinear optical response in photonic molecules. J. Chem. Phys. **127**, 154114 (2007)
66. E.J. Heller, The semiclassical way to molecular spectroscopy. Acc. Chem. Res. **14**, 368–375 (1981)
67. D.W. Silverstein, L. Jensen, Vibronic coupling simulations for linear and nonlinear optical processes: Theory. J. Chem. Phys. **136**, 064111 (2012)
68. T. Petrenko, F. Neese, Analysis and prediction of absorption band shapes, fluorescence band shapes, resonance Raman intensities, and excitation profiles using the time-dependent theory of electronic spectroscopy. J. Chem. Phys. **127**, 164319 (2007)
69. P.A. Weiss, D.W. Silverstein, L. Jensen, Non-condon effects on the doubly resonant sum frequency generation of Rhodamine 6G. J. Phys. Chem. Lett. **5**, 329–335 (2014)
70. F. Egidi, T. Giovannini, M. Piccardo, J. Bloino, C. Cappelli, V. Barone, Stereoelectronic, vibrational, and environmental contributions to polarizabilities of large molecular systems: A feasible anharmonic protocol. J. Chem. Theor. Comput. **10**, 2456–2464 (2014)
71. R. Zaleśny, R.W. Góra, J.M. Luis, W. Bartkowiak, On the particular importance of vibrational contributions to the static electrical properties of model linear molecules under spatial confinement. Phys. Chem. Chem. Phys. **17**, 21782–21786 (2015)

72. R. Zaleśny, M. Garcia-Borràs, R.W. Góra, M. Medved', J.M. Luis, On the physical origins of interaction-induced vibrational (hyper)polarizabilities. Phys. Chem. Chem. Phys. **18**, 22467–22477 (2016)

# Chapter 18
# Ab Initio Molecular Dynamics Study on Photoisomerization Reactions: Applications to Azobenzene and Stilbene

**Tetsuya Taketsugu and Yu Harabuchi**

**Abstract** Ab initio molecular dynamics (AIMD) approach to examine excited-state dynamics of ultrafast photoreactions was introduced, and its applications to photoisomerization of azobenzene and stilbene were described. AIMD with a surface hopping scheme was used to examine photoisomerization mechanism of nπ*-excited azobenzene. The rotation pathway was shown to be a preferred one compared with the inversion pathway. A small change in the NN stretching frequency of trans-azobenzene upon nπ* excitation was also reproduced by highly accurate calculations, which support the rotation pathway. To examine the photoreaction of ππ*-excited *cis*-stilbene, SF-TDDFT was used in AIMD simulations. It was shown that ππ*-excited *cis*-stilbene propagates primarily toward the twisted structural region due to dynamical effects, with partial branching to the 4,4-dihydrophenanthrene (DHP) structure. AIMD simulations were also carried out for ππ*-excited 1,1'-dimethylstilbene, and it was shown that the perpendicular structure around $S_1/S_0$ conical intersections has possibly the long lifetime.

**Keywords** Excited-state dynamics • Photoisomerization • Surface hopping

## 18.1 Introduction

Photochemistry provides a very broad frontier of quantum chemistry. Last 50 years a laser technology and spectroscopic experiments have clarified many aspects in photoreaction mechanism and dynamics. Especially, transient absorption and emission spectra have provided information of excited-state dynamics of molecules in detail, and sophisticated experiments have extended our understanding of the

T. Taketsugu (✉) · Y. Harabuchi
Department of Chemistry, Faculty of Science, Hokkaido University,
Sapporo 060-0810, Japan
e-mail: take@sci.hokudai.ac.jp

Y. Harabuchi
JST, PRESTO, 4-1-8 Honcho, Kawaguchi, Saitama 332-0012, Japan



M.J. Wójcik et al. (eds.), *Frontiers of Quantum Chemistry*,
https://doi.org/10.1007/978-981-10-5651-2_18

elementary reaction processes in electronically excited states. Of course, the entire picture of a photoreaction could not be drawn only from the experiments, and nowadays quantum chemical approach plays an essential role to understand the photoreaction mechanism. Owing to advanced ab initio wavefunction theory and density functional theory (DFT), vertical excitation energies can be evaluated with high accuracy, and global potential energy surfaces (PESs) in the excited states can be explored, which clarify the reaction pathways through a couple of excited states. Azobenzene and stilbene are basic textbook molecules in photochemistry. Both molecules show cis-trans photoisomerization, which leads to photoswitch functions in nanotechnology. The first excited state of stilbene is $\pi\pi^*$ state, while the first and second excited states of azobenzene are $n\pi^*$ and $\pi\pi^*$ states, respectively. In stilbene, the reaction pathway is restricted to a rotation about the central $C=C$ bond, while in azobenzene, the rotation and inversion pathways are possible. In this chapter, our recent theoretical studies on photoisomerization of azobenzene [1, 2] and stilbene [3, 4] were described.

Ab initio molecular dynamics (AIMD) approach has become a powerful theoretical tool to elucidate the ultrafast photoreaction process. AIMD is a classical trajectory method where atomic positions and velocities are propagated by solving Newton's equations of motion based on ab initio energy gradients [5–8]. The target of AIMD approach has been extended to photochemical reactions where non-radiative transitions play a significant role [9–11]. In our group, AIMD program code for excited-state dynamics was developed and applied to photochemical reactions in solutions [12–14] and to dissociative recombination reactions with the surface hopping scheme [15–17] based on a Tully's surface hopping scheme [18] in which the electronic wavefunction is expanded in terms of adiabatic eigenfunctions, and the expansion coefficients were developed by solving time-dependent Schrödinger equation using non-adiabatic coupling terms. Following Tully's fewest switches algorithm, the hopping probability between the electronic adiabatic states is evaluated from the electronic amplitudes and non-adiabatic coupling terms, and the hopping is invoked by comparing the hopping probability and a random number. In AIMD approach, all nuclear degrees of freedom are automatically included, and thus, a small time step is required to describe the fast atomic motions. If such fast motions are not important in a given reaction, a relatively large time step could be used by fixing those internal degrees of freedom through dynamics simulation. For this purpose, the RATTLE algorithm [19] was used in our AIMD simulation. A general AIMD code was developed in our group, which is named SPPR [20].

In AIMD simulations, choice of electronic structure theory is a key issue for reliability of dynamics. For electronically excited states, multi-configurational character and dynamic correlation effects are important factors. In ab initio wavefunction theory, we should employ multi-configurational-based theories since the target is not only the Franck–Condon region but also non-adiabatic regions (including conical intersection (CI) regions [21–26]) where more than two electronic states could approach each other. The mostly used multi-configurational wavefunction for multiple states is the state-averaged complete active space self-consistent field (SA-CASSCF) method. SA-CASSCF with the appropriate active space can describe the ground and excited electronic states in balance with qualitatively accuracy, but quantitative evaluation of the energetics is difficult because of no dynamic correlation effects included. To include dynamic correlation effects, one can move to multi-reference perturbation theory [27, 28] or multi-reference configuration interaction theory. The multi-reference methodology can reproduce the excited-state PESs with high accuracy, but it requires high computational costs. The alternative approaches, DFT and time-dependent DFT (TDDFT), are very attractive methods. The dynamic correlation effects are included as correlation functional, and thus, they can provide potential energy surfaces including electron correlation effects, with a relatively low cost. However, the usual linear response TDDFT calculations have a disadvantage for study of excited-state reactions [29]; TDDFT provides a discontinuous PES at the crossing point between the reference ground singlet state ($S_0$) and the first excited singlet state ($S_1$), as the reference state becomes the excited state after passing an intersystem crossing point. Furthermore, TDDFT cannot describe doubly excited electronic states. This limitation is serious for describing a process of rotations about double bonds, because the ground state and the doubly excited state cross each other. To overcome this limitation, the spin-flip (SF)-TDDFT method has been developed so that an open-shell triplet configuration is employed as the reference state and the ground and excited states are described as the response state [30–32]. In the SF-TDDFT method, five types of electronic states are obtained as solutions [33], i.e., the ground state, the open-shell singlet and triplet states of HOMO-LUMO single excitation, the singlet excited state of HOMO-LUMO double excitation, and spin-mixed states of singlet, triplet and quintet. In the recent AIMD simulation with SF-TDDFT [3], the target state was followed by monitoring orbital coefficients and configuration interaction coefficients along the trajectory, but such approach sometimes did not work in the region where the singlet and triplet states approach each other [3]. Very recently, a new approach, $T_{SF}$-index method [34], was proposed to trace the target state in $S_1/S_0$ minimum energy conical intersection (MECI) search calculations with SF-TDDFT.

In this chapter, we report applications of AIMD methodology to photoisomerization of azobenzene by SA-CASSCF method [1] and to photoisomerization of stilbene by SF-TDDFT methods [3, 4]. The theoretical results are compared with available experimental data, and the reaction pathways and dynamics for the respective photoisomerizations are discussed.

## 18.2 Photoisomerization of Azobenzene in $n\pi^*$ Excitation

### *18.2.1 Past Experimental and Theoretical Studies*

The photoisomerization of azobenzene has attracted a great deal of interests in both science and industry since this feature leads to its utilization for light-driven nano-scaled devices such as optical switches, and a number of experimental studies have been reported [35–43]. The earlier study on the absorption spectra indicated that the first and second excited states of both *cis*- and *trans*-azobenzene were attributed to $n\pi^*$ and $\pi\pi^*$ excitations, respectively [35, 37]. Quantum yields of the isomerizations for *cis*- and *trans*-azobenzene were reported to be different between $n\pi^*$ and $\pi\pi^*$ excitations [35–37]. Rau proposed that the photoisomerization of *trans*-azobenzene should follow different pathways in $n\pi^*$ and $\pi\pi^*$ excitations, through a comparison of the simple azobenzene and the sterically hindered azobenzenes [38, 39]; $n\pi^*$ excitation invokes the inversion of the NNC bond angle within a planar geometry, while $\pi\pi^*$ excitation invokes the rotation of two phenyl rings around the NN bond (Fig. 18.1). There have been a number of transient absorption experimental studies for $\pi\pi^*$ excitation of *trans*- and *cis*-azobenzene [41–43]. Tahara and co-workers [40–42] carried out the time-resolved Raman, fluorescence and absorption measurements with $\pi\pi^*$ excitation of *trans*-azobenzene in solution. They clarified that isomerization proceeds in $n\pi^*$ state after electronic relaxation, and proposed that the isomierization pathway is the same as that of direct $n\pi^*$ excitation. They also found that the NN stretching frequency shows only a slight decrease in the $S_1(n\pi^*)$ state [40].

There have been also a number of theoretical studies on the photoisomerization of azobenzene [1, 44–58]. First, Monti et al. [44] showed that in the $S_1(n\pi^*)$ state there is a large activation barrier along the rotation pathway while there is no barrier along the inversion pathway at the singly excitation configuration interaction (CIS) level, indicating that the photoisomerization of *trans*- and *cis*-azobenzene is likely to proceed via the inversion pathway in $n\pi^*$ excitation. In 1999, CASSCF method was employed to determine the rotation and inversion pathways between *trans*- and *cis*-azobenzene in the respective $S_1(n\pi^*)$ and $S_2(\pi\pi^*)$ states, followed by the multi-reference perturbation theory energy calculations along the CASSCF reaction pathways; it was concluded that the inversion is probably the preferred pathway for $n\pi^*$ excitation while the rotation is the preferred one for $\pi\pi^*$ excitation [45]. Ishikawa et al. [46] examined the two-dimensional PES of the $S_1(n\pi^*)$ state as a function of the CNNC torsion and NNC bond angles at the multi-reference single and double excitation configuration interaction (MR-SDCI) level, and proposed that the rotation pathway is preferred in the $S_1(n\pi^*)$ state based on the location of the $S_1$/$S_0$ conical intersection along the rotation pathway. After the Ishikawa's report, all

**Fig. 18.1** Scheme of rotation and inversion mechanism for *cis-trans* isomerization of azobenzene

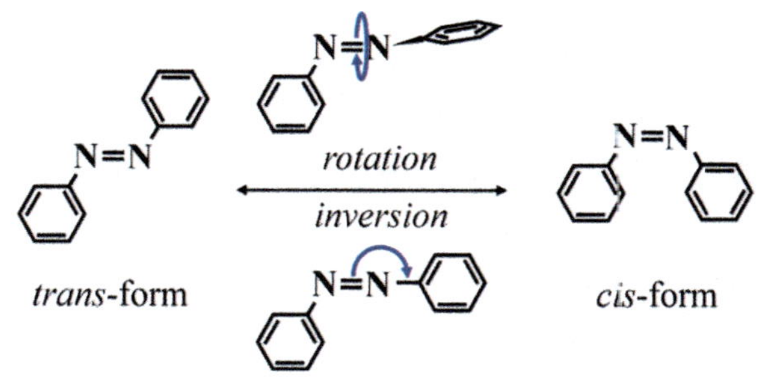

theoretical studies indicate that the rotation is the preferred pathway for $n\pi^*$ excitation in the gas phase [48, 53].

On-the-fly dynamics simulations were also performed for the photoisomerization of azobenzene on the basis of semi-empirical molecular orbital calculations with the surface hopping method [49, 52, 56, 58] and with the multiple spawning method [51]. Recently AIMD simulations at the CASSCF level [47, 48, 55, 57] and Car–Parrinello molecular dynamics simulations [54] were also performed for the photoisomerization of azobenzene in $n\pi^*$ excitation. In next section, we introduced a surface hopping AIMD simulation at SA-CASSCF level [1].

### *18.2.2 AIMD Study on Photoisomerization of Azobenzene*

Hence, we summarized a work of AIMD simulations for cis-trans photoisomerization of azobenzene in the lowest excited state $S_1(n\pi^*)$ reported by our group in 2009 [1]. In the AIMD simulations, the SA-CASSCF/STO-3G level was employed where $S_0$, $S_1(n\pi^*)$, $S_2(n\pi^*)$, $S_3(n^2\pi^{*2})$, and $S_4(\pi\pi^*)$ states were included with equal weight, and the active space was defined as six electrons in four orbitals (n, n, $\pi$, $\pi^*$). The RATTLE algorithm was applied to solve Newton's equations of motion with geometrical constraint that all CH bond lengths in two phenyl rings were fixed to the equilibrium distances in the ground state. All electronic structure calculations were performed by MOLPRO [59]. In total, 200 trajectories have been run starting from cis-azobenzene over 240 fs, while 100 trajectories have been run starting from trans-azobenzene over 3 ps. The initial atomic coordinates and velocities were determined by the canonical sampling ($T$ = 300 K) based on normal modes of vibration in the ground state. Among 200 trajectories for cis-azobenzene, the isomerization to trans-azobenzene occurs in 90 trajectories while the remaining 110 trajectories go back to the region of cis-azobenzene in the ground state, and thus, the quantum yield for cis to trans photoisomerization is estimated to be 0.45. For reactive trajectories, the reaction pathway can be classified into three patterns: (i) clockwise rotation, (ii) counterclockwise rotation, and (iii) counterclockwise ($S_1$) +clockwise ($S_0$) rotation. In all cases, only the central N–N part shows a drastic motion within 40 fs, while two phenyl rings do not move so much from the initial position because of a relatively heavy mass. The rotational direction of the N–N part is clockwise in the first pathway while it is counterclockwise in the second and third pathways. After the CNNC dihedral angle $d_{CNNC}$ reaches ~ 90° or -90°, surface hopping to $S_0$ occurs and both phenyl rings start to move toward a structure of trans-azobenzene. In the first and second pathways two phenyl rings keep the direction of motion after hopping to $S_0$, while in the third pathway they turn its direction in the ground state. Atomic motions along the AIMD trajectory do not correspond to the conventional rotation pathway where two phenyl rings rotate around the central NN bond. In the simple picture of rotation mechanism, a structural transformation accompanies a movement of two heavy phenyl rings in a long distance (rotation with a large moment of inertia), while in the pathway presented here, the rotation proceeds in two steps, i.e., (1) rotation of the central N–N part and (2) rotation of two phenyl rings with much shorter distance. Because two phenyl rings are deviated from the central C–N–N–C plane in cis-azobenzene, the initial rotation of NN part makes molecular structure close to the $S_1/S_0$ MECI point without a movement of phenyl rings. Then two phenyl rings rotate about the C–N bond, resulting in the structure of cis or trans form of azobenzene in the $S_0$ state. The similar reaction mechanism was also reported in dynamics simulations by Doltsinis et al. [54] and Thiel et al. [56].

In a typical trajectory starting from the $S_1(n\pi^*)$ state, surface hopping to the $S_0$ state occurs at around 40 fs. Before hopping to $S_0$, one NNC bond angle increases to 140° while the other decreases to 100°, and then they show fluctuations around the respective positions, indicating that this is not the inversion pathway. On the other hand, the CNNC dihedral angle immediately increases or decreases, and reaches to 90° or -90°, where surface hopping occurs. This feature indicates that the reaction proceeds along the rotation pathway, but as discussed above, this is not a conventional rotation pathway. It is interesting to note that both NNCC dihedral angles show twist motions in the direction opposite to change in the CNNC dihedral angle. The numbers of trajectories assigned to the respective patterns are 13, 64, and 13 for the clockwise rotation, counterclockwise rotation, and counterclockwise ($S_1$) +clockwise ($S_0$) rotation, respectively, and thus counterclockwise rotation is the most preferred pathway. We also performed the steepest descent path calculations from the Franck–Condon cis structure in the $S_1$ state, and found that it is the counterclockwise rotation pathway.

Here we discuss AIMD trajectories starting from *trans*-azobenzene. The reaction rate for trans-azobenzene is much slower than that for cis-azobenzene because it takes much longer time for trans-azobenzene to deviate from the planar structure. Thus, trajectories from trans-azobenzene were calculated over 3 ps. As the results, isomerization to cis-azobenzene occurs in 24 trajectories, while in 62 trajectories, the molecule goes back to the region of trans-azobenzene in the ground state. The remaining 14 trajectories continue to stay in the $S_1$ state during the simulation time of 3 ps. Thus, the quantum yield for trans to cis photoisomerization is estimated as 0.28. Compared to the reaction pathway in cis to trans isomerization, atomic motions along the trajectory show a conventional rotation pathway where two phenyl rings rotate around the NN bond. For the reactive trajectory, surface hopping to $S_0$ occurs at $t$ = 800–900 fs, and then CNNC dihedral angle starts to change drastically and the molecule transforms from trans to cis structure very quickly within 50 fs. For the unreactive trajectory, both NNCC dihedral angles show violent fluctuations but they do not correspond to the rotational motion of phenyl rings. Both phenyl rings keep their positions, while the central N atoms show a bicycle-pedal motion.

By fitting variations of $S_1$ population averaged over trajectories to exponential function, the lifetimes of $S_1$ for *cis*- and *trans*-azobenzenes are estimated as 0.06 and 1.86 ps, respectively. These values are in qualitative agreement with the experimental lifetimes, 0.17 and 2.6 ps, respectively. In this simulation, the quantum yield for cis to trans photoisomerization was calculated to be 0.45, which is almost twice as large as that for trans to cis photoisomerization, 0.28. These results are in good agreement with the experimental results.

### 18.2.3 Vibrational Spectroscopy of Trans-Azobenzene in $n\pi^*$ State

The excited-state PES exploration and AIMD simulations indicated that the $n\pi^*$-excited azobenzene prefers the rotation pathway. On the other hand, the transient Raman spectra showed that the NN stretching frequency of *trans*-azobenzene is almost unchanged in the $S_1(n\pi^*)$ state, indicating that the NN bond order does not change during the isomerization process [40]. In order to gain insight to the photoisomerization mechanism of *trans*-azobenzene, the NN stretching frequency in the $S_1$ state was investigated by theoretical calculations with several methodology including DFT, SA-CASSCF, and CASPT2. In SA-CASSCF calculations, the active space was 14 electrons in 12 orbitals involving two n orbitals and $\pi/\pi^*$ orbitals of nitrogens, and four sets of $\pi/\pi^*$ orbitals of phenyl rings. In CASPT2 calculations, the reference SA-CASSCF wavefunction was determined so that the active space includes two n orbitals and $\pi/\pi^*$ orbitals of nitrogens, and the lowest 17 orbitals are treated as frozen orbitals in the perturbation calculations. The rotation pathway was also calculated as a function of a CNNC torsion angle $d_{CNNC}$ and the inversion pathway as a function of NNC bond angle $a_{NNC}$ on the $S_1(n\pi^*)$ PES, by optimizing geometrical parameters other than the respective reaction coordinates. Then, the NN stretching frequency was calculated along the respective pathways by applying a projection technique [60] at the CASPT2 level. Sapporo-DZP basis sets [61] were employed in most calculations, and only anharmonic vibrational calculations with vibrational self-consistent field (VSCF) and correlation-corrected VSCF (cc-VSCF) methods were carried out with 6-31G*. DFT calculations were performed by GAMESS [62], while CASPT2 calculations were performed by MOLPRO program [59].

For the ground state of trans-azobenzene with $C_{2h}$ symmetry, the calculated geometrical parameters were in good agreements with the experimental values, while the NN stretching harmonic frequency was overestimated by *ca.* 80 $cm^{-1}$ relative to the experimental fundamental frequency at the CASPT2 level. The cc-VSCF calculations at the B3LYP/6-31G* level showed that the NN stretching harmonic frequency, 1561 $cm^{-1}$, was reduced to 1531 $cm^{-1}$ due to the anharmonicity effect. By using this value as the anharmonicity effect, the NN stretching fundamental frequency in the ground state of trans-azobenzene was estimated to be 1490 $cm^{-1}$ at the CASPT2 level, which is comparable to the experimental value, 1440 $cm^{-1}$.

For the $S_1(n\pi^*)$ state, geometry optimization was performed under the $C_{2h}$ symmetry constraint (referred to as $(S_{1\text{-}C2h})_{min}$). Normal mode analysis indicated that $(S_{1\text{-}C2h})_{min}$ has one imaginary frequency mode which corresponds to a torsional motion of the central -CNNC- ($A_u$) at DFT(CAM-B3LYP) (20*i* cm$^{-1}$) and CASPT2 (49*i* cm$^{-1}$) levels, although it has no imaginary frequency mode at the DFT (B3LYP) and SA-CASSCF levels. Thus, the planar structure of $(S_{1\text{-}C2h})_{min}$ is unstable with respect to the CNNC torsional motion. The comparison of geometrical parameters between $S_0$ and $S_1(n\pi^*)$ indicates that, by nπ* excitation, the NN bond length is almost unchanged while the NNC bond angle increases by *ca.* 14° and the CN bond length decreases by *ca.* 0.06 Å. These geometrical changes can be explained by the nature of molecular orbitals related to nπ* excitation. Figure 18.2 shows natural orbitals relevant to nπ* excitation, i.e., the lone pair n orbital of N atoms and the NN-π* anti-bonding orbital, at the $(S_0)_{min}$ structure determined by the SA-CASSCF(6,4) calculations. As shown in Fig. 18.2, the n orbital shows anti-bonding nature for the NN bond, while the π* orbital shows anti-bonding nature for the NN bond and bonding nature for two CN bonds. Thus, due to nπ* excitation, CN bonds should be strengthened while NN bond strength is not so affected. The extension in the NNC bond angle can be explained by the reduction of a repulsive force between bonding electron pairs (NN and CN bonds) and the lone pair of N atoms. The similar finding in geometrical changes of *trans*-azobenzene due to nπ* excitation was also reported at the semi-empirical OM2/MRCI level [58].

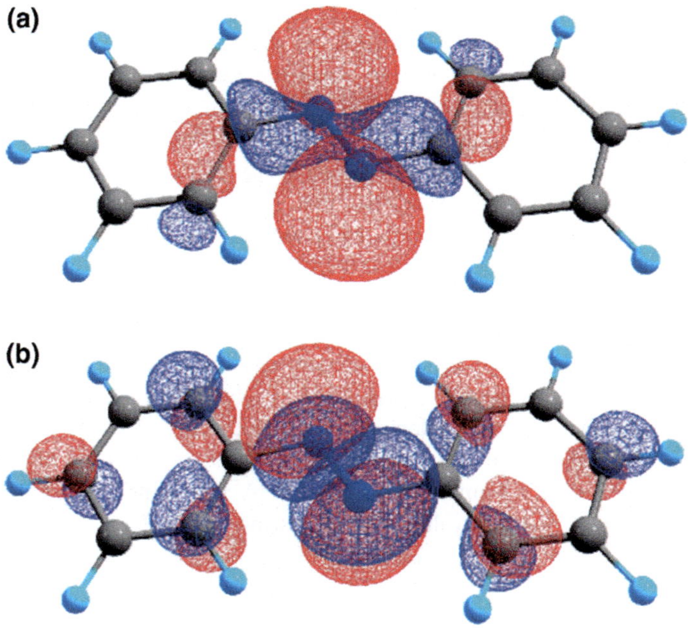


**Fig. 18.2** Natural orbitals corresponding to **a** lone pair orbital of nitrogen and **b** NN anti-bonding π* orbital of *trans*-azobenzene related to nπ* excitation determined by the SA-CASSCF(6,4) calculation at $(S_0)_{min}$. Reproduced from Ref. [2], with permission from AIP Publishing

Due to nπ* excitation, the experimental NN stretching frequency decreases by 12 $cm^{-1}$, while the calculated frequency increases by 116 $cm^{-1}$ (TDDFT(B3LYP)), 90 $cm^{-1}$ (TDDFT(CAM-B3LYP)), 76 $cm^{-1}$ (CASSCF), and 61 $cm^{-1}$ (CASPT2). By the cc-VSCF calculation at the TDDFT(B3LYP)/6-31G* level, the anharmonicity effect was estimated to be 29 $cm^{-1}$ (1679 → 1650 $cm^{-1}$) in the $S_1$ state, which is similar to the anharmonic effect in the ground state, and thus change of NN stretching fundamental frequency upon nπ* excitation should be similar to change in the harmonic frequency.

Before the appearance of ref [2], there was reported only one theoretical work on vibrational frequencies of *trans*-azobenzene in the $S_1$(nπ*) state [50], which employed state-specific CASSCF calculations to determine minimum energy structures and harmonic frequencies for the ground and $S_1$(nπ$^*$) states under the $C_{2h}$ symmetry restriction, with analytical gradient and analytical Hessian matrix, using the MOLCAS-5.2 program [63]. The calculated frequencies, multiplied by scaling factor 0.91 to take into account the effect of dynamic correlation and anharmonicity, were reported as 1440 $cm^{-1}$ ($S_0$) and 1438 $cm^{-1}$ ($S_1$), which were in good agreement with the experimental values. However, our SA-CASSCF frequencies were totally different from their results, and thus, the same calculations as ref [50] were performed by the state-specific CASSCF(14,12) and the atomic natural orbital (ANO) basis sets, using the MOLCAS-6.4 program [63]. Then, the NN stretching frequencies were calculated to be 1701 $cm^{-1}$ ($S_0$) and 1771 $cm^{-1}$ ($S_1$), which are reduced to 1548 $cm^{-1}$ and 1612 $cm^{-1}$, respectively, by scaling factor of 0.91, and this result is consistent with the CASSCF calculations with Sapporo-DZP basis sets [2].

The increase in the NN stretching frequency due to nπ* excitation can be explained by vibrational mixing concept [64]. The nπ* excitation strengthens CN bonds, leading to an increase of force constants in CN symmetric and anti-symmetric stretching modes. Then, the interaction between the CN symmetric stretching mode and the NN stretching mode of total symmetry representation is invoked, resulting in mixing of these two normal modes. Figure 18.3 shows mechanism of vibrational mixing of the NN stretching mode and the CN symmetric stretching mode due to nπ* excitation. As shown here, the NN stretching mode in the ground state shows almost a pure NN stretching motion while in the $S_1$(nπ*) state the corresponding mode contains a small contribution from a CN symmetric stretching motion; similarly, the original CN symmetric stretching mode contains a small contribution from NN stretching motion with anti-phase. As the result of this mixing, the frequency of the NN stretching mode increases from 1520 to 1581 $cm^{-1}$, while the frequency of the CN symmetric stretching mode decreases from 1235 to 1207 $cm^{-1}$.

At the CASPT2 level $C_{2h}$ minimum of *trans*-azobenzene in the $S_1$(nπ*) state has one imaginary frequency mode of $A_u$ which is directly related to a reaction coordinate of the rotation pathway, i.e., a torsion angle $d_{CNNC}$. Then, starting from $(S_{1\text{-}C2h})_{min}$, the rotation pathway was calculated by choosing $d_{CNNC}$ as a reaction coordinate and optimizing the other geometrical parameters by the CASPT2 method. Along the rotation pathway, vibrational frequencies were calculated with a

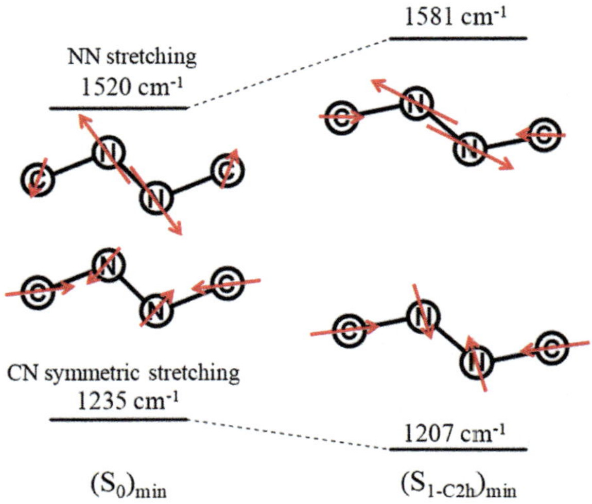


**Fig. 18.3** Scheme of the vibrational mixing between the NN stretching mode and the CN symmetric stretching mode due to nπ* excitation. Reproduced from Ref. [2], with permission from AIP Publishing

projection technique [64] at the CASPT2 level. The $S_1$ energy shows a gradual decrease until reaching $(S_1/S_0)_{CI}$ around $d_{CNNC} \sim 95°$. It is interesting to note that both $r_{NN}$ and $r_{CN}$ are almost unchanged along the rotation pathway. This result indicates that the NN bond order does not change while proceeding along the rotation pathway. The NN stretching frequency shows only a small decrease ($\sim$ 50 $cm^{-1}$) as $d_{CNNC}$ changes from 180° to 100°; this small decrease is consistent with an almost constant bond length of the central NN bond.

According to chemical intuition, the rotational motion about the central NN bond should weaken the NN bond because π-bonding orbital possibly breaks down. The calculated NN bond lengths and NN stretching frequencies, however, indicate that the NN bond is not weakened along the rotation pathway. Figure 18.4 shows natural orbitals in the active space of SA-CASSCF(6,4) wavefunction. The occupation numbers in these orbitals are also constant, $(n)^2(\pi)^2(n)^1(\pi^*)^1$. Each orbital almost keeps its shape (i.e., n, π, n, π*) along the rotation pathway against chemical intuition, and this should be the origin of the unchanged bond order of the central NN bond during the rotational motion.

The inversion pathway was also calculated as a function of a bond angle $a_{NNC1}$ where two carbon atoms of CNNC part are distinguished by $C_1$ and $C_2$. Along the inversion pathway, the $S_1$ energy increases by *ca.* 0.8 eV at $a_{NNC} \sim 180°$, the NN bond length decreases by 0.03 Å, and the NN stretching frequency increases by *ca.* 300 $cm^{-1}$. The increase in the adiabatic energy clearly indicates that *trans*-azobenzene prefers the rotation pathway in nπ* excitation. This finding can explain why the isomerization occurs through a rotation of the -N = N- fragment rather than a true torsion along the NN bond [1, 56].

Finally, we summarize significant results from the present calculations based on CASPT2 geometrical structures, vibrational frequencies, and reaction pathways. The energy variations along the reaction pathways indicate that the rotation pathway is preferred to the inversion pathway. Geometrical feature of $(S_{1\text{-}C2h})_{min}$ is

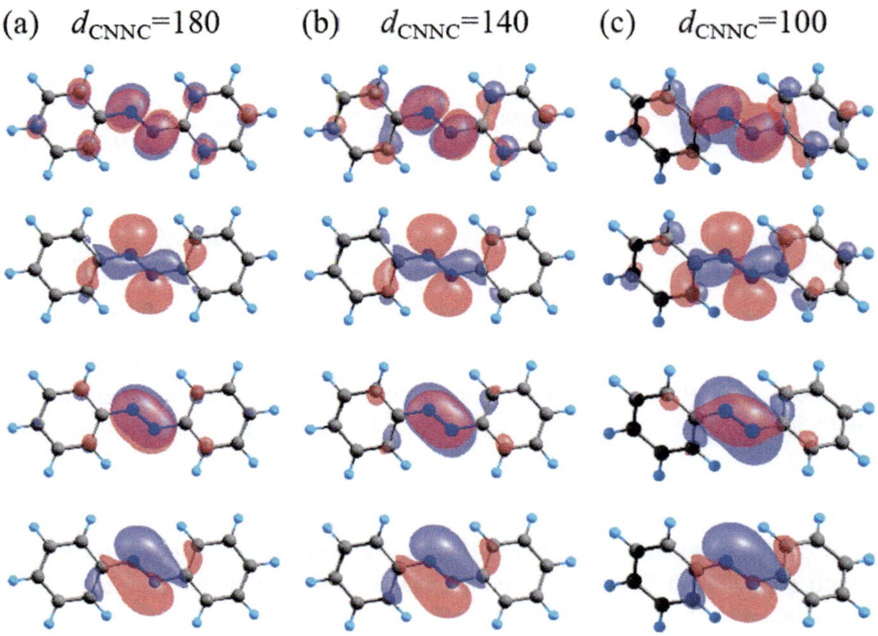


**Fig. 18.4** Set of natural orbitals in the active space of SA-CASSCF(6,4) at **a** $d_{CNNC}$ = 180° ($(S_{1\text{-}C2h})_{min}$), **b** 140°, and **c** 100°, along the rotation pathway [4]

shown to be minimum at the CASSCF level, while it changes to a first-order saddle point at the CASPT2 level, indicating that the planar structure of *trans*-azobenzene is unstable with respect to a torsional motion of the central CNNC dihedral angle. A shift of the NN stretching frequency due to nπ* excitation is relatively small, agreeing to the Raman spectra. A small shift in the NN stretching frequency is explained in terms of natural orbitals related to nπ* excitation and vibrational mixing between the NN stretching mode and the CN symmetric stretching mode. It was also shown that the NN stretching frequency shows a small decrease along the rotation pathway, and thus, the experimentally observed frequency could be assigned to the molecule under a structural transformation along the rotation pathway.

## 18.3 Photoisomerization of Cis-Stilbene in ππ* Excitation

### 18.3.1 Past Experimental and Theoretical Studies

Stilbene (SB) is a basic molecule that shows C = C photoisomerization between cis and trans forms. The photoisomerization mechanism of *cis*-SB has been widely studied both experimentally [65–80] and theoretically [76, 80–89]. Upon ππ*

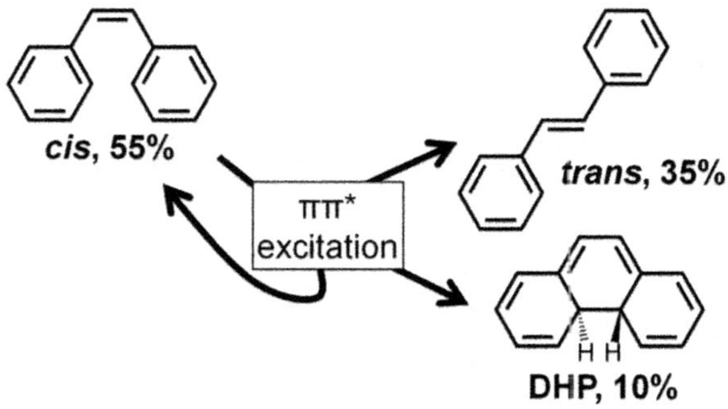


**Fig. 18.5** Scheme of photoisomerization of *cis*-stilbene

excitation, *cis*-SB can transform to trans-SB or 4,4-dihydrophenanthrene (DHP) with a time scale of 1.0 ps [66–70, 74, 77–79], and the branching ratio was reported as *cis* : *trans* : DHP = 55 : 35 : 10 as shown in Fig. 18.5 [66–68]. Steady state [75] and femtosecond time-resolved [79] fluorescence study revealed that there are two decay components with the lifetimes of 0.23 and 1.2 ps. The initial dynamics of $\pi\pi^*$-excited *cis*-SB was investigated by femtosecond Raman spectroscopy [76]. It showed a temporal change of the vibrational spectrum, which was assigned to the twisting motion of the central C = C bond through quantum chemical calculations [76].

Theoretical studies elucidated that $\pi\pi^*$-excited *cis*-SB evolves on the $S_1(\pi\pi^*)$ potential energy surface (PES) with the twisting motion of the central C = C bond [73, 81, 83, 87, 89]. Also, an accessible MECI point of $S_0$ and $S_1$ states (denoted as $S_1/S_0$-MECI) was located near the minimum on the $S_1$-PES, which corresponds to the C = C bond twisting structure (denoted as *twist*) [81, 83, 87, 89]. The molecular motion of the *cis-trans* photoisomerization has been predicted to be a 'hula-twist' motion, in which the central C = C bond rotates out of plane; the CH bonds remain out of the plane, while the other atoms reorient to remain coplanar [73]. Minezawa and Gordon investigated the reaction pathways in the relaxation process of $\pi\pi^*$-excited *cis*-SB by SF-TDDFT [30–32]. They located geometries of minima and $S_1/S_0$-MECIs for twist ($(S_1)_{twist\text{-min}}$ and $(S_1/S_0)_{twist}$) and DHP ($(S_1)_{\text{DHP-min}}$ and $(S_1/S_0)_{\text{DHP-MECI}}$) regions on the $S_1$-PES, and found that the photocyclization is in competition with the photoisomerization [89], which was later confirmed by the more sophisticated multi-reference method [90].

With respect to dynamics study for *cis*-SB, Berweger employed CIS/6-31G to generate a three-dimensional constrained PES, and performed dynamics simulations [82]. Dou et al. [85, 86, 88] performed semi-classical electron-radiation-ion dynamics simulations on the relaxation process of *cis*-SB in the $S_1$ state, and studied three dominant processes, i.e., *cis-trans* isomerization [86], *cis-cis* (no isomerization) [85], and *cis*-DHP photocyclization [88]. However, there has been no dynamics study which focused on the branching mechanism.

In 2012, Berndt et al. measured transient absorption spectra of 1,1′-dimethyl-stilbene (dmSB) in solution [91]. The lifetime of a transient band observed around 600 nm (CIS* band) was 0.2 ps, which is shorter than the corresponding CIS* band of *cis*-stilbene (0.92 ps). Another transient band, associated with the perpendicular conformation (denoted as P*), was also observed around 330 nm (P* band). It was found that the lifetime of the P* band of cis-dmSB is 19 ps in hexane and 2.9 ps in acetonitrile, which are much longer than the lifetime of the corresponding P* band of cis-stilbene [77]. They suggested that the longer lifetime of dmSB is attributable to the structural difference between the minimum and the conical intersection around P*. However, there has been no report which studied the relaxation process of dmSB in the $S_1$ state. Both the mechanism of changes in the decay time and the P* structure have never been revealed.

In the following sections, the relaxation mechanism and dynamics of $\pi\pi^*$-excited *cis*-SB and *cis*-dmSB were discussed by considering cis-trans photoisomerization and photocyclization to DHP and dmDHP, respectively. The substituent effect for the relaxation mechanism is discussed by comparing the excited-state PES and AIMD results for SB and dmSB. In addition, the origin of CIS* and P* bands in time-resolved absorption experiments is investigated.

### *18.3.2 Potential Energy Surface of the $\pi\pi^*$ State of SB and DmSB*

For SB [3] and dmSB [4], equilibrium and TS geometries in the ground and $\pi\pi^*$ states, as well as $S_1/S_0$-MECI geometries, were optimized at SF-TDDFT (BHHLYP)/6-31G(d) level. The intrinsic reaction coordinate (IRC) pathways and the steepest descent path (meta-IRC) from the Franck–Condon (FC) structure of *cis*-form in the $\pi\pi^*$ state were determined at the same level. All DFT calculations were performed with GAMESS [62].

Figure 18.6 shows geometries of cis-form minima in the ground and $\pi\pi^*$ states and $S_1/S_0$-MECI for SB and dmSB. Definitions of atom numbering are indicated for $(S_0)_{cis\text{-min}}$. Three minima in the $\pi\pi^*$ state, $(S_1)_{cis\text{-min}}$, $(S_1)_{\text{DHP-min}}$, and $(S_1)_{twist\text{-min}}$, were located for (a) SB and (b) dmSB. In both cases, there is one MECI in a DHP region ($(S_1/S_0)_{\text{DHP}}$), while there are two in a twist region $(S_1/S_0)_{twist\text{-}1}$ and $(S_1/S_0)_{twist\text{-}2}$. Geometries of $(S_1)_{twist\text{-min}}$, $(S_1/S_0)_{twist\text{-}1}$, and $(S_1/S_0)_{twist\text{-}2}$ are all characterized by a *twisted pyramidal* structure of the central CC bond. However, in the case of dmSB, the geometry of $(S_1)_{twist\text{-min}}$ is significantly different from those of $(S_1/S_0)_{twist\text{-}1}$ and $(S_1/S_0)_{twist\text{-}2}$. The dihedral angle $d_{\text{C3C1C2C7}}$ that characterizes a pyramidalization is 153.3° for $(S_1)_{twist\text{-min}}$, while $d_{\text{C3C1C2C7}}$ is 114.0° and 116.3° for $(S_1/S_0)_{twist\text{-}1}$ and $(S_1/S_0)_{twist\text{-}2}$, respectively, indicating that $(S_1)_{twist\text{-min}}$ shows a structural change between pyramidal and planar. This can be understood as the result of the steric hindrance between phenyl group and methyl group, and such a difference between minimum and MECI geometries was not reported for other molecules, e.g., ethylene, stylene, stilbene, and stiff-stilbene [29, 89, 90, 92]. It is

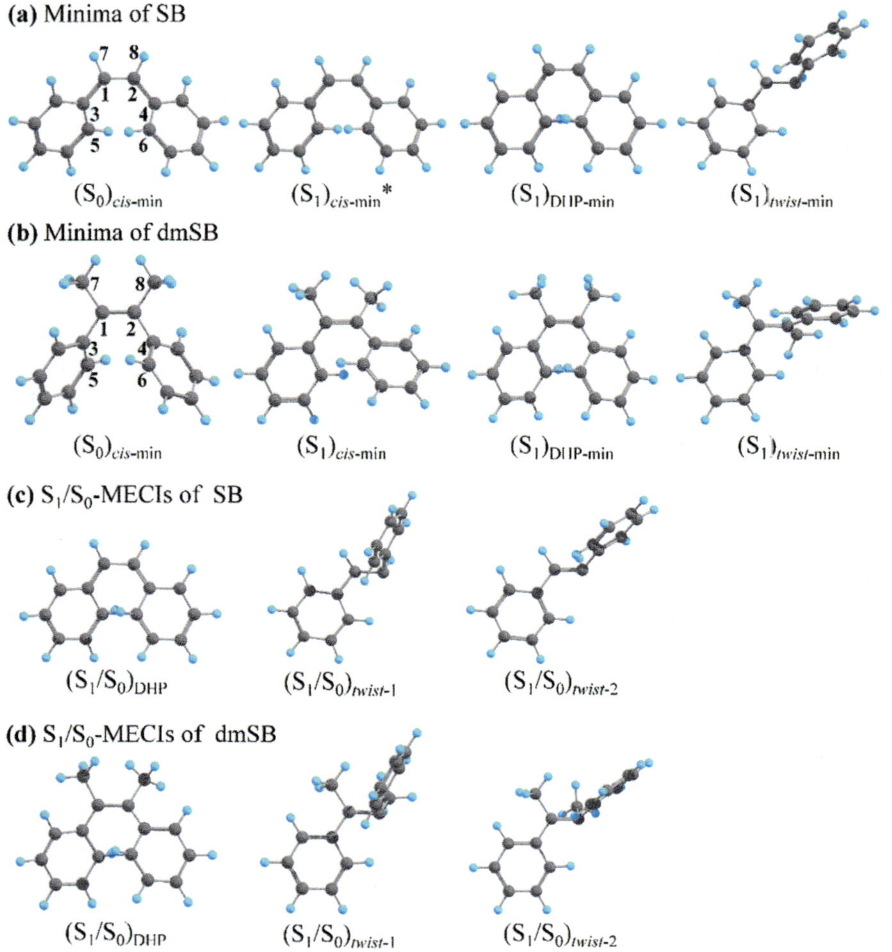


**Fig. 18.6** Geometries of minima in the $S_0$ and $S_1$ states for **a** SB and **b** dmSB; $S_1/S_0$-MECI geometries for **c** SB and **d** dmSB. Definitions of atom numbering are indicated for $S_0$-minimum of SB and dmSB

expected that the lifetime of the $S_1$ molecule in the twist region of dmSB should be longer than that of SB.

Figure 18.7 shows geometries and traces of reaction pathways on a two-dimensional configurational space in terms of $r_{C5C6}$ and $d_{C3C1C2C4}$ for (a) SB and (b) dmSB [4]. In both SB and dmSB, the meta-IRC pathways from $(S_1)_{FC}$ finally lead to a DHP-like geometry, $(S_1)_{DHP\text{-}min}$. In the case of SB, the meta-IRC from $(S_1)_{FC}$ passes through a *cis*-stilbene-like structure ($(S_1)_{cis\text{-}min}$) in the first stage, and the steepest descent path goes through a very flat region with a large reaction-path curvature. This result is interesting because previous studies predict that the reaction pathway leading from the FC region of *cis*-stilbene to the twisted

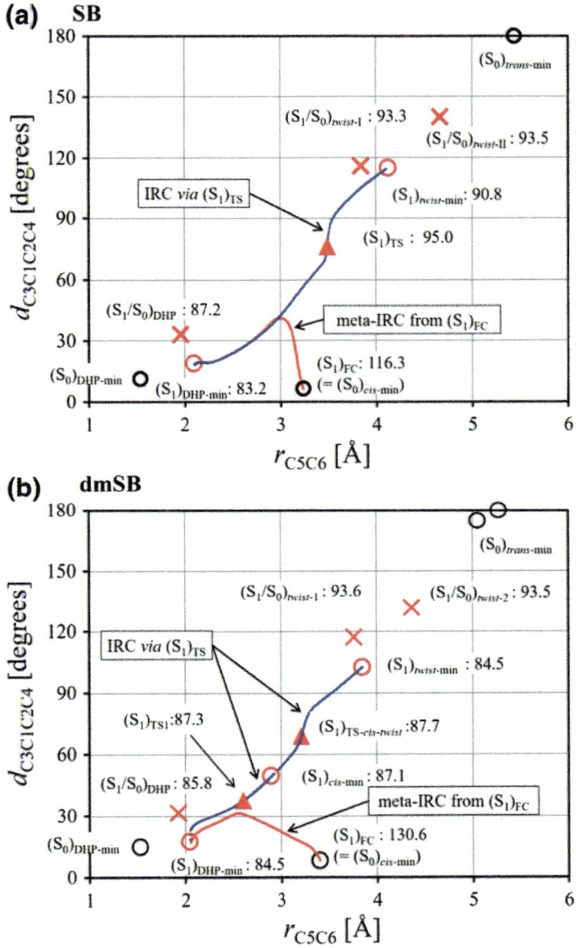


**Fig. 18.7** Plots of reaction pathways on the $S_1$-PES, projected on a two-dimensional configurational space in terms of $r_{C5C6}$ and $d_{C3C1C2C4}$, for **a** SB and **b** dmSB [4]. Minima in the ground state are denoted by a *black circle*, while minima, TS, and MECI in the $S_1(\pi\pi^*)$ state are denoted by a *red circle*, a *red triangle*, and a *red cross* mark, respectively. Meta-IRC from the FC point is denoted by a red solid line, while the IRC is denoted by a *blue dotted line*

structures is preferred [66–68]. It is suggested that branching for the twisted structure and the DHP structure occurs in this sharply curved region, and that dynamic effects may be important in determining the branching ratio of the products.

In the case of dmSB, the meta-IRC paths from the $(S_1)_{FC}$ of *cis*-dmSB approache $(S_1)_{TS1}$ between $(S_1)_{DHP\text{-}min}$ and $(S_1)_{cis\text{-}min}$, and enters a side of $(S_1)_{DHP\text{-}min}$. On the other hand, in SB case, there is no $(S_1)_{cis\text{-}min}$ and no $(S_1)_{TS1}$ (a kind of shoulder region with a very flat nature); the meta-IRC is initially oriented toward a twist direction, and turns its direction toward $(S_1)_{DHP\text{-}min}$. The difference in geometrical feature of the pathways suggests that, in *cis*-dmSB case, photocyclization is enhanced compared with *cis*-SB. Since $(S_1/S_0)_{DHP}$ point is located near $(S_1)_{DHP\text{-}min}$, the molecule entering the $(S_1)_{DHP\text{-}min}$ side would reach the $S_1/S_0$-CI region easily, resulting in a non-radiative transition to the ground state. Since the energy barriers for pathways from *cis*-dmSB to dmDHP or *twist*-dmSB are very low, the molecule near $(S_1)_{cis\text{-min}}$ region can easily enter DHP or *twist* regions.

The difference in the meta-IRC pathways for *cis*-dmSB and *cis*-SB was further investigated in the previous paper. In both dmSB and SB, $r_{C5C6}$ decreases gradually along the meta-IRC paths, and the dihedral angles, $d_{C3C1C2C4}$, first increase and then decrease almost simultaneously. The difference is the rate of increase in two dihedral angles (slow for dmSB and very fast for SB). It was concluded that a difference in mass of moving fragments, i.e., methyl group (dmSB) and hydrogen atoms (SB) caused the difference in meta-IRC [4].

### 18.3.3 Ab Initio Molecular Dynamics (AIMD) Simulations for SB and DmSB

For SB [3] and dmSB [4], the excited-state dynamics simulations were started upon excitation to the lowest $\pi\pi^*$ singlet state. Each trajectory in the excited state was terminated if the trajectory reached crossing regions between the ground and the first excited states, or if the simulation time reached the maximum (1.5 ps for stilbene and 1.0 ps for dmSB). As discussed in Sect. 18.3.2, there are two types of $S_1/S_0$-decay channel in the $\pi\pi^*$ state, i.e., DHP- and *twist*-sides. The branching dynamics toward DHP- and twist-sides in the $\pi\pi^*$ state of *cis*-SB and *cis*-dmSB were investigated based on the AIMD simulations with SF-TDDFT. 50 and 33 trajectories were calculated for SB and dmSB, respectively. It should be noted that the terminal points of the respective trajectories are not exactly the same as the $S_1/S_0$-MECI points; the $S_1/S_0$-crossing regions are distributed around the $S_1/S_0$-MECI points in configuration space.

In the case of SB, the 50 trajectories are divided into three groups: 13 trajectories (13/50) lead to the formation of DHP, 35 trajectories (35/50) undergo rotation via the torsion of the C1 = C2 bond, and two trajectories (2/50) first began torsional rotation, and then change to DHP. This means that the dynamically important product is different from the terminal geometry of the meta-IRC from $(S_1)_{FC}$, which indicates that this photobranching mechanism in the $\pi\pi^*$ state of SB is controlled by dynamics. Experimental quantum yields of the photoreaction cannot be discussed from the present dynamics simulations in a rigorous way, because the trajectories branch into three structures, i.e., DHP, *cis*-stilbene and *trans*-stilbene, after relaxing to the ground state. Interestingly, the calculated branching ratio (*trans*-stilbene : DHP = 35 : 13) is in good agreement with experimental data (*trans*-stilbene : DHP = 35 : 10), indicating that the dynamics simulations qualitatively reproduce the experimental quantum yield. In the case of dmSB, 33 trajectories are divided into four groups; trajectories entering the DHP-side and reaching the $S_1/S_0$-CI region (19/33), trajectories entering the DHP-side and moving to the twist-side (9/33), trajectories entering the twist-side (4/33), trajectories entering the twist-side and moving to the DHP-side (1/33). According to the rate of the destinations of trajectories (DHP : *twist* ~ 20 : 13), the photocyclization is favorable in the case of dmSB, which is opposite in the case of SB.

The difference in the branching ratio for SB and dmSB can be explained by the difference in reaction pathways [4]. In the case of SB, a dominant change in geometry along the meta-IRC path from $(S_1)_{FC}$ is an increase in torsional angle, $d_{C3C1C2C4}$, and thus, the torsional motion about the central CC bond receives a kinetic energy in the initial stage. Actually most trajectories run toward the twist-side initially, which results in a large branching ratio of twist-side for *cis*-SB. On the other hand, in the case of dmSB, most trajectories from $(S_1)_{FC}$ initially run toward the transition state between $(S_1)_{DHP\text{-}min}$ and $(S_1)_{cis\text{-}min}$. Thus, the trajectories which enter the DHP-side are dominant for *cis*-dmSB.

The calculated $S_1$-population decay of *cis*-SB and *cis*-dmSB in the $\pi\pi^*$ state is shown in Fig. 18.8. Here, the trajectories reaching $S_1/S_0$-crossing regions are regarded as those hopping to the ground state. The population decay for trajectories in the DHP-side is indicated by a red line, while the decay for those in the twist-side is indicated by a blue line. In the case of SB (Fig. 18.8a), 28/50 trajectories reached the $S_1/S_0$-crossing region before 1.0 ps. The population of the $\pi\pi^*$ state decreases to 0.44 at 1.0 ps. The time scale of the calculated population decay is in good agreement with the experimental decay of 1.2 ps [79]. The population decay for the trajectories leading to the *twist*-side (35 trajectories) and for the trajectories leading to DHP-side (13 trajectories) indicated that the lifetime for the DHP-side is relatively longer than that for the twist-side. An evidence from femotosecond time-resolved fluorescence spectra led to the proposal that the photoreaction process of *cis*-stilbene is a two-step mechanism [79]. The observed fluorescence wavelength, 420 nm [79], nearly coincides with the energy gap between the $S_0$ and $S_1$ states, 3.1 eV, for $(S_1)_{cis\text{-}min}$, while the energy difference between the $S_0$ and $S_1$ states in other regions along the IRC is too small to be observed. Thus, the two decay components in the experimental spectra (0.2 ps and 1.2 ps) were attributed to the molecular motion around $(S_1)_{cis\text{-}min}$ toward DHP-side and *twist*-side, respectively [3]. For the case of SB, most molecules go into the twist region due to the initial motion in the $S_1$ state, and it takes a relatively long time to enter the DHP side [3]. In the case of dmSB (Fig. 18.8b), the molecule at the DHP-side has a short lifetime while the molecule at the twist-side has a relatively long lifetime in the case of dmSB, which is just opposite to the case of dmSB. In the transient absorption spectra of *cis*-dmSB [91], the lifetime of the CIS* band at 600 nm and the P* band at 330 nm were reported to be 0.2 ps and 19 ps, respectively, in hexane. Based on the results of AIMD trajectories, the population of the CIS* state should increase initially and gradually decrease through a movement to the DHP side or the twist-side which occurred during the initial 0.2 ps. This time span is in very good agreement with the experimental lifetime of the CIS* band ($\tau = 0.2$ ps) [91]. On the other hand, most trajectories entering the twist-side stay around $(S_1)_{twist\text{-}min}$ for a long time (at least longer than 1 ps) without reaching the $S_1/S_0$-CI region. Thus, the molecule staying at the twist-side should have a long lifetime in the $S_1$ state, corresponding to the P* band observed in the experiment [91].

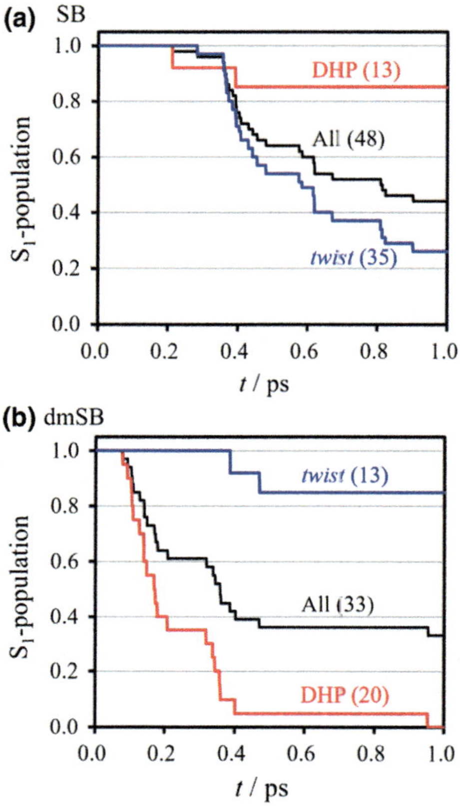


**Fig. 18.8** Decay of $S_1$-population as a function of time derived from AIMD simulations for **a** SB and **b** dmSB [4]. $S_1$-population for all the trajectories is *plotted in black*, while the decays for the trajectories staying at the DHP region and for those at the twist region are *plotted in red* and *in blue*, respectively

Next we discussed the experimental transient absorption spectra (Fig. 18.9b) based on AIMD simulations. In the reported time profile of the P* band [91], when normalized to the maximum amplitude of 1.0, the P* signal increases to ~ 0.8 in the first 0.2 ps, exhibits a shoulder feature at 0.2 ~ 0.3 ps, and increases again to 1.0 within 0.6 ps. Then, it gradually decreases, reflecting the lifetime of the P* state. It was also shown that the CIS* signal rapidly decays with a lifetime of 0.2 ps with a distinct shoulder feature observed at 0.2 ~ 0.3 ps. To understand these features seen in the time profiles of the CIS* and P* signals, the time variations of the $S_1$-population for the *cis*/DHP- and *twist*-sides in AIMD trajectories are analyzed. To distinguish the *cis*/DHP- and *twist*-sides based on geometrical parameters, the structures with $d_{C3C1C2C4} \leq 68.7°$ (value at $(S_1)_{TS2}$) are regarded as those located in the *cis*/DHP-side (red line in Fig. 18.9b), while the structures with $d_{C3C1C2C4} > 68.7°$ are regarded as those located in the *twist*-side (blue line in Fig. 18.9a). In AIMD simulations, all trajectories initially enter the cis region, corresponding to appearance of the transient absorption band in the 600-nm region (CIS* band). The $S_1$-population of the *cis*/DHP-side decreases with a lifetime of *ca.* 0.3 ps since some trajectories move out to the *twist*-side or reach the $S_1/S_0$-CI

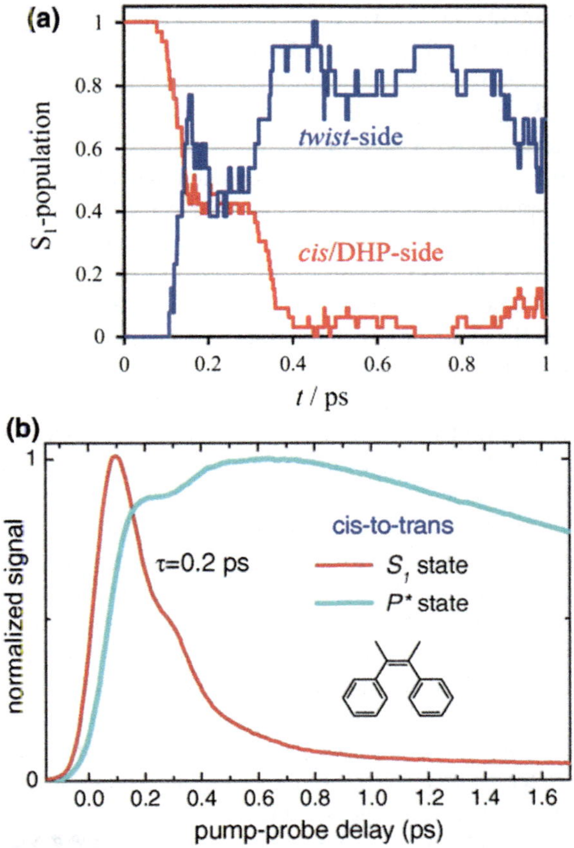


**Fig. 18.9** **a** Variations of the $S_1$-population from AIMD simulations for dmSB [4]. Rate of $S_1$-population in the *cis*/DHP-side ($d_{C3C1C2C4} \leq 68.7°$) is *plotted in red*, while that in twist-side ($d_{C3C1C2C4} > 68.7°$) is *plotted in blue*. The plot for twist-side is normalized so that the maximum value is equal to 1. **b** Decay of the CIS* band (*a red line*) and development of the P* band (*a cyan line*) for cis-dmSB (in acetonitrile). Reprinted from Ref. [91], Copyright (2017), with permission from Elsevier

regions within the DHP side (consistent with the experimental lifetime of the CIS* band, 0.2 ps). The $S_1$-population of the *twist*-side increases during 0.6 ps due to moving-in of the trajectories from the *cis*/DHP-side (in good agreement with the finite rise time in the P* band). In Fig. 18.9a, there is a shoulder feature in the plots of both *cis*/DHP- and twist-sides during 0.2 ~ 0.3 ps; these shoulders are related to trajectories which once move in the twist-side and then go back to the *cis*/DHP-side. This feature can explain the shoulder in the experimental time profiles of the CIS* and P* bands in Fig. 18.9b [91].

## 18.4 Conclusion

Recent progresses in electronic structure theory including ab initio wavefunction theory and DFT have made it possible to explore global excited-state PES involving non-adiabatic region quantitatively. This exploration leads to a new frontier in quantum chemistry. AIMD for photo reaction is possible once energy gradients for excited-state PES are available, and non-adiabatic transitions can be tractable if one gets non-adiabatic coupling terms. Dynamics simulations on the basis of accurate PES can provide a lot of valuable information that cannot be obtained from static calculations of the PES. Collaborations of AIMD approach and spectroscopic experiments can provide a detailed picture of photoreaction processes. In this chapter, we demonstrated the advanced applications of AIMD approach for photoisomerization reactions to elucidate mechanism and dynamics for ultrafast relaxation processes of $n\pi^*$-excited azobenzene and $\pi\pi^*$-excited stilbene.

Photoisomerization mechanism of $n\pi^*$-excited azobenzene was investigated by AIMD simulation with the RATTLE algorithm, based on the SA-CASSCF method. The photoisomerization from cis to trans proceeds much faster than that from trans to cis due to non-planarity of cis-form. It was clarified that cis to trans isomerization occurs via two-step rotation mechanism, accompanying rotations of the central NN part and two phenyl rings, and this process can be classified into two types, namely, clockwise and counterclockwise rotation pathways; on the other hand, trans to cis isomerization occurs via conventional rotation pathway where two phenyl rings rotate around the NN bond. The quantum yields for cis-trans photoisomerization were well reproduced. Experimentally, the transient Raman spectra showed that the NN stretching frequency of *trans*-azobenzene is almost unchanged in the $S_1(n\pi^*)$ state, indicating that the NN bond order does not change during the isomerization process. To examine this mechanism, a multi-reference second-order perturbation theory was applied to calculate equilibrium structures and vibrational frequencies of trans-azobenzene in the ground and $n\pi*$-excited states. It was found that the NN stretching frequency exhibits a slight increase at the minimum energy structure in the $n\pi*$ state, which is explained by the vibrational mixing of the NN stretching mode with the CN symmetric stretching mode. The NN stretching frequency was also calculated at several selected structures along the rotation and inversion pathways in the $n\pi*$ state; it was shown that the frequency decreases gradually along the rotation pathway while it increases by ca. 300 $cm^{-1}$ along the inversion pathway. The frequencies and energy variations along the respective pathways indicate that the rotation pathway is more consistent with the experimental observation of the NN stretching frequency in $n\pi^*$ excitation.

The first excited state of stilbene is $\pi\pi^*$ state, which is the most important electronic state to realize the photofunctionality of stilbene-like species. From theoretical viewpoint, the description of $\pi\pi^*$ state requires dynamic correlation effects in the electronic structure theory, and so SA-CASSCF is not a good choice for AIMD simulations. Recently, SF-TDDFT method was developed so that non-adiabatic region of $S_0$ and $S_1$ states can be described. Then, AIMD simulations with SF-TDDFT were carried out to examine the photoisomerization mechanism of $\pi\pi^*$-excited *cis*-stilbene. To follow the target state among nearly degenerate electronic states during the dynamics simulations, a state tracking method was developed. It was shown that the meta-IRC from the FC structure of *cis*-stilbene in the $\pi\pi^*$ state reached the $S_1$-minimum of DHP via a *cis*-stilbene-like structure on a very flat region of the $S_1$-PES. However, AIMD simulations showed that more trajectories reached CI in the twist region that is located between cis and trans forms, compared with CI in DHP region, and the branching ratio was very close to the experiment. The discrepancy between the meta-IRC and AIMD simulations can be understood from geometrical features of the excited-state PES. It was concluded that ππ*-excited *cis*-stilbene propagates primarily toward the twisted structural region due to dynamic effects, with partial branching to the DHP structural region via the flat-surface region around $(S_1)_{cis\text{-min}}$. AIMD simulations were also carried out for ππ*-excited cis-dmSB at the SF-TDDFT level to get insights into the substitution effects on the photoisomerization dynamics of stilbene. For cis-dmSB, the meta-IRC from the FC structure of *cis*-dmSB is oriented toward the DHP side, which is in contrast to the case of SB. The optimized geometries of minima and MECI suggested that molecules in the DHP region could easily decay to the ground state. On the other hand, $S_1/S_0$-MECI and $S_1$-minimum in the twist region have a relatively different geometry from each other, which is consistent with the experimental observation of the long lifetime of the perpendicular structure. AIMD simulations showed that more trajectories enter the well of the DHP side than the well of the twist-side, and that all of the trajectories going to the DHP-side reached the $S_1/S_0$-CI region with $\sim$ 0.2 ps on average, while very few trajectories reached $S_1/S_0$-CI even after 1 ps in the twist region.

**Acknowledgements** We are sincerely grateful to Dr. Yusuke Otani, Dr. Akira Nakayama, Dr. Takeshi Noro, Dr. Satoshi Maeda, Prof. Mark Gordon, Dr. Satoshi Takeuchi, and Dr. Tahei Tahara for collaborations and discussions on the target photoreaction dynamics. This work is supported by a grant from JSPS KAKENHI with Grant Number 26288001. A part of calculations was performed using the Research Center for Computational Science, Okazaki, Japan.

## References

1. Y. Ootani, K. Satoh, A. Nakayama, T. Noro, T. Taketsugu, J. Chem. Phys. **131**, 194306 (2009)
2. Y. Harabuchi, M. Ishii, A. Nakayama, T. Noro, T. Taketsugu, J. Chem. Phys. **138**, 064305 (2013)
3. Y. Harabuchi, K. Keipert, F. Zahariev, T. Taketsugu, M.S. Gordon, J. Phys. Chem. A **118**, 11987 (2014)
4. Y. Harabuchi et al., J. Phys. Chem. A **120**, 8804 (2016)
5. T. Taketsugu, M.S. Gordon, J. Phys. Chem. **99**, 8462 (1995)
6. T. Taketsugu, M.S. Gordon, J. Phys. Chem. **99**, 14597 (1995)
7. T. Taketsugu, M.S. Gordon, J. Chem. Phys. **103**, 10042 (1995)
8. M.S. Gordon, G. Chaban, T. Taketsugu, J. Phys. Chem. **100**, 11512 (1996)
9. T. Vreven et al., J. Am. Chem. Soc. **119**, 12687 (1997)
10. A.L. Kaledin, K. Morokuma, J. Chem. Phys. **113**, 5750 (2000)
11. M. Ben-Nun, J. Quenneville, T.J. Martinez, J. Phys. Chem. A **104**, 5161 (2000)
12. D. Kina, A. Nakayama, T. Noro, T. Taketsugu, M.S. Gordon, J. Phys. Chem. A **112**, 9675 (2008)
13. D. Kina et al., Int. J. Quantum Chem. **109**, 2308 (2009)
14. A. Nakayama, G. Arai, S. Yamazaki, T. Taketsugu, J Chem Phys **139**, 214304 (2013)
15. T. Taketsugu, A. Tajima, K. Ishii, T. Hirano, Astrophys. J. **608**, 323 (2004)

16. M. Kayanuma, T. Taketsugu, K. Ishii, Chem. Phys. Lett. **418**, 511 (2006)
17. M. Kayanuma, T. Taketsugu, K. Ishii, Theor. Chem. Acc. **120**, 191 (2007)
18. J.C. Tully, J. Chem. Phys. **93**, 1061 (1990)
19. H.C. Andersen, J. Comput. Phys. **52**, 24 (1983)
20. Y. Harabuchi et al., *SPPR, a developmental version at Hokkaido University* (Sapporo, Japan, 2016)
21. F. Bernardi, M. Olivucci, M.A. Robb, Chem. Soc. Rev. **25**, 321 (1996)
22. D.R. Yarkony, Acc. Chem. Res. **31**, 511 (1998)
23. A.L. Sobolewski, W. Domcke, C. Dedonder-Lardeux, C. Jouvet, Phys. Chem. Chem. Phys. **4**, 1093 (2002)
24. B.G. Levine, T.J. Martinez, Annu. Rev. Phys. Chem. **58**, 613 (2007)
25. T. Mori, S. Kato, Chem. Phys. Lett. **476**, 97 (2009)
26. S. Nanbu, T. Ishida, H. Nakamura, Chem. Sci. **1**, 663 (2010)
27. H. Nakano, J. Chem. Phys. **99**, 7983 (1993)
28. H. Nakano, Chem. Phys. Lett. **207**, 372 (1993)
29. B.G. Levine, C. Ko, J. Quenneville, T.J. Martinez, Mol. Phys. **104**, 1039 (2006)
30. Y.H. Shao, M. Head-Gordon, A.I. Krylov, J. Chem. Phys. **118**, 4807 (2003)
31. F. Wang, T. Ziegler, J. Chem. Phys. **121**, 12191 (2004)
32. N. Minezawa, M.S. Gordon, J. Phys. Chem. A **113**, 12749 (2009)
33. Y.A. Bernard, Y. Shao, A.I. Krylov, J. Chem. Phys. 136 (2012) 204103/1
34. S. Maeda, Y. Harabuchi, T. Taketsugu, K. Morokuma, J. Phys. Chem. A **118**, 12050 (2014)
35. G. Zimmerman, L.Y. Chow, U.J. Paik, J. Am. Chem. Soc. **80**, 3528 (1958)
36. D. Gegiou, K.A. Muszkat, E. Fischer, J. Am. Chem. Soc. **90**, 3907 (1968)
37. P. Bortolus, S. Monti, J. Phys. Chem. **83**, 648 (1979)
38. H. Rau, E. Luddecke, J. Am. Chem. Soc. **104**, 1616 (1982)
39. H. Rau In Photochromism: Molecules and Systems; Durr, H., Bounas-Laurent, H., Eds.; Elsevier: Amsterdam, 1990; Vol. 1, p 165
40. T. Fujino, T. Tahara, J. Phys. Chem. A **104**, 4203 (2000)
41. T. Fujino, S.Y. Arzhantsev, T. Tahara, J. Phys. Chem. A **105**, 8123 (2001)
42. T. Fujino, S.Y. Arzhantsev, T. Tahara, Bull. Chem. Soc. Jpn **75**, 1031 (2002)
43. C.W. Chang, Y.C. Lu, T.T. Wang, E.W.G. Diau, J. Am. Chem. Soc. **126**, 10109 (2004)
44. S. Monti, G. Orlandi, P. Palmieri, Chem. Phys. **71**, 87 (1982)
45. P. Cattaneo, M. Persico, Phys. Chem. Chem. Phys. **1**, 4739 (1999)
46. T. Ishikawa, T. Noro, T. Shoda, J. Chem. Phys. **115**, 7503 (2001)
47. T. Schultz et al., J. Am. Chem. Soc. **125**, 8098 (2003)
48. A. Cembran, F. Bernardi, M. Garavelli, L. Gagliardi, G. Orlandi, J. Am. Chem. Soc. **126**, 3234 (2004)
49. C. Ciminelli, G. Granucci, M. Persico, Chem. Eur. J. **10**, 2327 (2004)
50. L. Gagliardi, G. Orlandi, F. Bernardi, A. Cembran, M. Garavelli, Theo. Chem. Acc. **111**, 363 (2004)
51. A. Toniolo, C. Ciminelli, M. Persico, T.J. Martínez, J. Chem. Phys. **123**, 234308 (2005)
52. G. Granucci, M. Persico, Theor. Chem. Acc. **117**, 1131 (2007)
53. I. Conti, M. Garavelli, G. Orlandi, J. Am. Chem. Soc. **130**, 5216 (2008)
54. M. Boeckmann, N.L. Doltsinis, D. Marx, Angew. Chem. Int. Edit. **49**, 3382 (2010)
55. T. Cusati, G. Granucci, M. Persico, J. Am. Chem. Soc. **133**, 5109 (2011)
56. O. Weingart, Z. Lan, A. Koslowski, W. Thiel, J. Phys. Chem. Lett. **2**, 1506 (2011)
57. M. Pederzoli, J. Pittner, M. Barbatti, H. Lischka, J. Phys. Chem. A **115**, 11136 (2011)
58. J.A. Gamez, O. Weingart, A. Koslowski, W. Thiel, J. Chem. Theory Comput. **8**, 2352 (2012)
59. H.J. Werner et al. MOLPRO, version 2012, a package of ab initio programs, 2012
60. W.H. Miller, N.C. Handy, J.E. Adams, J. Chem. Phys. **72**, 99 (1980)
61. T. Noro, M. Sekiya, T. Koga, Theor. Chem. Acc. **131**, 1124 (2012)
62. M.W. Schmidt et al., J. Comput. Chem. **14**, 1347 (1993)
63. G. Karlström et al., Comput. Mater. Sci. **28**, 222 (2003)
64. T. Hirano, T. Taketsugu, Y. Kurita, The Journal of Physical Chemistry **98**, 6936 (1994)

65. W.M. Moore, D.D. Morgan, F.R. Stermitz, J. Am. Chem. Soc. **85**, 829 (1963)
66. K.A. Muszkat, E. Fischer, J. Chem. Soc. B (1967) 662
67. T. Wismonskiknittel, G. Fischer, E. Fischer, J. Chem. Soc., Perkin Trans. (1974) 1930
68. H. Petek et al., J. Phys. Chem. **94**, 7539 (1990)
69. D.C. Todd et al., J. Chem. Phys. **93**, 8658 (1990)
70. S.T. Repinec, R.J. Sension, A.Z. Szarka, R.M. Hochstrasser, J. Phys. Chem. **95**, 10380 (1991)
71. J.M. Rodier, X.P. Ci, A.B. Myers, Chem. Phys. Lett. **183**, 55 (1991)
72. S. Pedersen, L. Banares, A.H. Zewail, J. Chem. Phys. **97**, 8801 (1992)
73. W. Fuss, C. Kosmidis, W.E. Schmid, S.A. Trushin, Angew. Chem. Int. Edit. **43**, 4178 (2004)
74. K. Ishii, S. Takeuchi, T. Tahara, Chem. Phys. Lett. **398**, 400 (2004)
75. T. Nakamura, S. Takeuchi, N. Suzuki, T. Tahara, Chem. Phys. Lett. **465**, 212 (2008)
76. S. Takeuchi et al., Science **322**, 1073 (2008)
77. S.A. Kovalenko, A.L. Dobryakov, I. Ioffe, N.P. Ernsting, Chem. Phys. Lett. **493**, 255 (2010)
78. M. Sajadi, A.L. Dobryakov, E. Garbin, N.P. Ernsting, S.A. Kovalenko, Chem. Phys. Lett. **489**, 44 (2010)
79. T. Nakamura, S. Takeuchi, T. Taketsugu, T. Tahara, Phys. Chem. Chem. Phys. **14**, 6225 (2012)
80. A.L. Dobryakov, I. Ioffe, A.A. Granovsky, N.P. Ernsting, S.A. Kovalenko, J. Chem. Phys. **137**, 244505 (2012)
81. M.J. Bearpark et al., J. Phys. Chem. A **101**, 3841 (1997)
82. C.D. Berweger, W.F. van Gunsteren, F. Muller-Plathe, J. Chem. Phys. **108**, 8773 (1998)
83. Y. Amatatsu, Chem. Phys. Lett. **314**, 364 (1999)
84. V. Molina, M. Merchan, B.O. Roos, Spectrochimica Acta. Part A **55**, 433 (1999)
85. Y.S. Dou, R.E. Allen, Chem. Phys. Lett. **378**, 323 (2003)
86. Y.S. Dou, R.E. Allen, J. Chem. Phys. **119**, 10658 (2003)
87. J. Quenneville, T.J. Martinez, J. Phys. Chem. A **107**, 829 (2003)
88. Y.S. Dou, R.E. Allen, J. Mod. Opt. **51**, 2485 (2004)
89. N. Minezawa, M.S. Gordon, J. Phys. Chem. A **115**, 7901 (2011)
90. I.N. Ioffe, A.A. Granovsky, J. Chem. Theory Comput. **9**, 4973 (2013)
91. F. Berndt et al., Chem. Phys. Lett. **544**, 39 (2012)
92. Y. Amatatsu, Chem. Phys. Lett. **631–632**, 47 (2015)

# Chapter 19
# Density Functional Theoretical Studies on Chemical Enhancement of Surface-Enhanced Raman Spectroscopy in Electrochemical Interfaces

**Ran Pang, De-Yin Wu and Zhong-Qun Tian**

**Abstract** Since the surface-enhanced Raman scattering (SERS) observed in 1974, 40 years past, SERS spectroscopy are applied in several fields of chemistry, physics, biology, energy source and so on. However, the mechanism of SERS is not entirely clear. Density functional theory (DFT) calculations could give information at the molecular level and have been applied in the studies of electrochemical interfaces. In this paper, by combining DFT calculations and Raman scattering theory, we presented some results about SERS spectra of interfacial water and pyridine adsorbed on metal surfaces, including electrochemical adsorption, chemical reactions and photo-driven charge transfer processes. All these information helps us reveal the physical and chemical nature of SERS signals for studied molecules in complex electrochemical interfaces.

**Keywords** Density functional theory · Electrochemical SERS · Water · Pyridine · Photo-driven charge transfer

## 19.1 Introduction

Surface-enhanced Raman scattering (SERS) is an anomalous and special phenomenon observed when Raman spectroscopy was applied to surface science [1–3]. The first SERS spectrum was obtained from an electrochemical system of pyridine (Py) adsorbed on a rough silver electrode by Fleischmann et al. in 1974 [4]. Then Jeanmaire and Van Duyne found that the experimentally observed Raman intensity from an adsorbed Py should be 5–6 orders of magnitude greater than that in pure

R. Pang · D.-Y. Wu (✉) · Z.-Q. Tian
State Key Laboratory of Physical Chemistry of Solid Surfaces, Collaborative Innovation Center of Chemistry for Energy Materials, and Department of Chemistry, College of Chemistry and Chemical Engineering, Xiamen University, Xiamen 361005, China
e-mail: dywu@xmu.edu.cn



M.J. Wójcik et al. (eds.), *Frontiers of Quantum Chemistry*,
https://doi.org/10.1007/978-981-10-5651-2_19

liquid and aqueous solution [5]. Albreacht and Creighton also observed such an abnormal enhancement effect and proposed that the considerable enhancement was related to the resonance Raman scattering effect [6]. Because of the enormous enhancement, SERS spectra have provided rich information to understand the structures and the physical and chemical properties of adsorption processes and chemical reactions at interfaces or surfaces. So far the SERS has been developed to higher detection sensitivity up to the single-molecule level with high spatial resolution and ultrafast temporal resolution [7, 8]. The normal SERS technique is also extended to tip-enhanced Raman spectroscopy (TERS) [9–13], gap-mode SERS [14, 15], and shell-isolated nanoparticle-enhanced Raman spectroscopy (SHINERS) [16–18], so that the SERS spectroscopy was applied in several fields of chemistry, physics, biology, energy source and materials and so on.

The SERS spectroscopy provides vibrational information for understanding mechanistic and dynamic information on electrochemical interfaces at the molecular level. Meanwhile, the theoretical study of SERS develops with the enrichment of the experimental data [19–25]. Philpott suggested for the first time that the Raman signal could be enhanced similar to the resonance Raman scattering effect due to the surface plasmon resonance (SPR) state in 1975 [26]. In the case, the surface plasmon state was considered as the intermediate state of Raman scattering processes. To account for the enhancement of SERS signals, two types of enhancement mechanisms are considered, the electromagnetic field enhancement (EM) [27–29] and the chemical enhancement (CE) [30–32]. The former one is generally considered to have the main contribution to SERS signals from the surface plasmon resonance [23, 24, 33, 34], which depends on the optical property of the metal substrate and its roughness. In particular, for nanostructured surfaces considered as SERS substrates the enhancement is also a function of size, shape, and aggregation states of metal nanoparticles [35, 36]. The CE mechanism is a chemical effect, which strongly depends on the geometric and electronic structures of adsorbed molecules, the chemical bonding between molecules and metal surfaces, excited states related to the frequency of excitation lines, and chemical environments of interfaces such as the applied electrode potentials, the solvent effects, and the property of electrolytes [19, 25, 37–39]. Both chemical and physical enhancements can be influenced to some extent by changing the applied electrode potential, i.e., the Fermi level of metals, surface charge density, and dielectric constants of interfacial electrolyte in an electrochemical SERS (EC-SERS) system. This makes EC-SERS as one of the most complicated systems in the SERS.

Since the enhancement mechanisms in SERS are influenced by many factors, there are some difficulties to understand the cooperation effect between the two mechanisms [40]. Firstly, there are two distinctively different properties of electric fields, the frequency-dependent electromagnetic field and the static electric field co-existing in electrochemical systems [41]. The surface charge density of metal electrodes can be tuned by changing applied potentials to result in a shift in the SPR frequency [42]. Secondly, the change of applied electrode potentials results in the change of interaction (or bonding) between metal and adsorbates, the surface orientation of adsorbates and the structure of the double layer, which may cause the redistribution of the surface localized optical electric field [22, 37]. Thirdly, the materials of metal electrodes and solvent molecules are important components in electrochemical interfaces, which also lead to change the double layer structure and the optical property of the electrochemical system [43]. Finally, there may exist some unknown species and uncertain interfacial structures due to the polarization of applied potentials. This makes it difficult to analyze EC-SERS spectra by a simple comparison with normal Raman spectra of studied molecules in gas phase, liquid states, aqueous solution, and solid states.

Quantum chemical calculation plays a unique advantage for investigating the SERS spectra in electrochemical systems, even through there are many complicate factors to hinder completely understanding the SERS mechanism and electrochemical processes. Theoretical studies provide opportunities to obtain possible adsorption structures and chemical bonding implicitly in their vibrational spectra, as well as thermodynamic stability of interfacial structures. By predicting interfacial structures and corresponding vibrational spectra at different environments, it is very useful for us to explore the influence of different factors on molecular polarizability, such as double layer electric field, the chemical bonding in charged electrode surfaces, and surface adsorption orientation of probing molecules. Combining theoretical calculations and Raman scattering theory, it could give better assignment for Raman spectra of adsorption structures, and analyze the change of vibrational frequency and SERS intensities.

In this chapter, we presents our recent theoretical results of interfacial water and Py adsorbed on metal electrodes by using density functional theory calculations. To explore the CE of SERS in electrochemical interfaces, we focus on the change of vibrational frequency and SERS intensities in electrochemical adsorption, chemical reactions and photo-driven charge transfer (CT) processes. All these information helps us to reveal the physical and chemical nature of SERS signals for probing molecules in complex electrochemical interfaces.

## 19.2 Theory and Models

Density Functional theory (DFT) is one of the most popular and versatile computational methods of quantum chemistry used in physics, chemistry and material sciences. In this theory, the properties of a many-electron system can be determined by using a functional of electron density, which is the spatially dependent electron density, so that computational cost is relatively low when compared to traditional ab initio methods with the consideration of the electron correlation effect at different levels.

The DFT was developed from Hohenberg and Kohn [44] and Kohn and Sham [45] in 1960s, and shows that the total energy E of an atomic system can be expressed as a functional of its electron density r, i.e., E = E[r]. Based on the Kohn-Sham equation (Eq. 19.1), the energy can be further optimized for the electronic ground state density by using the first-order variations in the density.

$$\left[-\frac{\hbar^2}{2m}\nabla^2 + V_{eff}(r)\right]\psi(r) = \varepsilon_i\psi_i(r) \tag{19.1}$$

where

$$V_{eff}(r) = V_C(r) + \mu_{xc}[\rho(r)] \tag{19.2}$$

is an effective potential. The first term is the Coulombic electrostatic potential, that is given by

$$V_C(r) = -e^2 \sum\nolimits_\alpha \frac{Z_\alpha}{|r - R_\alpha|} + e^2 \frac{\rho(r')}{|r - r'|} dr' \tag{19.3}$$

which can also be calculated using Poisson equation,

$$\nabla^2 V_C(r) = -4\pi e^2 q(r) \tag{19.4}$$

where q(r) denotes the sum of the electronic charge distribution and the positive charge at atomic nuclei position $R_\alpha$. The second term in Eq. (19.2) is the exchange-correlation potential, which is give by

$$\mu_{xc}[\rho(r)] = \frac{\partial E_{xc}[\rho(r)]}{\partial \rho(r)} \tag{19.5}$$

Because the exchange-correlation functional energy are not known in the case, many approximations have been made as follows. Several different methods of exchange-correlation functionals have been developed for DFT calculations, for example, the local density approximation (LDA) [46, 47], the generalized gradient approximation (GGA) [48, 49], and hybrid density functional methods [50–52].

The LDA is a very successful and widely used exchange correlation functional of DFT energies. In the case the exchange correlation energy is considered to only depend on the local electron density of each volume element,

$$E_{xc}[\rho(r)] = \int \varphi(r)\varepsilon_{xc}[\rho(r)]dr \tag{19.6}$$

where $\varepsilon_{xc}[\rho(r)]$ is the exchange correlation energy per electron of a homogenous electron gas. It can be expressed as an analytic function of electron density under the exchange correlation potential $\mu_{xc}$. In our previous work, we compare the calculated results from the SVWN functional with others [53], and found the LDA functional can predict the good geometry but it fails to predict the vibrational frequency and Raman intensity. This was interpreted due to the defect from neglecting the non-local electron density.

The GGA functional was developed by Perdew and Wang (PW) from the first principle [49]. By combining the gradient expansion of the exchange correlation holes of a non uniform electron gas with a real space truncation to enforce constraints improved by properties of the physical exchange correlation hole.

$$E_{xc}^{GGA}[\rho(r)] = \int \rho(r)\varepsilon_{xc}[\rho(r), \nabla\rho(r)]dr \quad (19.7)$$

For example, the GGA was found to generally improve the prediction of the total energy, ionization energies, electron affinities of atoms, atomization energies of molecules, vibrational frequencies, and thermodynamic properties. Additionally, the GGA was thought to improve adsorption structures, adsorption energies, dissociation energies, and activation energy barriers for adsorbates on metal surfaces [54, 55].

Although the GGA improves the prediction of DFT calculations, ones found the accurate exchange energy used can further improve the results of GGA calculations. This is mainly due to the decrease of self-interaction errors of electrons. The representation approach is B3LYP, [50–52] a typical hybrid density functional method, which is widely used in molecular modeling calculations of vibrational spectra and chemical reactions.

So far the DFT approach has the following applications related to our studies in surface-enhanced Raman spectra of adsorbates on metal surfaces. DFT calculations can provide automatic optimization of the atomic geometry on the basis of evaluation of atomic forces. On the top of the optimization geometry, fundamental vibrational frequencies can be calculated from the Hessian matrix, i.e., force constant matrix at the harmonic or anharmonic approximation. Additionally, the function of DFT was extended to the treatment of excited states, providing the chance to predict the optical properties and spectral properties, in particular, Raman intensity with the consideration of the excitation wavelength from the coupled perturbation Hartree-Fock method.

Over the last decades, we carried out different DFT calculations to explore the CE in SERS of Py [39, 53, 56–63], water [43, 64–67], thiophenol [68, 69], p-aminothiophenol [68–71] and p, p'-dimercaptoazobenzene [69–72] on silver, gold, and copper surfaces of nanostructures. When DFT is applied in EC-SERS systems, our calculated results clearly showed that the hybrid DFT method can predict well the structural properties, the binding energies, the vibrational frequencies, and the Raman intensities of probing molecules [53, 62–66].

Metal clusters model was applied to simulate adsorption sites on metal electrodes of bulk and nanoparticles, that generally are considered as complex and unknown surface structures experimentally. We can assume that probing molecules prefers to adsorption on some surface sites, such as top, bridge, and hollow sites, with strongly localized chemical binding between probing molecules and surface metal atoms. On the basis of Boltzmann distribution, these stable adsorption structures could give good descriptions for adsorption of probing molecules on metal electrodes. For probing molecules like Py and water, the Pople's basis set 6-311 + G** [73] and the Dunning's basis set Aug-cc-pVTZ [74] were used for carbon, nitrogen, oxygen, and hydrogen atoms. For metal atoms, the electrons in the valence and internal shells were described by the basis set LANL2DZ and the corresponding relativistic effective core potentials, respectively [75].

In our simulated adsorption structures, different adsorption orientations are considered by checking characteristic groups, such as electrophilic and nucleophilic groups in adsorbed molecules. For metal electrodes, surface charges need to take into account, then different adsorption sites have also been considered by the analysis of symmetry matching, energy closing, and maximum orbital overlap from the view of chemical bonding [62, 76].

To simulate SERS spectra of a probing molecule adsorbed metal surfaces, Fig. 19.1 shows the schematic diagram for calculating the vibrational frequencies and Raman intensities in electrochemical interfaces. First, we consider the vibrational frequency and assignments of adsorbates at the electronic ground states. On top of the optimized structures, harmonic force constant matrix and vibrational frequencies are calculated at the same theoretical level. By using the force constant matrix of optimized structures, the scaled quantum mechanical force field (SQMF) procedure is used to do normal mode analysis [77, 78]. This procedure can effectively correct the defect in the theoretical methods, the basis set effect, and the anharmonic effect so that the present calculation reasonably reproduces experimentally observed frequencies. The scaled factors is dependent on different DFT methods and investigated systems [79, 80], for example, we used scaled factor of 0.981 for adsorption structures of Py calculated at the B3LYP/6-311 + G**(C, N, H)/LANL2DZ(Ag, Au, Cu) level [63]. As for interfacial water, two scaling factors of 0.927 for the B3LYP force constants of both O–H stretching coordinates and 0.963 for that of the HOH bending one. Compared to the molecule itself in the gas or liquid phases, there may exist some frequency shifts presenting the molecule-metal interactions in SERS conditions. For a better assignment of vibrational frequency shifts of the complexes, we also do normal mode analysis [77, 78].

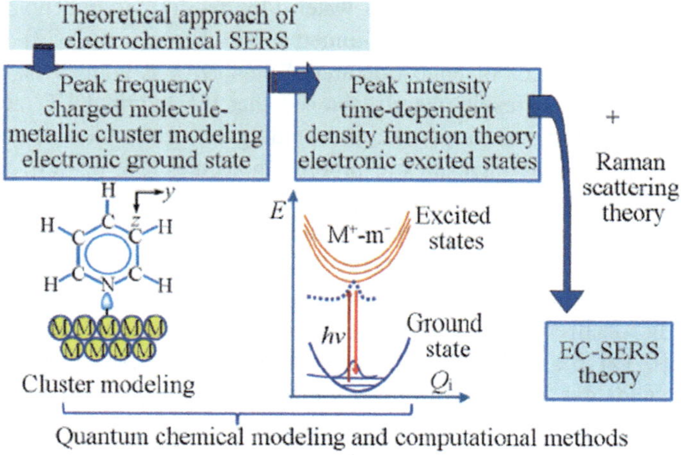


**Fig. 19.1** Schematic diagram of theoretical simulation for EC-SERS [81]. Copyright from the editorial office of chemical journal of Chinese universities

Next we obtain the absolute Raman intensity by calculating the differential Raman scattering cross section (DRSCS) at a given excitation wavelength. It needs to transform the atomic polarizability derivative tensor ($PD_X$) in the Cartesian coordinate to the polarizability derivative tensor ($PD_Q$) in the normal coordinate frame by using the following expression [82].

$$PD_Q = PD_X AL_S \tag{19.8}$$

where A and $L_s$ denote the transform matrix and the vibrational vector of the normal coordinate along isolated internal coordinators. Then the absolute Raman intensity of a given vibrational mode is [83, 84],

$$I_i^R = \frac{h}{8\pi^2 c\widetilde{\nu}_i} \cdot \frac{(\widetilde{\nu}_0 - \widetilde{\nu}_i)^4}{45\left[1 - \exp\left(-\frac{hc\widetilde{\nu}_i}{k_B T}\right)\right]} S_i \tag{19.9}$$

where

$$S_i = 45\overline{\alpha_i'}^2 + 7\gamma_i'^2 \tag{19.10}$$

$$\overline{\alpha_i'} = \frac{1}{3}\left(PD_{Q,i}^{XX} + PD_{Q,i}^{YY} + PD_{Q,i}^{ZZ}\right) \tag{19.11}$$

$$\begin{aligned}\gamma_i'^2 = \frac{1}{2}&\left[\left(PD_{Q,i}^{XX} - PD_{Q,i}^{YY}\right)^2 - \left(PD_{Q,i}^{YY} - PD_{Q,i}^{ZZ}\right)^2 + \left(PD_{Q,i}^{XX} - PD_{Q,i}^{ZZ}\right)^2\right] \\ &+ 3\left[\left(PD_{Q,i}^{YX}\right)^2 + \left(PD_{Q,i}^{ZX}\right)^2 + \left(PD_{Q,i}^{ZY}\right)^2\right]\end{aligned} \tag{19.12}$$

where $\tilde{\nu}_0$ and $\tilde{\nu}_i$ denote the frequency of the incident light and the vibrational frequency of the *i*th mode. The Raman scattering factor $S_i$ in Eq. 19.10 is calculated from the mean polarizability tensor derivative, $\overline{\alpha}_i'$, and the anisotropic polarizability derivative, $\gamma_i'^2$, of the *i*th vibrational mode, as seen in Eqs. 19.11 and 19.12. In this paper, the Raman intensity presented in simulated Raman spectra is in the DRSCS with the Lorentzian expansion in a line width of 10 $cm^{-1}$ at given laser line. The unit of $cm^2$ $sr^{-1}$ $mole^{-1}$ means the effective scattering cross section $cm^2$ per steradian per molecule.

Here, the polarizability is calculated by using two methods. The first one is the finite field method to consider the optical polarization effect. By using this method, ones can calculate the frequency-independent polarizability, that is also named as static polarizability. In the case, the photonic energy of Raman excitation light is much smaller than the transition energy of the first singlet excited state of studied systems. The second method is the coupled perturbation Hartree-Fock method, which is a frequency-dependent theoretical method to consider the influence of the light excitation [85, 86]. This method must be used when the photonic energy matches or approaches to the transition energy of the excited states of interest. The property of electronic transitions of low-lying excited states was obtained by using TD-DFT calculations. Then we obtain the preresonance Raman spectra, in which the Raman signals can be greatly enhanced.

Besides the influence of the light excitation, the Raman spectra also has changes in external field environments, such as the salvation effect and the interfacial electric-field effect [39, 66, 87–89]. In our work, we use two methods to considering the solvation effect on the structure fluctuation of probing molecules [66]. One is the explicit model by adding localized water molecules around the probing molecules. In this model, the hydrogen bonding (HB) interaction with surrounding water plays an important role in shaping the structure of clusters. Another method is using the polarizable continuum model (PCM) [90] or the solvation model of density (SMD) [91] models for the implicit solvent calculation of hydrated proton clusters in aqueous solution. Moreover, we also considered the influence of the external electric effect on the structure deformation and changes in polarizability tensors [66, 87].

## 19.3 SERS of Interfacial Water

Water is the most important solvent and plays vital roles in electrochemistry. Clarifying the interfacial water at the microscopic level can greatly improve our fundamental understanding of the electrode/electrolyte interface, which is still an eternal issue in electrochemistry and surface sciences [92–94]. Vibrational Raman spectroscopy has been widely used to characterize and analyze the interaction of water with solute molecules in chemistry and biochemistry [95–99]. In gas phases or pure liquid water, Raman signals of water O–H stretching modes are stronger than that of the bending mode [28, 100–102]. However, their relative Raman intensities can be changed significantly in halide electrolyte aqueous solutions [103–110], aqueous solutions containing hydrated electrons [111–118], and electrochemical interfaces [93, 119–124].

To understand the enhancement effect of hydrated electrons and halide ions in the Raman intensity of water molecules in the interface of metal electrode/halide electrolyte solutions, we investigated the binding interaction and the off-resonance Raman intensities for the intramolecular modes of water in $(H_2O)_3^-$, $H_2O\ldots X^-$ (X = F, Cl, Br, I) [65], and adsorbed on negative charged metal clusters $M_{10}$ (M = Au, Ag, Pt) [43]. Our calculated result proposes that the strong polarization of the proton acceptor causes a significant enhancement in Raman intensities of the HOH bending mode in some water molecules as proton donors [65].

### *19.3.1 Anionic Water Clusters*

As a special anion, hydrated electrons in electrochemical interfaces significantly enhance Raman intensities of librations and intramolecular modes of affected water molecules [112–118]. The enhancement factors in Raman intensities reach to ~2 × $10^3$ fold for the O–H stretch and ~3 × $10^5$ fold for the HOH bend by a resonance excitation at 683 nm [116]. Table 19.1 presents the calculated vibrational frequencies and the corresponding Raman intensities of the water trimer complexes. The most stable configurations are shown in Fig. 19.2. For neutral $(H_2O)_3$, one of three free O–H bonds adopts an anti configuration. The lower O–H stretching frequencies are associated with the HB O–H bonds, whereas the higher O–H stretching frequencies are associated with the free O–H bonds. Both B3LYP and MP2 methods concluded that the lowest O–H stretching frequency ($\nu_{s,1b}$) has the largest Raman intensity, such as 268.2 Å$^4$/amu for B3LYP and 276.2 Å$^4$/amu for MP2 with the aug-cc-pVTZ basis set. For the anion $(H_2O)_3^-$, all three free O–H bonds adopt a cis configuration, maximizing its dipole moment to stabilize the

**Table 19.1** Calculated vibrational frequencies (cm$^{-1}$) and Raman intensities ($I_R$ in Å$^4$/amu) of $(H_2O)_3$ and $(H_2O)_3^-$ [65–67]

| | Neutral | | | | Anion | | | |
|---|---|---|---|---|---|---|---|---|
| Mode | Freq [a] | $I_R^a$ | $I_R^b$ | $I_R^c$ | Freq [a] | $I_R^a$ | $I_R^b$ | $I_R^c$ |
| $\nu_{b,1}$ | 1608.0 | 1.2 | 1.1 | 1.1 | 1594.0 | 18125.7 | 18591.6 | 20988.4 |
| $\nu_{b,2}$ | 1610.1 | 1.1 | 1.1 | 1.3 | 1594.0 | 15867.5 | 23182.3 | 21258.0 |
| $\nu_{b,3}$ | 1633.7 | 1.3 | 1.2 | 1.1 | 1617.2 | 106774.6 | 136933.0 | 214875.8 |
| $\nu_{d1}$ | 3394.7 | 268.2 | 266.8 | 276.2 | 3375.0 | 33450.1 | 47452.4 | 29132.1 |
| $\nu_{d2}$ | 3457.1 | 34.0 | 34.4 | 27.6 | 3430.5 | 23138.9 | 22913.4 | 68332.3 |
| $\nu_{d3}$ | 3468.1 | 34.7 | 33.9 | 26.8 | 3431.2 | 23799.5 | 23101.3 | 55081.3 |
| $\nu_{f1}$ | 3720.7 | 52.5 | 51.6 | 47.8 | 3549.3 | 25700.9 | 25506.8 | 236198.6 |
| $\nu_{f2}$ | 3724.6 | 42.1 | 40.3 | 44.1 | 3549.5 | 25880.0 | 26122.5 | 183908.3 |
| $\nu_{f3}$ | 3725.8 | 99.1 | 99.8 | 76.7 | 3572.2 | 128495.0 | 222604.8 | 70502.4 |

[a]Scaled vibrational frequencies and Raman intensities are calculated at B3LYP/aug-cc-pVTZ
[b]B3LYP/aug-cc-pVQZ
[c]MP2/aug-cc-pVTZ

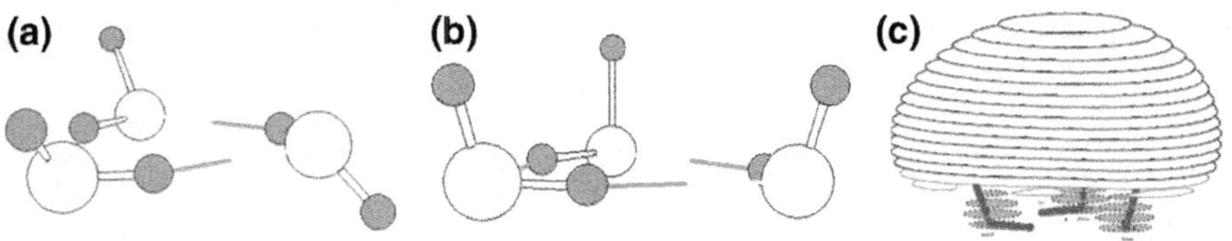


**Fig. 19.2** Optimized geometries of **a** $(H_2O)_3$ **b** $(H_2O)_3^-$ at the level of B3LYP/AVTZ. **c** the electron density plot of the highest occupied molecular orbital (HOMO) of $(H_2O)_3^-$ [65]. Copyright from the ACS Publications

excess electron [125]. The electron density distribution of HOMO plotted in Fig. 19.2c shows that the three free O–H groups are directed toward the excess electron. This results in the strongest Raman intensity associated with the free O–H stretching mode. As seen in Table 19.1, although B3LYP predicts that the highest O–H stretching frequency has the strongest Raman signal, MP2 predicts the strongest Raman signal comes from two near degenerate modes of the free O–H stretching vibrations. This discrepancy between two theoretical methods reflects a different response for the free O–H stretching vibrations to the polarization of the dipole-bound electron. Nevertheless, both methods conclude that the Raman intensity ratio between the strongest HOH bending and O–H stretching modes increases to 0.62 for B3LYP/aug-cc-pVQZ and 0.91 for MP2/aug-cc-pVTZ, as compared to 0.01 in the free water molecule.

### *19.3.2 Water-Halide Anionic Complexes*

Halide complexes are commonly used as electrolytes in electrochemical systems. Figure 19.3 shows equilibrium geometries of water-halide ($F^-$, $Cl^-$, $Br^-$, and $I^-$) complexes. All these complexes exhibit an asymmetric structure, in agreement with previous studies [126–132]. For these complexes, the binding energies (BE) in HO–H…$F^-$ is predicted to be 27.1 kcal/mol, which is almost twice larger than the values of 14.1 kcal/mol for $H_2O$…$Cl^-$, 11.0 kcal/mol for $H_2O$…$Br^-$, and 10.6 kcal/mol for $H_2O$…$I^-$. These calculated values are also in agreement with the measured disassociation energies of 26.2 kcal/mol for $H_2O$…$F^-$ and 14.9 kcal/mol for $H_2O$…$Cl^-$ [127, 130].

Table 19.2 presents scaled vibrational frequencies and Raman intensities of these four complexes. The results show that the HB O–H stretching frequencies ($\nu_5$) blue shift with increasing sizes of halide ions [127–133]. For example, the scaled vibrational frequencies of the $\nu_5$ mode at the present theoretical level red shifts to 2058.4 ($F^-$), 3175.3 ($Cl^-$), 3288.4 ($Br^-$), and 3367.3 cm$^{-1}$ ($I^-$), respectively. Their observed frequencies of the $\nu_5$ vibration are ~1523 ($F^-$), 3130 ($Cl^-$), 3267 ($Br^-$), and 3295 cm$^{-1}$ ($I^-$), respectively [127, 130, 134, 135]. The significant difference

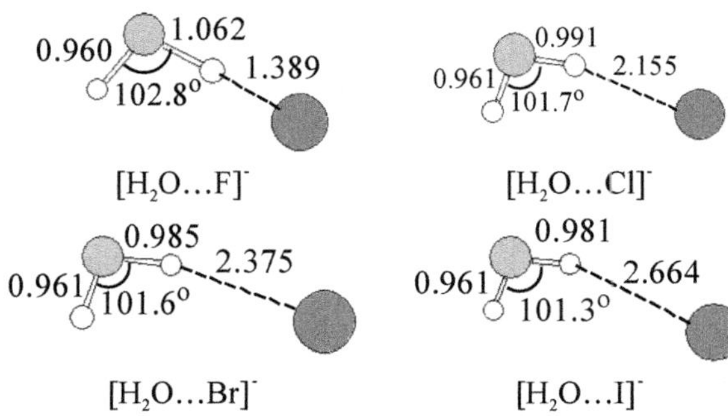


**Fig. 19.3** Optimal structures of $[H_2O…X]^-$ (X = F, Cl, Br, I) [65]. (Copyright from the ACS Publications.)

**Table 19.2** Calculated vibrational frequencies (ω: $cm^{-1}$) and Raman intensity ($I_R$: Å$^4$/amu) of $H_2O…X^-$ (X = F, Cl, Br, and I) [65]

| | $H_2O…F^-$ | | $H_2O…Cl^-$ | | $H_2O…Br^-$ | | $H_2O…I^-$ | |
|---|---|---|---|---|---|---|---|---|
| Mode | ω | $I_R$ | ω | $I_R$ | ω | $I_R$ | ω | $I_R$ |
| $\nu_1$ | 392.3 | 0.2 | 191.0 | 0.9 | 157.8 | 1.3 | 124.6 | 2.9 |
| $\nu_2$ | 570.0 | 2.1 | 359.0 | 2.2 | 314.3 | 2.0 | 250.1 | 26.3 |
| $\nu_3$ | 1144.1 | 4.8 | 729.2 | 6.9 | 658.6 | 7.9 | 603.3 | 104.0 |
| $\nu_4$ | 1668.4 | 1.7 | 1639.1 | 3.7 | 1634.1 | 5.9 | 1629.2 | 47.2 |
| $\nu_5$ | 2058.4 | 0.2 | 3175.3 | 222.8 | 3288.4 | 271.5 | 3367.3 | 477.2 |
| $\nu_6$ | 3704.3 | 87.8 | 3711.3 | 69.1 | 3709.7 | 70.9 | 3704.9 | 61.5 |

between the experimental and theoretical frequencies in $H_2O…F^-$ is mainly due to a large anharmonic effect for the O–H…$F^-$ structure [136]. On the other hand, the free O–H stretching ($\nu_6$) frequencies in the four complexes are around 3700 $cm^{-1}$, in agreement with the IR spectral peaks observed at 3690 ($H_2O…F^-$), 3698($H_2O…Cl^-$), 3690 ($H_2O…Br^-$), and 3710 $cm^{-1}$ ($H_2O…I^-$), respectively [127, 130]. Our normal mode analysis indicates that the $\nu_5$ and $\nu_6$ modes almost uncouple each other. Finally, the HOH…X stretching ($\nu_1$) mode has the lowest vibrational frequency among six fundamentals. The trend of changes in the $\nu_1$ frequencies is the same as the experimental observations [127, 130, 137].

The results presented in Table 19.2 show that the Raman intensities of the $\nu_4$ HOH bending and $\nu_5$ OH stretching vibrations increase in an order of $F^- < Cl^- < Br^- < I^-$. The enhancement of Raman intensities probably arises from two effects, the polarization effect and the CT of halide anions [127–132]. However, the latter effect may be not important and can be excluded due to the decrease in the amount of negative charge on the water molecule with an increase in the size of halide anions. From NBO analysis, we obtained the net transferred charges from halides to water molecules about 0.130e ($F^-$), 0.055e ($Cl^-$), 0.043e ($Br^-$), and 0.041e ($I^-$), respectively. For the polarization effect, it mainly originates from that the H-bonded proton of the water molecule may polarize the anions, yielding a larger induced dipole moment at the less polar O–H bond [106]. To quantitatively estimate the polarization effect, we calculated the isotropic and anisotropic polarizability derivatives with respect to the normal coordinates by using Eqs. 19.11 and 19.12, and the results are summarized in Table 19.3. For the $\nu_5$ mode, the HB interaction results in an increase in the anisotropic polarizability derivatives with the increasing size of halides, i.e., 3.76 ($F^-$), 22.84 ($Cl^-$), 29.76 ($Br^-$), and 29.38 Å$^4$/amu ($I^-$). For this mode, the isotropic polarizability derivatives are around 1 Å$^2$/amu$^{1/2}$ for X = F, Cl, and Br. It is more than doubled to 2.5 Å$^2$/amu$^{1/2}$ for X = I (see Table 19.3). Clearly, it is the anisotropic polarizability derivative that makes the Raman intensity of the $\nu_5$ mode larger for X = Cl, Br, and I than for X = F, whereas it is the isotropic polarizability derivative that makes the Raman intensity of this mode larger for X = I than for X = F, Cl, and Br. Hence, by combining these two polarization effects, we have the calculated Raman intensities

**Table 19.3** Isotropic and square anisotropic polarizability derivatives of $H_2O...X^-$ complexes[a] [65]

| Species | $H_2O...F^-$ | | $H_2O...Cl^-$ | | $H_2O...Br^-$ | | $H_2O...I^-$ | |
|---|---|---|---|---|---|---|---|---|
| Mode | $\bar{\alpha}'$ | $\gamma'^2$ | $\bar{\alpha}'$ | $\gamma'^2$ | $\bar{\alpha}'$ | $\gamma'^2$ | $\bar{\alpha}'$ | $\gamma'^2$ |
| $v_1$ | 0.03 | 0.02 | 0.14 | 0.00 | 0.17 | 0.01 | −0.20 | 0.14 |
| $v_2$ | 0.10 | 0.23 | 0.07 | 0.28 | 0.03 | 0.28 | 0.34 | 3.01 |
| $v_3$ | 0.00 | 0.68 | 0.00 | 0.99 | 0.00 | 1.13 | 0.00 | 14.84 |
| $v_4$ | 0.04 | 0.26 | 0.10 | 0.47 | 0.18 | 0.64 | −0.42 | 5.67 |
| $v_5$ | 0.93 | 3.76 | 1.18 | 22.84 | 1.18 | 29.76 | 2.45 | 29.38 |
| $v_6$ | 1.17 | 3.80 | 1.08 | 2.39 | 1.11 | 2.15 | 0.91 | 3.49 |

[a]The units of isotropic and square anisotropic polarizability derivatives are $Å^2/amu^{1/2}$ and $Å^4/amu$

which follow the trend of $I_R$ (0.2, $F^-$) ≪ $I_R$ (222.8, $Cl^-$) < $I_R$ (271.5, $Br^-$) < $I_R$ (477.2, $I^-$)

The Raman intensity of the bending mode of water strongly depends on the property of halide anions. As shown in Table 19.2, the bending mode in $H_2O...I^-$ exhibits stronger Raman intensity than the others. On the basis of Raman spectra of alkali halide aqueous solutions, Schultz and Hornig obtained the enhancement factors about 0.5, 1.9, 7.8, and 10.9 affected from $F^-$, $Cl^-$, $Br^-$, and $I^-$, respectively [7]. This is in agreement with our present results, such as 1.7, 3.7, 5.9 and 47.2. The results listed in Table 19.3 indicate that for the HOH bending vibration both absolute values of the isotropic and anisotropic polarizability derivatives increase from $F^-$ to $I^-$. Our results also indicated that the enhancement in the Raman intensity of the water bending vibration could be interpreted from the vibrational coupling between HOH and HOX bending coordinates. The former vibration borrows the polarizability derivative from the latter one. This plays an important role in the enhancement of the bonding interaction on the Raman intensity of the HOH bending mode in the $H_2O...I^-$ complex.

In summary, the HB interaction can induce significantly the vibrational frequency shifts and the relative Raman intensity changes when water molecules directly binding to anions. In this case, the large polarizability of proton acceptors in the hydrogen bond plays an important role.

### *19.3.3 Water Adsorbed on Noble Metal Cathodes*

To gain a deeper insight on water adsorbed on the silver, gold, and platinum electrodes, DFT calculations combined with metal cluster models were carried out [43]. Since the experimental potentials were more negative than the potential of zero charge, the electrode surfaces will carry negative charges. Figure 19.4 depicts the optimized structures for a water molecule interacting with negatively charged $Ag_{10}$, $Au_{10}$ and a hydrogen covered $Pt_{10}$ cluster, respectively. The modeling complexes indicate that the hydrogen bond is formed in the form of O–H...M for

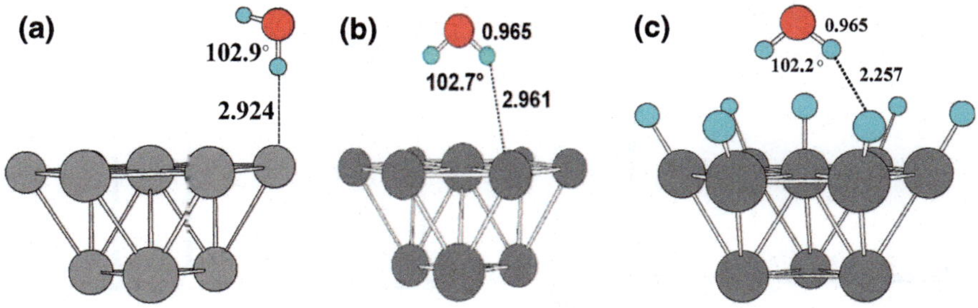


**Fig. 19.4** Optimized structures from cluster models of a water molecule adsorbed on negatively charged metal electrodes, which are mimicked by using $Ag_{10}^-$, $Au_{10}^-$, and $Pt_{10}H_7^-$complexes. **a** $[H_2O + Ag_{10}]^-$; **b** $[H_2O + Au_{10}]^-$; **c** $[H_2O + H_7Pt_{10}]^-$ [43]. Copyright from the Royal Society of Chemistry

silver and gold. For water adsorbed on the Pt electrode, the HB interaction is suggested for water adsorbed on the Pt electrode, i.e., the O–H…H–Pt hydrogen bond.

To understand the SERS signal of the bending vibration of water molecules, it is necessary to discuss the influence of the physical and chemical enhancements. In our previous report, we suggested that an enhancement of the local optical electric field at an adsorbed water molecule on the surface of the metal cluster from the penetration of the surface electronic tail into the solution and the high polarizability of the metal conduction electrons at the more negative potential [138, 139]. However, as mentioned in introduction, the physical enhancement is difficult to interpret the specific enhancement for the bending vibration. Meanwhile, the effect couldn't directly relate to the adsorption configuration of Raman probe molecules at that time.

Now the results of the DFT calculations presented in Fig. 19.5 show that the water molecule interacting through its H-end with a negatively charged metal cluster can result in a larger enhancement in the Raman intensity of the bending vibration than the stretching one. For a free water molecule, the calculated Raman scattering factor (RSF) values are 1.0, 101.3, and 26.5 Å$^4$/amu, comparable to the experimental data [65, 140, 141]. Accordingly, we may estimate the ratio of differential Raman cross-sections between the bending and symmetric stretching vibrations to be about 0.04 at an excitation line of 632.8 nm. Figure 19.5a shows that the relative Raman intensity in $H_2O$–Au is very similar to that of water in the gas phase or a pure liquid. However, charging a negative charge to $H_2O$–Au results in a significant enhancement in the Raman intensity of the bending vibration in $[H_2O–Au]^-$(Fig. 19.5b). Their RSFs of the bending and stretching vibrations calculated are 3.7 and 235.0 Å$^4$/amu for $H_2O$–Au, and 130.8 and 454.4 Å$^4$/amu for $[H_2O–Au]^-$. For water interacting with the negatively charged metal clusters, our calculated results also showed that the bending vibration was significantly enhanced to one or two orders compared with that of a free water molecule. For the O–H stretching vibrations, the biggest RSF magnitude is 2932.5 Å$^4$/amu in $[Ag_{10}–H_2O]^-$ (Fig. 19.5c), which is about 29 fold of

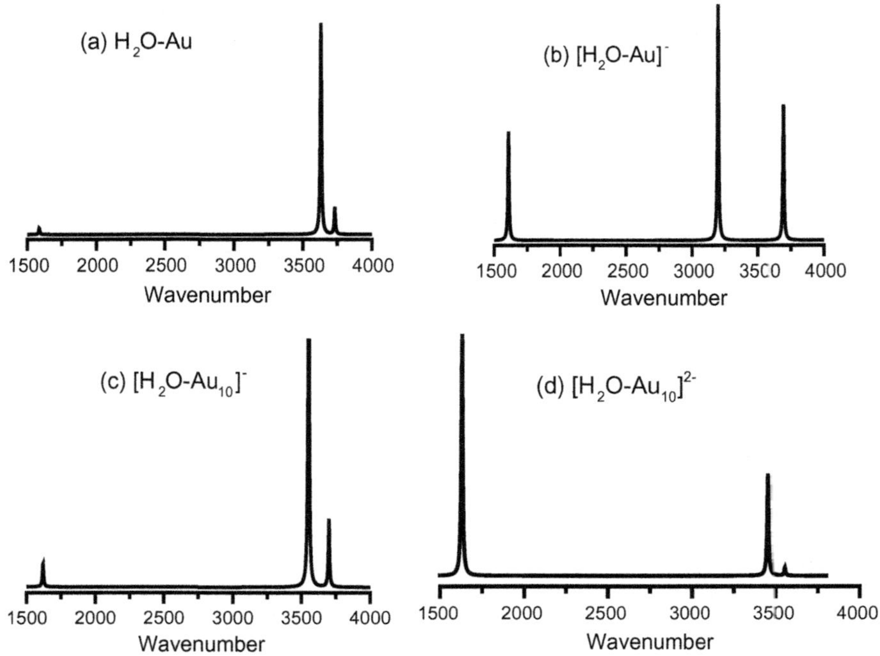


**Fig. 19.5** Theoretical simulation Raman spectra of **a** $H_2O$–Au; **b** $[H_2O–Au]^-$; **c** $[H_2O–Au_{10}]^-$; and **d** $[H_2O–Au_{10}]^{2-}$complexes. The parameters used are 632.8 nm for the excitation wavelength and 20 $cm^{-1}$ for the half-width at the half maximum height [43]. Copyright from the Royal Society of Chemistry

the RSF of free water. But it is smaller than ~590 fold calculated for the bending vibration. We have estimated the ratios of the DRSCS between the bending and stretching vibrations to be 0.78, 0.57, and 0.29 for $[Ag_{10}–H_2O]^-$, $[Au_{10}–H_2O]^-$, and $[Pt_{10}H_7–H_2O]^-$, respectively. Increasing the negative charge to −2 for $Au_{10}$ leads to the RSF of the bending vibration comparable to the stretching vibration (Fig. 19.5d). Although it is too large to doubt its reliability, it indicates that the bending vibration is very sensitive to the surface charge density of the electrode surface in the hydrogen bond configuration.

The present results proposed the changes of the relative Raman intensities are closely associated with the HB of the interfacial water molecules to the metal electrode surfaces. The enhancement in the Raman intensities of the bending vibrations of the interfacial water molecules can be considered as an indicator that the water binding to metal atoms through its H-end.

Copyright from the Royal Society of Chemistry.

Besides, we also considered the pH effect and the structure of adsorption of hydronium on noble metals. It is noted that Raman signals of hydrated proton is changing with the concentration ratio of acid and water [66]. Tian et al. proposed that in acidic solutions water molecules directly adsorbed on electrode surfaces, whereas the hydronium ion located at least at the second layer [94]. By using DFT, we calculated a interfacial structures of $H_3O^+(H_2O)_6$ adsorbed on silver and gold clusters, $H_{15}O_7$–$Ag_{19}$ and $H_{15}O_7$–$Au_{19}$ [64]. The size of hydrated proton selected is due to the structure of $(H_2O)_6$ is considered as the stable interfacial water layer on gold surfaces [67]. Strikingly, the relative Raman intensity of the bending vibration of water was significantly enhanced due to the pre-resonance Raman effect related to the excitation of surface bound electrons.

## 19.4 SERS of Pyridine on Metal Electrodes

Py is one of the most important probe molecules studied in EC-SERS. The adsorbed Py on metal surfaces is the model system for the theoretical study of the SERS mechanism [34, 39, 63, 142]. The adsorption orientation of Py was suggested to be upright or slightly tilted by the N-end of Py when the surface concentration reaches to a monolayer. Good quality SERS spectra can easily be observed from coinage metals (silver, gold, and copper) surfaces [5, 142, 143]. There are five intense peaks at 624, 1010, 1035, 1218, and 1598 $cm^{-1}$, which can be assigned to the totally symmetric $a_1$ modes under a $C_{2V}$ point group, named as asymmetric ring deformation ($\nu_{6a}$, the Wilson notation), ring breathing mode ($\nu_1$), symmetric ring triangular deformation ($\nu_{12}$), C–H in-plane bending mode ($\nu_{9a}$), and ring C–C stretching mode ($\nu_{8a}$), respectively [63]. Compared to the Py in the gas or liquid phases, however, $\nu_1$ and $\nu_{12}$ modes have the most strong normal Raman signals observed in the spectra [144, 145]. For these modes, the electromagnetic selection rule is not expected to provide an interpretation on the changes of their relative intensities [19, 24, 25]. It is the CE mechanism that is responsible for the change in relative Raman intensities in different electrochemical systems. Here we will discuss SERS spectra of Py from three aspects, i.e., the bonding interactions, vibrational frequency shifts, and the CE mechanism.

### *19.4.1 Bonding Interactions*

When Py adopts an upright configuration on metal electrode surfaces, it binds to the electrode surface through the lone-pair orbital of nitrogen that interacts with the conductance band of the surface metal. Molecular orbital theory reveals that the strength of the binding interaction depends on the property of the frontier molecular orbitals and the band structure of electrode materials. Although other adsorption configurations were also proposed in literature [146, 147], we pay attention on the lone-pair donation interaction here. Figure 19.6 shows the periodic trends of the predicted bond energies of Py interacting with monatomic metals, including

**Fig. 19.6** The periodic trend of the binding energies calculated at the B3LYP/6-311+G(d, p)(C, N, H)/LANL2DZ(M) level. The binding energies including the zero-point energy correction are also shown [60]. Copyright from the AIP Publishing

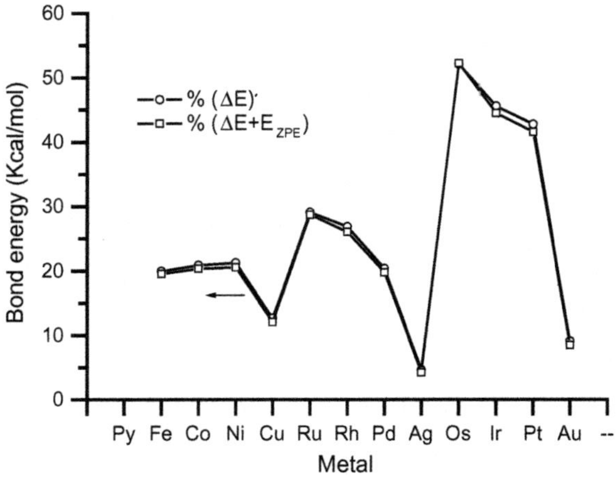


coinage metals and VIIIA transition-metals. For 3d transition metals, the bond energy increases from iron through cobalt to nickel, and then decreases to the minimum at copper. This is in good accordance with the bonding mechanism that the σ donation and the Pauli repulsion in the σ bonding space make a decisive contribution to the Nitrogen-Metal (N-M) bond energy. To make an efficient σ bonding, it is vital to reduce the σ-repulsion between the lone pair electrons on N of Py and the occupied electrons in the σ space of the metal moiety. Taking 3d transition metals, for example, three mechanisms including the 4s-4p polarization, the 4s-$3d_\sigma$ hybridization, and the $4s \rightarrow 3d_\sigma$ excitation are the probable contribution to the bonding between Py and the metals [60, 63, 148–150]. Due to the large energy difference between 4s and 4p orbitals, the latter two mechanisms are the leading interactions between 3d metals and Py [63]. Thus, for 3d metals the binding energy in M-Py follows the order of Fe < Co < Ni.

Unlike the 3d transition metals, the 4d and 5d transition metals exhibit a periodical trend that decreases monotonically from left to right across the periodic table. The same periodic trend has been observed for metal-halogen interactions [151]. The heat of adsorption follows the orders, Rh > Pd > Ag and Ir > Pt > Au. It was assumed that chemisorption arises from the interaction between the donor orbitals of the adsorbate and the vacancies in the metal d band [152, 153], such that the heat of the chemisorption is proportional to the amount of the metal d vacancy count [151]. We believe that a similar bonding mechanism is operating when Py binds to 4d and 5d transition metals. As a consequence, a metal surface with more electron vacancies in its d band will be capable of accepting greater adsorbate charge, resulting in a stronger and more covalent metal-adsorbate bond. Therefore, the adsorbate-metal bond is following the order $d^7 > d^8 > d^9 > d^{10}$. Similarly, the decrease of the vacancies in the d orbitals can be used to explain the order of the binding between Py and the 4d and the 5d metals [151–153]. On the basis of natural bonding orbital (NBO) analysis, the occupations in the 4d and the 5d orbitals are 7.47 for Ru and 6.50 for Os; 8.36 for Rh and 7.82 for Ir; 9.66 for Pd and 9.18 for Pt, respectively [60]. These results correlate well with the order of the binding energies in the same group such that Os > Ru, Ir > Rh and Pt > Pd. The above results show that the binding energies between Py and metals follow the order that Os > Ir > Pt > Ru > Rh > Ni > Co > Fe > Pd > Cu > Au > Ag. In absence of the efficient s-$d_\sigma$ hybridization, the interaction is the weakest for the binding between Py and the coinage metals. Therefore, the occupations in the s and the d orbitals play an important role in the binding between Py and the metals.

Although the single atomic model can give the reliable order in the binding energies between Py and metals, it is probably too simple for the coinage metal surfaces [63]. When considering the binding energies of Py interacting with IB metal clusters $M_n$ (n = 2–4; M = Ag, Au, Cu), our results showed that the binding interaction is significantly improved for describing the adsorption of pyridine on these metal surfaces [62]. Our results show that the binding energy from the metal cluster increases with respect to the single metal atomic case. This is due to the fact that these small metal clusters provide a good accepted orbital with the same symmetry and the matching energy for the lone pair orbital on the nitrogen of the Py. It is in favor of the chemical binding interaction of pyridine with surface metal atoms. In previous studies we in detail analyzed the binding interaction according to the orbital interaction. This is a special important factor to quantitively describe the adsorption property of Py on silver surfaces. Besides, our results agree well with the experimental values for Py adsorbed on copper and silver surfaces [154, 155]. This indicates that a small metal cluster model is probably suitable for a description of the localized chemical bonding of a Py molecule adsorbed on these metal surfaces.

### *19.4.2 Vibrational Frequency Shifts*

For transition-metal (VIII B elements) surfaces, Tian and co-workers devised a series of methods of electrochemical treatment of electrode surfaces and chemical synthesis of nanoparticles to directly record good-quality SERS spectra from pure transition metals [19, 156, 157]. They have systematically investigated SERS of Py from transition metal electrodes, such as iron, cobalt, nickel, ruthenium, rhodium, palladium, and platinum [158–168], and found that Py's SERS spectra on these metal surfaces was considerably different from that on the coinage metals. For example, SERS spectra of Py adsorbed on these transition metal surfaces showed that the signal of the $\nu_{12}$ vibration was weakened, but the Raman signals of the $\nu_{6a}$, $\nu_{9a}$, and $\nu_{8a}$ modes were enhanced significantly. And the Raman intensity of $\nu_{8a}$ at 1596 $cm^{-1}$ was comparable with that of the $\nu_1$ mode.

Through the analysis of bonding interaction, we found Py interacts with these transition metals much more strongly than the typical SERS substrates because of more s-$d_\sigma$ hybridization [60, 63]. Here, we present detailed theoretical analysis of the metal-dependent vibrational frequency shift for Py chemisorbed on the VIIIA and the IB metals. The frequency shifts of several intense bands in the SERS spectra are compared with the theoretical results to understand the nature of the

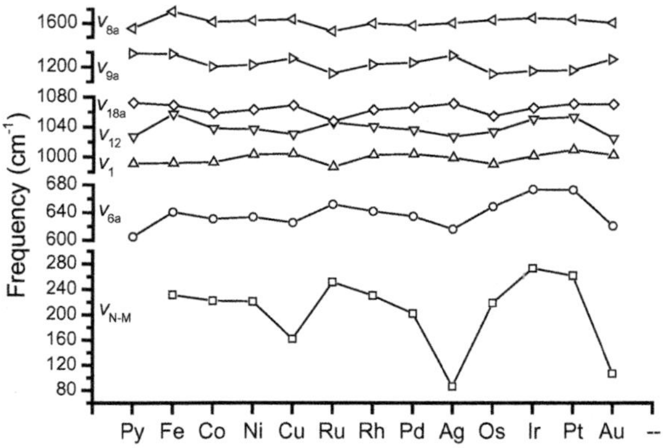


**Fig. 19.7** The dependence of the calculated vibrational frequencies of $\nu_1$, $\nu_{6a}$, $\nu_{8a}$, $\nu_{9a}$, $\nu_{12}$, $\nu_{18a}$ modes and the N-M stretching mode with the totally-symmetric property on the metal atoms in the Py-M complexes [60]. Copyright from the AIP Publishing

bonding between Py and metals (Fig. 19.7). Furthermore, we observed a linear dependence of the vibrational frequency of the $\nu_{6a}$ mode on the force constants of the Py N-metal bond (Fig. 19.8).

Figure 19.7 shows the vibrational frequencies of $\nu_1$, $\nu_{6a}$, $\nu_{12}$, $\nu_{18a}$, $\nu_{9a}$, $\nu_{8a}$ modes and the N-M stretching modes varying with the change of metals. As is shown in Fig. 19.7, the calculated orders of the $\nu_1$ frequencies are Cu > Ni > Co > Fe, Rh > Pd > Ag > Ru, and Pt > Ir > Au > Os. Compared with experimental SERS spectra, our calculations predicted well that the vibrational frequency of the $\nu_1$ mode was the largest on Cu among 3d metals, and that the vibrational frequency is the largest on Rh among 4d metals. And the vibrational frequency of the $\nu_1$ mode among the coinage metals followed the order of Cu > Au > Ag, in agreement with the experimental observations [53, 60, 63].

The vibrational frequencies of the N-M stretching modes locate within a range of 80–300 cm$^{-1}$. Among all complexes, the N-M stretching frequency is the largest for M = Ir, and the smallest for M = Ag. The periodic trends in the vibrational frequencies of the N-M stretches are Fe > Co ≈ Ni > Cu for the 3d metals, Ru > Rh > Pd > Ag for the 4d metals, and Ir > Pt > Os > Au for the 5d metals. Comparison of the binding energy mentioned above indicates that the order of the vibrational frequency of the N-M stretch is almost in conformity with that of the binding energies between Py and metals. We also found the periodic trend of the $\nu_{6a}$ mode mimics that of the $\nu_{N\text{-}M}$ mode in the M-Py systems in Fig. 19.7. Although the Py-M systems are too small to quantitatively describe the corresponding Py/M chemisorption systems, we anticipate that the finding that $\nu_{6a}$ and $\nu_{N\text{-}M}$ possess the same periodic trend still holds true for the Py/M chemisorption systems [53, 60, 63]. Thus we suggest that $\nu_{6a}$ is a better indicator for the strength of the M-Py bonding on the surfaces.

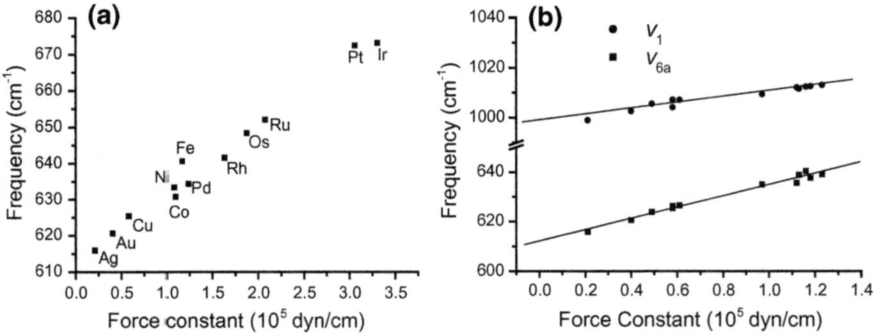


**Fig. 19.8** The calculated vibrational frequencies of the asymmetric ring deformation mode and the ring breathing mode varied with the force constants of the N-M stretching vibrations for the noble and transition metals. **a** Pyridine-Metal atom; [60] **b** Pyridine-Metal clusters (metals: copper, silver, and gold) [61]. Copyright from the AIP Publishing and ACS Publications

Figure 19.8 illustrates the linear dependence of the calculated $\nu_{6a}$ frequency on the force constants of the N-M stretching modes in the different systems of Py/metals. First, the $\nu_{6a}$ frequency depends on the N-M force constants in the Py-M complex with noble metals and transition metals, as shown in Fig. 19.8a. This is in agreement with the results mentioned above but the force constants vary in a range of 0.3–3.5 mdyn/Å. Second, the variations of the vibrational frequency on the force constants of N-M bonds in the size of different metal clusters including Cu, Ag, and Au, as shown in Fig. 19.8b. The bonding mechanism has been analyzed in our previous studies in detail [53, 60, 62, 63]. For the three metals, the vibrational frequency shifts of the $\nu_1$ and $\nu_{6a}$ modes depend on the strength of the bonding between Py and metal clusters. There are good linear relationships of $\nu_1$ and $\nu_{6a}$ frequencies on the N-M force constant. The $\nu_{6a}$ mode has about two fold of the slope of the $\nu_1$ mode, indicating that the former vibrational frequency blue-shifts more sensitive to the strength of the N-M bond than the latter one. It indicates that the coupling is stronger between the $\nu_{6a}$ mode and the N-M stretch than that between the $\nu_1$ mode and the N-M stretch [61]. Therefore, these results show the $\nu_{6a}$ mode is a good indicator for a measure of the strength of the Py-metal bond.

Finally, we also present the predicted frequencies of $\nu_{12}$, $\nu_{18a}$, $\nu_{9a}$ and $\nu_{8a}$ modes for Py interacting with metals. The $\nu_{12}$ mode corresponds to the symmetric ring deformation, whose periodic trend is close to those observed for the $\nu_{6a}$ and the $\nu_{\text{N-M}}$ modes. The $\nu_{18a}$ and $\nu_{8a}$ modes, on the other hand, are made up mainly from the symmetric stretching coordinates of C–N and C–C bonds. The $\nu_{9a}$ mode arises from the C–H in-plane bending motion. The periodic trends of these modes are more similar to the trend of the $\nu_1$ mode, although the change of the $\nu_{8a}$ frequency is small for the 5d metals (See Fig. 19.7). In summary, our results shows that the vibrational frequency shift can reflect the binding interaction between Py and different metals as mentioned above.

### 19.4.3 Chemical Enhancement

The CE effect depends on the specific mode for adsorbed molecules on electrochemical interfaces. This behavior in SERS spectra is closely associated with the adsorption interaction and the coordinate displacement related to the charge transfer states in the studied system [161, 169–172]. We first estimate the effect of the CT on the relative Raman intensity by using the method of Peticolas' formula [173]. In this formula, the relative Raman intensity of a given normal mode is proportional to the square of the displacement $\Delta Q_i$ between the potential energy minima along the $i$th normal mode at the excited and ground states. Thus, the strongest Raman band recorded in the SERS corresponds to the normal mode that has a large displacement of the potential energy minimum at the CT excited state with respect to the ground one. In this case, the approximate expressions can be written by [173]

$$I_i = K\Delta Q_i^2 \omega_i^2 \quad (19.13)$$

where K is a constant for a given incident exciting frequency and $\omega_i$ is the frequency of the $i$th normal mode. Here, K is defined the same as various $a_1$ normal modes. The relative intensity of each band is referred to the $\nu_1$ mode, which has been assigned an intensity of 100, as discussed in the literature [169]. In the present work, the displacements of the normal modes are calculated through the following transformation of the differences between the symmetric internal coordinates $\Delta R$ of both the excited state (CT state $^1B_1$) and the ground state ($^1A_1$).

$$\Delta Q = L^{-1}\Delta R \quad (19.14)$$

where $L^{-1}$ is the inversion of the vibrational normal mode matrix L, obtained by solving the Wilson's GF equation for the electronic ground state [174]. The changes of symmetric internal coordinates are calculated according to the relationships in references [62, 63]. Table 19.4 lists the calculated displacements of the normal coordinates and the relative Raman intensities.

Table 19.4 shows that the relative Raman intensities of the $\nu_{12}$ mode are sensitive to the interaction between Py and metals. For Py-$Cu_2$, the ratio of intensities between $\nu_{12}$ and $\nu_1$ modes are all very small for normal Raman intensities and the CT relative Raman intensities. From Table 19.4, one can obviously see that for Py-$Au_2$ the ratios slightly increase. The ratio between the DRSCS values for both modes increases to 0.18 while the CT ratio is 25/100 for both modes. These results agree well with the experimental SERS spectra [85, 146, 175]. The band intensity of the $\nu_{12}$ mode increases in the relative Raman intensity for Py adsorbed on gold than that on the Cu metal surfaces.

As for the relative Raman intensity of the both modes for Py on silver. It is known that the EM enhancement is significant in the silver substrate [25, 176]. But the EM mechanism should result in almost the same enhancement effect on the both $\nu_{12}$ and $\nu_1$ modes. Thus, the EM mechanism has a little influence on the spectral

**Table 19.4** The scaled theoretical frequencies (in $cm^{-1}$), the differential Raman scattering cross ($10^{-30}$ $cm^2$ $molecule^{-1}$ $sr^{-1}$), the coordinate displacement of the charge transfer states along the $a_1$ vibrational modes, and the relative Raman intensities of Py-$M_2$ systems [59]

| | Py-$Cu_2$ | | | | Py-$Ag_2$ | | | | Py-$Au_2$ | | | |
|---|---|---|---|---|---|---|---|---|---|---|---|---|
| | Freq | $I_R$ | $\Delta Q^a$ | $I_{CT}^b$ | Freq | $I_R$ | $\Delta Q^a$ | $I_{CT}^b$ | Freq | $I_R$ | $\Delta Q^a$ | $I_{CT}^b$ |
| $\nu_{6a}$ | 635.1 | 2.22(18.4) | 0.092 | 12 | 623.9 | 0.17(1.4) | −0.128 | 30 | 635.7 | 0.54(4.5) | 0.036 | 2 |
| $\nu_1$ | 1009.5 | 11.08(164.2) | 0.165 | 100 | 1005.6 | 3.38(50.1) | 0.144 | 100 | 1012.2 | 7.86(117.4) | 0.163 | 100 |
| $\nu_{12}$ | 1036.0 | 0.55(8.4) | 0.031 | 4 | 1030.8 | 3.02(46.2) | 0.048 | 12 | 1037.1 | 1.38(21.3) | 0.079 | 25 |
| $\nu_{9a}$ | 1214.0 | 1.57(29.6) | −0.143 | 109 | 1217.2 | 0.61(11.6) | −0.137 | 133 | 1215.2 | 0.73(13.8) | −0.162 | 142 |
| $\nu_{8a}$ | 1607.9 | 7.43(202.2) | −0.109 | 111 | 1605.4 | 1.03(27.9) | −0.110 | 149 | 1610.8 | 3.01(82.2) | −0.102 | 99 |

[a]The displacements of the normal coordinates in the $A_1$ species for the pyridine–$M_2$ system
[b]The relative Raman intensities of the normal modes of the pyridine–$M_2$ are calculated by use of the Eq. (19.13)

shapes of both the totally-symmetric modes or does not change their relative Raman intensities at all. Therefore, the asymmetry of both two bands on silver surfaces indicates the CT mechanism should be involved. Table 19.4 shows that the CT enhancement contributes to the Raman intensity of the $\nu_{12}$ mode smaller than $\nu_1$ modes under a resonance-like condition. The result is in a good agreement with the experimental spectra for Py on silver [85, 177]. A comparison of the normal mode displacement in Table 19.4 with that in literature implies that it is important for analysis of the relative Raman intensity to take into account the Py-metal bond in calculating the displacement of the normal modes between the CT state and the ground state [169, 178].

The above calculation is on the basis of the assumption that the displacement of the totally-symmetric mode is very small. For checking the reliability of the above result, we use the more general displaced harmonic oscillator model to calculate the relative Raman scattering cross-sections of both modes contributed by the CT excited state [179]. The Huang–Rhys factors of both modes are calculated in terms of the normal coordinate displacement listed in Table 19.4. The vibrational dephasing constants used for the $\nu_{12}$ mode and the $\nu_1$ mode are 2 and 3 $cm^{-1}$ in the gas phase, respectively. These values are in agreement with the experimental linewidths of both Raman bands. The excitation wavelength is 514.5 nm. The calculated results are shown in Fig. 19.9. When we use a small electronic dephasing constant, for example, 20 $cm^{-1}$, ones can see the relative intensities of both modes dependent on the resonance of the excitation light, (see Fig. 19.9a). However, in the metal surface there are large electronic dephasing constant in Raman scattering process. According to the experimental values [180], we use the electronic dephasing constant of 1880 $cm^{-1}$ in Fig. 19.9b. This value is the same order as those that Rubim et al. [181]. and Michaels et al. [182]. have used for Py and

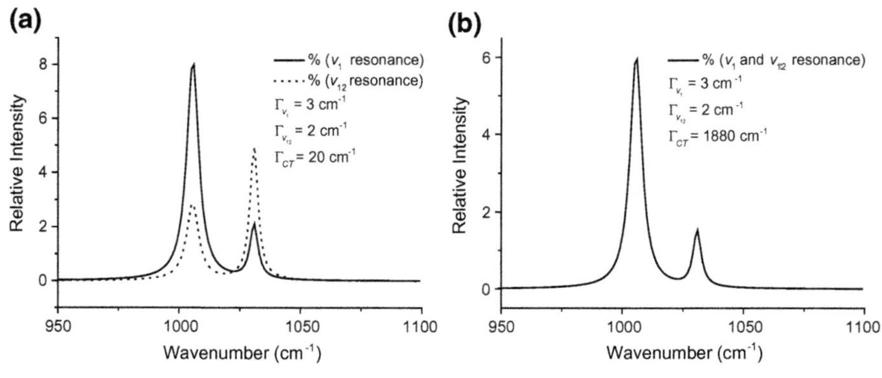


**Fig. 19.9** Effect of the CT excited state on resonance Raman spectra of the pyridine–silver complex. **a** The electronic dephasing constant $\Gamma_{CT} = 20\ cm^{-1}$. The solid line denotes the excitation energy corresponding to the energy of the 0–0 transition energy plus $\hbar\omega_{\nu 1}$; the dot line is the excitation energy corresponding to the energy of the 0–0 transition energy plus $\hbar\omega_{\nu 12}$. **b** The electronic dephasing constant $\Gamma_{CT} = 1880\ cm^{-1}$. The relative intensity is not sensitive to the change of the excitation energy [59]. Copyright from the Elsevier

Rhodamine 6G adsorbed on silver surfaces, respectively. From Fig. 19.9b, we obtain that the ratio of the integrated intensities of $\nu_{12}$ and $\nu_1$ modes is about 16/100. It is close to 12/100 in Table 19.4 for the Py-silver system.

From Table 19.4, one can see that the relative Raman intensities of the $\nu_{8a}$ and $\nu_{9a}$ modes are comparable to that of the $\nu_1$ mode, which agrees with the previous theoretical and experimental results [146, 169, 175, 178]. In the SERS experiments, three normal modes of $\nu_1$, $\nu_{8a}$, and $\nu_{9a}$ correspond to the intense Raman bands. It is worth to note that the relative Raman intensity of the $\nu_{6a}$ mode predicted by using the present method is smaller than the former three modes. This difference can be understood due to the small displacement of the normal modes in the Py-metal complexes.

More recently, the TD-DFT method has been used to predict the dependence of SERS spectra on the incident wavelengths for Py interacting with silver and gold clusters [34, 39, 183, 184]. It was found that the CT enhancement was about $10^3$ for a Py-$Ag_{20}$ complex, where the CT state was ascribed to a transition from a metal orbital to the 5π* orbital of Py. To yield the strongest enhancement, an UV light at 360 nm was chosen to calculate the SERS spectra of Py-$Ag_{20}$ [34, 184]. Similarly, for the Py-$Au_{20}$ complexes, resonance enhancements at 464 and 429 nm were estimated for vertical and surface complexes, respectively [183]. We believe that these incident photonic energies are too high such that interband transitions can be generally excited in the silver and gold clusters. In fact, the UV-SERS was only observed on the transition metals instead of the coinage metals so far [168]. Accordingly, we paid our attention on the influence of the charge-transfer mechanism on the dependence of the relative intensities of SERS spectra on the longer incident wavelengths in visible light.

Finally, we also considered the influence from the changing applied electrode potentials [19, 58, 87, 161]. Our calculated results showed that applied potentials can tune the Fermi level of metal electrodes so that the relative energy of the photo-driven charge transfer state can be matched with the photonic energy of excitation light in SERS measurements. The relative SERS intensities can be significantly changed due to the displacement effect of the totally-symmetric vibrations with applied electrode potentials [87]. This provides a direct evidence to understand the change of potential dependent SERS intensities observed in Py adsorbed on silver electrodes [5, 85].

## 19.5 Conclusion

DFT calculations could give information at the molecular level and plays a unique advantage for investigating the SERS mechanism and electrochemical processes. Combining theoretical calculation and Raman scattering theory, it could give better assignments for Raman spectra of adsorbed molecules, and analyze the change of vibrational frequency and SERS intensities. All these information reveals the physical and chemical nature of SERS signals for probing molecules in complex electrochemical interfaces. In this paper, we presents theoretical results of interfacial water and Py adsorbed on metal electrodes focusing on the CE mechanism of SERS. Our calculated results propose that the strong polarization of the proton acceptor in the hydrogen bond causes a significant enhancement in Raman intensities of the HOH bending mode in some water molecules as proton donors. As for Py adsorbed on different metal electrodes, the binding interaction and the charge transfer mechanism play an important role in the chemical enhancement of some typical modes in Py. The research is helpful to explore chemical adsorptions and chemical reactions of different probing molecules in electrochemical interfaces.

**Acknowledgements** National Natural Science Foundation of China (21533006, 21321062, and 21373712), National Key Basic Research Program of China (No. 2015CB932303), and Funds of State Key Laboratory of Physical Chemistry of Solid Surfaces.

## References

1. P.L. Stiles, J.A. Dieringer, N.C. Shah, R.P. Van Duyne, Annu. Rev. Anal. Chem. **1**, 601 (2008)
2. L. Jensen, C.M. Aikens, G.C. Schatz, Chem. Soc. Rev. **37**, 1061 (2008)
3. S.-Y. Ding, J. Yi, J.-F. Li, B. Ren, D.-Y. Wu, R. Panneerselvam, Z.-Q. Tian, Nat. Rev. Mater. **1**, 16021 (2016)
4. M. Fleischmann, P.J. Hendra, A.J. McQuillan, Chem. Phys. Lett. **26**, 163 (1974)
5. D.L. Jeanmaire, R.P. Van Duyne, J. Electroanal. Chem. Interfacial Electrochem. **84**, 1 (1977)
6. M.G. Albrecht, J.A. Creighton, J. Am. Chem. Soc. **99**, 5215 (1977)
7. S. Nie, S.R. Emory, Sci. **275**, 1102 (1997)
8. K. Kneipp, Y. Wang, H. Kneipp, L.T. Perelman, I. Itzkan, R.R. Dasari, M.S. Feld, Phys. Rev. Lett. **78**, 1667 (1997)
9. S. Zaleski, A.J. Wilson, M. Mattei, X. Chen, G. Goubert, M.F. Cardinal, K.A. Willets, R. P. Van Duyne, Acc. Chem. Res. **49**, 2023 (2016)
10. B. Pettinger, B. Ren, G. Picardi, R. Schuster, G. Ertl, Phys. Rev. Lett. **92**, 4 (2004)
11. Z. Liu, S.-Y. Ding, Z.-B. Chen, X. Wang, J.-H. Tian, J.R. Anema, X.-S. Zhou, D.-Y. Wu, B.-W. Mao, X. Xu, Nat. Commun. **2**, 305 (2011)
12. E. Bailo, V. Deckert, Chem. Soc. Rev. **37**, 921 (2008)
13. Z.-C. Zeng, S.-C. Huang, D.-Y. Wu, L.-Y. Meng, M.-H. Li, T.-X. Huang, J.-H. Zhong, X. Wang, Z.-L. Yang, B. Ren, J. Am. Chem. Soc. **137**, 11928 (2015)
14. L. Cui, B. Liu, D. Vonlanthen, M. Mayor, Y. Fu, J.-F. Li, T. Wandlowski, J. Am. Chem. Soc. **133**, 7332 (2011)
15. U. Jung, M. Müller, N. Fujimoto, K. Ikeda, K. Uosaki, U. Cornelissen, F. Tuczek, C. Bornholdt, D. Zargarani, R. Herges, J. Colloid Interface Sci. **341**, 366 (2010)
16. J.-F. Li, Y.-F. Huang, Y. Ding, Z.-L. Yang, S.-B. Li, X.-S. Zhou, F.-R. Fan, W. Zhang, Z.-Y. Zhou, D.-Y. Wu, Nat. **464**, 392 (2010)
17. C.-Y. Li, J.-C. Dong, X. Jin, S. Chen, R. Panneerselvam, A.V. Rudnev, Z.-L. Yang, J.-F. Li, T. Wandlowski, Z.-Q. Tian, J. Am. Chem. Soc. **137**, 7648 (2015)
18. J.-F. Li, Y.-J. Zhang, A.V. Rudnev, J.R. Anema, S.-B. Li, W.-J. Hong, P. Rajapandiyan, J. Lipkowski, T. Wandlowski, Z.-Q. Tian, J. Am. Chem. Soc. **137**, 2400 (2015)
19. Z.-Q. Tian, B. Ren, D.-Y. Wu, J. Phys. Chem. B **106**, 9463 (2002)
20. R. Chang, *Surface Enhanced Raman Scattering* (Springer Science & Business Media, 2013)
21. M. Moskovits, J. Raman Spectrosc. **36**, 485 (2005)

22. T.E. Madey, J.T. Yates Jr. *Vibrational Spectroscopy of Molecules on Surfaces,* vol. 1 (Springer Science & Business Media, 2013)
23. A. Campion, P. Kambhampati, Chem. Soc. Rev. **27**, 241 (1998)
24. M. Moskovits, Rev. Mod. Phys. **57**, 783 (1985)
25. A. Otto, I. Mrozek, H. Grabhorn, W. Akemann, J. Phys. Condens. Matter **4**, 1143 (1992)
26. M.R. Philpott, J. Chem. Phys. **62**, 1812 (1975)
27. S. Efrima, H. Metiu, J. Chem. Phys. **70**, 1602 (1979)
28. M. Moskovits, K. Michaelian, J. Chem. Phys. **69**, 2306 (1978)
29. F.W. King, R.P. Van Duyne, G.C. Schatz, J. Chem. Phys. **69**, 4472 (1978)
30. I. Pockrand, A. Otto, Solid State Commun. **35**, 861 (1980)
31. E. Burstein, Y. Chen, C. Chen, S. Lundquist, E. Tosatti, Solid State Commun. **29**, 567 (1979)
32. J.I. Gersten, R.L. Birke, J.R. Lombardi, Phys. Rev. Lett. **43**, 147 (1979)
33. J. Gersten, A. Nitzan, J. Chem. Phys. **73**, 3023 (1980)
34. L. Zhao, L. Jensen, G.C. Schatz, J. Am. Chem. Soc. **128**, 2911 (2006)
35. C.L. Haynes, A.D. McFarland, L. Zhao, R.P. Van Duyne, G.C. Schatz, L. Gunnarsson, J. Prikulis, B. Kasemo, M. Käll, J. Phys. Chem. B **107**, 7337 (2003)
36. J.P. Kottmann, O.J. Martin, D.R. Smith, S. Schultz, Chem. Phys. Lett. **341**, 1 (2001)
37. R.J. Gale, *Spectroelectrochemistry: Theory and Practice* (Springer Science & Business Media, 2012)
38. Selman, J.; Varma, R. *Techniques for Characterization of Electrodes and Electrochemical Processes* (Wiley, 1991)
39. D.-Y. Wu, X.-M. Liu, S. Duan, X. Xu, B. Ren, S.-H. Lin, Z.-Q. Tian, J. Phys. Chem. C **112**, 4195 (2008)
40. B. Pettinger, J. Chem. Phys. **85**, 7442 (1986)
41. D.-Y. Wu, J.-F. Li, B. Ren, Z.-Q. Tian, Chem. Soc. Rev. **37**, 1025 (2008)
42. A.H. Ali, C.A. Foss, J. Electrochem. Soc. **146**, 628 (1999)
43. J.-F. Li, Y.-F. Huang, S. Duan, R. Pang, D.-Y. Wu, B. Ren, X. Xu, Z.-Q. Tian, Phys. Chem. Chem. Phys. **12**, 2493 (2010)
44. P. Hohenberg, W. Kohn, Phys. Rev. **136**, B864 (1964)
45. W. Kohn, L.J. Sham, Phys. Rev. **140**, A1133 (1965)
46. J.P. Perdew, Y. Wang, Phys. Rev. B **45**, 13244 (1992)
47. L. Wilk, S. Vosko, M. Nusair, Can. J. Phys. **58**, 1200 (1980)
48. J.P. Perdew, K. Burke, M. Ernzerhof, Phys. Rev. Lett. **77**, 3865 (1996)
49. J.P. Perdew, J. Chevary, S. Vosko, K.A. Jackson, M.R. Pederson, D. Singh, C. Fiolhais, Phys. Rev. B **46**, 6671 (1992)
50. A.D. Becke, Phys. Rev. A **38**, 3098 (1988)
51. A.D. Becke, J. Chem. Phys. **98**, 5648 (1993)
52. C. Lee, W. Yang, R.G. Parr, Phys. Rev. B **37**, 785 (1988)
53. D.Y. Wu, M. Hayashi, Y.J. Shiu, K.K. Liang, C.H. Chang, Y.L. Yeh, S.H. Lin, J. Phys. Chem. A **107**, 9658 (2003)
54. W. Koch, M.C. Holthausen, *A Chemist's Guide to Density Functional Theory* (Wiley, 2015)
55. W. Schattke, M. Van Hove, Phys. J. **3**, 53 (2004)
56. D.Y. Wu, B. Ren, Z.Q. Tian, ChemPhysChem **7**, 619 (2006)
57. D.Y. Wu, B. Ren, Z.Q. Tian, Isr. J. Chem. **46**, 317 (2006)
58. D.Y. Wu, S. Duan, B. Ren, Z.Q. Tian, J. Raman Spectrosc. **36**, 533 (2005)
59. D.Y. Wu, M. Hayashi, S.H. Lin, Z.Q. Tian, Spectrochim. Acta Part A Mol. Biomol. Spectrosc. **60**, 137 (2004)
60. D.-Y. Wu, B. Ren, X. Xu, G.-K. Liu, Z.-L. Yang, Z.-Q. Tian, J. Chem. Phys. **119**, 1701 (2003)
61. D.Y. Wu, M. Hayashi, Y.J. Shiu, K.K. Liang, C.H. Chang, S.H. Lin, J. Chin. Chem. Soc. **50**, 735 (2003)
62. D.Y. Wu, M. Hayashi, C.H. Chang, K.K. Liang, S.H. Lin, J. Chem. Phys. **118**, 4073 (2003)
63. D.-Y. Wu, B. Ren, Y.-X. Jiang, X. Xu, Z.-Q. Tian, J. Phys. Chem. A **106**, 9042 (2002)

64. R. Pang, L.-J. Yu, D.-Y. Wu, B.-W. Mao, Z.-Q. Tian, Electrochim. Acta **101**, 272 (2013)
65. D.-Y. Wu, S. Duan, X.-M. Liu, Y.-C. Xu, Y.-X. Jiang, B. Ren, X. Xu, S.H. Lin, Z.-Q. Tian, J. Phys. Chem. A **112**, 1313 (2008)
66. R. Pang, L.-J. Yu, M. Zhang, Z.-Q. Tian, D.-Y. Wu, J. Phys. Chem. A **120**, 8273 (2016)
67. S. Duan, D.-Y. Wu, X. Xu, Y. Luo, Z.-Q. Tian, J. Phys. Chem. C **114**, 4051 (2010)
68. D.-Y. Wu, X.-M. Liu, Y.-F. Huang, B. Ren, X. Xu, Z.-Q. Tian, J. Phys. Chem. C **113**, 18212 (2009)
69. L.-B. Zhao, R. Huang, M.-X. Bai, D.-Y. Wu, Z.-Q. Tian, J Phys. Chem. C **115**, 4174 (2011)
70. L.-B. Zhao, Y.-F. Huang, X.-M. Liu, J.R. Anema, D.-Y. Wu, B. Ren, Z.-Q. Tian, Phys. Chem. Chem. Phys. **14**, 12919 (2012)
71. L.-B. Zhao, R. Huang, Y.-F. Huang, D.-Y. Wu, B. Ren, Z.-Q. Tian, J. Chem. Phys. **135**, 134707 (2011)
72. L.-B. Zhao, M. Zhang, Y.-F. Huang, C.T. Williams, D.-Y. Wu, B. Ren, Z.-Q. Tian, J. Phys. Chem. Lett. **5**, 1259 (2014)
73. L.A. Curtiss, K. Raghavachari, G.W. Trucks, J.A. Pople, J. Chem. Phys. **94**, 7221 (1991)
74. T.H. Dunning, J. Chem. Phys. **90**, 1007 (1989)
75. P.J. Hay, W.R. Wadt, J. Chem. Phys. **82**, 270 (1985)
76. R. Huang, H.-T. Yang, L. Cui, D.-Y. Wu, B. Ren, Z.-Q. Tian, J. Phys. Chem. C **117**, 23730 (2013)
77. P. Pulay, G. Fogarasi, F. Pang, J.E. Boggs, J Am. Chem. Soc. **101**, 2550 (1979)
78. Y. Xue, D. Xie, G. Yan, Int. J. Quantum Chem. **76**, 686 (2000)
79. Y. Bouteiller, J.-C. Gillet, G. Grégoire, J.P. Schermann, J. Phys. Chem. A **112**, 11656 (2008)
80. I. Alecu, J. Zheng, Y. Zhao, D.G. Truhlar, J. Chem. Theory Comput. **6**, 2872 (2010)
81. R. Pang, X. Jin, L.L. Zhao, S.-Y. Ding, D.-Y. Wu, Z.-Q. Tian, Chem. J. Chin. Univ. **36**, 2087 (2015)
82. G.B. Bacskay, S. Saebø, P.R. Taylor, Chem. Phys. **90**, 215 (1984)
83. B.A. Zilles, W.B. Person, J. Chem. Phys. **79**, 65 (1983)
84. H. Schrötter, H. Klöckner, *Raman Scattering Cross Sections in Gases and Liquids* (Springer, 1979)
85. J. Billmann, A. Otto, Solid State Commun. **44**, 105 (1982)
86. A. Bruckbauer, A. Otto, J. Raman Spectrosc. **29**, 665 (1998)
87. M.T. Lee, D.Y. Wu, Z.Q. Tian, S.H. Lin, J. Chem. Phys. **122**, 094719 (2005)
88. R. Huang, L.-B. Zhao, D.-Y. Wu, Z.-Q. Tian, J. Phys. Chem. C **115**, 13739 (2011)
89. L.-J. Yu, R. Pang, S. Tao, H.-T. Yang, D.-Y. Wu, Z.-Q. Tian, J. Phys. Chem. A **117**, 4286 (2013)
90. V. Barone, M. Cossi, J. Tomasi, J. Chem. Phys. **107**, 3210 (1997)
91. A.V. Marenich, C.J. Cramer, D.G. Truhlar, J. Phys. Chem. B **113**, 6378 (2009)
92. M. Futamata, D. Diesing, Vib. Spectrosc. **19**, 187 (1999)
93. A. Funtikov, S. Sigalaev, V. Kazarinov, J. Electroanal. Chem. Interfacial Electrochem. **228**, 197 (1987)
94. Z.Q. Tian, B. Ren, Y.X. Chen, S.Z. Zou, B.W. Mao, J. Chem. Soc. Faraday Transactions **92**, 3829 (1996)
95. M.J. Blandamer, M.F. Fox, Chem. Rev. **70**, 59 (1970)
96. S. Gopalakrishnan, D. Liu, H.C. Allen, M. Kuo, M.J. Shultz, Chem. Rev. **106**, 1155 (2006)
97. C.D. Cappa, J.D. Smith, K.R. Wilson, B.M. Messer, M.K. Gilles, R.C. Cohen, R.J. Saykally, J. Phys. Chem. B **109**, 7046 (2005)
98. C. Chatzidimitriou-Dreismann, U. Krieger, A. Möller, M. Stern, Phys. Rev. Lett. **75**, 3008 (1995)
99. G. Walrafen, *Water: A Comprehensive Treatise* (Plenum Press, New York, 1972) pp 151
100. J.S. Bartlett, K.J. Voss, S. Sathyendranath, A. Vodacek, Appl. Opt. **37**, 3324 (1998)
101. G.W. Faris, R.A. Copeland, Appl. Opt. **36**, 2686 (1997)
102. D.M. Carey, G.M. Korenowski, J. Chem. Phys. **108**, 2669 (1998)
103. W.R. Busing, D.F. Hornig, J. Phys. Chem. **65**, 284 (1961)
104. J.-W. Schultz, D. Hornig, The Journal of Physical Chemistry **65**, 2131 (1961)

105. G.E. Walrafen, J. Chem. Phys. **1962**, 36 (1035)
106. F. Rull, J. De Saja, J. Raman Spectrosc. **17**, 167 (1986)
107. R. Weston, Spectrochim. Acta **18**, 1257 (1962)
108. T.T. Wall, D.F. Hornig, J. Chem. Phys. **45**, 3424 (1966)
109. D. Liu, G. Ma, L.M. Levering, H.C. Allen, J. Phys. Chem. B **108**, 2252 (2004)
110. S. Burikov, T. Dolenko, P. Velikotnyi, A. Sugonyaev, V. Fadeev, Opt. Spectrosc. **98**, 235 (2005)
111. H. Telle, A. Laubereau, Optics Commun. **34**, 287 (1980)
112. H. Yui, T. Sawada, Phys. Rev. Lett. **85**, 3512 (2000)
113. M.J. Tauber, R.A. Mathies, J. Phys. Chem. A **105**, 10952 (2001)
114. M. Mizuno, T. Tahara, J. Phys. Chem. A **105**, 8823 (2001)
115. M.J. Tauber, R.A. Mathies, Chem. Phys. Lett. **354**, 518 (2002)
116. M. Mizuno, T. Tahara, J. Phys. Chem. A **107**, 2411 (2003)
117. M.J. Tauber, R.A. Mathies, J. Am. Chem. Soc. **125**, 1394 (2003)
118. M. Mizuno, S. Yamaguchi, T. Tahara, J. Phys. Chem. A **109**, 5257 (2005)
119. M. Fleischmann, P. Hendra, I. Hill, M. Pemble, J. Electroanal. Chem. Interfacial Electrochem. **117**, 243 (1981)
120. T. Chen, K. Smith, J. Owen, R. Chang, Chem. Phys. Lett. **108**, 32 (1984)
121. H. Kung, T. Chen, Chem. Phys. Lett. **130**, 311 (1986)
122. Y.-X. Chen, S.-Z. Zou, K.-Q. Huang, Z.-Q. Tian, J. Raman Spectrosc. **29**, 749 (1998)
123. Y.-X. Chen, Z.-Q. Tian, Chem. Phys. Lett. **281**, 379 (1997)
124. Y.X. Chen, A. Otto, J. Raman Spectrosc. **36**, 736 (2005)
125. J.M. Herbert, M. Head-Gordon, J. Am. Chem. Soc. **128**, 13932 (2006)
126. J.-H. Choi, K.T. Kuwata, Y.-B. Cao, M. Okumura, J. Phys. Chem. A **102**, 503 (1998)
127. W.H. Robertson, M.A. Johnson, Annu. Rev. Phys. Chem. **54**, 173 (2003)
128. P. Ayotte, G.H. Weddle, C.G. Bailey, M.A. Johnson, F. Vila, K.D. Jordan, J. Chem. Phys. **110**, 6268 (1999)
129. P. Ayotte, C.G. Bailey, G.H. Weddle, M.A. Johnson, J. Phys. Chem. A **102**, 3067 (1998)
130. K. Hiraoka, S. Mizuse, S. Yamabe, J. Phys. Chem. **92**, 3943 (1988)
131. S.S. Xantheas, J. Phys. Chem. **100**, 9703 (1996)
132. W.H. Thompson, J.T. Hynes, J. Am. Chem. Soc. **122**, 6278 (2000)
133. J. Kim, J.Y. Lee, K.S. Oh, J.M. Park, S. Lee, K.S. Kim, Phys. Rev. A **59**, R930 (1999)
134. X. Xu, W.A. Goddard, J. Phys. Chem. A **108**, 2305 (2004)
135. D.P. Schofield, J.R. Lane, H.G. Kjaergaard, J. Phys. Chem. A **111**, 567 (2007)
136. G.M. Chaban, S.S. Xantheas, R.B. Gerber, J. Phys. Chem. A **107**, 4952 (2003)
137. J. Craig, J. Raman Spectrosc. **33**, 191 (2002)
138. Y.X. Jiang, J.F. Li, D.Y. Wu, Z.L. Yang, B. Ren, J.W. Hu, Y.L. Chow, Z.Q. Tian, Chem. Commun. 4608 (2007)
139. B.I. Lundqvist, O. Gunnarsson, H. Hjelmberg, J.K. Norskov, Surf. Sci. **89**, 196 (1979)
140. W.F. Murphy, Mol. Phys. **33**, 1701 (1977)
141. W.F. Murphy, Mol. Phys. **36**, 727 (1978)
142. W.H. Yang, G.C. Schatz, J. Chem. Phys. **97**, 3831 (1992)
143. S. Angel, L. Katz, D. Archibald, L. Lin, D. Honigs, Appl. Spectrosc. **42**, 1327 (1988)
144. T. Klots, Spectrochim. Acta Part A Mol. Biomol. Spectrosc. **54**, 1481 (1998)
145. F.P. Urena, M.F. Gomez, J.L. González, E.M. Torres, Spectrochim. Acta Part A Mol. Biomol. Spectrosc. **59**, 2815 (2003)
146. J. Creighton, Surf. Sci. **173**, 665 (1986)
147. S. Su, Rong.; H.; Zhao, L.-B.; Wu, D.-Y.; Tian, Z.-Q. Acta Phys. Chim. Sin. **27**, 781 (2011)
148. P.J. Boussard, P.E. Siegbahn, M. Svensson, Chem. Phys. Lett. **231**, 337 (1994)
149. M. Rodgers, J. Stanley, R. Amunugama, J. Am. Chem. Soc. **122**, 10969 (2000)
150. R. Amunugama, M. Rodgers, J. Phys. Chem. A **105**, 9883 (2001)
151. S.A. Wasileski, M.J. Weaver, J. Phys. Chem. B **106**, 4782 (2002)
152. B. Hammer, J.K. Nørskov, Adv. Catal. **45**, 71 (2000)
153. S.S. Sung, R. Hoffmann, J. Am. Chem. Soc. **107**, 578 (1985)

154. J.-G. Lee, J. Ahner, J. Yates Jr., J. Chem. Phys. **114**, 1414 (2001)
155. M. Yang, T. Rockey, D. Pursell, H. Dai, J. Phys. Chem. B **105**, 11945 (2001)
156. Z.Q. Tian, B. Ren, Annu. Rev. Phys. Chem. **55**, 197 (2004)
157. Z.-Q. Tian, B. Ren, B.-W. Mao, J. Phys. Chem. B **101**, 1338 (1997)
158. P.G. Cao, J.L. Yao, B. Ren, B.W. Mao, R.A. Gu, Z.Q. Tian, Chem. Phys. Lett. **316**, 1 (2000)
159. R.A. Gu, P.G. Cao, J.L. Yao, B. Ren, Y. Xie, B.W. Mao, Z.Q. Tian, J. Electroanal. Chem. **505**, 95 (2001)
160. D.-Y. Wu, Y. Xie, B. Ren, J.-W. Yan, B.-W. Mao, Z.-Q. Tian, PhysChemComm **161**, 89, (2001)
161. Y. Xie, D.-Y. Wu, G.-K. Liu, Z.-F. Huang, B. Ren, J.-W. Yan, Z.-L. Yang, Z.-Q. Tian, J. Electroanal. Chem. **554**, 417 (2003)
162. Q.J. Huang, X.F. Lin, Z.L. Yang, J.W. Hu, Z.Q. Tian, J. Electroanal. Chem. **563**, 121 (2004)
163. B. Ren, Q.-J. Huang, W.-B. Cai, B.-W. Mao, F.-M. Liu, Z.-Q. Tian, J. Electroanal. Chem. **415**, 175 (1996)
164. J.S. Gao, Z.Q. Tian, Spectrochim. Acta Part a-Mol. Biomol. Spectrosc. **53**, 1595 (1997)
165. R.-A. Gu, X.-Y. Shen, G.-K. Liu, B. Ren, Z.-Q. Tian, J. Phys. Chem. B **108**, 17519 (2004)
166. B. Ren, X.F. Lin, J.W. Yan, B.W. Mao, Z.Q. Tian, J. Phys. Chem. B **107**, 899 (2003)
167. W.B. Cai, C.X. She, B. Ren, J.L. Yao, Z.W. Tian, Z.Q. Tian, J. Chem. Society-Faraday Trans. **94**, 3127 (1998)
168. B. Ren, X.-F. Lin, Z.-L. Yang, G.-K. Liu, R.F. Aroca, B.-W. Mao, Z.-Q. Tian, J. Am. Chem. Soc. **125**, 9598 (2003)
169. J.F. Arenas, I. López Tocón, J.C. Otero, J. Marcos, I. J. Phys. Chem. **100**, 9254 (1996)
170. J.F. Arenas, M.S. Woolley, I.L. Tocón, J.C. Otero, J.I. Marcos, J. Chem. Phys. **112**, 7669 (2000)
171. J.F. Arenas, J. Soto, I.L. Tocón, D.J. Fernández, J.C. Otero, J.I. Marcos, J. Chem. Phys. **116**, 7207 (2002)
172. D.C. Selmarten, J.T. Hupp, J. Chem. Soc. Faraday Trans. **92**, 3909 (1996)
173. W.L. Peticolas, D.P. Strommen, V. Lakshminarayanan, J. Chem. Phys. **73**, 4185 (1980)
174. S. Califano, *Vibrational states* (Wiley, 1976)
175. J.A. Creighton, C.G. Blatchford, M.G. Albrecht, J. Chem. Soc. Faraday Trans. 2 Mol. Chem. Phys. **75**, 790 (1979)
176. G. Abstreiter, M. Cardona, A. Pinczuk, *Light Scattering in Solids IV* (Springer, 1984) p. 5
177. L.A. Dick, A.D. McFarland, C.L. Haynes, R.P. Van Duyne, J. Phys. Chem. B **106**, 853 (2002)
178. I. López-Tocón, S. Centeno, J. Otero, J. Marcos, J. Mol. Struct. **565**, 369 (2001)
179. R. Islampour, M. Hayashi, S.H. Lin, Chem. Phys. Lett. **234**, 7 (1995)
180. M. Muniz-Miranda, Chem. Phys. Lett. **340**, 437 (2001)
181. J.C. Rubim, P. Corio, M.C. Ribeiro, M. Matz, J. Phys. Chem. **99**, 15765 (1995)
182. A.M. Michaels, J. Jiang, L. Brus, J. Phys. Chem. B **104**, 11965 (2000)
183. C.M. Aikens, G.C. Schatz, J. Phys. Chem. A **110**, 13317 (2006)
184. L. Jensen, L.L. Zhao, G.C. Schatz, J. Phys. Chem. C **111**, 4756 (2007)

# Chapter 20
# Advances in Anharmonic Methods and Their Applications to Vibrational Spectroscopies

**Krzysztof B. Beć, Justyna Grabska and Yukihiro Ozaki**

**Abstract** In this chapter, an overview of anharmonic time-independent approaches and their applications to vibrational spectroscopy will be presented. Attention to their significance for near-infrared (NIR) spectroscopy and studies on complex molecules in condensed phase will be paid. The benefits that are offered by NIR studies and particular difficulties that emerge in case of reproduction of NIR spectra due to the treatment of non-fundamental modes will be highlighted. A short introduction to available anharmonic methodologies, directly overviewed on the basis of the most recent reports in the field, will be presented. An exceptional possibility of elucidation of physicochemical properties, intermolecular interactions, hydrogen bonding, and solvent effects, through investigations of infrared and near-infrared modes, will be examined. Next, recent achievements allowing for accurate and efficient reproduction of experimental spectra in the entire NIR region will be presented. The perspectives for major advances approaching in the field of NIR spectroscopy due to the recent advances in anharmonic theoretical approaches will finally be deliberated.

**Keywords** Anharmonicity · Vibrational spectroscopy · IR and NIR spectroscopy

## 20.1 Introduction

Vibrational spectroscopy stands as an exceedingly rich source of structural and physicochemical information on the molecular level. Infrared (IR), near-infrared (NIR), and Raman spectra are sensitive and very specific, thus being able to provide

---

K.B. Beć (✉) · J. Grabska · Y. Ozaki (✉)
Department of Chemistry, School of Science and Technology,
Kwansei Gakuin University, 2-1 Gakuen, Sanda, Hyogo 669-1337, Japan
e-mail: krzysztof.bec@kwansei.ac.jp

Y. Ozaki
e-mail: ozaki@kwansei.ac.jp



M.J. Wójcik et al. (eds.), *Frontiers of Quantum Chemistry*,
https://doi.org/10.1007/978-981-10-5651-2_20

detailed data on molecular structure and intra- and intermolecular interactions [1, 2]. At the same time, the level of details of vibrational spectra seldom makes them easy to interpret, a fact which is even more notable in case of more complex molecules. This is why the aid which quantum chemistry can deliver to vibrational spectroscopy has been widely utilized since almost a century already [3]. In case of IR and Raman spectroscopy, this symbiosis has reached a truly outstanding level in the past 20 years or so, with countless results reported there so far. However, the last few decades have also seen an outstanding rise of NIR spectroscopy in various fields of applied sciences. It is otherwise a truly fortunate fact that most of the time the model of quantum harmonic oscillator approximates the vibrational properties of molecules quite well, considering its simplicity [4]. And although IR and Raman spectroscopies can undoubtedly gain from moving beyond the double-harmonic approximation, it is by no means necessary in their case [1, 5]. Obviously, the specific of NIR spectroscopy makes this entirely different, as it relies on non-fundamental transitions, thus inevitably requiring an anharmonic approach [6–8]. Although anharmonicity of molecular vibrations itself has been focused on since a long time, the connection between theoretical and applied research in the NIR has never reached a level which has been long known for IR spectroscopy. Investigations of entire NIR spectra based on theoretical studies still remain very rare compared to the number of similar studies in the spectral region of fundamental vibrations. However, we believe that in the forthcoming years, one will witness a remarkable expansion of NIR-related theoretical studies, mirroring the recent progress that has been achieved in the experimental NIR spectroscopy. The significance of the anharmonicity of molecular vibrations extends beyond pure interest of spectroscopy itself. Anharmonic effects influence, i.e., kinetics of chemical reactions [9–11], mechanisms of intermolecular relaxation of vibrational energy [12], for example, impacting stimulated Raman scattering effects [13], or have been evidenced to have a major influence on peptide binding and enzymatic reactions [14], thus having a major importance for general physical chemistry.

In this chapter, we will introduce the current state of knowledge in the field of time-independent anharmonic approaches. We will briefly explain the advantages that stem from utilization of NIR spectroscopy and the particular needs that applied research in this field has, which can be matched by modern implementations of anharmonic methods. Next, a selection of the most established anharmonic approaches will be briefly introduced. These deliberations will be based on key reports published in the field; both basic and particularly applied research will be overviewed with special attention paid to the most recent advances. Finally, we will attempt to foresee the future advances of anharmonic schemes, which we believe will be growing in importance in the forthcoming years.

## 20.2 An Overview of Anharmonic Approaches

### *20.2.1 Introductory Remarks*

We should briefly highlight the essential difference between the formalism standing behind theoretical IR and NIR spectroscopies. The calculation of IR modes can be performed within the double-harmonic approximation, in which both the harmonic frequencies and transition intensities can be obtained with relative ease [15]. Although this model remains very simplistic, it can offer adequate results with predictable error in most cases, which have led to the establishment of empirical scaling factors [16]. The determination of the scaling factors has become a major topic on itself with countless contributions, which have been initiated as early as in 1981 [17]. Scaled harmonic frequencies can, therefore, be accurate enough for general use in IR spectroscopy, as proved by an enormous number of research papers published so far. The limitations of the harmonic approximation may be evident for highly anharmonic modes, i.e., O–H or C–H stretching modes, for which the shape of vibrational surface deviates significantly from the harmonic simplification even for the bottom of the potential well [18]. But this again mostly concerns IR spectroscopy, while NIR spectroscopy may find almost no use from the harmonic approximation.

It is necessary here to explain the basic reasons for which NIR spectroscopy is presently growing in importance in various fields of applied sciences [6, 7]. It primarily benefits from simpler and more reliable construction of spectrometers and more convenient optical materials, resulting in an overall less-expensive laboratory equipment, factors which can frequently be deciding in applied studies. Because of the above reasons, a rapid development of portable spectrometers working in NIR wavelength region has been witnessed recently, bringing NIR spectroscopy outside laboratories for on-site measurements. NIR spectroscopy is also better suited for analysis of water-rich samples and aqueous solutions, which has nontrivial impact on its utilization in biochemical studies and in the industry. However, even from the point of view of basic research, NIR spectroscopy offers considerable advantages [8]. At a glance, it seems that the molecular information carried out in NIR spectra is somewhat duplicatory to the one obtained from fundamental modes. However, the NIR spectra-structure correlations [19, 20] differ from the ones established for IR spectra [1]. For example, the articulation of bands due to vibrations of functional groups and particularly X–H vibration is notably stronger in NIR spectra [6, 7]. As mentioned earlier, the anharmonicity of the molecular vibrations itself provides a major knowledge about the investigated system. Further, the complex mechanisms of mode coupling can be also explored in search of a deeper understanding of molecular vibrations and how they are influenced by various factors. Therefore, many open problems remain in this matter as for now. So far, the inherent complexity of vibrational spectra in the NIR region, which is considerably higher than in the IR, has been limiting our attempts to answer some of these fascinating questions. We believe that the future developments of anharmonic approaches may play a vital role in the future of vibrational spectroscopy.

### 20.2.2 Vibrational Self-consistent Field

Vibrational self-consistent field (VSCF) introduced in 1968 by Bowman [21] with further contributions by Carney et al. [22], Cohen et al. [23], and Gerber and Ratner [24] is a vibrational analogy to Hartree-Fock approach to determination of electronic structure. Within VSCF approximation, the solution to Schrödinger equation of $N$-dimensional system is defined as a Hartree product of one-mode functions

$$\Psi_S(Q) \approx \Phi_S(Q_1, \ldots, Q_M) = \prod_{m=1}^{M} \phi_{S_m}^{m}(Q_m) \tag{20.1}$$

The one-mode functions or 'modals,' $\phi_{S_m}^{m}$, by analogy to orbitals from HF approach for electronic structure [25], are expanded in a primitive basis, i.e., of harmonic oscillator functions. The minimization problem, therefore, becomes a set of coupled one-dimensional eigenvalue problems, which is solved self-consistently, giving in result the optimal modals. However, the analogy with electronic HF method breaks in one point that is the distinguishability of vibrational modes.

The amount of anharmonicity that is being covered within VSCF for a single mode is complete, and given a complete modal basis is used [26]. However, the mode-mode coupling for a given mode is treated by a mean-field approximation of the potential of other modes [21]. This introduces a substantial error in the systems with strongly coupled modes. Therefore, the exact anharmonicity stemming from mode coupling, the off-diagonal anharmonic terms, needs to be captured by anharmonic many-body approaches that extend beyond VSCF formalism. On the other hand, the common advantage of VSCF implementations is that no higher order properties other than Hessian are required, thus making VSCF calculations easily applicable.

### 20.2.3 Correlation-Corrected Vibrational Methods

There have been few improvements proposed so far addressing the major shortcoming in VSCF approach, which is an inadequate treatment of mode coupling. The most straightforward one, although not chronologically earliest one, is the correlation-corrected VSCF (cc-VSCF; or perturbation-corrected VSCF, PT2-VSCF), for which a foundation was laid in 1996 by Norris et al. [27]. Aiming at increasing the accuracy of VSCF calculations, they proposed Møller-Plesset perturbation theory for correcting the vibrational SCF approach, again in analogy to the treatment of electronic structure [28]. In the perturbative treatment, the perturbation Hamiltonian $V'$ is introduced as a difference between the full Hamiltonian and its part defined within VSCF that is the single-mode SCF

$$V' = H - \sum_k h_{\mathrm{SCF}} \tag{20.2}$$

Such formulated Hamiltonian is then used in Rayleigh-Schrodinger perturbation formalism. This approach approximates the off-diagonal anharmonicity effectively.

A different approach to a better approximation of the mode–mode interaction has been proposed in 1975 [29] by Whitehead and Handy, with numerous further contributions from various groups [30–32]. It is analogous to the configuration interaction (CI) method in the electronic structure theory. The core of vibrational configuration interaction (VCI) scheme is the diagonalization of the Hamiltonian in a basis of Hartree products (configurations) of one-mode functions, i.e., VSCF modals. The VCI wavefunction is defined as

$$\Psi_S^{\mathrm{VCI}}(Q_1, \ldots, Q_M) = \sum_S^{S'} C_S^{S'} \Phi_{S'}(Q_1, \ldots, Q_M) \tag{20.3}$$

In Eq. (20.3), the $K$ number is the total number of configurations. VCI coefficients $C_S^{S'}$ are then determined by a variational minimization of the energy. This is done by diagonalization of the Hamiltonian, and the resulting eigenvectors give VCI coefficients, while the eigenvalues give the corresponding energies. Thus, this approach determines variationally the best wavefunction within the basis set. If we assume a complete basis set and a full configuration space, this results in full VCI and exact energies of all states become available. The major disadvantage of VCI approach is the exponential scaling, $N^M$, where $N$ denotes the number of basis functions for each of $M$ nodes. Commonly, the basis set which is being used in VCI calculation consists of harmonic oscillator functions; eigenfunctions of a ground state VSCF can be used as well, for forming the configuration space. This gives much faster convergence, and therefore typically VCI step is being done after the initial VSCF calculations. Obviously, a full VCI is practically feasible for the smallest molecules only. A key importance in VCI implementations is held by the ways of reducing the configuration space. This can be obtained, i.e., by truncating the Eq. (20.3) to desired excitation level (i.e., VCIS for single excitations, VCISD for doubles, etc.); other solutions for reducing the computational cost of VCI calculations have been proposed over time [33].

Vibrational coupled cluster (VCC), proposed by Christiansen et al. [34, 35], is a different approach to anharmonicity beyond the VSCF mean-field formalism, again based on direct analogy to the coupled cluster method from the theory of electronic states. The exponential ansatz for VCC wavefunction is given as

$$|\mathrm{VCC}\rangle = \mathrm{e}^T |\Phi_{\mathrm{i}}\rangle \tag{20.4}$$

where $T$ denotes the cluster operator

$$T = \sum_{\mathrm{m} \in \mathrm{MCR}[T]} T^{\mathbf{m}} \tag{20.5}$$

and

$$T^{m} = \sum_{\mu^{\mathrm{m}}} \tau_{\mu}\mathbf{m}\tau_{\mu}\mathbf{m} \tag{20.6}$$

MCR[T] specifies the MCs within the excitation space, and $T^{\mathbf{m}}$ operator incorporates the excitation operator along with the corresponding amplitudes. Assuming that all possible MCs are included within MCR[T], we obtain a full excitation term, effectively obtaining a wavefunction equivalent to a full VCI wavefunction [26]. The inclusion of a full set of excitations is not feasible in practice, and the cluster operator is usually truncated. By discarding MCs above certain amount of modes, forms analogous to electronic CCS, CCSD, CCSDT, etc., can be derived. As modes, contrary to electrons, are distinguishable, more sophisticated approaches may be implemented in this case, for example, discarding weakly coupled modes. So far, VCC approach is on a relatively early stage of development; however, it has been already demonstrated that it is superior to VCI method at the corresponding excitation level [26].

There have been other proposals for an extension beyond VSCF level, such as the vibrational multi-configurational self-consistent field (VMCSCF) approach [36, 37]; these have remained rather of limited popularity so far. Both VCI and VCC solutions offer a considerable accuracy in reflecting anharmonic effects, provided that all prerequisites have been fulfilled. Due to exponential scaling, however, the practical applicability of both is severely limited to simple molecules. Currently, out of the post-VSCF approaches the PT2-VSCF method is one of the most used in applied studies, as it does not introduce overwhelming demand for computational resources.

### 20.2.4 Vibrational Second-Order Perturbation Theory

The techniques introduced above more or less rely on probing the potential energy surface, i.e., through a costly grid-based PES evaluation. The key idea behind vibrational second-order perturbation theory is that the quartic and quadratic force constants are being derived through one-dimensional numerical differentiation of the analytical Hessian, by employing displacements of the molecular structure from equilibrium geometry along normal coordinates. These terms are then used to define the vibrational Hamiltonian $\mathbf{H}_{\mathrm{vib}}$

$$\mathbf{H}_{\text{vib}} = \mathbf{H}_{\text{vib}}^{0} + \mathbf{H}_{\text{vib}}^{1} + \mathbf{H}_{\text{vib}}^{2} \tag{20.7}$$

The zero-order term $\mathbf{H}_{\text{vib}}^{0}$ is the harmonic one, the first-order term $\mathbf{H}_{\text{vib}}^{1}$ incorporate cubic components of the potential, and the second-order term $\mathbf{H}_{\text{vib}}^{2}$ includes all higher order components, respectively.

The VPT2 implementations offer significant advantage in computational efficiency; however, the derived vibrational energies and transition moments are typically very sensitive to the existence of tightly coupled modes, vibrational resonances in particular. The standard treatment of such cases often leads to erroneous results. The straightforward solution is an identification and subsequent removal of such tightly coupled terms, i.e., in the deperturbed VPT2 (DVPT2) route. However, this solution also removes a significant portion of anharmonicity captured in the model. Therefore, an improvement in the form of generalized VPT2 (GVPT2) approach was proposed in 2005 by Barone [38], in which the deperturbed tightly coupled terms are reintroduced through variational treatment. This way, the general advantages of VPT2 approach are preserved, while the existence of tightly coupled modes does not lead to singularity, provided that a well-defined criterion for near-degeneracies will be assured. Therefore, GVPT2 approach is suitable for treatment of relatively large molecular systems, for which the probability of tight coupling naturally increases; this is also augmented by the computational affordability of VPT2 methodology. This gives a wide potential applicability of GVPT2 calculations in the applied spectroscopy.

### 20.2.5 Other Approaches to Time-Independent Vibrational Schrödinger Equation

The schemes briefly introduced above are focused on capturing the anharmonicity within the entire vibrational degree of freedom. Both the diagonal, and off-diagonal anharmonicity matters there, and advanced and computationally demanding approaches have been proposed for better description of the latter. However, some application can see a substantial benefit from an accurate description of the diagonal anharmonicity of a selected mode. For the studies of higher overtones of modes with highly anharmonic potential, i.e., X–H stretching modes, a different approach, in form of solving the one-dimensional Schrodinger equation based on a large number of grid-based energy evaluations over displacement along a normal coordinate, may give very good results. The approach itself remains fairly straightforward; however, recently it has been used with good results in the applied NIR studies [18, 39–43].

The one-dimensional Schrödinger equation can be formulated as

$$H\psi_{\upsilon}(q) = \left[ -\frac{\hbar^2}{2\mu}\frac{\mathrm{d}^2}{\mathrm{d}q^2} + V(q) \right]\psi_{\upsilon}(q) = E_{\upsilon}\psi_{\upsilon}(q) \tag{20.8}$$

where $q$, $\mu$, and $V(q)$ denote, respectively, the normal coordinate, the reduced mass, and the vibrational potential of given molecular oscillator. For solving the Eq. (20.8), the Numerov approach can be utilized [44, 45] with further improvements [46, 47]; some advances have been noticed recently in this regard [48]. The route for reducing the cost of resolving the equation has been also reported very recently by Kuenzer et al. [49].

For calculation of single point energies, even for a relatively large grid, post-Hartree-Fock schemes can be utilized [40] yielding highly accurate results. This approach is capable of describing very well the nature of, i.e., N–H or X–H stretching vibrations, particularly when these are involved in a hydrogen bonding; the above statement holds, provided an adequate number of energy evaluations is being involved. It also works well, giving accurate wavenumbers and transition intensities, corresponding to fundamental and the first overtone but also higher overtone transitions. Therefore, it finds particular use in NIR studies of hydrogen bonding in condensed phase, or of subtle effects of interaction with apolar solvent molecules.

## 20.3 Applications

### *20.3.1 Exemplary Applications of VSCF Scheme*

Following a fairly straightforward scheme, VSCF, and PT2-VSCF approaches have been applied in various studies of basic and complex molecules, including biomolecules, where the fundamental vibrations were most often focused on [50–59]. The vibrational dynamics of water, and particularly the properties of bulk water and solvated ions, have always attracted an utmost attention. It should not come as a surprise that VSCF approach was routinely applied in this field as well. Water molecule, water dimer, and a solvated Cl anion have been a subject of PT2-VSCF study of Chaban et al. [50]. They have developed an algorithm utilizing, among others, the PT2-VSCF scheme. They have concluded that for these simple systems the vibrational part was determined to have a minor effect on the final accuracy. They have also stated that PT2-VSCF calculations offer an adequate accuracy for reproducing both the soft intermolecular and the stiff intramolecular modes in weakly bound clusters.

Chaban et al. then applied PT2-VSCF method in their investigation of a glycine molecule in the low-temperature matrix isolation and He droplets [51]. The calculations have been performed at MP2 level, and three low-energy isomers of glycine were taken into account. They concluded that anharmonic effects and mode coupling play a major role in the vibrational properties of glycine. However, the other very important conclusions have been drawn from the analysis of a reverse problem: the agreement between VSCF vibrational frequencies and the experimental ones has confirmed that the calculated potential energy surface resembles well the experimental one. Therefore, this reasoning proved that the ab initio PES obtained by Chaban et al. [51] has been much superior to the OPLS-AA potential, which was a state-of-the-art empirical potential at that time. The authors concluded that the inaccuracy of the empirical potential stems from its inability to describe in detail the effects of hydrogen bond, which are crucial in the case of one of the glycine conformers. The molecular and vibrational properties of glycine themselves have been investigated many times by anharmonic theoretical methods thereafter; these will be reviewed later in this chapter.

The topic of complexes of water with small biomolecules was further investigated with an aid of VSCF method by Gerber et al. [52]. Again, glycine, glycine–water complex, and also N-methylacetamide were chosen as the objects of the study. The anharmonic effects and mode coupling in particular have a significant impact on the vibrational properties of such kinds of molecular systems. Therefore, PT2-VSCF scheme was applied there, to take the advantage of its more accurate description of the off-diagonal anharmonicities. Gerber et al. [52] described quantitatively the amount of anharmonicity in vibrational modes of the systems they have investigated. They have found that the anharmonicity can become extremely large in case of soft modes, i.e., low-frequency partially torsional modes; in such cases, the anharmonic effects contribute almost equally to the harmonic effects. It was concluded that harmonic calculations are futile for such modes. On the other hand, in this study a disadvantage of VSCF approach was also acknowledged, which is the necessity of an extensive probing of PES for accurate results of the VSCF procedure. This obviously adds to the increased computational cost of the perturbative treatment of off-diagonal terms in the PT2-VSCF scheme. Various approaches for resolving the PES for further VSCF calculations were considered. As the general aim of Gerber et al. [52] has been complex biomolecules, the computational cost of this part is of high importance. In general MP2 or DFT calculations have offered adequate accuracy, while coupled cluster approaches were ruled out, due to an extensive cost. However, also empirical, QM/MM and semiempirical methods were discussed for generating PES of bigger molecular systems.

The impact of the way in which the PES is being determined has been recognized as significantly important for VSCF treatment of complex molecules. The opposite demands corresponding to this step have become obvious. From one point, a decrease in the cost of approximating PES has been deemed necessary for making VSCF-based computational schemes viable for treating molecular systems with practical significance to applied spectroscopy. At the same time, the accuracy of resolved vibrational properties should be considered. As demonstrated by Gregurick et al. in their study of peptides and peptide-water complexes [53], the quality of a description of the vibrational potential for selected modes has not always been perfect, leaving concerns about a viability of simplifications in this matter. The further developments of the PES approach for subsequent VSCF and VCI calculations have been also considered by Yagi et al. [54]. In their study based on $H_2O$ and $H_2CO$ molecules, also direct VSCF and VCI schemes were compared with standard ones. While numerous contributions into advancing this field were proposed by Yagi et al. [54], here we would like to stress one of their conclusions. It has been evidenced that the VSCF based on an exact PES, which includes all coordinate couplings, has a strong tendency to overestimate the energy. The workaround for decreasing the computational cost of determining the PES in the entire VSCF scheme was also discussed. The quartic force field (QFF) and modified Shepard interpolation (Int-PES) approaches were applied for approximating the PES. Yagi et al. have concluded that for simplest molecules the QFF solution allows for a good accuracy; but they have also admitted that this would likely not hold for more complex molecules with large amplitude vibrations. Moreover, this simplification results poorly for systems with strongly coupled modes, i.e., OH symmetric and antisymmetric stretching modes of water molecule. In the following report, Yagi et al. [55] further explored the viability of QFF approximation, with $n$-mode coupling representations of the quartic force field ($n$MR-QFF). This simplification only includes the terms up to $n$ number of mode couplings in the QFF. VSCF calculations with such an approximation of the PES can be done with a much lesser cost than the ones based on a $n$MR-PES route, as evidenced basing on results obtained for formaldehyde, ethylene, methanol, propyne, and benzene [55].

The particular significance of anharmonicity in biological systems, especially the effects introduced by hydrogen bonding, has resulted in a strong interest in employing VSCF approach for shedding light at these phenomena. As an example, VSCF methodology has been employed for studies of peptides in crystalline phase [56–58] and attempts have been made to extend the applications of VSCF method for proteins [59]. The topic of advancing VSCF scheme toward viability for the treatment of biomolecules continues, with recent report by Brauer et al. [60] on the anharmonic vibrational spectra of $\alpha$-D-glucose, $\beta$-D-glucose, and sucrose. The VSCF calculations were performed there in a hybrid manner. The harmonic part of the potential has been obtained on MP2 level, while the anharmonic part was treated either at Hartree-Fock AHF or semiempirical PM3 level. They have concluded that even the semiempirical approach is suitable for selected modes.

While VSCF method has been broadly adapted for anharmonic calculations of fundamental transitions, surprisingly the near-infrared modes have been studied less often. This is understandable, as the standard VSCF approach incorporates mean-field approximation of the potential does not describe the mode coupling well. Its perturbative augmented development PT2-VSCF, however, has been used with success for explanation of overtones and combination modes. Recently, Lutz et al. [61] have used PT2-VSCF in their applied analytical study of ethanol in gasoline, in which the anharmonic calculations have provided valuable insight into the origins of the observed NIR bands. However, not only IR and NIR spectroscopy have benefited from the advances in the VSCF theory. In the study of Brauer et al. reviewed above [60], a PT2-VSCF approach has been employed for obtaining the

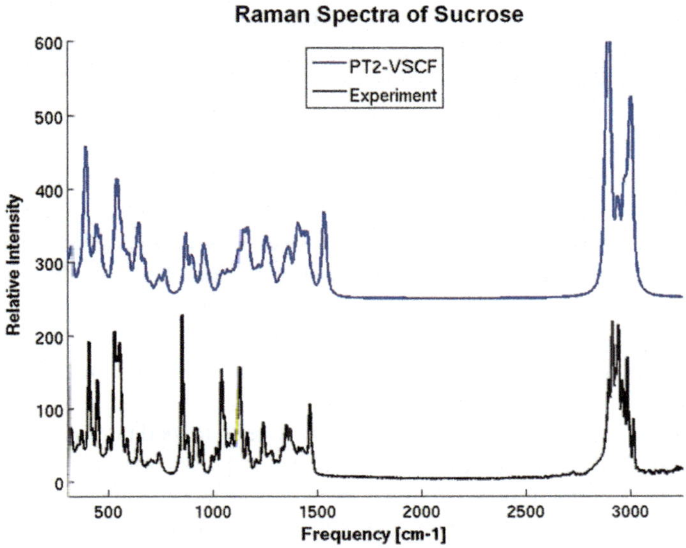


**Fig. 20.1** (*Left*) Raman spectrum of crystalline sucrose. VSCF-PT2 calculations for the hybrid HF/MP2 potential are compared with experiment. (Reproduced with permission from Ref. [60])

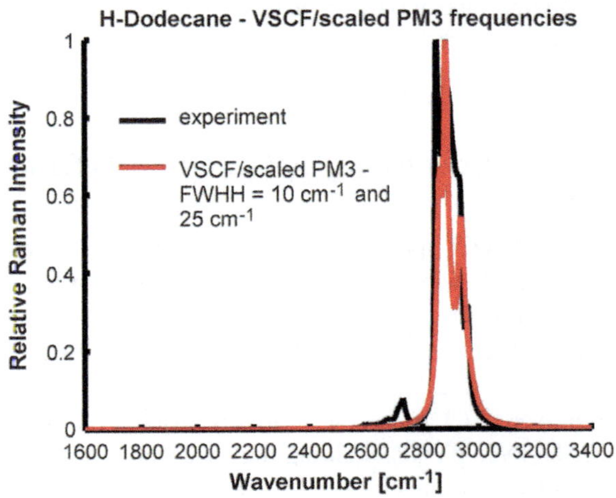


**Fig. 20.2** VSCF spectra of a non-deuterated dodecane isotopomer calculated with the improved PM3 method (*red curve*) compared to experiment (*black curve*). (Reproduced with permission from Ref. [62])

theoretical anharmonic Raman spectra of the studied systems (Fig. 20.1). VSCF scheme has as well been applied for an anharmonic modeling of Raman spectra of dodecane by Šebek et al. [62]. Good agreement with the experimental spectra, measured in liquid phase and at room temperature, has been obtained (Fig. 20.2). The VSCF calculations performed there have been based on a scaled PM3 semiempirical PES.

The examples discussed above demonstrate the capabilities of VSCF calculations in delivering good results for a variety of organic molecules, from very basic ones to heavy alkanes or biomolecules. On the other hand, VSCF technique has been also applied for a variety of other systems, including inorganic ones. For example, Mancera et al. have recently applied VSCF route of calculations for exploring the vibrations of gold and silver clusters [63]. They have employed VSCF and VCI methods, and they have also used an efficient way of approximation of the PES. By using periodic DFT with a plane–wave pseudopotential basis and the Voter–Chen potential, they have aimed for a reduction of the number of pair-couplings between modes. The conclusions from their study were that the anharmonicity of vibrations of Au and Ag clusters is rather small. However, their report has demonstrated the usefulness of their approach for the treatment of heavy atom systems.

As it should be stressed from the above overview, nowadays the VSCF scheme is mainly applied in its perturbatively augmented evolvement, PT2-VSCF (or cc-VSCF), as it offers a more adequate treatment of mode coupling. The standard approach was historically used with a success mainly in the studies of the diagonal, single-mode anharmonicity. However, the general VSCF scheme suffers from a costly PES sampling, and a huge effort has been made to overcome this difficulty. This is in fact one of the frontiers of advances in the VSCF scheme presently. The number of reported applications of VSCF routes in NIR spectroscopy has been so far limited, and with further developments of the theory, an increase in the future may be expected.

### *20.3.2 Exemplary Applications of VCI and VCC Methods*

Although VCI approach has been known for a long time, its exponential scaling prevented its wider use for molecules extending beyond a few atoms until recently, when the computational power has become more available. Here, we will briefly discuss a few recent studies which report on the applications of VCI scheme.

In their study on hydroxycarbene and formaldehyde, Koziol et al. [64] have employed VCI route on relatively high level of theory for electronic structure CCSD(T)/cc-pVQZ, yielding highly accurate results. As the *cis*- and *trans*-isomers of HCOH and HCOD exhibit strong anharmonicity of modes in the region below 4000 $cm^{-1}$, particularly due to mode coupling, with impact on IR intensities. Moreover, the existence of non-fundamental transitions in the IR region and the impact of Fermi resonances on complexity of IR spectra have been discussed therein. These factors may complicate the explanation of IR bands in the absence of anharmonic calculations (Fig. 20.3). They have stressed the fact that both VMP2 approach and obviously harmonic approximation in this case VCI route gave very good agreement, and the accuracy of obtained PES has given not only an insight into vibrational properties and spectra but also on the isomerization barriers. They have also compared the quality of results obtained with VSCF, VMP2, and VCI

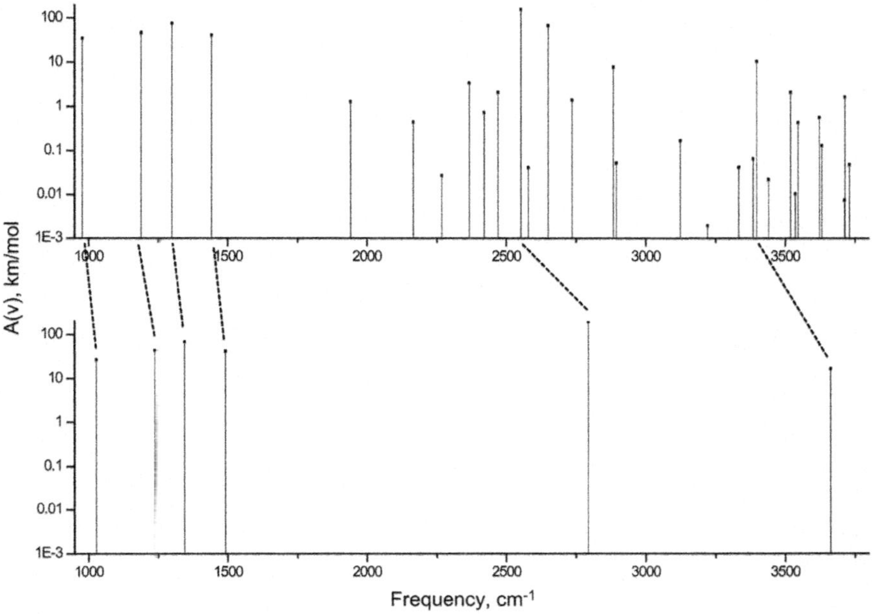


**Fig. 20.3** VCI (*top*) and harmonic (*bottom*) IR spectrum for cis-HCOH. (Reproduced with permission from Ref. [64])

calculations, with VCI ones coming clearly superior. The obtained error for VCI wavenumbers remained lower than 12 $cm^{-1}$ (1%), while in the case of transition intensities it held within 2–7%.

A good example of high powerful may be the VCI calculation on a high level of electronic theory has been demonstrated recently by Oschetzki et al. [65] in their study of azidoacetylene in gas phase and in low-temperature matrix isolation. It is a very interesting system as, as they have reported, the IR spectrum is dominated by anharmonic effects, mainly strong Fermi resonances (Fig. 20.4). The Fermi pair is the major feature in the vibrational spectrum of azidoacetylene, and fundamental mode in the same spectral range is reduced to a shoulder at the side of a very strong combination band. This case is perfect example of the strength of VCI approach, which is very well suited for treatment of tightly coupled modes, i.e., vibrational resonances, as VSCF fails at this, both standard route (caused by mean-field approximation) and perturbatively augmented one (due to denominator term sensitive to strong coupling). However, similar computational approach would be largely impractical for significantly more complex molecule. For these reasons, a considerable effort is constantly being made into making VCI implementations more affordable, i.e., adaptive VCI (A-VCI) [33].

A similar disadvantages limiting practical use of VCI route also hold for VCC method. It may change in the nearest future, as the VCC method is still considered to be at the early stage of its development [26]. However, it has been shown already

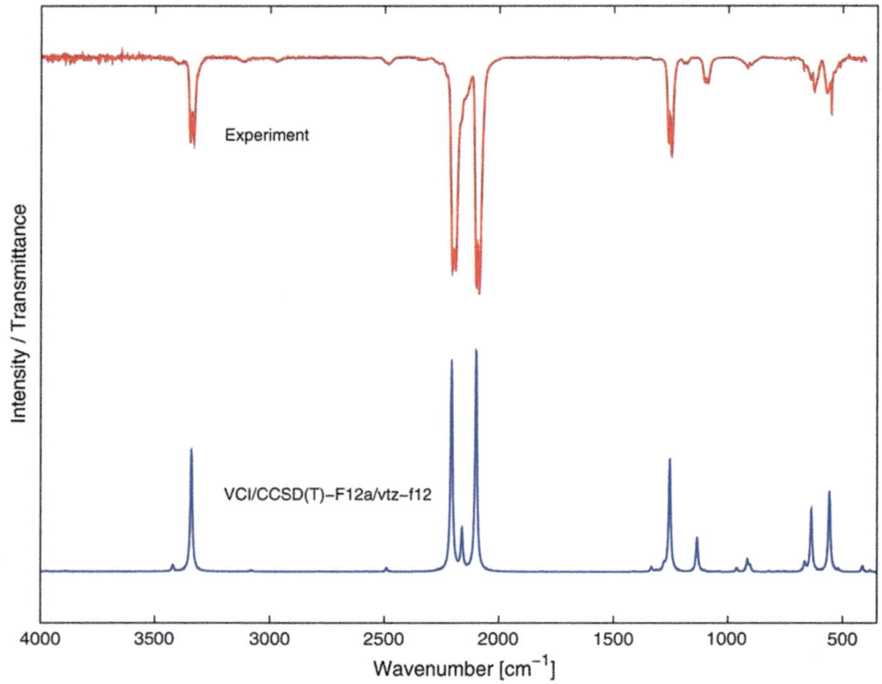


**Fig. 20.4** Comparison of a computed VCI spectrum and gas phase IR measurement (res. 2 cm$^{-1}$) for azidoacetylene. (Reproduced with permission from Ref. [65])

that it can provide better results than VCI method within the same computational cost. Recent report on VCC calculation of pyridine has been published by Godtliebsen and Christiansen [66] (Fig. 20.5). Their study focused on the implementation of band Lanczos procedure for VCI and the non-hermitian Lanczos algorithm for VCC. The IR and Raman spectra of pyridine reproduced with the use of these methods achieved great level of agreement with experimental data. Similarly, also pyridine-metal complexes, with emphasis on reproducing the Raman spectra, were studied as well.

### 20.3.3 Exemplary Applications of Vibrational Perturbation Theory

A variety of improvements over the standard VPT2 scheme was proposed over time [22, 67–69]; many of them were aimed at overcoming the sensitivity of the perturbative approach to near-degeneracies. However, here we would like to focus on the generalized VPT2 approach (GVPT2), mainly due to its tremendous popularity that it has gained in recent years. Since its introduction in 2005 [38], a substantial

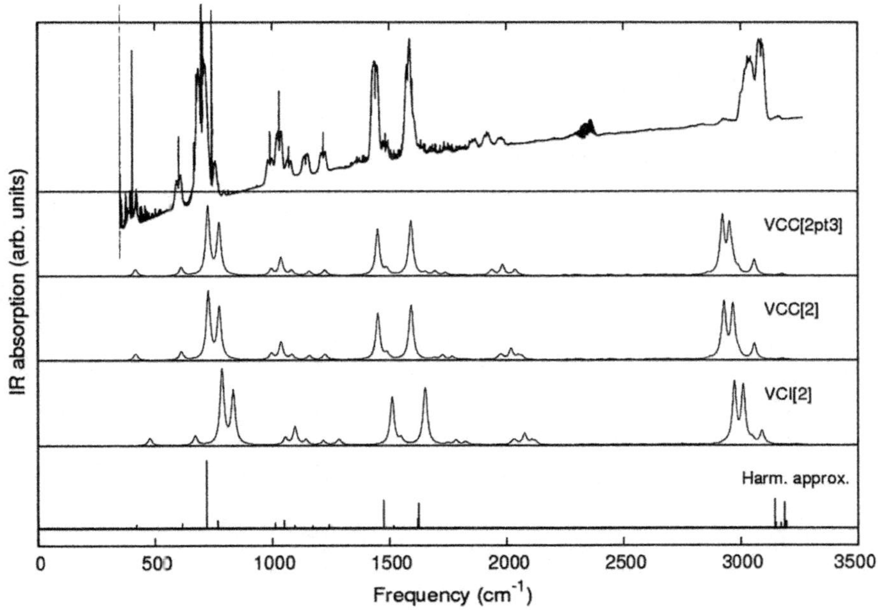


**Fig. 20.5** Pyridine IR spectra calculated from j = 3000 response functions for different vibrational wave functions. (Reproduced with permission from Ref. [66])

number of papers already reported on applications of GVPT2 route. Its subsequent implementation in the very popular Gaussian package from version 09 Rev. D.01 available since 2013 should have an effect in further increasing its widespread usage. The IR and Raman anharmonic spectra of formic acid, including the open and cyclic dimer, have been successfully reproduced by Barone et al. [70]. As presented in Fig. 20.6 the contributions of monomer and both dimeric forms into experimental data obtained in the low-temperature matrix isolation could be unequivocally identified.

GVPT2 calculations have been also extensively used for the studies of nucleobases, along with high-level harmonic calculations. A series of reports has recently emerged [71–74], which focused on IR and Raman spectra of these systems. Both monomeric species as well as hydrogen-bonded and stacked dimers were investigated. A reliable determination of stretching frequencies of C=O and N–H vibrations has been emphasized [74]. Similarly, high attention has been paid by Barone et al. to the conformational isomerism of glycine [75–78], where experimental vibrational data alone could not provide solid answers in this regard. The experimental data have been only able to deliver solid evidence on Ip conformer of glycine. This form was, therefore, used as a basis for validation of the theoretical model [76]. Barone et al. have succeeded in delivering highly accurate data about glycine, on the structure, rotational constants, conformational enthalpies, and vibrational levels, with 0.001 Å, 20 MHz, 1 kJ $\cdot$ mol$^{-1}$ and 10 cm$^{-1}$, respectively

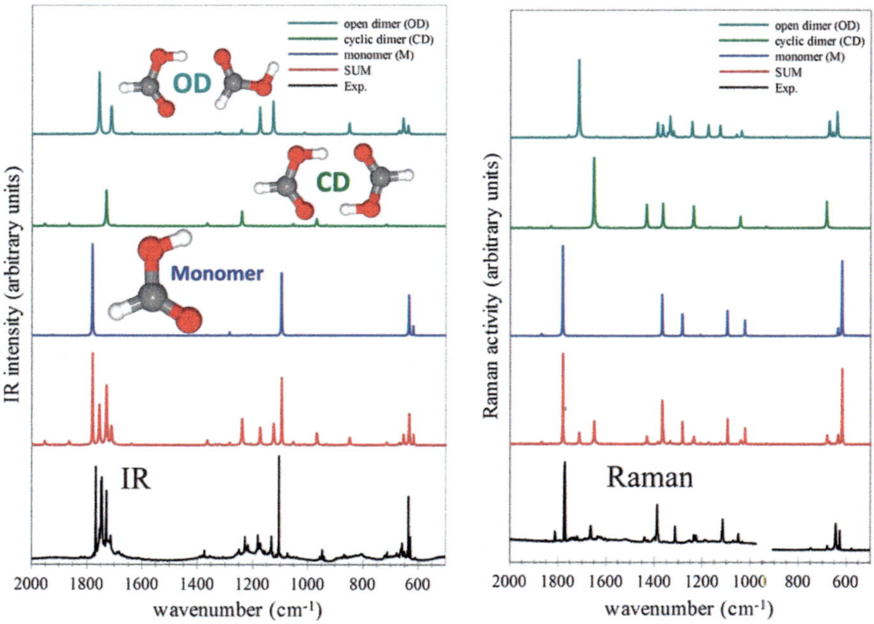


**Fig. 20.6** Computed anharmonic and experimental IR and Raman spectra of formic acid in the 2000–500 $cm^{-1}$ wavenumber range. For theoretical spectra, single contributions from the FA monomer (M) and its most stable dimers, cyclic dimer (CD) and open dimer (OD), and the overall spectra (SUM) obtained as the spectrum of the monomer complemented by the contributions (5–30%) from the CD and CO dimers are presented. Theoretical spectra line shape of monomer and dimers has been convoluted with Lorentzian distribution functions with HWHMs of 1 $cm^{-1}$ and 2 $cm^{-1}$, respectively. (Reproduced with permission from Ref. [70])

[76]. The anharmonic calculations within GVPT2 formalism allowed to compliment the limited information on structure, thermodynamics, and vibrational spectroscopy of glycine that could be derived from the experiment [75–78] (Fig. 20.7). The above studies have demonstrated the usefulness, robustness, and affordability of GVPT2 methodology and its capabilities, when paired with the experimental vibrational spectroscopy, for reflecting the physical chemistry, thermodynamics, intermolecular interactions, and vibrational properties of the biomolecules (Fig. 20.8).

The advantages of GVPT2 formalism make it very suitable for the treatment of various kinds of molecular systems. Yet, it has been applied to simple molecules and mostly in the region of fundamental vibrations; NIR investigations are still rather rare. Beć et al. [79] have recently reported very good agreement between GVPT2 and experimental NIR data of methanol, ethanol, and 1-propanol in diluted solution [79]. In their study of aliphatic alcohols, GVPT2 proved to be able to deliver highly accurate results, when employed within SCRF-CPCM solvent model of carbon tetrachloride at B2PLYP-D/SNST level of theory. Detailed NIR assignments were carried out (Fig. 20.9), and for example, the contributions into

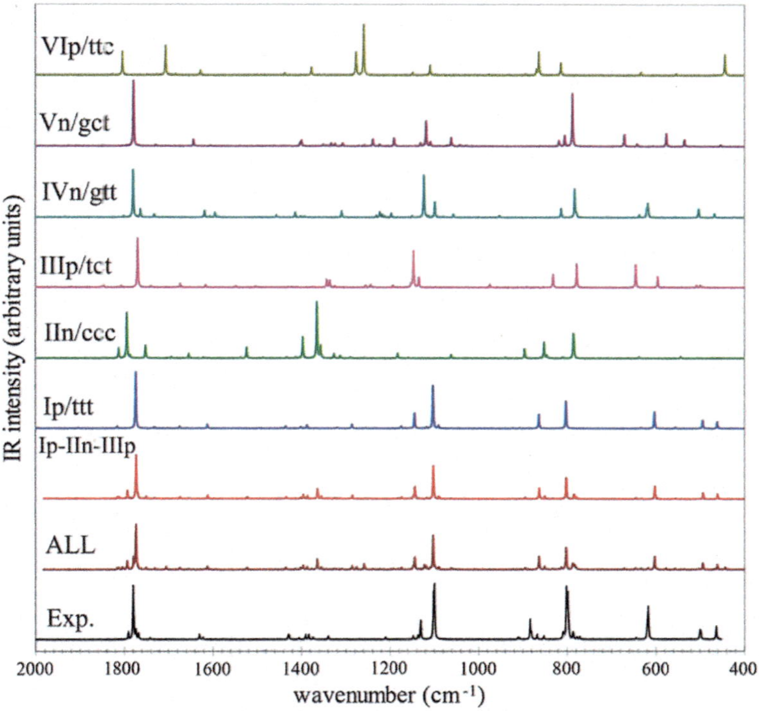


**Fig. 20.7** Best-estimated MI-IR spectra in the (2000–400 cm$^{-1}$) frequency region, for the main glycine isotopologue. Simulated theoretical spectra: single contributions from Ip/ttt, IIn/ccc, IIIp/tct, IVn/gtt, Vn/gct, and VIp/ttc, the sum of the Ip/ttt, IIn/ccc, and IIIp/tct (Ip–IIn–IIIp) contributions weighted for relative abundances (as computed in this work (T = 410 K), also assuming the conformational cooling IVn/gtt-Ip/ttt and Vn/gct-IIIp/tct), and the Ip–IIn–IIIp sum complemented by minor contributions (1%) from the IVn/gtt, Vn/gct, and VIp/tcc (ALL). Experimental IR spectra recorded in the low-temperature Ar matrix generated using the data of Table 5 of Ref. [14]. IR spectra line shapes (both theoretical and experimental) have been convoluted using Lorentzian functions with a half-width at half-maximum (HWHM) of 1 cm$^{-1}$. (Reproduced with permission from Ref. [75])

experimental NIR spectra arising from rotational isomers of ethanol and 1-propanol could be elucidated (Fig. 20.10). A theoretically supported analysis of experimental data of the series of basic alcohols revealed that NIR spectra carry a rich structural information about molecular system. This information is frequently difficult to elucidate and interpret, much more than the corresponding IR data.

Other recent studies using GVPT2 calculations have demonstrated its applicability to highly anharmonic vibrations and hydrogen-bonded systems. Fornaro et al. [80] have studied the hydrogen-bonding interactions in IR region of uracil, uracil dimers, and uracil-water system, by GVPT2 calculations performed at different levels of theory. Full B3LYP-D3, hybrid B3LYP-D3//B2PLYP (anharmonic//harmonic part), and ONIOM calculations, with different parts of the molecules treated

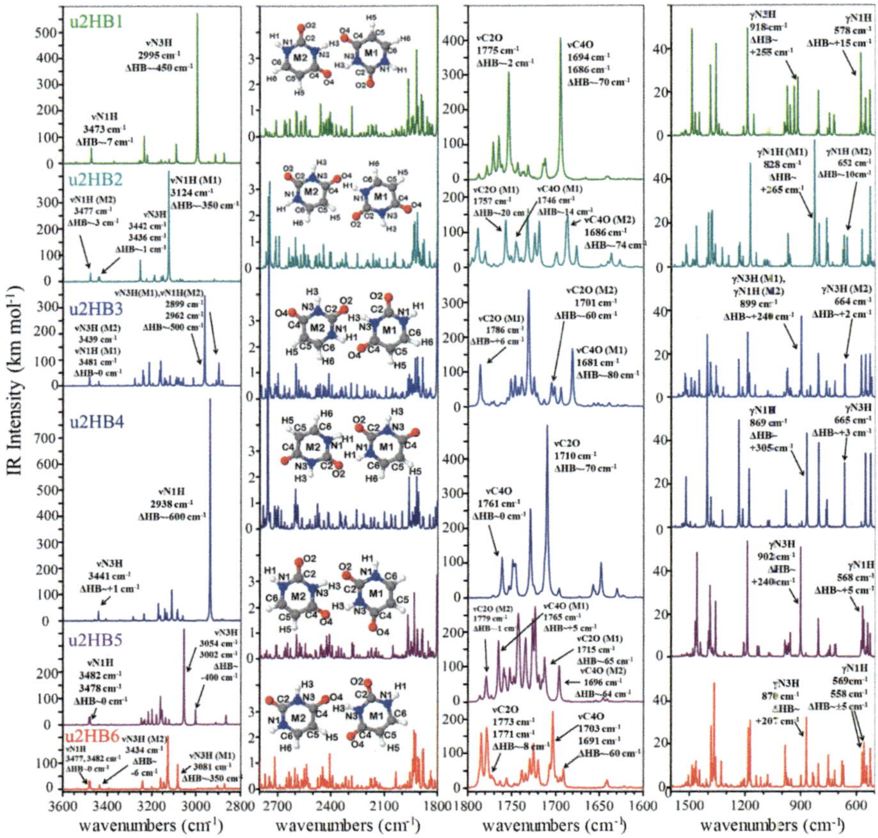


**Fig. 20.8** Anharmonic infrared spectra of the six different hydrogen-bonded uracil dimers computed with the B3LYP-D3/N07D method, along with the assignment of the NH and CO stretchings and NH bendings, and the corresponding shifts of the vibrational frequencies with respect to uracil monomer due to the formation of the hydrogen-bonding interactions (ΔHB). (Reproduced from Ref. [80] with permission)

on different levels of theory, were employed there. Good agreement with experimental IR spectra has been achieved, particularly in deriving the contributions stemming from non-fundamentals (Fig. 20.8). The study of Fornaro et al. [80] have demonstrated the ability of accurate prediction of IR patterns of hydrogen-bonded system by GVPT2 calculations. Their research proved that this approach remains viable for strongly anharmonic vibrations, including those of hydrogen-bonded bridge, resulting in the average error of 22 cm$^{-1}$ for the spectral region relevant to the bands affected by hydrogen bonding [80].

Beć et al. [81] have investigated the acetic acid in NIR region over wide range on concentrations ($CCl_4$), with particular interest in the double-hydrogen-bonded bridge and its influence on NIR spectra. The NIR data of acetic acid, and similarly other carboxylic acids, are difficult to interpret. By employing GVPT2 calculations,

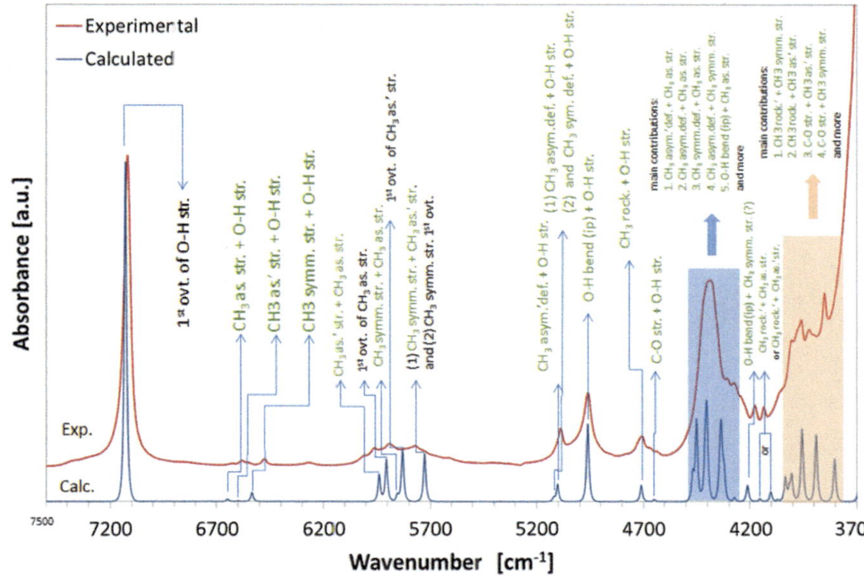


**Fig. 20.9** Examination of NIR spectra of low-concentration (0.005 M) methanol by the calculated NIR spectrum; GVPT2 calculations on CPCM-B2PLYP-D/SNST level of theory were used here. (Reprinted from Ref. [79] with permission)

some light could be shed at this phenomenon. The particular stability of cyclic dimer and the reasons for the prominent increase in the NIR baseline of carboxylic acids could be proposed, based on the results of theoretical study [81]. The hybrid approach, with calculation of harmonic part on the higher level of theory, i.e., on coupled cluster level for acetic acid monomer, was employed there as well.

While the reports on applications of GVPT2 method for basic molecules remain so far the most frequent ones, which is understandable as the framework itself is still rather new, by no means it is limited to such systems. The GVPT2 theory has been already used in studies on complex dielectric function and molar vibrational polarization of *tert*-butylmethyl ether in NIR and IR regions [82]. The affordability of GVPT2 approach enables also calculation of NIR spectra of increasingly more complex molecules, bringing new possibilities of applications in applied research. Recently, for example, GVPT2 resolved that NIR spectrum of rosmarinic acid has been reported in an applied analytical study of rosemary content, which has a growing phytopharmaceutical significance [83]. The GVPT2 calculations were performed by Kirchler et al. [83] with the use of DFT-B3LYP/N07D level of theory for determination of electronic structure. The agreement with the experimental NIR spectrum that they have obtained (Fig. 20.11) suggests that GVPT2 approach can succeed in delivering adequate results even for systems exceeding 180 electrons.

The GVPT2 scheme proves to have a capability of treating large molecules, hydrogen-bonded systems, and molecules in solution phase, all that while

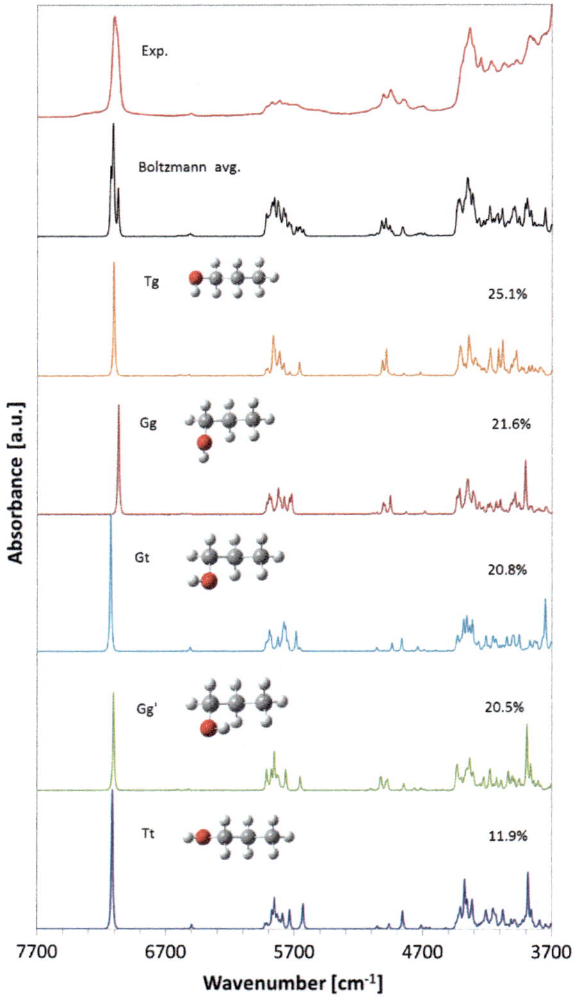


**Fig. 20.10** Experimental and calculated NIR spectra of 1-propanol. CPCM-B2PLYP-D/SNST-based spectra of conformational isomers and final (Boltzmann-averaged) predicted spectrum presented. Relative intensities of spectra of isomers are presented in accordance with their Boltzmann abundances (%). (Reprinted from Ref. [79] with permission)

remaining relatively straightforward in use. Therefore, these factors bring a possibility of applying the GVPT2 approach to an immensely large number of different systems which are in key interest of the experimental NIR spectroscopy. Accurate reproduction of NIR spectra provides invaluable contribution to experimental spectroscopy, which so far has frequently been forced to rely on ambiguous assignments. For the above reasons, we believe that in the nearest future one will

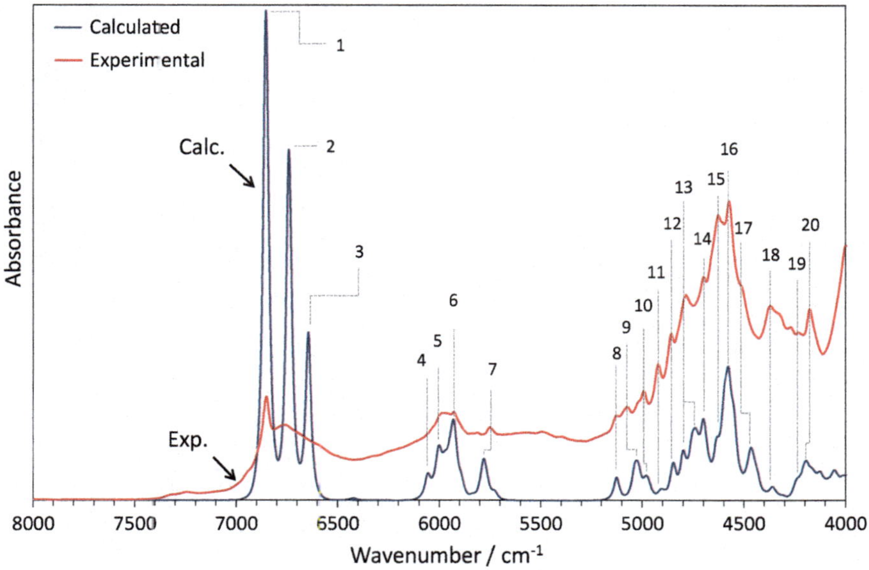


**Fig. 20.11** Experimental and theoretical NIR spectrum of rosmarinic acid obtained through fully anharmonic (GVPT2) DFT-B3LYP/N07D calculation. The experimental spectrum is a SNV spectrum normalized over 15 independent experimental datasets. The theoretical bandshapes were obtained with the application of Cauchy-Gauss product function [83]. Reproduced by permission of The Royal Society of Chemistry

witness a rapidly increasing number of applications of GVPT2 methodology, particularly in the field of applied NIR spectroscopy.

Since the general interest arising from the field of experimental spectroscopy is biased toward complex molecules, much attention has recently been paid to decrease the computational cost of VPT2 implementation, without sacrificing too much of its reliability [70, 71, 78]. The idea behind the reduced dimensionality VPT2 (RD-VPT2) approach is that most of the modes in a typical molecular system do not couple, or the coupling is small (Fig. 20.12). Therefore, such terms may be safely discarded. As the modes that couple usually fulfill a set of rules, i.e., arise from moieties located in the same part of the molecule and have similar frequencies, they can be easily located (Fig. 20.12). Only selected modes of interest, i.e., those that correspond to the experimental bands of high intensity, can be also treated anharmonically. The reduced dimensionality allows a much improved affordability of anharmonic treatment, as evidenced for chlorophyll (Fig. 20.13) [78].

Applications of GVPT2 approach to other kinds of vibrational spectroscopies, i.e., Raman or chiral spectroscopies VCD and ROA, have been also reported recently (Fig. 20.14) [70, 84–87]. For chiral spectroscopies, the availability of reliable anharmonic computational scheme is of crucial importance, as the anharmonicity affects the observed experimental bandshapes.

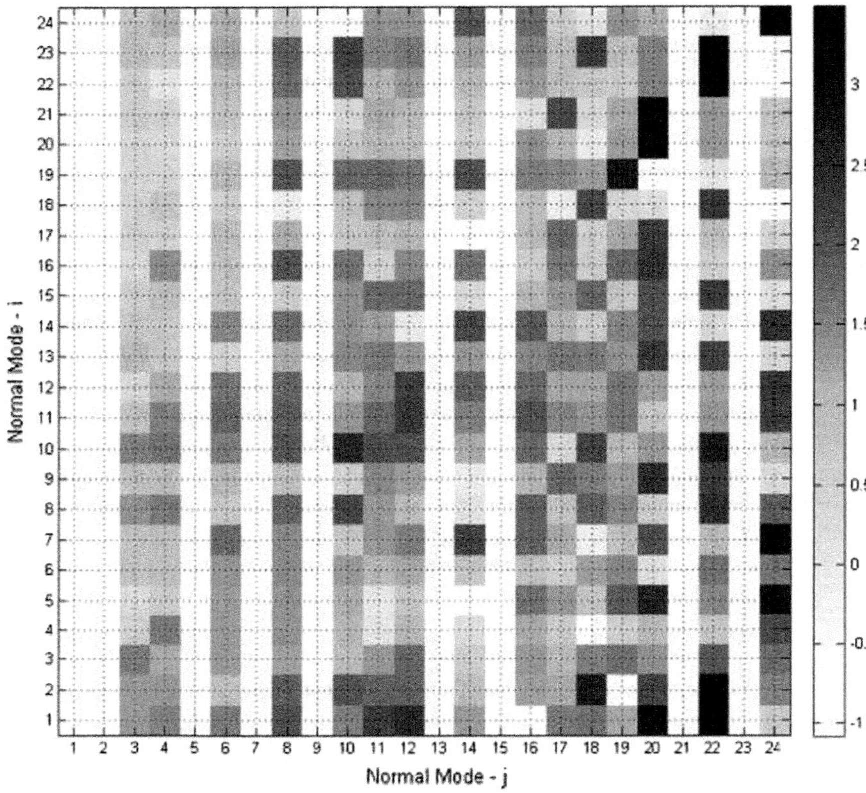


**Fig. 20.12** Graphical representation of the absolute value of the cubic force constants $K_{iij}$ for isolated glycine. A shade of *gray* is assigned depending on the value of $\log_{10}(|K_{iij}|)$, from *white* for the values lower than 0.1 to *black* for values above 3. (Reproduced from Ref. [78] with permission)

## 20.3.4 Miscellaneous Other Examples

NIR spectroscopy provides unique information about anharmonicity of X–H stretching vibrations. Due to typical pronunciation of bands originating from nonbonded X–H group [8], it is very well suited for studies of molecular systems in which both bonded and nonbonded moieties coexist. Moreover, overtones of different order may be investigated based on experimental data, and overall the quality and richness of structural information gained from NIR spectroscopy of such systems is considerable. The general anharmonic approaches, such as VSCF, VCI, or VPT2, not always can offer adequate aid for analysis of such data. Partially because of the theory, but mainly due to ways of implementation these cannot provide desired accuracy nor efficiency. Examples of recent applications of this approach can be pointed out, i.e., in the work of Yabushita group [88–91]. In their

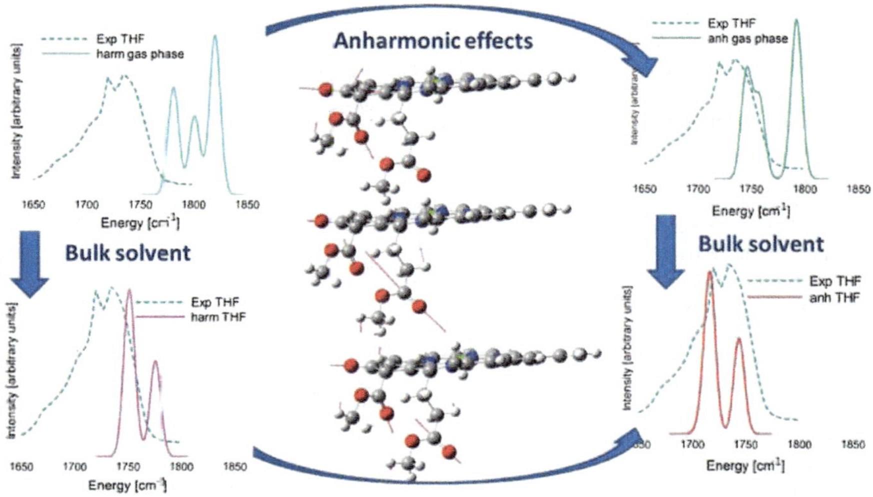


**Fig. 20.13** Vibrational harmonic and anharmonic spectra of the chlorophyll-*a* cation in the gas phase and in tetrahydrofuran solution. (Reproduced from Ref. [78] with permission)

studies of different kinds of X–H stretching vibrations, i.e., C–H stretching [88] and O–H stretching [89–91], they have focused on investigation of dipole transition moments. They have been in particular able to successfully reproduce the subtle dependencies on transition intensities of fundamental and overtone modes. Furthermore, Futami et al. [18, 39–43] have reported a series of studies focused on NIR modes, in which experimental data have been explained by employing the 1D Schrödinger time-independent equation, resolved basing on dense grid of energy evaluations over normal coordinate of selected vibrations.

The vibrational effects of hydrogen bonding in solution have been investigated by Futami et al. [39]. In their study of pyrrole-pyridine complex they particularly focused on the fundamental and first overtone transitions of N–H stretching mode. It has been demonstrated that upon formation of pyrrole-pyridine hydrogen-bonded complex the intensity of N–H stretching first overtone band diminishes significantly. To explain this phenomenon observed experimentally, they employed the approach of solving 1D Schrödinger equation with a dense potential energy surface grid. They have succeeded in following the changes in the vibrational potential, vibrational energies, wavefunctions, and transition dipole moment of N–H stretching mode upon formation of hydrogen bonding between molecules of pyrrole and pyridine (Fig. 20.15). This is a notably representative example of a highly anharmonic vibration, which is even further extended upon formation of the hydrogen-bonded complex. The change in the vibrational potential curve and the overlap of wavefunctions would alone suggest an increase of the relative (nonbonded/bonded) intensity of the first overtone transition vs. fundamental transition in this case. However, the experimentally observed phenomenon could be explained by a prominent decrease of the dipole transition moment of $\nu$N–H mode. This can be

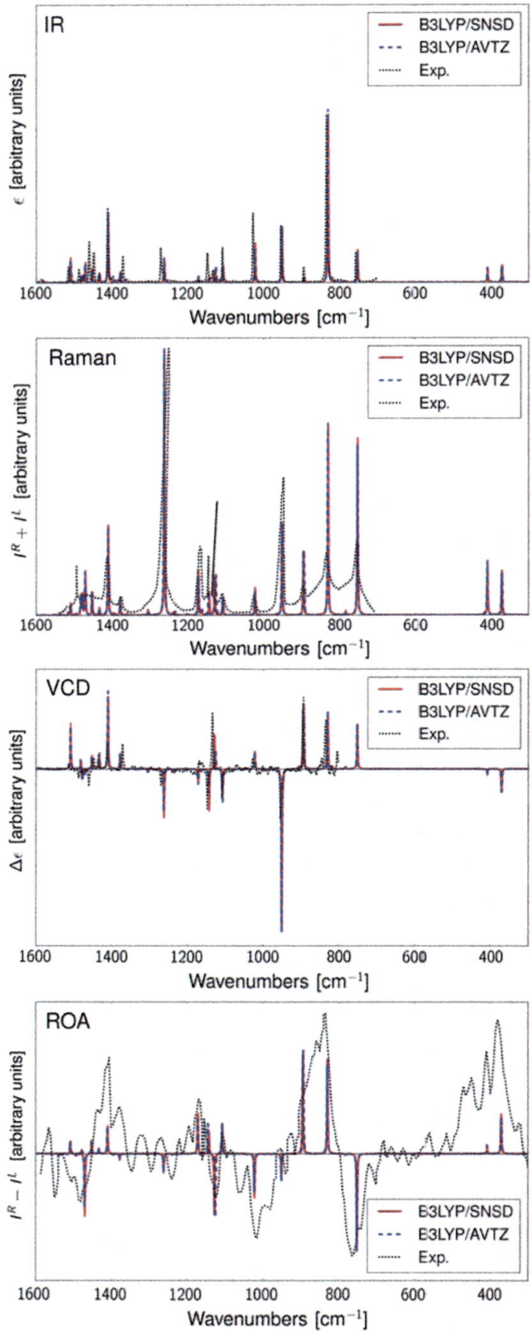


**Fig. 20.14** Fully anharmonic IR, Raman, VCD, and ROA spectra of (*R*)-methyloxirane compared to their experimental counterparts measured in the low-temperature Ar matrix (IR, VCD) or gas phase (Raman, ROA). Vibrational wavenumbers have been computed at the 'cheapCC'/B3LYP level in conjunction with B3LYP intensities. B3LYP computations have been performed with the SNSD and aug-cc-pVTZ (AVTZ) basis sets. All spectra have been convoluted by means of Lorentzian distribution functions with FWHM of 2 $cm^{-1}$. (Reproduced from Ref. [85] with permission)

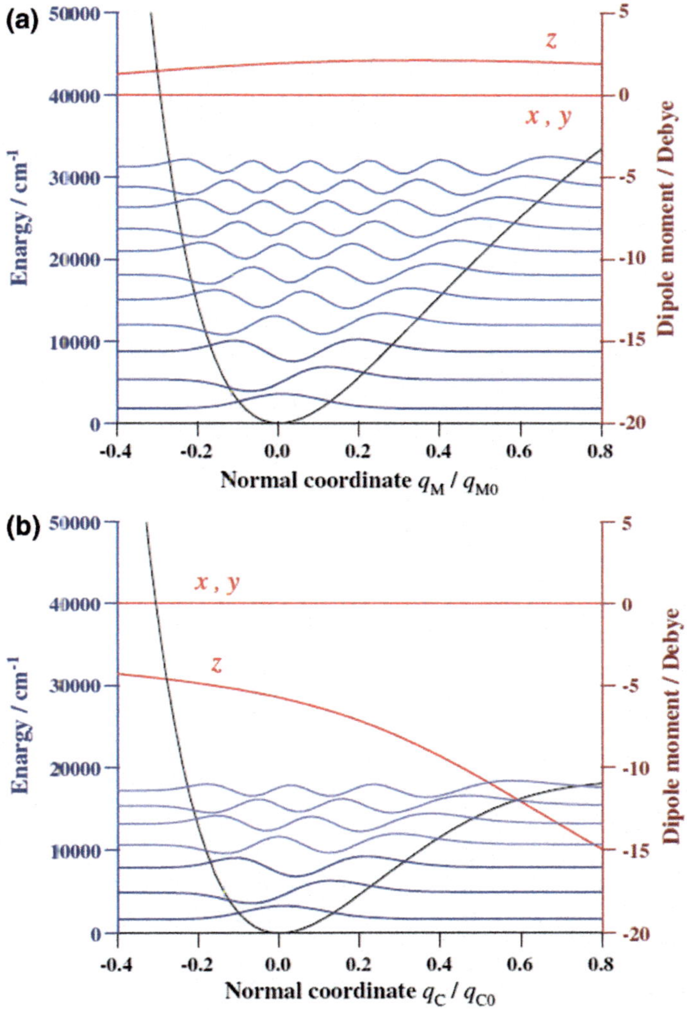


**Fig. 20.15** Vibrational wave functions and dipole moment functions along potential energy curves of the NH stretching mode of **a** pyrrole monomer and **b** pyrrole–pyridine complex calculated at the DFT//B3LYP/6-311 ++G(3df,3pd) level. Symbols $q_{M0}$ and $q_{C0}$ denote units of the normal coordinates for the NH stretching mode in the monomer and the complex, respectively. (Reproduced from Ref. [18] with permission)

clearly noticed for the complex (Fig. 20.15 b), as opposed to nonbonded pyrrole molecule (Fig. 20.15 b). Futami et al. [39], therefore, concluded that the dipole moment function plays the major role in the experimentally observed decrease of overtone intensity of $\nu$N–H mode of pyrrole upon formation of hydrogen-bonded complex with pyridine. This research was also further continued by investigations of

**Table 20.1** Observed and calculated wavenumbers ($cm^{-1}$) and relative absorption intensities of the fundamentals and the first overtones of NH stretching bands of pyrrole. The intensities are normalized against those of bands of $CCl_4$ solutions of pyrrole. (Reproduced form Ref. [18] with permission)

| | | First overtone | | | | Fundamental | | | | Anharmonicity constant $\chi$ | |
|---|---|---|---|---|---|---|---|---|---|---|---|
| | | Obs. | | Calc. | | Obs. | | Calc. | | Obs. | Calc. |
| solvent | $\varepsilon_0$ | $\nu$ | int. | $\nu$ | int. | $\nu$ | int. | $\nu$ | int. | | |
| $CCl_4$ | 2.2 | 6855.6 | 1 | 6932.5 | 1 | 3496.6 | 1 | 3533.2 | 1 | −68.8 | −66.9 |
| $CHCl_3$ | 4.8 | 6837.3 | 1.11 | 6919.4 | 1.1 | 3486 | 1.4 | 3526.6 | 1.35 | −67.3 | −66.8 |
| $CH_2Cl_2$ | 8.9 | 6822.4 | 1.13 | 6913.8 | 1.13 | 3477.2 | 1.65 | 3523.7 | 1.52 | −66 | −66.8 |

other hydrogen-bonded systems, such as methanol-pyridine [43].The conclusion that can be drawn from these studies is that the diagonal anharmonicity plays a key role in the NIR spectra of hydrogen-bonded N–H···N complexes.

Another good example of applying nonstandard approaches is research reported by Futami et al. on molecule–solvent interaction observed in NIR spectra [18]. Through solving of 1D Schrödinger equation with a dense PES grid, they have managed to explore solvent dependence of absorption intensities and wavenumbers of the fundamental and first overtone modes of N–H stretching vibration of pyrrole molecule. The NIR experimental data collected in the study with the use of different solvents revealed a distinct and consistent shift of both $\nu_{0\rightarrow 1}$NH and $\nu_{0\rightarrow 2}$NH bands of pyrrole. In the theoretical study, it was concluded that dielectric constant $\varepsilon$ of the solvent plays a key role in the changes in NIR spectrum observed experimentally. They have successfully reproduced the observed patterns in their calculated data, as presented in Table 20.1. With an increasing dielectric constant $\varepsilon$ of the solvent (2.2, 4.8, and 8.9 for respectively: $CCl_4$, $CHCl_3$, and $CH_2Cl_2$), the frequency of the $\nu_{0\rightarrow 1}$NH and $\nu_{0\rightarrow 2}$NH transition decreases. Further, the corresponding absorption intensities increase in this case as well, and the increase itself is higher for the fundamental transition compared to the first overtone transition. Moreover, these studies have revealed that the changes in vibrational properties of pyrrole due to the interaction with solvent are clearly different than these upon formation of hydrogen-bonded complex with pyridine. For the complex the intensity of fundamental transition of NH stretching vibration increases, while the intensity for the first overtone decreases. On the opposite, the changes in transition intensities for $\nu_{0\rightarrow 1}$NH and $\nu_{0\rightarrow 2}$NH introduced by the interaction with solvent both happen in the same direction; both fundamental and first overtone transition either increase or decrease in the intensity, depending on dielectric constant of the respectful solvent.

## 20.4 Summary and Future Perspectives

Fairly straightforward calculation of vibrational frequencies and intensities within double-harmonic approximation has had its golden age, and countless reports with calculated harmonic IR modes have been published to date. However, a lack of description of anharmonicity, which is the real state of molecular vibrations, remains the major drawback of the harmonic approximation. One of the most common workaround, allowing at least for a general correction of calculated fundamental frequencies, was the introduction of the empirical scaling factors. A rough idea of how important anharmonic effects truly are is the fact that on the recently published list [92] of the most cited papers ever in The Journal of Physical Chemistry the fifth place holds a paper by Scott and Radom [16] on the evaluation of scaling factors. Obviously, anharmonicity is of great importance for vibrational spectroscopies. It remains of an absolutely crucial significance to NIR spectroscopy, which relies on the transitions that exist purely because of the anharmonicity. But also the anharmonic nature of intra- and intermolecular vibrations itself provides rich information on the structure, thermodynamics, and other properties of molecular system. For these reasons, we believe that the interest of both quantum chemists and spectroscopists will be turning even more into anharmonic computational methods.

In this chapter, we briefly introduced the most popular anharmonic approaches. We have highlighted their advantages and disadvantages and the recent progress which has been made. The vibrational self-consistent field (VSCF) approach often struggles because of an inadequate description of mode coupling. It has, therefore, led to an emergence of its perturbatively augmented derivative, PT2-VSCF or cc-VSCF. The evidence of its successful application to complex molecules has been reported, although often only selected modes could be reproduced perfectly. Much further advanced description of the interaction between modes can be offered by vibrational configuration interaction and vibrational coupled cluster approaches. These methods have proved to be able to deliver outstanding results, even for tightly coupled modes, or strong Fermi resonances, in which cases the IR spectrum appears to be entirely different from its harmonic approximation. However, these advanced methods are limited by an overwhelming computational cost, and their applications to complex molecules remain rare. This may change in the future, as both a considerable effort is being put into making VCI and VCC approaches more affordable, and the availability of computational power is ever increasing. On this background, the recent developments in vibrational perturbation theory seem promising. Particularly, the emergence of generalized VPT2 have allowed for very efficient, yet reasonably accurate calculation of vibrational spectra. The capability of GVPT2 scheme to be applied to various molecular systems, including fairly complex ones, and to reproduce IR, NIR, Raman, VCD, and ROA spectra with a good accuracy will likely lead to a notable increase of its popularity in the area of applied spectroscopy. Presently, the field of anharmonic calculations is strongly evolving, meeting an ever-increasing demand from the experimental vibrational spectroscopy. We believe that this trend will notably intensify in the forthcoming years.

## References

1. F. Siebert, P. Hildebrandt. Chapter 2. *Theory of Infrared Absorption and Raman Spectroscopy in Vibrational Spectroscopy in Life Science*. (Wiley, 2008)
2. J.M. Chalmers, P.R. Griffiths (eds.) *Handbook of Vibrational Spectroscopy*, (Wiley, 2002)
3. D.M. Dennison, Rev. Mod. Phys. **3**, 280 (1931)
4. A. Beiser, *Perspectives of Modern Physics*. (McGraw-Hill, 1969)
5. J. Coates, Interpretation of infrared spectra a practical approach, in *Encyclopedia of Analytical Chemistry* (Wiley, 2006)
6. H.W. Siesler, Y. Ozaki, S. Kawata, H.M. Heise (eds.), *Near-Infrared Spectroscopy* (Wiley-VCH, Weinheim, 2002)

7. Y. Ozaki, W.F. McClure, A.A. Christy (eds.), *Near-Infrared Spectroscopy in Food Science and Technology* (Wiley-Interscience, Hoboken, NJ, USA, 2007)
8. M.A. Czarnecki, Y. Morisawa, Y. Futami, Y. Ozaki, Chem. Rev. **115**, 9707 (2015)
9. J. Bron, Can. J. Chem. **53**, 3069 (1975)
10. J.A. Sanson, M.-L. Sanchez, J.C. Corchado, J. Phys. Chem. A **110** 589 (2006)
11. J. Zheng, P. Seal, D.G. Truhlar, Chem. Sci. **4**, 200 (2013)
12. J.A. Stride, P.H. Dallin, U.A. Jayasooriya, J. Chem. Phys. **119**, 2747–2752 (2003)
13. W. Fang, Z. Li, G. Qu, A. Cao, Z. Men, Sun C. Appl. Phys. B **107**, 145–149 (2012)
14. W. Ebeling, A. Netrebko, Y.M. Romanovsky effects of coupling and resonances on transition rates in (eds.) by W. Ebeling, L. Schimansky-Geier, Y.M. Romanovsky *Stochastic Dynamics of Reacting Biomolecules*. (World Scientific, 2002)
15. P.L. Polavarapu. Vibrational Spectra: Principles and Applications with Emphasis on Optical Activity. (Elsevier, 1998)
16. A.P. Scott, L. Radom, J. Phys. Chem. **100**, 16502 (1996)
17. J.A. Pople, H.B. Schlegel, R. Krishnan, D.J. DeFrees, J.S. Binkley, M.J. Frisch, R.A. Whiteside, R.F. Hout, W.J. Hehre, Int. J. Quantum Chem., Quantum Chem. Symp. **15**, 269 (1981)
18. Y. Futami, Y. Ozaki, Y. Hamada, M.J. Wójcik, Y. Ozaki, Chem. Phys. Lett. **482**, 320 (2009)
19. R.F. Goddu, D.A. Delker, Anal. Chem. **32**, 140 (1960)
20. L.G. Weyer, S.-C. Lo. Chapter: Spectra-structure correlations in the near-infrared. in eds. by J. M. Chalmers, P.R. Griffiths *Handbook of Vibrational Spectroscopy*. vol. 3. (Wiley, 2006)
21. J.M. Bowman, J. Chem. Phys. **68**, 608 (1978)
22. G.D. Carney, L.L. Sprandel, C.W. Kern, Adv. Chem. Phys. **37**, 305 (1978)
23. M. Cohen, S. Greita, R.D. McEarchran, Chem. Phys. Lett. **60**, 445 (1979)
24. R.B. Gerber, M.A. Ratner, Chem. Phys. Lett. **68**, 195 (1979)
25. A. Szabo, N.S. Ostlund, *Modern Quantum Chemistry* (Dover Publishing, Mineola, New York, 1996)
26. P. Seidler, O. Christiansen. Chapter 18. Vibrational Coupled Cluster Theory. in *Recent Progress in Coupled Cluster Methods* eds. by P. Carsky, J. Pladus, J. Pittner. (Springer, 2010)
27. L.S. Norris, M.A. Ratner, A.E. Roitberg, R.B. Gerber, J. Chem. Phys. **105**, 11261 (1996)
28. J.S. Binkley, J.A. Pople, Int. J. Quant. Chem. **9**, 229 (1975)
29. R.J. Whitehead, N.C. Handy, J. Mol. Spec. **55**, 356 (1975)
30. J.M. Bowman, K. Christoffel, F. Tobin, J. Phys. Chem. **83**, 905 (1979)
31. M.A. Ratner, V. Buch, R.B. Gerber, Chem. Phys. **53**, 345 (1980)
32. T.C. Thompson, D.G. Truhlar, Chem. Phys. Lett. **75**, 87 (1980)
33. R. Garnier, M. Odunlami, V. Le Bris, D. Bégué, I. Baraille, O. Coulaud. J. Chem. Phys. **144**, 204123 (2016)
34. O. Christiansen, J. Chem. Phys. **120**, 2140 (2004)
35. O. Christiansen, J. Chem. Phys. **120**, 2149 (2004)
36. F. Culot, J. Liévin, Theor. Chim. Acta **89**, 227 (1994)
37. H.D. Meyer, Wiley Interdiscip. Rev. Comput. Mol. Sci. **2**, 351 (2012)
38. V. Barone, J. Chem. Phys. **122**, 14108 (2005)
39. Y. Futami, Y. Ozaki, Y. Hamada, M.J. Wójcik, Y. Ozaki, J. Phys. Chem. A **115**, 1194 (2011)
40. Y. Futami, Y. Morisawa, Y. Ozaki, Y. Hamada, M.J. Wojcik, Y. Ozaki, J. Mol. Struct. **1018**, 102 (2012)
41. Y. Chen, Y. Morisawa, Y. Futami, M.A. Czarnecki, H.-S. Wang, Y. Ozaki, J. Phys. Chem. A **118**, 2576 (2014)
42. Y. Futami, Y. Ozaki, Y. Hamada, Y. Ozaki, Vib. Spectrosc. **72**, 124 (2014)
43. Y. Futami, Y. Ozaki, Y. Ozaki, Phys. Chem. Chem. Phys. **18**, 5580 (2016)
44. B.V. Numerov, Publications de l'Observatoire Astrophysique Central de Russie **2**, 188 (1923)
45. B.V. Numerov, Mon. Not. R. Astron. Soc. **84**, 592 (1924)
46. B.R. Johnson, J. Chem. Phys. **67**, 4086 (1977)
47. B.R. Johnson, J. Chem. Phys. **69**, 4678 (1978)
48. M. Pillai, J. Goglio, T.G. Walke, Am. J. Phys. **80**, 1017 (2012)

49. U. Kuenzer, J.-A. Soraru, T.S. Hofer, Phys. Chem. **18**, 31521 (2016)
50. G.M. Chaban, J.O. Jung, R.B. Gerber, J. Chem. Phys. **111**, 1823 (1999)
51. G.M. Chaban, J.O. Jung, R.B. Gerber, J. Phys. Chem. A **104**, 10035 (2000)
52. R.B. Gerber, B. Brauer, S.K. Gregurick, G.M. Chaban, Phys. Chem. Comm. **5**, 142 (2002)
53. S.K. Gregurick, E. Fredj, R. Elber, R.B. Gerber, J. Phys. Chem. B **101**, 8595 (1997)
54. K. Yagi, T. Taketsugu, K. Hirao, M.W. Schmidt, M.S. Gordon, J. Chem. Phys. **121**, 1383 (2004)
55. K. Yagi, T. Taketsugu, K. Hirao, M.S. Gordon, J. Chem. Phys. **113**, 1005 (2000)
56. R. Balu, K. Siegrist, D. Plusquellic, E. Heilweil, S.K. Gregurick, Biophys. J. **160A** (2007)
57. D. Shemesh, J. Mullin, M.S. Gordon, R.B. Gerber, Chem. Phys. **347**, 218 (2008)
58. H. Zhang, K. Siegrist, D.F. Plusquellic, S.K. Gregurick, J. Am. Chem. Soc. **130**, 17846 (2008)
59. A. Roitberg, R.B. Gerber, R. Elber, M.A. Ratner, Science **268**, 1319 (1995)
60. B. Brauer, M. Pincu, V. Buch, I. Bar, J.P. Simons, R.B. Gerber, J. Phys. Chem. A **115**, 5859 (2011)
61. O.M.D. Lutz, G.K. Bonn, B.M. Rode, C.W. Huck, Anal. Chim. Acta **826**, 61 (2014)
62. J. Sebek, L. Pele, E.O. Potma, R.B. Gerber, Phys. Chem. Chem. Phys. **13**, 12724 (2011)
63. L.A. Mancera, D.M. Benoit, Phys. Chem. Chem. Phys. **18**, 529 (2016)
64. L. Koziol, Y. Wang, B.J. Braams, J.M. Bowman, A.I. Krylov, J. Chem. Phys. **128**, 204310 (2008)
65. D. Oschetzki, X. Zeng, H. Beckers, K. Banert, G. Rauhut, Phys. Chem. Chem. Phys. **15**, 6719 (2013)
66. I.H. Godtliebsen, O. Christiansen, Phys. Chem. Chem. Phys. **15**, 10035 (2013)
67. S. Califano, *Vibrational States* (Wiley, London, 1976)
68. H.H. Nielsen, Rev. Mod. Phys. **23**, 90 (1951)
69. A.B. McCoy, E.L. Sibert III, J. Chem. Phys. **92**, 1893 (1990)
70. V. Barone, M. Biczysko, J. Bloino, Phys. Chem. Chem. Phys. **16**, 1759 (2014)
71. T. Fornaro, I. Carnimeo, M. Biczysko, J. Phys. Chem. A **119**, 5313 (2015)
72. T. Fornaro, M. Biczysko, S. Monti, V. Barone, Phys. Chem. Chem. Phys. **16**, 10112 (2014)
73. C. Puzzarini, M. Biczysko, V. Barone, J. Chem. Theory Comput. **7**, 3702 (2011)
74. T. Fornaro, M. Biczysko, J. Bloino, V. Barone, Phys. Chem. Chem. Phys. **18**, 8479 (2016)
75. V. Barone, M. Biczysko, J. Bloino, C. Puzzarini, Phys. Chem. Chem. Phys. **15**, 10094 (2013)
76. V. Barone, M. Biczysko, J. Bloino, C. Puzzarini, J. Chem. Theory Comput. **9**, 1533 (2013)
77. V. Barone, M. Biczysko, J. Bloino, C. Puzzarini, Phys. Chem. Chem. Phys. **15**, 1358 (2013)
78. V. Barone, M. Biczysko, J. Bloino, M. Borkowska-Panek, I. Carnimeo, P. Panek, Int. J. Quant. Chem. **112**, 2185 (2012)
79. K.B. Beć, Y. Futami, M.J. Wójcik, Y. Ozaki, Phys. Chem. Chem. Phys. **18**, 13666 (2016)
80. T. Fornaro, D. Burini, M. Biczysko, V. Barone, J. Phys. Chem. A **119**, 4224 (2015)
81. K.B. Beć, Y. Futami, M.J. Wójcik, T. Nakajima, Y. Ozaki, J. Phys. Chem. A **120**, 6170 (2016)
82. K.B. Beć, J. Grabska, C.W. Huck, Y. Ozaki, J.P. Hawranek, J. Mol. Liq. **224**, 1189 (2016)
83. C.G. Kirchler, C.K. Pezzei, K.B. Beć, S. Mayr, Y. Ozaki, C.W. Huck, Analyst **8374**, 837404 (2016)
84. V. Barone, M. Biczysko, J. Bloino, C. Puzzarini, J. Chem. Phys. **141**, 034107 (2014)
85. J. Bloino, M. Biczysko, V. Barone, J. Phys. Chem. A **119**, 11862 (2015)
86. C. Cappelli, J. Bloino, F. Lipparini, V. Barone, J. Phys. Chem. Lett. **3**, 1766 (2012)
87. J. Bloino, V. Barone, J. Chem. Phys. **136**, 124108 (2012)
88. K. Takahashi, M. Sugawara, S. Yabushita, J. Phys. Chem. A **106**, 2676 (2002)
89. K. Takahashi, M. Sugawara, S. Yabushita, J. Phys. Chem. A **107**, 11092 (2003)
90. K. Takahashi, M. Sugawara, S. Yabushita, J. Phys. Chem. A **109**, 4242 (2005)
91. H. Takahashi, S. Yabushita, J. Phys. Chem. A **117**, 5491 (2013)
92. http://pubs.acs.org/page/vi/jphyschem120-anniversary.html